PROCEEDINGS OF
SYMPOSIA IN PURE MATHEMATICS
Volume XI

ENTIRE FUNCTIONS AND RELATED PARTS OF ANALYSIS

American Mathematical Society
PROVIDENCE, RHODE ISLAND
1968

Proceedings of the Symposium in Pure Mathematics
of the American Mathematical Society
Held at the University of California, San Diego
La Jolla, California
June 27–July 22, 1966

Sponsored by The National Science Foundation under Grant GP–5069

Edited by
Jacob Korevaar,
(Coordinating editor)
S. S. Chern
Leon Ehrenpreis
W. H. J. Fuchs
L. A. Rubel

Library of Congress Catalog Number 68–10458

Preface

The present volume is a direct outcome of the Society's thirteenth Summer Research Institute, the topic of which was Entire Functions and Related Parts of Analysis.

The Institute was organized by a committee consisting of A. Beurling, R. P. Boas (chairman), L. Ehrenpreis, W. H. J. Fuchs, J. Korevaar and L. A. Rubel. It was held from June 27 to July 22, 1966, at the University of California, San Diego, in La Jolla, with the undersigned as chairman. The primary objective of the Institute was to improve communication between the specialists in the different parts of the subject.

In order to acquaint the classically oriented participants with the new ideas and methods coming from various more abstract branches of mathematics, the following sections were set up:

Complex Analysis from the point of view of Mapping of Manifolds. In this section, organized by S. S. Chern, lectures were presented by E. Bishop, S. S. Chern (four), F. W. Gehring (two), B. Rodin, L. Sario (two), W. Stoll (two), and H. Wu.

Fourier Methods, a section organized by L. Ehrenpreis and dealing primarily with applications of complex Fourier analysis. Lectures were given by L. Coburn (two), L. Ehrenpreis (three), J. Kazdan (two), D. J. Newman (two), S. M. Shah, and P. van Goethem.

Function Spaces. This section, organized by L. A. Rubel, was devoted to methods derived from functional analysis and algebra. Lectures were preeented by N. L. Alling, L. de Branges, J. J. Kelleher (two), H. J. Landau, B. Logan, L. A. Rubel (two), and B. A. Taylor (two).

The Institute participants of more modern orientation had ample opportunity to become acquainted with developments and problems in many active classical areas. Most of the papers dealing with classical topics were presented in the section on

Distribution of Values. In this section, organized by W. H. J. Fuchs, lectures were given by J. Clunie, E. F. Collingwood, H. Delange, W. H. J. Fuchs, D. Gaier, T. H. Ganelius, A. Gray, F. Gross, W. Hayman (two), A. Hyllengren, W. Al-Katifi, J. Korevaar (two), P. Lelong, B. Lepson, T. L. McCoy, P. Noverraz, C. Pommerenke, and A. Schumitzky.

Other lecturers on relatively classical topics included E. J. Akutowicz, A. Dvoretzky, P. Koosis, G. Labelle, A. L. Shields, and H. Wilf. During the final week of the Institute, L. Ehrenpreis organized a successful series of problem sessions.

All in all, the seventy-three participants in the Institute (including twenty-two from abroad) provided more than sixty formal lectures and a number of informal ones. Approximately half the speakers wrote up lecture notes which were distributed in offset form shortly after the Institute disbanded.

The Institute had a brief continuation at the Houston meeting of the Society, with a day-long session on January 25, 1967. The program included papers by E. F. Beckenbach, S. S. Chern, A. Edrei, W. H. J. Fuchs, J. Korevaar and S. Hellerstein, P. Malliavin, L. Nachbin and C. P. Gupta, L. A. Rubel, and W. Stoll.

All but one of the thirty-four articles in these Proceedings have at least one author who participated in the Institute. Approximately twenty of the papers correspond to a lecture given in the Institute; two major papers (by W. Stoll and B. A. Taylor) present solutions to a problem raised during the sessions.

For this publication, the editors decided against an organization by sections as in the Institute. No manuscripts were available for a majority of the lectures; the dividing lines between the sections had never been very clear; and several of the manuscripts did not fit into any of the original categories. For these reasons, the articles are simply arranged alphabetically by (first) author.

The volume ends with a problem section edited by L. Ehrenpreis. It provides a record of the problem sessions of the Institute, containing both the problems and a number of solutions.

On behalf of the organizing committee for the 1966 Summer Institute and the editorial committee for these Proceedings, the undersigned wishes to acknowledge the financial support received from the National Science Foundation under grant GP 5069, the hospitality provided by the University of California, San Diego, and the generous and able assistance supplied by the executive director of the Society, Dr. G. L. Walker, and his staff.

JACOB KOREVAAR
La Jolla, 1967

Contents

The Invariant Subspaces of H^1 and an Extension of Schwarz's Lemma

EDWIN J. AKUTOWICZ

Introduction. Let H^p ($p \geqq 1$) denote as usual the Banach space of all functions $f(z)$ holomorphic in the unit disc $|z| < 1$ such that the integral [1]

$$\frac{1}{2\pi} \int |f(re^{i\vartheta})|^p d\vartheta$$

remains bounded as r tends to 1. The norm in the space H^p is the L^p-norm of the boundary function $\lim_{r\to 1} f(re^{i\vartheta})$ denoted $f(e^{i\vartheta})$:

$$\|f\|_{H^p} = \|f\|_p = \left\{\frac{1}{2\pi}\int |f(e^{i\vartheta})|^p d\vartheta\right\}^{1/p}.$$

We shall assume certain of the principal facts about these spaces to be known, such as the factorisation into inner and outer factors, representation by the Cauchy integral over the boundary values, etc. (for details consult K. Hoffman [**6**]).

For $f \in H^1$, the inner factor of f shall be denoted by I_f and the outer factor by $O_f : f = I_f O_f$. The Blaschke factor of f shall be denoted by B_f and the singular factor by $\Sigma_f : I_f = B_f \Sigma_f$.

The closed, nonnull, linear subspaces S of H^1 which are invariant under multiplication by z (briefly, invariant subspaces) have been described as follows by deLeeuw and Rudin [**7**] (also Helson [**4**]):

THEOREM. *Every invariant subspace S of H^1 is of the form*

$$S = s \cdot H^1,$$

where s is an inner function essentially uniquely determined by the subspace S (i.e., to within a constant factor of modulus unity). [2]

The proof suggested (Rudin [**8**, p. 432]) would give the generating element s

[1] Here, and in the sequel, an unadorned integral sign refers to integration over the interval $-\pi \leqq \vartheta \leqq \pi$.

[2] This is of course also true for H^p, $p > 1$, but these cases are simpler.

directly as the greatest common divisor of the inner functions occurring in the factorisation of elements of the invariant subspace S.

In the first part of the present note we observe that the invariant subspaces S of H^1 are of the form

$$S = z^n \cdot \sigma \cdot H^1,$$

where σ is the unique inner factor of all elements λ of the invariant subspace $z^{-n}S$ which realise the best approximation to the constant 1,

$$(*) \qquad \| 1 - \lambda \|_1 = \inf_{f \in z^{-n}S} \| 1 - f \|_1 ,$$

and where n is the largest integer such that every function of the subspace S vanishes at the origin to order n. This is obviously consistent with the situation in H^2, where the unique best approximant coincides with the generator. But in H^1 the phenomena are essentially different. Thus for the invariant subspace $z \cdot H^1$, the function z does not realise the best approximation (*).

In §2 we shall connect the extremal problem (*) for an invariant subspace S of H^1 with a conjugate extremal problem in the space B of all bounded holomorphic functions in the exterior of the unit disc and vanishing at infinity. It is well known for Banach spaces that a minimum such as (*) coincides with a maximum in the dual Banach space. This relationship has been investigated by S. Havinson and by G. Tumarkin in a series of papers (cf. [**5**], which contains further references; cf. also [**2**], [**9**]). Morally, the dual space of H^1 should be B; this is not the case, but B does nevertheless play the role of the dual in the present considerations. We use here a method which was shown to us by L. Carleson a number of years ago. This consists in the examination of the coefficient of $1/z$ in the Taylor expansion about the point at infinity of certain functions arising by approximating the H^1 problem (*) by a similar problem in H^p and letting p tend to 1. Such arguments occur in Carleson's notes [**3**]. We thus obtain the following theorem.

THEOREM 1. *For every invariant subspace S of H^1 ($S = s \cdot H^1$, s inner), there subsists the equality*

$$\inf_{f \in S} \| 1 - f \|_1 = \sup_{h \in B; \|h\|_\infty \leqq 1} \left| \frac{1}{2\pi} \int hs \, d\vartheta \right| .$$

In the special case $S = z \cdot H^1$, the supremum on the right is precisely the extremal condition of the classical lemma of Schwarz. Cf. [**1**] for an extension of Schwarz's lemma in another direction.

I am very much obliged to L. Carleson, whose criticisms of earlier, faulty versions of this note have been extremely useful.

1. A characterisation of the invariant subspaces of H^1. In this section we shall give a proof of

THEOREM 2. *Every invariant subspace S of H^1 is essentially uniquely of the form*

$$S = z^n \cdot \sigma \cdot H^1,$$

for some $n = 0, 1, \cdots$ and some inner function σ which does not vanish at the origin. Furthermore, the function σ is the unique inner factor $\sigma = I_\lambda$ of all solutions λ of the extremal problem

$$\inf_{f \in z^{-n} \cdot S} \| 1 - f \|_1 = \| 1 - \lambda \|_1 . \tag{1.1}$$

Let us first suppose that not every element of the subspace S vanishes at the origin, i.e., $n = 0$. Consider then the extremum of (1.1) for such S. Let λ be any element of S for which the lower bound (1.1) is attained. Such an element exists because S is closed. It follows by making a variation that λ must satisfy[3]

$$\int \phi \operatorname{sgn}(1 - \bar{\lambda}) d\vartheta = 0 \tag{1.2}$$

for every $\phi \in S$. Hence λ cannot be identically zero as then (1.2) would imply that every $\phi \in S$ vanished at the origin. Replacing ϕ by $e^{in\vartheta}\phi$ ($n = 0, 1, \cdots$) in (1.2), we obtain by the F. and M. Riesz theorem that

$$\phi \operatorname{sgn}(1 - \bar{\lambda}) = g \quad \text{a.e. on } |z| = 1, \tag{1.3}$$

where g is the boundary function of a function (likewise denoted by g) in H^1. Hence, in particular,

$$\lambda = \operatorname{sgn}(1 - \lambda) \cdot g_0, \qquad g_0 \in H^1 . \tag{1.4}$$

Therefore

$$I_\lambda = \operatorname{sgn}(1 - \lambda) \cdot I_{g_0},$$

I_λ, I_{g_0} being the inner factors of λ, g_0. It follows by squaring that

$$\bar{\lambda} I_\lambda^2 = \lambda I_{g_0}^2 - I_{g_0}^2 + I_\lambda^2 \equiv h \in H^1,$$

and hence that

$$\bar{O}_\lambda I_\lambda = I_h O_\lambda, \quad \text{a.e. on } |z| = 1 . \tag{1.5}$$

From (1.5) we shall draw the conclusion that the inner factor I_λ of λ divides every element of S. It is false that the outer factor O_λ is necessarily trivial.

Let us take a nontrivial element $\phi \in S \cap H^\infty$; such ϕ exists according to the Rudin-deLeeuw result cited earlier. It then follows from (1.3) that

$$\phi^2 \bar{\lambda} = \phi^2 + (1 - \lambda) g^2 \equiv \psi . \tag{1.6}$$

---

[3] For a complex number c, $\operatorname{sgn} c = c / |c|$.

Here g is a bounded function,

(1.7)
$$\sup_{|z|<1} |g(z)| \leqq \|\phi\|_\infty.$$

For

$$\left\{\frac{1}{2\pi}\int |g(re^{i\vartheta})|^p d\vartheta\right\}^{1/p} \leqq \left\{\frac{1}{2\pi}\int |g(e^{i\vartheta})|^p d\vartheta\right\}^{1/p} = \left\{\frac{1}{2\pi}\int |\phi(e^{i\vartheta})|^p d\vartheta\right\}^{1/p}$$

which yields (1.7) upon passing to the limit $p = \infty$.

Therefore $\psi \in H^1$. Now, in view of (1.5) and (1.6),

$$(\phi^2/I_\lambda)\cdot(I_h O_\lambda/I_\lambda) = \psi, \quad \text{a.e. on } |z| = 1.$$

Hence, by the uniqueness of inner factors,

(1.8)
$$I_\lambda^2 I_\psi = I_\phi^2 I_h.$$

Cancelling a possible common factor of I_λ and I_h, (1.8) becomes

(1.9)
$$\tilde{I}_\lambda I_\lambda I_\psi = I_\phi^2 \tilde{I}_h,$$

where $\tilde{I}_\lambda$ and $\tilde{I}_h$ are relatively prime. Therefore I_λ and $\tilde{I}_h$ are relatively prime and (1.9) implies that I_λ must divide I_ϕ^2 and hence $I_\phi : I_\phi = I_\lambda I$ (I inner). This proves that

$$\phi = I_\lambda I O_\phi \in I_\lambda \cdot H^\infty.$$

Thus

$$S \cap H^\infty \subset I_\lambda \cdot H^\infty.$$

But H^∞ is dense in H^1, so that passing to the closure in H^1 we obtain

$$S \cap H^1 = S \subset I_\lambda \cdot H^1.$$

As the opposite inclusion is obvious[4], we have

$$S = I_\lambda \cdot H^1$$

in the case that not all functions of S vanish at the origin.

On the other hand, if z^n is the highest power of z dividing all elements of S, it is easy to verify that $S = z^n \cdot S_0$, where S_0 is an invariant subspace and not all elements of S_0 vanish at the origin.

Knowing that $S_0 = I_0 \cdot H^1$, I_0 inner, the uniqueness of I_0 follows immediately. Theorem 2 is completely proved.

2. **The conjugate extremal problem.** We shall exploit the fact that the invariant subspaces of H^1 are characterised as in the deLeeuw-Rudin theorem (Introduction) in order to prove the equality of Theorem 1. Instead of using antiholomorphic

[4] Using the fact that $O \cdot H^1 = H^1$ for any outer function O, which follows at once from the deLeeuw-Rudin theorem.

functions in the unit disc, we shall take the corresponding holomorphic functions in the exterior $|z| > 1$. Recall that B is the space of functions holomorphic and bounded in the exterior of the unit disc and vanishing at infinity. Put

$$p = \inf_{f \in S} \|1 - f\|_1 \text{ and } c = \sup_{h \in B; \|h\|_\infty \leqq 1} \left| \frac{1}{2\pi} \int hs\, d\vartheta \right|,$$

where $S = s \cdot H^1$. Observe first that

$$p = \inf_{\phi \in \bar{H}^1} \|1 - \bar{s}\phi\|_1,$$

and that for every $\phi \in \bar{H}^1$ and every $g \in B$ we have

$$\int \phi g\, d\vartheta = 0,$$

so that

$$\frac{1}{2\pi} \int gs\, d\vartheta = \frac{1}{2\pi} \int (1 - \bar{s}\phi) gs\, d\vartheta,$$

and hence

$$c \leqq p. \tag{2.1}$$

In order to prove the opposite inequality, we shall construct a particular function $h \in B$, $\|h\|_\infty \leqq 1$, using a method of Carleson's. Consider for $q > 1$ the minimum

$$p_q = \inf_{\phi \in \bar{H}^q} \|1 - \bar{s}\phi\|_q = \inf_{\phi \in \bar{H}^q} \|s - \phi\|_q = \inf_{\phi = s + \beta;\, \beta \in \bar{H}^q} \|\phi\|_q,$$

which is attained for a unique $\phi_q = s + \beta_q, \beta_q \in \bar{H}_q$:

$$p_q = \left\{ \frac{1}{2\pi} \int |\phi_q|^q\, d\vartheta \right\}^{1/q}.$$

If we make a variation $\phi_q + \eta\beta$, $\beta \in \bar{H}^q$, η complex, we find that the extremal element ϕ_q must satisfy

$$\int |\phi_q|^{q-2} \bar{\phi}_q \beta\, d\vartheta = 0 \tag{2.2}$$

for all β of the above form. Define the function

$$\Phi_q(z) = \frac{1}{2\pi} \int |\phi_q|^{q-2} \bar{\phi}_q s \frac{d\vartheta}{z - e^{i\vartheta}}, \qquad |z| > 1.$$

Since we can take $\beta = \beta_q = \phi_q - s$ in (2.2), we find that

$$\frac{1}{2\pi} \int |\phi_q|^{q-2} \bar{\phi}_q s\, d\vartheta = \frac{1}{2\pi} \int |\phi_q|^q d\vartheta,$$

which shows that the coefficient of $1/z$ in the Taylor expansion of $\Phi_q(z)$ about the point at infinity is equal to p_q^q. If we take the normalised family of functions $\psi_q = \phi_q/p_q$, we obtain a family of measures

$$d\mu_q = |\psi_q|^{q-2}\bar{\psi}_q s \frac{d\vartheta}{2\pi}$$

which satisfy

$$\int |d\mu_q| = \frac{1}{2\pi}\int |\psi_q|^{q-1} d\vartheta \leqq \left\{\frac{1}{2\pi}\int |\psi_q|^q d\vartheta\right\}^{(q-1)/q} = 1.$$

Hence a subsequence $d\mu_{q_\nu}$ will converge weakly to a measure $d\mu$, $\int|d\mu| \leqq 1$ as $q_\nu \to 1$. Hence in particular

$$\lim_{\nu\to\infty} \Phi_{q_\nu}(z) = \int \frac{d\mu(\vartheta)}{z - e^{i\vartheta}}, \qquad |z| > 1. \tag{2.3}$$

Since we have, according to (2.2),

$$\frac{1}{2\pi}\int |\psi_q|^{q-2}\bar{\psi}_q e^{-in\vartheta} d\vartheta = \int e^{-in\vartheta}\bar{s}\, d\mu_q(\vartheta) = 0, \tag{2.4}$$

$n = 0, 1, \cdots$, we also must have

$$\int e^{-in\vartheta}\bar{s}\, d\mu(\vartheta) = 0, \qquad n = 0, 1, \cdots.$$

Therefore it follows by the F. and M. Riesz theorem that

$$d\mu = hs\, d\vartheta/2\pi,$$

where $h \in \bar{H}^1$ and vanishes at infinity. Actually h satisfies $\|h\|_\infty \leqq 1$. For observe that, according to (2.4), $|\psi_q|^{q-2}\bar{\psi}_q$ is the boundary function of a function $h_q \in \bar{H}^r$, $1/r + 1/q = 1$. The weak convergence as measures of $h_q\, d\vartheta$ to $h\, d\vartheta$ then implies by the Cauchy integral representation that $h_q \to h$ uniformly on compact sets of $|z| > 1$. Now for any $R > 1$,

$$\left\{\frac{1}{2\pi}\int |h_q(Re^{i\vartheta})|^r d\vartheta\right\}^{1/r} \leqq \left\{\frac{1}{2\pi}\int |h_q(e^{i\vartheta})|^r d\vartheta\right\}^{1/r} = \left\{\frac{1}{2\pi}\int |\psi_q|^q d\vartheta\right\}^{1/r} = 1,$$

which easily yields the conclusion that $\|h\|_\infty \leqq 1$.

Now the coefficient of $1/z$ in the expansion of the function $p_{q_\nu}^{1-q_\nu}\Phi_{q_\nu}(z)$ is precisely p_{q_ν}, which on the one hand clearly decreases to a limit at least as large as $p_1 = p$ as $q_\nu \to 1$, and on the other hand, since

$$\lim_{\nu\to\infty} \Phi_{q_\nu}(z) = \int \frac{d\mu(\vartheta)}{z - e^{i\vartheta}} = \frac{1}{2\pi}\int \frac{hs}{z - e^{i\vartheta}} d\vartheta,$$

the coefficient must tend to

$$\frac{1}{2\pi}\int hs\,d\vartheta,$$

whence $p = p_1 \leqq \lim_{q \to 1} p_q \leqq c$. The proof of Theorem 1 is thus completed.

References

1. E. J. Akutowicz, *Schwarz's lemma in the Hardy class* H^1, Rend. Circ. Mat. Palermo Ser. (2), **8** (1959), 185–191.

2. R. C. Buck, "Linear spaces and approximation theory" in *On numerical approximation*, Univ. of Wisconsin Press, Madison, Wis., 1956.

3. L. Carleson, *Selected problems in exceptional sets*, van Nostrand, Princeton, N. J., 1962.

4. H. Helson, *Invariant subspaces*, Academic Press, New York, 1964.

5. S. Havinson, *The theory of extremal problems for bounded analytic functions satisfying additional conditions inside the domain*, Uspehi Mat. Nauk **18** (1963), 25–98. (Russian).

6. K. Hoffman, *Banach spaces of analytic functions*, Prentice-Hall, Englewood Cliffs, N. J., 1962.

7. K. deLeeuw and W. Rudin, *Extreme points and extremal problems in* H^1, Pacific J. Math. **8** (1958), 467–485.

8. W. Rudin, *The closed ideals in an algebra of analytic functions*, Canad. J. Math. **9** (1957), 426–434.

9. W. W. Rogosinski and H. S. Shapiro, *On certain extremum problems for analytic functions*, Acta Math. **80** (1953), 287–318.

Université de Montpellier
Montpellier, France

The Valuation Theory of Meromorphic Function Fields[1]

NORMAN L. ALLING[2]

Introduction. There follows an extended version of the lecture notes I used for my talk on the valuation theory of meromorphic function fields over noncompact Riemann surfaces, at the 1966 Summer Institute at La Jolla. In view of the professed aim of the Institute, and since my subject was far removed from the classical theory of entire functions (being on the edge of pure algebra), I have tried to weave in bits of algebra along the way, as they come up. If I seem to be patronizing to the reader, please excuse me; it has been my effort to inform the reader not steeped in algebra about a nearly complete theory in algebraic analysis.

With the exception of some of the results in §5, all of the results have appeared in the literature, although a few minor technical improvements have been effected here and there in this manuscript. A Bibliography is appended which represents virtually all the relevant papers I know of. Since many of the contributors to our knowledge of this subject did not know the existing literature very well, there was a great deal of duplication of work. In addition, the development of this relatively small bit of mathematics seems to me like a vignette of the history of our science, which should at least be interesting to students. In view of this, and since the theory is now very nearly complete for the full ring of analytic functions $A(X)$ on a noncompact (connected) Riemann surface X, and for its quotient field $F(X)$, some historical notes seem in order. These lie between §6 and the Bibliography.

§1 deals with the beginnings of the ideal theory of $A(X)$, and with an interesting algebraic-geometric application of this theory; the Chevalley-Kakutani-Bers Theorem (1.1). In §2 the residue class fields of $A(X)$ are analyzed. §3 contains a quick sketch of valuation theory and of the notion of localization needed further on. The valuation theory of $F(X)$ is dealt with in §4. In §5 we return to the ideal theory of $A(X)$, which is very nearly completely resolved. The last section, §6, deals with Iss'sa's Theorem and corollaries.

[1] This research was supported in part by grant NSF GP 5775.

[2] During the course of 1966 Summer Institute, I was in part supported by the National Science Foundation, via the Institute.

ACKNOWLEDGEMENTS. Acknowledging my debt to the mathematicians mentioned in the Bibliography, by dint of their placement there, does not sufficiently take note of the help and encouragement which Henriksen, Royden, and Schilling gave me. Nor does it do justice to Hausdorff, who invented η_1-sets **[H₁]**; or to Krull who invented valuation theory and from whose *Allgemeine Bewertungstheorie*, J. Reine Angew. Math. **167** (1932), 160–196, I learned valuation theory; or to H. Hahn, whose *Über die nichtarchimedischen Grössensysteme* S.-B. der K. Akad. Wiss. (Vienna) **116** (1907), 601–653, first suggested the connection between order and valuation, by inventing both Hahn groups and formal power series fields, Hahn having been motivated by a question of Hilbert's.

1. **Ideal theory.** Let X and Y be two connected noncompact Riemann surfaces and let θ be a one-to-one conformal mapping of X onto Y. Clearly θ^*, the adjoint of θ on $A(Y)$, is a C-isomorphism onto $A(X)$. Having noted this, the converse naturally comes to mind.

THEOREM 1.1 (CHEVALLEY-KAKUTANI **[CK]**, BERS **[B₂]**). *X and Y are conformally equivalent if and only if $A(Y)$ and $A(X)$ are C-isomorphic.*

To prove this it is necessary to identify Y, given only $A(Y)$ as an algebra over C; thus we are led to a study of the ideals, and particularly the maximal ideals, of A.

In many ways A is a very tractable ring. It is an integral domain which is integrably closed, semisimple, and is an algebra over the field of complex numbers. (For a general reference to algebra see e.g., **[ZS₁]**.) Since elements in A may have an infinite number of zeros, the ideal theory of A would appear to be nonfinitistic. The generalized Weierstrass (product) Theorem **[F]** can be used to show that A is non-Noetherian. Given $f \in A$ let $Z(f) \equiv \{x \in X : f(x) = 0\}$. Clearly $Z(f)$ is either X itself or its a discrete subset of X. Conversely, by the Weierstrass Theorem, given any discrete subset D of X, there exists $f \in A$ such that $D = Z(f)$. The multiplicative group $U(A)$ of invertible elements of A, called the *group of units* of A, cannot of course meet any proper ideal, and in the case under consideration is $\{f \in A : Z(f) = \varnothing\}$. The following lemma reduces the study of finitely generated ideals to principal ideals. Since A is non-Noetherian, there are still some ideals left to analyze, however.

LEMMA 1.2 (HELMER **[H₂]**). *Let $f_1, \cdots, f_n \in A$. There exists $d \in A$ such that $(f_1, \cdots, f_n) = (d)$.*

Having proved this in case $n = 2$, the rest follows. This case is easily handled by invoking the Weierstrass and Mittag-Leffler Theorems.

COROLLARY 1.3 (HELMER **[H₂]**). *Let $f, g \in A$. $Z(f) \cap Z(g) = \varnothing$ if and only if there exists $a, b \in A$ such that $af + bg = 1$: i.e., $(f, g) = (1)$.*

Let S be a subset of A and let $Z(S) \equiv \{Z(f) : f \in S\}$. $Z(A) \equiv \Delta = \{X\}$

$\cup \{D \subset X: D$ discrete$\}$. By a *Δ-filter* on X is meant a nonempty subset δ of Δ such that the following conditions hold.

(1.a) $\varnothing \notin \delta$.

(1.b) U and V in δ implies $U \cap V \in \delta$.

(1.c) $U \in \delta$ and $V \in \Delta$ such that $U \subset V$ implies $V \in \delta$.

(If the set of all subsets of X plays the role of Δ above, δ is called a *filter on X*.)

THEOREM 1.4 (HENRIKSEN [$\mathbf{H}_3$]). *If I is a proper ideal in A, then $Z(I)$ is a Δ-filter. Conversely, if δ is a Δ-filter, then $Z^{-1}(\delta)$ is a proper ideal in A. Further, $I \subset Z^{-1}Z(I)$, and equality need not hold. (If equality does occur, I is called a* Z-ideal.) *Thus,* specm A, *the set of all maximal ideals of A, is in bijective correspondence under Z with δX, the set of all* Δ-ultrafilters (= *maximal Δ filters) on X.*

Having shown that maximal ideals in A correspond naturally with Δ-ultrafilters, let us consider two examples of such objects.

EXAMPLE 1.1. Let $x \in X$ and let $\delta_x \equiv \{D \in \Delta : x \in D\}$. Clearly $\delta_x \in \delta X$. Further the map $x \to \delta_x$ is an injection of X into δX. Let $M_x \equiv Z^{-1}(\delta_x) \in$ specm A. Note that $0 \to M_x \to A \to C \to 0$ is an exact sequence of C-algebras and C-homomorphisms, which in future we will express by saying that the sequence is *C-exact*. (Thus M_x is injected into A by means of the identity map, and for all $f \in A$ there exists a unique $\lambda (= f(x))$ such that $f - \lambda \in M_x$.) Note that $\bigcap_{D \in \delta} D = \{x\}$.

EXAMPLE 1.2. Let $D \equiv \{x_n : n \in N\}$ be an infinite discrete subset of X. Given $m \in N$, let $D_m \equiv \{x_n : n \in N$ and $n \geqq m\}$. $(D_m)_{m \in N}$ satisfies (1.a) and (1.b), and as a consequence is called a *Δ-filter base*. Note that $\bigcap_{m \in N} D_m = \varnothing$. By Zorn's Lemma there exists an ultrafilter μ on D that contains $(D_m)_{m \in N}$. (μ is by no means unique.) It is easily seen that an ultrafilter μ on D is a filter on D such that for all $D' \subset D$, either $D' \in \mu$ or $D - D' \in \mu$. Furthermore, any such filter is an ultrafilter on D. Let $\delta \equiv \{E \in \Delta : E \cap D \in \mu\}$. It is easy to see that $\delta \in \delta X$. Since $\bigcap_{E \in \delta} E \subset \bigcap_{m \in N} D_m$, which is empty, $\bigcap_{E \in \delta} E = \varnothing$.

Let us now return to the general case of an element $\delta \in \delta X$. It will be called *fixed* or *free* in case $\bigcap_{D \in \delta} D$ is nonempty or is empty. The first example, δ_x, is fixed and the second is free. Let δ be fixed. Since δ is a Δ-*ultra*filter, $\bigcap_{D \in \delta} D$ cannot contain two distinct points; thus it is of the form $\{x\}$, and $\delta = \delta_x$. Assume now that δ is free and let D be a discrete subset of δ. Let $\mu \equiv \{E \cap D : E \in \delta\}$. μ is easily seen to be an ultrafilter on D and $\delta = \{E \in \Delta : E \cap D \in \mu\}$. Thus we see that Examples 1.1 and 1.2 generate all Δ-ultrafilters on X.

$M \equiv Z^{-1}(\delta)$ will be called *fixed* or *free* in case δ is fixed or free. Let $M_x \equiv Z^{-1}(\delta_x)$. This ideal is, of course, $\{f \in A : f(x) = 0\}$. Now let δ be free. M is easily seen to be $\{f \in A$: there exists $D \in \delta$ such that $f|D = 0\}$. δ is a directed set, directed down, and thus we can think of convergence occurring along δ.

In this language, M consists of all $f \in A$ that are eventually identically zero on δ.

Let $x \in X$, and note that due to the analyticity of the elements of A at x, M_x is principal, being generated by an element of A with only one zero on X and that zero of order one at x. Now let δ be free. If $M(= Z^{-1}(\delta))$ is not principal, then by Helmer's Lemma (1.2), M is not finitely generated. Since, as we remarked earlier, A is not Noetherian this is possible. Let $f \in M$ and let $Z(f) \equiv D$ be partitioned into nonempty sets D_1 and D_2. Since δ is a Δ-*ultra*filter either D_1 or D_2 is in δ; assume that $D_1 \in \delta$. Let $f_1 \in A$ such that $Z(f_1) = D_1$. Then $f_1 \in M$ and yet $f_1 \notin (f)$, which shows that for all $f \in M$, (f) is a proper subset of M. Hence M is infinitely generated. Thus we have proved the following.

PROPOSITION 1.5 (HENRIKSEN [$\mathbf{H_3}$]). *Let M be a maximal ideal in A. M is fixed if and only if it is principal.*

When the notion of a fixed ideal was introduced it was not *a priori* a concept that depended only on the ring structure of A. Now we find that the notion is an algebraic one. Following Bers, a sketch of a proof of (1.1) will be given.

PROOF OF (1.1). Let ϕ be a C-isomorphism of $A(Y)$ onto $A(X)$. ϕ^*, the adjoint of ϕ, is an injection of specm $A(X)$ onto specm $A(Y)$. Clearly M is principal if and only if $\phi(M)$ is principal; thus by (1.5), ϕ^* maps the fixed maximal ideals of $A(X)$ onto the fixed maximal ideals of $A(Y)$: i.e., ϕ^* may be regarded as an injection θ of X onto Y. The weak topology on X, making all $f \in A(X)$ continuous, is just the original topology on X, and θ is clearly a homeomorphism relative to these weak topologies. The conformal structure on X can be recovered using generators of the principal maximal ideals to serve as local uniformizers. Since ϕ takes generators to generators, θ is easily seen to be conformal, proving the theorem.

As will be seen in the next section there is another way of distinguishing between fixed and free ideals on algebraic grounds.

2. **Residue class fields.** Let M be a maximal ideal in A. Recall that if M is fixed, $M = M_x$ for some $x \in X$, and

$$0 \to M_x \to A \xrightarrow{x} C \to 0$$

is C-exact, where $v_x(f) = f(x)$ for all $f \in A$: i. e., given $f \in A$, there exists a unique $\lambda \in C$ such that $f - \lambda \in M_x$. What, then, is the situation if $M(= Z^{-1}(\delta))$ is free? Let v and K be defined so that

$$0 \to M \to A \xrightarrow{v} K \to 0$$

is C-exact, and let C denote the image in K of the constant maps. There is another way of representing K that is very useful. Let D be a discrete element in δ and let $\mu \equiv \{E \cap D : E \in \delta\}$. As remarked earlier, μ is an ultrafilter on D. Let ρ map

A into C^D, the set of all maps of D into C as follows: for $f \in A$ let $\rho(f) = f|D$, the restriction of f to D. By the generalized Weierstrass and Mittag-Leffler Theorems, A maps onto C^D. Let $M' \equiv \{f \in C^D$: there exists $U \in \mu$ such that $f \mid U = 0\}$. It is easy to see that M' is an ideal in C^D. Let ξ be the canonical homomorphism of C^D onto $L \equiv C^D/M'$. Since the kernel of $\xi\rho$ is M, K and L are naturally C-isomorphic. Let C denote the image of the constant functions in L. The image of $f \in C^D$ in L is λ in C if and only if there exists $U \in \mu$ such that $f|U = \lambda$; thus it is evident that $K \neq C$. In fact K is a very large extension of C. However the following holds.

THEOREM 2.1 ((HENRIKSEN [$\mathbf{H}_3$]). *K is an algebraically closed proper extension of C.*

Since K and L are C-isomorphic, it suffices to prove these statements for L. Let $f \in C^D$ have a different value at each point of D; then $\xi(f) \notin C$, proving the second, assertion. To prove the first let $p(t) \equiv t^n + a_1 t^{n-1} + \cdots + a_n \in L[t]$, and let f_i be a preimage of a_i in C^D. Given $x \in D$, let $p_x(t) \equiv t^n + f_1(x)t^{n-1} + \cdots + f_n(x) \in C[t]$. Since C is algebraically closed, $p_x(t)$ has a root $f(x) \in C$. Clearly $f \in C^D$ and $p(\xi(f)) = 0$, proving the theorem.

COROLLARY 2.2 (HENRIKSEN [$\mathbf{H}_3$]). *Let M be a maximal ideal in A. The image of the constant functions in A maps onto A/M if and only if M is fixed.*

(2.2) can be used to give a variant of Bers' proof of (1.1), given in §1.

As mentioned before, $f \in M$ if and only if f is eventually zero on δ: i.e., there exists $D \in \delta$ such that $f|D = 0$. From this, one immediately sees that $\nu(f) = \lambda \in C$ if and only if f is eventually λ on δ. Clearly this is a trivial kind of convergence along the directed set δ. Let us now investigate a deeper form of convergence along δ. Let $\mathfrak{D}' \equiv \{f \in A$: there exists $D \in \delta$ such that $f|D$ is bounded$\}$, and let $f \in \mathfrak{D}'$. Since δ is a Δ-*ultra*filter, $\lim_{D \in \delta} f(D)$ always exists in C: i.e., there exists a unique $\lambda \in C$ such that for all $\varepsilon > 0$ there exists $E \in \delta$ such that for all $x \in E$, $|f(x) - \lambda| < \varepsilon$. This may be seen as follows: let $f \in \mathfrak{D}'$. By definition there exists a discrete set $D \in \delta$ such that $f|D$ is bounded. Let n be an integer such that $|f(x)| < n$ for all $x \in D$. Let the disc of radius n be partitioned into subsets $U_1, \cdots, U_m$ each of which has diameter less than ε. Let $D_j \equiv f^{-1}(U_j) \cap D$, for $j = 1, \cdots, m$, and notice that $D_1, \cdots, D_m$ partition D. Since δ is a Δ-ultrafilter, one and only one D_j is in δ, showing that the limit does indeed exist. The following more sophisticated argument shows the same thing. $\mu \equiv \{E \cap D : E \in \delta\}$ may be identified with a point of the Stone-Čech compactification of D; thus each bounded function on D converges along μ, and thus along δ. In this setting it is natural to regard $f \in C^D$ as a mapping of D into the Riemann sphere Σ. Since the latter is compact, $\lim_{U \in \mu} f(U)$ always exists in Σ. Let it be denoted by $\alpha'(f)$. Note: the direct argument above can be used to show that if $f \notin \mathfrak{D}'$, then $\alpha'(f)$ exists and is ∞; thus $\mathfrak{O}'$ is the preimage under α' of C. Clearly $\alpha' \mid \mathfrak{O}'$ is a C-homorphism onto C whose kernel $\mathfrak{M}'$ equals the set of all $f \in A$ that converge to zero along δ. All this is by way of motivating the following: let $\mathfrak{O}$ be the image of $\mathfrak{O}'$ in $K(\equiv A/M)$. Since $M \subset \mathfrak{M}'$, α' induces a C-homomorphism α of $\mathfrak{O}$ onto C.

PROPOSITION 2.3 [A_3]. *$\mathfrak{O}$ is a valuation ring of K: i.e., let $a \in K - \mathfrak{O}$, then $1/a \in \mathfrak{O}$.*

PROOF. Let $f \in A$ such that $v(f) = a$. Since $a \neq 0$ there exists $D_0 \in \delta$ such that $f \mid D_0$ does not assume zero. There exists $g \in A$, without zeros, such that $g \mid D_0 = f \mid D_0$. Then of course $v(g) = a$, $h \equiv 1/g \in A$, and $v(h) = 1/a$. Let n be an integer. Since $f \notin \mathfrak{O}'$ there exists $D_n \in \delta$ such that $|f \mid D_n| \geqq n$. Clearly $|h \mid D_0 \cap D_n| \leqq 1/n$ for each n, and $D_0 \cap D_n \in \delta$, completing the argument.

Before deriving some corollaries from (2.3), let me recall the essentials of valuation theory and the process of localization for you, in case they are not readily at hand for you.

3. Valuation and localization theories recalled. Let K/k be a field extension. Throughout this section two parallel sets of definitions will be given, one of which will be treated parenthetically. A proper subring $\mathfrak{O}$ of K is called a *valuation ring of* K (of K/k) if $a \in K - \mathfrak{O}$ implies $1/a \in \mathfrak{O}$ (and $k \subset \mathfrak{O}$). Let $U(\mathfrak{O})$ denote the group of units of $\mathfrak{O}$. $\mathfrak{M} \equiv \mathfrak{O} - U(\mathfrak{O})$ turns out to be the maximal ideal of $\mathfrak{O}$; thus $\mathfrak{O}$ is a *local ring*. Let α be the canonical homomorphism of $\mathfrak{O}$ onto $\mathfrak{O}/\mathfrak{M} \equiv \Omega$, the *residue class field of* $\mathfrak{O}$. (α is a k-homomorphism and Ω, an extension of k.) Let α extend to K by mapping $K - \mathfrak{O}$ to ∞. α is called a *place of K (over k) onto* Ω. The following example is important, and may prove useful to the reader.

EXAMPLE 3.1. Let F be the field of meromorphic functions on X, let $x \in X$, and let $\mathfrak{O}_x \equiv \{f \in F : f \text{ is regular at } x\}$. $\mathfrak{O}_x$ is then a valuation ring of F over C onto C, and $\alpha(f) = f(x)$ for all $f \in \mathfrak{O}_x$. The maximal ideal $\mathfrak{M}_x$ of $\mathfrak{O}_x$ is $\{f \in \mathfrak{O}_x : f(x) = 0\}$.

REMARK. This example suggests that a place corresponds to the evaluation of functions at a certain place on a "surface". The next notion, that of a valuation, allows one to analyze the ideal theory of a valuation ring.

Let K^* denote the multiplicative group of nonzero elements of K. Since the group of units of $\mathfrak{O}$, $U(\mathfrak{O})$, is a subgroup of K^* we may form the factor group G. Let V be the canonical homomorphism and let G, an Abelian group, be written additively. Thus the following sequence of groups and group homomorphisms is exact:

$$1 \to U(\mathfrak{O}) \to K^* \overset{V}{\to} G \to 0.$$

According to the characteristic property of valuation rings, $a \in \mathfrak{O}$ or $1/a \in \mathfrak{O}$; both occurring if and only if $a \in U(\mathfrak{O})$. Let $G(\geqq 0) \equiv V(\mathfrak{O} \cap K^*)$. Since $\mathfrak{O}$ is a ring, $G(\geqq 0)$ is closed under addition. Using the characteristic property of a valuation, in the form in which it appears above, letting $G(\geqq 0)$ play the role of the set of nonnegative elements of G defines the structure of a totally (= linearly) ordered group on G. Let V extend to K by taking 0 to ∞, an ideal element greater than all $g \in G$. Let $\infty + g = \infty$ for all $g \in G$ and let $\infty + \infty = \infty$. V satisfies the following *strong triangle law*: $V(a \pm b) \geqq \min V(a), V(b)$ for all $a, b \in K$. From this it follows that if $V(a) \neq V(b)$ equality holds in this triangle law. V is called a *valuation of* K (*over* k), with *value group* G.

The value group in Example 3.1 is the additive group Z of integers, a most important value group in analysis, geometry, and number theory.

Valuation rings are intrinsic (just as are maximal ideals in rings) and give rise to places and valuations, which are uniquely determined up to isomorphism. Conversely, places and valuations give rise to unique valuation rings. Valuation rings, places, and valuations, each emphasize one particular aspect of the situation, and each has its special uses.

Let V be the valuation of K (over k) associated with $\mathfrak{O}$ and let G be its value group. Given an ideal I in $\mathfrak{O}$, $i \in I$, and $g \in G$, such that $V(i) \leqq g$, then there exists $j \in \mathfrak{O}$ such that $V(j) = g$. Hence $V(j/i) \geqq 0$, showing that $j/i \in \mathfrak{O}$. Since I is an ideal, $j(=(j/i)i) \in I$, proving that $V(I)$ is a nonvoid upper interval in $\bar{G}(\geqq 0) \equiv \{g \in G \cup \{\infty\} : g \geqq 0\}$. Conversely, given such an upper interval G' in $\bar{G}(\geqq 0)$, let $J \equiv V^{-1}(G')$. Then it is easy to see that J is an ideal. Further, $I = V^{-1}V(I)$; thus the correspondence is a bijection. Clearly this set of upper intervals of $\bar{G}(\geqq 0)$ is totally ordered under inclusion; thus the set of all ideals of $\mathfrak{O}$ is similarly ordered. Other pleasant correspondences exist. For example, $I = (f)$, for some $f \in \mathfrak{O}$, if and only if $V(f)$ is the least element of $V(I)$. Hence I is principal if and only if $V(I)$ has a least element. Since the value group occurring in Example 3.1 is Z, all upper intervals of $\bar{G}(\geqq 0)$ have least elements, giving another proof that $\mathfrak{O}_x$ is a principal ideal ring.

Recall that a proper ideal I is prime if $\mathfrak{O} - I$ is closed under multiplication, or equivalently if $\mathfrak{O}/I$ is an integral domain. Thus I is prime if and only if $G(\geqq 0) - V(I)$ is nonempty and closed under addition (where $G(\geqq 0) \equiv \{g \in G : g \geqq 0\}$). This can be made even more useful with the aid of the following definition: let G' be an upper interval in $\bar{G}(\geqq 0)$ and let $G(G') \equiv \{g \in G : |g| < G'\}$. Clearly a proper ideal I of $\mathfrak{O}$ is prime if and only if $G(V(I))$ is a convex subgroup of G. Since the only convex subgroups of Z are $\{0\}$ and Z, $\mathfrak{O}_x$ has two prime ideals, namely $\mathfrak{M}_x$ and $\{0\}$. spec associates with a ring the set of all proper prime ideals of it. As noted above, spec $\mathfrak{O}$ is totally ordered under inclusion. $P \to G(V(P))$ is an order reversing mapping of spec $\mathfrak{O}$ into Γ, which we will define to be the set of all convex subgroups of G, ordered by inclusion. The totally ordered set Γ is known as the *rank of* G. If $\Gamma - \{0\}$ is finite, its cardinal number n is an important invariant of G; in this case G is said to be of *rank* n. For example, rank one groups are merely Archimedean, and as such can be mapped isomorphically onto a subgroup of the reals.

For some totally ordered groups G, the principal rank is of particular importance. Let $g \in G$ and let $W(g)$ be the smallest convex subgroup of G that contains g. Such groups are called *principal convex subgroups*. Then $W(g \pm h) \leqq W(g) \cup W(h)$ for all g and $h \in G$; thus W possesses the additive properties of a valuation, where the order on the "values" is here reversed. Let $\Pi \equiv \{W(g): g \in G, g \neq 0\}$ be called the *principal rank of* G. If G is of finite rank $\Pi = \Gamma - \{0\}$. Since every convex subgroup of G is the union of principal convex subgroups, Γ is order isomorphic to the set of all lower intervals of Π; this is a good deal like the Dedekind completion of Π. It is easy to construct examples for which $\Pi \neq \Gamma - \{0\}$.

Localization occurs in the following context. For the moment let A be an integral domain, let F be its quotient field, and let M be a maximal ideal in A; then $A - M$ is closed under multiplication. (It is this last fact that is essential. For example, let P be a prime ideal. Then $A - P$ is also closed under multiplication.) Let $A_M \equiv \{a/b \in F: a \in A \text{ and } b \in A - M\}$. One can easily check to see that A_M is indeed a ring. Let M^e, the extension of M to A_M, be MA_M ($\equiv \{m_1 f_1 + \cdots + m_n f_n : m_i \in M,\ f_i \in A_M, \text{ and } n \text{ an integer}\}$). Then M^e is *the* maximal ideal of A_M and hence A_M is a *local ring*. A_M is said to have been obtained from A be means of *localization of* A *at* M. Let A again be $A(X)$, the ring of analytic functions on X. Referring to the notation of Example 3.1, $A_{M_x} = \mathfrak{O}_x$. Two inferences can be drawn. First, A_M can be a valuation ring. Secondly, in the example above, localization is related to local behaviour of functions at x, in the sense of analysis.

EXAMPLE 3.2. Let $A = C[x, y]$, the ring of polynomials in two variables over the complex field. Let $M \equiv \{f(x, y) \in A: f(0, 0) = 0\}$. M is a maximal ideal and A_M is, of course, a local ring. Since neither x/y nor y/x is in A_M, it is not a valuation ring, however.

4. **The valuation theory of** $F(X)$. Having recalled the notions of valuation theory (in §3), let us first compute the value group of $\mathfrak{O} \subset K$ ($\equiv A/M$, M free) defined in §2. Recall that K is C-isomorphic to $L \equiv C^D/M'$, where D is an infinite discrete subset of X and $M' = \{f \in C^D$: there exists $U \in \mu$ such that $f | U = 0\}$, μ being a free ultrafilter on D derived as follows: let $Z(M) = \delta$ and let $\mu \equiv \{E \cap D$: $E \in \delta\}$. Under the natural isomorphism between K and L, the image of $\mathfrak{O}$ in L is $\{\phi(f) \in L$: there exists $f \in C^D$ and there exists $U \in \mu$ such that $f | U$ is bounded$\}$. Let $\mathfrak{O}$ also denote this valuation ring in L. Thus it suffices to compute the value group of this valuation ring.

This can be done relatively easily by considering R^D, the ring of all maps of D into R and let L_R denote its image in L. Clearly L_R is an extension of R,

$L_R(i) = L$, and $\mathfrak{O}_R \equiv \mathfrak{O} \cap L_R$ is a valuation ring of L_R. Let L_R be ordered as follows: clearly under pointwise ordering R^D is a lattice ordered ring. Let the image of $R^D(\geqq 0)$ $(\equiv \{f \in R^D : f \geqq 0\})$ be $L_R(\geqq 0)$. From this one sees that $L_R(\geqq 0)$ is closed under addition, multiplication, and multiplication by nonnegative real scalars; thus L_R becomes a partially ordered field. The remarkable thing is that L_R is actually totally (= linearly) ordered. This is however just a reflection of the fact that μ is an ultrafilter; for let $a \in L_R$ and let f be a preimage in R^D. Let $U \equiv \{x \in D : f(x) \geqq 0\}$ and let $V \equiv \{x \in D : f(x) < 0\}$. Obviously $U \cup V = D$ and $U \cap V = \varnothing$. Since μ is an ultrafilter on D, either $U \in \mu$ or $V \in \mu$. Assume for the moment that $U \in \mu$. Let $g \in R^D$ such that $g|U = f|U$ and $g|V = 0$. Then $g \geqq 0$ and the image of g in L_R is a, showing that $a \geqq 0$. If $V \in \mu$, let $h \in R^D$ such that $h|V = -f|V$ and $h|U = 0$; then the image of h is $-a$ and $-a \geqq 0$, proving that L_R is indeed a totally ordered field. It is not difficult to show that given a countable set in L_R, a set of preimages can be found in R^D which are order isomorphic to the original set. (One proceeds by induction and uses the lattice structure on R^D. See e.g., [**GJ**, 13.5].) Using this, it then becomes evident that given countable subsets S_1 and S_2 in L_R such that $S_1 < S_2$ (i.e., for all $s_1 \in S_1$ and $s_2 \in S_2, s_1 < s_2$), there exists $s \in L_R$ such that $S_1 < \{s\} < S_2$: i.e., L_R is an η_1-*set* (See, e.g., [**GJ**, 13.6] for details). These sets were introduced by Hausdorff in his Gründzuge in 1914 [$\mathbf{H_1}$]. Returning to valuation theory, note that $\mathfrak{O}_R = \{a \in L_R$: there exists an integer n such that $|a| \leqq n\}$; thus $\mathfrak{O}_R$ is related to the order on L_R. L_R is a totally ordered group. As such W (defined in §3) maps L_R^* onto Π, the principal rank of L_R. Let Π be given an addition as follows: let $W(a) + W(b) = W(ab)$, for all a, $b \in L_R$. Then Π is a value group which is naturally antiorder-isomorphic to G. It is straightforward to verify that if a totally ordered group is an η_1-set, then so is its principal rank [$\mathbf{A_2}$]. Hence, G and its principal rank Π are η_1-sets.

THEOREM 4.1 [$\mathbf{A_3}$]. *Let M be a free maximal ideal in A, let $K \equiv A/M$, and let $\mathfrak{O}$ be the image in K of $\{f \in A$: there exists $D \in \delta$ $(\equiv Z(M))$ such that $f|D$ is bounded$\}$. Then $\mathfrak{O}$ is a valuation ring whose value group G is divisible and is an η_1-set; thus the transcendence degree of K over C is $2^{\aleph_0}$.*

We have shown that $\mathfrak{O}$ is a valuation ring and noted that its value group is an η_1-set. Since K is algebraically closed (2.1), K^* is certainly divisible. Since G is a homomorphic image of K^*, it too is divisible. Hence G is a vector space over the field Q of rational numbers. Since G is of power at least $2^{\aleph_0}$ [$\mathbf{H_1}$, **GJ**] its Q-dimension is at least also the power of the continuum. Using the triangle law it is easy to see that elements in K whose values are linearly independent over Q are algebraically independent over C. Thus the transcendence degree of K over C is at least the power of the continuum. Since K has only a continuum of elements, this transcendence degree is bounded above by the power of the continuum, proving the theorem.

The following theorem has great impact on our state of knowledge of the ideal theory of A (see §5).

THEOREM 4.2 [$\mathbf{A}_3$]. *Let M be a maximal ideal of A. Then the localized ring A_M is a valuation ring whose value group is isomorphic to the integers if and only if M is fixed. Assume that M is free; then the value group H of A_M is a nondivisible group with a least positive element, which is a near η_1-set.*

Before sketching the proof let us fill in the missing definition. H is said to be a *near η_1-set* if given countable sets H_1 and H_2 of H such that $H_1 \leqq H_2$ (i.e. for all $h_i \in H_i$, $h_1 \leqq h_2$), there exists $h \in H$ such that $H_1 \leqq \{h\} \leqq H_2$. Clearly an η_1-set is a near η_1-set. H is an example of a near η_1-set which is not an η_1-set.

PROOF. Let $f \in F - A_M$. Using the general Weierstrass Theorem we can find $a, b \in A$ such that the following conditions hold: $f = a/b$, and a and b have no common zeros (i.e., $(a, b) = (1)$). Since $f \notin A_M$, $b \in M$; thus $Z(b) \in \delta \equiv Z(M)$. Since $(a, b) = (1)$, $Z(a) \cap Z(b) = \varnothing$ ((1.3) (Helmer)). As a consequence $Z(a) \notin \delta$, $a \notin M$, and hence $1/f = b/a \in A_M$, proving that A_M is a valuation ring. (Note: this proof, which differs from the one given in [$\mathbf{A}_3$], is a generalization of the proof that if A is a unique factorization domain and m is an irreducible element in A then $A_{(m)}$ is a valuation ring.)

Clearly if M is fixed ($= M_x$ for some $x \in X$), $A_M = \mathfrak{O}_x$ (Example 3.1), showing that its value group is isomorphic to the integers. Assume now that M is free. Let $\delta \equiv Z(M)$ and let D be a discrete element in δ. Given $f \in K^*$, let $\mathfrak{d}(f) \in Z^X$ be defined as follows: $\mathfrak{d}(f)(x) = V_x(f)$, where V_x is the valuation of $\mathfrak{O}_x$. Clearly $\mathfrak{d}$ is a homomorphism of K^*, the multiplicative group of nonzero elements of K, into Z^X whose kernel is $U(A)$ the group of units of A. Let r be the restriction map of Z^X onto Z^D; $r\mathfrak{d}$ is surjective. Let $\mu = \{E \cap D : E \in \delta\}$, and recall that μ is an ultrafilter on D. Let $N \equiv \{c \in Z^D$: there exists $U \in \mu$ such that $c \mid U = 0\}$. Clearly N is a subgroup of Z^D; let s be the conical homomorphism of Z^D onto $Z^D/N \equiv H$, and let the positive elements of Z^D, under the pointwise order, give H a partial ordering. It is easy to check that the kernel of $V \equiv sr\mathfrak{d}$ is $U(A_M)$ and that the image of $A_M - \{0\}$ is $H(\geqq 0)$. Thus H is a value group A_M. The properties cited for H can now virtually be read off, using the fact that $H = Z^D/N$, and that μ is an ultrafilter, completing our sketch of the proof.

Before proceeding to the corollary to (4.2), let us find an element $v(f)$ in K^* such that $V(f)$ will be the least positive element of H. Let $v(f) \in K^*$ be such that $V_x(f) = 1$ for all $x \in D$. Then f has this property.

COROLLARY 4.3 [$\mathbf{A}_3$]. *The set of all proper ideals of A_M is totally ordered by inclusion, its order type being the same as the set of proper upper intervals of $\bar{H}(\geqq 0)$. The set of all prime ideals of A_M,* spec A_M, *is totally ordered by inclusion, its order type being order anti-isomorphic to the rank of H. The* *principal rank of H is an η_1-set to which a smallest element has been adjoined.*

Note: an η_1-set is necessarily of power at least $2^{\aleph_0}$. Hausdorff [$\mathbf{H}_1$] showed that two η_1-sets of power $\aleph_1$ are isomorphic. A similar result holds for divisible Abelian groups that are η_1-sets of power $\aleph_1$ [$\mathbf{A}_1$]. The existence of such an object of course implies the continuum hypothesis, an hypothesis which seems daily less plausible as a "natural" axiom in set theory.

5. **Application of valuation theory to the ideal theory of A.** Since A_M is a valuation ring (4.2) its ideals are totally ordered by inclusion (4.3). Further we have a grip on the order type of this set and of spec A_M, the set of all prime ideals of A_M (4.3). Now we want to study proper ideals I in A. Let $I_M \equiv IA_M (\equiv \{i_1a_1 + \cdots + i_na_n; i_j \in I, a_j \in A_M$, and n an integer$\})$; this is an ideal in A_M. Given an ideal J of A_M let J^c (the c is for "contracted") be $J \cap A$. This is an ideal of A. $P \to P^c$ is a one-to-one mapping of spec A_M onto $\{Q \in \operatorname{spec} A; Q \subset M\}$ and $Q \to Q_M$ is its inverse; thus we have the following description of the prime ideals of A.

THEOREM 5.1 [A_3]. spec A, *the set of all prime ideals of* A, *is* $\bigcup_{M \in \operatorname{specm} A} (\operatorname{spec} A_M)^c$, *the sets sans* $\{0\}$ *making up the union being disjoint.* $(\operatorname{spec} A_M)^c$ *consists of two points if M is fixed, and if M is free it is a totally ordered set described in* (4.3).

The object of any ideal theory is to give a complete description of any ideal, in an informative way. For example in the ring of integers the ideal theoretic version of the fundamental theorem of arithmetic provides such a description Let us now introduce a bit of algebraic-geometric language. Let I be an ideal in A and let $\mathfrak{v}(I) \equiv \{M \in \operatorname{specm} A : I \subset M\}$, and think of this as the *variety of I*. Let specm $A \equiv \mathfrak{X}$ be topologized by making $\{\mathfrak{v}(I)$: I an ideal in $A\}$ the closed sets of $\mathfrak{X}$. One can readily check that $\mathfrak{X}$ is a compact T_1-space. The following is well known.

LEMMA 5.2. *Let I be a proper ideal in A; then* $I = \bigcap_{M \in \mathfrak{v}(I)} (I_M)^c$.

PROOF. Let $J \equiv \bigcap_{M \in \mathfrak{v}(I)} (I_M)^c$. Clearly $I \subset J$. To prove the other inequality let $f \in A - I$, and let $B = \{b \in A : bf \in I\}$. B is a proper ideal of A which contains I. By Zorn's Lemma, it is contained in a maximal ideal $M \in \mathfrak{v}(I)$. Clearly $f \notin I_M$, so $f \notin J$. This shows that $A - I \subset A - J$, and hence that $J \subset I$, proving the lemma.

LEMMA 5.3. *Let* $M_1, \cdots, M_n$ *be distinct points of* specm A, *and let* $G_1, \cdots, G_n$ *be the value groups of the valuation rings* $A_{M_1}, \cdots, A_{M_n}$, *the valuations being denoted by* $V_1, \cdots, V_n$ *respectively. Let* $g_i \in G_i$, $g_i \geqq 0$, *be given for* $i = 1, \cdots, n$. *There exists* $a \in A$ *such that* $V_i(a) = g_i$, *for all* $i = 1, \cdots, n$.

This is of course a good deal like the theorem on independence for algebraic curves.

PROOF. Let $\delta_i \equiv Z(M_i)$, $i = 1, \cdots, n$. Since the M's are distinct so are the δ's (1.4). For each pair (i,j), $i \neq j$, there exists a discrete subset D_{ij} of X such that it is in δ_i but not δ_j. Let $D_i \equiv \bigcap_{j \neq i} D_{ij}$. (Here it is understood that j is a free variable and takes on all values in $\{1, \cdots, n\} - \{i\}$, and that i is fixed.) Since each such $D_{ij} \in \delta_i$, $D_i \in \delta_i$. For $j \neq i$, $D_i \in \delta_j$ implies $D_{ij} \in \delta_j$, which is a contradiction, showing that $D_i \notin \delta_j$. Note that $E_i \equiv D_i - \bigcup_{j \neq i} D_j \in \Delta$. Suppose for a moment that $E_i \notin \delta_i$. Since $D_i \in \delta_i$, $D_i - E_i \in \Delta$, since $\{D \cap D_i : D \in \delta_i\}$ is an ultrafilter on D_i, and since $E_i \notin \delta_i$; $D_i - E_i \in \delta_i$. Clearly $D_i - E_i = \bigcup_{j \neq i}(D_j \cap D_i)$. Since δ_i is a Δ-*ultra*filter, at least one of $D_j \cap D_i \in \delta_i$, $j \neq i$. Since $D_j \cap D_i \subset D_j$ and δ_i is a Δ-filter, $D_j \in \delta_i$, which is a contradiction. What we have achieved then is the existence of $E_i \in \delta_i$ for $i = 1, \cdots, n$ such that $E_i \cap E_j = \emptyset$ if $i \neq j$. Let $a_i \in A$ such that $V(a_i) = g_i$, $i = 1, \cdots, n$. Let $a \in A$ such that the zeros of a on E_i agree with those of a_i on E_i. Then a has the requisite properties, proving the lemma.

The generalization of the Weierstrass (product) theorem to X may be viewed as a generalization of (5.3) to the infinite case in which each M_n is fixed such that $\{M_n\}_{n \in N}$ is a discrete subset of X (using the usual identification). That (5.3) does not admit an unrestricted generalization to the countably infinite case can be seen in view of the following example.

EXAMPLE 5.1. Let M be a free maximal ideal and let D be a discrete subset of X in δ. Let $D = (x_n)_{n \in N}$. Let $M_n \equiv M_{x_n}$, let $g_n \equiv 0$ for all n, and let $g \in G$, the value group of A_M, be greater than zero. Let $a \in A$ such that $V_{x_n}(a) = 0$, then $V(a) = 0$, V being the valuation of A_M.

Of course the fact that M is in the closure of $\{M_{x_n}\}$ in specm A is cardinal in the example above.

LEMMA 5.4. *Let I be an ideal in A that is contained in a maximal ideal M, and let $J \equiv (I_M)^c$. Then $J = \{a \in A: V(a) \in V(I)\}$ and J is contained in a unique maximal ideal, namely M, where V is the valuation of A_M. Further, if I is contained in a unique maximal ideal M, then $I = J$.*

PROOF. Since A_M is a valuation ring $I_M = \{a \in A_M: V(a) \in V(I)\}$ [**ZS**$_2$, p. 40]. Since $V(I) = V(I_M)$, the equality holds. Clearly $J \subset M$. Let N be another element in specm A. By (5.3) there exists $a \in J$ which is not in N, showing that M is the unique maximal ideal that contains J. Finally if I is contained in a unique maximal ideal, M, then $\mathfrak{v}(I)$ (the variety of I) is $\{M\}$. By (5.2) $I = (I_M)^c$, proving the lemma.

This motivates the following definition: an ideal J of A is called *primary* (*for* M) if J is contained in a unique maximal ideal (M).

REMARK. This definition is not the same as the classical notion of a primary ideal, and yet it seems to share enough of the properties of classical primary ideals in this context to deserve the name primary. This notion has also proved useful in certain Banach algebras of analytic functions.

Let $(I_M)^c$ be called the *primary component of I*, for all $M \in \mathfrak{v}(I)$. By a *primary decomposition for I* will be meant a set of primary ideals $(J_t)_{t \in T}$ in A such that $I = \bigcap_{t \in T} J_t$. From (5.2) and (5.4) we deduce the following.

THEOREM 5.5. *Let I be a proper ideal of A. There exists a primary decomposition for I, namely* $((I_M)^c)_{M \in \mathfrak{v}(I)}$.

The notion of the radical of an ideal is an important one. Recall that if I is an ideal then its *radical*, $I^{1/2}$, is defined to be $\{a \in A$: there exists an integer n for which $a^n \in I\}$. It is easy to see that $I^{1/2}$ is again an ideal (using the binomial theorem), and that $I \subset I^{1/2}$. Using Zorn's Lemma it is easy to see that $I^{1/2}$ is the intersection of the prime ideals that contain it. Let J be primary for M. By (5.2) and (5.4), $J = \{a \in A: V(a) \in V(J)\}$. Recall (§3) that $G(V(J))$ is defined to be $\{g \in G: |g| < V(J)\}$, and that J is prime if and only if the symmetric interval $G(V(J))$ is a subgroup of G (the value group of A_M). Clearly $V(I^{1/2}) - \{\infty\}$ $= \{g \in G: g > 0$ and there exists n such that $ng \in V(I)\}$. From this one easily sees that $G(V(I^{1/2}))$ is a convex subgroup of G; thus $((I_M)^c)^{1/2}$ is prime. One can also see that $G(V(I^{1/2}))$ is the largest convex subgtroup of G contained in $G(V(I))$. An ideal I is said to admit a *prime decomposition* if there exists a set of prime ideals $(P_t)_{t \in T}$ of A such that $I = \bigcap_{t \in T} P_t$.

THEOREM 5.6. *Let I be a proper ideal in A.* $I^{1/2}$ *admits a prime decomposition, namely* $I^{1/2} = \bigcap_{M \in \mathfrak{v}(I)} ((I_M)^c)^{1/2}$.

The equality is the result of a standard compacteness argument, which will be given for the sake of completeness, and the fact that $((I_M)^c)^{1/2}$ is prime, as was noted above.

PROOF. Let $L \equiv \bigcap_{M \in \mathfrak{v}(I)} ((I_M)^c)^{1/2}$. Given $f \in I^{1/2}$ there exists, by definition, an integer n such that $f^n \in I$. By (5.5), $f^n \in (I_M)^c$ for all $M \in \mathfrak{v}(I)$, showing that $f \in ((I_M)^c)^{1/2}$ for all such M and hence $f \in L$. Conversely, if $M \in \mathfrak{v} - \mathfrak{v}(I)$, $((I_M)^c)^{1/2} = A$, showing that $L = \bigcap_{M \in \mathfrak{x}} ((I_M)^c)^{1/2}$. Let $f \in L$. Given $M \in \mathfrak{v}$ ($\equiv$ specm A), there exists an integer n such that $f^n \in (I_M)^c$: i.e., there exists $a \in I$ and $b \in A - M$ such that $f^n = a/b$. But $\mathfrak{v}$ is compact, thus there exists $b_1, \cdots, b_k \in A$, $a_1, \cdots, a_k \in I$ and integers n_i such that $b_i f^{n_i} = a_i$ for $i = 1, \cdots, k$ and $(b_1, \cdots, b_k) = (1)$. Possibly changing the a_i's we can further assume that $n_i = m$ for all i. Since $(b_1, \cdots, b_k) = (1)$, there exist $c_1, \cdots, c_k \in A$ such that $1 = c_1 b_1 + \cdots + c_k b_k$ and hence $f^m = c_1 b_1 f^m + \cdots + c_k b_k f^m \in I$, showing that $f \in I^{1/2}$, proving the theorem.

The notion of a Z-ideal was introduced in (1.4). Recall that I is a *Z-ideal* if $I = Z^{-1}Z(I)$. It was pointed out in (1.4) that all maximal ideals are Z-ideals. Let $(M_t)_{t \in T}$ be a nonempty set of maximal ideals and let $I \equiv \bigcap_{t \in T} M_t$. I may easily be seen to be a Z-ideal. Any proper id eal I of A that has such a decomposition will be said to admit a *maximal decomposition*.

THEOREM 5.7. *Let I be a proper ideal of A. I is a Z-ideal if and only if I admits a maximal decomposition; in particular $I = \bigcap_{M \in \mathfrak{v}(I)} M$ is such a decomposition.*

PROOF. Let I be a proper Z-ideal of A. If $I = 0$, $\mathfrak{v}(I) = \operatorname{specm} A$. Since A is obviously semisimple the equality above holds. Assume now that $I \neq 0$. Let $a \in I - \{0\}$, and let $M \in \mathfrak{v}(I)$. Let $D \equiv Z(a)$, let V be the valuation of A_M, and let G be its value group. Let $g \in G$, $g > 0$. There exists $b \in A$ having zero set D such that $V(b) = g$ (see the proof of (4.2) and use the Weierstrass Theorem). Since $Z(b) = Z(a)$, and since I is a Z-ideal, $b \in I$. Thus we have shown that $V(I) = G(>0) \cup \{\infty\}$, and hence by (5.4), $(I_M)^c = M$, proving the theorem.

We have shown that under various hypotheses certain decompositions exist. Following classical ideal theory, one is lead to inquire into the uniqueness of these decompositions. The following example shows that uniqueness fails in each case.

EXAMPLE 5.2. Let us continue using the notation of Example 5.1. Let $I \equiv \bigcap_{n \in N} M_n$; thus I is a Z-ideal. Clearly I is a radical ideal (i.e., $I = I^{1/2}$). Let $f \in I$ and note that $D \subset Z(f)$, showing that $Z(f) \in \delta$ and hence $f \in M$; thus $I = I \cap M$.

Of course it is the topology on specm A that accounts for this, since M is in the closure of $(M_n)_{n \in N}$ in specm$A$.

Let us examine Example 5.2 more closely in an effort to see how much information is necessary to determine I. First note that since I is an intersection of maximal ideals, I is a Z-ideal. Further, any maximal ideal in the closure of $(M_n)_{n \in N}$ can play the role of M. Thus $I = \bigcap_{n \in N} M_n$ is determined by a dense subset $(M_n)_{n \in N}$ of the closed set $\mathfrak{v}(I)$. This suggests the following theorem.

THEOREM 5.8. *Let I be a proper ideal of A that admits a maximal decomposition. $I = \bigcap_{M \in \mathfrak{Y}} M$ if and only if $\mathfrak{Y}$ is dense in $\mathfrak{v}(I)$.*

PROOF. Let $I = \bigcap_{M \in \mathfrak{Y}} M$. The sets $(\mathfrak{v}(f))_{f \in A}$ for a base for the topology on $\mathfrak{v}$ ($\equiv$ specm A). Thus it suffices to show that $\mathfrak{v}(I) \subset \mathfrak{v}(f)$ if and only if $\mathfrak{Y} \subset \mathfrak{v}(f)$. Let $\mathfrak{v}(I) \subset \mathfrak{v}(f)$. Then $f \in \bigcap_{M \in \mathfrak{v}(I)} M$, which by (5.7) is I. Since $I = \bigcap_{M \in \mathfrak{Y}} M$, $f \in M$ for all $M \in \mathfrak{Y}$: i.e., $\mathfrak{Y} \subset \mathfrak{v}(f)$. Now let $\mathfrak{Y} \subset \mathfrak{v}(f)$. Then $f \in \bigcap_{M \in \mathfrak{Y}} M$, which is I. By (5.7), $I = \bigcap_{M \in \mathfrak{v}(I)} M$, showing that $\mathfrak{v}(I) \subset \mathfrak{v}(f)$, proving one implication.

Conversely, let $\mathfrak{Y}$ be a dense subset of $\mathfrak{v}(I)$ and let $J \equiv \bigcap_{M \in \mathfrak{Y}} M$. Since $\mathfrak{Y} \subset \mathfrak{v}(I)$ and since $I = \bigcap_{M \in \mathfrak{v}(I)} M$ (5.7), $I \subset J$. Let $f \in J$. Since $\mathfrak{Y} \subset \mathfrak{v}(f)$ and since $\mathfrak{v}(I) \subset \bar{\mathfrak{Y}}$, $\mathfrak{v}(I) \subset \mathfrak{v}(f)$; thus $f \in \bigcap_{M \in \mathfrak{v}(I)} M$, which by (5.7) is I, proving that $J \subset I$, completing the proof.

6. Iss'sa's Theorem and corollaries.

THEOREM 6.1 (ISS'SA [I]). *Let X and Y be noncompact Riemann surfaces, X and Y are conformally equivalent if and only if $F(Y)$ and $F(X)$ are C-isomorphic.*

This theorem, long stated and only recently proved by Iss'sa, is an analogy of the Chevalley-Kakutani-Bers Theorem (1.1). Clearly the only point at issue is the statement that a C-isomorphism ϕ of $F(Y)$ onto $F(X)$ gives rise to a conformal map θ of X onto Y. If we knew that $\phi | A(Y)$ was an isomorphism onto $A(X)$ we could apply (1.1) and find θ. Given only the extension $F(Y)/C$ there is a fairly obvious way to try to extract $A(Y)$ from $F(Y)$, namely the following: let $\mathrm{Riem}_Z F(Y) \equiv \{\mathfrak{O}: \mathfrak{O}$ a valuation ring of $F(Y)$ over C onto C whose value group is $Z\}$. Think of this as the formal Riemann surface associated with $F(Y)$. Given $y \in Y$, $\mathfrak{O}_y \in \mathrm{Riem}_Z F(Y)$, and $y \to \mathfrak{O}_y$ is clearly a one-to-one map of Y into $\mathrm{Riem}_Z F(Y)$. Is it surjective? If so the problem is solved, since $A(Y)$ would then be the intersection of the valuation rings in $\mathrm{Riem}_Z F(Y)$, and since ϕ clearly maps $\mathrm{Riem}_Z F(Y)$ onto $\mathrm{Riem}_Z F(X)$. Hence what is asserted is that given $F(Y)/C$, there is something intrinsic about $A(Y)$: i.e., $A(Y)$ is the ring of "integers" of $F(Y)$. Note: we know that $C \subset A(Y)$, that the latter is integrally closed in $F(Y)$, and that $F(Y)$ is the smallest subfield of itself that contains $A(Y)$. However this is not in general enough to make such a ring intrinsic, as the following example shows.

EXAMPLE 6.1. Note that the polynomial ring $C[z]$ is integrally closed in $C(z)$, contains the constants, and has $C(z)$ as its quotient field; but the same is true of $C[az + b/cz + d]$, $ad - bc \neq 0$. Hence $C[z]$ is in no sense singled out as the distinguished subring of $C(z)$, having the properties listed above. Finally note that $\mathrm{Riem}_Z C(z)$ is the sphere, not the plane, the latter being specm $C[z]$, and that the intersection of the rings in $\mathrm{Riem}_Z C(z)$ is C, not $C[z]$.

More encouraging and of greater analytic interest is the following example.

EXAMPLE 6.2. Let Y be a noncompact Riemann surface that is the interior of a compact bordered Riemann surface. Let $B(Y)$ be the ring of bounded analytic functions on Y and let $K(Y)$ be its quotient field in $F(Y)$. (These turn out to be the functions of *bounded characteristic* on Y.) In 1958 Royden [$\mathbf{R}_2$] showed that $y \to \mathfrak{O}_y$ maps Y *onto* $\mathrm{Riem}_Z K(Y)$; thus (6.1) holds if we change F to K in its statement. Royden's proof uses the fact that $B(Y)$ is a Banach algebra, under the sup norm, in an essential way.

Iss'sa uses two lemmas to establish the result. The first is technical but straightforward. The second is elementary but inspired.

LEMMA 6.2 (ISS'SA [I]). *Let Y be the complex plane and let $\mathfrak{O} \in \mathrm{Riem}_Z F(Y)$; then the identity map z is in $\mathfrak{O}$.*

It is convenient to use the ring of p-adic integers in the proof of (6.2). Briefly this is the following. Let Q denote the rational number field and let p be a prime number. Given $r \in Q^*$, there exist integers a and b such that $r = a/b$. $a = a_0 p^e$ and $b = b_0 p^f$, where a_0 and p, and b_0 and p are relatively prime. Let $v_p(r) = e - f$; then v_p is a valuation on Q, called the *p-adic valuation*, with integer value group. Let $\| r \| = e^{-v_p(r)}$, and note that Q can be completed in this norm. Let Q_p be the completion. v_p extends to a valuation v_p on Q_p. Let Z_p denote its valuation ring. This is the ring of *p-adic integers*. p is an element of value 1 in Q_p, thus it plays the role of a uniformizing variable. Let M_p be the maximal ideal of Z_p. Then it is easily seen that Z_p/M_p is the field with p elements. Let the integers $0, 1, \cdots, p-1$ serve as coset representatives for this residue class field. Each $a \in Q_p$ may thus be represented as a formal pseudo-Laurent series $\sum_{n=k}^{\infty} a_n p^n$, where $a_n \in \{0, 1, \cdots, p-1\}$.

PROOF OF (6.2). Let V be the valuation of $\mathfrak{O}$. Assume for a moment that $V(z) < 0$, V being the valuation of $\mathfrak{O}$. Then $\gamma \equiv -V(z) \in Z^*$ and γZ_p is not contained in Z. Choose $\alpha \in Z_p$ such that $\alpha\gamma \notin Z$. α may be written as $\sum_{n=0}^{\infty} a_n p^n$ where $a_n \in \{0, 1, \cdots, p-1\}$. Note: $\alpha \notin Z$, thus an infinite number of the a_n's are nonzero. By the classical Weierstrass Theorem there exists $h \in A(Y)$ such that $V_n(h) = a_n p^n$ for all nonnegative integers n. Let $h_n \equiv h/\prod_{j=0}^{n-1}(z-j)^{a_j p^j}$ for all such n, and let $V(h) \equiv \delta \in Z$. Since $V(z) < 0$ and $C \subset \mathfrak{O}$, $V(j) \geqq 0$; thus $V(z-j) = \min V(z), V(j) = V(z)$. As a consequence $V(h_n) = \delta + (\sum_{j=0}^{n-1} a_j p^j)\gamma$. Clearly the zero set of h_n is $\{m \in Z\colon m \geqq n\}$, and $V_m(h_n) = a_m p^m$ for all integers $m \geqq n$. Let $t_n \in A$ have $\{m \in Z\colon m \geqq n\}$ as zero set, and let $V_m(t_n) = a_m p^{m-n}$ for all $m \geqq n$. Let $k_n \equiv h_n/t_n^{p^n} \in F(Y)$. Now $V_m(k_n) = a_m p^m - p^n V(t_n) = a_m p^m - p^n a_m p^{m-n} = 0$, for all integers $m \geqq n$. We conclude that k_n has no zeros on Y and is thus a unit in $A(Y)$. Since Y, the plane, is simply connected, every unit of $A(Y)$ is an exponential: i.e., there exists $f_n \in A(Y)$ such that $k_n = e^{f_n}$. Let $b_n \equiv e^{f_n/p^n}$ and note that $b_n^{p^n} = k_n$. Let $g_n \equiv b_n t_n$ and observe that $g_n^{p^n} = h_n$ for all n. Thus, $\delta + (\sum_{j=0}^{n-1} a_j p^j)\gamma = V(h_n)$ is an integer which is divisible by p^n for all n. In Z_p this means that $\delta = \alpha\gamma$, which is a contradiction, since α was chosen so that $\alpha\gamma \notin Z$, and yet $\delta \in Z$.

LEMMA 6.3 (ISS'SA [I]). *Let $\mathfrak{O} \in \mathrm{Riem}_Z F(Y)$; then $A(Y) \subset \mathfrak{O}$.*

PROOF. Let $f \in A(X) - C$. f may be regarded as an analytic mapping of X into the plane C. Then the adjoint f^* of f is a C-isomorphism of $F(C)$ into $F(X)$, which we will view as an identification. $\mathfrak{O} | F(C) \equiv \mathfrak{O}'$ is a subring of $F(C)$ which contains C. Were $F(C) = \mathfrak{O}'$, then $f \in \mathfrak{O}'$ and $V(f) \geqq 0$. Assume $\mathfrak{O}'$ is a proper subring of $F(C)$; then it is a valuation ring of $F(C)$ over C onto C. Let G be its value group. Since the value group of $\mathfrak{O}$ is Z, the integers, G is isomorphic to a nontrivial subgroup of Z: i.e., G is isomorphic to Z. Thus $\mathfrak{O}' \in \mathrm{Riem}_Z F(C)$. By (6.2) $V'(z) \geqq 0$, V' being the valuation of $\mathfrak{O}'$. But in $F(Y)$, $z(f) = f^*(z)$, which is merely f. Therefore $V(f) = V'(z) \geqq 0$, proving the lemma.

PROOF OF (6.1). Let ϕ be a C-isomorphism of $F(Y)$ onto $F(X)$ and let θ be the induced bijection of $\mathrm{Riem}_Z F(X)$ onto $\mathrm{Riem}_Z F(Y)$. By (6.3) $A(Y) = \bigcap_{\mathfrak{O} \in \mathrm{Riem}_Z F(Y)} \mathfrak{O}$;

thus $\phi| A(Y)$ is a C-isomorphism onto $A(X)$. By the Chevalley-Kakutani-Bers Theorem (1.1), X and Y are conformally equivalent, proving the theorem.

Theorem (6.1) is very reassuring for it tells us that algebraic equivalence and analytic equivalence are equivalent. A natural outgrowth of this is the following.

THEOREM 6.4 [$\mathbf{A}_6$]. *Let X and Y be noncompact Riemann surfaces. Let θ be a C-isomorphism of $F(Y)$ into $F(X)$ such that $F(X)/\theta F(Y)$ is a finite algebraic extension (of degree n). Then there exists an analytic n to 1 mapping ϕ of X onto Y. Conversely, given such a map ϕ let θ be the adjoint of ϕ on $F(Y)$. Then θ has the properties described above. Finally θ and ϕ are in bijective correspondence.*

The proof of this relies on (6.1) and standard ideas in the theory of algebraic function fields, essentially due to Riemann. (6.4) is obviously a generalization of (6.1).

Let me raise a final question. Let X be the interior of a compact bordered Riemann surface: e.g., an open annulus. Given only the extension $F(X)/C$ find an "effective" means of computing the first Betti number g of X: e.g., find a way of defining $\boldsymbol{d}_2(X)$, the space of meromorphic differentials on X of the second kind given *only* $F(X)/C$. If this can be done then $g = \dim_C \boldsymbol{d}_2(X)/dF(X)$. The following methods seems too removed from $F(X)$ to be classified as being effective: form $\mathrm{Riem}_Z F(X)$, give it the weak topology, and then compute the rank of $H_1(X, Z)$.

We know from algebraic geometry that differentials are dual objects to derivations, so one is tempted to consider the space of all derivations of $F(X)$, trivial on C, but this space is enormous and seemingly almost devoid of interest. Each $\mathfrak{O} \in \mathrm{Riem}_Z F(X)$ induces an $\mathfrak{O}$-adic topology on $F(X)$, under which it is a topological field. Let $\boldsymbol{D}(X)$ be the space of all derivations of $F(X)$, trivial on C, that are continuous in the $\mathfrak{O}$-adic topology for all $\mathfrak{O} \in \mathrm{Riem}_Z F(X)$; and let $\boldsymbol{d}(X) = \mathrm{Hom}_{F(X)}(\boldsymbol{D}(X), F(X))$. It is fairly easily seen that $\boldsymbol{d}(X)$ is naturally the space of meromorphic differentials on X. From this one can naturally extract $\boldsymbol{d}_2(X)$, the differentials with no logarithmic singularities, and then obtain the following C-exact sequence:

$$0 \to C \to F(X) \overset{d}{\to} \boldsymbol{d}_2(X) \to \check{H}^1(X, C) \to 0$$

the cohomology group being the first Čech cohomology group with constant coefficients. For details see [$\mathbf{A}_7$].

Historical notes. Although the fact that the set of all entire functions $A(C)$ is a ring under pointwise operations has been known for a very long time, the first paper dealing with the ring theory of this ring seems to be Helmer's [$\mathbf{H}_2$], written in 1940. Although several subjects are considered by Helmer, including the ring of functions of finite order, it is largely the observation that every finitely generated ideal of $A(C)$ is principal for which this paper is remembered. This indicated that the ideal theory of the finitely generated ideals was trivial and, since the ring is not Noetherian, the rest of its ideal theory would be very infinitistic. It should be remembered that at the time most algebra was very finitistic.

In 1942 Chevalley and Kakutani **[CK]** proved that if X and Y are domains in the complex plane, such that for each boundary point there exists a bounded analytic function on the domain that does not admit an analytic continuation through the point in question, then X and Y are conformally equivalent if and only if their rings of bounded analytic functions are C-isomorphic. This manuscript was apparently never published. Kakutani refers to it **[K]**, and so does Bers **[B_2]**, who apparently learned about the result from Chevalley. According to Kakutani **[K]**, he and Chevalley also proved the corresponding result for $A(X)$ in 1942. (Since I have never seen the manuscript in question, I cannot personally vouch for these assertions.)

Schilling's report to the Society in 1946 **[S]** noted that all closed ideals of the ring of entire functions, under the topology of uniform convergence on compacta, are fixed. Since all fixed maximal ideals are closed, the two classes coincide. A curious slip occurs in §4 of this paper, in which it is stated that all nonzero prime ideals of $A(C)$ are maximal. Later Henriksen **[H_4]** noted that given a free maximal ideal M, there exist at least $2^{\aleph_1}$ nonmaximal prime ideals in it. $A(C)_M$ is a valuaring **[A_3]**, a consequence of the Weierstrass (product) Theorem **[F]**; thus for all nonzero prime ideals to be maximal, the value group of $A(C)_M$ would have to be Archimedean, which is not the case. Schilling goes on to study finite algebraic extensions of $F(C)$ and the corresponding surface covering the plane. This very suggestive notion lay unexploited until Royden employed it in 1958. Schilling's paper is one of the most highly algebraic accounts of the subject to be given.

Bers **[B_2]** proved in 1946 that two plane domains X and Y that possess boundary points are conformally equivalent if and only if $A(Y)$ and $A(X)$ are $Q(i)$-isomorphic. Bers states that only after his paper was completed did he become aware of the Chevalley-Kakutani manuscript. Bers based his proof on ideal theory and on the existence of an injection in the rings. Since the latter does not in general exist, Bers' proof needed to be modified to achieve the general result.

Henriksen's dissertation **[H_3]** (1952) on the ideal theory of $A(C)$ was deeply influenced by Hewitt's paper **[H_6]** on rings of continuous real-valued functions on noncompact spaces, which appeared in 1948. Rather than using the zero set of a function, Henriksen utilized its "algebraic" zero set in which multiplicity was noted. This slightly vague notion was used by him to great effect. In **[H_3]** we find a concrete realization of specm $A(C)$ in the terms of Δ-ultrafilters and a proof that $A(C)/M$ is always algebraically closed (cf., (1.4), (2.1), and (2.2)). In his second paper on this subject [$\mathbf{H}_4$] (1953), Henriksen writes that Kaplansky had told him in a conversation about the error in Schilling's paper, noted above. Henriksen goes on to obtain a great deal of information about the prime ideal of $A(C)$, without using valuation techniques. In [$\mathbf{H}_5$] Henriksen considers the ring of entire functions of finite order. The results are fragmentary and indicate that this ring is much more subtle than is $A(C)$.

Kakutani's clear account of the subject, presented at the Michigan Conference on Functions of a Complex Variable in 1953, appeared in 1955 [**K**]. In it the Chevalley-Kakutani-Bers Theorem is proved for plane domains, for which (in the case of bounded functions) each boundary point has an essential singularity. Apparently Kakutani only learned of Henriksen's paper at a late stage in the preparation of his manuscript. (Lack of communication has been widespread in this small field.) Fixed and free ideals and the correspondence between maximal ideals and Δ-ultrafilters is here established in its present form. It is these ideas that are used by Kakutani to prove the theorem. Kakutani's paper closes with a discussion of the algebra of bounded analytic functions on plane domains X. Some of the basic devices used in this study seem to have appeared here for the first time. For example, the notion of a fiber in the maximal ideal space over a boundary point of its planar domain appears here.

In December of 1953 Rudin [$\mathbf{R}_4$] gave a new proof of the Chevalley-Kukutani-Bers Theorem. Rudin states that according to Kakutani, his proof is an improvement on the proof that appeared in the Chevalley-Kakutani manuscript. This proof of Rudin's is based on function theory rather than ideal theory, or Banach algebras. In the course of the proof Rudin shows that a ring isomorphism of $A(Y)$ onto $A(X)$, which carries i to i, is necessarily complex linear. Two years later an abstract of Rudin's appeared [$\mathbf{R}_3$] in which he states a generalization of the Chevalley-Kakutani-Bers Theorem (1.1) to noncompact Riemann surfaces X, using Florack's generalizations of the Weierstrass product theorem to X [**F**], which appeared in 1948. (Of course with this result and with her generalization of the Mittag-Leffler Theorem to X, the whole theory of A and F instantly generalizes to X. However, it was still many years before all papers were written at this level of generality.) Now it is clear, using Florack's Theorem [**F**], that for all $x \in X$, M_x is principal. Rudin states that the converse is true, which it is, without any indication of proof. By analyzing the free ideals this may of course be seen (1.5). (This was known to Henriksen in 1951 in case $X = C$.)

Unlike some of his predecessors, Royden was aware of the literature as he wrote his paper on the subject in 1956 [$\mathbf{R}_1$]. Using Florack's Theorems, cited above, Royden points out that Helmer's theorems go over to the general case and implicitly that Henriksen's theorems do also· Royden's proof of the Chevalley-Kakutani-Bers Theorem is based on ideal theory and in particular uses the bijective correspondence between the points of the surface and the maximal principal ideals. The existence of an injective function, which Bers used, no longer works; thus the ideal theory won an undisputed place for itself in the theory. The second subject of many of these papers, that of the ring $B(X)$, is here dealt with most incisively. Let $K(X)$ be the quotient field of $B(X)$. Let $\mathfrak{O}$ be a valuation ring of K over C whose value group is the integers. Royden showed that $B(X) \subset \mathfrak{O}$; thus the extension of the Chevalley-Kakutani-Bers theorem for K, for appropriate domains, is proved.

Banaschewski [$\mathbf{B}_1$] proceeded as follows: the divisor map $\mathfrak{d}$ takes $A(C)$ to a lattice-ordered semigroup W. The ideal theory of $A(C)$ is reflected in the ideal theory of W, where he chose to make his analysis. Actually, Banaschewski abstracts the setting and thus implicitly generalizes to $A(X)$. In this abstracts setting, Banaschewski showed that the appropriate generalizations of the ideal theoretic results of Helmer and Henriksen go over. He went on to obtain some of the results stated in §5 on general ideal theory of $A(X)$.

In Royden's 1958 paper [$\mathbf{R}_2$] the duality between covering maps of $\bar{X}$ onto $\bar{Y}$ and extensions of meromorphic function rings was again introduced, having lain dormant in this field since Schilling's paper (1946). However in view of Ahlfors' result of 1950 (which proves Lemma 2), one could show that if $\bar{X}$ is any compact bordered Riemann surface, then there exists an n-to-1 analytic map of $\bar{X}$ onto $\bar{D}$, the closed unit disc. Royden used this to show again that each discrete rank one valuation of $K(X)$ (the functions of finite characteristic on X) over C onto C corresponds to an x in X, thus proving an analogy of the Chevalley-Kakutani-Bers Theorem for K once more. Note: this paper also has an essential slip. In the second to last line of Lemma 3, for analytic read meromorphic over $\bar{X}$. A counter example is not, as far as I know, known. However, the proof Royden had in mind certainly does not work.

In 1963 Nakai [**N**] showed that in the Chevalley-Kakutani-Bers Theorem for arbitrary X one need only assume that $A(Y)$ and $A(X)$ are *ring* isomorphic and that $i \to i$ in order to make its adjoint conformal. (I am indebted to Professor Chern for pointing out this reference to me.) See also [$\mathbf{R}_4$].

Having been present in vestigial form for years (e.g., in Henriksen [$\mathbf{H}_4$] and Banaschewski [$\mathbf{B}_1$]), the valuation theory of $F(X)$ became manifest in [$\mathbf{A}_3$]. Since a summary of these results appears here, little more need be said of them in these notes. Interestingly, there is a strong formal resemblance between the results in [$\mathbf{A}_3$] and those in [$\mathbf{A}_5$] on rings of continuous integer-valued functions and non-standard arithmetic.

Several authors have treated the rings $A(X)$ and $B(X)$ as part of the same subject. Using ideas of Royden's [$\mathbf{R}_2$], I was able to lift Carleson's very deep corona theorem for the disc to the interior of a compact bordered Riemann surface in 1963 [$\mathbf{A}_4$]. These results appeared *in extenso* in [$\mathbf{A}_6$] and were further developed in [$\mathbf{A}_7$]; however most of the results lie more properly in the theory of Banach algebias of analytic functions, a large and active field not closely related to entire functions.

Hej Iss'sa proved a general result [I] in 1964, which has (6.1) as a special case, thus settling a question which had vexed many of the workers in this field for years.

Going back to the plane, there is still much work to be done on rings of entire functions which satisfy some sort of growth condition, and on their quotient fields. The sort of difficulties that arise in dealing with the corona question of bounded analytic functions on the open disc arise in this study. Further, the theory of cohomology with bounds may be applicable here. In general the questions are much more subtle and analytic than the questions treated in this paper. Kelleher and Taylor read papers at La Jolla which dealt with such problems. Nevertheless, many problems remain unsolved.

Finally, subscripts have been written after the papers in the Bibliography which are mentioned in these historical notes and which lie in the mainstream of this research: e.g., $[\mathbf{H}_2]_1$ refers to Helmer's paper $[\mathbf{H}_2]$, as the first to be written.

BIBLIOGRAPHY[3]

$[A_1]$ N. L. Alling, *On ordered divisible groups*, Trans. Amer. Math. Soc. **94** (1960), 498–514.

$[A_2]$ ———, *A characterization of Abelian η_α-groups in terms of their natural valuation*, Proc. Nat. Acad. Sci. U.S.A. **47** (1961), 711–713.

$[A_3]_{15}$ ———, *The valuation theory of meromorphic function fields over open Riemann surfaces*. Acta Math. **110** (1963), 79–96.

$[A_4]_{16}$ ———, *A proof of the corona conjecture for finite open Riemann surfaces*, Bull. Amer. Math. Soc. **70** (1964), 110–112.

$[A_5]$ ———, *Rings of continuous integer valued functions and nonstandard arithmetic*, Trans. Amer. Math. Soc. **118** (1965), 498–525.

$[A_6]$ ———, *Extensions of meromorphic function rings over non-compact Riemann surfaces*. I, Math. Z. **89** (1965), 273–299.

$[A_7]$ ———, *Extensions of meromorphic function rings over non-compact Riemann surfaces*. II, Math. Z. **93** (1966), 345–394.

$[\mathbf{B}_1]_{12}$ B. Banaschewski, *Zur Idealtheorie der ganzen Funktionen*, Math. Nachr. **19** (1958), 136–160.

$[\mathbf{B}_2]_4$ L. Bers, *On rings of analytic functions*, Bull. Amer. Math. Soc. **54** (1948), 311–315.

$[\mathbf{CK}]_2$ C. Chevalley and S. Kakutani, an unpublished manuscript on rings of analytic functions on planar domains, 1942. See Historical Notes for further details.

[F] H. Florack, *Reguläre und meromorphe Funktionen auf nicht geschlossenen Riemannschen Flächen*, Math. Inst. Univ. Münster No. 1 (1948).

[GJ] L. Gillman and M. Jerison, *Rings of continuous functions*, Van Nostrand, Princeton, N. J., 1960.

$[\mathbf{H}_1]$ F. Hausdorff, *Grundzüge der Mengenlehre*, Leipzig, 1914.

$[\mathbf{H}_2]_1$ O. Helmer, *Divisibility properties of integral functions*, Duke Math. J. **6** (1940), 345–356.

$[\mathbf{H}_3]_5$ M. Henriksen, *On the ideal structure of the ring of entire functions*, Pacific J. Math. **2** (1952), 179–184.

[3] The subscripts following some of the square brackets above refer to the chronological sequence of the principal papers in this field: e.g., by $[A_3]_{15}$ it is meant that $[A_3]$ is the 15th paper in this field. Each such paper is discussed in the Historical Notes.

[H4]6 ———, *On the prime ideals of the ring of entire functions*, Pacific J. Math. **3** (1953), 711–720.

[H5]7 ———, *On rings of entire functions of finite order*, J. Indian Math. Soc. **17** (1953), 59–61.

[H6] E. Hewitt, *Rings of real-valued continuous functions*. I, Trans. Amer. Math. Soc. **64** (1948), 54–99.

[I]17 H. Iss'sa, *On meromorphic function fields on a Stein variety*, Ann. of Math. **83** (1966), 34–46.

[K]8 S. Kakutani, "Rings of analytic functions," in *Lectures on functions of a complex variable*, edited by W. Kaplan et. al., Univ. of Mich., 1955, 71–83.

[N]14 M. Nakai, *On rings of analytic functions on Riemann surfaces*, Proc. Japan. Acad. **39** (1963), 79–84.

[R1]11 H. Royden, *Rings of analytic and meromorphic functions*, Trans. Amer. Math. Soc. **83** (1956), 269–276.

[R2]13 ———, *Rings of meromorphic functions*, Proc. Amer. Math. Soc. **9** (1958), 959–965.

[R3]10 W. Rudin, *An algebraic characterization of conformal equivalence*, Bull. Amer. Soc. **61** (1955), 543.

[R4]9 ———, *Some theorems on bounded analytic functions*, Trans. Amer. Math. Soc. **83** (1956), 333–342.

[S]3 O. F. G. Schilling, *Ideal theory on open Riemann surfaces*, Bull. Amer. Math. Soc. **52** (1946), 945–963.

[ZS1] O. Zariski and P. Samuel, *Commutative algebra*. I, Van Nostrand, Princeton, N. J. 1958.

[ZS2] ———, ———, *Commutative algebra*. II, Van Nostrand, Princeton N. J., 1960.

UNIVERSITY OF ROCHESTER
ROCHESTER, NEW YORK

Further Results on Factorization of Entire Functions

I. N. BAKER
F. GROSS

I. **Introduction**. An entire function $h(z) = f(g(z))$ is said to have $f(z)$ and $g(z)$ as left and right factors respectively, provided that they are both nonlinear and entire. $h(z)$ is said to be prime (pseudo-prime) if every factorization of the above form implies that one of f or g is linear (a polynomial).

In [4] Rosenbloom stated without proof that $e^z + z$ is prime. This was proved in [2]. It was also shown in [2] that every entire function of the form $e^z + P(z)$, where $P(z)$ is a nonconstant polynomial is pseudoprime.

In the present note we shall study a more general class of entire functions of the form $F(z) = H(z) + g(z)$, where $H(z)$ is a transcendental entire periodic function and $g(z)$ is an entire function of order less than 1. One of the consequences of this study will be that every function of the form $e^z + P(z)$ ($P(z)$ a nonconstant polynomial) is prime.

II. **Periodic functions** mod g. Let g be an entire function of order less than 1.

DEFINITION. An entire function $F(z)$ is said to be periodic mod g, (with period τ) if $F(z+\tau) - F(z) = g(z)$.

LEMMA 1 (BOREL [1]). *Let $a_i(z)$ be an entire function of order ρ, let $g_i(z)$ also be entire and let $g_j(z) - g_i(z)$ $(i \neq j)$ be a transcendental function or polynomial of degree higher than ρ, then*

$$\sum_{i=1}^{n} a_i(z)e^{g_i(z)} = a_0(z)$$

holds only when

$$a_0(z) = a_1(z) = \cdots = a_n(z) = 0.$$

Note. Though this was first stated by Borel, the first complete proof was given by R. Nevanlinna [3].

THEOREM 1. *Every right factor of a function h (h periodic* mod g *with period τ) is of the form*

$$l(z) = H_1(z) + g_2(z)e^{H_2(z)+az}$$

where $H_i(z)$, $i = 1, 2$, *are periodic with period* τ, a *is a constant and* g_2 *is an entire function of order less than or equal to* 1. *If* g *is a polynomial then* g_2 *is also.*

PROOF. For convenience we assume that $\tau = 1$. Suppose $h = f(l)$. From the hypotheses of the theorem we have

$$f(l(z+1)) - f(l(z)) = g(z) \tag{1}$$

and

$$f(l(z+2)) - f(l(z)) = g(z+1) + g(z). \tag{2}$$

It follows from (1) and (2) that the order of convergence of the zeros of $l(z+i) - l(z)$, $i = 1, 2$, is less than 1.

We may write

$$l(z+1) - l(z) = g_1(z)e^{\alpha(z)}, \tag{3}$$

where α is entire and $g_1(z)$ is of order less than 1. Also

$$l(z+2) - l(z) = g_1(z+1)e^{\alpha(z+1)} + g_1(z)e^{\alpha(z)}. \tag{4}$$

Since the order of the zeros of the left side of (4) is less than 1, we may apply Lemma 1 to conclude that $\alpha(z+1) = \alpha(z) + c$, c a constant. Thus $\alpha(z) = H_2(z) + cz$, where $H_2(z+1) = H_2(z)$. Using this fact (3) becomes

$$l(z+1) = l(z) + g_1(z)\exp(H_2(z) + cz).$$

There exists an entire function $g_2(z)$ of order $\leqq 1$ such that $e^c g_2(z+1) - g_2(z) = g_1(z)$ (see Whittaker [5], p.22). One can easily verify that when $g(z)$ (and hence $g_1(z)$) is a polynomial then so is $g_2(z)$. It follows that $l(z)$ has the desired form.

It is likely that g_2 is actually of order < 1.

Note that when g is a polynomial not every function l in Theorem 1 is actually possible.

$$l(z) = H_1(z) + \exp(H_2(z) + az),$$

for example, is never a right factor of h. To prove this, we first observe that it may be assumed without any loss of generality that $|e^a| \leqq 1$.

Suppose that

$$f(H_1(z) + e^{an}\exp(H_2(z) + az)) - f(H_1(z) + \exp(H_2(z) + az)) = \sum_{i=1}^{n-1} Q(z+i).$$

Then for a fixed z, the left side of this equation is a bounded function, while the right side is not.

THEOREM 2. *If P is any nonlinear polynomial and g is any transcendental entire function, then P(g) is not periodic* mod *any nonconstant polynomial.*

PROOF. Suppose that $P(g(z+1)) - P(g(z)) = L(z)$, $L(z)$ a nonconstant polynomial. By virtue of Theorem 1, g is of the form

$$g(z) = H_1(z) + Q(z)\exp(H_2(z) + az),$$

where $H_i(z+1) = H_i(z)$, $i = 1, 2$, Q is a polynomial and a is a constant.

$$g(z+n) - g(z) = [Q(z+n)e^{an} - Q(z)]\exp(H_2(z) + az)$$

and

$$P(g(z+n)) - P(g(z)) = L(z) + L(z+1) + \cdots + L(z+n-1) \tag{5}$$

so that

$$P(g(z) + (Q(z+n)e^{an} - Q(z))\exp(H_2(z) + az) = P(g(z)) + L(z) + L(z+1) + \cdots + L(z+n-1). \tag{6}$$

If $|e^a| < 1$ then at $z = 0$ the right side of (6) approaches infinity with n while the left side is bounded. A similar argument is valid when $|e^a| > 1$, simply replace z by $-z$ in the expression for $g(z)$.

Thus $|e^a| = 1$. Let s, t and r be the degrees of P, Q and L respectively. When $z = 0$, the absolute value of the right side of (6) is greater than

$$\lambda_1(n/2)(n/2)^r$$

and less than

$$\lambda_2(n-1)^r(n-1), \ \lambda_i, \qquad (i = 1, 2) \text{ positive constants.}$$

Thus $st = r + 1$.

If $\alpha(z) = \exp(H_2(z) + az)$ is a constant, then

$$P(H_1(z) + \alpha Q(z+1)) - P(H_1(z) + \alpha Q(z)) = L(z). \tag{7}$$

For large R, the maximum modulus of the left side of (7) is greater than $\gamma \max_{|z|=R}|H_1(z)|^{s-1}$, for some $\gamma > 0$. Since H_1 must be transcendental, this, of course, is impossible.

Assume that $\alpha(z)$ is not constant. One can easily verify that for every z the right side of (6) is less in absolute value than λn^{r+1}, for some $\lambda > 0$ indedendent of z and all sufficiently large $n > N(z)$. We may assume that the leading coefficients of P and Q respectively are λ_P and 1. Choose z_0 such that

$$|\lambda_P| \ |\alpha(z_0)|^s > 2\lambda.$$

For any $\varepsilon > 0$, the left side of (6) is greater than

$$(1-\varepsilon)|\lambda_P||\alpha(z_0)|^s n^{st} > (1-\varepsilon)2\lambda n^{r+1}$$

for sufficiently large $n > N(z_0, \varepsilon)$. Thus we would have for any $\varepsilon > 0$

$$\lambda n^{r+1} > (1-\varepsilon)2\lambda n^{r+1}$$

or sufficiently large n. This is impossible, of course, and our proof is complete,

THEOREM 3. *If P is a nonconstant polynomial of degree greater than 2 and f is transcendental entire, then $F = f(P)$ is not periodic* mod *any polynomial.*

PROOF. Assume that $F(z+i) - F(z) = g(z)$; $g(z)$ a polynomial (and $i = (-1)^{1/2}$). We may write

$$F(z) = H(z) + g_1(z),$$

where $H(z)$ is periodic with period i and $g_1(z)$ is a polynomial. Since $F(z)$ is transcendental, it follows that for sufficiently large r

(8) $$M_F(r) > r^2 M_g(2r),$$

where $M_F(r) = \max_{|z|=r}|F(z)|$. Also for all positive n

(9) $$F(z+ni) - F(z) = g(z) + g(z+i) + \cdots + g(z+(n-1)i).$$

From (8) and (9) we get, for every $\varepsilon > 0$ and sufficiently large $r > R(\varepsilon)$,

(10) $$(1-\varepsilon)M_F(r) < M_{F(z+ni)}(r) < (1+\varepsilon)M_F(r)$$

for all n with $|n| < r$.

Let z_0 ($|z_0| = r$) be a point such that

$$|F(z_0)| = M_F(r).$$

If r is sufficiently large, then for a sufficiently small $\delta > 0$ there is a point z' with $|z'| \approx |z_0|$ (i.e. $(1-\varepsilon)|z'| < |z_0| < (1+\varepsilon)|z'|$) and $\pi + \delta < \arg z' < 2\pi - \delta$ such that

$$f(p(z_0)) = f(p(z')).$$

Thus

(11) $$\begin{aligned} M_F(r) &= |f(p(z'))| = |F(z')| \\ &= |F(z'+ni) - g(z') - g(z'+i) - \cdots - g(z'+n-1)i)|, \end{aligned}$$

where n is chosen so that $z' + ni = x + iy$ and $|y| < 1/2$. The right side of (11) is less than

$$M_F(x+1) + 2rM_g(2r) < M_F(x+1) + (2/r)M_F(r).$$

Hence, using (11),

$$(1 - 2/r)M_F(r) < M_F(1 + |z'|(\cos\delta)) < M_F(\gamma r) \tag{12}$$

for some γ with $\cos\delta < \gamma < 1$.

One can easily show using the Hadamard convexity theorem that for any $\varepsilon > 0$, $M_F(\gamma r) < \varepsilon M_F(r)$ for sufficiently large r. Thus (12) implies

$$(1 - 2/r)M_F(r) < \varepsilon M_F(r)$$

for sufficiently large r. This is impossible, of course, and our theorem follows.

THEOREM 4. *An entire function h of exponential type which is periodic* mod *some nonconstant polynomial is either prime or of the form $h = f((z+c)^2)$, where f is entire and c is a constant.*

PROOF. We know [2] that h is pseudoprime. By virtue of Theorems 2 and 3, the factorizations given in the theorem are the only possible ones.

The function $\cos z + z^2 = h(z^2)$, h entire, illustrates that Theorem 4 is sharp.

THEOREM 5. *Let $P(z)$ be any nonconstant polynomial and let $a(\neq 0)$ and b be any constants. Then $h(z) = e^{az+b} + P(z)$ is prime.*

PROOF. By virtue of Theorem 4, if $h(z)$ is not prime then

$$h(z) = f((z+c)^2)$$

or

$$h(z - c) = f(z^2)\,.$$

Since

$$e^{az+b'} + P(z-c) = h(z-c),$$

b' a constant, we have, for some integer $k > 0$, that $e^{ka}e^{az+b'}$ is an even function.

This of course, is not the case and our proof is complete.

It is reasonable to expect that Theorem 5 remains valid, when $P(z)$ is replaced by any nonconstant entire function of order less than 1.

Along similar lines we prove

THEOREM 6. *If $H(z)$ is periodic of finite lower order and a is a nonzero constant, then $h(z) = H(z) + az$ is prime.*

PROOF. We may assume that H has period 1.

Suppose $h(z) = f(g(z)), f$ and g entire.

Then

$$f(g(z+n)) - f(g(z)) = an. \tag{13}$$

Since $H(z)$ is of finite lower order, one easily concludes from (13) that

$$g(z+n) - g(z) = e^{\alpha(z)+cz} \cdot (e^{cn} - 1), \tag{14}$$

where $\alpha(z)$ is entire and periodic with period 1 and c is a constant with $e^c \neq 1$.

Hence (13) and (14) yield

$$f\left| g(z) + e^{\alpha(z)+cz}\, (e^{cn} - 1)\right| = f(g(z)) + an. \tag{15}$$

We may assume without any loss of generality that $|e^c| \leqq 1$.

For a fixed z the left side of (15), as n goes to infinity, is either finite or has a limit point, while the right side of (15) is an infinite discrete set.

Subsequent to the completion of this paper the authors discovered an extension of Whittaker's results due to C. T. Rajagopal and A. R. Reddy; see Proc. Cambridge Philos. Soc. **61** (1965) 673–677 and Ibid. **63** (1967) 1091–1092. From this extension it follows that g_2 in Theorem 1 is of order less than 1.

REFERENCES

1. E. Borel, *Sur les zéros des fonctions entières*, Acta Math. **20** (1897).
2. F. Gross, *On the factorization of meromorphic functions*, Lecture Notes Math. Inst. On Entire Functions and Related Parts of Analysis, La Jolla, Calif., June 1966.
3. R. Nevanlinna, *Le théorème de Picard-Borel et la théorie des fonctions méromorphes*, Paris, 1929.
4. P. C. Rosenbloom, *The fix-points of entire functions*, Medd. Lunds Univ. Mat. Sem., Tome Suppl. (1952), 186–192.
5. J. M. Whittaker, *Interpolatory function theory*, Stechert-Hafner, New York, 1964.

IMPERIAL COLLEGE
LONDON, ENGLAND, AND
BELLCOMM. INC.
WASHINGTON, D. C.

An Introduction to the Theory of Meromorphic Minimal Surfaces

E. F. BECKENBACH*

1. **Introduction.** A surface S is said to be given in terms of isothermic parameters (u, v) if and only if the representation

$$S: x_j = x_j(u, v), \qquad j = 1, 2, 3, \tag{1}$$

for $(u, v) \in D$, where D is a finite domain of definition, is such that

$$E = G = \lambda(u, v), \qquad F = 0, \tag{2}$$

where

$$E = \sum_{j=1}^{3} \left(\frac{\partial x_j}{\partial u}\right)^2, \qquad F = \sum_{j=1}^{3} \left(\frac{\partial x_j}{\partial u}\right)\left(\frac{\partial x_j}{\partial v}\right), \qquad G = \sum_{j=1}^{3} \left(\frac{\partial x_j}{\partial v}\right)^2. \tag{3}$$

Such a representation is conformal, or angle-preserving, except at points where $\lambda(u, v) = 0$.

According to a theorem of Weierstrass [**12**, p. 27], a necessary and sufficient condition that a surface S, given in terms of isothermic parameters, be minimal is that the coordinate functions be harmonic, that is, that for all $(u, v) \in D$ the functions $x_j(u, v)$, $j = 1, 2, 3$, satisfy the equation

$$\Delta x_j(u, v) = 0, \tag{4}$$

where Δ denotes the Laplace operator,

$$\Delta = \partial^2/\partial u^2 + \partial^2/\partial v^2. \tag{5}$$

Then in any simply connected part of D, the functions given by (1) are the real parts of analytic functions of a complex variable,

$$x_j = \Re f_j(w), \qquad w = u + iv, \tag{6}$$

and (2) is equivalent to

---

* The preparation of this paper was sponsored in part by the National Science Foundation Grant GP-5279.

(7) $$\sum_{j=1}^{3} \left[\frac{df_j(w)}{dw}\right]^2 = 0.$$

If, in an isothermic representation (1) of a minimal surface S, one of the coordinate functions is identically zero, say $x_3(u,v) \equiv 0$, then the map lies on a plane, and either $x_1(u,v) + ix_2(u,v)$ or $x_2(u,v) + ix_1(u,v)$ is an analytic function of the complex variable $w = u + iv$. Then $x_1(u,v)$ and $x_2(u,v)$ are said to form a *couple of conjugate harmonic functions.* By analogy, the coordinate functions of any minimal surface in isothermic representation are called a *triple of conjugate harmonic functions* [6].

The analogy here indicated between analytic functions of a complex variable and isothermic representations of minimal surfaces has often been noted, and since the time of Weierstrass it has served as a guiding principle in the study of minimal surfaces. It is the purpose of the present paper to give a brief introduction to notions and methods [2]–[6] whereby this analogy can be pursued in the direction of the classical theory of meromorphic functions of a complex variable.

2. **Meromorphic minimal surfaces.** Let the real-valued function $x(u,v)$ be harmonic for (u,v) in a deleted circular neighborhood $\mathscr{U}_\varepsilon^*(P_0)$ of a point $P_0:(u_0,v_0)$, that is, for (u,v) satisfying

$$0 < (u-u_0)^2 + (v-v_0)^2 < \varepsilon^2.$$

Then $x(u,v)$ can be represented [11, p. 692] in $\mathscr{U}_\varepsilon^*(P_0)$ by a series of the form

$$x(u,v) = c\log r + \sum_{k=-\infty}^{\infty} r^k(a_k\cos k\theta + b_k\sin k\theta),$$

where (r,θ) are polar coordinates with pole P_0:

$$u - u_0 = r\cos\theta, \qquad v - v_0 = r\sin\theta.$$

The constant b_0 is arbitrary; throughout this paper, we shall assume that it has been assigned the value 0. Otherwise, the constants c, a_k $(k = 0, \pm 1, \pm 2, \cdots)$, and b_k $(k = \pm 1, \pm 2, \cdots)$ are uniquely determined by the function $x(u,v)$.

We then have, for $\omega = w - w_0 = (u+iv) - (u_0 + iv_0)$,

$$x(u,v) = \Re[c\log\omega + f(\omega)], \qquad \omega = r(\cos\theta + i\sin\theta),$$

where

$$f(\omega) = \sum_{k=-\infty}^{\infty} (a_k - ib_k)\omega^k$$

is an analytic function of ω in $\mathscr{U}_\varepsilon^*(P_0)$.

By (7), three such functions,

$$
\begin{aligned}
x_j(u,v) &= c_j \log r + \sum_{k=-\infty}^{\infty} r^k (a_{j,k} \cos k\theta + b_{j,k} \sin k\theta) \\
&= \Re[c_j \log \omega + f_j(\omega)], \qquad j = 1,2,3,
\end{aligned} \tag{8}
$$

harmonic in $\mathscr{U}_\varepsilon^*(P_0)$, are a triple of conjugate harmonic functions there if and only if

$$
\sum_{j=1}^{3} \left[c_j \omega^{-1} + \frac{df_j(\omega)}{d\omega} \right]^2 = 0. \tag{9}
$$

Computing the coefficients of ω^k in (9) for each k, $k = 0, \pm 1, \pm 2, \cdots$, we find that (9) holds if and only if

$$
2k \sum_{j=1}^{3} a_{j,k} c_j + \sum_{l=-\infty}^{\infty} l(k-l) \sum_{j=1}^{3} (a_{j,l} a_{j,k-l} - b_{j,l} b_{j,k-l}) = 0, \qquad k = \pm 1, \pm 2 \cdots; \tag{10}
$$

$$
\sum_{j=1}^{3} c_j^2 - \sum_{l=-\infty}^{\infty} l^2 \sum_{j=1}^{3} (a_{j,l} a_{j,-l} - b_{j,l} b_{j,-l}) = 0; \tag{11}
$$

and

$$
k \sum_{j=1}^{3} b_{j,k} c_j + \sum_{l=-\infty}^{\infty} l(k-l) \sum_{j=1}^{3} a_{j,l} b_{j,k-l} = 0, \quad k = 0, \pm 1, \pm 2, \cdots. \tag{12}
$$

Thus (10), (11), *and* (12) *are necessary and sufficient conditions for the functions* (8) *to be a triple of conjugate harmonic functions.*

If the functions (8) are a triple of conjugate harmonic functions in $\mathscr{U}_\varepsilon^*(P_0)$, that is, if the functions (8) are the coordinate functions of a minimal surface S in isothermic representation for $(u,v) \in \mathscr{U}_\varepsilon^*(P_0)$, then we shall say that P_0 is an *isolated singular point* of S.

If for an infinitude of negative indices l we have

$$
\sum_{j=1}^{3} (a_{j,l}^2 + b_{j,l}^2) \neq 0, \tag{13}
$$

then we say that the isolated singularity of S at P_0 is *essential*, otherwise that it is *nonessential*.

If the isolated singularity of S at P_0 is nonessential, and the lowest index $l = t$ for which (13) holds is negative, then we say that S has a *pole of order* $|t|$ at P_0. By definition, then, the poles of S are isolated.

We note by (11) that if (13) does not hold for any negative value of l, that is, if

$$
a_{j,l} = b_{j,l} = 0, \qquad l = -1, -2, \cdots,
$$

then

(14) $$\sum_{j=1}^{3} c_j^2 = 0,$$

or $c_1 = c_2 = c_3 = 0$. Hence a minimal surface given in isothermic representation by functions $x_j(u, v)$ *cannot have a singularity that is merely logarithmic.*

If (13) does not hold for any $l < 0$, then we say that S has a *removable singularity* at P_0. If in this case we adjoin to S the point

(15) $$\boldsymbol{a}_0 = (a_{1,0}, a_{2,0}, a_{3,0})$$

corresponding to P_0, then the functions (8) give an isothermic map of the neighborhood $\mathscr{U}_\varepsilon(P_0)$ that is, of the set of values (u, v) satisfying

$$(u - u_0)^2 + (v - v_0)^2 < \varepsilon^2,$$

onto the extended surface, which we again denote by S. We then say that S is *regular* at P_0.

If S is regular at P_0, then either each $x_j(u, v) \equiv a_{j,0}$, and S reduces to a point, or there is a lowest positive index $l = t$ for which (13) holds. In the former case, we say that S is a *constant minimal surface.* In the latter case, we say that S has an ***a**-point of order* t at P_0; in particular, if $\boldsymbol{a} = (0, 0, 0)$ then we say that S has a *zero of order* t at P_0.

If S has a pole of order $-t > 0$ or an $\boldsymbol{a}$-point of order $t > 0$ at P_0, then for $k = 2t$, (10) and (12) reduce respectively to

$$t^2 \sum_{j=1}^{3} (a_{j,t}^2 - b_{j,t}^2) = 0$$

and

$$t^2 \sum_{j=1}^{3} a_{j,t} b_{j,t} = 0,$$

so that, by (13),

(16) $$\sum_{j=1}^{3} a_{j,t}^2 = \sum_{j=1}^{3} b_{j,t}^2 \neq 0, \qquad \sum_{j=1}^{3} a_{j,t} b_{j,t} = 0.$$

If S has a pole of order $-t > 0$ at P_0, then by (8) and (16) for $r \to 0$ we have

$$\sum_{j=1}^{3} [x_j(u, v)]^2$$

(17) $$= r^{2t} \left(\sum_{j=1}^{3} a_{j,t}^2 \cos^2 t\theta + 2 \sum_{j=1}^{3} a_{j,t} b_{j,t} \cos t\theta \sin t\theta + \sum_{j=1}^{3} b_{j,t}^2 \sin^2 t\theta \right) + o(r^{2t})$$

$$= r^{2t} \sum_{j=1}^{3} a_{j,t}^2 + o(r^{2t}).$$

Similarly, if S has an $\boldsymbol{a}$-point of order $t > 0$ at P_0, then

$$\sum_{j=1}^{3} [x_j(u,v) - a_{j,0}]^2 = r^{2t} \sum_{j=1}^{3} a_{j,t}^2 + o(r^{2t}). \tag{18}$$

By (16) we thus see that if S is a nonconstant minimal surface, then *not only the poles but also the $\boldsymbol{a}$-points of S are isolated* [**4**].

If, except for poles, S is a regular minimal surface given in isothermic representation by (1) for (u,v) in a finite domain D, then we say that S is a *meromorphic minimal surface for (u,v) in D*. In particular, if D is the entire finite plane, then we say simply that S is a *meromorphic minimal surface.* If D is the entire finite plane and S has no poles in D, then we say that S is an *entire minimal surface.*

3. **Examples.** The functions

$$\begin{aligned} x_1 &= \Re(1/w + w) = (1/r + r)\cos\theta, \\ x_2 &= \Re i(1/w - w) = (1/r + r)\sin\theta, \\ x_3 &= \Re(2\log w) = 2\log r, \qquad w = r(\cos\theta + i\sin\theta), \end{aligned} \tag{19}$$

are the coordinate functions of a meromorphic minimal surface (actually a catenoid) in isothermic representation. Its single pole in the finite plane is at the origin and is of order 1.

The minimal surface of Enneper [**10**, p. 221] is given in isothermic representation by

$$\begin{aligned} x_1 &= \Re(3w - w^3) = 3r\cos\theta - r^3\cos 3\theta, \\ x_2 &= \Re i(3w + w^3) = -3r\sin\theta - r^3\sin 3\theta, \\ x_3 &= \Re(3w^2) = 3r^2\cos 2\theta, \qquad w = r(\cos\theta + i\sin\theta). \end{aligned} \tag{20}$$

This is an entire minimal surface. Its single zero is at the origin and is of order 1.

The relations

$$\begin{aligned} x_1 &= \Re(\log w - \tfrac{1}{2}w^2) = \log r - \tfrac{1}{2}r^2\cos 2\theta, \\ x_2 &= \Re i(\log w + \tfrac{1}{2}w^2) = -i\theta - \tfrac{1}{2}r^2\sin 2\theta, \\ x_3 &= \Re(2w) = 2r\cos\theta, \qquad w = r(\cos\theta + i\sin\theta), \end{aligned} \tag{21}$$

give an isothermic representation of a minimal surface S with a singularity of a different sort at the origin. The second of the relations (21) is not a (single-valued) function of $w = u + iv$, however, so that this surface is not included in the class of surfaces presently under consideration.

The subclass of meromorphic minimal surfaces (1) for which the c_j are restricted to have the value 0 in the representation (8) for each P_0 is somewhat more tractable than the unrestricted class. The restriction is not needed for the validity of the results of this paper, however, and accordingly we shall not make it here.

4. **Formulas.** In this section, we shall develop some formulas that will be needed later. These are concerned with the differential geometry of meromorphic minimal surfaces.

Let the functions (8) be the coordinate functions of a nonconstant meromorphic minimal surface S in isothermic representation in $\mathscr{U}_\varepsilon^*(P_0)$, and let τ denote the lowest index l for which

$$\sum_{j=1}^{3} a_{j,l}^2 \neq 0. \tag{22}$$

By (16) and the fact that we have chosen $b_{j,0} = 0$, $j = 1, 2, 3$, τ is also the lowest index for which (13) holds. Then (8) can be written as

$$\begin{aligned} x_j(u,v) &= c_j \log r + \sum_{k=\tau}^{\infty} r^k(a_{j,k}\cos k\theta + b_{j,k}\sin k\theta) \\ &= \Re[c_j \log \omega + f_i(\omega)], \qquad j = 1, 2, 3, \end{aligned} \tag{23}$$

where

$$\sum_{j=1}^{3} a_{j\,\tau}^2 \neq 0 \tag{24}$$

and

$$f_j(\omega) = \sum_{k=\tau}^{\infty} (a_{j,k} - ib_{j,k})\omega^k. \tag{25}$$

If $\tau = 0$, let t denote the lowest positive index for which (13) holds; if $\tau \neq 0$, let $t = \tau$. Then $t \geqq \tau$, with inequality if and only if $\tau = 0$. Recalling that (14) holds if $\tau \geqq 0$, we see that:

If $\tau < 0$, then $t = \tau$ and S has a pole of order $-t$ at P_0.

If $\tau = 0$, then $t > \tau$ and S has an $\boldsymbol{a}$-point ($a \neq 0$) of order t at P_0.

If $\tau > 0$, then $t = \tau$ and S has a zero of order t at P_0.

A computation analogous to that which led to (18) yields

$$\sum_{j=1}^{3} [x_j(u,v)]^2 = r^{2\tau} \sum_{j=1}^{3} a_{j,\tau}^2 + o(r^{2\tau}). \tag{26}$$

From (25) and the definition of t, by differentiation we obtain

$$\frac{\partial x_j}{\partial u} - i\frac{\partial x_j}{\partial v} = \frac{c_j}{\omega} + \sum_{k=t}^{\infty} k(a_{j,k} - ib_{j,k})\omega^{k-1}. \tag{27}$$

Equating real parts, and equating imaginary parts, in (27), from (2), (3), (14), and (16) we obtain, by a computation,

$$\lambda(u,v) = t^2 r^{2t-2} \sum_{j=1}^{3} a_{j,t}^2 + O(r^{2t-1}). \tag{28}$$

It follows from (16) and (28) (cf. [**4**]) that, *for a nonconstant meromorphic minimal surface S given in terms of isothermic parameters (u,v), the zeros and infinities of the area-deformation ratio $\lambda(u,v)$ are isolated.*

At points where $\lambda(u,v) \neq 0$ and $\lambda(u,v) \neq \infty$, S has a tangent plane. The direction cosines $X_j(u,v)$ of its normal are given [**10**, p. 147] by

$$X_j(u,v) = \left(\frac{\partial x_k(u,v)}{\partial u}\frac{\partial x_l(u,v)}{\partial v} - \frac{\partial x_l(u,v)}{\partial u}\frac{\partial x_k(u,v)}{\partial v}\right)\Big/\lambda(u,v), \tag{29}$$

where j, k, $l = 1,2,3$ in cyclic order. For the functions (23), let $\varepsilon > 0$ be so small that $\lambda(u,v) \neq 0$ and $\lambda(u,v) \neq \infty$ in $\mathscr{U}_\varepsilon^*(P_0)$. Then for $(u,v) \in \mathscr{U}_\varepsilon^*(P_0)$, from (27), (28), and (29) we obtain, by a computation,

$$\begin{aligned} X_j(u,v) &= \frac{t^2 r^{2t-2}(a_{k,t}b_{l,t} - a_{l,t}b_{k,t}) + O(r^{2t-1})}{t^2 r^{2t-2} \sum_{q=1}^{3} a_{q,t}^2 + O(r^{2t-1})} \\ &= (a_{k,t}b_{l,t} - a_{l,t}b_{k,t}) \Big/ \sum_{q=1}^{3} a_{q,t}^2 + O(r), \end{aligned} \tag{30}$$

From (30), we see that

$$\lim_{r\to 0} X_j(u,v) = (a_{k,t}b_{l,t} - a_{l,t}b_{k,t}) \Big/ \sum_{q=1}^{3} a_{q,t}^2, \tag{31}$$

j, k, $l = 1,2,3$ in cyclic order, even if $\lambda(u_0,v_0) = 0$ or $\lambda(u_0,v_0) = \infty$. We take this limiting value as the definition of $X_j(u_0,v_0)$ if $\lambda(u_0,v_0) = 0$ or $\lambda(u_0,v_0) = \infty$. With this extended definition, we see by (16) that *a nonconstant meromorphic minimal surface S given in terms of isothermic parameters (u,v) in a domain D has a continuous unit normal vector function*

$$X(u,v) = (X_1(u,v),\ X_2(u,v),\ X_3(u,v))$$

throughout D.

5. The distance function. The next formula we shall develop is fundamental for the present investigation. It is an expression [**6**] for the Laplacian of the logarithm of the distance function

$$\left\{\sum_{j=1}^{3} [x_j(u,v)]^2\right\}^{\frac{1}{2}} \tag{32}$$

for a meromorphic minimal surface given in isothermic representation.

For the isothermic coordinate functions (23) of a meromorphic minimal surface S, let $\varepsilon > 0$ be so small that in $\mathscr{U}_\varepsilon^*(P_0)$ the distance function (32) and the area-deformation ratio $\lambda(u,v)$ have no zeros or infinities. They might or might not vanish or be infinite at P_0.

Using vector notation, for $(u,v) \in \mathscr{U}_\varepsilon^*(P_0)$ we obtain, by a computation,

$$(33)\quad \Delta \log (\boldsymbol{x} \cdot \boldsymbol{x})^{\frac{1}{2}} = \frac{(\boldsymbol{x} \cdot \boldsymbol{x})(\boldsymbol{x} \cdot \Delta \boldsymbol{x} + \boldsymbol{x}_u \cdot \boldsymbol{x}_u + \boldsymbol{x}_v \cdot \boldsymbol{x}_v) - 2[(\boldsymbol{x} \cdot \boldsymbol{x}_u)^2 + (\boldsymbol{x} \cdot \boldsymbol{x}_v)^2]}{(\boldsymbol{x} \cdot \boldsymbol{x})^2},$$

where the subscripts indicate partial differentiation. By (2) and (3) we have

$$\boldsymbol{x}_u \cdot \boldsymbol{x}_u = \boldsymbol{x}_v \cdot \boldsymbol{x}_v = \lambda,$$

and from (4) we obtain

$$\Delta \boldsymbol{x} = \boldsymbol{0},$$

so that (33) reduces to

$$(34)\qquad \Delta \log(\boldsymbol{x} \cdot \boldsymbol{x})^{\frac{1}{2}} = 2[\lambda(\boldsymbol{x} \cdot \boldsymbol{x}) - (\boldsymbol{x} \cdot \boldsymbol{x}_u)^2 - (\boldsymbol{x} \cdot \boldsymbol{x}_v)^2]/(\boldsymbol{x} \cdot \boldsymbol{x})^2 .$$

Since $\lambda(u,v) \neq 0$ in $\mathscr{U}_\varepsilon^*(P_0)$, $\boldsymbol{x}_u$ and $\boldsymbol{x}_v$ are nonnull vectors there. Further, these vectors lie in the tangent plane to S and therefore are perpendicular to the unit normal vector $\boldsymbol{X}(u,v)$; and since $F = 0$, they are perpendicular to each other. Accordingly, for any $(u,v) \in \mathscr{U}_\varepsilon^*(P_0)$, there are scalars a, b, c such that

$$(35)\qquad \boldsymbol{x} = a\boldsymbol{x}_u + b\boldsymbol{x}_v + c\boldsymbol{X}.$$

From (35), we obtain

$$\begin{aligned} \boldsymbol{x} \cdot \boldsymbol{x} &= a^2\lambda + b^2\lambda + c^2, \quad \boldsymbol{x} \cdot \boldsymbol{x}_u = a\lambda, \\ \boldsymbol{x} \cdot \boldsymbol{x}_v &= b\lambda, \quad \boldsymbol{x} \cdot \boldsymbol{X} = c. \end{aligned}$$

Hence (34) can be written as

$$\begin{aligned} \Delta \log(\boldsymbol{x} \cdot \boldsymbol{x})^{\frac{1}{2}} &= 2(a^2\lambda^2 + b^2\lambda^2 + c^2\lambda^2 - a^2\lambda^2 - b^2\lambda^2)/(\boldsymbol{x} \cdot \boldsymbol{x}) \\ &= 2c^2\lambda/(\boldsymbol{x} \cdot \boldsymbol{x})^2, \end{aligned}$$

or, finally,

$$(36)\qquad \Delta \log(\boldsymbol{x} \cdot \boldsymbol{x})^{\frac{1}{2}} = 2(\boldsymbol{x} \cdot \boldsymbol{X})^2\lambda/(\boldsymbol{x} \cdot \boldsymbol{x})^2 .$$

In obtaining an estimate of the behavior of the right-hand member of (36) as $r \to 0$, we can use the expressions (26) and (28) for $\boldsymbol{x} \cdot \boldsymbol{x}$ and λ, respectively. For $\boldsymbol{x} \cdot \boldsymbol{X}$, by (23) and (30) we have

$$(37)\qquad \boldsymbol{x} \cdot \boldsymbol{X} = \frac{r^\tau \sum_{j,k,l} (a_{j,\tau}\cos\tau\theta + b_{j,\tau}\sin\tau\theta)(a_{k,t}b_{l,t} - a_{l,t}b_{k,t})}{\sum_{q=1}^{3} a_{q,t}^2} + o(r^\tau),$$

where j, k, $l = 1, 2, 3$ in cyclic order in the sum in the numerator.

If $\tau = 0$, then by (16), (24), (26), (28), and (37) we have

$$\lim_{r\to 0} (\boldsymbol{x}\cdot\boldsymbol{x})^2 = \left(\sum_{j=1}^{3} a_{j,0}^2\right)^2 > 0,$$

$$\lim_{r\to 0} \lambda = t^2 \sum_{j=1}^{3} a_{j,t}^2 \lim_{r\to 0} r^{2t-2} \geqq 0,$$

with equality if and only if $t > 1$, and

$$\lim_{r\to 0} (\boldsymbol{x}\cdot\boldsymbol{X})^2 = \left[\sum_{j,k,l} a_{j,0}(a_{k,t}b_{l,t} - a_{l,t}b_{k,t}) \Big/ \sum_{q=1}^{3} a_{q,t}^2\right]^2 \geqq 0.$$

Since each of these three limits exists and is finite, and the limit of $(\boldsymbol{x}\cdot\boldsymbol{x})^2$ is not 0, it follows from (36) that $\Delta\log(\boldsymbol{x}\cdot\boldsymbol{x})^{\frac{1}{2}}$ is continuous in this case.

If $\tau \neq 0$, then $t = \tau$, and the sum in the numerator of the first term in the right-hand member of (37) vanishes identically. We then have, by (23) and (30),

$$\boldsymbol{x}\cdot\boldsymbol{X} = O(r^{\tau+1}) \tag{38}$$

unless

$$\tau = -1 \quad\text{and}\quad \sum_{j=1}^{3} c_j^2 \neq 0. \tag{39}$$

In the exceptional case (39), we have

$$\boldsymbol{x}\cdot\boldsymbol{X} = O(\log r). \tag{40}$$

If (38) holds, then from (24), (26), (28), (36), and (38), we get

$$\begin{aligned}\Delta\log(\boldsymbol{x}\cdot\boldsymbol{x})^{\frac{1}{2}} &= [O(r^{\tau+1})]^2 O(r^{2\tau-2}) \Big/ \left[r^{2\tau}\sum_{j=1}^{3} a_{j,\tau}^2 + o(r^{2\tau})\right]^2 \\ &= O(r^{2\tau+2+2\tau-2-4\tau}) \qquad (41) \\ &= O(1),\end{aligned}$$

so that $\Delta\log(\boldsymbol{x}\cdot\boldsymbol{x})^{\frac{1}{2}}$ is bounded in the neighborhood of P_0. It is not necessarily continuous at P_0, however. For example, for the minimal surface of Enneper (20), we have

$$\Delta\log(\boldsymbol{x}\cdot\boldsymbol{x})^{\frac{1}{2}} = 2\cos^2 2\theta + o(r),$$

so that the limiting behavior of $\Delta\log(\boldsymbol{x}\cdot\boldsymbol{x})^{\frac{1}{2}}$ depends on the limiting behavior of θ.

In the exceptional case (39), in place of (41) we have, by (24) and (40),

$$\begin{aligned}\Delta\log(\boldsymbol{x}\cdot\boldsymbol{x})^{\frac{1}{2}} &= [O(\log r)]^2 O(r^{-4}) \Big/ \left[r^{-2}\sum_{j=1}^{3} a_{j,-1}^2 + o(r^{-2})\right]^2 \\ &= O[r^{-4+4}(\log r)^2] \qquad (42) \\ &= O[(\log r)^2].\end{aligned}$$

For example, for the catenoid (19) we have

$$\Delta \log (\boldsymbol{x} \cdot \boldsymbol{x})^{\frac{1}{2}} = 8(\log r)^2[1 + o(1)].$$

Thus in the exceptional case (39), $\Delta \log (\boldsymbol{x} \cdot \boldsymbol{x})^{\frac{1}{2}}$ becomes infinite as $r \to 0$. As we shall see in §7, however, it does not become infinite too rapidly for the applications we shall be making.

6. **Notation.** Let $w = u + iv$; let $A_r(w_0)$ denote the closed circular disc $|w - w_0| \leqq r$; and let $\partial A_r(w_0)$ denote the boundary, $|w - w_0| = r$, of $A_r(w_0)$.

Let the functions (1) be the coordinate functions of a nonconstant meromorphic minimal surface S in isothermic representation for (u, v) in a finite domain D, and for a given fixed $R > 0$ let $A_R(0)$ be contained in D.

If S has a zero at the origin, let s_0 denote the order of this zero; otherwise, let $s_0 = 0$. Similarly, if S has a pole at the origin, let n_0 denote the order of this pole, and otherwise let $n_0 = 0$. Then at least one of s_0 and n_0 must be equal to 0, and both might be equal to 0.

Let s_0 be denoted by $n(0,0; S)$, and n_0 by $n(0, \infty; S)$. Then by (23) and the definition of τ, if τ_0 is the value of τ for the functions (1) when P_0 is the origin, we have

$$\boldsymbol{x} \cdot \boldsymbol{x} = r^{2\tau_0} \sum_{j=1}^{3} a_{j,\tau_0}^2 + o(r^{2\tau}), \qquad r = |w|, \tag{43}$$

with

$$\sum_{j=1}^{3} a_{j,\tau_0}^3 \neq 0 \tag{44}$$

and

$$\tau_0 = s_0 - n_0 = n(0,0; S) - n(0, \infty; S). \tag{45}$$

For any ρ, $0 < \rho \leqq R$, there can be only a finite number of zeros and poles of S in $A_\rho(0)$, since the zeros and poles of a nonconstant meromorphic minimal surface are isolated. In the punctured disc $0 < |w| \leqq \rho$, let the zeros of S be at the points

$$w = z_1, z_2, \cdots, z_k, \qquad k = k(\rho) \geqq 0,$$

with

$$0 < |z_1| \leqq |z_2| \leqq \cdots \leqq |z_k| \leqq \rho,$$

and let the poles be at

$$w = p_1, p_2, \cdots, p_l, \qquad l = l(\rho) \geqq 0,$$

with

$$0 < |p_1| \leqq |p_2| \leqq \cdots \leqq |p_l| \leqq \rho.$$

Let the orders of these zeros and poles be, respectively, $s_1, s_2, \cdots, s_k$ and $n_1, n_2, \cdots, n_l$, and denote the sum of the orders of the zeros and poles of S in $A_\rho(0)$ by $n(\rho, 0; S)$ and $n(\rho, \infty; S)$, respectively:

$$n(\rho, 0; S) = s_0 + s_1 + s_2 + \cdots + s_k, \tag{46}$$

$$n(\rho, \infty; S) = n_0 + n_1 + n_2 + \cdots + n_l. \tag{47}$$

Then we have $n(\rho, 0; S) \geqq 0$, with $s_j > 0$ for $j = 1, 2, \cdots, k$, and similarly $n(\rho, \infty; S) \geqq 0$, with $n_q > 0$ for $q = 1, 2, \cdots, l$.

7. **An application of Green's theorem.** Now let ρ, $0 < \rho \leqq R$, be chosen so that S does not have a zero or pole on $\partial A_\rho(0)$. Since there are only a finite number of zeros and poles in $A_\rho(0)$, we can choose $r > 0$ so small that the $k + l + 1$ closed circular discs $A_r(0)$, $A_r(z_j)$, $j = 1, 2, \cdots, k$, and $A_r(p_q)$, $q = 1, 2, \cdots, l$, are disjoint from one another and interior to $A_\rho(0)$. Let Ω_r denote the domain interior to the circle $\partial A_\rho(0)$ and exterior to the $k + l + 1$ circles $\partial A_r(0)$, $\partial A_r(z_j)$, and $\partial A_r(p_q)$.

In Ω_r, the function

$$g(u, v) = \log (\mathbf{x} \cdot \mathbf{x})^{\frac{1}{2}} \tag{48}$$

has continuous derivatives of all orders. Hence we can apply Green's theorem to $g(u, v)$ in Ω_r:

$$\int_{\partial\Omega_r} \frac{\partial g(u, v)}{\partial \nu} ds = \iint_{\Omega_r} \Delta g(u, v)\, dA, \tag{49}$$

where ν refers to the normal directed outwardly from Ω_r.

By the definition of Ω_r, we have

$$\begin{aligned} \int_{\partial\Omega_r} \frac{\partial g(u, v)}{\partial \nu} ds = & \int_{\partial A_\rho(0)} \frac{\partial g}{\partial \rho} \rho\, d\theta - \int_{\partial A_r(0)} \frac{\partial g}{\partial r} r\, d\theta \\ & - \sum_{j=1}^{k} \int_{\partial A_r(z_j)} \frac{\partial g}{\partial r} r\, d\theta - \sum_{q=1}^{l} \int_{\partial A_r(p_q)} \frac{\partial g}{\partial r} r\, d\theta. \end{aligned} \tag{50}$$

From (23), (43), and (45), by a computation we find that on $\partial A_r(0)$ we have

$$\partial g / \partial r = \tau_0 / r + o(r^{-1}) = (s_0 - n_0)/r + o(r^{-1}),$$

so that

$$\begin{aligned} \int_{\partial A_r(0)} \frac{\partial g}{\partial r} r\, d\theta &= 2\pi\tau_0 + o(1) \\ &= 2\pi(s_0 - n_0) + o(1). \end{aligned}$$

Similarly, on the $\partial A_r(z_j)$ and $\partial A_r(p_q)$ we have, respectively,

$$\int_{\partial A_r(z_j)} \frac{\partial g}{\partial r} r\, d\theta = 2\pi s_j + o(1)$$

and

$$\int_{\partial A_r(pq)} \frac{\partial g}{\partial r} r\, d\theta = -2\pi n_q + o(1).$$

By (50), (46), and (47) we accordingly have

$$\begin{aligned}\int_{\partial\Omega_r} \frac{\partial g(u,v)}{\partial v} ds &= \int_{\partial A_\rho(0)} \frac{\partial g}{\partial \rho} \rho\, d\theta - 2\pi \sum_{j=0}^{k} s_j + 2\pi \sum_{q=0}^{l} n_q + o(1) \\ &= \int_{\partial A\ (0)} \frac{\partial g}{\partial \rho} \rho\, d\theta - 2\pi n(\rho, 0; S) + 2\pi n(\rho, \infty; S) + o(1),\end{aligned}$$

whence

$$\lim_{r\to 0} \int_{\partial\Omega_r} \frac{\partial g(u,v)}{\partial v} ds = \int_{\partial A_\rho(0)} \frac{\partial \sigma}{\partial \rho} \rho\, d\theta - 2\pi n(\rho, 0; S) + 2\pi n(\rho, \infty; S). \tag{51}$$

By (41) and (42), for any $A_r(w) \subset D$ we have

$$\iint_{A_r(w)} \Delta g(u,v)\, dA = \int_0^{2\pi} \int_0^r O(1)\sigma\, d\sigma\, d\theta,$$

or at worst, in the exceptional case (39),

$$\iint_{A_r(w)} \Delta g(u,v)\, dA = \int_0^{2\pi} \int_0^r O[(\log \sigma)^2]\sigma\, d\sigma d\theta.$$

Since $\sigma O(1) = o(1)$ and $\sigma O[(\log \sigma)^2] = o(1)$, in either case we have

$$\lim_{r\to 0} \iint_{A_r(w)} \Delta g(u,v)\, dA = 0.$$

Therefore, by the definition of Ω_r,

$$\lim_{r\to 0} \iint_{\Omega_r} \Delta g(u,v)\, dA = \iint_{A_\rho(0)} \Delta g(u,v)\, dA. \tag{52}$$

From (49), (51), and (52), we obtain

$$\int_{\partial A_\rho(0)} \frac{\partial g(u,v)}{\partial \rho} \rho\, d\theta - 2\pi n(\rho, 0; S) + 2\pi n(\rho, \infty; S) = \iint_{A_\rho(0)} \Delta g(u,v)\, dA. \tag{53}$$

8. **An integration.** Equation (53) yields

$$\frac{1}{2\pi} \int_{\partial A_\rho(0)} \frac{\partial g(u,v)}{\partial \rho} d\theta - \frac{n(\rho, 0; S) - n(\rho, \infty; S)}{\rho} = \frac{1}{2\pi\rho} \iint_{A_\rho(0)} \Delta g(u,v)\, dA,$$

which can be rewritten as

$$\frac{1}{2\pi}\int_0^{2\pi}\frac{dg}{\partial\rho}d\theta-\frac{n(0,0;S)-n(0,\infty;S)}{\rho}-\frac{n(\rho,0;S)-n(0,0;S)}{\rho}$$

$$+\frac{n(\rho,\infty;S)-n(0,\infty;S)}{\rho}=\frac{1}{2\pi\rho}\iint_{A_\rho(0)}\Delta g\,dA. \tag{54}$$

Noting that, by (45),

$$-\frac{n(0,0;S)-n(0,\infty;S)}{\rho}=\frac{-\tau_0}{\rho}=\frac{\partial}{\partial\rho}\log\rho^{-\tau_0}$$

$$=\frac{1}{2\pi}\int_0^{2\pi}\frac{\partial}{\partial\rho}\log\rho^{-\tau_0}\,d\theta,$$

we can apply (48) to rewrite (54) as

$$\frac{1}{2\pi}\int_0^{2\pi}\frac{\partial}{\partial\rho}\log[(\boldsymbol{x}\cdot\boldsymbol{x})^{\frac{1}{2}}\rho^{-\tau_0}]\,d\theta-\frac{n(\rho,0;S)-n(0,0;S)}{\rho}$$

$$+\frac{n(\rho,\infty;S)-n(0,\infty;S)}{\rho}=\frac{1}{2\pi\rho}\iint_{A_\rho(0)}\Delta g\,dA. \tag{55}$$

By (41) and (42), for any r, $0<r\leqq R$, the right-hand member of (55) can be integrated from 0 to r with respect to ρ. The numerators in the second and third terms on the left vanish in an interval $0\leqq\rho\leqq r_0$, $r_0>0$, so these terms also can be integrated. For the first term in the left-hand member of (55), by (43) we have

$$\int_\varepsilon^r\left\{\frac{1}{2\pi}\int_0^{2\pi}\frac{\partial}{\partial\rho}\log[(\boldsymbol{x}\cdot\boldsymbol{x})^{\frac{1}{2}}\rho^{-\tau_0}]\,d\theta\right\}d\rho$$

$$=\frac{1}{2\pi}\int_0^{2\pi}\left\{\int_\varepsilon^r\frac{\partial}{\partial\rho}\log[(\boldsymbol{x}\cdot\boldsymbol{x})^{\frac{1}{2}}\rho^{-\tau_0}]\,d\rho\right\}d\theta$$

$$=\frac{1}{2\pi}\int_0^{2\pi}\log[(\boldsymbol{x}\cdot\boldsymbol{x})^{\frac{1}{2}}\rho^{-\tau_0}]\Big|_\varepsilon^r d\theta$$

$$=\frac{1}{2\pi}\int_{\partial A_r(0)}\log(\boldsymbol{x}\cdot\boldsymbol{x})^{\frac{1}{2}}d\theta-\tau_0\log r-\log\left(\sum_{j=1}^3 a_{j,\tau_0}^2\right)^{\frac{1}{2}}+o(1),$$

so that

$$\lim_{\varepsilon\to 0}\int_\varepsilon^r\left\{\frac{1}{2\pi}\int_0^{2\pi}\frac{\partial}{\partial\rho}\log[(\boldsymbol{x}\cdot\boldsymbol{x})^{\frac{1}{2}}\rho^{-\tau_0}]\,d\theta\right\}d\rho$$

$$=\frac{1}{2\pi}\int_{\partial A_r(0)}\log(\boldsymbol{x}\cdot\boldsymbol{x})^{\frac{1}{2}}\,d\theta-n(0,0;S)\log r$$

$$+\,n(0,\infty;S)\log r-\log\left(\sum_{j=1}^3 a_{j,\tau_0}^2\right)^{\frac{1}{2}}.$$

Hence (55) yields

$$\frac{1}{2\pi}\int_{\partial A_r(0)} \log(\boldsymbol{x}\cdot\boldsymbol{x})^{\frac{1}{2}}\,d\theta - \int_0^r \frac{n(\rho,0;S)-n(0,0;S)}{\rho}\,d\rho - n(0,0;S)\log r$$

$$(56)\qquad + \int_0^r \frac{n(\rho,\infty;S)-n(0,\infty;S)}{\rho}\,d\rho + n(0,\infty;S)\log r$$

$$-\log\left(\sum_{j=1}^{3} a_{j,\tau_0}^2\right)^{\frac{1}{2}} = \frac{1}{\pi}\int_0^r\left[\frac{1}{2\rho}\iint_{A_\rho(0)} \Delta g\,dA\right]d\rho.$$

If for a nonnegative function ϕ the function $\log^+\phi$ is defined by

$$\begin{aligned}\log^+\phi &= \log\phi \quad \text{for} \quad \phi \geqq 1,\\ &= 0 \qquad\ \ \text{for} \quad 0 \leqq \phi < 1,\end{aligned}$$

then we have the identity

$$(57)\qquad \log\phi = \log^+\phi - \log^+(1/\phi).$$

By (36), (48), and (57), (56) can accordingly be written as

$$\frac{1}{2\pi}\int_{\partial A_r(0)} \log^+(\boldsymbol{x}\cdot\boldsymbol{x})^{\frac{1}{2}}\,d\theta$$

$$+ \int_0^r \frac{n(\rho,\infty;S)-n(0,\infty;S)}{\rho}\,d\theta + n(0,\infty;S)\log r$$

$$(58)\qquad = \frac{1}{2\pi}\int_{\partial A\ (0)} \log^+(\boldsymbol{x}\cdot\boldsymbol{x})^{-\frac{1}{2}}\,d\theta$$

$$+ \int_0^r \frac{n(\rho,0;S)-n(0,0;S)}{\rho}\,d\rho + n(0,0;S)\log r$$

$$+ \int_0^r\left[\frac{1}{\rho}\iint_{A_\rho(0)} \frac{(\boldsymbol{x}\cdot\boldsymbol{X})^2}{\pi(\boldsymbol{x}\cdot\boldsymbol{x})^2}\lambda\,dA\right]d\rho + \log\left(\sum_{j=1}^{3} a_{j,\tau_0}^2\right)^{\frac{1}{2}}.$$

9. **The proximity, enumerative, and visibility functions.** For the meromorphic minimal surface S given in isothermic representation by the functions (1), and for any given finite

$$\boldsymbol{a} = (a_1, a_2, a_3),$$

consider the surface

$$(59)\qquad S - \boldsymbol{a}\colon\ x_j = x_j(u,v) - a_j, \qquad j = 1,2,3,$$

for $(u,v) \in D$. This again is a nonconstant meromorphic minimal surface in isothermic representation.

Applying (58) to $S - \boldsymbol{a}$, we obtain

$$\begin{aligned}
&\frac{1}{2\pi}\int_{\partial A_r(0)} \log^+[(\boldsymbol{x}-\boldsymbol{a})\cdot(\boldsymbol{x}-\boldsymbol{a})]^{\frac{1}{2}}\,d\theta \\
&\qquad + \int_0^r \frac{n(\rho,\infty;S-\boldsymbol{a}) - n(0,\infty;S-\boldsymbol{a})}{\rho}\,d\rho + n(0,\infty;S-\boldsymbol{a})\log r \\
&= \frac{1}{2\pi}\int_{\partial A_r(0)} \log^+[(\boldsymbol{x}-\boldsymbol{a})\cdot(\boldsymbol{x}-\boldsymbol{a})]^{-\frac{1}{2}}\,d\theta \\
&\qquad + \int_0^r \frac{n(\rho,0;S-\boldsymbol{a}) - n(0,0;S-\boldsymbol{a})}{\rho}\,d\rho + n(0,0;S-\boldsymbol{a})\log r \qquad (60)\\
&\qquad + \int_0^r\left[\frac{1}{\rho}\iint_{A_\rho(0)} \frac{[(\boldsymbol{x}-\boldsymbol{a})\cdot \boldsymbol{X}]^2}{\pi[(\boldsymbol{x}-\boldsymbol{a})\cdot(\boldsymbol{x}-\boldsymbol{a})]^2}\lambda\,dA\right]d\rho \\
&\qquad + \log\left(\sum_{j=1}^{3} a_{j,\tau_{\boldsymbol{a}}}^2\right)^{\frac{1}{2}},
\end{aligned}$$

where $\tau_{\boldsymbol{a}}$ is the value of τ for the functions (59) when P_0 is the origin.

Since the poles of S and the poles of $S - \boldsymbol{a}$ occur at the same (u, v)-points, we have

$$n(\rho,\infty;S) = n(\rho,\infty;S-\boldsymbol{a}). \tag{61}$$

Since the zeros of the surface $S - \boldsymbol{a}$ are the $\boldsymbol{a}$-points of the surface S, the function $n(\rho,\boldsymbol{a};S)$, defined by

$$n(\rho,\boldsymbol{a};S) = n(\rho,0;S-\boldsymbol{a}), \tag{62}$$

gives an expression of the number of $\boldsymbol{a}$-points of the surface S in $|w| \leqq \rho$.

If the map of $A_\rho(0)$ on S is projected from $\boldsymbol{a}$ on the unit sphere with center at $\boldsymbol{a}$, then the function $h(\rho,\boldsymbol{a};S)$, defined by

$$h(\rho,\boldsymbol{a};S) = \iint_{A_\rho(0)} \frac{[(\boldsymbol{x}-\boldsymbol{a})\cdot\boldsymbol{X}]^2}{\pi[(\boldsymbol{x}-\boldsymbol{a})\cdot(\boldsymbol{x}-\boldsymbol{a})]^2}\lambda\,dA, \tag{63}$$

gives an expression of the area of this projection, except that the element of area is weighted by

$$(1/\pi)\left|\cos(\boldsymbol{x}-\boldsymbol{a},\ \mathbf{X})\right|.$$

Thus $h(\rho,\boldsymbol{a}: S)$ gives a measure of the visibility of the surface S for $|w| \leqq \rho$, as viewed from $\boldsymbol{a}$.

In particular, we have

$$h(0,\boldsymbol{a};S) = 0. \tag{64}$$

Since

$$\lim_{a \to \infty} h(\rho, a; S) = 0,$$

we define $h(\rho, \infty; S)$ by

$$h(\rho, \infty; S) = 0. \tag{65}$$

By analogy with the Nevanlinna theory [9] of meromorphic functions of a complex variable, let us define a *proximity function* (*Schmiegungsfunktion*) for S by

$$\begin{aligned} m(r, \infty; S) &= \frac{1}{2\pi} \int_{\partial A_r(0)} \log^+ (x \cdot x)^{\frac{1}{2}} d\theta, \\ m(r, a; S) &= \frac{1}{2\pi} \int_{\partial A_r(0)} \log^+ [(x - a) \cdot (x - a)]^{-\frac{1}{2}} d\theta, \end{aligned} \tag{66}$$

and an *enumerative function* (*Anzahlfunktion*) for S by

$$\begin{aligned} N(r, \infty; S) &= \int_0^r \frac{n(\rho, \infty; S) - n(0, \infty; S)}{\rho} d\rho + n(0, \infty; S) \log r, \\ N(r, a; S) &= \int_0^r \frac{n(\rho, a; S) - n(0, a; S)}{\rho} d\rho + n(0, a; S) \log r, \end{aligned} \tag{67}$$

for a finite.

To these we now adjoin a *visibility function* (*Sichtbarkeitsfunktion*) *to* S, defined by

$$\begin{aligned} H(r, \infty; S) &= 0, \\ H(r, a; S) &= \int_0^r \frac{h(\rho, a; S)}{\rho} d\rho. \end{aligned} \tag{68}$$

By (64) and (65), the definition (68) of H is quite analogous to the definition (67) of N, with h in place of n.

Substituting from (68), (69), and (70) into (58) and (60), and using (61) and (62), we obtain, respectively,

$$\begin{aligned} &m(r, \infty; S) + N(r, \infty; S) + H(r, \infty; S) \\ &\quad = m(r, 0; S) + N(r, 0; S) + H(r, 0; S) + \log \left(\sum_{j=1}^{3} a_{j,\tau_0}^2 \right)^{\frac{1}{2}} \end{aligned} \tag{69}$$

and

$$\begin{aligned} &m(r, \infty; S - a) + N(r, \infty; S) + H(r, \infty; S) \\ &\quad = m(r, a; S) + N(r, a; S) + H(r, a; S) + \log \left(\sum_{j=1}^{3} a_{j,\tau_a}^2 \right)^{\frac{1}{2}}. \end{aligned} \tag{70}$$

10. The Nevanlinna characteristic function. It is well known that for any two nonnegative numbers, say γ_1 and γ_2 with $\gamma_1 \geqq \gamma_2 \geqq 0$, we have

$$\log^+(\gamma_1 + \gamma_2) \leqq \log^+\gamma_1 + \log^+\gamma_2 + \log 2. \tag{71}$$

Thus if $\gamma_1 \geqq 1$ then

$$\begin{aligned}\log^+(\gamma_1 + \gamma_2) = \log(\gamma_1 + \gamma_2) \leqq \log 2\gamma_1 = \log\gamma_1 + \log 2 \\ \leqq \log^+\gamma_1 + \log^+\gamma_2 + \log 2,\end{aligned}$$

while if $\gamma_1 < 1$ then

$$\log^+(\gamma_1 + \gamma_2) \leqq \log 2 = \log^+\gamma_1 + \log^+\gamma_2 + \log 2.$$

Hence (71) holds in any case.

By the triangle inequality, we have

$$[(\boldsymbol{x} - \boldsymbol{a}) \cdot (\boldsymbol{x} - \boldsymbol{a})]^{\frac{1}{2}} \leqq (\boldsymbol{x} \cdot \boldsymbol{x})^{\frac{1}{2}} + (\boldsymbol{a} \cdot \boldsymbol{a})^{\frac{1}{2}}$$

and

$$\begin{aligned}(\boldsymbol{x} \cdot \boldsymbol{x})^{\frac{1}{2}} &= [(\boldsymbol{x} - \boldsymbol{a} + \boldsymbol{a}) \cdot (\boldsymbol{x} - \boldsymbol{a} + \boldsymbol{a})]^{\frac{1}{2}} \\ &\leqq [(\boldsymbol{x} - \boldsymbol{a}) \cdot (\boldsymbol{x} - \boldsymbol{a})]^{\frac{1}{2}} + (\boldsymbol{a} \cdot \boldsymbol{a})^{\frac{1}{2}}.\end{aligned}$$

Therefore, by (66) and (71), we have

$$m(r, \infty; S - \boldsymbol{a}) \leqq m(r, \infty; S) + \log^+(\boldsymbol{a} \cdot \boldsymbol{a})^{\frac{1}{2}} + \log 2 \tag{72}$$

and

$$m(r, \infty; S) \leqq m(r, \infty; S - \boldsymbol{a}) + \log^+(\boldsymbol{a} \cdot \boldsymbol{a})^{\frac{1}{2}} + \log 2. \tag{73}$$

From (72) and (73), we obtain

$$m(r, \infty; S) - m(r, \infty; S - \boldsymbol{a}) = B(r, \boldsymbol{a}; S), \tag{74}$$

with

$$|B(r, \boldsymbol{a}; S)| \leqq \log^+(\boldsymbol{a} \cdot \boldsymbol{a})^{\frac{1}{2}} + \log 2. \tag{75}$$

Substituting for $m(r, \infty; S - \boldsymbol{a})$ from (74) into (70), we obtain

$$\begin{aligned}m(r, \infty; S) + N(r, \infty; S) + H(r, \infty; S) \\ = m(r, \boldsymbol{a}; S) + N(r, \boldsymbol{a}; S) + H(r, \boldsymbol{a}; S) + C(r, \boldsymbol{a}; S),\end{aligned} \tag{76}$$

where

$$C(r, \boldsymbol{a}; S) = B(r, \boldsymbol{a}; S) + \log\left(\sum_{j=1}^{3} a_{j,\tau_{\boldsymbol{a}}}^2\right)^{\frac{1}{2}}. \tag{77}$$

By (75) and (77), we have

$$(78)\qquad |C(r,\boldsymbol{a};S)| \leqq \left|\log\left(\sum_{j=1}^{3} a_{j,\tau_a}^2\right)^{\frac{1}{2}}\right| + \log^+(\boldsymbol{a}\cdot\boldsymbol{a})^{\frac{1}{2}} + \log 2.$$

We define the *total affinity* of S to $\boldsymbol{a}$ in $|w| \leqq r$, or the *affinity function* for S, by

$$(79)\qquad \mathfrak{A}(r,\boldsymbol{a};S) = m(r,\boldsymbol{a};S) + N(r,\boldsymbol{a};S) + H(r,\boldsymbol{a};S).$$

In particular, we call the total affinity of S to ∞ the *Nevanlinna characteristic function* of S and denote it by $T(r;S)$, so that

$$(80)\qquad \begin{aligned} T(r;S) = \mathfrak{A}(r,\infty;S) &= m(r,\infty;S) + N(r,\infty;S) + H(r,\infty;S) \\ &= m(r,\infty;S) + N(r,\infty;S). \end{aligned}$$

The first fundamental theorem of R. Nevanlinna **[9]**, **[8]**, **[14]** concerning meromorphic functions of a complex variable is generalized by the inequality (78) to meromorphic minimal surfaces:

THEOREM 1. *If the functions*

$$x_j = x_j(u,v), \qquad j = 1,2,3,$$

for $u^2 + v^2 < \infty$, *are the coordinate functions of a nonconstant meromorphic minimal surface* S *in isothermic representation, then for each finite* $\boldsymbol{a}$ *the total affinity* $\mathfrak{A}(r,\boldsymbol{a};S)$ *of* S *to* $\boldsymbol{a}$ *satisfies*

$$T(r;S) = \mathfrak{A}(r,\boldsymbol{a};S) + C(r,\boldsymbol{a};S),$$

where $T(r;S)$ *is the Nevanlinna characteristic function of* S *and* $C(r,\boldsymbol{a};S)$ *is a bounded function of* r *for each* $\boldsymbol{a}$:

$$|C(r,\boldsymbol{a};S)| \leqq \left|\log\left(\sum_{j=1}^{3} a_{j,\tau_a}^2\right)^{\frac{1}{2}}\right| + \log^+(\boldsymbol{a}\cdot\boldsymbol{a})^{\frac{1}{2}} + \log 2.$$

Thus S has essentially the same affinity for all points $\boldsymbol{a}$ in space, in the sense that for any two given points $\boldsymbol{a}$ and $\boldsymbol{b}$ the difference $\mathfrak{A}(r,\boldsymbol{a};S) - \mathfrak{A}(r,\boldsymbol{b};S)$ is a bounded function of r:

$$|\mathfrak{A}(r,\boldsymbol{a};S) - \mathfrak{A}(r,\boldsymbol{b};S)|$$

$$\leqq \left|\log\left(\sum_{j=1}^{3} a_{j,\tau_a}^2\right)^{\frac{1}{2}}\right| + \left|\log\left(\sum_{j=1}^{3} a_{j,\tau_b}^2\right)^{\frac{1}{2}}\right| + \log^+(\boldsymbol{a}\cdot\boldsymbol{a})^{\frac{1}{2}} + \log^+(\boldsymbol{b}\cdot\boldsymbol{b})^{\frac{1}{2}} + \log 4.$$

This is true in particular if S lies on a plane (the complex-variable case), even if the point $\boldsymbol{a}$, or the point $\boldsymbol{b}$, or both are not on the plane.

11. A four-dimensional inversion. In the four-dimensional (x_1, x_2, x_3, x_4)-space, let $\mathscr{S}_0$ be the sphere having center $P_0\colon (\kappa_1, \kappa_2, \kappa_3, \kappa_4)$ and radius δ_0. Then the points $P\colon (x_1, x_2, x_3, x_4)$ and $P'\colon (x_1, x_2, x_3, x_4)$ are inverses of each other [7, p. 121] with respect to $\mathscr{S}_0$ if and only if

$$(81) \qquad (x_j' - \kappa_j)/(x_j - \kappa_j) = \delta'/\delta \quad \text{and} \quad \delta'\delta = \delta_0^2, \qquad j = 1, 2, 3, 4,$$

where δ and δ' are the four-dimensional Euclidean distances P_0P and P_0P', respectively.

In particular, for $(\kappa_1, \kappa_2, \kappa_3, \kappa_4) = (0, 0, 0, 1)$ and $\delta_0 = 1$, we have

$$\frac{\delta'}{\delta} = \frac{\delta'\delta}{\delta^2} = \frac{\delta_0^2}{\delta^2} = \frac{1}{\delta^2} = \frac{1}{x_1^2 + x_2^2 + x_3^2 + (x_4 - 1)^2},$$

whence (81) yields

$$\begin{aligned} x_j' &= x_j/(x_1^2 + x_2^2 + x_3^2 + (x_4 - 1)^2), \qquad j = 1, 2, 3, \\ x_4' &= (x_1^2 + x_2^2 + x_3^2 + x_4(x_4 - 1))/(x_1^2 + x_2^2 + x_3^2 + (x_4 - 1)^2). \end{aligned}$$

Under this inversion, the coordinates of the image of a point $(x_1, x_2, x_3, 0)$ on the hyperplane $x_4 = 0$ are given by

$$(82) \qquad \begin{aligned} x_j' &= x_j/(x_1^2 + x_2^2 + x_3^2 + 1), \qquad j = 1, 2, 3, \\ x_4' &= (x_1^2 + x_2^2 + x_3^2)/(x_1^2 + x_2^2 + x_3^2 + 1). \end{aligned}$$

The coordinates (82) satisfy

$$x_1'^2 + x_2'^2 + x_3'^2 + (x_4' - 1/2)^2 = (1/2)^2,$$

so that the image of $(x_1, x_2, x_3, 0)$ lies on the sphere $\mathscr{S}$ with center $(0, 0, 0, \frac{1}{2})$ and radius $\frac{1}{2}$. In fact, the stereographic projection of the three-dimensional hyperspherical "surface" $\mathscr{S}$ from its "north" pole $(0, 0, 0, 1)$ onto the hyperplane $x_4 = 0$ tangent to $\mathscr{S}$ at its "south" pole $(0, 0, 0, 0)$ coincides with the mapping of $\mathscr{S}$ onto this hyperplane under inversion in $\mathscr{S}_0$.

We shall henceforth call the hyperplane $x_4 = 0$ the (x_1, x_2, x_3)-space. For two points

$$P\colon (x_1, x_2, x_3) \quad \text{and} \quad Q\colon (y_1, y_2, y_3)$$

in the (x_1, x_2, x_3)-space, the line segment joining their images

$$P'\colon (x_1', x_2', x_3', x_4') \quad \text{and} \quad Q'\colon (y_1', y_2', y_3', y_4')$$

under the above inversion in $\mathscr{S}_0$ is a chord of the sphere $\mathscr{S}$. From (82), we obtain

$$\begin{aligned}
\text{distance}\,(P', Q') &= [(x_1' - y_1')^2 + (x_2' - y_2')^2 + (x_3' - y_3')^2 + (x_4' - y_4')^2]^{\frac{1}{2}} \\
&= \frac{[(x_1 - y_1)^2 + (x_2 - y_2)^2 + (x_3 - y_3)^2]^{\frac{1}{2}}}{(x_1^2 + x_2^2 + x_3^2 + 1)^{\frac{1}{2}}(y_1^2 + y_2^2 + y_3^2 + 1)^{\frac{1}{2}}} \\
&= \frac{[(\boldsymbol{x} - \boldsymbol{y}) \cdot (\boldsymbol{x} - \boldsymbol{y})]^{\frac{1}{2}}}{(\boldsymbol{x} \cdot \boldsymbol{x} + 1)^{\frac{1}{2}}(\boldsymbol{y} \cdot \boldsymbol{y} + 1)^{\frac{1}{2}}} \qquad (83) \\
&= \chi(P, Q).
\end{aligned}$$

We shall call $\chi(P,Q)$ the *chordal distance* of P and Q. Since P' and Q' both lie on the sphere $\mathscr{S}$ of radius $\frac{1}{2}$, we have

$$\chi(P, Q) \leqq 1 \tag{84}$$

for all P, Q in the (x_1, x_2, x_3)-space.

Noting that

$$\lim_{P \to \infty} \chi(P, Q) = \frac{1}{(\boldsymbol{y} \cdot \boldsymbol{y} + 1)^{\frac{1}{2}}},$$

we define $\chi(\infty, Q)$ by

$$\chi(\infty, Q) = 1/(\boldsymbol{y} \cdot \boldsymbol{y} + 1)^{\frac{1}{2}}. \tag{85}$$

In §12, below, we shall use $\chi(P,Q)$ in defining an alternative proximity function for meromorphic minimal surfaces in isothermic representation.

12. **The spherical characteristic.** The definition (66) of the proximity function $m(r, \boldsymbol{a}; S)$ for a meromorphic minimal surface S in isothermic representation is such that only points of S at distance no more than 1 from $\boldsymbol{a}$ for $\boldsymbol{a}$ finite, and at distance no less than 1 from the origin for $\boldsymbol{a}$ infinite, have any effect on the value of $m(r, \boldsymbol{a}; S)$.

Since all points P and Q in the (x_1, x_2, x_3)-space have chordal distance $\chi(P, Q) \leqq 1$, all points of S affect the value of the proximity function if this function is redefined with the chordal metric in place of the Euclidean metric. Following the complex-variable lead of Ahlfors [**1**] and Shimizu [**13**], for a meromorphic minimal surface S in isothermic representation, we use (83) and (85) to define a *spherical proximity function* $m°(r, \boldsymbol{a}; S)$ in terms of $\log 1/\chi(P, Q)$ as follows:

$$\begin{aligned}
m°(r, \infty; S) &= \frac{1}{2\pi} \int_{\partial A_r(0)} \log (\boldsymbol{x} \cdot \boldsymbol{x} + 1)^{\frac{1}{2}} d\theta, \\
m°(r, \boldsymbol{a}; S) &= \frac{1}{2\pi} \int_{\partial A_r(0)} \log \frac{(\boldsymbol{x} \cdot \boldsymbol{x} + 1)^{\frac{1}{2}}(\boldsymbol{a} \cdot \boldsymbol{a} + 1)^{\frac{1}{2}}}{[(\boldsymbol{x} - \boldsymbol{a}) \cdot (\boldsymbol{x} - \boldsymbol{a})]} \, d\theta,
\end{aligned} \tag{86}$$

for $\boldsymbol{a}$ finite.

From (57), (66), and (86), for $\boldsymbol{a}$ finite we obtain

$$(87)\qquad m^{\circ}(r,\boldsymbol{a};S)=m^{\circ}(r,\infty;S)+m(r,\boldsymbol{a};S)-m(r,\infty;S-\boldsymbol{a})+\log(\boldsymbol{a}\cdot\boldsymbol{a}+1)^{\frac{1}{2}}.$$

Substituting from (87) into (72), we get

$$(88)\qquad \begin{aligned} &m^{\circ}(r,\infty;S)+N(r,\infty;S)+H(r,\infty;S)\\ &\quad=m^{\circ}(r,\boldsymbol{a};S)+N(r,\boldsymbol{a};S)+H(r,\boldsymbol{a};S)\\ &\qquad+\log\left(\sum_{j=1}^{3}a_{j,\tau_a}\right)-\log(\boldsymbol{a}\cdot\boldsymbol{a}+1)^{\frac{1}{2}}. \end{aligned}$$

We can write (88) as

$$(89)\qquad \begin{aligned} &m^{\circ}(r,\infty;S)+N(r,\infty;S)+H(r,\infty;S)+C(\infty;S)\\ &\quad=m^{\circ}(r,\boldsymbol{a};S)+N(r,\boldsymbol{a};S)+H(r,\boldsymbol{a};S)+C(\boldsymbol{a};S), \end{aligned}$$

where the constants $C(\infty;S)$ and $C(\boldsymbol{a};S)$ are such that in (89) both the left-hand member and the right-hand member $\to 0$ as $r\to 0$. Namely, if there is not a pole of S at $r=0$ then we have

$$(90)\qquad \begin{aligned} C(\infty;S)&=-\log\left(\sum_{j=1}^{3}a_{j,0}^{2}+1\right)^{\frac{1}{2}},\\ C(\boldsymbol{a};S)&=-\log\left(\sum_{j=1}^{3}a_{j,0}^{2}+1\right)^{\frac{1}{2}}+\log\left(\sum_{j=1}^{3}a_{j,\tau_a}^{2}\right)-\log(\boldsymbol{a}\cdot\boldsymbol{a}+1)^{\frac{1}{2}}, \end{aligned}$$

while if there is a pole of S at $r=0$ then $a_{j,\tau_a}=a_{j,\tau}$, $j=1,2,3$, is independent of $\boldsymbol{a}$ and we have

$$(91)\qquad \begin{aligned} C(\infty;S)&=-\log\left(\sum_{j=1}^{3}a_{j,\tau}^{2}\right)^{\frac{1}{2}},\\ C(\boldsymbol{a};S)&=-\log(\boldsymbol{a}\cdot\boldsymbol{a}+1)^{\frac{1}{2}}. \end{aligned}$$

We define the *spherical affinity* of S to $\boldsymbol{a}$ in $|w|\leqq r$, or the *spherical affinity function* for S, by

$$(92)\qquad \mathfrak{A}^{\circ}(r,\boldsymbol{a};S)=m^{\circ}(r,\boldsymbol{a};S)+N(r,\boldsymbol{a};S)+H(r,\boldsymbol{a};S)+C(\boldsymbol{a};S).$$

In particular, we call the spherical affinity of S to ∞ the *spherical characteristic function* of S and denote it by $T^{\circ}(r;S)$:

$$(93)\qquad \begin{aligned} T^{\circ}(r;S)&=m^{\circ}(r,\infty;S)+N(r,\infty;S)+H(r,\infty;S)+C(\infty;S)\\ &=m^{\circ}(r,\infty;S)+N(r,\infty;S)+C(\infty;S). \end{aligned}$$

Substituting $\mathfrak{A}^{\circ}(r,\boldsymbol{a};S)$ and $T^{\circ}(r;S)$ in (89), we have the following generalization of the spherical form of the first fundamental theorem of Nevanlinna to meromorphic minimal surfaces:

THEOREM 2. *If the functions*

$$x_j = x\,(u, v), \qquad j = 1, 2, 3,$$

for $u^2 + v^2 < \infty$, *are the coordinate functions of a nonconstant meromorphic minimal surface S in isothermic representation, then for each finite* $\boldsymbol{a}$ *the spherical affinity* $\mathfrak{A}^\circ(r, \boldsymbol{a}; S)$ *of S to* $\boldsymbol{a}$ *satisfies*

$$T^\circ(r; S) = \mathfrak{A}^\circ(r, \boldsymbol{a}; S), \tag{94}$$

where $T^\circ(r; S)$ *is the spherical characteristic function of S.*

It follows from (66), (71), and (86) that

$$m(r, \infty; S) \leqq m^\circ(r, \infty; S) \leqq m(r, \infty; S) + \log 2.$$

Therefore, by (80) and (93), *the difference* $T^\circ(r; S) - T(r, S)$ *is a bounded function of* r.

13. **Convexity properties.** For $u^2 + v^2 = |w|^2 < \infty$, let the functions (1) be the coordinate functions of a nonconstant meromorphic minimal surface S in isothermic representation. For $\boldsymbol{a}$ finite or infinite, let the $\boldsymbol{a}$-points of S in $0 < |w| \leqq r$ be at points $w = w_j$, $j = 1, 2, \cdots, k$, of moduli $r_1, r_2, \cdots, r_k$, and let the orders of these $\boldsymbol{a}$-points be α_j, $j = 1, 2, \cdots, k$. Then

$$n(r, \boldsymbol{a}; S) = \alpha_0 + \alpha_1 + \cdots + \alpha_k,$$

where $\alpha_0 \geqq 0$ is the order of the $\boldsymbol{a}$-point, if any, at $w = 0$. Evaluating the integral in (67), we obtain

$$N(r, \boldsymbol{a}; S) = \log\left(r^{n(r, \boldsymbol{a}; S)} / r_1^{\alpha_1} r_2^{\alpha_2} \cdots r_k^{\alpha_k}\right). \tag{95}$$

From either (67) or (95), we see that $N(r, \boldsymbol{a}; S)$ is a continuous function of $\log r$. Further, by differentiating either (67) or (95) we get

$$dN(r, \boldsymbol{a}; S) / d \log r = n(r, \boldsymbol{a}; S)$$

except at the points of discontinuity of $n(r, \boldsymbol{a}; S)$. Accordingly, since $n(r, \boldsymbol{a}; S)$ is a nondecreasing, nonnegative function of $\log r$, we have the following result:

The function $N(r, \boldsymbol{a}; S)$ *is a nondecreasing, piecewise-linear convex function of* $\log r$.

Similarly, for $\boldsymbol{a}$ finite, by differentiating (68) we obtain

$$dH(r, \boldsymbol{a}; S) / d \log r = h(r, \boldsymbol{a}; S).$$

By (63), we have

$$h(r, \boldsymbol{a}; S) \geqq 0,$$

with equality for $r > 0$ if and only if S is a plane surface and $\boldsymbol{a}$ lies in the plane. Further, by (68), we have $H(r, \boldsymbol{a}; S) > 0$ for $r > 0$ if $h(r, \boldsymbol{a}; S) \not\equiv 0$. Hence we have the following result:

The function $H(r,\boldsymbol{a};S)$ vanishes identically if $\boldsymbol{a}$ is infinite or if S is a plane surface and $\boldsymbol{a}$ lies in the plane. Otherwise, $H(r,\boldsymbol{a};S)$ is a positive, increasing, strictly convex function of $\log r$.

To determine the behavior of $T^\circ(r;S)$, we integrate (94) with respect to $\boldsymbol{a}$ over the three-dimensional hyperspherical "surface" $\mathscr{S}$ and divide by the content $\mathscr{V} = 2\pi^2 r^3$ of $\mathscr{S}$ to obtain

$$\frac{1}{\mathscr{V}}\iiint_{\mathscr{S}} T^\circ(r;S)d\mathscr{V}_{\boldsymbol{a}} = \frac{1}{\mathscr{V}}\iiint_{\mathscr{S}} m^\circ(r,\boldsymbol{a};S)d\mathscr{V}_{\boldsymbol{a}} + \frac{1}{\mathscr{V}}\iiint_{\mathscr{S}} N(r,\boldsymbol{a};S)\,d\mathscr{V}_{\boldsymbol{a}} \tag{96}$$
$$+ \frac{1}{\mathscr{V}}\iiint_{\mathscr{S}} H(r,\boldsymbol{a};S)\,d\mathscr{V}_{\boldsymbol{a}} + \frac{1}{\mathscr{V}}\iiint_{\mathscr{S}} C(\boldsymbol{a};S)\,d\mathscr{V}_{\boldsymbol{a}}.$$

The integrand $T^\circ(r;S)$ of the integral in the left-hand member of (96) does not vary with $\boldsymbol{a}$, and accordingly the value of this integral is $T^\circ(r;S)$.

The first integral in the right-hand member of (96) is

$$\frac{1}{\mathscr{V}}\iiint_{\mathscr{S}} m^\circ(r,\boldsymbol{a};S)\,d\mathscr{V}_{\boldsymbol{a}} = \frac{1}{\mathscr{V}}\iiint_{\mathscr{S}}\left[\frac{1}{2\pi}\int_{\partial A_r(0)} \log\frac{1}{\chi(\boldsymbol{x},\boldsymbol{a})}\,d\theta\right]d\mathscr{V}_{\boldsymbol{a}} \tag{97}$$
$$= \frac{1}{2\pi}\int_{\partial A_r(0)}\left[\frac{1}{\mathscr{V}}\iiint_{\mathscr{S}} \log\frac{1}{\chi(\boldsymbol{x},\boldsymbol{a})}\,d\mathscr{V}_{\boldsymbol{a}}\right]d\theta,$$

and here by geometric symmetry the inner integral in the last expression is the same for all $\boldsymbol{x}\in\mathscr{S}$. Accordingly, we can replace $\chi(\boldsymbol{x},\boldsymbol{a})$ by $\chi(\boldsymbol{x}(0,0),\boldsymbol{a})$ in (97). Except at $\boldsymbol{a} = \boldsymbol{x}(0,0)$, by (90) and (91) we have

$$C(\boldsymbol{a};S) = -\log(1/\chi(\boldsymbol{x}(0,0),\boldsymbol{a})).$$

Therefore, the sum of the first and fourth integrals in the right-hand member of (96) is 0.

The function $N(r,\boldsymbol{a};S)$ has value 0 for all $\boldsymbol{a}$ not on S. It therefore has value 0 almost everywhere on $\mathscr{S}$. Hence the second integral in the right-hand member of (96) has value 0.

It follows, accordingly, that (96) can be written as

$$T^\circ(r;S) = \frac{1}{\mathscr{V}}\iiint_{\mathscr{S}} H(r,\boldsymbol{a};S)\,d\mathscr{V}_{\boldsymbol{a}}.$$

By (78), then, we have

$$T^\circ(r;S) = \frac{1}{\mathscr{V}}\iiint_{\mathscr{S}}\left[\int_0^r \frac{h(\rho,\boldsymbol{a};S)}{\rho}d\rho\right]d\mathscr{V}_{\boldsymbol{a}} \tag{98}$$
$$= \int_0^r\left[\frac{1}{\mathscr{V}}\iiint_{\mathscr{S}} h(\rho,\boldsymbol{a};S)\,d\mathscr{V}_{\boldsymbol{a}}\right]\frac{d\rho}{\rho}$$
$$= \int_0^r \mathscr{P}(\rho;S)\frac{d\rho}{\rho},$$

where the function $\mathscr{P}(\rho; S)$ can be given a quasi-geometric interpretation, as indicated in the discussion of $h(\rho, \boldsymbol{a}; S)$ in §9.

From (98) we obtain

$$dT^{\circ}(r; S)/d\log r = \mathscr{P}(\rho; S),$$

which is a positive and strictly increasing function of $\log r$. Hence we have the following result:

The function $T^{\circ}(r; S)$ *is a positive, increasing, strictly convex function of* $\log r$.

14. **Order and type.** It follows from (98) that for any nonconstant meromorphic minimal surface S in isothermic representation, we have

$$T^{\circ}(r; S) > \mathscr{P}(1; S)\log r$$

for $r > 1$, and hence

$$\liminf_{r\to\infty} \frac{T^{\circ}(r; S)}{\log r} > 0. \tag{99}$$

It can be shown [3] that the left-hand member of (99) has a finite value c if and only if S is, apart from possible logarithmic terms, a rational minimal surface, as defined below in §15. In that case, c is a positive integer; namely, c is the *degree* of S.

In conformity with the analogous definition for meromorphic functions of a complex variable, we now define the *order* $\rho(S)$ of a meromorphic minimal surface S by

$$\rho(S) = \limsup_{r\to\infty} \frac{\log T^{\circ}(r; S)}{\log r}. \tag{100}$$

If $0 < \rho(S) < \infty$, we define the *type* $\tau(S)$ of the surface S by

$$\tau(S) = \limsup_{r\to\infty} r^{-\rho(S)} T^{\circ}(r; S). \tag{101}$$

A further investigation of problems of order and type of nonplane entire and meromorphic minimal surfaces, similar to kindred problems for functions of a complex variable, might be fruitful.

15. **Rational minimal surfaces.** Let $f(w)$ be a rational function of the complex variable w,

$$f(w) = \frac{p(w)}{q(w)} = \frac{a_0 + a_1 w + \cdots + a_n w^n}{b_0 + b_1 w + \cdots + b_m w^m}, \qquad a_n b_m \neq 0,$$

where the polynomials $p(w)$ and $q(w)$ have no linear factor in common, that is, where $p(w)$ and $q(w)$ are relatively prime polynomials,

$$[p(w), q(w)] = 1.$$

Then the *degree* of $f(w)$ is the maximum of the degree of $p(w)$ and the degree of $q(w)$:

$$\deg f(w) = \max[\deg p(w), \deg q(w)] = \max(m, n).$$

A *rational minimal surface in isothermic representation* is a meromorphic minimal surface in isothermic representation for which the harmonic coordinate functions

$$x_j = x_j(u, v) = \Re f_j(w)$$

are rational functions of u, v,

$$x_j = x_j(u, v) = P_j(u, v)/Q_j(u, v), \qquad j = 1, 2, 3. \tag{102}$$

If (102) holds, then we have

$$\begin{aligned} f_j(w) &= 2x_j(w/2, w/2i) + \text{const} \\ &= 2P_j(w/2, w/2i)/Q_j(w/2, w/2i) + \text{const}, \end{aligned}$$

so that

$$f_j(w) = p_j(w)/q_j(w), \qquad j = 1, 2, 3;$$

that is, the $f_j(w)$ are rational functions of w. Conversely, if it is given that the coordinate functions of a meromorphic minimal surface in isothermic representation are the real parts of rational functions of w,

$$x_j = x_j(u, v) = \Re f_j(w) = \Re(p_j(w)/q_j(w)), \qquad j = 1, 2, 3,$$

then

$$\begin{aligned} x_j &= x_j(u, v) = \Re(p_j(w)/q_j(w)) = \Re(p_j(w)\overline{q_j(w)}/|q_j(w)|^2) \\ &= \Re(p_j(w)\overline{q_j(w)})/|q_j(w)|^2 = P_j(u, v)/Q_j(u, v), \end{aligned}$$

where $P_j(u, v)$ and $Q_j(u, v)$ are polynomials in u, v. Hence an equivalent definition of a rational minimal surface in isothermic representation is that it is a meromorphic minimal surface in isothermic representation for which the coordinate functions $x_j(u,v)$ are the real parts of rational functions of w:

$$x_j = x_j(u, v) = \Re f_j(w) = \Re p_j(w)/q_j(w), \qquad j = 1, 2, 3. \tag{103}$$

If the $x_j(u, v)$ are given by (103), then the $p_j(w)/q_j(w)$, $j = 1, 2, 3$, can be expressed with a common polynomial denominator, so that, with a slight adjustment of notation, we have

$$x_j = x_j(u, v) = \Re\, p_j(w)/q(w), \qquad j = 1, 2, 3, \tag{104}$$

and it is no loss of generality to assume that $p_1(w)$, $p_2(w)$, $p_3(w)$, and $q(w)$ are relatively prime:

$$[p_1(w), p_2(w), p_3(w), q(w)] = 1. \tag{105}$$

When a rational minimal surface S is given in a representation (104) for which (105) holds, we say that S is given in *standard representation* and that the *degree* of S is the maximum of the degrees of $p_1(w)$, $p_2(w)$, $p_3(w)$, $q(w)$:

$$\deg S = \max[\deg p_1(w), \deg p_2(w), \deg p_3(w), \deg q(w)].$$

For example, the rational minimal surfaces of Eng [3], given by

$$x_1 = \Re f_1(w) = \Re \int^w (1-\zeta^2)\left[\frac{a}{(1-\zeta)^4} + \frac{b}{(i-\zeta)^4} + \frac{c}{\zeta^4}\right] d\zeta,$$

$$x_2 = \Re f_2(w) = \Re \int^w i(1+\zeta^2)\left[\frac{a}{(1-\zeta)^4} + \frac{b}{(i-\zeta)^4} + \frac{c}{\zeta^4}\right] d\zeta,$$

$$x_3 = \Re f_3(w) = \Re \int^w 2\zeta\left[\frac{a}{(1-\zeta)^4} + \frac{b}{(i-\zeta)^4} + \frac{c}{\zeta^4}\right] d\zeta,$$

where a, b, and c are nonzero constants, are of degree 9, although each $f_j(w)$ is of degree 8 since

$$f_1(w) = \frac{p_1(w)}{q_1(w)} = \frac{p_1(w)}{(1-w)^2(i-w)^3w^3},$$

$$f_2(w) = \frac{p_2(w)}{q_2(w)} = \frac{p_2(w)}{(1-w)^3(i-w)^2w^3},$$

$$f_3(w) = \frac{p_3(w)}{q_3(w)} = \frac{p_3(w)}{(1-w)^3(i-w)^3w^2},$$

with $(p_j, q_j) = 1$, $\deg p_j \leqq 7$, and

$$\max_{j=1,2,3} (\deg p_j, \deg q_j) = 8.$$

A standard representation here is

$$x_1 = \Re f_1(w) = \Re(1-w)p_1(w)/q(w),$$

$$x_2 = \Re f_2(w) = \Re(i-w)p_2(w)/q(w),$$

$$x_3 = \Re f_3(w) = \Re w p_3(w)/q(w),$$

where $q(w) = (1-w)^3(i-w)^3w^3$.

If S is a rational minimal surface given in standard representation by (104), and if $q(w) \equiv 1$, then we say that S is a *polynomial minimal surface* or an *entire rational minimal surface.* For example, the minimal surface of Enneper (20) is a polynomial minimal surface of degree 3.

16. **The fundamental theorem of algebra.** It is well known that on the extended complex w-plane, that is, on the Riemann sphere, a rational function $f(w)$ of degree p takes on each complex value, counting multiplicities and including ∞, exactly p times.

In this section, we shall indicate how the above result, *if suitably reformulated*, extends to rational minimal surfaces and to *all points of* 3-*dimensional* (actually n-dimensional) *space*. Since the result also includes the fundamental theorem of algebra, it follows that the fundamental theorem of algebra, in the present context, can be considered as applying to all points in space, not merely to the points of the complex plane on which the map of the Riemann sphere by the polynomial function lies. Actually, the result holds more generally, for meromorphic minimal surfaces given in isothermic representation by functions of the form

$$x_j = \Re f_j(w) = \Re[p_j(w)/q(w) + c_j \log w], \qquad c_j \text{ real}, \qquad j = 1,2,3,$$

where the $p_j(w)$ and $q(w)$ are rational functions of w.

To illustrate the result, let us consider the catenoid (19), whose coordinate functions can be represented as

$$\begin{aligned} x_1 &= x_1(u,v) = \Re(1+w^2)/w, \\ x_2 &= x_2(u,v) = \Re i(1-w^2)/w, \\ x_3 &= x_3(u,v) = \Re(2\log w). \end{aligned} \tag{106}$$

Because of the term w^2 in the formulas for $x_1(u,v)$ and $x_2(u,v)$, we say that this surface S is of degree 2. For it, we have

$$\begin{aligned} m^\circ(r,\infty;S) &= \log[(r+1/r)^2 + 4(\log r)^2]^{\frac{1}{2}}, \\ N(r,\infty;S) &= \log\, r, \\ H(r,\infty;S) &= 0, \\ C(\infty;S) &= 0, \end{aligned}$$

so that

$$T^\circ(r;S) = \log[(r+1/r)^2 + 4(\log r)^2]^{\frac{1}{2}} + \log r$$

and

$$\lim_{r\to\infty} \frac{T^\circ(r;S)}{\log r} = 2. \tag{107}$$

As was pointed out in §3, S has a single pole in the finite plane; it is at the origin and is of order 1. In the extended plane, however, S has a second pole, of order 1, at $w = \infty$. Hence *the point at* ∞ in the (x_1, x_2, x_3)-space is taken on exactly twice on S. We express this fact by writing

$$n(\infty,\infty;S) = 2.$$

Since

$$h(r, \infty; S) = 0$$

for all r, we have

$$\lim_{r \to \infty} h(r, \infty; S) = 0;$$

we express this by writing

$$h(\infty, \infty; S) = 0.$$

Hence we have

$$n(\infty, \infty; S) + h(\infty, \infty; S) = 2.$$

On S, we have

$$(\boldsymbol{x} \cdot \boldsymbol{x})^{\frac{1}{2}} = [(r + 1/r)^2 + 4(\log r)^2]^{\frac{1}{2}} \geqq 2,$$

so that no point in $(\boldsymbol{x} \cdot \boldsymbol{x})^{\frac{1}{2}} < 2$ is on S. In particular, the origin is not on S, and we write

(108) $$n(\infty, \boldsymbol{0}; S) = 0.$$

Further, we have

$$m^{\circ}(r, \boldsymbol{0}; S) = \log \frac{[(r + 1/r)^2 + 4(\log r)^2 + 1]^{1/2}}{[(r + 1/r)^2 + 4(\log r)^2]^{1/2}} \underset{r \to \infty}{\to} 0,$$

$$N(r, \boldsymbol{0}; S) = 0, \qquad C(\boldsymbol{0}; S) = 0,$$

so that

$$\lim_{r \to \infty} \frac{\mathfrak{A}^{\circ}(r, \boldsymbol{0}; S)}{\log r} = \lim_{r \to \infty} \frac{H(r, \boldsymbol{0}; S)}{\log r}.$$

From (63), (94), and (107), then,

$$\lim_{r \to \infty} \left(\left(\int_0^r \frac{h(\rho, \boldsymbol{0}; S)}{\rho \, d\rho} \right) \Big/ \log r \right) = 2;$$

therefore, since $h(\rho, \boldsymbol{0}; S)$ is a nondecreasing function of ρ,

$$\lim_{r \to \infty} h(r, \boldsymbol{0}; S) = 2.$$

We write this as

$$h(\infty, \boldsymbol{0}; S) = 2,$$

so that, by (108),

$$n(\infty, \boldsymbol{0}; S) + h(\infty, \boldsymbol{0}; S) = 2.$$

In a similar way, for any finite $\boldsymbol{a}$ we have

$$\lim_{r\to\infty} m^{\circ}(r,\boldsymbol{a};S)=0$$

so that

$$\lim_{r\to\infty}\frac{N(r,\boldsymbol{a};S)+H(r,\boldsymbol{a};S)}{\log r}=2,$$

whence

$$n(\infty,\boldsymbol{a};S)+h(\infty,\boldsymbol{a};S)=2.$$

For points $\boldsymbol{a}$ not on S, such as $(0,0,0)$, we have

$$n(\infty,\boldsymbol{a};S)=0, \qquad h(\infty,\boldsymbol{a};S)=2,$$

while for points $\boldsymbol{a}$ on S, such as $(2,0,0)$, we have

$$n(\infty,\boldsymbol{a};S)=1, \qquad h(\infty,\boldsymbol{a};S)=1;$$

but for *all* points a in space, including ∞, we have

$$n(\infty,\boldsymbol{a};S)+h(\infty,\boldsymbol{a};S)=2.$$

The above example illustrates the following result:

THEOREM 3. *If the functions*

$$x_j=x_j(u,v), \qquad j=1,2,3,$$

are the coordinate functions of a rational minimal surface S (logarithmic terms might be present) of degree p in standard representation, then for each $\boldsymbol{a}$, *including* $\boldsymbol{a}=\infty$, *in the* (x_1,x_2,x_3)*-space, we have*

$$n(\infty,\boldsymbol{a};S)+h(\infty,\boldsymbol{a};S)=p.$$

The function $h(r,\boldsymbol{a};S)$ is positive, and is an increasing function of r, unless either $\boldsymbol{a}=\infty$ or S is a plane surface and $\boldsymbol{a}$ lies on S. In these exceptional cases, we have $h(r,\boldsymbol{a};S)\equiv 0$. This fact yields the following result:

COROLLARY. *Under the hypothesis of Theorem* 3, *we have*

$$n(\infty,\boldsymbol{a};S)\leqq p,$$

with equality if and only if either $\boldsymbol{a}=\infty$ *or S is a plane surface and* $\boldsymbol{a}$ *lies on S.*

The corollary includes the fundamental theorem of algebra. A further investigation of the possible values of $n(\infty\ \boldsymbol{a};S)$ relative to p, for nonplane rational minimal surfaces S, might lead to interesting considerations.

17. **Plane meromorphic maps viewed from space.** For a plane meromorphic map S, and for points $\boldsymbol{a}_1^*, \boldsymbol{a}_2^*, \cdots, \boldsymbol{a}_k^*$, $k>2$, *in the plane*, the second fundamental theorem of Nevanlinna for meromorphic functions of a complex variable yields

$$\sum_{j=1}^{k} m(r, \boldsymbol{a}_j^*; S) < 2T(r; S) - N_1(r; S) + O[\log r] + O[\log T(r; S)], \tag{109}$$

where $N_1(r; S)$ is an enumerative function for multiple values on the map, as $r \to \infty$ outside an open set Δ_k such that

$$\int_{\Delta_k} r^k dr < \infty.$$

Now if the points $\boldsymbol{a}_1, \boldsymbol{a}_2, \cdots, \boldsymbol{a}_k$ are not necessarily on the plane, then their respective orthogonal projections $a_1^*, a_2^*, \cdots, a_k^*$ on the plane are at least as close to each point of the plane as are $a_1, a_2 \cdots, a_k$, and accordingly their proximity-function values are at least as great:

$$m(r, a_j; S) \leqq m(r\ a_j^*; S), \qquad i = 1, 2, \cdots, k.$$

Therefore (109), which holds for $a_1^*, a_2^*, \cdots, a_k^*$, holds also for $a_1, a_2, \cdots, a_k$.

From the above observation, defect relations, Picard's theorem, and so on, can be discussed for plane meromorphic maps from a space viewpoint. Since a treatment of the second fundamental theorem for meromorphic minimal surfaces in general will appear in a separate publication, however, we shall not here pursue these matters further.

References

1. L. V. Ahlfors, *Beiträge der meromorphen Funktionen*, Seventh Congr. Math. Scand., Oslo, 1929.
2. E. F. Beckenbach, *Meromorphic plane maps viewed from space*, (to appear).
3. E. F. Beckenbach and F. H. Eng, *Rational minimal surfaces*, (to appear).
4. E. F. Beckenbach and J. W. Hahn, *Triples of conjugate harmonic functions and minimal surfaces*, Duke Math. J. **2** (1936), 698–704.
5. E. F. Beckenbach and G. A. Hutchinson, *Meromorphic minimal surfaces*, Bull. Amer. Math. Soc. **68** (1962), 519–522.
6. E. F. Beckenbach and T. Radó, *Subharmonic functions and minimal surfaces*, Trans. Amer. Math. Soc. **35** (1933), 648–661.
7. L. R. Ford, *Automorphic functions*, McGraw-Hill, New York, 1929.
8. Einar Hille, *Analytic function theory*, Vol. 2, Ginn, New York, 1962.
9. R. Nevanlinna, *Zur Theorie der meromorphen Funktionen*, Acta Math. **46** (1925), 1–99.
10. Barrett O'Neill, *Elementary differential geometry*, Academic Press, New York, 1966.
11. W. F. Osgood, *Lehrbuch der Funktionentheorie*, Vol. 1, Teubner, Leipzig, 1928.
12. Tibor Radó, *On the problem of Plateau*, Springer-Verlag, Berlin, 1933.
13. T. Shimizu, *On the theory of meromorphic functions*, Japan. J. Math. **6** (1929), 119–171.
14. M. Tsuji, *Potential theory in modern function theory*, Maruzen, Tokyo, 1959.

University of California
Los Angeles, California

Entire Functions Related to the Dirichlet Divisor Problem

N. A. BOWEN
A. J. MACINTYRE

1. Consider the functions introduced by af Hallstrom [7].

$$c(z) = \cos z \cdot \cos\tfrac{1}{2}z \cdot \cos\tfrac{1}{3}z \cdots. \tag{1}$$

$$s(z) = \left(\frac{1}{z}\sin z\right)\left(\frac{2}{z}\sin\frac{z}{2}\right)\left(\frac{3}{z}\sin\frac{z}{3}\right)\cdots. \tag{2}$$

Evidently

$$c(z) = \prod_{m=1}^{\infty}\prod_{n=1}^{\infty}\left(1 - \frac{4z^2}{m^2(2n-1)^2\pi^2}\right) = \prod_{N=1}^{\infty}\left(1 - \frac{4z^2}{N^2\pi^2}\right)^{\delta(N)} \tag{3}$$

where $\delta(N)$ is the number of odd divisors of N.

Similarly

$$s(z) = \prod_{m=1}^{\infty}\prod_{n=1}^{\infty}\left(1 - \frac{z^2}{m^2n^2\pi^2}\right) = \prod_{N=1}^{\infty}\left(1 - \frac{z^2}{N^2\pi^2}\right)^{d(N)} \tag{4}$$

where $d(N)$ is the number of divisors of N. The two functions are related. We may note the formulae [**12**, p. 182]

$$c(z) = \left(\frac{1}{2z}\sin 2z\right)\left(\frac{3}{2z}\sin\frac{2z}{3}\right)\left(\frac{5}{2z}\sin\frac{2z}{5}\right)\cdots \tag{5}$$

and [7, p. 164]

$$s(2z) = s(z)c(z). \tag{6}$$

We investigate the asymptotic behavior of these functions in the complex plane in rather more detail than by Hallstrom [7, p.171] and along the real axis we find the order of magnitude of $c(z)$ correct to the exponential factor [cf. **12**].

From the behavior of $s(z)$ and $c(z)$ along the imaginary axis follow by Tauberian theorems ([**2**; **3**; **4**; **11**; **13**, p. 271; **14**; **17**]—only the last reference explicitly deals with sufficiently general growth order) some results about the zeros, indeed the relations

(7) $$n(r, s(z)) \sim 2(r/\pi)\log r$$

and

(8) $$n(r\ c(z)) \sim 2(r/\pi)\log r$$

are so determined. But as shown by Tjan [**16**] and Anderson [**1**] we cannot extend (7) and (8) to complete asymptotic expansions as a Tauberian consequence of asymptotic expansions for $\log|s(iy)|$ and $\log|c(iy)|$. Indeed comparison of our expansions with the known results for the divisor problem provides a partial alternative to their work. However it is to be hoped that further investigation of the functions $s(z)$ and $c(z)$ will provide new information about the divisor problem.

In §2 we apply Dirichlet's estimates for $d(1) + \cdots + d(N)$ to $s(z)$ and $c(z)$. In §3 we take up an independent investigation of the asymptotic behavior of the functions. §4 contains calculations of some relevant integrals and §5 deals with $c(z)$ on the real axis.

2. **Implications of Dirichlet's estimate.** Dirichlet [**10**, pp. 262, 270] gave the approximation

(9) $$\begin{aligned} D(x) &= d(1) + d(2) + \cdots + d([x]) \\ &= x\log x + (2\gamma - 1)x + O(x^{\frac{1}{2}}). \end{aligned}$$

From this result and the representation

(10) $$\begin{aligned} \log s(iy) &= \int_0^\infty \log(1 + y^2/t^2\pi^2)\, dD(t) \\ &= 2y^2 \int_1^\infty \frac{D(t)\, dt}{t(t^2\pi^2 + y^2)} \end{aligned}$$

it appears that

(11) $$\begin{aligned} \log s(iy) &= 2y^2 \int_1^\infty \frac{\log t + (2\gamma - 1)}{t^2\pi^2 + y^2}\, dt + O(y^{\frac{1}{2}}) \\ &= \frac{2y}{\pi} \int_{\pi/y}^\infty \frac{\log y + \log u - \log\pi + (2\gamma - 1)}{1 + u^2}\, du + O(y^{\frac{1}{2}}). \end{aligned}$$

Noting that

$$\int_0^\infty \frac{\log u}{1 + u^2}\, du = 0,$$

it appears that

(12) $$\begin{aligned} \log s(iy) = \frac{2y}{\pi}(\log y - \log\pi + 2\gamma - 1)\left(\frac{\pi}{2} + O(1/y)\right) \\ - \frac{2y}{\pi} \int_0^{\pi/y} \frac{\log u\, du}{1 + u^2} + O(y^{\frac{1}{2}}). \end{aligned}$$

This gives

(13) $$\log s(iy) = y\log y + (2\gamma - 1 - \log\pi)y + O(y^{\frac{1}{2}}),$$

and since $c(iy) = s(2iy)/s(iy)$ there follows

(14) $$\log c(iy) = y\log y + (2\gamma - 1 + 2\log 2 - \log\pi)y + O(y^{\frac{1}{2}}).$$

The exponent $\frac{1}{2}$ in the error term in (9) has been reduced [**10**, p. 270] but cannot be reduced to as little as $\frac{1}{4}$ [**6, 8, 10**, p. 272]. In the next section we show that the terms $O(y^{\frac{1}{2}})$ in (13) and (14) can be replaced by precise asymptotic estimates.

3.1. *Asymptotic behavior of* $\log c(iy)$.

(15) $$\begin{aligned}\log c(iy) &= \int_0^\infty \log\cosh(y/t)\,d[t] = y\int_1^\infty \tanh(y/t)[t]\,dt/t^2\\ &= y\int_0^y \tanh u\,\frac{du}{u} - \int_1^\infty y\tanh(y/t)(t-[t])\,dt/t^2.\end{aligned}$$

It is easily seen that the second integral is less than

$$\int_0^y \tanh u\,du$$

and that

(16) $$\log c(iy) = y\log y + O(y).$$

To obtain a more precise result set

(17) $$\begin{aligned}N(x) &= \int_1^x (t-[t])\,dt/t^2\\ &= \log x - \left(1 + \frac{1}{2} + \frac{1}{3} + \cdots + \frac{1}{[x]} - \frac{[x]}{x}\right)\\ &= \log x - \log[x] - \gamma + [x]/x + O(1/x)\\ &= -\gamma + 1 + O(1/x).\end{aligned}$$

We now observe that

(18) $$\begin{aligned}y\int_1^\infty \tanh(y/t)\,dN(t) &= y^2\int_1^\infty \frac{N(t)\,dt}{t^2\cosh^2(y/t)} \sim y^2\int_1^\infty \frac{(1-\gamma)\,dt}{t^2\cosh^2(y/t)}\\ &= (1-\gamma)y\int_0^y \frac{du}{\cosh^2 u}\\ &\sim (1-\gamma)y.\end{aligned}$$

As we shall prove in §4

$$\int_0^y \tanh u\, du/u = \log y + (\gamma + 2\log 2 - \log \pi) + O(1), \tag{19}$$

so that

$$\log c(iy) = y\log y + (2\gamma - 1 + 2\log 2 - \log \pi)y + o(y). \tag{20}$$

This result is less complete than (14) but can be improved by considering $N(x)$ in more detail. Instead of (17) note that [**5**, p. 324]

$$\begin{aligned} N(x) &= \log x + [x]/x - (\log [x] + \gamma + 1/2[x]) + O(1/x^2) \\ &= 1 - \gamma - 1/2x + O(1/x^2). \end{aligned} \tag{21}$$

The term $O(1/x^2)$ contributes to $\log c(iy)$ an amount which is

$$O\left(y^2 \int_1^\infty dt/t^4 \cosh^2(y/t)\right) = O(1/y) \int_0^y u^2 du/\cosh^2 u = O(1/y).$$

The term $-1/2x$ contributes

$$\begin{aligned} &-(y^2/2)\int_1^\infty dt/t^3 \cosh^2(y/t) \\ &= -\frac{1}{2}\int_0^y u\, du/\cosh^2 u = -\frac{1}{2}\log 2 + O(1/y). \end{aligned} \tag{22}$$

Hence[1]

$$\log c(iy) = y\log y + (2\gamma - 1 + 2\log 2 - \log \pi)y + \tfrac{1}{2}\log 2 + O(1/y). \tag{23}$$

3.2. *Asymptotic behavior of* $\log s(iy)$.

$$\begin{aligned} \log s(iy) &= \int_1^\infty (\log\sinh(y/t) - \log(y/t))\, d[t] \\ &= y\int_1^\infty (\coth(y/t) - t/y)[t]\, dt/t^2 \\ &= y\int_1^\infty (\coth(y/t) - t/y)dt/t - y\int_1^\infty (\coth(y/t) - t/y)\, dN(t) \\ &= y\int_0^y (\coth u - 1/u)du/u + y^2\int_1^\infty (1/\sinh^2(y/t) - t^2/y^2)N(t)dt/t^2 . \end{aligned} \tag{24}$$

[1] The result (23) can also be obtained by using the Euler-Maclaurin summation formula, but the working is a little tedious.

The first of the last two integrals may be integrated by parts giving

$$y \int_0^y (1/\sinh^2 u - 1/u^2) \log u \, du + y(\coth y - 1/y) \log y$$

$$= y \int_0^\infty (1/\sinh^2 u - 1/u^2) \log u \, du$$

$$\text{(25)} \qquad + y \int_y^\infty \log u \, du/u^2 + (y-1)\log y + O(ye^{-2y} \log y)$$

$$= y(\gamma - \log \pi) + \log y + 1 + (y-1)\log y$$

$$+ \; O(ye^{-2y} \log y).$$

The evaluation of $\int_0^\infty (1/\sinh^2 u - 1/u^2) \log u \, du$ is given in §4.

When the expression $(1-\gamma) - 1/2t + O(1/t^2)$ for $N(t)$ is substituted in the last integral of (24) the term $O(1/x^2)$ contributes

$$O\left(y^2 \int_1^\infty (1/\sinh^2(y/t) - t^2/y^2)\, dt/t^4\right)$$

$$\text{(26)} \qquad = O\left(y^2 \int_0^y (1/\sinh^2 u - 1/u^2) u^2 \, du/y^3\right)$$

$$= O(1).$$

The other two terms give

$$(1-\gamma)y^2 \int_0^y (1/\sinh^2 u - 1/u^2)\, du/y$$

$$\text{(27)} \qquad = (1-\gamma) y \int_0^\infty (1/\sinh^2 u - 1/u^2)\, du + (1-\gamma) + O(ye^{-2y})$$

and

$$\text{(28)} \qquad -\frac{1}{2} y^2 \int_0^y (1/\sinh^2 u - 1/u^2) u \, du/y^2 = \frac{1}{2} \log y - \frac{1}{2} C + O(ye^{-2y})$$

where

$$\text{(29)} \qquad C = \int_0^1 (1/\sinh^2 u - 1/u^2) u \, du + \int_1^\infty u \, du/\sinh^2 u.$$

Combining these results we find[2]

$$\text{(30)} \qquad \log s(iy) = y \log y + y(2\gamma - 1 - \log \pi) + \tfrac{1}{2} \log y + O(1).$$

---

[2] There the Euler-Maclaurin summation formula again applies and it appears that the $O(1)$ in (30) may be replaced by $\log 2\pi + O(y^{-1} (\log y)^2)$.

4.1. *Asymptotic behavior of* $\int_0^y \tanh u\, du/u$. Integration by parts gives

$$\begin{aligned}\int_0^y \tanh u\, du/u &= \tanh y \log y - \int_0^y \log u\, du/\cosh^2 u \\ &= \tanh y \log y + \int_y^\infty \log u\, du/\cosh^2 u - \int_0^\infty \log u\, du/\cosh^2 u.\end{aligned} \tag{31}$$

This formula gives a very satisfactory indication of the asymptotic behavior of the integral as y tends to infinity apart from the question of the value of the constant

$$\int_0^\infty \log u\, du/\cosh^2 u. \tag{32}$$

An independent investigation of the asymptotic behavior provides us with the value of this integral. Consider

$$\frac{1}{2}\int_{-c}^{c} \tanh x\, dx/x, \qquad c > 0, \tag{33}$$

as a contour integral in the complex plane and deform the contour into the polygonal path of the three segments

$$x = -c,\ \ 0 \leqq y \leqq n\pi; \qquad y = n\pi,\ \ -c \leqq x \leqq c; \qquad x = c,\ \ n\pi \geqq y \geqq 0.$$

The integrand has poles at $z = (k + \frac{1}{2})\pi i$ and hence

$$\begin{aligned}&\frac{1}{2}\int_{-c}^{c} \tanh x\, dx/x \\ &= \frac{1}{2}\int_0^{n\pi} \tanh(-c+iy) i\,dy/(-c+iy) + \frac{1}{2}\int_{-c}^{c} \tanh(x) dx/(x+in\pi) \\ &- \frac{1}{2}\int_0^{n\pi} \tanh(c+iy) i\,dy/(c+iy) + \pi i\left\{\frac{1}{\pi i/2} + \frac{1}{3\pi i/2} + \cdots + \frac{1}{(2n-1)\pi i/2}\right\},\end{aligned} \tag{34}$$

and

$$\begin{aligned}\int_0^c \tanh x\, dx/x &= \int_0^c x \tanh x\, dx/(x^2 + n^2\pi^2) - \int_0^{n\pi} y\, dy/(c^2 + y^2) \\ &+ 2\left\{1 + \frac{1}{3} + \frac{1}{5} + \cdots + 1/(2n-1)\right\} + R.\end{aligned} \tag{35}$$

Here we have set

$$\begin{aligned}\tanh(c+iy) &= 1 - (2/(e^{2c+2iy}+1)), \\ \tanh(-c+iy) &= -\tanh(c-iy) = -1 + (2/(e^{2c-2iy}+1))\end{aligned}$$

so that

$$|R| \leqq 2\pi n / c(e^{2c} - 1) .$$

If we set $x = n\pi u$ in the first integral and $n = [c/\pi]$ we find for the first two terms on the right-hand side of (35)

$$\int_0^1 \frac{u\,(\tanh(n\pi u) - 1)\,du}{1 + u^2} + \int_{n\pi}^c y\,dy/(c^2 + y^2) + O(1/c)$$

and this expression is easily seen to tend to zero as c tends to infinity. Hence as c tends to infinity we have

$$\begin{aligned} &\int_0^c \tanh x\,dx/x - \log c \\ &\quad = 2\{1 + \tfrac{1}{3} + \cdots + 1/(2n-1)\} - \log n\pi + o(1) \\ &\quad = 2\{1 + \tfrac{1}{2} + \cdots + 1/2n\} - \{1 + \tfrac{1}{2} + \cdots + 1/n\} - \log n - \log \pi + o(1) \qquad (36) \\ &\quad = 2\{\log 2n + \gamma\} - \{\log n + \gamma\} - \log n - \log \pi + o(1) \\ &\quad = 2\log 2 + \gamma - \log \pi + o(1). \end{aligned}$$

We infer that

$$\int_0^\infty \frac{\log u\,du}{\cosh^2 u} = \log \pi - \gamma - 2\log 2. \qquad (37)$$

4.2. *An alternative method for the evaluation of* (32) *is as follows.*

$$\begin{aligned} I(x) &= \int_0^\infty \frac{u^x}{\cosh^2 u}\,du = \int_0^\infty \frac{4u^x e^{-2u}}{(1 + e^{-2u})^2}\,du \\ &= 4 \int_0^\infty u^x\{e^{-2u} - 2e^{-4u} + 3e^{-6u} \cdots\}\,du \\ &= 4\Gamma(x+1)\{1/2^{x+1} - 2/4^{x+1} + 3/6^{x+1} \cdots\} \qquad (38) \\ &= \frac{2}{2^x}\Gamma(x+1)\{1 - 1/2^x + 1/3^x - \cdots\} \\ &= \frac{2}{2^x}\cdot\Gamma(x+1)\zeta(x)(1 - 2/2^x). \end{aligned}$$

The inversion of summation and integration is certainly valid for $x > 1$ and the result valid by analytic continuation for $x > -1$. Hence

$$\begin{aligned} I'(0) &= \int_0^\infty \frac{\log u\,du}{\cosh^2 u} = -2\Gamma'(1)\zeta(0) - 2\Gamma(1)\zeta'(0) \\ &\qquad + \Gamma(1)\zeta(0)\{-2\log 2 + 4\log 4\} \qquad (39) \\ &= -2\Gamma'(1)\zeta(0) - 2\Gamma(1)\zeta'(0) + 6\Gamma(1)\zeta(0)\log 2. \end{aligned}$$

Since $\zeta(0) = -1/2$ [**15**, p. 19] and $\zeta'(0) = -(1/2)\log 2\pi$ (ibid, p. 20)

(40) $$\int_0^\infty \frac{\log u\, du}{\cosh^2 u} = \Gamma'(1) + \log 2\pi - 3\log 2.$$

But $\Gamma'(1) = -\gamma$[**5**, p. 507] so we have

(41) $$\int_0^\infty \frac{\log u\, du}{\cosh^2 u} = \log \pi - \gamma - 2\log 2.$$

4.3. *Two related integrals.* First note the simple relations

$$\begin{aligned}\int_\alpha^\beta &(1/\sinh^2 u - 1/\cosh^2 u - 1/u^2)\log u\, du \\ &= \int_\alpha^\beta (4/\sinh^2 2u - 1/u^2)\log u\, du \\ &= \int_{2\alpha}^{2\beta} (2/\sinh^2 v - 2/v^2)(\log v - \log 2)\, dv \qquad (42)\\ &= 2\int_{2\alpha}^{2\beta} (1/\sinh^2 v - 1/v^2)\log v\, dv \\ &\quad - 2\log 2(\coth 2\alpha - \coth 2\beta - 1/2\alpha + 1/2\beta).\end{aligned}$$

Now with

$$A = \int_0^\infty \log u\, du/\cosh^2 u, \qquad B = \int_0^\infty (1/\sinh^2 u - 1/u^2)\log u\, du,$$

let α tend to zero and β to infinity whereupon (42) gives

(43) $$B - A = 2B - 2\log 2(-1), \qquad A + B = -2\log 2.$$

We have determined A to be $\log \pi - \gamma - 2\log 2$ so it appears that

(44) $$\int_0^\infty (1/\sinh^2 u - 1/u^2)\log u\, du = \gamma - \log \pi.$$

The relation (41) appears in Bieren de Hahn's tables [Table 257, no. 4] but we have been unable to locate the original source of the relation (44).

5. The expansions obtained in §3 for $\log c(z)$ and $\log s(z)$ with $z = iy$ on the imaginary axis can be extended into the complex plane. For the first term this follows from the general theory [**2, 3, 4, 11, 13, 17**] when extended to proninate orders as in [**17**]. But this general theory gives no indication of the behavior of the functions along the real axis. The question arose in the problem of D. J. Newman [**12**]. The detailed behavior of $c(z)$ and $s(z)$ along the real axis must reflect the somewhat erratic distribution of the number of factors of the integers. However we find that outside intervals $|x - n| < \delta$ near the integers

(45) $$\log c(x) \sim -\tfrac{1}{2}\pi x, \qquad \log s(z) \sim -\tfrac{1}{2}\pi x.$$

To obtain this result we start from the known integral [**5**, p. 516, Example 12]

$$\int_0^\infty \log\cos^2 u\, du/u^2 = -\pi. \tag{46}$$

From this it follows that for $x > 0$

$$\int_1^\infty \log|\cos(x/t)|\,dt = -\tfrac{1}{2}\pi x + o(x). \tag{47}$$

We propose to compare this integral with the similar Stieltjes integral

$$\log|c(x)| = \int_1^\infty \log|\cos(x/t)|\,d[t]. \tag{48}$$

It will be more convenient to work with positive quantities. Let us refer to

$$\int_1^\infty \log|\sec(x/t)|\,dt = \tfrac{1}{2}\pi x + o(x) \tag{49}$$

as "the integral" and

$$-\log|c(x)| = \int_1^\infty \log|\sec(x/t)|\,d[t] \tag{50}$$

as "the sum". With x a large positive parameter the graph of $\log|\sec(x/t)|$ appears as in the following diagram.

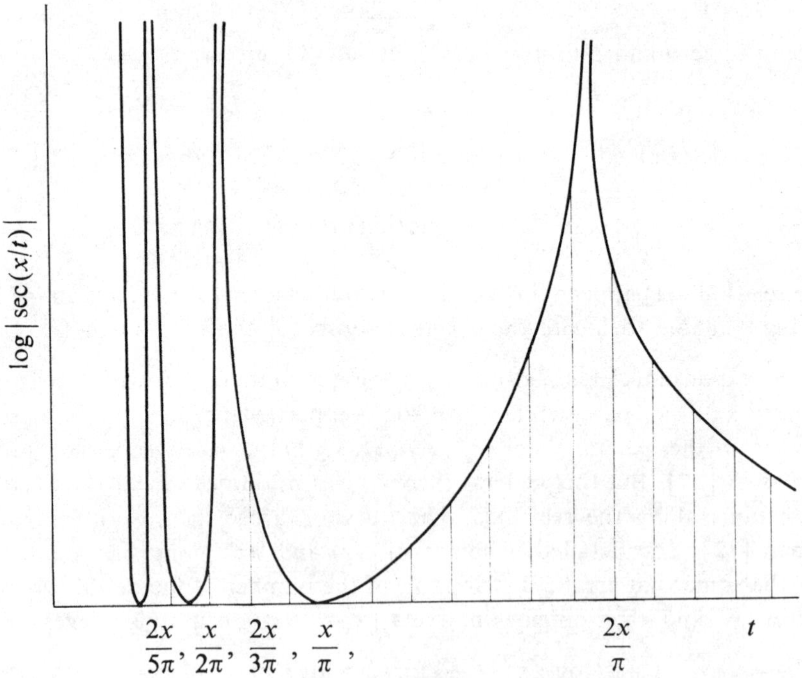

FIGURE 1.

The diagram illustrates the graph of $\log|\sec(x/t)|$ as a function of t for $t > 1$. The vertical lines are segments of $t = 1$, $t = 2$, $t = 3, \cdots$, from the t axis up to the curve.

In this diagram "the integral" is represented by the area under the curve and "the sum" by the sum of the vertical segments. Since the curve has infinities at the values

(52) $$t = 2x/\pi, \qquad 2x/3\pi, \cdots, 2x/(2p+1)\pi, \quad p = [(2x - \pi)/2\pi],$$

we cannot expect "the sum" to approximate "the integral" unless this set of values all differ appreciably from integers. We do not however have to consider p different conditions. For if $2x/n\pi$ differs from an integer m by a small amount η then $2x/\pi$ differs from mn by $n\eta$ and hence from the nearest integer by at most $n\eta$. Hence if $2x/\pi$ differs from the nearest integer by at least δ then

$$\delta \leqq n\eta, \qquad \eta \geqq \delta/n.$$

If we then make the restriction that for some integer s

(53) $$s + \delta \leqq 2x/\pi \leqq s + 1 - \delta, \qquad 0 < \delta < \tfrac{1}{2}$$

we are assured that $2x/n\pi$ differs from the nearest integer by at least δ/n. We shall impose the restriction (53) on x from now on. δ will appear then in our results.

Over relatively large values of t we may then expect sums and integrals to be approximately equal. For relatively small values of t, however, several infinities of $\log|\sec(x/t)|$ may fall into the same interval $n \leqq t \leqq n+1$ and the relation between sum and integral will there be erratic. We must thus prove that the contribution of some suitable range $1 \leqq t \leqq X = X(x)$ to both sum and integral is relatively small. Consider first the integral

$$\int_1^X \log|\sec(x/t)|\,dt = x\int_{x/X}^x \log|\sec u|\,du/u^2.$$

We take $0 < X < x$. The integral is less than εx provided x/X and x are sufficiently large, that is

(54) $$\int_1^X \log|\sec(x/t)|\,dt < \varepsilon x$$

provided $x > x_0(t)$ and $x/X > x_0(t)$. It will be convenient to leave the corresponding estimate for the sum until the end of the discussion.

Next consider the question of approximate equality of sum and integral in the range $X_0 \leqq t \leqq x$. We shall require that the intervals of the form $2x/(2k+1)\pi \leqq t \leqq 2x/(2k-1)\pi$ between successive infinities are reasonably long so as to contain several integer values of t. If we restrict k to be no greater than $[x^{1/2}/4]$, then the relevant length

(55) $$4x/(4k^2 - 1)\pi > 16/\pi > 5,$$

and the interval contains at least four integers. The values of t then allowed are

(56) $$t \geqq 2x/(2[x^{\frac{1}{2}}/4] + 1)\,\pi = 2x/N\pi.$$

For t in this range we describe the areas over intervals $n \leqq t \leqq n+1$ between the t axis and the curve $\log|\sec(x/t)|$ as peaks if the interval contains an infinity of the curve. The areas between successive peaks will be called valleys. One valley extends to infinity. The area of a valley is less than the corresponding sum but exceeds the sum less its two extreme values [**9**, Theorem 154, p. 111]. Let A be the area between the curve $\log|\sec(x/t)|$ and the t axis for $t > 2x/N\pi$. Let S be the corresponding sum. Let $A(p)$ and $S(p)$ be the respective contributions associated with the peaks. $S(p)$ will contain two terms for each peak after the first. The remarks just made lead to the inequality

(57) $$S - S(p) + A(p) < A < S + A(p).$$

The value $X = 2x/N\pi$ clearly makes

(58) $$\int_1^X \log|\sec(x/t)|\,dt = o(x).$$

It remains for us to prove that $A(p)$, $S(p)$ and

(59) $$\int_1^X \log|\sec(x/t)|\,d[t]$$

are all $o(x)$ and (45) is established.

First let us note that in the range $1 \leqq t \leqq X$ fall at most $2x/(2[x^{\frac{1}{2}}/4] - 1)\pi$ or $O(x^{\frac{1}{2}})$ integers. In the range $X \leqq t \leqq x$ fall at most $[x^{\frac{1}{2}}/4]$ peaks.

Now consider the Stieltjes integral (59). The contribution at $t = n$ is

$$\log|\sec(x/n)|.$$

But x/n differs from the nearest multiple (and a priori the nearest odd multiple) of $\pi/2$ by at least $\delta\pi/2n$. Hence

(60) $$\log|\sec(x/n)| \leqq \log|\operatorname{cosec}(\delta\pi/2n)| \leqq \log|n/\delta|.$$

Hence the integral (59) is $O(x^{\frac{1}{2}}\log x)$. The inequality (60) also applies to each of the two contributions to $S(p)$ associated with all but the last peak and also to the single contribution associated with the last peak. The maximum value of $t = n$ concerned is less than $2x/\pi + 1$. Again the number of terms is $O(x^{\frac{1}{2}})$ and S_p is $O(x^{\frac{1}{2}}\log x)$. We have finally to consider $A(p)$. The area associated with a single peak falling between $t = n$ and $t = n + 1$ is

$$\int_n^{n+1} \log|\sec(x/t)|\,dt = x\int_{x/(n+1)}^{x/n} \log|\sec u|\,du/u^2$$

$$\leqq (n+1)^2 x^{-1} \int_{x/(n+1)}^{x/n} \log|\sec u|\,du.$$

For a given length l $(<\frac{1}{2}\pi)$ the integral

$$\int_v^{v+l} \log|\sec u|\,du$$

has a maximum

$$2\int_0^{(1/2)l} \log|\operatorname{cosec} u|\,du < 2\int_0^{(1/2)l} \log(\pi/2u)du = l(1+\log(\pi/l)),$$

$$l = x/n(n+1).$$

Hence the contribution of a single peak to $A(p)$ is

$$O(1+\log(\pi/l)) = O(\log x).$$

This completes the proof of (45) which follows from (57) in the case of $\log c(x)$. For $\log s(x)$ there is no significant change in the argument. In place of (46) we start with the integral

$$\int_0^\infty (\log\sin u - \log u)\,du/u^2.$$

The value of this integral follows from the two following results.

$$\int_0^\infty \log\sin^2 x \,\frac{dx}{a^2+x^2} = \frac{\pi}{a}\log\frac{1}{2}(1-e^{-2a}),$$

$$\int_0^\infty \log x^2 \,\frac{dx}{a^2+x^2} = \frac{\pi}{a}\log a.$$

Of these the first is given by Bromwich [**5**, p. 516, Example 12, (2)] and the second is easily obtained by calculus of residues.

We then argue as follows.

$$\int_0^\infty (\log\sin^2 x - \log x^2)\,dx/(a^2+x^2)$$

$$= \frac{\pi}{a}\log\frac{1-e^{-2a}}{2a} = \frac{\pi}{a}\log\left(\frac{2a-(2a)^2/2!+\cdots}{2a}\right)$$

$$\to -\pi \quad \text{as } a\to 0.$$

We can now infer the second assertion (45).

The first of the two results in (45) was announced by the second author[3] at the 1958 International Congress, Edinburgh.

REFERENCES

1. J. M. Anderson, *Integral functions and Tauberian theorems*, Duke Math. J. **32** (1965), 597–606.

2. R. P. Boas, *Entire functions*, Academic Press, New York, 1954.

3. N. A. Bowen, *A function theory proof of Tauberian theorems on integral functions*, Quart. J. Math. Oxford Ser. **19** (1948), 90–100.

4. N. A. Bowen and A. J. Macintyre, *Some theorems on integral functions with negative zeros*, Trans. Amer. Math. Soc. **70** (1951) 114–126.

5. T. J. I'a Bromwich, *Introduction to the theory of infinite series* (2), London, 1926.

6. K. S. Ganghadaran, *Two classical lattice point problems*. Proc. Cambridge Philos. Soc. **57** (1961) 699–721.

7. G. af Hallstrom, *Zwei Beispiele ganzer Funktionen mit algebraischen Hochstindex einer Stellensorte*, Math. Z. **47** (1941) 161–174.

8. G. H. Hardy, *On Dirichlet's divisor problem*, Proc. London Math. Soc. **15** (1916) 1–25.

9. G. H. Hardy, J. E. Littlewood and G. Pólya, *Inequalities*, Univ. Press, Cambridge, 1934.

10. G. H. Hardy and E. M. Wright, *An introduction to the theory of numbers*, Clarendon Press, Oxford, 1938.

11. M. H. Heins, *Entire functions with bounded minimum modulus. Subharmonic function analogs*, Ann. of Math. (2) **49** (1948) 200–213.

12. D. J. Newman, American Math. Monthly, problem 4615, 1954 p. 718 and 1956 p. 182 (solution by N. Fine).

13. E. C. Titchmarsh, *Theory of functions*, Oxford 1932.

14. ———, *On integral functions with real negative zeros*, Proc. London Math. Soc. (2) **26** (1926), 185–200.

15. ———, *The theory of the Riemann zeta-function*, Oxford, at the Clarendon Press, 1951.

16. M. M. Tjan, *On an application of the Carlemann-Subhanulo Tauberian Theorem*, Izv. Akad. Nauk. UzSSR Ser. Fiz.-Mat. Nauk. (1963) no. 3, 18–20.

17. G. Valiron, Ann. Fac. Sci. Univ. Toulouse (3) **5** (1913), 117–257.

ROYAL UNIVERSITY OF MALTA
MALTA
UNIVERSITY OF CINCINNATI
CINCINNATI, OHIO

[3] A. J. Macintyre's death on August 4, 1967, after a short illness, has caused a severe shock to his family and many friends, as well as a great loss to mathematics. I should like to record here that (i) he did not see the printer's proofs of this paper, (ii) the whole of §5 above is due to him. N. A. Bowen

The Expansion Theorem for Hilbert Spaces of Entire Functions[1]

LOUIS DE BRANGES

A fundamental problem is to determine the invariant subspaces of an everywhere defined and bounded transformation and to write the transformation as an integral in terms of invariant subspaces. Analogous problems exist also for unbounded or partially defined transformations once the meaning of "invariant subspace" is clarified. We now study closed symmetric transformations H having the property that $H - w$ has a bounded inverse for every complex number w. The main result is an integral representation of the transformation in terms of invariant subspaces when sufficiently many invariant subspaces exist and a complete continuity hypothesis is satisfied. The method of proof uses a canonical model of the transformation in a Hilbert space whose elements are entire functions. The result is given as an eigenfunction expansion which generalizes the Paley-Wiener theorem and Fourier transformation. The present treatment of the expansion differs from previous work [1] in the use of vector valued entire functions. The following notation is convenient for vector and operator valued functions.

Let $\mathscr{C}$ be a fixed Hilbert space which is used as a coefficient space. By a vector we always mean an element of this space. By an operator we mean a bounded transformation of vectors into vectors. If b is a vector, $\bar{b}$ is the linear functional on vectors defined by the inner product $\bar{b}a = \langle a, b\rangle$ for every vector a. If a and b are vectors, $a\bar{b}$ is the operator defined by $(a\bar{b})c = a(\bar{b}c)$ for every vector c. Standard conventions are used for operator inequalities. The absolute value symbol is used for the norm of a vector and for the operator norm of an operator. The adjoint of an operator A is denoted $\bar{A}$. A vector valued function $f(z)$ is said to be analytic in a region if the complex valued function $\bar{c}f(z)$ is analytic in the region for every vector c. The function $f(z)$ is said to be of bounded type in the region if $\bar{c}f(z)$ is of bounded type in the region for every vector c. A vector valued function $f(z)$, which is of bounded type in the upper half plane $y > 0$, is said to be of nonpositive

[1] Research supported by the Alfred P. Sloan Foundation and the National Science Foundation.

The paper revises and extends an unpublished manuscript of the author (Institute for Advanced Study, 1959). He is indebted to James Rovnyak for reading and criticizing the work.

mean type in the half plane if $\bar{c}f(z)$ is of nonpositive mean type in the half plane for every vector c, or in other words if

$$\lim_{y\to+\infty} e^{-\varepsilon y}\bar{c}f(iy)=0$$

for every $\varepsilon>0$. The analogous concept of mean type is used in the lower half plane $y<0$. Similar conventions are used for an operator valued function $g(z)$, the related complex valued functions being $\bar{b}g(z)\,a$ for all vectors a and b.

Let $A(z)$ and $B(z)$ be operator valued entire functions such that $B(z)\bar{A}(\bar{z}) = A(z)\bar{B}(\bar{z})$, $A(z)-iB(z)$ has invertible values in the upper half plane, and $A(z)+iB(z)$ has invertible values in the lower half plane. Assume that $1-A(z)$ and $B(z)$ have completely continuous values and that $K(z,z)\geqq 0$ for all complex z where

$$K(w,z)=[B(z)\bar{A}(w)-A(z)\bar{B}(w)]/[\pi(z-\bar{w})].$$

Let $\mathscr{H}(A,B)$ be the set of vector valued entire functions $F(z)$ such that

$$\|F(t)\|^2=\int_{-\infty}^{+\infty}|[A(t)-iB(t)]^{-1}F(t)|^2\,dt<\infty,$$

$[A(z)-iB(z)]^{-1}F(z)$ is of bounded type and of nonpositive mean type in the upper half plane, and $[A(z)+iB(z)]^{-1}F(z)$ is of bounded type and of nonpositive mean type in the lower half plane. Then $\mathscr{H}(A,B)$ is a Hilbert space which contains $K(w,z)c$ as a function of z for every vector c and complex number w. If $F(z)$ is in $\mathscr{H}(A,B)$, then $\bar{c}F(w)=\langle F(t),K(w,t)c\rangle$ for every c and w.

If $\mathscr{H}(A,B)$ is a given space, multiplication by z in $\mathscr{H}(A,B)$ is the transformation defined by $F(z)\to zF(z)$ whenever $F(z)$ and $zF(z)$ are in $\mathscr{H}(A,B)$. It is a closed symmetric transformation whose deficiency indices do not exceed the dimension of the coefficient space $\mathscr{C}$, and multiplication by $z-w$ has a bounded (partially defined) inverse for every complex number w. Multiplication by z in $\mathscr{H}(A,B)$ has an extension T which is a closed relation with spectrum below the real axis such that

$$i(T^*-\bar{w})^{-1}-i(T-w)^{-1}-i(w-\bar{w})(T^*-\bar{w})^{-1}(T-w)^{-1}$$

is nonnegative and completely continuous when w lies above the real axis. It is known [2] that any closed symmetric transformation which has these properties is unitarily equivalent to multiplication by z in a space $\mathscr{H}(A,B)$.

Since multiplication by z in $\mathscr{H}(A,B)$ is a partially defined transformation and since it does not have everywhere defined resolvents, extensions of the transformation are needed in the formulation of the invariant subspace concept. Let $S(z)$ be an operator valued entire function, which has at least one invertible value, such that values of $1-S(z)$ are completely continuous operators. It is known [2] that $S(z)$ has invertible values in the complex plane except at isolated points. If $S(z)$ has an invertible value at a point w, then the transformation

$$F(z)\to[F(z)-S(z)S(w)^{-1}F(w)]/(z-w)$$

is an extension of division by $z - w$ provided that $[F(z) - S(z)S(w)^{-1}F(w)]/(z - w)$ belongs to $\mathcal{H}(A, B)$ whenever $F(z)$ belongs to $\mathcal{H}(A, B)$.

THEOREM 1. *Let $\mathcal{H}(A, B)$ be a given space and let $S(z)$ be an operator valued entire function such that $1 - S(z)$ has completely continuous values. Assume that $S(z)$ and $A(z) - iB(z)$ have invertible values at a point w. A necessary and sufficient condition that*

$$\{[A(z) - iB(z)] - S(z)S(w)^{-1}[A(w) - iB(w)]\}c/(z - w)$$

belong to $\mathcal{H}(A, B)$ for every vector c is that

$$\int_{-\infty}^{+\infty} (1 + t^2)^{-1} |[A(t) - iB(t)]^{-1}S(t)c|^2 \, dt < \infty$$

for every vector c, that $[A(z) - iB(z)]^{-1}S(z)$ be of bounded type and of nonpositive mean type in the upper half plane, and that $[A(z) + iB(z)]^{-1}S(z)$ be of bounded type and of nonpositive mean type in the lower half plane. In this case

$$F(z) \to [F(z) - S(z)S(w)^{-1}F(w)]/(z - w)$$

is an everywhere defined and bounded transformation in $\mathcal{H}(A, B)$.

Note that these conditions do not depend on w. The same conclusion holds on the hypothesis that $A(z) + iB(z)$ has an invertible value at w. We now characterize such functions $S(z)$ in terms of matrices

$$M(z) = \begin{pmatrix} A(z) & B(z) \\ C(z) & D(z) \end{pmatrix}$$

whose entries are operator valued analytic functions of z. The following conventions are used for matrix valued analytic functions. Let $\mathcal{C}^2$ be the Hilbert space of pairs

$$\begin{pmatrix} u \\ v \end{pmatrix}$$

of elements of $\mathcal{C}$ in the norm

$$\left\| \begin{pmatrix} u \\ v \end{pmatrix} \right\|^2 = |u|^2 + |v|^2.$$

For each pair $\begin{pmatrix} u \\ v \end{pmatrix}$, let

$$\begin{pmatrix} u \\ v \end{pmatrix}^- = (\bar{u}, \bar{v})$$

be the linear functional on $\mathcal{C}^2$ defined by the matrix product

$$\begin{pmatrix} u \\ v \end{pmatrix}^- \begin{pmatrix} a \\ b \end{pmatrix} = (\bar{u}, \bar{v}) \begin{pmatrix} a \\ b \end{pmatrix} = \bar{u}a + \bar{v}b.$$

By a matrix we always mean a 2×2-matrix having operator entries. The norm and adjoint of a matrix are defined with respect to its action as a transformation in $\mathscr{C}^2$. Thus the adjoint $\bar{M}$ of the matrix

$$M = \begin{pmatrix} A & B \\ C & D \end{pmatrix} \quad \text{is} \quad \bar{M} = \begin{pmatrix} \bar{A} & \bar{C} \\ \bar{B} & \bar{D} \end{pmatrix} .$$

Much use is made of the matrix

$$I = \begin{pmatrix} 0 & -1 \\ 1 & 0 \end{pmatrix}.$$

Matrix valued analytic functions are used to construct auxiliary Hilbert spaces whose elements are pairs of vector valued analytic functions.

THEOREM 2. *Let $S(z)$ be an operator valued entire function, which has an invertible value, such that the values of $1 - S(z)$ are completely continuous operators. Let $M(z)$ be a matrix valued function which is defined and analytic in the region of analyticity of $S(z)^{-1}$. Assume that $M(z)$ has at least one invertible value, that*

$$M(z)I\bar{M}(\bar{z}) = S(z)I\bar{S}(\bar{z})$$

and that

$$[M(z)I\bar{M}(z) - S(z)I\bar{S}(z)]/(z - \bar{z}) \geqq 0$$

for all complex z. If the values of $M(z)I - I$ are completely continuous matrices, then there exists a Hilbert space $\mathscr{H}_S(M)$ whose elements are pairs

$$\begin{pmatrix} F_+ (z) \\ F_- (z) \end{pmatrix}$$

of vector valued functions, defined and analytic in the region of analyticity of $S(z)^{-1}$, such that

$$\frac{M(z)I\bar{M}(w) - S(z)I\bar{S}(w)}{2\pi(z - \bar{w})} \begin{pmatrix} u \\ v \end{pmatrix}$$

belongs to $\mathscr{H}_S(M)$ for all choices of vectors u and v and for all complex numbers w, and such that the identity

$$\begin{pmatrix} u \\ v \end{pmatrix}^- \begin{pmatrix} F_+(w) \\ F_-(w) \end{pmatrix} = \left\langle \begin{pmatrix} F_+ (t) \\ F_- (t) \end{pmatrix}, \frac{M(t)I\bar{M}(w) - S(t)I\bar{S}(w)}{2\pi(t - \bar{w})} \begin{pmatrix} u \\ v \end{pmatrix} \right\rangle$$

holds for every element

$$\begin{pmatrix} F_+(z) \\ F_-(z) \end{pmatrix}$$

of $\mathscr{H}_S(M)$.

The space $\mathscr{H}_S(M)$ is also used when the values of $M(z)I - I$ are not completely continuous matrices.

THEOREM 3. *Let $S(z)$ be an operator valued entire function, which has at least one invertible value, such that the values of $1 - S(z)$ are completely continuous operators. Let $M(z)$ be a matrix valued function which is defined and analytic in the region of analyticity of $S(z)^{-1}$. Assume that $M(z)$ has at least one invertible value, that*

$$M(z)I\bar{M}(\bar{z}) = S(z)I\bar{S}(\bar{z}),$$

and that

$$[M(z)I\bar{M}(z) - S(z)I\bar{S}(z)]/(z - \bar{z}) \geqq 0$$

for all complex z. Assume that there exists a Hilbert space $\mathcal{H}_S(M)$ whose elements are pairs

$$F(z) = \begin{pmatrix} F_+(z) \\ F_-(z) \end{pmatrix}$$

of vector valued functions, defined and analytic in the region of analyticity of $S(z)^{-1}$, such that

$$\frac{M(z)I\bar{M}(w) - S(z)I\bar{S}(w)}{2\pi(z - \bar{w})} \begin{pmatrix} u \\ v \end{pmatrix}$$

belongs to $\mathcal{H}_S(M)$ for all choices of vectors u and v and for all complex numbers w, and such that the identity

$$\begin{pmatrix} u \\ v \end{pmatrix}^- \begin{pmatrix} F_+(w) \\ F_-(w) \end{pmatrix} = \left\langle \begin{pmatrix} F_+(t) \\ F_-(t) \end{pmatrix}, \frac{M(t)I\bar{M}(w) - S(t)I\bar{S}(w)}{2\pi(t - \bar{w})} \begin{pmatrix} u \\ v \end{pmatrix} \right\rangle$$

holds for every element

$$\begin{pmatrix} F_+(z) \\ F_-(z) \end{pmatrix}$$

of $\mathcal{H}_S(M)$. The space is uniquely determined by $S(z)$ and $M(z)$. The pair $[F(z) - S(z)S(w)^{-1}F(w)]/(z - w)$ belongs to $\mathcal{H}_S(M)$ whenever $F(z)$ belongs to $\mathcal{H}_S(M)$ and w is not a singularity of $S(z)^{-1}F(z)$. The identity

$$\begin{aligned} &2\pi\bar{G}(\beta)\bar{S}(\beta)^{-1}IS(\alpha)^{-1}F(\alpha) \\ &\quad = \langle F(t), [G(t) - S(t)S(\beta)^{-1}G(\beta)]/(t - \beta)\rangle \\ &\qquad - \langle [F(t) - S(t)S(\alpha)^{-1}F(\alpha)]/(t - \alpha), G(t)\rangle \\ &\qquad + (\alpha - \bar{\beta})\langle [F(t) - S(t)S(\alpha)^{-1}F(\alpha)]/(t - \alpha), [G(t) - S(t)S(\beta)^{-1}G(\beta)]/(t - \beta)\rangle \end{aligned}$$

holds for all elements $F(z)$ and $G(z)$ of $\mathcal{H}_S(M)$ when α and β are not singularities of $S(z)^{-1}$.

The space $\mathscr{H}_S(M)$ is used to characterize the operator valued entire functions $S(z)$ associated with any given space $\mathscr{H}(A,B)$.

THEOREM 4. *Let $\mathscr{H}(A,B)$ be a given space, and let $S(z)$ be an operator valued entire function, which has an invertible value, such that the values of $1 - S(z)$ are completely continuous operators. A necessary and sufficient condition that*

$$\{[A(z) - iB(z)] - S(z)S(w)^{-1}[A(w) - iB(w)]\}c/(z - w)$$

belong to $\mathscr{H}(A,B)$ for every vector c when $S(w)$ is an invertible operator is that a space $\mathscr{H}_S(M)$ exist,

$$M(z) = \begin{pmatrix} A(z) & B(z) \\ C(z) & D(z) \end{pmatrix},$$

for some operator valued analytic functions $C(z)$ and $D(z)$.

The space $\mathscr{H}_S(M)$ is denoted $\mathscr{H}(M)$ when $S(z) = 1$. The space is used to determine the isometric inclusions of spaces $\mathscr{H}(A,B)$.

THEOREM 5. *Let $\mathscr{H}(A(a),B(a))$ and $\mathscr{H}(M(a,b))$ be given spaces. Let $A(b,z)$ and $B(b,z)$ be the operator valued entire functions defined by*

$$(A(b,z), B(b,z)) = (A(a,z), B(a,z))M(a,b,z).$$

If $1 - A(b,z)$ and $B(b,z)$ have completely continuous values, a space $\mathscr{H}(A(b),B(b))$ exists. The space $\mathscr{H}(A(a),B(a))$ is contained in the space $\mathscr{H}(A(b),B(b))$, and the inclusion does not increase norms. If there is no nonzero constant

$$\begin{pmatrix} u \\ v \end{pmatrix}$$

in $\mathscr{H}(M(a,b))$ such that $A(a,z)u + B(a,z)v$ belongs to $\mathscr{H}(A(a),B(a))$, then $\mathscr{H}(A(a),B(a))$ is contained isometrically in $\mathscr{H}(A(b), B(b))$. The transformation

$$\begin{pmatrix} F_+(z) \\ F_-(z) \end{pmatrix} \to 2^{\frac{1}{2}}[A(a,z)F_+(z) + B(a,z)F_-(z)]$$

is an isometry of $\mathscr{H}(M(a,b))$ onto the orthogonal complement of $\mathscr{H}(A(a),B(a))$ in $\mathscr{H}(A(b),B(b))$. The function $[A(a,z) - iB(a,z)]^{-1}[A(b,z) - iB(b,z)]$ has no real singularities. Let $S(z)$ be an operator valued entire function, which has an invertible value at a point w, such that the values of $1 - S(z)$ are completely continuous operators. If

$$\{[A(a,z) - iB(a,z)] - S(z)S(w)^{-1}[A(a,w) - iB(a,w)]\}c/(z - w)$$

belongs to $\mathscr{H}(A(a),B(a))$ for every vector c, then

$$\{[A(b,z) - iB(b,z)] - S(z)S(w)^{-1}[A(b,w) - iB(b,w)]\}c/(z - w)$$

belongs to $\mathscr{H}(A(b),B(b))$ for every vector c.

The theory of the space $\mathscr{H}(A, B)$ is related to the perturbation theory of selfadjoint transformations. A fundamental problem is to study the relation between the spectra of two selfadjoint transformations H_+ and H_- in the same Hilbert space when the difference $(H_+ - w)^{-1} - (H_- - w)^{-1}$ of resolvents is completely continuous for some, and hence all, nonreal numbers w. A canonical model of the perturbation situation has been developed in previous work [3]. Let $\phi(z)$ be an operator valued function, which is analytic in the upper and lower half planes such that $\phi(\bar{z}) = -\bar{\phi}(z)$. Assume that $\operatorname{Re}\phi(z) \geqq 0$ for $y > 0$. Then there exists a unique Hilbert space $\mathscr{L}(\phi)$ whose elements are vector valued functions, analytic in the upper and lower half planes, with this property: For every vector c and nonreal number w, $\frac{1}{2}[\phi(z) + \bar{\phi}(w)]c/[\pi i(\bar{w} - z)]$ belongs to $\mathscr{L}(\phi)$ as a function of z, and the identity

$$\bar{c}F(w) = \langle F(t), \tfrac{1}{2}[\phi(t) + \bar{\phi}(w)]c/[\pi i(\bar{w} - t)]\rangle_{\mathscr{L}(\phi)}$$

holds for every $F(z)$ in $\mathscr{L}(\phi)$. The space $\mathscr{L}(\phi)$ is interesting because of a selfadjoint relation H in the space such that

$$(H - w)^{-1} : F(z) \to [F(z) - F(w)]/(z - w)$$

for every nonreal number w. A canonical model of the perturbation situation is obtained using a pair of spaces, $\mathscr{L}(\phi_+)$ and $\mathscr{L}(\phi_-)$, related so that $\phi_+(z)\phi_-(z) = 1 = \phi_-(z)\phi_+(z)$. The transformation $F(z) \to \phi_-(z)F(z)$ is then an isometry of $\mathscr{L}(\phi_+)$ onto $\mathscr{L}(\phi_-)$. A perturbation situation can therefore be constructed from the underlying selfadjoint relations H_+ and H_- in $\mathscr{L}(\phi_+)$ and $\mathscr{L}(\phi_-)$. It is known that the general perturbation problem can be reduced to this situation. The problem is equivalent to the study of the relation between the singularities of the reciprocal functions $\phi_+(z)$ and $\phi_-(z)$. The inverse problem of scattering theory [2] is the problem of determining all spaces $\mathscr{H}(M)$ which are related to a given space $\mathscr{L}(\phi)$ in such a way that the transformation

$$\begin{pmatrix} F_+(z) \\ F_-(z) \end{pmatrix} \to 2^{\frac{1}{2}}[i\phi(z)F_+(z) + F_-(z)]$$

is an isometry of $\mathscr{H}(M)$ into $\mathscr{L}(\phi)$. We give a condition for a space $\mathscr{H}_S(M)$ to be related in a similar way to a space $\mathscr{L}(\phi)$.

THEOREM 6. *If $\mathscr{H}_S(M)$ and $\mathscr{L}(\phi)$ are given spaces and if the values of $M(z)I - S(z)I$ are completely continuous matrices, a necessary and sufficient condition that*

$$\begin{pmatrix} F_+(z) \\ F_-(z) \end{pmatrix} \to 2^{\frac{1}{2}}[i\phi(z)S(z)^{-1}F_+(z) + S(z)^{-1}F_-(z)]$$

be a transformation of $\mathscr{H}_S(M)$ into $\mathscr{L}(\phi)$ which is bounded by 1 is that

$$\phi(z)S(z)^{-1}\{[A(z)-iB(z)]-[A(z)+iB(z)]W(z)\}$$
$$= S(z)^{-1}\{[D(z)+iC(z)]+[D(z)-iC(z)]W(z)\}$$

for some operator valued function $W(z)$ which is analytic and bounded by 1 in the upper half plane. If $S(z)^{-1}$ has no real singularities, the orthogonal complement in $\mathscr{H}_S(M)$ of the set on which the transformation is isometric consists of elements of the form

$$M(z)\begin{pmatrix} u \\ v \end{pmatrix}$$

where u and v are vectors such that $\bar{u}v = \bar{v}u$.

By a theorem of von Neumann, every completely continuous transformation in a Hilbert space of dimension greater than 1 has a proper invariant subspace. We now use an integral representation of such a transformation, due to Brodskii [**4, 5**], and an equivalent multiplicative representation of its characteristic operator function. An expository account of the theory of characteristic operator functions and its relation to spaces $\mathscr{H}(M)$ appears in previous work [**6, 7, 8**].

A Hilbert space $L_m^2(a,b)$ is associated with any continuous, nondecreasing function

$$m(t) = \begin{pmatrix} \alpha(t) & \beta(t) \\ \bar{\beta}(t) & \gamma(t) \end{pmatrix}$$

whose values are selfadjoint matrices, $a \leqq t \leqq b$. A pair

$$f(t) = \begin{pmatrix} f_+(t) \\ f_-(t) \end{pmatrix}$$

of vector valued functions defined in (a,b) is said to be piecewise constant if there exists a partition $a = t_0 < t_1 < \cdots < t_r = b$ of (a,b) such that $f(t)$ is constant in each interval (t_{k-1}, t_k). The piecewise constant functions are a vector space over the complex numbers. If $f(t)$ and $g(t)$ belong to the space, there exists a partition $a = t_0 < t_1 < \cdots < t_r = b$ of (a,b) such that $f(t)$ and $g(t)$ are constant in each interval (t_{k-1}, t_k). The definition

$$\langle f(t), g(t)\rangle_m = \sum_{k=1}^{r} g(\tfrac{1}{2}t_{k-1}+\tfrac{1}{2}t_k)^-[m(t_k)-m(t_{k-1})]f(\tfrac{1}{2}t_{k-1}+\tfrac{1}{2}t_k)$$

does not depend on the choice of partition. The expression $\langle f,g\rangle_m$ is linear in f for each fixed g, it satisfies the symmetry condition $\langle g,f\rangle_m = \langle f,g\rangle_m^-$, and $\langle f,g\rangle_m$ is nonnegative when $f = g$. The Hilbert space $L_m^2(a,b)$ is obtained by identifying any two piecewise constant functions $f(t)$ and $g(t)$ such that $\langle g-f, g-f\rangle_m = 0$ and completing the resulting inner product space. The function notation

$$\langle f,g\rangle_m = \int_a^b \bar{g}(t)dm(t)f(t)$$

is used for general elements of $L_m^2(a,b)$ although such elements need not be represented by functions.

The m-derivative in the interval (a,b) of a pair

$$f(t) = \begin{pmatrix} f_+(t) \\ f_-(t) \end{pmatrix}$$

of vector valued functions is defined to be the unique element

$$g(t) = \begin{pmatrix} g_+(t) \\ g_-(t) \end{pmatrix}$$

of $L_m^2(a,b)$, if one exists, such that $f(c) - f(a) = \int_a^c dm(t)g(t)$ whenever $a \leqq c \leqq b$. The integral is taken in the sense that

$$\sum_{k=1}^{r} \bar{h}(\tfrac{1}{2}t_{k-1} + \tfrac{1}{2}t_k)[f(t_k) - f(t_{k-1})] = \int_a^b \bar{h}(t)dm(t)g(t)$$

whenever

$$h(t) = \begin{pmatrix} h_+(t) \\ h_-(t) \end{pmatrix}$$

is a piecewise constant pair of vector valued functions and $a = t_0 < t_1 < \cdots < t_r = b$ is a partition of (a,b) such that $h(t)$ is constant in each interval (t_{k-1}, t_k). A pair

$$g(t) = \begin{pmatrix} g_+(t) \\ g_-(t) \end{pmatrix}$$

of vector valued functions defined for $a \leqq t \leqq b$ is said to determine an element of $L_m^2(a,b)$ if there exists an m-differentiable function

$$f(t) = \begin{pmatrix} f_+(t) \\ f_-(t) \end{pmatrix}$$

such that $f(c) - f(a) = \int_a^c dm(t)g(t)$ whenever $a \leqq c \leqq b$ for some interpretation of the integral given by the context. For example, it may be a norm limit of Stieltjes sums. In such cases we identify $g(t)$ with the element of $L_m^2(a,b)$ which is the m-derivative of $f(t)$.

THEOREM 7. *Let $\mathcal{H}(M)$ be a given space such that $M(0)$ is the identity matrix and $M'(0)I$ is completely continuous. Then there exists a family $\{\mathcal{H}(M(a,b))\}$ of spaces, $0 \leqq a \leqq b \leqq 1$, such that $M(a,c,z) = M(a,b,z)M(b,c,z)$ when $a \leqq b \leqq c$, such that $M(0,1,z) = M(z)$ identically, and such that $M(a,b,0)$ is the identity matrix for all a and b, and there exists a continuous, matrix valued function*

$$m(t) = \begin{pmatrix} \alpha(t) & \beta(t) \\ \bar{\beta}(t) & \gamma(t) \end{pmatrix}$$

such that $M'(a,b,0)I = m(b) - m(a)$ *whenever* $a \leqq b$. *The increments in* $m(t)$ *are nonnegative, completely continuous matrices, and the integral equation*

$$M(a,b,z)I\bar{M}(a,b,w) - I = (z - \bar{w}) \int_a^b M(a,t,z)dm(t)\bar{M}(a,t,w)$$

holds for all complex z *and* w *when* $a \leqq b$, *the integral being a norm limit of Stieltjes sums. The function pair*

$$\bar{M}(0,t,w) \begin{pmatrix} u \\ v \end{pmatrix}$$

determines an element of $L^2_m(0,1)$ *for every pair of vectors* u *and* v *and for every complex number* w. *If*

$$f(t) = \begin{pmatrix} f_+(t) \\ f_-(t) \end{pmatrix}$$

belongs to $L^2_m(0,1)$, *its eigentransform*

$$F(z) = \begin{pmatrix} F_+(z) \\ F_-(z) \end{pmatrix}$$

is the unique pair of vector valued entire functions such that

$$2\pi \begin{pmatrix} u \\ v \end{pmatrix}^- \begin{pmatrix} F_+(w) \\ F_-(w) \end{pmatrix} = \int_0^1 \begin{pmatrix} u \\ v \end{pmatrix}^- M(0,t,w)dm(t) \begin{pmatrix} f_+(t) \\ f_-(t) \end{pmatrix}$$

for all vectors u *and* v *and for all complex numbers* w. *The eigentransform of an element of* $L^2_m(0,1)$ *always belongs to* $\mathcal{H}(M(0,1))$, *and the identity*

$$2\pi \| F(t) \|^2_{\mathcal{H}(M(0,1))} = \int_0^1 \bar{f}(t)dm(t)f(t)$$

holds whenever $f(t)$ *is orthogonal to the kernel of the eigentransformation. The Stieltjes integral* $\int_0^c m(t)Idm(t)$ *converges in norm when* $0 \leqq c \leqq 1$, *and*

$$Im(t) \begin{pmatrix} u \\ v \end{pmatrix}$$

determines an element of $L^2_m(0,1)$ *for every pair of vectors* u *and* v. *Let* $g(t)$ *belong to* $L^2_m(0,1)$, *and let* $G(z)$ *be its eigentransform in* $\mathcal{H}(M(0,1))$. *Then there exists a corresponding element* $f(t)$ *of* $L^2_m(0,1)$ *such that*

$$\int_c^1 dm(t)f(t) = \int_c^1 [m(c) - m(t)]Idm(t)g(t)$$

when $0 \leqq c \leqq 1$, *it is orthogonal to the kernel of the eigentransformation, and its eigentransform is* $F(z) = [G(z) - G(0)]/z$.

Let

$$m(t) = \begin{pmatrix} \alpha(t) & \beta(t) \\ \bar{\beta}(t) & \gamma(t) \end{pmatrix}$$

be a continuous, nondecreasing function whose values are selfadjoint matrices. The function is said to be stable in an interval (a,b) under the following conditions:

(1) *The matrix $m(b) - m(a)$ is completely continuous.*

(2) *The Stieltjes integral $\int_0^c m(t)Idm(t)$ converges in norm when $a \leqq c \leqq b$, and*

$$Im(t) \begin{pmatrix} u \\ v \end{pmatrix}$$

determines an element of $L_m^2(a, b)$ for every pair of vectors u and v.

(3) *There exists a completely continuous transformation $g(t) \to f(t)$ in $L_m^2(a, b)$ such that*

$$\int_c^b dm(t)f(t) = \int_c^b [m(c) - m(t)]Idm(t)g(t)$$

when $a \leqq c \leqq b$.

Stability is a necessary and sufficient condition for the existence of a family of spaces $\mathscr{H}(M)$ related to $m(t)$ as in Theorem 7.

THEOREM 8. *Let*

$$m(t) = \begin{pmatrix} \alpha(t) & \beta(t) \\ \bar{\beta}(t) & \gamma(t) \end{pmatrix}$$

be a continuous, nondecreasing function of t, $0 \leqq t \leqq 1$, whose values are selfadjoint matrices. If m(t) is stable in (0,1), then there exists a unique family $\{\mathscr{H}(M(a, b))\}$ of spaces, $0 \leqq a \leqq b \leqq 1$, such that $M(a, c, z) = M(a, b, z)M(b, c, z)$ when $a \leqq b \leqq c$, such that $M(a, b, 0)$ is the identity matrix for all a and b, and such that $M'(a, b, 0)I = m(b) - m(a)$ when $a \leqq b$.

The structure problem for any given space $\mathscr{H}(A(b), B(b))$ is the problem of finding all spaces $\mathscr{H}(A(a), B(a))$ such that

$$(A(b, z), B(b, z)) = (A(a, z), B(a, z))M(a, b, z)$$

for some space $\mathscr{H}(M(a, b))$. The existence of nontrivial subspaces is not known for a general space $\mathscr{H}(A, B)$. The expansion theorem for Hilbert spaces of entire functions is a decomposition of any given space, which is possible when sufficiently many subspaces exist and when a complete continuity condition is satisfied. The following statement of the expansion generalizes the results of previous work [1]. The expansion uses the space $L_m^2(0, b)$ when $m(t)$ has a possible discontinuity at the origin. The space is defined as before except that only those piecewise constant functions are used which are square integrable with respect to $m(t)$.

THEOREM 9. *Let $\{\mathcal{H}(A(t), B(t))\}$ be a given family of spaces, $t > 0$, such that $A(t,0) = 1$ and $B(t,0) = 0$ for every t. Let*

$$m(t) = \begin{pmatrix} \alpha(t) & \beta(t) \\ \bar{\beta}(t) & \gamma(t) \end{pmatrix}$$

be a continuous, nondecreasing function of $t > 0$ whose values are selfadjoint matrices, and which is stable in any interval (a,b), $0 < a < b < \infty$. Le $\{\mathcal{H}(M(a,b))\}$ be the unique family of spaces, $0 < a < b < \infty$, such that $M(a,c,z) = M(a,b,z)M(b,c,z)$ when $a \leqq b \leqq c$, such that $M(a,b,0)$ is the identity matrix for all a and b, and such that $M'(a,b,0)I = m(b) - m(a)$ when $a \leqq b$. If

$$(A(b,z), B(b,z)) = (A(a,z), B(a,z))M(a,b,z)$$

whenever $a \leqq b$, then the integral equation

$$(A(b,z), B(b,z))I(A(b,w), B(b,w))^- - (A(a,z), B(a,z))I(A(a,w), B(a,w))^-$$

$$= (z - \bar{w}) \int_a^b (A(t,z), B(t,z))dm(t)(A(t,w), B(t,w))^-$$

holds for all complex z and w when $a \leqq b$, the integral being a norm limit of Stieltjes sums. If

$$\lim_{a \searrow 0} [B(a,z)\bar{A}(a,w) - A(a,z)\bar{B}(a,w)] = 0$$

for all complex z and w, then the function pair $(A(t,w), B(t,w))^- c$ determines an element of $L^2_m(0,b)$ for every vector c and every complex number w when $0 < b < \infty$. If

$$f(t) = \begin{pmatrix} f_+(t) \\ f_-(t) \end{pmatrix}$$

belongs to $L^2_m(0,b)$, its eigentransform $F(z)$ is the unique vector valued entire function such that

$$\pi \bar{c} F(w) = \int_0^b \bar{c}(A(t,w), B(t,w))dm(t)f(t)$$

for every vector c and complex number w. The eigentransform of an element of $L^2_m(0,b)$ always belongs to $\mathcal{H}(A(b), B(b))$, and the identity

$$\pi \| F(t) \|^2_{\mathcal{H}(A(b),B(b))} = \int_0^b \bar{f}(t)dm(t)f(t)$$

holds whenever $f(t)$ is orthogonal to the kernel of the eigentransformation.

Let $g(t)$ *be an element of* $L_m^2(0,b)$, *and let* $G(z)$ *be its eigentransform in* $\mathscr{H}(A(b),B(b))$. *If* $G(0)=0$ *and if*

$$\int_a^b dm(t)f(t) = \int_a^b [m(a)-m(t)]I\,dm(t)g(t)$$

for $0<a<b$, *then* $f(t)$ *determines an element of* $L_m^2(0,b)$ *which is orthogonal to the kernel of the eigentransformation, and its eigentransform is given by* $F(z)=G(z)/z$.

Proof of Theorem 1. If

$$\{[A(z)-iB(z)]-S(z)S(w)^{-1}[A(w)-iB(w)]\}c/(z-w)$$

belongs to $\mathscr{H}(A,B)$ for every vector c, then

$$\{1-[A(z)-iB(z)]^{-1}S(z)S(w)^{-1}[A(w)-iB(w)]\}c/(z-w)$$

is of bounded type and of nonpositive mean type in the upper half plane,

$$[A(z)+iB(z)]^{-1}\{[A(z)-iB(z)]-S(z)S(w)^{-1}[A(w)-iB(w)]\}c/(z-w)$$

is of bounded type and of nonpositive mean type in the lower half plane, and

$$\int_{-\infty}^{+\infty}|\{1-[A(t)-iB(t)]^{-1}S(t)S(w)^{-1}[A(w)-iB(w)]\}c/(t-w)|^2\,dt < \infty$$

for every vector c. Since $A(z)-iB(z)$ and $S(z)$ have invertible values at w and since $[A(z)+iB(z)]^{-1}[A(z)-iB(z)]$ is bounded by 1 in the lower half plane, it follows that $[A(z)-iB(z)]^{-1}S(z)c$ is of bounded type and of nonpositive mean type in the upper half plane, $[A(z)+iB(z)]^{-1}S(z)c$ is of bounded type and of nonpositive mean type in the lower half plane, and

$$\int_{-\infty}^{+\infty}(1+t^2)^{-1}|[A(t)-iB(t)]^{-1}S(t)c|^2\,dt<\infty$$

for every vector c. On the other hand, if these conditions are satisfied, it is easily verified that

$$\{[A(z)-iB(z)]-S(z)S(w)^{-1}[A(w)-iB(w)]\}c/(z-w)$$

belongs to $\mathscr{H}(A,B)$ for every vector c and that $[F(z)-S(z)S(w)^{-1}F(w)]/(z-w)$ belongs to $\mathscr{H}(A,B)$ whenever $F(z)$ belongs to $\mathscr{H}(A,B)$. Boundedness of the transformation $F(z)\to[F(z)-S(z)S(w)^{-1}F(w)]/(z-w)$ is obtained by the closed graph theorem.

Proof of Theorem 2. The proof generalizes the proof of Theorem 4 of **[2]**. We show that the positive-definiteness inequality

$$\sum \begin{pmatrix} u_i \\ v_i \end{pmatrix}^- \frac{M(w_i)I\bar{M}(w_k) - S(w_i)I\bar{S}(w_k)}{2\pi(w_i - \bar{w}_k)} \begin{pmatrix} u_k \\ v_k \end{pmatrix} \geqq 0$$

holds whenever $w_1, \cdots, w_r$ is a finite set of points in the plane, and $u_1, \cdots, u_r$ and $v_1, \cdots, v_r$ are corresponding vectors. The proof uses the choice of a real reference point where $S(z)$ has an invertible value. For definiteness we assume that this point is the origin. The identity

$$S(z)^{-1}M(z)I\bar{M}(\bar{z})\bar{S}(\bar{z})^{-1} = I$$

implies the identity

$$\bar{M}(\bar{z})\bar{S}(\bar{z})^{-1}IS(z)^{-1}M(z) = I$$

at all points where $S(z)^{-1}M(z)$ has an invertible value. Since we assume that $M(z)$ has an invertible value, the last identity holds in a nonempty open set. It is valid by analytic continuation in the region of analyticity of $S(z)^{-1}$. We can now write

$$[M(z)I\bar{M}(w) - S(z)I\bar{S}(w)]/[2\pi(z - \bar{w})]$$
$$= \frac{S(z) - M(z)I\bar{M}(0)\bar{S}(0)^{-1}I}{2} \; \frac{\phi(z) + \bar{\phi}(w)}{\pi i(\bar{w} - z)} \; \frac{\bar{S}(w) - IS(0)^{-1}M(0)I\bar{M}(w)}{2}$$

where

$$\phi(z) = [1 - S(z)^{-1}M(z)I\bar{M}(0)\bar{S}(0)^{-1}I]^{-1}[1 + S(z)^{-1}M(z)I\bar{M}(0)\bar{S}(0)^{-1}I]iI$$

is a matrix valued function which is defined and analytic in the set where $\frac{1}{2}[1 - S(z)^{-1}M(z)I\bar{M}(0)\bar{S}(0)^{-1}I]$ has invertible values. This set is nonempty because the function has the identity matrix as its value at the origin. Since the values of $\frac{1}{2}[1 + S(z)^{-1}M(z)I\bar{M}(0)\bar{S}(0)^{-1}I]$ are matrices of completely continuous operators, it follows that $\frac{1}{2}[1 - S(z)^{-1}M(z)I\bar{M}(0)\bar{S}(0)^{-1}I]$ has invertible values at all except isolated points in the complex plane. The hypothesis that $[M(z)I\bar{M}(z) - S(z)I\bar{S}(z)]/(z - \bar{z}) \geqq 0$ for all complex z implies that $[\phi(z) + \bar{\phi}(z)]/[\pi i(\bar{z} - z)] \geqq 0$ at all points where $\phi(z)$ is defined. The hypothesis that $M(z)I\bar{M}(\bar{z}) = S(z)I\bar{S}(\bar{z})$ implies that $\phi(z) = -\bar{\phi}(\bar{z})$. By the Poisson representation of a function which is analytic and has a nonnegative real part in the upper half plane, the singularities of $\phi(z)$ are a real sequence (t_n), and there exists a nonnegative matrix P and a sequence of nonnegative matrices (P_n) such that

$$\frac{\phi(z) + \bar{\phi}(w)}{\pi i(\bar{w} - z)} = P + \sum \frac{P_n}{(t_n - z)(t_n - \bar{w})}$$

whenever z and w are not singularities of the function. It follows that

$$\sum \begin{pmatrix} u_i \\ v_i \end{pmatrix}^- \frac{\phi(w_i) + \bar{\phi}(w_k)}{\pi i(\bar{w}_k - w_i)} \begin{pmatrix} u_k \\ v_k \end{pmatrix} \geqq 0$$

whenever $w_1, \cdots, w_r$ are nonreal numbers and $u_1, \cdots, u_r$ and $v_1, \cdots, v_r$ are corresponding vectors. The desired positive-definiteness inequality for $[M(z)I\bar{M}(w) - S(z)I\bar{S}(w)]/(z - \bar{w})$ is obtained by a change of variable. The existence of the space $\mathscr{H}_S(M)$ follows as in the proof of Lemma 11 of [9].

Proof of Theorem 3. The proof generalizes the proof of Theorem 4 of [2]. We omit the proof that $\mathscr{H}_S(M)$ is uniquely determined by $S(z)$ and $M(z)$ since the argument is essentially the same as one used in the proof of Lemma 11 of [9]. Since we assume that $S(z)^{-1}M(z)I\bar{M}(\bar{z})\bar{S}(\bar{z})^{-1} = I$ and since $M(z)$ has an invertible value, the identity $\bar{M}(\bar{z})\bar{S}(\bar{z})^{-1}IS(z)^{-1}M(z) = I$ holds in the common region of analyticity of $S(z)^{-1}$ and $\bar{S}(\bar{z})^{-1}$. It follows that $M(z)$ has invertible values in the common region of analyticity of $S(z)^{-1}$ and $\bar{S}(\bar{z})^{-1}$. Since we assume that

$$\frac{M(z)I\bar{M}(w) - S(z)I\bar{S}(w)}{2\pi(z - \bar{w})} \begin{pmatrix} u \\ v \end{pmatrix} = \frac{M(z) - S(z)S(\bar{w})^{-1}M(\bar{w})}{2\pi(z - \bar{w})} I\bar{M}(w) \begin{pmatrix} u \\ v \end{pmatrix}$$

belongs to $\mathscr{H}_S(M)$ for all pairs of vectors u and v and for all complex numbers w, the pair

$$\frac{M(z) - S(z)S(w)^{-1}M(w)}{z - w} \begin{pmatrix} u \\ v \end{pmatrix}$$

belongs to $\mathscr{H}_S(M)$ for all vectors u and v when w is not a singularity of $S(z)^{-1}$ or of $\bar{S}(\bar{z})^{-1}$, and the identity

$$2\pi \begin{pmatrix} u \\ v \end{pmatrix}^- \bar{M}(w)\bar{S}(w)^{-1}IS(\bar{w})^{-1} \begin{pmatrix} F_+(w) \\ F_-(w) \end{pmatrix}$$
$$= \left\langle \begin{pmatrix} F_+(t) \\ F_-(t) \end{pmatrix}, \frac{M(t) - S(t)S(w)^{-1}M(w)}{t - w} \begin{pmatrix} u \\ v \end{pmatrix} \right\rangle$$

holds for every element

$$F(z) = \begin{pmatrix} F_+(z) \\ F_-(z) \end{pmatrix}$$

of $\mathscr{H}_S(M)$. If we let

$$G(z) = \begin{pmatrix} G_+(z) \\ G_-(z) \end{pmatrix} = \frac{M(z) - S(z)S(w)^{-1}M(w)}{z - w} \begin{pmatrix} u \\ v \end{pmatrix}$$

for any fixed u, v, and w, then

$$(\beta - w)[G(z) - S(z)S(w)^{-1}G(w)]/(z - w)$$
$$= \frac{M(z) - S(z)S(\beta)^{-1}M(\beta)}{z - \beta} \begin{pmatrix} u \\ v \end{pmatrix} - \frac{M(z) - S(z)S(w)^{-1}M(w)}{z - w} \begin{pmatrix} u \\ v \end{pmatrix}$$

belongs to $\mathcal{H}_S(M)$, and $[G(z) - S(z)S(w)^{-1}G(w)]/(z-w)$ belongs to $\mathcal{H}_S(M)$ if $\beta \neq w$. Let $F(z)$ be an element of $\mathcal{H}_S(M)$ such that $[F(z) - S(z)S(\alpha)^{-1}F(\alpha)]/(z-\alpha)$ belongs to $\mathcal{H}_S(M)$ for some number α which is not a singularity of $S(z)^{-1}$. Then

$$\begin{aligned}
&\langle F(t), [G(t) - S(t)S(\beta)^{-1}G(\beta)]/(t-\beta)\rangle \\
&\quad - \langle [F(t) - S(t)S(\alpha)^{-1}F(\alpha)]/(t-\alpha), G(t)\rangle \\
&\quad + (\alpha - \bar\beta)\langle [F(t) - S(t)S(\alpha)^{-1}F(\alpha)]/(t-\alpha), [G(t) - S(t)S(\beta)^{-1}G(\beta)]/(t-\beta)\rangle \\
&= (\bar\beta - \bar w)^{-1}\left\langle \begin{pmatrix} F_+(t) \\ F_-(t)\end{pmatrix}, \frac{M(t) - S(t)S(\beta)^{-1}M(\beta)}{t-\beta}\begin{pmatrix} u \\ v\end{pmatrix}\right\rangle \\
&\quad - (\bar\beta - \bar w)^{-1}\left\langle \begin{pmatrix} F_+(t) \\ F_-(t)\end{pmatrix}, \frac{M(t) - S(t)S(w)^{-1}M(w)}{t-w}\begin{pmatrix} u \\ v\end{pmatrix}\right\rangle \\
&\quad - \left\langle \begin{pmatrix} [F_+(t) - S(t)S(\alpha)^{-1}F_+(\alpha)]/(t-\alpha) \\ [F_-(t) - S(t)S(\alpha)^{-1}F_-(\alpha)]/(t-\alpha)\end{pmatrix}, \frac{M(t) - S(t)S(w)^{-1}M(w)}{t-w}\begin{pmatrix} u \\ v\end{pmatrix}\right\rangle \\
&\quad + \frac{\alpha - \bar\beta}{\bar\beta - \bar w}\left\langle \begin{pmatrix} [F_+(t) - S(t)S(\alpha)^{-1}F_+(\alpha)]/(t-\alpha) \\ [F_-(t) - S(t)S(\alpha)^{-1}F_-(\alpha)]/(t-\alpha)\end{pmatrix}, \right. \\
&\qquad\qquad\qquad\qquad \left. \frac{M(t) - S(t)S(\beta)^{-1}M(\beta)}{t-\beta}\begin{pmatrix} u \\ v\end{pmatrix}\right\rangle \\
&\quad - \frac{\alpha - \bar\beta}{\bar\beta - \bar w}\left\langle \begin{pmatrix} [F_+(t) - S(t)S(\alpha)^{-1}F_+(\alpha)]/(t-\alpha) \\ [F_-(t) - S(t)S(\alpha)^{-1}F_-(\alpha)]/(t-\alpha)\end{pmatrix}, \right. \\
&\qquad\qquad\qquad\qquad \left. \frac{M(t) - S(t)S(w)^{-1}M(w)}{t-w}\begin{pmatrix} u \\ v\end{pmatrix}\right\rangle \\
&= (\bar\beta - \bar w)^{-1}2\pi\begin{pmatrix} u \\ v\end{pmatrix}^- \bar M(\beta)\bar S(\beta)^{-1}IS(\bar\beta)^{-1}\begin{pmatrix} F_+(\bar\beta) \\ F_-(\bar\beta)\end{pmatrix} \\
&\quad - (\bar\beta - \bar w)^{-1}2\pi\begin{pmatrix} u \\ v\end{pmatrix}^- \bar M(w)\bar S(w)^{-1}IS(\bar w)^{-1}\begin{pmatrix} F_+(\bar w) \\ F_-(\bar w)\end{pmatrix} \\
&\quad - 2\pi\begin{pmatrix} u \\ v\end{pmatrix}^- \bar M(w)\bar S(w)^{-1}IS(\bar w)^{-1}\begin{pmatrix} [F_+(\bar w) - S(\bar w)S(\alpha)^{-1}F_+(\alpha)]/(\bar w - \alpha) \\ [F_-(\bar w) - S(\bar w)S(\alpha)^{-1}F_-(\alpha)]/(\bar w - \alpha)\end{pmatrix} \\
&\quad + \frac{\alpha - \bar\beta}{\bar\beta - \bar w}2\pi\begin{pmatrix} u \\ v\end{pmatrix}^- \bar M(\beta)\bar S(\beta)^{-1}IS(\bar\beta)^{-1}\begin{pmatrix} [F_+(\bar\beta) - S(\bar\beta)S(\alpha)^{-1}F_+(\alpha)]/(\bar\beta - \alpha) \\ [F_-(\bar\beta) - S(\bar\beta)S(\alpha)^{-1}F_-(\alpha)]/(\bar\beta - \alpha)\end{pmatrix} \\
&\quad - \frac{\alpha - \bar\beta}{\bar\beta - \bar w}2\pi\begin{pmatrix} u \\ v\end{pmatrix}^- \bar M(w)\bar S(w)^{-1}IS(\bar w)^{-1}\begin{pmatrix} [F_+(\bar w) - S(\bar w)S(\alpha)^{-1}F_+(\alpha)]/(\bar w - \alpha) \\ [F_-(\bar w) - S(\bar w)S(\alpha)^{-} F_-(\alpha)]/(\bar w - \alpha)\end{pmatrix} \\
&= 2\pi\bar G(\beta)\bar S(\beta)^{-1}IS(\alpha)^{-1}F(\alpha).
\end{aligned}$$

It follows that the identity for difference quotients,

$$\langle F(t), [G(t) - S(t)S(\beta)^{-1}G(\beta)]/(t-\beta)\rangle$$
$$- \langle [F(t) - S(t)S(\alpha)^{-1}F(\alpha)]/(t-\alpha), G(t)\rangle$$
$$+ (\alpha - \bar{\beta})\langle [F(t) - S(t)S(\alpha)^{-1}F(\alpha)]/(t-\alpha), [G(t) - S(t)S(\beta)^{-1}G(\beta)]/(t-\beta)\rangle$$
$$= 2\pi \bar{G}(\beta)\bar{S}(\beta)^{-1}IS(\alpha)^{-1}F(\alpha),$$

holds whenever $F(z)$ and $G(z)$ are finite sums of elements of $\mathscr{H}_S(M)$ of the form

$$\frac{M(z) - S(z)S(w)^{-1}M(w)}{z - w}\begin{pmatrix} u \\ v \end{pmatrix}$$

for some vectors u and v and for some number w different from α and β if α, β, and w are not singularities of $S(z)^{-1}$ or of $\bar{S}(\bar{z})^{-1}$.

The transformation $R(w)\colon F(z) \to [F(z) - S(z)S(w)^{-1}F(w)]/(z-w)$ is densely defined in $\mathscr{H}_S(M)$. We show that it is bounded when w is not real. If $F(z)$ is a finite sum of elements of $\mathscr{H}_S(M)$ of the form

$$\frac{M(z) - S(z)S(\alpha)^{-1}M(\alpha)}{z - \alpha}\begin{pmatrix} u \\ v \end{pmatrix}$$

for some vectors u and v and for some number α different from w, α and w not singularities of $S(z)^{-1}$ or of $\bar{S}(\bar{z})^{-1}$, then $F(z)$ is in the domain of $R(w)$, as we have seen. By the identity for difference quotients,

$$\| F(t) + (w - \bar{w})[F(t) - S(t)S(w)^{-1}F(w)]/(t-w) \|^2$$
$$= \| F(t) \|^2 + 2\pi(\bar{w} - w)\bar{F}(w)\bar{S}(w)^{-1}IS(w)^{-1}F(w).$$

But if $F(z)$ is in $\mathscr{H}_S(M)$ and if u and v are vectors, then

$$\left| \begin{pmatrix} u \\ v \end{pmatrix}^- \begin{pmatrix} F_+(w) \\ F_-(w) \end{pmatrix} \right|^2 = \left| \left\langle \begin{pmatrix} F_+(t) \\ F_-(t) \end{pmatrix}, \frac{M(t)I\bar{M}(w) - S(t)I\bar{S}(w)}{2\pi(t - \bar{w})}\begin{pmatrix} u \\ v \end{pmatrix} \right\rangle \right|^2$$
$$\leqq \| F \|^2 \begin{pmatrix} u \\ v \end{pmatrix}^- \frac{M(w)I\bar{M}(w) - S(w)I\bar{S}(w)}{2\pi(w - \bar{w})}\begin{pmatrix} u \\ v \end{pmatrix}$$

by the Schwarz inequality. It follows that

$$| S(w)^{-1}F(w) |^2 \leqq \| F \|^2 \left| \frac{S(w)^{-1}M(w)I\bar{M}(w)\bar{S}(w)^{-1} - I}{2\pi(w - \bar{w})} \right|$$

when w is not a singularity of $S(z)^{-1}$. So the transformation $1 + (w - \bar{w})R(w)$ is bounded and

$$\| 1 + (w - \bar{w})R(w) \|^2 \leqq 1 + | S(w)^{-1}M(w)I\bar{M}(w)\bar{S}(w)^{-1} - I |$$

when w is not a singularity of $S(z)^{-1}$. The transformation $R(w)$ is density defined and bounded when w is not real and is not a singularity of $S(z)^{-1}$ or of $\bar{S}(\bar{z})^{-1}$.

Since the transformation clearly has a closed graph, it is everywhere defined for these values of w. Since $(\alpha-\beta)R(\alpha)R(\beta)=R(\alpha)-R(\beta)$ whenever $R(\alpha)$ or $R(\beta)$ is everywhere defined and since $R(w)$ is bounded independently of w in a neighborhood of any nonreal singularity of $\bar{S}(\bar{z})^{-1}$ which is not a singularity of $S(z)^{-1}$, the transformation $R(w)$ is everywhere defined and bounded whenever w is a nonreal number which is not a singularity of $S(z)^{-1}$. But if w is a real number which is not a singularity of $S(z)^{-1}$ and if $\varepsilon>0$ is so small that $S(z)^{-1}$ has no singularities for $(z-w)\leqq\varepsilon$ then $(1-e^{2i\theta})R(w+\varepsilon e^{i\theta})$ is bounded independently of θ, $0\leqq\theta\leqq 2\pi$, and

$$2\pi R(w)=\int_0^{2\pi}(1-e^{2i\theta})R(w+\varepsilon e^{i\theta})d\theta$$

by Cauchy's formula. It follows that $R(w)$ is everywhere defined and bounded also for real numbers w which are not singularities of $S(z)^{-1}$. The identity for difference quotients holds by continuity whenever α and β are not singularities of $S(z)^{-1}$.

Proof of Theorem 4, the sufficiency. We show that the transformation

$$\begin{pmatrix}F_+(z)\\F_-(z)\end{pmatrix}\to 2^{1/2}F_+(z)$$

is a partial isometry of $\mathscr{H}_S(M)$ onto $\mathscr{H}(A,B)$. If

$$\begin{pmatrix}F_+(z)\\F_-(z)\end{pmatrix}=\frac{M(z)I\bar{M}(w)-S(z)I\bar{S}(w)}{2\pi(z-\bar{w})}\begin{pmatrix}c\\0\end{pmatrix}$$

for some vector c, then

$$F_+(z)=\frac{B(z)\bar{A}(w)-A(z)\bar{B}(w)}{2\pi(z-\bar{w})}c$$

is the corresponding element of $\mathscr{H}(A,B)$. For all complex numbers α and β and for all vectors a and b,

$$\begin{aligned}&\left\langle\frac{M(t)I\bar{M}(\beta)-S(t)I\bar{S}(\beta)}{2\pi(t-\bar{\beta})}\begin{pmatrix}b\\0\end{pmatrix},\frac{M(t)I\bar{M}(\alpha)-S(t)I\bar{S}(\alpha)}{2\pi(t-\bar{\alpha})}\begin{pmatrix}a\\0\end{pmatrix}\right\rangle_{\mathscr{H}_S(M)}\\&=\begin{pmatrix}a\\0\end{pmatrix}^-\frac{M(\alpha)I\bar{M}(\beta)-S(\alpha)I\bar{S}(\beta)}{2\pi(\alpha-\bar{\beta})}\begin{pmatrix}b\\0\end{pmatrix}\\&=\bar{b}\,\frac{B(\alpha)\bar{A}(\beta)-A(\alpha)\bar{B}(\beta)}{2\pi(\alpha-\bar{\beta})}a\\&=2\left\langle\frac{B(t)\bar{A}(\beta)-A(t)\bar{B}(\beta)}{2\pi(t-\bar{\beta})}b,\frac{B(t)\bar{A}(\alpha)-A(t)\bar{B}(\alpha)}{2\pi(t-\bar{\alpha})}a\right\rangle_{\mathscr{H}(A,B)}.\end{aligned}$$

It follows that if

$$\begin{pmatrix} F_+(z) \\ F_-(z) \end{pmatrix}$$

is a finite sum of elements

$$\frac{M(z)I\bar{M}(w) - S(z)I\bar{S}(w)}{2\pi(z - \bar{w})}\begin{pmatrix} c \\ 0 \end{pmatrix}$$

of $\mathscr{H}_S(M)$, then $F_+(z)$ belongs to $\mathscr{H}(A,B)$ and

$$\left\| \begin{pmatrix} F_+(t) \\ F_-(t) \end{pmatrix} \right\|^2_{\mathscr{H}_S(M)} = 2\| F_+(t) \|^2_{\mathscr{H}(A,B)}.$$

The same conclusion follows by continuity whenever

$$\begin{pmatrix} F_+(z) \\ F_-(z) \end{pmatrix}$$

belongs to the closed span of such special elements of $\mathscr{H}_S(M)$. On the other hand, if

$$\begin{pmatrix} F_+(z) \\ F_-(z) \end{pmatrix}$$

is an element of $\mathscr{H}_S(M)$ which is orthogonal to such special functions, then

$$\bar{c}F_+(w) = \left\langle \begin{pmatrix} F_+(t) \\ F_-(t) \end{pmatrix}, \frac{M(t)I\bar{M}(w) - S(t)I\bar{S}(w)}{2\pi(t - \bar{w})} \begin{pmatrix} c \\ 0 \end{pmatrix} \right\rangle_{\mathscr{H}_S(M)}$$

vanishes for all vectors c and for all complex numbers w. It follows that $F_+(z)$ vanishes identically. The transformation

$$\begin{pmatrix} F_+(z) \\ F_-(z) \end{pmatrix} \to 2^{1/2}F_+(z)$$

is therefore a partial isometry of $\mathscr{H}_S(M)$ into $\mathscr{H}(A,B)$. Since the range of the partial isometry is a closed subspace of $\mathscr{H}(A,B)$ which contains

$$[B(z)\bar{A}(w) - A(z)\bar{B}(w)]c/[\pi(z - \bar{w})]$$

for every vector c and every complex number w, it is $\mathscr{H}(A,B)$.

Since the expression

$$\frac{M(z)I\bar{M}(w) - S(z)I\bar{S}(w)}{2\pi(z - \bar{w})}\begin{pmatrix} u \\ v \end{pmatrix} = \frac{M(z) - S(z)S(\bar{w})^{-1}M(\bar{w})}{2\pi(z - \bar{w})}I\bar{M}(w)\begin{pmatrix} u \\ v \end{pmatrix}$$

belongs to $\mathscr{H}_S(M)$ for every pair of vectors u and v and for every complex number w, and since $M(w)$ is an invertible matrix when $S(w)$ and $S(\bar{w})$ are invertible operators,

$$\frac{M(z) - S(z)S(\bar{w})^{-1}M(\bar{w})}{z - \bar{w}} \begin{pmatrix} u \\ v \end{pmatrix}$$

belongs to $\mathscr{H}_S(M)$ for every pair of vectors u and v when w is not a singularity of $S(z)^{-1}$ or of $\bar{S}(\bar{z})^{-1}$. It follows that $[A(z) - S(z)S(\bar{w})^{-1}A(\bar{w})]u/(z - \bar{w})$ and $[B(z) - S(z)S(\bar{w})^{-1}B(\bar{w})]v/(z - \bar{w})$ belong to $\mathscr{H}(A, B)$ for every pair of vectors u and v when w is not a singularity of $S(z)^{-1}$ or of $\bar{S}(\bar{z})^{-1}$, and this implies that

$$\{[A(z) - iB(z)] - S(z)S(\bar{w})^{-1}[A(\bar{w}) - iB(\bar{w})]\}c/(z - \bar{w})$$

belongs to $\mathscr{H}(A, B)$ for every vector c.

PROOF OF THEOREM 4, THE NECESSITY. The proof generalizes the proof of necessity for Theorem 27 of [**1**]. The construction requires the choice of a point on the real axis where $S(z)$ and $A(z)$ have invertible values. For definiteness we assume that this point is the origin. Note that the hypotheses of Theorem 1 are satisfied when $S(z) = A(z) - iB(z)$ or when $S(z) = A(z) + iB(z)$. It follows that $S(z) = A(z)$ also satisfies the hypotheses of the theorem. Since $A(0)$ is an invertible operator, $[F(z) - A(z)A(0)^{-1}F(0)]/z$ belongs to $\mathscr{H}(A, B)$ whenever $F(z)$ belongs to $\mathscr{H}(A, B)$. We show that the identity

$$0 = \langle F(t), [G(t) - A(t)A(0)^{-1}G(0)]/t\rangle - \langle [F(t) - A(t)A(0)^{-1}F(0)]/t, G(t)\rangle$$

holds for all elements $F(z)$ and $G(z)$ of $\mathscr{H}(A, B)$. Since the vector span of the functions $K(w, z)c$, c in $\mathscr{C}$ and $w \neq 0$, is dense in $\mathscr{H}(A, B)$, it is sufficient to verify the identity in the case that $G(z) = K(w, z)c$ for some vector c and some nonzero number w. Since

$$[K(w, z) - A(z)A(0)^{-1}K(w, 0)]/z = [K(w, z) - K(0, z)\bar{A}(0)^{-1}\bar{A}(w)]/\bar{w},$$

we then have

$$\begin{aligned} &\langle F(t), [G(t) - A(t)A(0)^{-1}G(0)]/t\rangle \\ &\quad = \langle F(t), [K(w, t) - K(0, t)\bar{A}(0)^{-1}\bar{A}(w)]c/\bar{w}\rangle \\ &\quad = \bar{c}[F(w) - A(w)A(0)^{-1}F(0)]/w \end{aligned}$$

and also

$$\begin{aligned} &\langle [F(t) - A(t)A(0)^{-1}F(0)]/t, G(t)\rangle \\ &\quad = \langle [F(t) - A(t)A(0)^{-1}F(0)]/t, K(w, t)c\rangle \\ &\quad = \bar{c}[F(w) - A(w)A(0)^{-1}F(0)]/w. \end{aligned}$$

The desired identity follows.

By Theorem 1 we know that

$$\{[A(z) - iB(z)] - S(z)S(0)^{-1}[A(0) - iB(0)]\}c/z$$

belongs to $\mathscr{H}(A, B)$ for every vector c. The proof of Theorem 1 will show that

$$\{[A(z) + iB(z)] - S(z)S(0)^{-1}[A(0) + iB(0)]\}c/z$$

belongs to $\mathscr{H}(A, B)$ for every vector c. It follows that $[A(z) - S(z)S(0)^{-1}A(0)]c/z$ belongs to $\mathscr{H}(A, B)$ for every vector c. Since $A(0)$ and $S(0)$ are invertible operators, $[S(z) - A(z)A(0)^{-1}S(0)]c/z$ belongs to $\mathscr{H}(A, B)$ for every vector c. Since the transformation $c \to [S(z) - A(z)A(0)^{-1}S(0)]c/z$ of $\mathscr{C}$ into $\mathscr{H}(A, B)$ has a closed graph, it is bounded. The transformation therefore has a bounded adjoint which takes $\mathscr{H}(A, B)$ into $\mathscr{C}$. From this we can conclude that there exists a bounded transformation $F(z) \to \tilde{F}(0)$ of $\mathscr{H}(A, B)$ into $\mathscr{C}$ such that

$$\pi \bar{c} S(0)^{-1} \tilde{F}(0) = \langle F(t), [S(t) - A(t)A(0)^{-1}S(0)]c/t \rangle$$

for every vector c. It follows that the identity

$$\begin{aligned} \pi \tilde{G}(0)^{-} \bar{S}(0)^{-1} S(0)^{-1} F(0) - \pi \bar{G}(0) \bar{S}(0)^{-1} S(0)^{-1} \tilde{F}(0) \\ = \langle F(t), [G(t) - S(t)S(0)^{-1}G(0)]/t \rangle \\ - \langle [F(t) - S(t)S(0)^{-1}F(0)]/t, G(t) \rangle \end{aligned}$$

holds for all elements $F(z)$ and $G(z)$ of $\mathscr{H}(A, B)$.

We now define a linear transformation $F(z) \to \tilde{F}(w)$ of $\mathscr{H}(A, B)$ into $\mathscr{C}$ for every complex number w in the region of analyticity of $S(z)^{-1}$. If $F(z)$ belongs to $\mathscr{H}(A, B)$, consider the function $G(z) = F(z) + w[F(z) - S(z)S(w)^{-1}F(w)]/(z - w)$, which belongs to $\mathscr{H}(A, B)$ by Theorem 1. Note that $G(0) = S(0)S(w)^{-1}F(w)$. Define $\tilde{F}(w)$ so that $\tilde{G}(0) = S(0)S(w)^{-1}\tilde{F}(w)$. Since

$$R(w)\colon F(z) \to [F(z) - S(z)S(w)^{-1}F(w)]/(z - w)$$

is a bounded transformation of $\mathscr{H}(A, B)$ into itself, and since $S(0)$ and $S(w)$ are invertible operators, $F(z) \to \tilde{F}(w)$ is a bounded transformation of $\mathscr{H}(A, B)$ into $\mathscr{C}$. The resolvent identity $(\alpha - \beta)R(\alpha)R(\beta) = R(\alpha) - R(\beta)$ is easily verified when α and β are in the region of analyticity of $S(z)^{-1}$. The transformation $R(w)$ is analytic as a function of w by general properties of resolvents. It follows that $\tilde{F}(z)$ is analytic in the region of analyticity of $S(z)^{-1}$ whenever $F(z)$ is in $\mathscr{H}(A, B)$. If $F(z)$ is in $\mathscr{H}(A, B)$ and if $G(z) = [F(z) - S(z)S(\beta)^{-1}F(\beta)]/(z - \beta)$ for some number β in the region of analyticity of $S(z)^{-1}$, then

$$\begin{aligned} (\alpha - \beta)\{G(z) + \alpha[G(z) - S(z)S(\alpha)^{-1}G(\alpha)]/(z - \alpha)\} \\ = (\alpha - \beta)[F(z) - S(z)S(\beta)^{-1}F(\beta)]/(z - \beta) \\ + \alpha[F(z) - S(z)S(\alpha)^{-1}F(\alpha)]/(z - \alpha) \\ - \alpha[F(z) - S(z)S(\beta)^{-1}F(\beta)]/(z - \beta) \\ = F(z) + \alpha[F(z) - S(z)S(\alpha)^{-1}F(\alpha)]/(z - \alpha) \\ - F(z) - \beta[F(z) - S(z)S(\beta)^{-1}F(\beta)]/(z - \beta). \end{aligned}$$

It follows that $\tilde{G}(z) = [\tilde{F}(z) - S(z)S(\beta)^{-1}\tilde{F}(\beta)]/(z-\beta)$.

If $F(z)$ and $G(z)$ are in $\mathscr{H}(A,B)$ and if α and β belong to the region of analyticity of $S(z)^{-1}$, then

$$F_1(z) = F(z) + \alpha[F(z) - S(z)S(\alpha)^{-1}F(\alpha)]/(z-\alpha)$$

and

$$G_1(z) = G(z) + \beta[G(z) - S(z)S(\beta)^{-1}G(\beta)]/(z-\beta)$$

belong to $\mathscr{H}(A,B)$ and

$$[F_1(z) - S(z)S(0)^{-1}F_1(0)]/z = [F(z) - S(z)S(\alpha)^{-1}F(\alpha)]/(z-\alpha)$$
$$[G_1(z) - S(z)S(0)^{-1}G_1(0)]/z = [G(z) - S(z)S(\beta)^{-1}G(\beta)]/(z-\beta)$$

for all complex numbers α and β. It follows that

$$\begin{aligned}
&\langle F(t), [G(t) - S(t)S(\beta)^{-1}G(\beta)]/(t-\beta)\rangle \\
&\quad - \langle [F(t) - S(t)S(\alpha)^{-1}F(\alpha)]/(t-\alpha), G(t)\rangle \\
&\quad + (\alpha - \bar{\beta})\langle [F(t) - S(t)S(\alpha)^{-1}F(\alpha)]/(t-\alpha), \\
&\qquad\qquad [G(t) - S(t)S(\beta)^{-1}G(\beta)]/(t-\beta)\rangle \\
&= \langle F_1(t), [G_1(t) - S(t)S(0)^{-1}G_1(0)]/t\rangle \\
&\quad - \langle [F_1(t) - S(t)S(0)^{-1}F_1(0)]/t, G_1(t)\rangle \\
&= \pi\tilde{G}_1(0)^{-}\bar{S}(0)^{-1}S(0)^{-1}F_1(0) - \pi\bar{G}_1(0)\bar{S}(0)^{-1}S(0)^{-1}\tilde{F}_1(0) \\
&= \pi\tilde{G}(\beta)^{-}\bar{S}(\beta)^{-1}S(\alpha)^{-1}F(\alpha) - \pi\bar{G}(\beta)\bar{S}(\beta)^{-1}S(\alpha)^{-1}\tilde{F}(\alpha).
\end{aligned}$$

This is the general identity for difference quotients.

Let $\mathscr{H}$ be the Hilbert space of pairs

$$\begin{pmatrix} F(z) \\ \tilde{F}(z) \end{pmatrix}$$

consisting of an element $F(z)$ of $\mathscr{H}(A,B)$ and corresponding analytic function $\tilde{F}(z)$ in the norm

$$\left\| \begin{pmatrix} F(t) \\ \tilde{F}(t) \end{pmatrix} \right\|^2 = 2\|F(t)\|^2.$$

Then the transformation

$$\begin{pmatrix} F(z) \\ \tilde{F}(z) \end{pmatrix} \to \begin{pmatrix} F(w) \\ \tilde{F}(w) \end{pmatrix}$$

of $\mathscr{H}$ into $\mathscr{C}^2$ is continuous whenever w is not a singularity of $S(z)^{-1}$. It follows that there exists a matrix valued function $L(w, z)$, for every number w which is not a singularity of $S(z)^{-1}$, such that

$$L(w,z)\begin{pmatrix} u \\ v \end{pmatrix}$$

belongs to $\mathscr{H}$ for every pair of vectors u and v and such that

$$\begin{pmatrix} u \\ v \end{pmatrix}^- \begin{pmatrix} F(w) \\ \tilde{F}(w) \end{pmatrix} = \left\langle \begin{pmatrix} F(t) \\ \tilde{F}(t) \end{pmatrix}, L(w,t)\begin{pmatrix} u \\ v \end{pmatrix} \right\rangle$$

for every element

$$\begin{pmatrix} F(z) \\ \tilde{F}(z) \end{pmatrix}$$

of $\mathscr{H}$. If

$$\begin{pmatrix} F(z) \\ \tilde{F}(z) \end{pmatrix} \quad \text{and} \quad \begin{pmatrix} G(z) \\ \tilde{G}(z) \end{pmatrix}$$

are in $\mathscr{H}$, then

$$\begin{aligned} 2\pi \begin{pmatrix} S(0)^{-1}G(0) \\ S(0)^{-1}\tilde{G}(0) \end{pmatrix}^- I \begin{pmatrix} S(0)^{-1}F(0) \\ S(0)^{-1}\tilde{F}(0) \end{pmatrix} &= \left\langle \begin{pmatrix} F(t) \\ \tilde{F}(t) \end{pmatrix}, \begin{pmatrix} [G(t) - S(t)S(0)^{-1}G(0)]/t \\ [\tilde{G}(t) - S(t)S(0)^{-1}\tilde{G}(0)]/t \end{pmatrix} \right\rangle \\ &\quad - \left\langle \begin{pmatrix} [F(t) - S(t)S(0)^{-1}F(0)]/t \\ [\tilde{F}(t) - S(t)S(0)^{-1}\tilde{F}(0)]/t \end{pmatrix}, \begin{pmatrix} G(t) \\ \tilde{G}(t) \end{pmatrix} \right\rangle. \end{aligned}$$

When

$$\begin{pmatrix} G(z) \\ \tilde{G}(z) \end{pmatrix} = L(w,z)\begin{pmatrix} u \\ v \end{pmatrix}$$

for some vectors u and v and for some number w which is not a singularity of $S(z)^{-1}$, the identity reads

$$\begin{aligned} 2\pi \begin{pmatrix} u \\ v \end{pmatrix}^- \bar{L}(w,0)\bar{S}(0)^{-1} I S(0)^{-1} \begin{pmatrix} F(0) \\ \tilde{F}(0) \end{pmatrix} &= \left\langle \begin{pmatrix} F(t) \\ \tilde{F}(t) \end{pmatrix}, \begin{pmatrix} [G(t) - S(t)S(0)^{-1}G(0)]/t \\ [\tilde{G}(t) - S(t)S(0)^{-1}\tilde{G}(0)]/t \end{pmatrix} \right\rangle \\ &\quad - \begin{pmatrix} u \\ v \end{pmatrix}^- \begin{pmatrix} [F(w) - S(w)S(0)^{-1}F(0)]/w \\ [\tilde{F}(w) - S(w)S(0)^{-1}\tilde{F}(0)]/w \end{pmatrix}. \end{aligned}$$

An equivalent identity is

$$- \left\langle \begin{pmatrix} F(t) \\ \tilde{F}(t) \end{pmatrix}, \ 2\pi L(0,t)\bar{S}(0)^{-1} I S(0)^{-1} L(w,0) \begin{pmatrix} u \\ v \end{pmatrix} \right\rangle$$

$$= \left\langle \begin{pmatrix} F(t) \\ \tilde{F}(t) \end{pmatrix}, \ \frac{L(w,t) - S(t)S(w)^{-1} L(w,0)}{t} \begin{pmatrix} u \\ v \end{pmatrix} \right\rangle$$

$$- \left\langle \begin{pmatrix} F(t) \\ \tilde{F}(t) \end{pmatrix}, \ \frac{L(w,t) - L(0,t)\bar{S}(0)^{-1}\bar{S}(w)}{\bar{w}} \begin{pmatrix} u \\ v \end{pmatrix} \right\rangle .$$

By the arbitrariness of

$$\begin{pmatrix} F(z) \\ \tilde{F}(z) \end{pmatrix} \text{ and } \begin{pmatrix} u \\ v \end{pmatrix},$$

we can conclude that

$$- 2\pi L(0,z)\bar{S}(0)^{-1} I S(0)^{-1} L(w,0)$$
$$= [L(w,z) - S(z)S(0)^{-1} L(w,0)]/z$$
$$- [L(w,z) - L(0,z)\bar{S}(0)^{-1}\bar{S}(w)]/\bar{w}.$$

The identity can be written

$$2\pi(z - \bar{w})S(z)^{-1} L(w,z)\bar{S}(w)^{-1}$$
$$= 2\pi z S(z)^{-1} L(0,z)S(0)^{-1}$$
$$- 2\pi \bar{w}\bar{S}(0)^{-1} L(w,0)\bar{S}(w)^{-1}$$
$$+ 2\pi z S(z)^{-1} L(0,z)\bar{S}(0)^{-1} I 2\pi \bar{w} S(0)^{-1} L(w,0)\bar{S}(w)^{-1}.$$

Define $M(z) = S(z)S(0)^{-1}M(0) - 2\pi z L(0,z)\bar{S}(0)^{-1} I M(0)$ where

$$M(0) = \begin{pmatrix} A(0) & B(0) \\ 0 & S(0)\bar{S}(0)\bar{A}(0)^{-1} \end{pmatrix}.$$

Then $M(z)$ has an invertible value at the origin,

$$S(z)^{-1} M(z) M(0)^{-1} S(0) = 1 - 2\pi z S(z)^{-1} L(0,z)\bar{S}(0)^{-1} I,$$

and

$$2\pi(z - \bar{w})S(z)^{-1} L(w,z)\bar{S}(w)^{-1}$$
$$= [1 - 2\pi z S(z)^{-1} L(0,z)\bar{S}(0)^{-1} I] I_1 [1 + 2\pi \bar{w} I S(0)^{-1} L(w,0)\bar{S}(w)^{-1}] - I.$$

Since $M(0) I \bar{M}(0) = S(0) I \bar{S}(0)$, the identity simplifies to

$$L(w,z) = [M(z) I \bar{M}(w) - S(z) I \bar{S}(w)]/[2\pi(z - \bar{w})]$$

when w is not a singularity of $S(z)^{-1}$. Since $L(w,z)$ is continuous at $\bar{w}$, $M(\bar{w})I\bar{M}(w) = S(\bar{w})I\bar{S}(w)$. The space $\mathscr{H}$ is therefore a space $\mathscr{H}_S(M)$.

If

$$M(z) = \begin{pmatrix} A_1(z) & B_1(z) \\ C_1(z) & D_1(z) \end{pmatrix},$$

then

$$\frac{M(z)I\bar{M}(w) - S(z)I\bar{S}(w)}{2\pi(z-\bar{w})}\begin{pmatrix} 1 \\ 0 \end{pmatrix}$$

$$= \begin{pmatrix} [B_1(z)\bar{A}_1(w) - A_1(z)\bar{B}_1(w)]/[2\pi(z-\bar{w})] \\ [S(z)\bar{S}(w) - A_1(z)\bar{D}_1(w) + B_1(z)\bar{C}_1(w)]/[2\pi(z-\bar{w})] \end{pmatrix},$$

$$\frac{M(z)I\bar{M}(w) - S(z)I\bar{S}(w)}{2\pi(z-\bar{w})}\begin{pmatrix} 0 \\ 1 \end{pmatrix}$$

$$= \begin{pmatrix} [S(z)\bar{S}(w) - A_1(z)\bar{D}_1(w) + B_1(z)\bar{C}_1(w)]/[2\pi(z-\bar{w})] \\ [D_1(z)\bar{C}_1(w) - C_1(z)\bar{D}_1(w)]/[2\pi(z-\bar{w})] \end{pmatrix}.$$

By the construction of $\mathscr{H}_S(M)$, and by the definition of $L(w,z)$,

$$[B_1(z)\bar{A}_1(w) - A_1(z)\bar{B}_1(w)]c/[\pi(z-\bar{w})]$$

and

$$[S(z)\bar{S}(w) - A_1(z)\bar{D}_1(w) + B_1(z)\bar{C}_1(w)]c/[\pi(z-\bar{w})]$$

belong to $\mathscr{H}(A,B)$ for every vector c and

$$\bar{c}F(w) = \left\langle F(t), \frac{B_1(t)\bar{A}_1(w) - A_1(t)\bar{B}_1(w)}{\pi(t-\bar{w})}c \right\rangle$$

$$\bar{c}\tilde{F}(w) = \left\langle F(t), \frac{S(t)\bar{S}(w) - A_1(t)\bar{D}_1(w) + B_1(t)\bar{C}_1(w)}{\pi(t-\bar{w})}c \right\rangle$$

for every $F(z)$ in $\mathscr{H}_S(M)$. Since

$$\bar{c}F(0) = \left\langle F(t), \frac{B(t)\bar{A}(0) - A(t)\bar{B}(0)}{\pi t}c \right\rangle$$

$$\bar{c}\tilde{F}(0) = \left\langle F(t), \frac{S(t)\bar{S}(0) - A(t)A(0)^{-1}S(0)\bar{S}(0)}{\pi t}c \right\rangle$$

for every vector c,

$$B_1(z)\bar{A}_1(0) - A_1(z)\bar{B}_1(0) = B(z)\bar{A}(0) - A(z)\bar{B}(0)$$

$$A_1(z)\bar{D}_1(0) - B_1(z)\bar{C}_1(0) = A(z)A(0)^{-1}S(0)\bar{S}(0).$$

Since $A_1(0) = A(0)$, $B_1(0) = B(0)$, $C_1(0) = 0$, and $D_1(0) = S(0)\bar{S}(0)\bar{A}(0)^{-1}$, we obtain $A_1(z) = A(z)$ and $B_1(z) = B(z)$. This completes the proof of the theorem.

PROOF OF THEOREM 5. The proof generalizes the proof of Theorem 34 of **[1]**. The function $S(z) = A(a,z) - iB(a,z)$ obviously satisfies the hypotheses of Theorem 1 when $A(z) = A(a,z)$ and $B(z) = B(a,z)$. The hypotheses of Theorem 2 are satisfied with

$$\frac{M(a,z)I\bar{M}(a,w) - S(z)I\bar{S}(w)}{2\pi(z-\bar{w})} = \frac{B(a,z)\bar{A}(a,w) - A(a,z)\bar{B}(a,w)}{2\pi(z-\bar{w})}\begin{pmatrix}1 & -i\\ i & 1\end{pmatrix}.$$

If

$$\begin{pmatrix}F_+(z)\\ F_-(z)\end{pmatrix}$$

is an element of $\mathscr{H}_S(M(a))$ which is a finite sum of elements of the form

$$\frac{M(a,z)I\bar{M}(a,w) - S(z)I\bar{S}(w)}{2\pi(z-\bar{w})}\begin{pmatrix}u\\ v\end{pmatrix}$$

then $F_-(z) = iF_+(z)$. Since such sums are dense in $\mathscr{H}_S(M(a))$, the same formula holds for every element

$$\begin{pmatrix}F_+(z)\\ F_-(z)\end{pmatrix}$$

of $\mathscr{H}_S(M(a))$. By the proof of sufficiency for Theorem 4, the transformation

$$\begin{pmatrix}F_+(z)\\ F_-(z)\end{pmatrix} \to 2^{\frac{1}{2}}F_+(z)$$

is a partial isometry of $\mathscr{H}_S(M(a))$ onto $\mathscr{H}(A(a),B(a))$. Since the transformation has no nonzero element in its kernel, it is an isometry.

We show that there is no nonzero element

$$L(z) = \begin{pmatrix}L_+(z)\\ L_-(z)\end{pmatrix}$$

of $\mathscr{H}(M(a,b))$ such that $M(a,z)L(z)$ belongs to $\mathscr{H}_S(M(a))$. It is easily verified that the set of such elements $L(z)$ is a Hilbert space in the norm

$$\|L(t)\|_{\mathscr{L}}^2 = \|M(a,t)L(t)\|_{\mathscr{H}(M(a))}^2 + \|L(t)\|_{\mathscr{H}(M(a,b))}^2,$$

that $[L(z) - L(w)]/(z-w)$ belongs to $\mathscr{L}$ whenever $L(z)$ belongs to $\mathscr{L}$, for every complex number w, and that the identity

$$\begin{aligned}0 = &\langle F(t), [G(t) - G(\beta)]/(t - \beta)\rangle_{\mathscr{L}} \\ &- \langle [F(t) - F(\alpha)]/(t - \alpha), G(t)\rangle_{\mathscr{L}} \\ &+ (\alpha - \bar{\beta})\langle [F(t) - F(\alpha)]/(t - \alpha), [G(t) - G(\beta)]/(t - \beta)\rangle_{\mathscr{L}}\end{aligned}$$

holds for all elements $F(z)$ and $G(z)$ of $\mathscr{L}$ and all complex numbers α and β. As in Problems 98 and 99 of [**1**], there exists a matrix valued entire function $\Phi(z)$ such that $\Phi(z) = -\bar{\Phi}(\bar{z})$, such that

$$\frac{\Phi(z) + \bar{\Phi}(w)}{\pi i(\bar{w} - z)} \begin{pmatrix} u \\ v \end{pmatrix}$$

belongs to $\mathscr{L}$ for every pair of vectors u and v and for all complex numbers w, and such that the identity

$$\begin{pmatrix} u \\ v \end{pmatrix}^- \begin{pmatrix} L_+(w) \\ L_-(w) \end{pmatrix} = \left\langle \begin{pmatrix} L_+(t) \\ L_-(t) \end{pmatrix}, \frac{\Phi(t) + \bar{\Phi}(w)}{\pi i(\bar{w} - t)} \begin{pmatrix} u \\ v \end{pmatrix} \right\rangle$$

holds for every element $L(z)$ of $\mathscr{L}$. If u and v are vectors,

$$\begin{pmatrix} u \\ v \end{pmatrix}^- \Phi(z) \begin{pmatrix} u \\ v \end{pmatrix}$$

is an entire function whose real part is nonnegative in the upper half plane and zero on the real axis. By the Poisson representation of a function which is positive and harmonic in a half plane, the entire function is linear. By the arbitrariness of u and v, $\Phi(z)$ is a linear function. It follows that $[\Phi(z) + \bar{\Phi}(w)]/[\pi i(\bar{w} - z)]$ is a constant and that the elements of $\mathscr{L}$ are constants. But if

$$\begin{pmatrix} u \\ v \end{pmatrix}$$

is a constant in $\mathscr{L}$,

$$M(a,z) \begin{pmatrix} u \\ v \end{pmatrix} = \begin{pmatrix} A(a,z)u + B(a,z)v \\ -B(a,z)u + A(a,z)v \end{pmatrix}.$$

Since this function belongs to $\mathscr{H}_S(M(a))$,

$$i[A(a,z)u + B(a,z)v] = -B(a,z)u + A(a,z)v.$$

Since $[A(a,z) - iB(a,z)](u + iv) = 0$ and since $A(a,z) - iB(a,z)$ has an invertible value, $u + iv = 0$. Since $\bar{u}v = vu$ by the identity for difference quotients in $\mathscr{H}(M(a,b))$, $u = v = 0$. This completes the proof that $\mathscr{H}(M(a,b))$ contains no nonzero element.

By the proof of Theorem 34 of [**1**], a space $\mathscr{H}_S(M(b))$ exists, the space $\mathscr{H}_S(M(a))$ is contained isometrically in $\mathscr{H}_S(M(b))$, and the transformation

$$\begin{pmatrix} F_+(z) \\ F_-(z) \end{pmatrix} \to M(a, z)\begin{pmatrix} F_+(z) \\ F_-(z) \end{pmatrix}$$

takes $\mathcal{H}(M(a,b))$ isometrically onto the orthogonal complement of $\mathcal{H}_S(M(a))$ in $\mathcal{H}_S(M(b))$. Since $M(b,z)I\bar{M}(b,\bar{z}) = S(z)I\bar{S}(\bar{z})$, $A(b,z)$ and $B(b,z)$ are operator valued entire functions such that $A(b,z)\bar{B}(b,\bar{z}) = B(b,z)\bar{A}(b,\bar{z})$. The matrix inequality

$$[M(b,z)I\bar{M}(b,z) - S(z)I\bar{S}(z)]/(z-\bar{z}) \geqq 0$$

implies the operator inequality

$$[B(b,z)\bar{A}(b,z) - A(b,z)\bar{B}(b,z)]/(z-\bar{z}) \geqq 0.$$

By hypothesis $1 - A(b,z)$ and $B(b,z)$ have completely continuous values. To show that $A(b,z) - iB(b,z)$ has an invertible value at a point w on or above the the real axis, we need only show that there is no nonzero vector c in the kernel of $\bar{A}(b,w) + i\bar{B}(b,w)$. Since

$$[A(b,w) + iB(b,w)][\bar{A}(b,w) - i\bar{B}(b,w)]$$
$$\leqq [A(b,w) - iB(b,w)][\bar{A}(b,w) + i\bar{B}(b,w)],$$

the vector c is then in the kernel of $\bar{A}(b,w) - i\bar{B}(b,w)$, and hence also in the kernel of $\bar{A}(b,w)$ and $\bar{B}(b,w)$. Since

$$\begin{pmatrix} \bar{A}(b,w) \\ \bar{B}(b,w) \end{pmatrix} = \bar{M}(a,b,w)\begin{pmatrix} \bar{A}(a,w) \\ \bar{B}(a,w) \end{pmatrix}$$

and since $M(a,b,\bar{w})I\bar{M}(a,b,w) = I$, c is then in the kernel of $\bar{A}(a,w)$ and $\bar{B}(a,w)$ and hence in the kernel of $\bar{A}(a,w) + i\bar{B}(a,w)$. It follows that $A(b,w) - iB(b,w)$ is an invertible operator if $A(a,w) - iB(a,w)$ is an invertible operator. If $A(a,w) - iB(a,w)$ is not an invertible operator, we can still conclude that $[A(a,z) - iB(a,z)]^{-1}[A(b,z) - iB(b,z)]$ has an analytic continuation at the point w. A similar argument will show that $A(b,z) + iB(b,z)$ has an invertible value at every point on or below the real axis where $A(a,z) + iB(a,z)$ has an invertible value, and that $[A(a,z) + iB(a,z)]^{-1}[A(b,z) + iB(b,z)]$ has an analytic continuation at points where $A(a,z) + iB(a,z)$ does not have an invertible value. We have now verified the conditions for the existence of a space $\mathcal{H}(A(b),B(b))$. By the proof of sufficiency for Theorem 4, the transformation

$$\begin{pmatrix} F_+(z) \\ F_-(z) \end{pmatrix} \to 2^{1/2}F_+(z)$$

is a partial isometry of $\mathcal{H}_S(M(b))$ onto $\mathcal{H}(A(b),B(b))$.

If $F(z)$ is in $\mathcal{H}(A(a),B(a))$, then

$$\begin{pmatrix} F(z) \\ iF(z) \end{pmatrix}$$

is in $\mathscr{H}_S(M(a))$ and hence in $\mathscr{H}_S(M(b))$, and $F(z)$ is in $\mathscr{H}(A(b), B(b))$. Since

$$\| F(t) \|^2_{\mathscr{H}(A(b),B(b))} \leqq \frac{1}{2}\left\| \begin{pmatrix} F(t) \\ iF(t) \end{pmatrix} \right\|^2_{\mathscr{H}_S(M(b))} = \frac{1}{2}\left\| \begin{pmatrix} F(t) \\ iF(t) \end{pmatrix} \right\|^2_{\mathscr{H}_S(M(a))} = \| F(t) \|^2_{\mathscr{H}(A(a),B(a))},$$

the inclusion of $\mathscr{H}(A(a), B(a))$ in $\mathscr{H}(A(b), B(b))$ does not increase norms. If we replace $S(z) = A(a,z) - iB(a,z)$ by a general operator valued entire function, which has an invertible value, such that $1 - S(z)$ has completely continuous values, and such that

$$\{[A(a,z) - iB(a,z)] - S(z)S(w)^{-1}[A(a,w) - iB(a,w)]\}c/(z - w)$$

belongs to $\mathscr{H}(A(a),B(a))$ for every vector c, then a space $\mathscr{H}_S(M(a))$ still exists by Theorem 4 for some operator valued analytic functions $C(a, z)$ and $D(a, z)$. The above argument will show that a space $\mathscr{H}_S(M(b))$ exists, $M(b, z) = M(a,z)M(a,b,z)$. By Theorem 4,

$$\{[A(b,z) - iB(b,z)] - S(z)S(w)^{-1}[A(b,w) - iB(b,w)]\}c/(z - w)$$

belongs to $\mathscr{H}(A(b), B(b))$ for every vector c.

PROOF OF THEOREM 6. The proof generalizes the proof of Theorem 13 of **[2]**. Assume that the transformation

$$T\colon \begin{pmatrix} F_+(z) \\ F_-(z) \end{pmatrix} \to 2^{1/2}[i\phi(z)S(z)^{-1}F_+(z) + S(z)^{-1}F_-(z)]$$

takes $\mathscr{H}_S(M)$ into $\mathscr{L}(\phi)$ and is bounded by 1. Then the adjoint T^* of T takes $\mathscr{L}(\phi)$ into $\mathscr{H}_S(M)$ and is bounded by 1. If

$$\begin{pmatrix} F_+(z) \\ F_-(z) \end{pmatrix}$$

is in $\mathscr{H}_S(M)$, then

$$\begin{aligned} &\langle 2^{1/2}[i\phi(t)S(t)^{-1}F_+(t) + S(t)^{-1}F_-(t)], [\phi(t) + \bar{\phi}(w)]c/[\pi i(\bar{w} - t)]\rangle_{\mathscr{L}(\phi)} \\ &\quad = 2^{1/2}\bar{c}[i\phi(w)S(w)^{-1}F_+(w) + S(w)^{-1}F_-(w)] \\ &\quad = \left\langle \begin{pmatrix} F_+(t) \\ F_-(t) \end{pmatrix}, 2^{1/2}\frac{M(t)I\bar{M}(w) - S(t)I\bar{S}(w)}{2\pi(t - \bar{w})}\begin{pmatrix} - i\bar{S}(w)^{-1}\bar{\phi}(w)c \\ \bar{S}(w)^{-1}c \end{pmatrix}\right\rangle_{\mathscr{H}_S(M)} \end{aligned}$$

for every vector c when w is not a singularity of $S(z)^{-1}$. By the definition of the adjoint,

$$T^*\colon \frac{\phi(z) + \bar{\phi}(w)}{\pi i(\bar{w} - z)}c \to 2^{1/2}\frac{M(z)I\bar{M}(w) - S(z)I\bar{S}(w)}{2\pi(z - \bar{w})}\begin{pmatrix} - i\bar{S}(w)^{-1}\bar{\phi}(w)c \\ \bar{S}(w)^{-1}c \end{pmatrix}.$$

Since T is bounded by 1,

$$\bar{c}\,\frac{\phi(w)+\bar{\phi}(w)}{\pi i(\bar{w}-w)}\,c = \left\|\frac{\phi(t)+\bar{\phi}(w)}{\pi i(\bar{w}-t)}\,c\right\|^2_{\mathcal{L}(\phi)}$$

$$\geq \left\| 2^{1/2}\,\frac{M(t)I\bar{M}(w)-S(t)I\bar{S}(w)}{2\pi(t-\bar{w})}\begin{pmatrix}-i\bar{S}(w)^{-1}\bar{\phi}(w)c\\ \bar{S}(w)^{-1}c\end{pmatrix}\right\|^2_{\mathcal{H}_S(M)}$$

$$\geq (i\bar{c}\phi(w)S(w)^{-1},\bar{c}S(w)^{-1})\,\frac{M(w)I\bar{M}(w)-S(w)I\bar{S}(w)}{\pi(w-\bar{w})}\begin{pmatrix}-i\bar{S}(w)^{-1}\bar{\phi}(w)c\\ \bar{S}(w)^{-1}c\end{pmatrix}.$$

By the arbitrariness of c,

$$[\phi(w)+\bar{\phi}(w)]/[\pi i(\bar{w}-w)]$$

$$\geq (i\phi(w)S(w)^{-1},S(w)^{-1})\,\frac{M(w)I\bar{M}(w)-S(w)I\bar{S}(w)}{\pi(w-\bar{w})}\begin{pmatrix}-i\bar{S}(w)^{-1}\bar{\phi}(w)\\ \bar{S}(w)^{-1}\end{pmatrix}.$$

When w is in the upper half plane, the inequality can be written

$$\{\phi(w)S(w)^{-1}[A(w)+iB(w)]+S(w)^{-1}[D(w)-iC(w)]\}$$

$$\times\{[\bar{A}(w)-i\bar{B}(w)]\bar{S}(w)^{-1}\bar{\phi}(w)+[\bar{D}(w)+i\bar{C}(w)]\bar{S}(w)^{-1}\}$$

$$\geq \{\phi(w)S(w)^{-1}[A(w)-iB(w)]-S(w)^{-1}[D(w)+iC(w)]\}$$

$$\times\{[\bar{A}(w)+i\bar{B}(w)]\bar{S}(w)^{-1}\bar{\phi}(w)-[\bar{D}(w)-i\bar{C}(w)]\bar{S}(w)^{-1}\}.$$

Since we assume that the values of $M(z)I-S(z)I$ are completely continuous matrices, and since the values of $1-S(z)$ are completely continuous operators, the values of $M(z)I-I$ are completely continuous matrices. By the proof of Theorem 5 the existence of a space $\mathcal{H}_S(M)$ implies the existence of a space $\mathcal{H}(A,B)$. It follows that $A(z)+iB(z)$ has invertible values at all except isolated points. The function

$$1-[S(z)^{-1}D(z)-iS(z)^{-1}C(z)][S(z)^{-1}A(z)+iS(z)^{-1}B(z)]^{-1}$$

is analytic at all except isolated points in the complex plane, and it has completely continuous values. Since the existence of a space $\mathcal{H}_S(M)$ implies that

$$[I-\bar{M}(\bar{z})\bar{S}(\bar{z})^{-1}IS(\bar{z})^{-1}M(\bar{z})]/(z-\bar{z})\geq 0$$

and that

$$\mathrm{Re}[S(z)^{-1}D(z)-iS(z)^{-1}C(z)][S(z)^{-1}A(z)+iS(z)^{-1}B(z)]^{-1}\geq 1$$

for $y\leq 0$, the inequality

$$\mathrm{Re}[S(z)^{-1}D(z)-iS(z)^{-1}C(z)][S(z)^{-1}A(z)+iS(z)^{-1}B(z))]^{-1}\geq \tfrac{1}{2}$$

holds for some point above the real axis. Since $\operatorname{Re}\phi(z) \geqq 0$ for $y > 0$, the function

$$\phi(z) + [S(z)^{-1}D(z) - iS(z)^{-1}C(z)][S(z)^{-1}A(z) + iS(z)^{-1}B(z)]^{-1}$$

has an invertible value in the upper half plane. Since $\operatorname{Re}\phi(z) \geqq 0$ for $y > 0$, the function $1 + \phi(z)$ has invertible values in the upper half plane, and the function

$$1 + [1 + \phi(z)]^{-1}\{[S(z)^{-1}D(z) - iS(z)^{-1}C(z)][S(z)^{-1}A(z) + iS(z)^{-1}B(z)]^{-1} - 1\}$$

has an invertible value in the upper half plane. Since the values of the function differ from the identity operator by completely continuous operators, the function has invertible values at all except isolated points in the upper half plane. It follows that the function

$$\phi(z)[S(z)^{-1}A(z) + iS(z)^{-1}B(z)] + [S(z)^{-1}D(z) - iS(z)^{-1}C(z)]$$

has invertible values at all except isolated points in the upper half plane. The function

$$\begin{aligned} W(z) = {} & \{\phi(z)S(z)^{-1}[A(z) + iB(z)] + S(z)^{-1}[D(z) - iC(z)]\}^{-1} \\ & \times \{\phi(z)S(z)^{-1}[A(z) - iB(z)] - S(z)^{-1}[D(z) + iC(z)]\} \end{aligned}$$

is analytic at all except isolated points in the upper half plane, and $W(z)\bar{W}(z) \leqq 1$. It follows that $W(z)$ has an analytic continuation in the half plane and is bounded by 1. The required form of $\phi(z)$ follows.

Conversely, suppose that $\phi(z)$ is of this form for some operator valued function $W(z)$ which is analytic and bounded by 1 in the upper half plane. As in the proof of Theorem 13 of [**2**], the inequality

$$\sum \bar{c}_i \frac{1 - W(w_i)\bar{W}(w_k)}{2\pi i(\bar{w}_k - w_i)} c_k \geqq 0$$

holds for all finite sets of points $w_1, \cdots, w_r$ in the upper half plane and corresponding vectors $c_1, \cdots, c_r$. This implies the existence of a transformation of $\mathscr{L}(\phi)$ into $\mathscr{H}_S(M)$ which is bounded by 1, such that

$$\frac{\phi(z) + \bar{\phi}(w)}{\pi i(\bar{w} - z)} c \to 2^{1/2} \frac{M(z)I\bar{M}(w) - S(z)I\bar{S}(w)}{2\pi(z - \bar{w})} \begin{pmatrix} -i\bar{S}(w)^{-1}\bar{\phi}(w)c \\ \bar{S}(w)^{-1}c \end{pmatrix}$$

for every vector c when w is in the upper half plane. The adjoint transformation of $\mathscr{H}_S(M)$ into $\mathscr{L}(\phi)$, which is the transformation

$$\begin{pmatrix} F_+(z) \\ F_-(z) \end{pmatrix} \to 2^{1/2}[i\phi(z)S(z)^{-1}F_+(z) + S(z)^{-1}F_-(z)],$$

is then bounded by 1. We determine the set on which the transformation is isometric.

Consider the elements of the space $\mathscr{H}_S(M) = \mathscr{H}_S(M(a))$ in a new norm, the b-norm, defined by

$$\left\| \begin{pmatrix} F_+(t) \\ F_-(t) \end{pmatrix} \right\|_b^2 = \frac{1}{2} \left\| \begin{pmatrix} F_+(t) \\ F_-(t) \end{pmatrix} \right\|_{\mathscr{H}_S(M(a))}^2 + \| i\phi(t)S(t)^{-1}F_+(t) + S(t)^{-1}F_-(t) \|_{\mathscr{L}(\phi)}^2.$$

Since the inequality

$$\tfrac{1}{2} \| F(t) \|_{\mathscr{H}_S(M(a))}^2 \leqq \| F(t) \|_b^2 \leqq \| F(t) \|_{\mathscr{H}_S(M(a))}^2.$$

holds for every element $F(z)$ of $\mathscr{H}_S(M(a))$, the space $\mathscr{H}_S(M(a))$ is a Hilbert space in the b-norm. It is easily verified that the transformation $F(z) \to F(w)$ of $\mathscr{H}_S(M(a))$ into $\mathscr{C}^2$ is continuous in the b-norm and that the identity for difference quotients of Theorem 3 holds in the b-norm. By the proof of necessity for Theorem 4, $\mathscr{H}_S(M(a))$ is equal to a space $\mathscr{H}_S(M(b))$ in the b-norm.

If

$$F(z) = \begin{pmatrix} F_+(z) \\ F_-(z) \end{pmatrix}$$

is a pair of vector valued functions such that $M(a,z)F(z)$ belongs to $\mathscr{H}_S(M(b))$, define

$$\| F(t) \|_{ab}^2 = \sup [\, \| M(a,t)F(t) + G(t) \|_{\mathscr{H}_S(M(b))}^2 - \| G(t) \|_{\mathscr{H}_S(M(a))}^2]$$

where the supremum is taken over all elements $G(z)$ of $\mathscr{H}_S(M(a))$. It is easily verified that the set of such pairs $F(z)$ such that $\| F(t) \|_{ab} < \infty$ is a Hilbert space $\mathscr{H}(a,b)$ in the ab-norm. The elements of $\mathscr{H}(a,b)$ are analytic in the common region of analyticity of $S(z)^{-1}$ and $\bar{S}(\bar{z})^{-1}$. The transformation $F(z) \to F(w)$ of of $\mathscr{H}(a,b)$ into $\mathscr{C}^2$ is continuous when w is not a singularity of $S(z)^{-1}$ or of $\bar{S}(\bar{z})^{-1}$. The function $[F(z) - F(w)]/(z-w)$ belongs to $\mathscr{H}(a,b)$ whenever $F(z)$ belongs to $\mathscr{H}(a,b)$ if w is not a singularity of $S(z)^{-1}$ or of $\bar{S}(\bar{z})^{-1}$. The identity for difference quotients

$$\begin{aligned} 2\pi \bar{G}(\beta) I F(\alpha) &= \langle F(t), [G(t) - G(\beta)]/(t-\beta) \rangle_{ab} \\ &\quad - \langle [F(t) - F(\alpha)]/(t-\alpha), G(t) \rangle_{ab} \\ &\quad + (\alpha - \bar{\beta}) \langle [F(t) - F(\alpha)]/(t-\alpha), [G(t) - G(\beta)]/(t-\beta) \rangle_{ab} \end{aligned}$$

holds for all elements $F(z)$ and $G(z)$ of $\mathscr{H}(a,b)$ when α and β are not singularities of $S(z)^{-1}$ or of $\bar{S}(\bar{z})^{-1}$. Let $\mathscr{L}$ be the space $\mathscr{H}(a,b)$ considered in the norm

$$\| F(t) \|_{\mathscr{L}}^2 = \| F(t) \|_{ab}^2 + \| M(a,t)F(t) \|_{\mathscr{H}_S(M(a))}^2.$$

It is easily verified that $\mathscr{L}$ is a Hilbert space and that the identity

$$0 = \langle F(t), [G(t) - G(\beta)]/(t - \beta) \rangle - \langle [F(t) - F(\alpha)]/(t - \alpha), G(t) \rangle$$
$$+ (\alpha - \bar{\beta}) \langle [F(t) - F(\alpha)]/(t - \alpha), [G(t) - G(\beta)]/(t - \beta) \rangle$$

holds in the metric of $\mathscr{L}$ when α and β are not singularities of $S(z)^{-1}$ or of $\bar{S}(\bar{z})^{-1}$. Since we assume that $S(z)$ has invertible values on the real axis, the proof of Theorem 5 will show that the elements of $\mathscr{L}$, and hence of $\mathscr{H}(a,b)$, are constants. The identity for difference quotients in $\mathscr{H}(a,b)$ implies that $\bar{u}v = \bar{v}u$ whenever

$$\begin{pmatrix} u \\ v \end{pmatrix}$$

belongs to $\mathscr{H}(a,b)$.

By the proof of Theorem 8 of [9], every element $F(z)$ of $\mathscr{H}(a,b)$ has a unique minimal decomposition

$$F(z) = G(z) + M(a,z) \begin{pmatrix} u \\ v \end{pmatrix}$$

with $G(z)$ in $\mathscr{H}_S(M(a))$ and

$$\begin{pmatrix} u \\ v \end{pmatrix}$$

in $\mathscr{H}(a,b)$, such that

$$\| F(t) \|_{\mathscr{H}_S(M(b))}^2 = \| G(t) \|_{\mathscr{H}_S(M(a))}^2 + \left\| \begin{pmatrix} u \\ v \end{pmatrix} \right\|_{ab}^2.$$

If $F(z)$ is orthogonal to

$$M(a,z) \begin{pmatrix} u \\ v \end{pmatrix}$$

in the metric of $\mathscr{H}_S(M(b))$, then

$$\left\| \begin{pmatrix} u \\ v \end{pmatrix} \right\|_{\mathscr{H}(a,b)}^2 = \left\langle F(t), M(a,t) \begin{pmatrix} u \\ v \end{pmatrix} \right\rangle_{\mathscr{H}_S(M(b))} = 0$$

and

$$\| F(t) \|_{\mathscr{H}_S(M(a))} = \| F(t) \|_{\mathscr{H}_S(M(b))}$$
$$= \| 2^{1/2} [i\phi(t) S(t)^{-1} F_+(t) + S(t)^{-1} F_-(t)] \|_{\mathscr{L}(\phi)}.$$

PROOF OF THEOREM 7. The theorem is true trivially when $M(z)$ is a constant. The matrices $M(a,b,z)$ are all constant in this case. By Theorem III of [6], the matrix $M'(0)I$ is nonnegative, and it is nonzero if $M(z)$ is not a constant. Since we assume that $M'(0)I$ is completely continuous, there exists an orthogonal sequence of elements

$$\begin{pmatrix} u_r \\ v_r \end{pmatrix} \text{ of } \mathscr{C}^2, \ r = 0,1,2,\cdots,$$

having limit zero at infinity, such that

$$M'(0)I = \sum_{r=0}^{\infty} \begin{pmatrix} u_r \\ v_r \end{pmatrix} \begin{pmatrix} u_r \\ v_r \end{pmatrix}^- .$$

Let (p_r) be the choice of a sequence of positive numbers such that $\sum_{r=0}^{\infty} p_r[\,|u_r|^2 + |v_r|^2]^2 = 1$. Let t be a given number, $0 \leqq t \leqq 1$. By the proof of Theorem XV of [7], there exist spaces $\mathscr{H}(M(t))$ and $\mathscr{H}(M(t,1))$ such that the values of $M(t,z)$ and $M(t,1,z)$ at the origin are the identity matrix, such that $M(1,z) = M(t,z)M(t,1,z)$, and such that

$$\sum_{r=0}^{\infty} p_r \begin{pmatrix} u_r \\ v_r \end{pmatrix}^- M'(t,0)I \begin{pmatrix} u_r \\ v_r \end{pmatrix} = t.$$

It is easily seen from the proof of the theorem that these factors can be chosen so that $M(b,z) = M(a,z)M(a,b,z)$ for some space $\mathscr{H}(M(a,b))$ when $a \leqq b$. If $m(a) = M'(a,0)I$ when $0 \leqq a \leqq 1$, then $m(a)$ is a nonnegative matrix and $M'(a,b,0)I = m(b) - m(a)$ is a nonnegative matrix when $a \leqq b$. Since $m(a) \leqq m(1)$ where $m(1) = M'(0)I$ is completely continuous by hypothesis, $m(a)$ is completely continuous for every a. Since

$$\sum_{r=0}^{\infty} p_r \begin{pmatrix} u_r \\ v_r \end{pmatrix}^- [m(b) - m(a)] \begin{pmatrix} u_r \\ v_r \end{pmatrix} = b - a,$$

$m(t)$ is a continuous function of t.

Let $\kappa(a,b)$ be the bound of the transformation $F(z) \to [F(z) - F(0)]/z$ in $\mathscr{H}(M(a,b))$. We show that $\kappa(a,b) \leqq \kappa(a,c)$ whenever $a \leqq b \leqq c$. By Theorem XII of [7], the most general element $L(z)$ of $\mathscr{H}(M(a,b))$ is of the form $L(z) = F(z) + M(a,b,z)G(z)$ with $F(z)$ in $\mathscr{H}(M(a,b))$, $G(z)$ in $\mathscr{H}(M(b,c))$, and

$$\| F(t) + M(a,b,t)G(t) \|^2_{\mathscr{H}(M(a,c))}$$

$$= \| F(t) \|^2_{\mathscr{H}(M(a,b))} + \| G(t) \|^2_{\mathscr{H}(M(b,c))}.$$

Since $G(z)$ is an element of $\mathscr{H}(M(b,c))$ such that $M(a,b,z)G(z)$ belongs to $\mathscr{H}(M(a,b))$, the proof of Theorem 5 will show that $G(z)$ is a constant. On the other hand, the function

$$[L(z)-L(0)]/z = [F(z)-F(0)]/z$$
$$+ [M(a,z)-M(a,0)]G(0)/z + M(a,z)[G(z)-G(0)]/z$$

belongs to $\mathcal{H}(M(a,b))$, and it is easily verified that

$$\|[L(t)-L(0)]/t\|^2_{\mathcal{H}(M(a,c))}$$
$$= \|[F(t)-F(0)]/t + [M(a,t)-M(a,0)]G(0)/t\|^2_{\mathcal{H}(M(a,b))}$$
$$+ \|[G(t)-G(0)]/t\|^2_{\mathcal{H}(M(b,c))}.$$

Since $G(z)$ is a constant, the identity states that $[L(z)-L(0)]/z$ has the same norm in $\mathcal{H}(M(a,b))$ as in $\mathcal{H}(M(a,c))$. Since the inclusion of $\mathcal{H}(M(a,b))$ in $\mathcal{H}(M(a,c))$ does not increase norms, we obtain the inequality

$$\|[L(t)-L(0)]/t\|_{\mathcal{H}(M(a,b))} = \|[L(t)-L(0)]/t\|_{\mathcal{H}(M(a,c))}$$
$$\leqq \kappa(a,c)\|L(t)\|_{\mathcal{H}(M(a,c))}$$
$$\leqq \kappa(a,c)\|L(t)\|_{\mathcal{H}(M(a,b))}.$$

By the arbitrariness of $L(z)$, $\kappa(a,b) \leqq \kappa(a,c)$.

We show that $\kappa(b,c) \leqq \kappa(a,c)$ whenever $a \leqq b \leqq c$. It follows from the identity for difference quotients in $\mathcal{H}(M(a,c))$ that the adjoint of the transformation $F(z) \to [F(z)-F(0)]/z$ in $\mathcal{H}(M(a,c))$ is the transformation $F(z) \to [F(z)-M(a,c,z)F(0)]/z$. Since a transformation and its adjoint have the same bound, $\kappa(a,c)$ is the bound of the transformation $F(z) \to [F(z) - M(a,c,z)F(0)]/z$ in $\mathcal{H}(M(a,c))$. For the same reason, $\kappa(b,c)$ is the bound of the transformation $F(z) \to [F(z)-M(b,c,z)F(0)]/z$ in $\mathcal{H}(M(b,c))$. If $L(z)$ is in $\mathcal{H}(M(b,c))$, then $M(a,b,z)L(z)$ is in $\mathcal{H}(M(a,c))$ and can be written $M(a,b,c)L(z) = F(z) + M(a,b,z)G(z)$ with $F(z)$ in $\mathcal{H}(M(a,b))$, $G(z)$ in $\mathcal{H}(M(b,c))$, and

$$\|M(a,b,t)L(t)\|^2_{\mathcal{H}(M(a,c))} = \|F(t)\|^2_{\mathcal{H}(M(a,b))} + \|G(t)\|^2_{\mathcal{H}(M(b,c))}.$$

Since $L(z)-G(z)$ is an element of $\mathcal{H}(M(b,c))$ such that $M(a,b,z)[L(z)-G(z)]$ belongs to $\mathcal{H}(M(a,b))$, it is a constant. It follows that $F(z) = M(a,b,z)c$ for some vector c. It is easily verified that

$$[M(a,b,z)L(z) - M(a,c,z)L(0)]/z$$
$$= [F(z)-M(a,b,z)F(0)]/z + M(a,b,z)[F(0)-M(a,c,z)F(0)]/z$$
$$+ M(a,b,z)[G(z)-M(b,c,z)G(0)]/z$$

belongs to $\mathcal{H}(M(a,c))$ and that

$$\begin{aligned}\|[M(a,b,t)L(t) - M(a,c,t)L(0)]/t\|^2_{\mathcal{H}(M(a,c))} \\ = \|[F(t) - M(a,b,t)F(0)]/t\|^2_{\mathcal{H}(M(a,b))} \\ + \|[F(0) - M(a,c,t)F(0)]/t \\ + [G(t) - M(b,c,t)G(0)]/t\|^2_{\mathcal{H}(M(b,c))}.\end{aligned}$$

Since $F(z) = M(a,b,z)c$ for some vector c, the identity states that the norm of $M(a,b,z)[L(z) - M(b,c,z)L(0)]/z$ in $\mathcal{H}(M(a,c))$ is equal to the norm of $[L(z)-M(b,c,z)L(0)]/z$ in $\mathcal{H}(M(b,c))$. Since the transformation $L(z) \to M(a,b,z)L(z)$ of $\mathcal{H}(M(b,c))$ into $\mathcal{H}(M(a,c))$ does not increase norms, the inequality

$$\begin{aligned}\|[L(t) - M(b,c,t)L(0)]/t\|_{\mathcal{H}(M(b,c))} \\ = \|[M(a,b,t)L(t) - M(a,c,t)L(0)]/t\|_{\mathcal{H}(M(a,c))} \\ \leqq \kappa(a,c)\|M(a,b,t)L(t)\|_{\mathcal{H}(M(a,c))} \\ \leqq \kappa(a,c)\|L(t)\|_{\mathcal{H}(M(b,c))}\end{aligned}$$

holds for every element $L(z)$ of $\mathcal{H}(M(b,c))$. By the arbitrariness of $L(z)$, $\kappa(b,c) \leqq \kappa(a,c)$. It follows that $\kappa(b,c) \leqq \kappa(a,d)$ whenever $a \leqq b \leqq c \leqq d$.

If $a < 1$ we show that there is no nonzero element $F(z)$ of $\mathcal{H}(M(a,1))$, which belongs to $\mathcal{H}(M(a,b))$ whenever $b > a$, such that $\|F(t)\|_{\mathcal{H}(M(a,b))} = \|F(t)\|_{\mathcal{H}(M(a,1))}$ whenever $b > a$. To see this, observe that the set $\mathcal{M}(a)$ of such elements $F(z)$ is a closed subspace of $\mathcal{H}(M(a,1))$ which contains $[F(z) - F(0)]/z$ whenever it contains $F(z)$. On the other hand, if $F(z)$ belongs to $\mathcal{M}(a)$ and if $b > a$, then

$$2\pi\bar{c}F(0) = \langle F(t), [M(a,b,t)I - I]c/t\rangle$$

for every vector c. By the Schwarz inequality,

$$\begin{aligned}2\pi|\bar{c}F(0)|^2 \leqq \|F(t)\|^2_{\mathcal{H}(M(a,b))}\bar{c}[m(b) - m(a)]c \\ \leqq \|F(t)\|^2_{\mathcal{H}(M(a,1))}\bar{c}[m(b) - m(a)]c.\end{aligned}$$

Since $b > a$ is arbitrary and since $m(t)$ is continuous at a, $\bar{c}F(0) = 0$ By the arbitrariness of c, $F(0) = 0$. Since $[F(z) - F(0)]/z$ belongs to $\mathcal{M}(a)$ whenever $F(z)$ belongs to $\mathcal{M}(a)$, $F(z) = 0$. A similar argument will show that if $b > 0$, there is no nonzero element $F(z)$ of $\mathcal{H}(M(0,b))$ which is of the form $F(z) = M(a,b,z)G_a(z)$ for every $a < b$ with $\|F(t)\|_{\mathcal{H}(M(a,1))} = \|G_a(t)\|_{\mathcal{H}(M(a,b))}$.

We show that $\lim \kappa(a,b) = 0$ as $b \searrow a$ for every $a < 1$. Argue by contradiction, assuming that for some fixed a there exists an $\varepsilon > 0$ such that $\kappa(a,b) > \varepsilon$ for every $b > a$. Then for every $b > a$ there exists an element $F_b(z)$ of $\mathcal{H}(M(a,b)$. with $\|F_b(t)\|_{\mathcal{H}(M(a,1))} = 1$ and $\|[F_b(t) - F_b(0)]/t\|_{\mathcal{H}(M(a,1))} \geqq \varepsilon$. In this case the functions $F_b(z)$ can be chosen so that there exists an element $F(z)$ of $\mathscr{H}(M(a,1))$ such that

$$\langle F(t), G(t)\rangle_{\mathscr{H}(M(a,1))} = \lim_{b \searrow a} \langle F_b(t), G(t)\rangle_{\mathscr{H}(M(a,1))}$$

for every element $G(z)$ of $\mathscr{H}(M(a,1))$. Since $m(1)-m(a)$ is completely continuous, the transformation $F(z) \to [F(z) - F(0)]/z$ in $\mathscr{H}(M(a,1))$ is completely continuous. It follows that $[F(z) - F(0)]/z = \lim [F_b(z) - F_b(0)]/z$ in the metric of $\mathscr{H}(M(a,1))$ as $b \searrow a$. Since

$$\| [F_b(t) - F_b(0)]/t \|_{\mathscr{H}(M(a,1))} \geqq \varepsilon$$

for every b,

$$\| [F(t) - F(0)]/t \|_{\mathscr{H}(M(a,1))} \geqq \varepsilon.$$

But we have seen that

$$\| [F(t) - F(0)]/t \|_{\mathscr{H}(M(a,b))} = \| [F(t) - F(0)]/t \|_{\mathscr{H}(M(a,1))}$$

when $b > a$, which contradicts the results of the preceeding paragraph. We can therefore conclude that $\lim \kappa(a,b) = 0$ as $b \searrow a$. A similar argument will show that $\lim \kappa(a,b) = 0$ as $a \nearrow b$ for every $b > 0$. If now $\varepsilon > 0$ is given, there exists a $\delta > 0$ such that $\kappa(a,b) \leqq \varepsilon$ whenever $b - a \leqq \delta$.

If $0 \leqq a \leqq b \leqq 1$ the transformation $F(z) \to M(a,b,z)F(z)$ takes $\mathscr{H}(M(a,b))$ into $\mathscr{H}(M(0,1))$ and is bounded by 1. The adjoint transformation takes $\mathscr{H}(M(0,1))$ into $\mathscr{H}(M(a,b))$ and is bounded by 1. If $F(z)$ is in $\mathscr{H}(M(0,1))$, let $F(a,b,z)$ be the element of $\mathscr{H}(M(a,b))$ which is obtained from $F(z)$ under the adjoint transformation. Let $0 = h_0 < h_1 < \cdots < h_r = 1$ be any partition of $(0,1)$. By the theory of minimal decompositions, Theorem XII of [7] and Theorem 4 in the appendix of [**2**],

$$F(z) = \sum_{k=1}^{r} M(0,h_{k-1},z)F(h_{k-1},h_k,z)$$

and

$$\| F(t) \|^2_{\mathscr{H}(M(0,1))} = \sum_{k=1}^{r} \| F(h_{k-1},h_k,t) \|^2_{\mathscr{H}(M(h_{k-1},h_k))}.$$

The minimal decomposition of $[F(z) - F(0)]/z$ is

$$[F(z) - F(0)]/z = \sum_{k=1}^{r} M(0,h_{k-1},z)\{[M(h_{k-1},h_k,z) - 1]F(h_k,1,0)/z + [F(h_{k-1},h_k,z) - F(h_{k-1},h_k,0)]/z\}$$

where

$$\sum_{k=1}^{r} \| [F(h_{k-1}, h_k, t) - F(h_{k-1}, h_k, 0]/t \|^2_{\mathscr{H}(M(h_{k-1}, h_k))}$$

$$\leqq \sum_{k=1}^{r} \kappa(h_{k-1}, h_k)^2 \| F(h_{k-1}, h_k, t) \|^2_{\mathscr{H}(M(h_{k-1}, h_k))}$$

$$\leqq \| F(t) \|^2_{\mathscr{H}(M(0,1))} \max \kappa\, (h_{k-1}, h_k)^2$$

$$\leqq \varepsilon^2 \| F(t) \|^2_{\mathscr{H}(M(0,1))}$$

when the mesh of the partition is so small that $\kappa(h_{k-1}, h_k) \leqq \varepsilon$ for every k. It follows that

$$[F(z) - F(0)]/z = \lim \sum_{k=1}^{r} M(0, h_{k-1}, z)\; [M(h_{k-1}, h_k, z) - 1] F(h_k, 1, 0)/z$$

in the metric of $\mathscr{H}(M(0,1))$ as the mesh of the partition goes to zero. Also if $F(z)$ and $G(z)$ are in $\mathscr{H}(M(0,1))$

$$\langle [F(t) - F(0)]/t, [G(t) - G(0)]/t \rangle_{\mathscr{H}(M(0,1))}$$

$$= \sum_{k=1}^{r} \Big\langle \frac{[M(h_{k-1}, h_k, t) - 1] F(h_k, 1, 0)}{t} + \frac{F(h_{k-1}, h_k, t) - F(h_{k-1}, h_k, 0)}{t},$$

$$\frac{[M(h_{k-1}, h_k, t) - 1] G(h_k, 1, 0)}{t} + \frac{G(h_{k-1}, h_k, t) - G(h_{k-1}, h_k, 0)}{t} \Big\rangle_{\mathscr{H}(M(h_{k-1}, h_k))}$$

and

$$\langle [F(t) - F(0)]/t, [G(t) - G(0)]/t \rangle_{\mathscr{H}(M(0,1))}$$

$$= \lim \sum_{k=1}^{r} \langle [M(h_{k-1}, h_k, t) - 1] F(h_k, 1, 0)/t,$$

$$[M(h_{k-1}, h_k, t) - 1] G(h_k, 1, 0)/t \rangle_{\mathscr{H}(M(h_{k-1}, h_k))}$$

$$= 2\pi \lim \sum_{k=1}^{r} \bar{G}(h_k, 1, 0)[m(h_k) - m(h_{k-1})] F(h_k, 1, 0)$$

as the mesh of the partition goes to zero. If we choose

$$G(z) = \frac{M(0,1,z) I \bar{M}(0,1,w) - I}{2\pi(z - \bar{w})} \begin{pmatrix} u \\ v \end{pmatrix}$$

for some pair of vectors u and v and for some nonzero number w, then

$$G(a,b,z) = \frac{M(a,b,z) I \bar{M}(a,b,w) - I}{2\pi(z - \bar{w})} \bar{M}(0,a,w) \begin{pmatrix} u \\ v \end{pmatrix}$$

and

$$\bar{w}[F(z) - F(0)]/z$$

$$= \frac{M(0,1,z)I\bar{M}(0,1,w) - I}{2\pi(z - \bar{w})} \begin{pmatrix} u \\ v \end{pmatrix} - \frac{M(0,1,z)I - I}{2\pi z} \bar{M}(0,1,w) \begin{pmatrix} u \\ v \end{pmatrix}.$$

A routine calculation will now show that

$$M(0,1,z)I\bar{M}(0,1,w) - I$$

$$= \lim (z - \bar{w}) \sum_{k=1}^{r} M(0,h_k,z)[m(h_k) - m(h_{k-1})]\bar{M}(0,h_k,w)$$

in the "operator" norm, uniformly for z and w in any bounded set, as the mesh of the partition goes to zero.

If $F(z)$ belongs to $\mathscr{H}(M(0,1))$ and if $a \leq b \leq c$, then

$$F(a,c,z) = F(a,b,z) + M(a,b,z)F(b,c,z).$$

It follows that $F(a,b,0) = F(0,b,0) - F(0,a,0)$ when $a \leq b$. We show that $F(0,t,0)$ is always an m-differentiable function of t in $(0,1)$ and that $2\pi F(0,c,0) = \int_0^c dm(t)f(t), 0 < c < 1$, for an element $f(t)$ of $L_m^2(0,1)$ such that

$$2\pi \| F(t) \|^2_{\mathscr{H}(M(0,1))} = \int_0^1 \bar{f}(t)\,dm(t)f(t).$$

It is sufficient to make the verification in the case that

$$F(z) = \frac{M(0,1,z)I\bar{M}(0,1,w) - I}{2\pi(z - \bar{w})} \begin{pmatrix} u \\ v \end{pmatrix}$$

for some vectors u and v and for some nonzero number w since the finite sums of such special functions are dense in $\mathscr{H}(M(0,1))$. In this case

$$F(a,b,z) = \frac{M(a,b,z)I\bar{M}(a,b,w) - I}{2\pi(z - \bar{w})} \begin{pmatrix} u \\ v \end{pmatrix}.$$

It follows from the integral equation for $M(z)$ that

$$f(t) = \bar{M}(0,t,w) \begin{pmatrix} u \\ v \end{pmatrix}$$

determines an element of $L_m^2(0,1)$ which is the m-derivative of

$$-\frac{I\bar{M}(0,t,w) - I}{\bar{w}} \begin{pmatrix} u \\ v \end{pmatrix}$$

and that

$$2\pi \| F(t) \|^2_{\mathscr{H}(M(0,1))} = \begin{pmatrix} u \\ v \end{pmatrix}^- \frac{M(0,1,w)I\bar{M}(0,1,w) - I}{w - \bar{w}} \begin{pmatrix} u \\ v \end{pmatrix}$$

$$= \int_0^1 \begin{pmatrix} u \\ v \end{pmatrix}^- M(0,t,w)\,dm(t)\bar{M}(0,t,w) \begin{pmatrix} u \\ v \end{pmatrix} = \int_0^1 \bar{f}(t)\,dm(t)f(t)\,.$$

This completes the verification for the special choice of $F(z)$. A similar argument applies when $F(z)$ is a finite sum of such special functions. Note that $F(z)$ is always the eigentransform of $f(t)$ since

$$2\pi \begin{pmatrix} u \\ v \end{pmatrix}^- F(w) = 2\pi \left\langle F(t), \frac{M(0,1,t)I\bar{M}(0,1,w) - I}{2\pi(t - \bar{w})} \begin{pmatrix} u \\ v \end{pmatrix} \right\rangle_{\mathscr{H}(M(0,1))}$$

$$= \left\langle f(t), \bar{M}(0,t,w) \begin{pmatrix} u \\ v \end{pmatrix} \right\rangle_{L^2_m(0,1)}$$

$$= \int_0^1 \begin{pmatrix} u \\ v \end{pmatrix}^- M(0,t,w)\,dm(t)f(t)$$

for every pair of vectors u and v and for all complex numbers w. It is clear that the eigentransformation is isometric on the closed span of the elements

$$\bar{M}(0,t,w) \begin{pmatrix} u \\ v \end{pmatrix} \quad \text{of} \quad L^2_m(0,1)$$

and that the transformation annihilates the orthogonal complement of these elements.

Consider now elements $F(z)$ and $G(z)$ of $\mathscr{H}(M(0,1))$ such that $F(z) = [G(z) - G(0)]/z$. Let $f(t)$ and $g(t)$ be elements of $L^2_m(0,1)$ whose eigentransforms are $F(z)$ and $G(z)$. If $f(t)$ is orthogonal to the kernel of the eigentransformation, we show that

$$\int_c^1 \begin{pmatrix} u_1 \\ v_1 \end{pmatrix}^- dm(t)f(t) = \int_c^1 \begin{pmatrix} u_1 \\ v_1 \end{pmatrix}^- [m(c) - m(t)]I\,dm(t)\,g(t)$$

for every pair of vectors u_1 and v_1 when $0 \leqq c \leqq 1$. An explicit proof is given only in the case that there exists a number a, $0 < a < 1$, such that

$$g(t) = \bar{M}(0,t,w) \begin{pmatrix} u \\ v \end{pmatrix}$$

for $0 \leqq t \leqq a$ and $g(t) = 0$ for $t > a$ for some vectors u and v and for some nonzero number w. The general case follows because the finite sums of such special functions are dense in $L^2_m(0,1)$. In this special case we have

$$G(z) = \frac{M(0,a,z)I\bar{M}(0,a,w) - I}{2\pi(z - \bar{w})} \begin{pmatrix} u \\ v \end{pmatrix},$$

$$\bar{w}F(z) = \frac{M(0,a,z)I\bar{M}(0,a,w) - I}{2\pi(z - \bar{w})} \begin{pmatrix} u \\ v \end{pmatrix} - \frac{M(0,a,z)I - I}{2\pi z} \bar{M}(0,a,w) \begin{pmatrix} u \\ v \end{pmatrix}.$$

Since $F(z)$ has the same norm in $\mathscr{H}(M(0,a))$ as in $\mathscr{H}(M(0,1))$,

$$\bar{w}f(t) = \bar{M}(0,t,w)\begin{pmatrix} u \\ v \end{pmatrix} - \bar{M}(0,a,w)\begin{pmatrix} u \\ v \end{pmatrix}$$

for $0 \leqq t \leqq a$ and $f(t) = 0$ for $t > a$. The desired identity is obtained by substituting $z = 0$ in the identity

$$\int_c^a M(0,t,z)\,dm(t)\bar{M}(0,t,w) - \int_c^a M(0,t,z)\,dm(t)\bar{M}(0,a,w)$$

$$= \bar{w}\int_c^a \frac{M(0,t,z) - M(0,c,z)}{z}\,dm(t)\bar{M}(0,t,w),$$

which follows from the integral equation for $M(z)$.

Proof of Theorem 8. If $m(t)$ is stable in $(0,1)$, there exists a completely continuous transformation T in $L^2_m(0,1)$ such that the identity

$$\int_c^1 dm(t)f(t) = \int_c^1 [m(c) - m(t)]I\,dm(t)\,g(t)$$

holds for $0 \leqq c \leqq 1$ whenever $T: g \to f$. We show that the identity

$$\int_c^a dm(t)I[m(t) - m(a)] = \int_c^a [m(c) - m(t)]I\,dm(t)$$

holds whenever $0 \leqq c \leqq a \leqq 1$. The hypotheses imply that each integral exists as a norm limit of Stieltjes sums. If $c = t_0 < t_1 < \cdots < t_r = a$ is any partition of the integral (c,a), then

$$\sum_{k=1}^{r} [m(t_k) - m(t_{k-1})]I[m(t_k) - m(a)]$$

$$= \sum_{k=1}^{r} [m(c) - m(t_{k-1})]I[m(t_k) - m(t_{k-1})].$$

The desired identity is obtained in the limit as the mesh of the partition goes to zero since

$$\lim \sum_{k=1}^{r} [m(t_k) - m(t_{k-1})]I[m(t_k) - m(t_{k-1})] = 0.$$

The identity is used to compute the action of T on special functions. If

$$g(t) = \begin{pmatrix} u \\ v \end{pmatrix}$$

is a constant for $0 \leqq t \leqq a$ and if $g(t) = 0$ for $t > a$ for some number $a, 0 \leqq a \leqq 1$, then $T: g \to f$ where

$$f(t) = I[m(t) - m(a)] \begin{pmatrix} u \\ v \end{pmatrix}$$

for $0 \leqq t \leqq$ a and $f(t) = 0$ for $t > a$. From this we see that the identity

$$\langle Tf, g\rangle_{L_m^2(0,1)} - \langle f, Tg\rangle_{L_m^2(0,1)} = \int_0^1 \bar{g}(s)dm(s)I \int_0^1 dm(t)f(t)$$

holds for all elements f and g of $L_m^2(0,1)$. It is sufficient to verify the identity in the case that $f(t)$ is constant in some interval $(0,a)$, $0 < a < 1$, and vanishes in $(a,1)$, since the finite sums of such special functions are dense in $L_m^2(0,1)$. For the same reason it is sufficient to verify the identity when $g(t)$ is constant in some interval $(0,b)$, $0 < b < 1$, and vanishes in $(b,1)$. By symmetry we can assume that $a \leqq b$. The desired identity now follows from the identity

$$\int_0^a [m(b) - m(t)]Idm(t) - \int_0^a dm(t)I[m(t) - m(a)]$$
$$= [m(b) - m(0)]I[m(a) - m(0)]$$

which is equivalent to an identity obtained earlier in the proof.

For every number a, $0 \leqq a \leqq 1$, let $\mathscr{M}(a)$ be the closure in $L_m^2(0, 1)$ of the piecewise constant functions $f(t)$ which vanish for $t > a$. Then $\mathscr{M}(a)$ is an invariant subspace for T. For every $a < 1$, $\mathscr{M}(a)$ is the intersection of the spaces $\mathscr{M}(b)$ with $b > a$. For every $b > 0$, $\mathscr{M}(b)$ is the closure of the union of the spaces $\mathscr{M}(a)$ with $a < b$. The existence of such a continuous family of invariant subspaces implies that T has no nonzero spectrum. Let J be the transformation of $L_m^2(0,1)$ into $\mathscr{C}$ defined by $Jf = \int_0^1 dm(t)f(t)$. If $f(t)$ is in $L_m^2(0,1)$, define a corresponding vector valued entire function $F(z)$ by $2\pi F(z) = J(1 - zT)^{-1} f$ for all complex z. It is easily verified that the entire function corresponding to $T(1 - wT)^{-1} f$ is then $[F(z) - F(w)]/(z - w)$ for every complex number w. The set $\mathscr{H}$ of such entire functions $F(z)$ is a Hilbert space in the norm defined by $2\pi \| F(t) \|^2 = \int_0^1 \bar{f}(t) dm(t) f(t)$ whenever $f(t)$ is orthogonal to all elements $g(t)$ of $L_m^2(0,1)$ such that $G(z) = 0$. The transformation $F(z) \to F(0)$ of $\mathscr{H}$ into $\mathscr{C}$ is continuous by construction, and the special identity for difference quotients

$$2\pi \bar{G}(0) I F(0) = \langle [F(t) - F(0)]/t, G(t)\rangle - \langle F(t), [G(t) - G(0)]/t\rangle$$

holds for all elements $F(z)$ and $G(z)$ of $\mathscr{H}$. The general identity for difference quotients follows by a routine calculation. By Theorem III of [**6**], the space $\mathscr{H}$

is equal isometrically to a space $\mathcal{H}(M(0,1))$ for some operator valued entire function $M(0,1,z)$ such that $M(0,1,0)$ is the identity matrix. Since

$$\frac{M(0,1,z)\,I - I}{2\pi z}\begin{pmatrix} u \\ v \end{pmatrix}$$

belongs to $\mathcal{H}(M(0,1))$ for every pair of vectors u and v, and since

$$2\pi\left\langle F(t), \frac{M(0,1,t)I - I}{2\pi t}\begin{pmatrix} u \\ v \end{pmatrix}\right\rangle = 2\pi\begin{pmatrix} u \\ v \end{pmatrix}^{-} F(0) = \int_0^1 \begin{pmatrix} u \\ v \end{pmatrix}^{-} dm(t)f(t)$$

for every $F(z)$ in $\mathcal{H}(M(0,1))$, and since

$$\begin{pmatrix} u \\ v \end{pmatrix}$$

is orthogonal to every element $g(t)$ of $L_m^2(0,1)$ such that $G(z) = 0$, we can conclude that

$$J(1 - zT)^{-1}\begin{pmatrix} u \\ v \end{pmatrix} = \frac{M(0,1,z)I - I}{z}\begin{pmatrix} u \\ v \end{pmatrix}$$

and that

$$2\pi\left\| \frac{M(0,1,t)I - I}{2\pi t}\begin{pmatrix} u \\ v \end{pmatrix}\right\|^2 = \int_0^1 \begin{pmatrix} u \\ v \end{pmatrix}^{-} dm(t)\begin{pmatrix} u \\ v \end{pmatrix}.$$

It follows that $M'(0,1,0)I = m(1) - m(0)$.

If $0 \leqq a \leqq 1$ consider the set of vector valued entire functions $F(z)$ corresponding to elements $f(t)$ of $\mathcal{M}(a)$. The above argument will show that these entire functions are a Hilbert space $\mathcal{H}(M(0,a))$ with norm defined so that

$$2\pi\|F(t)\|^2_{\mathcal{H}(M(0,a))} = \int_0^1 \bar{f}(t)dm(t)f(t)$$

whenever $f(t)$ is orthogonal to all elements $g(t)$ of $\mathcal{M}(a)$ such that $G(z) = 0$. If $M(0,a,z)$ is chosen so that $M(0,a,0)$ is the identity matrix, then $M'(0,a,0)I = m(a) - m(0)$. Since $\mathcal{H}(M(0,a))$ is contained in $\mathcal{H}(M(0,b))$ when $a \leqq b$ and since the inclusion does not increase norms,

$$M(0,b,z) = M(0,a,z)M(a,b,z)$$

for some space $\mathcal{H}(M(a,b))$. The value of $M(a, b, z)$ at the origin is the identity matrix and $M'(a,b,0)I = m(b) - m(a)$. Since $M(a,c,z) = M(a,b,z)M(b,c,z)$ when $a \leqq b \leqq c$, the hypotheses of Theorem 7 are satisfied. Uniqueness of the construction follows from the integral equation for $M(z)$, which holds by Theorem 7.

PROOF OF THEOREM 9. By Theorem 7,

$$M(a,b,z)I\bar{M}(a,b,w) - I = (z - \bar{w})\int_a^b M(a,t,z)dm(t)\bar{M}(a,t,w)$$

when $0 < a < b < \infty$, where the integral is a norm limit of Stieltjes sums. Since

$$(A(t,z), B(t,z)) = (A(a,z), B(a,z))\, M(a,t,z)$$

when $a \leqq t \leqq b$,

$$\begin{aligned}(A(b,z), B(b,z))\, I(A(b,w), B(b,w))^- &- (A(a,z), B(a,z))\, I(A(a,w), B(a,w))^- \\ &= (A(a,z), B(a,z))[M(a,b,z)I\bar{M}(a,b,w) - I](A(a,w), B(a,w))^- \\ &= (z - \bar{w})\int_a^b (A(a,z), B(a,z))M(a,t,z)dm(t)\bar{M}(a,t,w)(A(a,w), B(a,w))^- \\ &= (z - \bar{w})\int_a^b (A(t,z), B(t,z))\, dm(t)(A(t,w), B(t,w))^-\end{aligned}$$

for all complex z and w, where the integral is again a norm limit of Stieltjes sums. If

$$\lim_{a \searrow 0}\, [B(a,z)\bar{A}(a,w) - A(a,z)\bar{B}(a,w)] = 0$$

for all complex z and w, it follows that $(A(t,w), B(t,w))^- c$ determines an element of $L^2_m(0, b)$ for every vector c and that

$$\begin{aligned}&\bar{u}[B(b,z)\bar{A}(b,w) - A(b,z)\bar{B}(b,w)]c/(z - \bar{w}) \\ &\qquad = \int_0^b \bar{u}(A(t,z), B(t,z))\, dm(t)(A(t,w), B(t,w))^- c\end{aligned}$$

for all vectors c and u and for all complex numbers z and w. If $f(t)$ is a finite sum of elements $(A(t,w), B(t,w))^- c$ of $L^2_m(0, b)$, then it is easily verified from this identity that its eigentransform $F(z)$, defined by

$$\pi F(z) = \int_0^b (A(t,z), B(t,z))\, dm(t) f(t)$$

belongs to $\mathscr{H}(A(b), B(b))$ and that

$$\pi \| F(t) \|^2_{\mathscr{H}(A(b), B(b))} = \int_0^b \bar{f}(t)\, dm(t) f(t).$$

The same conclusion follows by continuity whenever $f(t)$ belongs to the closed span of such special functions. On the other hand, the eigentransform $F(z)$ of

$f(t)$ clearly vanishes identically if $f(t)$ is orthogonal to such special functions.

Let a be a fixed number, $0 < a < b$. By the proof of Theorem 5 and by the proof of Theorem XII of [7], every element $F(z)$ of $\mathcal{H}(A(b), B(b))$ has a unique minimal decomposition

$$F(z) = F(a,z) + (A(a,z), B(a,z))F(a,b,z)$$

with $F(a,z)$ in $\mathcal{H}(A(a), B(a))$, $F(a,b,z)$ in $\mathcal{H}(M(a,b))$, and

$$\| F(t) \|^2_{\mathcal{H}(A(b),B(b))} = \| F(t) \|^2_{\mathcal{H}(A(a),B(a))} + \tfrac{1}{2} \| F(a,b,t) \|^2_{\mathcal{H}(M(a,b))}.$$

If $F(z)$ is the eigentransform of the element $f(t)$ of $L^2_m(0,b)$ which is orthogonal to the kernel of the eigentransformation, we show that

$$\pi F(a,z) = \int_0^a (A(t,z), B(t,z))\, dm(t) f(t),$$

$$\pi F(a,b,z) = \int_a^b M(a,t,z) dm(t) f(t).$$

It is sufficient to establish the formulas in the case that

$$F(z) = [B(b,z)\bar{A}(b,w) - A(b,z)\bar{B}(b,w)]c/[\pi(z - \bar{w})]$$

for some vector c and complex number w since the finite sums of such special functions are dense in $\mathcal{H}(A(b), B(b))$. In this case $f(t) = (A(t,w), B(t,w))^{-}c$, and the minimal decomposition is obtained with

$$F(a,z) = \frac{B(a,z)\bar{A}(a,w) - A(a,z)\bar{B}(a,w)}{\pi(z - \bar{w})} c,$$

$$F(a,b,z) = \frac{M(a,b,z) I \bar{M}(a,b,w) - I}{\pi(z - \bar{w})} (\bar{A}(a,w), \bar{B}(a,w))c.$$

The required formulas now follow from the integral equation for $M(z)$.

Let a be a fixed number, $0 < a < b$, and consider any element $G(z)$ of $\mathcal{H}(A(b), B(b))$ which is the eigentransform of an element $g(t)$ of $L^2_m(0,b)$. Let $F(z) = [G(z) - A(a,z)G(0)]/z$, and let $f(t)$ be the element of $L^2_m(0,b)$ which has $F(z)$ as its eigentransform and which is orthogonal to all elements of $L^2_m(0,b)$ whose eigentransforms vanish identically. We show that

$$\int_a^b dm(t) f(t) = \int_a^b [m(a) - m(t)] I dm(t) g(t).$$

Let

$$G(z) = G(a,z) + (A(a,z), B(a,z))G(a,b,z)$$

be the minimal decomposition of $G(z)$ as an element of $\mathcal{H}(A(b), B(b))$ with $G(a,z)$

in $\mathscr{H}(A(a), B(a))$ and $G(a, b, z)$ in $\mathscr{H}(M(a, b))$. Then the minimal decomposition

$$F(z) = F(a, z) + (A(a, z), B(a, z)) F(a, b, z)$$

of $F(z)$ as an element of $\mathscr{H}(A(b), B(b))$ with $F(a, z)$ in $\mathscr{H}(A(a), B(a))$ and $F(a, b, z)$ in $\mathscr{H}(M(a, b))$ is obtained with

$$\begin{aligned} F(a, b, z) &= [G(a, b, z) - G(a, b, 0)]/z, \\ F(a, z) &= [G(a, z) - A(a, z) G(a, 0)]/z \\ &\quad + [(A(a, z), B(a, z)) - (A(a, 0), B(a, 0))] F(a, b, 0)/z \end{aligned}$$

where

$$F(a, b, z) = \int_a^b M(a, t, z) dm(t) f(t)$$

and

$$G(a, b, z) = \int_a^b M(a, t, z) dm(t) g(t).$$

The desired formula follows from Theorem 7 since $f(t)$ is orthogonal to all elements $h(t)$ of $L_m^2(a, b)$ such that $\int_a^b M(a, t, z) dm(t) h(t) = 0$.

Appendix on square summable power series

The theory of Hilbert spaces of entire functions presented in the body of the paper depends on factorization properties of operator valued entire functions $F(z)$ such that $1 - F(z)$ has completely continuous values. If such a function has an invertible value, it will have invertible values at all except isolated points in the complex plane. The values of w such that $F(w)$ fails to have an invertible value are analogous to the zeros of a complex valued entire function and have been used **[2]** to give an operator generalization of the Weierstrass factorization. Difficult unsolved problems surround the factorization properties of operator valued entire functions when no complete continuity hypothesis is made. As in the theory of complex valued functions, a half plane rather than the full plane is often the most natural region in which to study factorization. The unit disk, which is conformally equivalent to a half plane, is now used instead because of technical advantages lying with power series expansions.

Let $\mathscr{C}(z)$ be the Hilbert space of power series $f(z) = \sum a_n z^n$ with vector coefficients such that $\|f(z)\|^2 = \sum |a_n|^2 < \infty$. Let $B(z)$ be a power series with operator coefficients such that $B(z) f(z)$ belongs to $\mathscr{C}(z)$ whenever $f(z)$ belongs to $\mathscr{C}(z)$ and such that $\|B(z) f(z)\| \leq \|f(z)\|$ for every $f(z)$ in $\mathscr{C}(z)$. The B-norm of an element $f(z)$ of $\mathscr{C}(z)$ is defined by

$$\|f(z)\|_B^2 = \sup[\|f(z) + B(z) g(z)\|^2 - \|g(z)\|^2]$$

where the supremum is taken over all elements $g(z)$ of $\mathscr{C}(z)$. By $\mathscr{H}(B)$ we mean the set of all power series in $\mathscr{C}(z)$ which have finite B-norm. The space $\mathscr{H}(B)$ is a Hilbert space in the B-norm, and its inclusion in $\mathscr{C}(z)$ does not increase norms. If $f(z)$ is in $\mathscr{H}(B)$, then $[f(z)-f(0)]/z$ is in $\mathscr{H}(B)$ and

$$\|[f(z)-f(0)]/z\|_B^2 \leqq \|f(z)\|_B^2 - |f(0)|^2.$$

It is known [2] that the transformation $f(z) \to [f(z)-f(0)]/z$ in a space $\mathscr{H}(B)$ is a canonical model of transformations in Hilbert space which are bounded by 1. The study of factors of $B(z)$ is closely related to the study of invariant subspaces of the transformation $f(z) \to [f(z)-f(0)]/z$.

The principal complication which arises in the theory of the space $\mathscr{H}(B)$ is due to the fact that the space need not be contained isometrically in $\mathscr{C}(z)$. A new Hilbert space $\mathscr{L}(\phi)$ is needed in the study of the space $\mathscr{H}(B)$ in this case. A space $\mathscr{L}(\phi)$ is associated with any power series $\phi(z)$, having operator coefficients which converges in the unit disk and represents a function $\phi(w)$ such that $\operatorname{Re}\phi(w) \geqq 0$ for $|w|<1$. The space $\mathscr{L}(\phi)$ is a space of power series with vector coefficients which converge in the unit disk. The series $\frac{1}{2}[\phi(z)+\bar{\phi}(w)]c/(1-z\bar{w})$ belongs to $\mathscr{L}(\phi)$ for every vector c when $|w|<1$ and the identity

$$\bar{c}L(w) = \langle L(z), \tfrac{1}{2}[\phi(z)+\bar{\phi}(w)]c/(1-z\bar{w})\rangle_{\mathscr{L}(\phi)}$$

holds for every element $L(z)$ of $\mathscr{L}(\phi)$. The series $[L(z)-L(0)]/z$ belongs to $\mathscr{L}(\phi)$ whenever $L(z)$ belongs to $\mathscr{L}(\phi)$, the transformation $L(z) \to [L(z)-L(0)]/z$ is bounded, and its adjoint is isometric.

If $\mathscr{H}(B)$ is a given space which is not contained isometrically in $\mathscr{C}(z)$, it contains a nonzero element of the form $B(z)L(z)$ where $L(z)$ is in $\mathscr{C}(z)$. Let $\mathscr{L} = \mathscr{L}_B$ be the set of elements $L(z)$ of $\mathscr{C}(z)$ such that $B(z)L(z)$ belongs to $\mathscr{H}(B)$. The overlapping space $\mathscr{L}$ is a Hilbert space in the overlapping norm

$$\|L(z)\|_{\mathscr{L}}^2 = \|L(z)\|^2 + \|B(z)L(z)\|_B^2.$$

It is known [2] that the space $\mathscr{L}$ is equal isometrically to a space $\mathscr{L}(\phi)$. However not every space $\mathscr{L}(\phi)$ is the overlapping space of a space $\mathscr{H}(B)$. A method of Lowdenslager [10] characterizes those spaces which are overlapping spaces.

THEOREM 1. *A necessary and sufficient condition that a space $\mathscr{L}(\phi)$ be the overlapping space of a space $\mathscr{H}(B)$ is that a space $\mathscr{L}(\psi)$ exist, $\psi(z) = 1-\phi(z)$, and that the polynomials which belong to $\mathscr{L}(\psi)$ be dense in $\mathscr{L}(\psi)$. In this case $B(z)$ can be chosen so that the orthogonal complement in $\mathscr{H}(B)$ of the elements of the form $B(z)L(z)$, where $L(z)$ is in $\mathscr{L}(\phi)$, contains $zf(z)$ whenever it contains $f(z)$.*

The theory of overlapping spaces gives a factorization theorem for operator valued functions which are analytic and bounded by 1 in the unit disk.

THEOREM 2. *Let $\mathscr{H}(B)$ and $\mathscr{H}(C)$ be given spaces with overlapping spaces $\mathscr{L}(\phi)$ and $\mathscr{L}(\theta)$ such that $\mathscr{L}(\theta)$ is contained in $\mathscr{L}(\phi)$ and such that the inclusion does not increase norms. Assume that the orthogonal complement of the range of multiplication by $C(z)$ in $\mathscr{C}(z)$ contains $zf(z)$ whenever it contains $f(z)$. Then $B(z) = A(z)C(z)$ where $\mathscr{H}(A)$ exists and multiplication by $A(z)$ annihilates the orthogonal complement of the range of multiplication by $C(z)$ in $\mathscr{C}(z)$.*

If $\mathscr{H}(A)$, $\mathscr{H}(B)$, and $\mathscr{H}(C)$ are spaces such that $B(z) = A(z)C(z)$, then $\mathscr{H}(A)$ is contained in $\mathscr{H}(B)$ and the inclusion does not increase norms. Of particular interest are factorizations which determine isometric inclusions of the corresponding spaces since $\mathscr{H}(A)$ is then a closed invariant subspace of the transformation $f(z) \to [f(z) - f(0)]/z$ in $\mathscr{H}(B)$.

THEOREM 3. *Under the hypotheses of Theorem 2, assume that $\mathscr{L}(\theta)$ is contained isometrically in $\mathscr{L}(\phi)$, that $\mathscr{L}(\theta)$ is not contained isometrically in $\mathscr{C}(z)$, and that $\mathscr{L}(\phi - \theta)$ is not the closed span of the polynomials it contains. Then $\mathscr{H}(A)$ is contained isometrically in $\mathscr{H}(B)$, it contains a nonzero element, and it does not fill $\mathscr{H}(B)$.*

We now give conditions for the existence of such a subspace $\mathscr{L}(\theta)$ of $\mathscr{L}(\phi)$.

THEOREM 4. *Let $\mathscr{H}(B)$ be a given space with overlapping space $\mathscr{L}(\phi)$ such that $\| L(z) \|_{\mathscr{L}(\phi)} = \| L(z) \|$ whenever $L(z)$ is a polynomial in $\mathscr{L}(\phi)$. Assume that there exists a nonzero space $\mathscr{L}(\psi)$ contained in $\mathscr{L}(\phi)$ and in $\mathscr{L}(1 - \phi)$ in such a way that the inclusions do not increase norms. Then there exists a space $\mathscr{L}(\theta)$ contained isometrically in $\mathscr{L}(\phi)$ such that the polynomials which belong to $\mathscr{L}(1 - \theta)$ are dense in $\mathscr{L}(1 - \theta)$, such that $\mathscr{L}(\theta)$ is not contained isometrically in $\mathscr{C}(z)$, and such that the polynomials which belong to $\mathscr{L}(\phi - \theta)$ are not dense in $\mathscr{L}(\phi - \theta)$.*

THEOREM 5. *Let $\mathscr{H}(B)$ be a given space with overlapping space $\mathscr{L}(\phi)$ such that the elements of $\mathscr{H}(B)$ of the form $B(z)L(z)$ with $L(z)$ in $\mathscr{L}(\phi)$ are dense in $\mathscr{H}(B)$. A necessary and sufficient condition that multiplication by $B^*(z)$ be isometric in $\mathscr{C}(z)$ is that there exist no nonzero space $\mathscr{L}(\psi)$ contained in $\mathscr{L}(\phi)$ and in $\mathscr{L}(1 - \phi)$ in such a way that the inclusions do not increase norms.*

A consequence of the factorization theorem is an existence theorem for invariant subspaces, due to Nagy and Foiaş **[11]**, who obtain it by a different method.

THEOREM 6. *Let T be a transformation of a Hilbert space $\mathscr{H}$ into itself which is bounded by 1. Assume that the dimension of $\mathscr{H}$ is greater than 1 and that there exist elements u and v of $\mathscr{H}$ such that $\lim \| T^n u \| \neq 0$ and $\lim \| T^{*n} v \| \neq 0$ as $n \to \infty$. Then there exists a closed subspace $\mathscr{M}$ of $\mathscr{H}$, other than the zero subspace and the full space, which is an invariant subspace for T.*

Difficulties arise in factorization if multiplication by $B^*(z)$ is isometric in $\mathscr{C}(z)$. We now construct examples of spaces $\mathscr{H}(B)$ having this property.

THEOREM 7. *Let $B(0)$ and S be given operators such that*

$$1 + S\bar{B}(0)B(0)\bar{S} = B(0)\bar{B}(0) + S\bar{S}.$$

Assume that there exists no nonzero element $f(z)$ of $\mathscr{C}(z)$ such that $[S\bar{B}(0) - z]f(z) = S\bar{B}(0)f(0)$. Then there exists a space $\mathscr{H}(B)$ corresponding to

$$B(z) = [1 - zS\bar{B}(0)]^{-1}[B(0) - zS],$$

and multiplication by $B^*(z)$ is isometric in $\mathscr{C}(z)$. The space $\mathscr{H}(B^*)$ is contained isometrically in $\mathscr{C}(z)$ and is the set of elements $f(z)$ of $\mathscr{C}(z)$ such that $[S - zB(0)]f(z) = Sf(0)$. The transformation $f(z) \to [1 - zS\bar{B}(0)]^{-1}Sf(0)$ is a partial isometry of $\mathscr{H}(B^*)$ onto $\mathscr{H}(B)$. The overlapping space $\mathscr{L}(\phi)$ of $\mathscr{H}(B)$ is the set of elements $f(z)$ of $\mathscr{C}(z)$ such that $[B(0) - Sz]f(z) = B(0)f(0)$ and such that there exists a corresponding element $g(z)$ of $\mathscr{H}(B^*)$ such that $B(0)f(0) = Sg(0)$. If $g(z)$ is chosen orthogonal to all elements $h(z)$ of $\mathscr{H}(B^*)$ such that $Sh(0) = 0$, then

$$\|f(z)\|^2_{\mathscr{L}(\phi)} = \|f(z)\|^2 + \|g(z)\|^2.$$

Invariant subspaces determine factors of $B(z)$ if multiplication by $B^*(z)$ is isometric in $\mathscr{C}(z)$.

THEOREM 8. *Let $\mathscr{H}(B)$ be a given space such that multiplication by $B^*(z)$ is isometric in $\mathscr{C}(z)$. If $\mathscr{M}$ is a closed invariant subspace for the transformation $f(z) \to [f(z) - f(0)]/z$ in $\mathscr{H}(B)$, then there exist spaces $\mathscr{H}(A)$ and $\mathscr{H}(C)$ such that $B(z) = A(z)C(z)$, $\mathscr{H}(A)$ is contained isometrically in $\mathscr{H}(B)$ and coincides with $\mathscr{M}$, multiplication by $C^*(z)$ is isometric in $\mathscr{C}(z)$, $f(z) \to A(z)f(z)$ is an isometric transformation of $\mathscr{H}(C)$ onto the orthogonal complement of $\mathscr{H}(A)$ in $\mathscr{H}(B)$, the overlapping space $\mathscr{L}(\theta)$ of $\mathscr{H}(C)$ is contained isometrically in the overlapping space $\mathscr{L}(\phi)$ of $\mathscr{H}(B)$, and there is no nonzero constant in $\mathscr{L}(\theta)$.*

We determine the form of these factors when $B(z)$ satisfies the hypotheses of Theorem 7.

THEOREM 9. *Let $B(0)$ and S be given operators such that*

$$1 + S\bar{B}(0)B(0)\bar{S} = B(0)\bar{B}(0) + S\bar{S}.$$

Assume that there exists no nonzero element $f(z)$ of $\mathscr{C}(z)$ such that $[S\bar{B}(0) - z]f(z) = S\bar{B}(0)f(0)$. If

$$B(z) = [1 - zS\bar{B}(0)]^{-1}[B(0) - zS]$$

and if $B(z) = A(z)C(z)$ as in Theorem 8, then

$$A(z) = [1 - zS\bar{B}(0)]^{-1}[A(0) - zS\bar{C}(0)],$$

$$[A(0) - zS\bar{C}(0)]C(z) = B(0) - zS,$$

and

$$1 + S\bar{B}(0)\,B(0)\bar{S} \;=\; A(0)\,\bar{A}(0) + S\bar{C}(0)\,C(0)\bar{S}.$$

The theory of square summable power series is used also in the perturbation theory of unitary transformations. If U and V are unitary transformations in the same Hilbert space $\mathscr{H}$, the wave limit Wf is defined by $Wf = \lim_{r\to\infty} \bar{U}^r V^r f$ whenever the limit exists. The domain of W is a closed subspace of $\mathscr{H}$ which reduces V and its range is a closed subspace of $\mathscr{H}$ which reduces U. Since W is isometric and since the identity $UWf = WVf$ holds whenever f is in the domain of W, the restriction of V to the domain of W is unitarily equivalent to the restriction of U to the range of W. A fundamental problem is to determine the domain and range of W when U and V are given transformations. A canonical model of the perturbation situation has been introduced in previous work [3], where unitary equivalence is obtained without the use of wave limits. The following variant of the canonical model is useful in computing wave limits. Note that the set of elements f of $\mathscr{H}$ such that $V^r f = U^r f$ for every $r = 1, 2, 3, \cdots$, is a closed subspace of $\mathscr{H}$ which is invariant under U and V. Since the wave limit exists and is the identity transformation on this subspace, we can restrict the study of wave limits to the orthogonal complement of the subspace.

THEOREM 10. *Let U and V be unitary transformations in the same Hilbert space $\mathscr{H}$. If the dimension of the range of $V - U$ does not exceed the dimension of the coefficient space $\mathscr{C}$, then there exist partially isometric transformations S of $\mathscr{H}$ onto a space $\mathscr{L}(\phi)$ and T of $\mathscr{H}$ onto a space $\mathscr{L}(\psi)$ with these properties:*

(1) The kernel of S is the set of elements f of $\mathscr{H}$ such that $U^n f = V^n f$ for every $n = 1, 2, 3, \cdots$. Whenever $S: h \to f(z)$, then $SU: h \to [f(z) - f(0)]/z$.

(2) The kernel of T is the set of elements f of $\mathscr{H}$ such that $U^n f = V^n f$ for every $n = 1, 2, 3, \cdots$. Whenever $T: h \to g(z)$, then $TV: h \to [g(z) - g(0)]/z$.

(3) The functions $\phi(z)$ and $\psi(z)$ are related by

$$\phi(z)\psi(z) \;=\; 1 \;=\; \psi(z)\phi(z).$$

Whenever h is in $\mathscr{H}$ and $S: h \to f(z)$ and $T: h \to g(z)$, then $g(z) = \psi(z) f(z)$ and $f(z) = \phi(z) g(z)$.

A perturbation situation is obtained from any pair of spaces $\mathscr{L}(\phi)$ and $\mathscr{L}(\psi)$ such that $\phi(z)$ and $\psi(z)$ are reciprocals in the algebra of formal power series with operator coefficients.

THEOREM 11. *If $\mathscr{L}(\phi)$ and $\mathscr{L}(\psi)$ are given spaces such that*

$$\phi(z)\,\psi(z) \;=\; 1 \;=\; \psi(z)\,\phi(z),$$

then there exists a Hilbert space $\mathscr{H}$ and unitary transformations U and V in $\mathscr{H}$ which are related to $\mathscr{L}(\phi)$ and $\mathscr{L}(\psi)$ as in Theorem 10. Let $\phi(z) = \sum \phi_n z^n$.

If h is in $\mathscr{H}$ and if $Th = g(z) = \sum a_n z^n$ in $\mathscr{L}(\psi)$, then $S\bar{U}^r V^r h = f_r(z)$ for every $r = 1,2,3,\cdots$, where

$$f_r(z) = -\sum_{n=0}^{\infty}\sum_{k=0}^{r-1-n} \bar{\phi}_k a_{k+n} z^n + \phi(z)\sum_{n=r}^{\infty} a_n z^n.$$

A necessary and sufficient condition that h be in the domain of W is that $f_\infty(z) = \lim f_r(z)$ exists in the metric of $\mathscr{L}(\phi)$ as $r \to \infty$. In this case $SWh = f_\infty(z)$.

PROOF OF THEOREM 1. If $\mathscr{L}(\phi)$ is the overlapping space of a space $\mathscr{H}(B)$, then $\mathscr{L}(\phi)$ is contained in $\mathscr{C}(z)$ and the inclusion does not increase norms. If $g(z)$ is in $\mathscr{C}(z)$, define its one-norm to be

$$\|g(z)\|_1^2 = \sup[\|f(z)+g(z)\|^2 - \|f(z)\|_{\mathscr{L}(\phi)}^2]$$

where the supremum is taken over all elements $f(z)$ of $\mathscr{L}(\phi)$. It is easily verified that the elements of $\mathscr{C}(z)$ which have finite one-norm are a Hilbert space $\mathscr{L}_1$ in the one-norm. Theorem 4 in the appendix of [2] has an obvious generalization for the one-norm. If $f(z)$ is in $\mathscr{L}(\phi)$, if $g(z)$ is in $\mathscr{L}_1$, and if $h(z) = f(z) + g(z)$, then

$$\|h(z)\|^2 \leqq \|f(z)\|_{\mathscr{L}(\phi)}^2 + \|g(z)\|_1^2.$$

Every element $h(z)$ of $\mathscr{C}(z)$ has a unique minimal decomposition $h(z) = f(z) + g(z)$ with $f(z)$ in $\mathscr{L}(\phi)$ and $g(z)$ in $\mathscr{L}_1$, such that equality holds in the last inequality. The element $f(z)$ of $\mathscr{L}(\phi)$ is obtained from $h(z)$ under the adjoint of the inclusion of $\mathscr{L}(\phi)$ in $\mathscr{C}(z)$. The element $g(z)$ of $\mathscr{L}_1$ is obtained from $h(z)$ under the adjoint of the inclusion of $\mathscr{L}_1$ in $\mathscr{C}(z)$. If $f(z)$ is in $\mathscr{L}(\phi)$, then

$$\langle f(z),(1-z\bar{w})^{-1}c\rangle = \bar{c}f(w) = \langle f(z),\tfrac{1}{2}[\phi(z)+\bar{\phi}(w)]c/(1-z\bar{w})\rangle_{\mathscr{L}(\phi)}$$

for every vector c when $|w|<1$. So $\frac{1}{2}[\phi(z)+\bar{\phi}(w)]c/(1-z\bar{w})$ is the element of $\mathscr{L}(\phi)$ obtained from $(1-z\bar{w})^{-1}c$ under the adjoint of the inclusion of $\mathscr{L}(\phi)$ in $\mathscr{C}(z)$. It follows that

$$(1-z\bar{w})^{-1}c = \tfrac{1}{2}[\phi(z)+\bar{\phi}(w)]c/(1-z\bar{w}) + \tfrac{1}{2}[\psi(z)+\bar{\psi}(w)]c/(1-z\bar{w})$$

is the minimal decomposition of $(1-z\bar{w})^{-1}c$ in $\mathscr{C}(z)$ with $\frac{1}{2}[\phi(z)+\bar{\phi}(w)]c/(1-z\bar{w})$ in $\mathscr{L}(\phi)$ and $\frac{1}{2}[\psi(z)+\bar{\psi}(w)]c/(1-z\bar{w})$ in $\mathscr{L}_1$, and that the identity

$$\bar{c}g(w) = \langle g(z),(1-z\bar{w})^{-1}c\rangle = \langle g(z),\tfrac{1}{2}[\psi(z)+\bar{\psi}(w)]c/(1-z\bar{w})\rangle_1$$

holds for every element $g(z)$ of $\mathscr{L}_1$. In other words, a space $\mathscr{L}(\psi)$ exists and is equal isometrically to $\mathscr{L}_1$.

Let $h(z)$ be in $\mathscr{C}(z)$ and let $h(z) = f(z) + g(z)$ be its minimal decomposition with $f(z)$ in $\mathscr{L}(\phi)$ and $g(z)$ in $\mathscr{L}(\psi)$. By the proof of Theorem 7 in the appendix of [2],

$$\|B(z)h(z)\|^2 = \|h(z)\|^2 - \|f(z)\|_{\mathscr{L}(\phi)}^2 = \|g(z)\|_{\mathscr{L}(\psi)}^2.$$

Let V be the transformation of $\mathscr{C}(z)$ into $\mathscr{L}(\psi)$ which is the adjoint of the inclusion of $\mathscr{L}(\psi)$ in $\mathscr{C}(z)$. Since the inclusion is one-to-one, the adjoint V has a dense range in $\mathscr{L}(\psi)$. By the last identity, there exists a unique isometric transformation U of $\mathscr{L}(\psi)$ into $\mathscr{C}(z)$ such that $UV: h(z) \to B(z)h(z)$ for every $h(z)$ in $\mathscr{C}(z)$. Let $R(0)$ be the transformation $g(z) \to [g(z) - g(0)]/z$ in $\mathscr{L}(\psi)$ and let $R(0)^*$ be its adjoint in $\mathscr{L}(\psi)$. The adjoint in $\mathscr{C}(z)$ of the transformation $h(z) \to [h(z) - h(0)]/z$ is the transformation $h(z) \to zh(z)$. The inclusion of $\mathscr{L}(\psi)$ in $\mathscr{C}(z)$ clearly commutes with the transformation $u(z) \to [u(z) - u(0)]/z$. Let $h(z)$ be in $\mathscr{C}(z)$ and let $V: h(z) \to g(z)$. If $R(0)^*: g(z) \to u(z)$, then $V: zh(z) \to u(z)$. In this case $U: g(z) \to B(z)h(z)$ and $U: u(z) \to zB(z)h(z)$. The same conclusion follows by continuity for every element $g(z)$ of $\mathscr{L}(\psi)$. That is, if $R(0)^*: g(z) \to u(z)$ and if $U: g(z) \to v(z)$, then $U: u(z) \to zv(z)$. Since there is no nonzero element of $\mathscr{C}(z)$ in the range of multiplication by z^n for every $n = 1,2,3,\cdots$, there is no nonzero element of $\mathscr{L}(\psi)$ in the range of $R(0)^{*n}$ for every $n = 1,2,3,\cdots$. Since $R(0)^*$ is an isometry, it follows that the union of the kernels of the transformations $R(0)^n$, $n = 1,2,3,\cdots$, is dense in $\mathscr{L}(\psi)$. Since the kernel of $R(0)^{n+1}$ is the set of polynomials of degree at most n which belong to $L(\psi)$, the polynomials which belong to $\mathscr{L}(\psi)$ are dense in $\mathscr{L}(\psi)$.

Conversely, assume that a space $\mathscr{L}(\psi)$ exists and that the polynomials which belong to $\mathscr{L}(\psi)$ are dense in $\mathscr{L}(\psi)$. If $w_1, \cdots, w_r$ are numbers in the unit disk and if $c_1, \cdots, c_r$ are corresponding vectors, then

$$\begin{aligned} &\| \Sigma \tfrac{1}{2}[\phi(z) + \bar{\phi}(w_k)]c_k/(1 - z\bar{w}_k) \|^2_{\mathscr{L}(\phi)} \\ &\quad = \Sigma \tfrac{1}{2}\bar{c}_i[\phi(w_i) + \bar{\phi}(w_k)]c_k/(1 - w_i\bar{w}_k) \\ &\quad = \Sigma \bar{c}_i(1 - w_i\bar{w}_k)^{-1}c_k - \Sigma \tfrac{1}{2}\bar{c}_i[\psi(w_i) + \bar{\psi}(w_k)]c_k/(1 - w_i\bar{w}_k) \\ &\quad = \| \Sigma(1 - z\bar{w}_k)^{-1}c_k \|^2 - \| \Sigma \tfrac{1}{2}[\psi(z) + \bar{\psi}(w_k)]c_k/(1 - z\bar{w}_k) \|^2_{\mathscr{L}(\psi)} \\ &\quad \leqq \| \Sigma(1 - z\bar{w}_k)^{-1}c_k \|^2. \end{aligned}$$

Since the span of the elements $(1 - z\bar{w})^{-1}c$ is dense in $\mathscr{C}(z)$, there exists a transformation T of $\mathscr{C}(z)$ into $\mathscr{L}(\phi)$, which is bounded by 1, such that

$$T: (1 - z\bar{w})^{-1}c \to \tfrac{1}{2}[\phi(z) + \bar{\phi}(w)]c/(1 - z\bar{w})$$

for every vector c when $|w| < 1$. The adjoint transformation T^* takes $\mathscr{L}(\phi)$ into $\mathscr{C}(z)$ and is bounded by 1. If $T: f(z) \to h(z)$, then

$$\begin{aligned} \bar{c}h(w) &= \langle h(z), (1 - z\bar{w})^{-1}c \rangle \\ &= \langle g(z), \tfrac{1}{2}[\phi(z) + \bar{\phi}(w)]c/(1 - z\bar{w}) \rangle_{\mathscr{L}(\phi)} = \bar{c}g(w). \end{aligned}$$

By the arbitrariness of c and w, $h(z) = g(z)$, and T^* is the inclusion of $\mathscr{L}(\phi)$ in $\mathscr{C}(z)$. We have shown that $\mathscr{L}(\phi)$ is contained in $\mathscr{C}(z)$ and that the inclusion does not increase norms. If $\mathscr{L}_1$ is constructed as before, the previous argument will show that $\mathscr{L}_1$ is equal isometrically to $\mathscr{L}(\psi)$.

The dimension of the space of constants which belong to $\mathscr{L}(\psi)$ is obviously no larger than the dimension of the coefficient space $\mathscr{C}$. So there exists an isometric transformation S which takes the space of constants in $\mathscr{L}(\psi)$ into $\mathscr{C}$. Extend S so as to be a partially isometric transformation of $\mathscr{L}(\psi)$ into $\mathscr{C}$ which annihilates the orthogonal complement of the constants in $\mathscr{L}(\psi)$. Then

$$U: g(z) \to \sum_{n=0}^{\infty} [SR(0)^n g] z^n$$

is a transformation of $\mathscr{L}(\psi)$ into $\mathscr{C}(z)$ which is isometric on polynomials. Since we assume that the polynomials which belong to $\mathscr{L}(\psi)$ are dense in $\mathscr{L}(\psi)$, U is an isometry of $\mathscr{L}(\psi)$ into $\mathscr{C}(z)$. If $R(0)^*: g(z) \to u(z)$ in $\mathscr{L}(\psi)$ and if $U: g(z) \to v(z)$, then $U: u(z) \to zv(z)$. Define V as before to be the adjoint of the inclusion of $\mathscr{L}(\psi)$ in $\mathscr{C}(z)$. Then V is a transformation of $\mathscr{C}(z)$ into $\mathscr{L}(\psi)$ which is bounded by 1. If $V: h(z) \to g(z)$ and if $R(0)^*: g(z) \to u(z)$, then $V: zh(z) \to u(z)$. The composed transformation UV takes $\mathscr{C}(z)$ into itself, it is bounded by 1, and it commutes with multiplication by z. By Lemma 1 in the appendix of **[2]**, UV is multiplication by $B(z)$ for some space $\mathscr{H}(B)$.

If $h(z)$ is in $\mathscr{C}(z)$ and if $h(z) = f(z) + g(z)$ is its minimal decomposition with $f(z)$ in $\mathscr{L}(\phi)$ and $g(z)$ in $\mathscr{L}(\psi)$, then

$$\| B(z)h(z) \|^2 = \| g(z) \|^2_{\mathscr{L}(\psi)} = \| h(z) \|^2 - \| f(z) \|^2_{\mathscr{L}(\phi)}.$$

When $h(z) = (1 - z\bar{w})^{-1}c$ for some vector c and number w, $|w| < 1$, $f(z) = \frac{1}{2}[\phi(z) + \bar{\phi}(w)]c/(1 - z\bar{w})$. Since an isometric transformation preserves inner products, the identity

$$\langle B(z)(1 - z\bar{\alpha})^{-1}a, B(z)(1 - z\bar{\beta})^{-1}b \rangle = \langle (1 - z\bar{\alpha})^{-1}a, (1 - z\bar{\beta})^{-1}b \rangle$$
$$- \langle \tfrac{1}{2}[\phi(z) + \bar{\phi}(\alpha)]a/(1 - z\bar{\alpha}), \tfrac{1}{2}[\phi(z) + \bar{\phi}(\beta)]b/(1 - z\bar{\beta}) \rangle_{\mathscr{L}(\phi)}$$

holds for all vectors a and b when $|\alpha| < 1$ and $|\beta| < 1$. The identity implies that

$$\tfrac{1}{2}\bar{b}[\phi(\beta) + \bar{\phi}(\alpha)]a = \bar{b}c - (1 - \beta\bar{\alpha})\langle B(z)(1 - z\bar{\alpha})^{-1}a, B(z)(1 - z\bar{\beta})^{-1}b \rangle.$$

The same formula clearly holds when $\phi(z)$ is replaced by $\theta(z)$ where $\mathscr{L}(\theta)$ is the overlapping space of $\mathscr{H}(B)$. By the arbitrariness of a and b, $\theta(\beta) + \bar{\theta}(\alpha) = \phi(\beta) + \bar{\phi}(\alpha)$. So $\phi(z) - \theta(z)$ is a constant whose real part is zero. Since the addition of such a constant does not change the corresponding space, $\mathscr{L}(\phi)$ is the overlapping space of $\mathscr{H}(B)$.

By construction the range of U is the set of all elements of $\mathscr{C}(z)$ whose coefficients lie in the range of S. Since the inclusion of $\mathscr{L}(\phi)$ in $\mathscr{C}(z)$ is one-to-one, the adjoint transformation V has a dense range in $\mathscr{L}(\psi)$. It follows that the range of multiplication by $B(z)$ in $\mathscr{C}(z)$ is a dense subspace of the range of U. The orthogonal complement of the range of multiplication by $B(z)$ in $\mathscr{C}(z)$ therefore contains $zf(z)$ whenever it contains $f(z)$.

If $h(z)$ is in $\mathscr{C}(z)$, let $h(z) = f(z) + B(z)g(z)$ be its minimal decomposition with $f(z)$ in $\mathscr{H}(B)$ and $g(z)$ in $\mathscr{C}(z)$. If $u(z)$ is in $\mathscr{C}(z)$,

$$\langle g(z), u(z)\rangle = \langle h(z), B(z)u(z)\rangle$$

by Theorem 4 in the appendix of [2]. It follows that $h(z)$ is orthogonal to the range of multiplication by $B(z)$ in $\mathscr{C}(z)$ if, and only if, $g(z) = 0$, which is the case if, and only if, $h(z)$ belongs to $\mathscr{H}(B)$ and $\| h(z) \|_B = \| h(z) \|$. If $L(z)$ is in $\mathscr{L}(\phi)$,

$$\langle f(z), B(z)L(z)\rangle_B = \langle g(z), L(z)\rangle = \langle h(z), B(z)L(z)\rangle .$$

If $h(z)$ is orthogonal to the range of multiplication by $B(z)$ in $\mathscr{C}(z)$, then $h(z) = f(z)$ is orthogonal in $\mathscr{H}(B)$ to $B(z)L(z)$ whenever $L(z)$ is in $\mathscr{L}(\phi)$. If $h(z)$ is in $\mathscr{H}(B)$ and is orthogonal in $\mathscr{H}(B)$ to $B(z)L(z)$ whenever $L(z)$ is in $\mathscr{L}(\phi)$, then $g(z)$ is in $\mathscr{L}(\phi)$,

$$\begin{aligned} &\langle g(z), g(z)\rangle + \langle B(z)g(z), B(z)g(z)\rangle_B \\ &= \langle f(z), B(z)g(z)\rangle_B + \langle B(z)g(z), B(z)g(z)\rangle_B \\ &= \langle h(z), B(z)g(z)\rangle_B = 0, \end{aligned}$$

and $g(z) = 0$.

Proof of Theorem 2. By the proof of Theorem 1, $\mathscr{L}(1-\phi)$ is contained in $\mathscr{C}(z)$ and is the set of elements $f(z)$ of $\mathscr{C}(z)$ such that

$$\| f(z) \|^2_{\mathscr{L}(1-\phi)} = \sup [\| f(z) + g(z) \|^2 - \| g(z) \|^2_{\mathscr{L}(\phi)}]$$

where the supremum is taken over all elements $g(z)$ of $\mathscr{L}(\phi)$. The elements of $\mathscr{L}(1-\theta)$ are characterized in a similar way in terms of $\mathscr{L}(\theta)$. Since we assume that $\mathscr{L}(\theta)$ is contained in $\mathscr{L}(\phi)$ and that the inclusion does not increase norms, it follows that $\mathscr{L}(1-\phi)$ is contained in $\mathscr{L}(1-\theta)$ and that the inclusion does not increase norms. If $h(z)$ is in $\mathscr{C}(z)$, let $f(z)$ be the element of $\mathscr{L}(1-\phi)$ obtained from $h(z)$ under the adjoint of the inclusion of $\mathscr{L}(1-\phi)$ in $\mathscr{C}(z)$. Let $g(z)$ be the element of $\mathscr{L}(1-\theta)$ obtained from $h(z)$ under the adjoint of the inclusion of $\mathscr{L}(1-\theta)$ in $\mathscr{C}(z)$. Then $f(z)$ is the element of $\mathscr{L}(1-\phi)$ obtained from $g(z)$ under the adjoint of the inclusion of $\mathscr{L}(1-\phi)$ in $\mathscr{L}(1-\theta)$. It follows that $\| f(z) \|_{\mathscr{L}(1-\phi)} \leqq \| g(z) \|_{\mathscr{L}(1-\theta)}$. Since $\| B(z)h(z) \| = \| f(z) \|_{\mathscr{L}(1-\phi)}$ and $\| C(z)h(z) \| = \| g(z) \|_{\mathscr{L}(1-\theta)}$ by the proof of Theorem 1, we can conclude that $\| B(z)h(z) \| \leqq \| C(z)h(z) \|$. Let T be the unique transformation of $\mathscr{C}(z)$ into itself, which is bounded by 1 and annihilates the orthogonal complement of the range of multiplication by $C(z)$ in $\mathscr{C}(z)$, such that $T: C(z)h(z) \to B(z)h(z)$ whenever $h(z)$ is in $\mathscr{C}(z)$. If $f(z)$ is in the closure of the range of multiplication by $C(z)$ in $\mathscr{C}(z)$ and if $T: f(z) \to g(z)$, then $T: zf(z) \to zg(z)$. Since the orthogonal complement of the range of multiplication by $C(z)$ is invariant under multiplication by z by hypothesis, T commutes with multiplication by z in $\mathscr{C}(z)$. By Lemma 1 in the appendix of **[2]**, T is multiplication by $A(z)$ for some space $\mathscr{H}(A)$. Since $B(z)h(z) = A(z)C(z)h(z)$ for every $h(z)$ in $\mathscr{C}(z)$ by construction, $B(z) = A(z)C(z)$.

Proof of Theorem 3. If $L(z)$ is in $\mathscr{L}(\theta)$, then $C(z)L(z)$ is in $\mathscr{H}(C)$, $B(z)L(z) = A(z)C(z)L(z)$ is in $\mathscr{H}(B)$, and

$$\begin{aligned} \| B(z)L(z) \|_B^2 &= \| L(z) \|_{\mathscr{L}(\phi)}^2 - \| L(z) \|^2 \\ &= \| L(z) \|_{\mathscr{L}(\theta)}^2 - \| L(z) \|^2 = \| C(z)L(z) \|_C^2. \end{aligned}$$

It follows that $\| A(z)f(z) \|_B = \| f(z) \|_C$ whenever $f(z)$ belongs to the closed span of those elements of $\mathscr{H}(C)$ which are of the form $B(z)L(z)$ with $L(z)$ in $\mathscr{L}(\theta)$. Since $A(z)f(z) = 0$ whenever $f(z)$ belongs to $\mathscr{H}(C)$ and is orthogonal to $C(z)L(z)$ whenever $L(z)$ is in $\mathscr{L}(\theta)$, multiplication by $A(z)$ is a partially isometric transformation of $\mathscr{H}(C)$ into $\mathscr{H}(B)$. We show that there is no nonzero element of $\mathscr{H}(A)$ of the form $A(z)f(z)$ with $f(z)$ in $\mathscr{H}(C)$. For then $f(z)$ could be chosen so that $\| A(z)f(z) \|_B = \| f(z) \|_C$ and $A(z)f(z) = 0 + A(z)f(z)$ would be a minimal decomposition of $A(z)f(z)$ as an element of $\mathscr{H}(B)$ with 0 in $\mathscr{H}(A)$ and $f(z)$ in $\mathscr{H}(C)$. Since $f(z)$ is an element of $\mathscr{H}(C)$ such that $A(z)f(z)$ belongs to $\mathscr{H}(A)$,

$$\langle f(z), f(z) \rangle_C = \langle 0, -A(z)f(z) \rangle_A = 0$$

and $f(z)$ vanishes identically. By Theorem 4 in the appendix of **[2]**, $\mathscr{H}(A)$ is contained isometrically in $\mathscr{H}(B)$. If $L(z)$ is an element of $\mathscr{L}(\theta)$ such that $\| L(z) \|_{\mathscr{L}(\theta)} > \| L(z) \|$ then $\| B(z)L(z) \|_B = \| C(z)L(z) \|_C \neq 0$ and $B(z)L(z)$ is orthogonal in $\mathscr{H}(B)$ to $\mathscr{H}(A)$. Since we assume that $\mathscr{L}(\theta)$ is not contained isometrically in $\mathscr{C}(z)$, $\mathscr{H}(A)$ does not fill $\mathscr{H}(B)$. It remains to show that $\mathscr{H}(A)$ contains a nonzero element.

Let $\mathscr{L}$ be the orthogonal complement of $\mathscr{L}(\theta)$ in $\mathscr{L}(\phi)$. If c is a vector and if $|w| < 1$, then $\frac{1}{2}[\phi(z) + \bar{\phi}(w)]c/(1 - z\bar{w})$ and $\frac{1}{2}[\theta(z) + \bar{\theta}(w)]c/(1 - z\bar{w})$ belong to $\mathscr{L}(\phi)$ and

$$\begin{aligned} \bar{c}L(w) &= \langle L(z), \tfrac{1}{2}[\phi(z) + \bar{\phi}(w)]c/(1 - z\bar{w}) \rangle_{\mathscr{L}(\phi)} \\ &= \langle L(z), \tfrac{1}{2}[\theta(z) + \bar{\theta}(w)]c/(1 - z\bar{w}) \rangle_{\mathscr{L}(\theta)} \end{aligned}$$

for every $L(z)$ in $\mathscr{L}(\theta)$. It follows that $\frac{1}{2}[\phi(z) + \bar{\phi}(w)]c/(1 - z\bar{w}) - \frac{1}{2}[\theta(z) + \bar{\theta}(w)]c/(1 - z\bar{w})$ belongs to $\mathscr{M}$. If $L(z)$ is in $\mathscr{M}$, then

$$\begin{aligned} &\langle L(z), \tfrac{1}{2}[\phi(z) + \bar{\phi}(w)]c/(1 - z\bar{w}) - \tfrac{1}{2}[\theta(z) + \bar{\theta}(w)]c/(1 - z\bar{w}) \rangle_{\mathscr{L}(\phi)} \\ &= \langle L(z), \tfrac{1}{2}[\phi(z) + \bar{\phi}(w)]c/(1 - z\bar{w}) \rangle_{\mathscr{L}(\phi)} = \bar{c}L(w). \end{aligned}$$

The space $\mathscr{M}$ is therefore a space $\mathscr{L}(\phi - \theta)$ in the metric of $\mathscr{L}(\phi)$.

In the remainder of the proof we argue by contradiction, assuming that $\mathcal{H}(A)$ contains no nonzero element. Since $[A(z) - A(0)]c/z$ belongs to $\mathcal{H}(A)$ for every vector c, $A(z) = A(0)$ is then a constant. Since $[1 - A(z)\bar{A}(0)]c$ belongs to $\mathcal{H}(A)$ for every vector c, $A(0)$ is an isometric operator. Consider any element $h(z)$ of $\mathcal{H}(C)$ such that $\| h(z) \|_C = \| A(z)h(z) \|_B$. Then there exists a unique element $g(z)$ of $\mathcal{C}(z)$ such that

$$h(z) = [h(z) - C(z)g(z)] + C(z)g(z)$$

is a minimal decomposition of $h(z)$ as an element of $\mathcal{C}(z)$ with $h(z) - C(z)g(z)$ in $\mathcal{H}(C)$ and $g(z)$ in $\mathcal{C}(z)$. Since $h(z)$ belongs to $\mathcal{H}(C)$, $g(z)$ belongs to $\mathcal{L}(\theta)$ and

$$\begin{aligned} \| h(z) \|^2 &= \| h(z) - C(z)g(z) \|_C^2 + \| g(z) \|^2 \\ &= \| h(z) \|_C^2 - 2\,\mathrm{Re}\,\langle h(z), C(z)g(z)\rangle_C + \| C(z)g(z) \|_C^2 + \| g(z) \|^2 \\ &= \| h(z) \|_C^2 - 2\| g(z) \|^2 + \| g(z) \|_{\mathcal{L}(\theta)}^2. \end{aligned}$$

If $L(z)$ is in $\mathcal{L}(\theta)$, then

$$\begin{aligned} &\langle h(z), C(z)L(z)\rangle_C \\ &\quad= \langle h(z) - C(z)g(z), C(z)L(z)\rangle_C + \langle C(z)g(z), C(z)L(z)\rangle_C \\ &\quad= \langle g(z), L(z)\rangle + \langle C(z)g(z), C(z)L(z)\rangle \\ &\quad= \langle g(z), L(z)\rangle_{\mathcal{L}(\theta)}. \end{aligned}$$

So $g(z)$ is the element of $\mathcal{L}(\theta)$ obtained from $h(z)$ under the adjoint of the transformation $L(z) \to C(z)L(z)$ of $\mathcal{L}(\theta)$ into $\mathcal{H}(C)$. For the same reason, there is a unique element $f(z)$ of $\mathcal{C}(z)$ such that

$$A(z)h(z) = [A(z)h(z) - B(z)f(z)] + B(z)f(z)$$

is a minimal decomposition of $A(z)h(z)$ as an element of $\mathcal{C}(z)$ with $A(z)h(z) - B(z)f(z)$ in $\mathcal{H}(B)$ and $f(z)$ in $\mathcal{C}(z)$. The element $f(z)$ belongs to $\mathcal{L}(\phi)$ and

$$\| A(z)h(z) \|^2 = \| A(z)h(z) \|_A^2 - 2\| f(z) \|^2 + \| f(z) \|_{\mathcal{L}(\phi)}^2.$$

Since $f(z)$ is the element of $\mathcal{L}(\phi)$ obtained from $h(z)$ under the adjoint of the transformation $L(z) \to B(z)L(z)$ of $\mathcal{L}(\phi)$ into $\mathcal{H}(B)$, $g(z)$ is the element of $\mathcal{L}(\theta)$ obtained from $f(z)$ under the adjoint of the inclusion of $\mathcal{L}(\theta)$ in $\mathcal{L}(\phi)$. Since $\mathcal{L}(\theta)$ is contained isometrically in $\mathcal{L}(\phi)$, $g(z)$ is the projection of $f(z)$ in $\mathcal{L}(\theta)$. It follows that $f(z) - g(z)$ is the projection of $f(z)$ in $\mathcal{L}(\phi - \theta)$ and that

$$\begin{aligned} \| f(z) - g(z) \|_{\mathcal{L}(\phi-\theta)}^2 &= \| f(z) \|_{\mathcal{L}(\phi)}^2 - \| g(z) \|_{\mathcal{L}(\theta)}^2 \\ &= \| A(z)h(z) \|^2 + 2\| f(z) \|^2 - \| h(z) \|^2 - 2\| g(z) \|^2. \end{aligned}$$

Construct a sequence $(h_n(z))$ of elements of $\mathcal{H}(C)$ inductively by $h_0(z) = h(z)$ and $h_{n+1}(z) = [h_n(z) - h_n(0)]/z$. Then $h_n(z)$ belongs to $\mathcal{H}(C)$ for every n and $\| h_n(z) \|_C = \| A(z)h_n(z) \|_B$. For every n there exists a unique element $g_n(z)$ of $\mathcal{C}(z)$ such that

$$h_n(z) = [h_n(z) - C(z)g_n(z)] + C(z)g_n(z)$$

is a minimal decomposition of $h_n(z)$ as an element of $\mathcal{C}(z)$ with $h_n(z) - C(z)g_n(z)$ in $\mathcal{H}(C)$ and $g_n(z)$ in $\mathcal{C}(z)$. For every n there exists a unique element $f_n(z)$ of $\mathcal{C}(z)$ such that

$$A(z)h_n(z) = [A(z)h_n(z) - B(z)f_n(z)] + B(z)f_n(z)$$

is a minimal decomposition of $A(z)h_n(z)$ as an element of $\mathcal{C}(z)$ with $A(z)h_n(z) - B(z)f_n(z)$ in $\mathcal{H}(B)$ and $f_n(z)$ in $\mathcal{C}(z)$. By construction, $f_0(z) = f(z)$ and $g_0(z) = g(z)$. By the proof of Lemma 5 of [9], $f_{n+1}(z) = [f_n(z) - f_n(0)]/z$ and $g_{n+1}(z) = [g_n(z) - g_n(0)]/z$ for every n. Since

$$\begin{aligned} &\| f_n(z) - g_n(z) \|^2_{\mathcal{L}(\phi-\theta)} \\ &\quad = \| A(z)h_n(z) \|^2 + 2\| f_n(z) \|^2 - \| h_n(z) \|^2 - 2\| g_n(z) \|^2 \end{aligned}$$

where $\lim \| h_n(z) \| = 0$, $\lim \| A(z)h_n(z) \| = 0$, $\lim \| f_n(z) \| = 0$, and $\lim \| g_n(z) \| = 0$ as $n \to \infty$ by the definition of the norm in $\mathcal{C}(z)$, we obtain

$$\lim_{n\to\infty} \| f_n(z) - g_n(z) \|_{\mathcal{L}(\phi-\theta)} = 0.$$

It follows that $f(z) - g(z)$ belongs to the closed span of the polynomials which belong to $\mathcal{L}(\phi - \theta)$. But $f(z)$ is an arbitrary element of $\mathcal{L}(\phi)$ in the range of the adjoint of the transformation $L(z) \to B(z)L(z)$ of $\mathcal{L}(\phi)$ into $\mathcal{H}(B)$. The closure of the range of the adjoint is the orthogonal complement of the set of elements $L(z)$ of $\mathcal{L}(\phi)$ such that $\| L(z) \|_{\mathcal{L}(\phi)} = \| L(z) \|$. Let $\mathcal{M}$ be the set of all elements $f(z)$ of $\mathcal{L}(\phi)$ with projection $g(z)$ in $\mathcal{L}(\theta)$ such that $f(z) - g(z)$ belongs to the closed span of the polynomials in $\mathcal{L}(\phi - \theta)$. Then $\mathcal{M}$ is a closed subspace of $\mathcal{L}(\phi)$ whose orthogonal complement is contained in the set of elements $L(z)$ of $\mathcal{L}(\phi)$ such that $\| L(z) \|_{\mathcal{L}(\phi)} = \| L(z) \|$. Since the orthogonal complement of $\mathcal{M}$ is invariant under the transformation $f(z) \to [f(z) - f(0)]/z$, it contains no nonzero element. It follows that $\mathcal{L}(\phi - \theta)$ is the closed span of the polynomials it contains, which is contrary to hypothesis. We must therefore grant that $\mathcal{H}(A)$ contains a nonzero element.

PROOF OF THEOREM 4. We show that $\mathcal{L}(\psi)$ is orthogonal to all polynomials in $\mathcal{L}(\phi)$. Every element $h(z)$ of $\mathcal{C}(z)$ has a unique minimal decomposition $h(z) = f(z) + g(z)$ with $f(z)$ in $\mathcal{L}(\phi)$ and $g(z)$ in $\mathcal{L}(1 - \phi)$ such that

$$\| h(z) \|^2 = \| f(z) \|^2_{\mathcal{L}(\phi)} + \| g(z) \|^2_{\mathcal{L}(1-\phi)}.$$

If $L(z)$ belongs to $\mathcal{L}(\phi)$ and to $\mathcal{L}(1 - \phi)$, then

$$\langle f(z), L(z)\rangle_{\mathscr{L}(\phi)} = \langle g(z), L(z)\rangle_{\mathscr{L}(1-\phi)}.$$

If $h(z)$ is a polynomial in $\mathscr{L}(\phi)$, then $\| h(z) \|_{\mathscr{L}(\phi)} = \| h(z) \|$ by hypothesis, and the minimal decomposition is obtained with $f(z) = h(z)$ and $g(z) = 0$. If $L(z)$ is in $\mathscr{L}(\psi)$, it belongs to $\mathscr{L}(\phi)$ and $\mathscr{L}(1-\phi)$ by hypothesis, and

$$\langle h(z), L(z)\rangle_{\mathscr{L}(\phi)} = \langle f(z), L(z)\rangle_{\mathscr{L}(\phi)} = \langle g(z), L(z)\rangle_{\mathscr{L}(1-\phi)} = 0.$$

It follows in particular that $\mathscr{L}(\psi)$ contains no nonzero constant and that the transformation $L(z) \to [L(z) - L(0)]/z$ in $\mathscr{L}(\psi)$ is unitary. It is well known that a unitary transformation has reducing subspaces in which the spectrum of the restricted transformation is an arbitrary closed subset of the spectrum of the given transformation. By Theorem 8 in the appendix of [**2**], any reducing subspace is a space $\mathscr{L}(\psi_1)$ in the metric of $\mathscr{L}(\psi)$. We can therefore assume with no loss of generality that the spectrum of the transformation $L(z) \to [L(z) - L(0)]/z$ in $\mathscr{L}(\psi)$ is properly contained in the spectrum of the transformation $L(z) \to [L(z) - L(0)]/z$ considered in that portion of $\mathscr{L}(\phi)$ which is orthogonal to all polynomials in $\mathscr{L}(\phi)$. In this case there exists a nonzero space $\mathscr{L}(\phi_1)$ contained isometrically in $\mathscr{L}(\phi)$ such that $\mathscr{L}(\phi_1)$ is orthogonal to all polynomials in $\mathscr{L}(\phi)$ and the spectrum of the transformation $L(z) \to [L(z) - L(0)]/z$ in $\mathscr{L}(\phi_1)$ lies at a positive distance from the spectrum of the transformation $L(z) \to [L(z) - L(0)]/z$ in $\mathscr{L}(\psi)$. Let $\rho_0(w)$ be the distance from a point w to the spectrum of the transformation $L(z) \to [L(z) - L(0)]/z$ in $\mathscr{L}(\psi)$ and let $\rho_1(w)$ be the distance from w to the spectrum of the transformation $L(z) \to [L(z) - L(0)]/z$ in $\mathscr{L}(\phi_1)$. If $f(z)$ is in $\mathscr{L}(\psi)$ and if $\rho_0(w) > 0$, then $[f(z) - f(w)]/(z - w)$ is in $\mathscr{L}(\psi)$ and

$$\| [f(z) - f(w)]/(z - w) \|_{\mathscr{L}(\psi)} \leqq \| f(z) \|_{\mathscr{L}(\psi)} / \rho_0(w).$$

If $g(z)$ is in $\mathscr{L}(\phi_1)$ and if $\rho_1(w) > 0$, then $[g(z) - g(w)]/(z - w)$ is in $\mathscr{L}(\phi_1)$ and

$$\| [g(z) - g(w)]/(z - w) \|_{\mathscr{L}(\phi_1)} \leqq \| g(z) \|_{\mathscr{L}(\phi_1)} / \rho_1(w).$$

The expression

$$\langle [f(z) - f(w)]/(z - w), g(z)\rangle_{\mathscr{L}(\phi)} = \langle f(z), [g(z) - g(\bar{w}^{-1})]/(1 - z\bar{w})\rangle_{\mathscr{L}(\phi)}$$

is an analytic function $\lambda(w)$ of w in the region $\rho_0(w) > 0$ and $\rho_1(1/\bar{w}) > 0$. From the expression on the left we see that $\lambda(w)$ has an analytic continuation in the region $\rho_0(w) > 0$. From the expression on the right we see that $\lambda(w)$ has an analytic continuation in the region $\rho_1(1/\bar{w}) > 0$. It is therefore an entire function. An obvious use of the Schwarz inequality will show that $\lim \lambda(w) = 0$ as $|w| \to \infty$. By Liouville's theorem, $\lambda(w)$ vanishes identically. It follows that $\mathscr{L}(\psi)$ is orthogonal to $\mathscr{L}(\phi_1)$ in the metric of $\mathscr{L}(\phi)$. Since the orthogonal complement of $\mathscr{L}(\psi)$ in $\mathscr{L}(\phi)$ is a reducing subspace of the transformation $L(z) \to [L(z) - L(0)]/z$ in $\mathscr{L}(\phi)$, it is a space $\mathscr{L}(\theta)$ in the metric of $\mathscr{L}(\phi)$. By construction $\mathscr{L}(\theta)$ contains all polynomials in $\mathscr{L}(\phi)$ but is not the closed span of the polynomials it contains. It follows that $\mathscr{L}(\theta)$ is not contained isometrically in $\mathscr{C}(z)$.

Since the space $\mathcal{L}(\phi - \theta)$ is the orthogonal complement of $\mathcal{L}(\theta)$ in $\mathcal{L}(\phi)$, it contains no nonzero polynomial. Since $\mathcal{L}(\phi - \theta)$ contains $\mathcal{L}(\psi)$, it contains a nonzero element. Let $\mathcal{M}$ be the closed span in $\mathcal{L}(1 - \theta)$ of the polynomials which belong to $\mathcal{L}(1 - \theta)$. Since

$$1 - \theta(z) = [1 - \phi(z)] + [\phi(z) - \theta(z)]$$

where $\mathcal{L}(1 - \phi)$ and $\mathcal{L}(\phi - \theta)$ exist, $\mathcal{L}(1 - \phi)$ is contained in $\mathcal{L}(1 - \theta)$ and the inclusion does not increase norms. Since $\mathcal{L}(1 - \phi)$ is the closed span of the polynomials it contains by Theorem 1, $\mathcal{M}$ contains $\mathcal{L}(1 - \phi)$. Since $\mathcal{L}(\psi)$ is contained in $\mathcal{L}(1 - \phi)$, $\mathcal{M}$ contains $\mathcal{L}(\psi)$. Since $\mathcal{L}(\psi)$ is dense in $\mathcal{L}(\phi - \theta)$ and the inclusion of $\mathcal{L}(\phi - \theta)$ in $\mathcal{L}(1 - \theta)$ does not increase norms, $\mathcal{M}$ contains $\mathcal{L}(\phi - \theta)$. Since every element of $\mathcal{L}(1 - \theta)$ is the sum of an element of $\mathcal{L}(1 - \phi)$ and an element of $\mathcal{L}(\phi - \theta)$, $\mathcal{M}$ contains every element of $\mathcal{L}(1 - \theta)$.

PROOF OF THEOREM 5. Let $R(0)$ be the transformation $f(z) \to [f(z) - f(0)]/z$ in $\mathcal{H}(B)$ and let $R(0)^*$ be its adjoint in $\mathcal{H}(B)$. By Theorems 12 and 13 in the appendix of [**2**], multiplication by $B^*(z)$ is isometric in $\mathcal{C}(z)$ if, and only if, $\lim \| R(0)^{*n} f \|_B = 0$ as $n \to \infty$ for every $f(z)$ in $\mathcal{H}(B)$. Let $S(0)$ be the transformation $L(z) \to [L(z) - L(0)]/z$ in $\mathcal{L}(\phi)$ and let $S(0)^*$ be its adjoint in $\mathcal{L}(\phi)$. If $f(z)$ is in $\mathcal{H}(B)$ there exists a unique element $g(z)$ of $\mathcal{L}(\phi)$ such that

$$f(z) = [f(z) - B(z)g(z)] + B(z)g(z)$$

is a minimal decomposition of $f(z)$ as an element of $\mathcal{C}(z)$ with $f(z) - B(z)g(z)$ in $\mathcal{H}(B)$ and $g(z)$ in $\mathcal{C}(z)$. By the proof of Lemma 5 of [**9**], the transformation $f(z) \to g(z)$ of $\mathcal{H}(B)$ into $\mathcal{L}(\phi)$ so defined commutes with the difference-quotient transformation. As in the proof of Theorem 3, the adjoint of the transformation $f(z) \to g(z)$ is the transformation $L(z) \to B(z)L(z)$ of $\mathcal{L}(\phi)$ into $\mathcal{H}(B)$. It follows that if $L(z)$ is in $\mathcal{L}(\phi)$ and if $S(0)^{*n}: L(z) \to L_n(z)$, then $R(0)^{*n}: B(z)L(z) \to B(z)L_n(z)$ for every $n = 1, 2, 3, \cdots$. If multiplication by $B^*(z)$ is isometric in $\mathcal{C}(z)$, then

$$\lim_{n \to \infty} \| B(z) L_n(z) \|_B = 0.$$

Since $S(0)^*$ is isometric in $\mathcal{L}(\phi)$, it follows that

$$\lim_{n \to \infty} \| L_n(z) \|_{\mathcal{L}(\phi)} = \| L(z) \|.$$

Conversely, if this last formula holds for every element $L(z)$ of $\mathcal{L}(\phi)$, then $\lim \| R(0)^{*n} f \|_B = 0$ as $n \to \infty$ for every element $f(z)$ of $\mathcal{H}(B)$ of the form $f(z) = B(z)L(z)$ with $f(z)$ in $\mathcal{L}(\phi)$. Since we assume that such elements of $\mathcal{H}(B)$ are dense in $\mathcal{H}(B)$, the same formula holds for every $f(z)$ in $\mathcal{H}(B)$, and multiplication by $B^*(z)$ is isometric in $\mathcal{C}(z)$. As in the proof of Theorem 10 in the appendix of [**2**], we define $\mathcal{E}(\phi)$ as the set of pairs $(f(z), g(z))$ with $f(z)$ in $\mathcal{L}(\phi)$ and $g(z) = \sum a_n z^n$ a power series with vector coefficients such that

$z^n f(z) + a_0 z^{n-1} + \cdots + a_{n-1}$ belongs to $\mathcal{L}(\phi)$ for every $n = 1,2,3,\cdots$, and such that the sequence of norms

$$\| z^n f(z) + a_0 z^{n-1} + \cdots + a_{n-1} \|_{\mathcal{L}(\phi)}$$

is bounded. Then $\mathcal{E}(\phi)$ is a Hilbert space in the norm

$$\| (f(z), g(z)) \|_{\mathcal{E}(\phi)} = \lim_{n \to \infty} \| z^n f(z) + a_0 z^{n-1} + \cdots + a_{n-1} \|_{\mathcal{L}(\phi)}.$$

If $f(z)$ is in $\mathcal{L}(\phi)$, there exists a unique power series $\tilde{f}(z)$ such that $(f(z), \tilde{f}(z))$ belongs to $\mathcal{E}(\phi)$ and

$$\| (f(z), \tilde{f}(z)) \|_{\mathcal{E}(\phi)} = \| f(z) \|_{\mathcal{L}(\phi)}.$$

The transformation $(f(z), g(z)) \to ([f(z) - f(0)]/z, zg(z) + f(0))$ is unitary in $\mathcal{E}(\phi)$, and $\mathcal{E}(\phi)$ is the smallest closed invariant subspace of the transformation which contains $(f(z), \tilde{f}(z))$ whenever $f(z)$ is in $\mathcal{L}(\phi)$. If $f(z)$ is in $\mathcal{E}(\phi)$ and if $\tilde{f}(z) = \sum a_n z^n$, then

$$S(0)^{*n} : f(z) \to z^n f(z) + a_0 z^{n-1} + \cdots + a_{n-1}.$$

It follows that multiplication by $B^*(z)$ is isometric in $\mathcal{C}(z)$ if, and only if,

$$\| (f(z), \tilde{f}(z)) \|^2_{\mathcal{E}(\phi)} = \| f(z) \|^2 + \| \tilde{f}(z) \|^2$$

for every element $f(z)$ of $\mathcal{L}(\phi)$. In this case the identity

$$\| (f(z), g(z)) \|^2_{\mathcal{E}(\phi)} = \| f(z) \|^2 + \| g(z) \|^2$$

holds for every element $(f(z), g(z))$ of $\mathcal{E}(\phi)$. But $\mathcal{E}(1)$ is the Hilbert space of pairs $(f(z), g(z))$ of elements of $\mathcal{C}(z)$ in the norm

$$\| (f(z), g(z)) \|^2_{\mathcal{E}(1)} = \| f(z) \|^2 + \| g(z) \|^2.$$

So multiplication by $B^*(z)$ is isometric in $\mathcal{C}(z)$ if, and only if, $\mathcal{E}(\phi)$ is contained isometrically in $\mathcal{E}(1)$. It is easily verified that $\mathcal{E}(1 - \phi)$ is the set of elements $(u(z), v(z))$ of $\mathcal{E}(1)$ such that

$$\sup [\| (u(z), v(z)) + (f(z), g(z)) \|^2_{\mathcal{E}(1)} - \| (f(z), g(z)) \|^2_{\mathcal{E}(\phi)}] < \infty$$

where the supremum is taken over all elements $(f(z), g(z))$ of $\mathcal{E}(\phi)$. The supremum is equal to $\| (u(z), v(z)) \|^2_{\mathcal{E}(1-\phi)}$. It follows from the obvious generalization of Theorem 4 in the appendix of [**2**] that $\mathcal{E}(\phi)$ is contained isometrically in $\mathcal{E}(1)$ if, and only if, the intersection of $\mathcal{E}(\phi)$ and $\mathcal{E}(1 - \phi)$ contains no nonzero element. If $\mathcal{L}(\psi)$ is a space contained in $\mathcal{L}(\phi)$ and in $\mathcal{L}(1 - \phi)$ in such a way that the inclusions do not increase norms, then $\mathcal{E}(\psi)$ is contained in $\mathcal{E}(\phi)$ and in $\mathcal{E}(1 - \phi)$ in such a way that the inclusions do not increase norms. Thus if multiplication by $B^*(z)$ is isometric in $\mathcal{C}(z)$, $\mathcal{E}(\psi)$ and hence $\mathcal{L}(\psi)$ contains no nonzero element. On the other hand, if multiplication by $B^*(z)$ is not isometric in $\mathscr{C}(z)$, consider the set $\mathscr{E}$ of elements $(f(z), g(z))$ of $\mathscr{E}(\phi)$, which also belong to $\mathscr{E}(1-\phi)$, in the norm

$$\|(f(z), g(z))\|_{\mathscr{E}}^2 = \|(f(z), g(z))\|_{\mathscr{E}(\phi)}^2 + \|(f(z), g(z))\|_{\mathscr{E}(1-\phi)}^2.$$

Then $\mathscr{E}$ contains a nonzero element, $([f(z)-f(0)]/z, zg(z)+f(0))$ belongs to $\mathscr{E}$ whenever $(f(z), g(z))$ belongs to $\mathscr{E}$, and the transformation $(f(z), g(z)) \rightarrow ([f(z)-f(0)]/z, zg(z)+f(0))$ is unitary in $\mathscr{E}$. Let $\mathscr{L}$ be the set of power series $f(z)$ such that $(f(z), g(z))$ belongs to $\mathscr{E}$ for some corresponding power series $g(z)$. Then $\mathscr{L}$ is a Hilbert space in the unique norm which makes the transformation $(f(z), g(z)) \rightarrow f(z)$ a partial isometry of $\mathscr{E}$ onto $\mathscr{L}$. The space $\mathscr{L}$ is contained in $\mathscr{L}(\phi)$ and in $\mathscr{L}(1-\phi)$ and the inclusions do not increase norms. The series $[L(z) - L(0)]/z$ belongs to $\mathscr{L}$ whenever $L(z)$ belongs to $\mathscr{L}$, the transformation $L(z) \rightarrow [L(z) - L(0)]/z$ is bounded in $\mathscr{L}$, and its adjoint is isometric. By Theorem 8 in the appendix of [**2**], $\mathscr{L}$ is equal isometrically to a space $\mathscr{L}(\psi)$. If multiplication by $B^*(z)$ is not isometric in $\mathscr{C}(z)$, then $\mathscr{E}$, and hence $\mathscr{L}(\psi)$, contains a nonzero element.

PROOF OF THEOREM 6. The required invariant subspace is given by the spectral theorem for unitary transformations if T is unitary. If T is an isometric transformation which is not unitary, then the range of T is a closed subspace, other than the zero subspace and the full space, which is invariant. If T is not an isometric transformation, then the set of elements f of the Hilbert space such that $\|T^n f\| = \|f\|$ for every $n = 1, 2, 3, \cdots$, is a closed invariant subspace for T which is not the full space. It is the required invariant subspace if it contains a nonzero element. If $\mathscr{H}$ is an inseparable Hilbert space and if f is a nonzero element of $\mathscr{H}$, then the closed span of f, Tf, $T^2 f, \cdots$, is a closed invariant subspace which contains a nonzero element and which does not contain the full space. We can therefore assume without loss of generality that $\mathscr{H}$ is a separable Hilbert space and that there exists no nonzero element f of $\mathscr{H}$ such that $\|T^n f\| = \|f\|$ for every $n = 1, 2\ 3, \cdots$. Choose the coefficient space $\mathscr{C}$ to be a separable Hilbert space of infinite dimension. By Theorems 1 and 6 in the appendix of [**2**], T is unitarily equivalent to the transformation $f(z) \rightarrow [f(z) - f(0)]/z$ in a space $\mathscr{H}(B)$ such that the identity

$$\|[f(z) - f(0)]/z\|_B^2 = \|f(z)\|_B^2 - |f(0)|^2$$

holds for every element $f(z)$ of $\mathscr{H}(B)$. In what follows we assume that T is the transformation $f(z) \rightarrow [f(z) - f(0)]/z$ in such a space $\mathscr{H}(B)$.

Let $\mathscr{L}(\phi)$ be the overlapping space of $\mathscr{H}(B)$. By Lemma 4 in the appendix of [**2**], $B(z)c = 0$ whenever c is a constant in $\mathscr{L}(\phi)$. Since $[L(z) - L(0)]/z$ belongs to $\mathscr{L}(\phi)$ whenever $L(z)$ belongs to $\mathscr{L}(\phi)$, it follows that $B(z)L(z) = 0$ whenever $L(z)$ is a polynomial in $\mathscr{L}(\phi)$. It follows that $\|L(z)\|_{\mathscr{L}(\phi)} = \|L(z)\|$ whenever

$L(z)$ is a polynomial in $\mathscr{L}(\phi)$. Since we assume that $\lim \| T^n u \|_B \neq 0$ as $n \to \infty$ for some element u of $\mathscr{H}(B)$, the space $\mathscr{H}(B)$ is not contained isometrically in $\mathscr{C}(z)$. The set of elements $f(z)$ of $\mathscr{H}(B)$ such that $\|f(z)\|_B = \|f(z)\|$ is a closed subspace of $\mathscr{H}(B)$ which contains $[f(z) - f(0)]/z$ whenever it contains $f(z)$. It is the required invariant subspace if it contains a nonzero element. In what follows we assume that there exists no nonzero element $f(z)$ of $\mathscr{H}(B)$ such that $\|f(z)\|_B = \|f(z)\|$. By the proof of Theorem 1, the elements of $\mathscr{H}(B)$ of the form $B(z)L(z)$ with $L(z)$ in $\mathscr{L}(\phi)$ are dense in $\mathscr{H}(B)$ and the range of multiplication by $B(z)$ is dense in $\mathscr{C}(z)$. Since we assume that $\lim \| R(0)^{*n} v \|_B \neq 0$ as $n \to \infty$ for some element $v(z)$ of $\mathscr{H}(B)$, multiplication by $B^*(z)$ is not isometric in $\mathscr{C}(z)$. By Theorem 5 there exists a nonzero space $\mathscr{L}(\psi)$ contained in $\mathscr{L}(\phi)$ and in $\mathscr{L}(1-\phi)$ in such a way that the inclusions do not increase norms. By Theorem 4 there exists a space $\mathscr{L}(\theta)$ contained isometrically in $\mathscr{L}(\phi)$ such that the polynomials which belong to $\mathscr{L}(1-\theta)$ are dense in $\mathscr{L}(1-\theta)$, such that $\mathscr{L}(\theta)$ is not contained isometrically in $\mathscr{C}(z)$, and such that $\mathscr{L}(\phi - \theta)$ is not the closed span of the polynomials it contains. By Theorem 1, $\mathscr{L}(\theta)$ is the overlapping space of a space $\mathscr{H}(C)$ such that the orthogonal complement in $\mathscr{H}(C)$ of the elements of the form $B(z)L(z)$, with $L(z)$ in $\mathscr{L}(\phi)$, contains $zf(z)$ whenever it contains $f(z)$. By the proof of Theorem 1, the orthogonal complement in $\mathscr{C}(z)$ of the range of multiplication by $C(z)$ contains $zf(z)$ whenever it contains $f(z)$. By Theorem 2, $B(z) = A(z)C(z)$ where $\mathscr{H}(A)$ exists and multiplication by $A(z)$ annihilates the orthogonal complement of the range of multiplication by $C(z)$ in $\mathscr{C}(z)$. By Theorem 3, $\mathscr{H}(A)$ is contained isometrically in $\mathscr{H}(B)$, it contains a nonzero element, and it does not fill $\mathscr{H}(B)$. It is the required invariant subspace.

PROOF OF THEOREM 7. We show that multiplication by $B^*(z)$ is isometric on the range of multiplication by $1 - zB(0)\bar{S}$ in $\mathscr{C}(z)$. If $f(z)$ is in $\mathscr{C}(z)$, then

$$B^*(z)[1 - zB(0)\bar{S}]f(z) = [\bar{B}(0) - z\bar{S}]f(z)$$

is in $\mathscr{C}(z)$. So all we need show is that

$$\langle [1 - zB(0)\bar{S}]f(z), [1 - zB(0)\bar{S}]g(z) \rangle = \langle [\bar{B}(0) - z\bar{S}]f(z), [\bar{B}(0) - z\bar{S}]g(z) \rangle$$

whenever $f(z)$ and $g(z)$ are in $\mathscr{C}(z)$. By linearity and continuity, it is sufficient to obtain the identity when $f(z) = az^r$ and $g(z) = bz^s$ for some vectors a and b. By symmetry it is sufficient to consider the case $r \geqq s$. Since multiplication by z is isometric in $\mathscr{C}(z)$, a common factor of z can be cancelled in inner products. The problem reduces to the case $s = 0$. Since the identity is true trivially when $r > 1$, we need only verify the case $r = 0$

$$\langle [1 - zB(0)\bar{S}]a, [1 - zB(0)\bar{S}]b \rangle = \langle [\bar{B}(0) - z\bar{S}]a, [\bar{B}(0) - z\bar{S}]b \rangle$$

and the case $r = 1$

$$\langle [1 - zB(0)\bar{S}]za, [1 - zB(0)\bar{S}]b \rangle = \langle [\bar{B}(0) - z\bar{S}]za, [\bar{B}(0) - z\bar{S}]b \rangle.$$

The first identity reduces to

$$\bar{b}a + \bar{b}S\bar{B}(0)B(0)\bar{S}a = \bar{b}B(0)\bar{B}(0)a + \bar{b}S\bar{S}a$$

and is true by hypothesis. The second identity is true trivially. So multiplication by $B^*(z)$ is isometric on the closure of the range of multiplication by $1 - zB(0)\bar{S}$ in $\mathscr{C}(z)$. But if $h(z)$ is an element of $\mathscr{C}(z)$ which is orthogonal to the range of multiplication by $1 - zB(0)\bar{S}$, then

$$0 = \langle h(z), [1 - zB(0)\bar{S}]z^n c \rangle$$

for every vector c, $n = 0,1,2,3\cdots$. If $h(z) = \sum a_n z^n$, then

$$0 = \bar{c}b_n - \bar{c}S\bar{B}(0)b_{n+1}.$$

By the arbitrariness of c, $b_n = S\bar{B}(0)b_{n+1}$. It follows that

$$[S\bar{B}(0) - z]h(z) = S\bar{B}(0)h(0).$$

Since we assume that such an element of $\mathscr{C}(z)$ must vanish identically, multiplication by $1 - zB(0)\bar{S}$ has a dense range in $\mathscr{C}(z)$, and multiplication by $B^*(z)$ is isometric in $\mathscr{C}(z)$. By Lemma 2 in the appendix of [2], $B^*(z)$ converges to a function which is bounded by 1 in the unit disk. It follows that $B(z)$ converges to a function which is bounded by 1 in the unit disk and that a space $\mathscr{H}(B)$ exists.

Since multiplication by $B^*(z)$ is isometric in $\mathscr{C}(z)$, the space $\mathscr{H}(B^*)$ is contained isometrically in $\mathscr{C}(z)$ and coincides with the orthogonal complement of the range of multiplication by $B^*(z)$ in $\mathscr{C}(z)$. Since the range of multiplication by $1 - zB(0)\bar{S}$ is dense in $\mathscr{C}(z)$, the range of multiplication by $B^*(z)$ is the closure of the range of multiplication by $\bar{B}(0) - z\bar{S}$ in $\mathscr{C}(z)$. It follows that an element $f(z)$ of $\mathscr{C}(z)$ is in $\mathscr{H}(B^*)$ if, and only if,

$$0 = \langle f(z), [\bar{B}(0) - z\bar{S}]z^n c \rangle$$

for every vector c, $n = 0,1,2,\cdots$. If $f(z) = \sum a_n z^n$, this condition reads

$$0 = \bar{c}B(0)a_n - \bar{c}Sa_{n+1}.$$

By the arbitrariness of c, $f(z)$ is in $\mathscr{H}(B^*)$ if, and only if,

$$B(0)a_n = Sa_{n+1}$$

for every n, or in other words if, and only if,

$$[S - zB(0)]f(z) = Sf(0).$$

By Theorem 5 in the appendix of [2], there exists a transformation $f(z) \to g(z)$ of $\mathscr{H}(B)$ into $\mathscr{H}(B^*)$, which is bounded by 1, such that

$$\bar{c}g(w) = \langle f(z), [B(z) - B(\bar{w})]c/(z - \bar{w}) \rangle_B$$

for every vector c when $|w|<1$. Since multiplication by $B^*(z)$ is isometric in $\mathscr{C}(z)$, there is no nonzero vector c such that $B^*(z)c$ belongs to $\mathscr{H}(B^*)$. By Theorem 13 in the appendix of [2], the transformation of $\mathscr{H}(B)$ into $\mathscr{H}(B^*)$ is an isometry. By Theorem 5 in the appendix of [2], the adjoint transformation of $\mathscr{H}(B^*)$ into $\mathscr{H}(B)$ is of the form $f(z)\to\tilde{f}(z)$ where

$$\bar{c}\tilde{f}(w) = \langle f(z),[B^*(z)-\bar{B}(w)]c/(z-\bar{w})\rangle$$

for every vector c when $|w|<1$. The adjoint of the isometry is a partial isometry which takes $\mathscr{L}(B^*)$ onto $\mathscr{H}(B)$. The identity

$$[B^*(z)-\bar{B}(w)]/(z-\bar{w}) = -[1-B^*(z)B(0)]\bar{S}[1-\bar{w}B(0)\bar{S}]^{-1}$$

is obtained by a straightforward calculation when $|w|<1$. It implies that

$$\bar{c}\,\tilde{f}(w) = -\bar{c}[1-wS\bar{B}(0)]^{-1}Sf(0)$$

for every vector c when $|w|<1$. By the arbitrariness of c and w,

$$\tilde{f}(z) = -[1-zS\bar{B}(0)]^{-1}Sf(0).$$

A necessary and sufficient condition that an element $f(z)$ of $\mathscr{C}(z)$ belong to the overlapping space $\mathscr{L}(\phi)$ of $\mathscr{H}(B)$ is that

$$B(z)f(z) = [1-zS\bar{B}(0)]^{-1}Sg(0)$$

for some element $g(z)$ of $\mathscr{H}(B^*)$, or equivalently that

$$[B(0)-zS]f(z) = B(0)f(0) = Sg(0).$$

In this case $g(z)$ can be chosen orthogonal to all elements $h(z)$ of $\mathscr{H}(B^*)$ such that $Sh(0)=0$ and

$$\begin{aligned}\|f(z)\|^2_{\mathscr{L}(\phi)} &= \|f(z)\|^2+\|[1-zS\bar{B}(0)]^{-1}Sg(0)\|^2_B\\ &= \|f(z)\|^2+\|g(z)\|^2.\end{aligned}$$

Proof of Theorem 8. Since $\mathscr{M}$ is an invariant subspace for the transformation $R(0):f(z)\to[f(z)-f(0)]/z$ in $\mathscr{H}(B)$, its orthogonal complement is an invariant subspace for the adjoint $R(0)^*$ of $R(0)$. By Theorem 5 in the appendix of [2], there exists a transformation $f(z)\to\tilde{f}(z)$ of $\mathscr{H}(B)$ into $\mathscr{H}(B^*)$ such that

$$\bar{c}\,\tilde{f}(w) = \langle f(z),[B(z)-B(\bar{w})]c/(z-\bar{w})\rangle_B$$

for every vector c when $|w|<1$. If $R(0)^*:f(z)\to g(z)$, then $\tilde{g}(z)=[\tilde{f}(z)-\tilde{f}(0)]/z$. Since multiplication by $B^*(z)$ is isometric in $\mathscr{C}(z)$, there is no nonzero vector c such that $B^*(z)c$ belongs to $\mathscr{H}(B^*)$. By Theorem 13 in the appendix of [2], the transformation $f(z)\to\tilde{f}(z)$ is an isometry of $\mathscr{H}(B)$ into $\mathscr{H}(B^*)$. By the proof of Theorem 14 in the appendix of [2], the orthogonal complement of the range of the transformation is the set of elements $L(z)$ of $\mathscr{H}(B^*)$ such that $z^nL(z)$ belongs to $\mathscr{H}(B^*)$ for every $n = 1,2,3,\cdots$. The transformation maps the orthogonal complement of $\mathscr{M}$ into a closed subspace of $\mathscr{C}(z)$ which contains $[f(z) - f(0)]/z$ whenever it contains $f(z)$. This subspace is therefore a space $\mathscr{H}(C^*)$ for some power series $C^*(z)$ with operator coefficients such that multiplication by $C^*(z)$ is a partially isometric transformation in $\mathscr{C}(z)$. By construction, $\mathscr{H}(C^*)$ is contained isometrically in $\mathscr{H}(B^*)$. By Theorem 3 in the appendix of [**2**], $B^*(z) = C^*(z)A^*(z)$ for some space $\mathscr{H}(A^*)$. Since multiplication by $B^*(z)$ is isometric in $\mathscr{C}(z)$ and since multiplication by $C^*(z)$ does not increase norms, multiplication by $A^*(z)$ is isometric in $\mathscr{C}(z)$. Multiplication by $C^*(z)$ is isometric on the range of multiplication by $A^*(z)$. Since multiplication by $C^*(z)$ is a partially isometric transformation in $\mathscr{C}(z)$, the kernel of the transformation is the orthogonal complement of the set on which the transformation is isometric. Since the set on which multiplication by $C^*(z)$ is isometric is invariant under multiplication by z, the kernel of multiplication by $C^*(z)$ is invariant under the transformation $f(z) \to [f(z) - f(0)]/z$. It follows that multiplication by $C^*(z)$ annihilates every coefficient of $f(z)$ if multiplication by $C^*(z)$ annihilates $f(z)$. The kernel of multiplication by $C^*(z)$ is contained in $\mathscr{H}(A^*)$. Let $\mathscr{H}_0$ be the set of elements of $\mathscr{H}(A^*)$ which are orthogonal to the kernel of multiplication by $C^*(z)$. Then $\mathscr{H}_0$ is a closed subspace of $\mathscr{C}(z)$ which contains $[f(z) - f(0)]/z$ whenever it contains $f(z)$. The coefficients of elements of $\mathscr{H}_0$ are orthogonal to all vectors in the kernel of multiplication by $C^*(z)$. The space $\mathscr{H}_0$ is therefore equal isometrically to a space $\mathscr{H}(A_0{}^*)$ where multiplication by $A_0{}^*(z)$ is a partially isometric transformation in $\mathscr{C}(z)$ and where multiplication by $A_0(z)$ and by $A_0{}^*(z)$ annihilate the kernel of multiplication by $C^*(z)$. It follows from Theorem 4 in the appendix of [**2**] that the space $\mathscr{H}(C^*A_0{}^*)$ is equal isometrically to the space $\mathscr{H}(C^*A^*)$. It follows that $C^*(z)A^*(z) = C^*(z)A_0{}^*(z)S_0$ where S_0 is an isometric operator whose range is orthogonal to all vectors c in the kernel of multiplication by $C^*(z)A_0{}^*(z)$. The range of S_0 is then orthogonal to all vectors c in the kernel of multiplication by $C^*(z)$. The dimension of the set of vectors c which are orthogonal to the kernel of multiplication by $C^*(z)$ is therefore equal to the dimension of $\mathscr{C}$. So there exists an isometric operator S whose range is the set of vectors c such that $\| C^*(z)c \| = \| c \|$. Multiplication by $C^*(z)S$ is isometric in $\mathscr{C}(z)$ and $\mathscr{H}(C^*S)$ is equal isometrically to $\mathscr{H}(C^*)$. Since we can replace $C^*(z)$ by $C^*(z)S$ without changing the previous constructions, we can assume $C^*(z)$ chosen so that multiplication by $C^*(z)$ is isometric in $\mathscr{C}(z)$. By construction $\mathscr{H}(C^*)$ is orthogonal to all elements $f(z)$ of $\mathscr{H}(B^*)$ such that $z^n f(z)$ belongs to $\mathscr{H}(B^*)$ for every $n = 1,2,3,\cdots$. By Theorem 4 in the appendix of [**2**], any such element $f(z)$ is of the form $f(z) = C^*(z)g(z)$ where $g(z)$ is an element of $\mathscr{H}(A^*)$ such that $z^n g(z)$ belongs to $\mathscr{H}(A^*)$ for every $n = 1,2,3,\cdots$. It follows from Theorems 5 and 13 in the appendix of [**2**] that there is no nonzero vector c such that $C(z)c$ belongs to $\mathscr{H}(C)$.

We show that there is no nonzero element $h(z)$ of $\mathscr{H}(C)$ such that $A(z)h(z)$

belongs to $\mathcal{H}(A)$. For then there exists an element $f(z)$ of $\mathcal{H}(C^*)$ such that

$$\bar{c}h(w) = \langle f(z), [C^*(z) - \bar{C}(w)]c/(z - \bar{w})\rangle$$

for every vector c when $|w| < 1$. There exists an element $g(z)$ of $\mathcal{H}(A^*)$ such that

$$\bar{c}A(w)h(w) = \langle g(z), [A^*(z) - \bar{A}(w)]c/(z - \bar{w})\rangle$$

for every vector c when $|w| < 1$, and $g(z)$ can be chosen orthogonal to every element $L(z)$ of $\mathcal{H}(A^*)$ such that $z^n L(z)$ belongs to $\mathcal{H}(A^*)$ for every $n = 1, 2, 3, \cdots$. Then $f(z) - C^*(z)g(z)$ is an element of $\mathcal{H}(B^*)$ which is orthogonal to every element $L(z)$ of $\mathcal{H}(B^*)$ such that $z^n L(z)$ belongs to $\mathcal{H}(B^*)$ for every $n = 1, 2, 3, \cdots$. Since

$$\begin{aligned} &\langle f(z) - C^*(z)g(z), [B^*(z) - \bar{B}(w)]c/(z - \bar{w})\rangle \\ &\quad = \langle f(z), [C^*(z) - \bar{C}(w)]\bar{A}(w)c/(z - \bar{w})\rangle - \langle g(z), [A^*(z) - \bar{A}(w)]c/(z - \bar{w})\rangle \\ &\quad = \bar{c}A(w)h(w) - \bar{c}A(w)h(w) = 0 \end{aligned}$$

for every vector c when $|w| < 1$, $f(z) - C^*(z)g(z)$ vanishes identically. This implies that $f(z)$, $g(z)$, and $h(z)$ vanish identically. By Theorem 4 in the appendix of [2], $\mathcal{H}(A)$ is contained isometrically in $\mathcal{H}(B)$, and multiplication by $A(z)$ is an isometric transformation of $\mathcal{H}(C)$ onto the orthogonal complement of $\mathcal{H}(A)$ in $\mathcal{H}(B)$.

If $L(z)$ belongs to the overlapping space $\mathcal{L}(\theta)$ of $\mathcal{H}(C)$, then $C(z)L(z)$ belongs to $\mathcal{H}(C)$ and

$$\| L(z) \|^2_{\mathcal{L}(\theta)} = \| L(z) \|^2 + \| C(z)L(z) \|^2_C.$$

Since $B(z)L(z) = A(z)C(z)L(z)$ belongs to $\mathcal{H}(B)$ and since

$$\| B(z)L(z) \|_B = \| C(z)L(z) \|_C,$$

we find that $L(z)$ is an element of $\mathcal{C}(z)$ such that $B(z)L(z)$ belongs to $\mathcal{H}(B)$ and

$$\| L(z) \|^2_{\mathcal{L}(\phi)} = \| L(z) \|^2 + \| B(z)L(z) \|^2_B = \| L(z) \|^2_{\mathcal{L}(\theta)}.$$

So $\mathcal{L}(\theta)$ is contained isometrically in $\mathcal{L}(\phi)$.

PROOF OF THEOREM 9. By Theorem 8, $[f(z) - f(0)]/z = Pf(z)$ for every $f(z)$ in $\mathcal{H}(B)$ where $P = S\bar{B}(0)$. If

$$K(w, z) = [1 - A(z)\bar{A}(w)]/(1 - z\bar{w}),$$

then $K(w, z)c$ belongs to $\mathcal{H}(A)$ for every vector c when $|w| < 1$ and

$$\bar{c}f(w) = \langle f(z), K(w, z)c\rangle_A$$

for every $f(z)$ in $\mathcal{H}(A)$. Since $\mathcal{H}(A)$ is contained in $\mathcal{H}(B)$,

$$[K(w,z) - K(w,0)]/z = \bar{w}K(w,z) - [A(z) - A(0)]\bar{A}(w)/z = PK(w,z),$$

from which we obtain the identity

$$(1 - zP)A(z)\bar{A}(w) = z\bar{w} - zP + (1 - z\bar{w})A(0)\bar{A}(w).$$

Since $[A(z) - A(0)]c/z$ belongs to $\mathscr{H}(A)$ for every vector c,

$$\{[A(z) - A(0)]/z - A'(0)\}/z = P[A(z) - A(0)]/z.$$

It follows that $(1 - zP)A(z) = A(0) - zT$ where $T = PA(0) - A'(0)$. Substitution now gives the identity $z - \bar{P} = A(z)[z\bar{A}(0) - \bar{T}]$ where $P = T\bar{A}(0)$. The identity

$$[1 - zT\bar{A}(0)][z - A(0)\bar{T}] = [A(0) - zT][z\bar{A}(0) - \bar{T}]$$

implies that

$$1 + S\bar{B}(0)B(0)\bar{S} = 1 + T\bar{A}(0)A(0)\bar{T} = A(0)\bar{A}(0) + T\bar{T}.$$

Since $B(z) = A(z)C(z)$,

$$[A(0) - zS\bar{C}(0)]C(z) = B(0) - zS.$$

PROOF OF THEOREM 10. By hypothesis the dimension of the range of $1 - U^*V = U^*(U - V)$ does not exceed the dimension of the coefficient space $\mathscr{C}$. It follows that there exists a bounded transformation J of $\mathscr{H}$ into $\mathscr{C}$, with adjoint J^* taking $\mathscr{C}$ into $\mathscr{H}$, such that

$$1 - U^*V = 2J^*\psi(0)J$$

for some invertible operator $\psi(0)$ and such that the ranges of $\psi(0) + \bar{\psi}(0)$ and $\bar{\psi}(0)J$ are orthogonal to the kernel of J^*. Define

$$Sh = \sum JU^nhz^n \quad \text{and} \quad Th = \sum \psi(0)JV^nhz^n$$

for h in $\mathscr{H}$. The kernel of S is the set of elements h of $\mathscr{H}$ such that $UU^nh = VU^nh$ for every $n = 0, 1, 2, \cdots$. This is the set of elements f of $\mathscr{H}$ such that $U^nf = V^nf$ for every $n = 1, 2, 3, \cdots$. If

$$Sh = \sum JU^nhz^n = f(z),$$

then

$$SUh = \sum JU^{n+1}hz^n = [f(z) - f(0)]/z.$$

This completes the proof of (1), and (2) has a similar proof. The range of S is a Hilbert space $\mathscr{L}$ in the unique norm which makes S a partial isometry. The transformation $f(z) \to f(0)$ of $\mathscr{L}$ into $\mathscr{C}$ is continuous by construction. Since the kernel of S is an invariant subspace for U, the set on which S is isometric is an invariant subspace for U^*. It follows that the transformation $f(z) \to [f(z) - f(0)]/z$ has an isometric adjoint in $\mathscr{L}$. By Theorem 8 in the appendix of **[2]**, $\mathscr{L}$ is equal isometrically to a space $\mathcal{L}(\phi)$. For every vector c, $\frac{1}{2}[\phi(z)+\bar{\phi}(0)]c$ belongs to $\mathcal{L}$ and

$$\bar{c}Sh = \bar{c}f(0) = \langle f(z), \tfrac{1}{2}[\phi(z)+\bar{\phi}(0)]c\rangle_{\mathcal{L}(\phi)}$$

for every element $f(z) = Sh$ of $\mathcal{L}(\phi)$. It follows that

$$SJ^*c = \tfrac{1}{2}[\phi(z)+\bar{\phi}(0)]c \quad \text{and} \quad JJ^*c = \tfrac{1}{2}[\phi(0)+\bar{\phi}(0)]c$$

for every vector c. But $1-2J^*\psi(0)J = U^*V$ is a unitary transformation in $\mathcal{H}$. Since

$$[1-2J^*\bar{\psi}(0)J][1-2J^*\psi(0)J] = 1,$$
$$J^*[\psi(0)+\bar{\psi}(0)]J = 2J^*\bar{\psi}(0)JJ^*\psi(0)J.$$

Since the ranges of $\psi(0)+\bar{\psi}(0)$ and $\bar{\psi}(0)J$ are orthogonal to the kernel of J^*,

$$\psi(0)+\bar{\psi}(0) = 2\bar{\psi}(0)JJ^*\psi(0) = \bar{\psi}(0)[\phi(0)+\bar{\phi}(0)]\psi(0).$$

Since $\psi(0)$ is an invertible operator,

$$\psi(0)^{-1}+\bar{\psi}(0)^{-1} = \phi(0)+\bar{\phi}(0).$$

Since the addition to $\phi(z)$ of an operator with vanishing real part does not alter the corresponding space, we can choose $\phi(z)$ so that $\phi(0)\psi(0) = 1 = \psi(0)\phi(0)$. Since $\phi(0)$ is an invertible operator, there exists a power series $\psi(z)$ with operator coefficients such that $\phi(z)\psi(z)=1=\psi(z)\phi(z)$. If h is in $\mathcal{H}$ and if $Sh=f(z)$, then

$$S(1-U^*V)h = 2SJ^*\psi(0)Jh = [\phi(z)+\bar{\phi}(0)]\psi(0)f(0),$$
$$S(U-V)h = SU(1-U^*V)h = [\phi(z)-\phi(0)]\psi(0)f(0)/z,$$
$$SVh = SUh - S(U-V)h = [f(z)-\phi(z)\psi(0)f(0)]/z.$$

It follows that $Th = \sum\phi(0)JV^nh = \psi(z)f(z)$. By an argument earlier in the proof, the range of T is a space $\mathcal{L}(\theta)$ in the unique norm which makes T a partial isometry. So multiplication by $\psi(z)$ is an isometry of $\mathcal{L}(\phi)$ onto $\mathcal{L}(\theta)$. It follows that

$$\tfrac{1}{2}[\psi(z)+\bar{\psi}(0)]c = \tfrac{1}{2}\psi(z)[\phi(z)+\bar{\phi}(0)]\bar{\psi}(0)c$$

belongs to $\mathcal{L}(\theta)$ for every vector c and that

$$\langle g(z), \tfrac{1}{2}[\psi(z)+\bar{\psi}(0)]c\rangle_{\mathcal{L}(\theta)} = \langle f(z), \tfrac{1}{2}[\phi(z)+\bar{\phi}(0)]\bar{\psi}(0)c\rangle_{\mathcal{L}(\phi)}$$
$$= \bar{c}\psi(0)f(0) = \bar{c}g(0) = \langle g(z), \tfrac{1}{2}[\theta(z)+\bar{\theta}(0)]c\rangle_{\mathcal{L}(\theta)}$$

for every element $g(z) = \psi(z)f(z)$ of $\mathcal{L}(\theta)$. By the arbitrariness of c and $g(z)$, $\psi(z)+\bar{\psi}(0) = \theta(z)+\bar{\theta}(0)$. Since $\psi(z)-\theta(z) = \bar{\theta}(0)-\bar{\psi}(0)$ is an operator whose real part is zero, we can choose $\theta(z)$ so that $\theta(z) = \psi(z)$.

PROOF OF THEOREM 11. We show that the transformation $f(z)\to\phi(z)f(z)$ is an isometry of $\mathcal{L}(\psi)$ onto $\mathcal{L}(\phi)$. If $f(z)$ is an element of $\mathcal{L}(\psi)$ of the form

$f(z) = \frac{1}{2}[\psi(z) + \bar{\psi}(w)]\bar{\phi}(w)c/(1 - z\bar{w})$ for some vector c and number w, $|w| < 1$, then

$$\phi(z)f(z) = \tfrac{1}{2}[\phi(z) + \bar{\phi}(w)]c/(1 - z\bar{w})$$

belongs to $\mathscr{L}(\phi)$. For all vectors a and b and numbers α and β, $|\alpha| < 1$ and $|\beta| < 1$,

$$\begin{aligned}&\langle \tfrac{1}{2}[\psi(z) + \bar{\psi}(\beta)]\bar{\phi}(\beta)b/(1 - z\bar{\beta}), \tfrac{1}{2}[\psi(z) + \bar{\psi}(\alpha)]\bar{\phi}(\alpha)a/(1 - z\bar{\alpha})\rangle_{\mathscr{L}(\psi)} \\ &\quad = \tfrac{1}{2}\bar{a}\phi(\alpha)[\psi(\alpha) + \bar{\psi}(\beta)]\bar{\phi}(\beta)b/(1 - \alpha\bar{\beta}) \\ &\quad = \langle \tfrac{1}{2}[\phi(z) + \bar{\phi}(\beta)]b/(1 - z\bar{\beta}), \tfrac{1}{2}[\phi(z) + \bar{\phi}(\alpha)]a/(1 - z\bar{\alpha})\rangle_{\mathscr{L}(\phi)}.\end{aligned}$$

It follows that the transformation $f(z) \to \phi(z)f(z)$ is isometric on finite sums of elements of $\mathscr{L}(\psi)$ of the form $\frac{1}{2}[\psi(z) + \bar{\psi}(w)]\bar{\phi}(w)c/(1 - z\bar{w})$. Since such sums are dense in $\mathscr{L}(\psi)$, the transformation $f(z) \to \phi(z)f(z)$ is an isometry of $\mathscr{L}(\psi)$ into $\mathscr{L}(\phi)$. Since the range of the transformation is a closed subspace of $\mathscr{L}(\phi)$ which contains $\frac{1}{2}[\phi(z) + \bar{\phi}(w)]c/(1 - z\bar{w})$ for every vector c when $|w| < 1$, it is the full space.

By the proof of Theorem 10 in the appendix of [**2**], the pairs

$$\begin{aligned}&(\tfrac{1}{2}[\phi(z) + \bar{\phi}(w)]c/(1 - z\bar{w}),\ \tfrac{1}{2}[\phi^*(z) - \bar{\phi}(w)]c/(z - \bar{w})),\\ &(\tfrac{1}{2}[\phi(z) - \phi(\bar{w})]c/(z - \bar{w}),\ \tfrac{1}{2}[\phi^*(z) + \phi(\bar{w})]c/(1 - z\bar{w}))\end{aligned}$$

belong to the extension space $\mathscr{E}(\phi)$ of $\mathscr{L}(\phi)$ for every vector c when $|w| < 1$ and the identities

$$\bar{c}f(w) = \langle (f(z), g(z)), (\tfrac{1}{2}[\phi(z) + \bar{\phi}(w)]c/(1 - z\bar{w}), \tfrac{1}{2}[\phi^*(z) - \bar{\phi}(w)]c/(z - \bar{w}))\rangle_{\mathscr{E}(\phi)}$$

$$\bar{c}g(w) = \langle (f(z), g(z)), (\tfrac{1}{2}[\phi(z) - \phi(\bar{w})]c/(z - \bar{w}), \tfrac{1}{2}[\phi^*(z) + \phi(\bar{w})]c/(1 - z\bar{w}))\rangle_{\mathscr{E}(\phi)}$$

hold for every element $(f(z), g(z))$ of $\mathscr{E}(\phi)$. The analogous identities are valid in the extension space $\mathscr{E}(\psi)$ of $\mathscr{L}(\psi)$. An argument similar to the one just given for $\mathscr{L}(\phi)$ and $\mathscr{L}(\psi)$ will show that the transformation

$$(f(z), g(z)) \to (\phi(z)f(z), -\phi^*(z)g(z))$$

is an isometry of $\mathscr{E}(\psi)$ onto $\mathscr{E}(\phi)$. The transformation

$$(f(z), g(z)) \to ([f(z) - f(0)]/z, zg(z) + f(0))$$

is unitary in $\mathscr{E}(\phi)$ and in $\mathscr{E}(\psi)$. The transformation $(f(z), g(z)) \to f(z)$ is a partial isometry of $\mathscr{E}(\phi)$ into $\mathscr{L}(\phi)$ and of $\mathscr{E}(\psi)$ into $\mathscr{L}(\psi)$. Let $\mathscr{H}$ be a Hilbert space which admits an isometry A of $\mathscr{H}$ onto $\mathscr{E}(\phi)$. Let B be the induced isometry of $\mathscr{H}$ onto $\mathscr{E}(\psi)$ defined by $Bh = (\psi(z)f(z), -\psi^*(z)g(z))$ if $Ah = (f(z), g(z))$. Let U be the unitary transformation in $\mathscr{H}$ which corresponds to the transformation $(f(z), g(z)) \to ([f(z) - f(0)]/z, zg(z) + f(0))$ in $\mathscr{E}(\phi)$. Let V be the unitary transformation in $\mathscr{H}$ which corresponds to the transformation defined by the same formula in $\mathscr{E}(\psi)$. Define $Sh = f(z)$ whenever $Ah = (f(z), g(z))$ and $Th = f(z)$ whenever $Bh = (f(z), g(z))$. It is easily seen that the three conditions of Theorem 10 are satisfied.

If h is in $\mathscr{H}$ and if $Th = g(z) = \sum a_n z^n$, then $Bh = (g(z), h(z))$ for some power series $h(z)$ and

$$BV^r h = \left(\sum_{n=0}^{\infty} a_{n+r} z^n, \; z^r h(z) + \sum_{n=0}^{r-1} a_{r-1-n} z^n \right).$$

It follows that

$$AV^r h = \left(\phi(z) \sum_{n=0}^{\infty} a_{n+r} z^n, \; -\phi^*(z) z^r h(z) - \phi^*(z) \sum_{n=0}^{r-1} a_{r-1-n} z^n \right).$$

From this we find that

$$S\bar{U}^r V^r h = \phi(z) \sum_{n=0}^{\infty} a_{n+r} z^{n+r} - \sum_{n=0}^{\infty} \sum_{k=0}^{r-1-n} \bar{\phi}_k a_{k+n} z^n.$$

The theorem follows.

References

1. L. de Branges, *Hilbert spaces of entire functions*, Prentice-Hall, Englewood Cliffs, New Jersey, 1968.

2. L. de Branges and J. Rovnyak, "Canonical models in quantum scattering theory" in *Perturbation theory and its applications in quantum mechanics*, C. Wilcox editor, Wiley, New York, 1966.

3. L. de Branges, *Perturbations of self-adjoint transformations*, Amer. J. Math. **84** (1962), 543–560.

4. M. S. Brodskii, *On the triangular representation of some operators with completely continuous imaginary part*, Dokl. Akad. Nauk SSSR **137** (1960), 1271–1274 = Soviet Math. Dokl. **1** (1961), 952–955.

5. ———, *On the multiplicative representation of some analytic operator functions*, Dokl. Akad. Nauk SSSR **138** (1961), 751–754 = Soviet Math. Dokl. **2** (1961), 695–698.

6. L. de Branges, *Some Hilbert spaces of analytic functions*. I, Trans. Amer. Math. Soc. **106** (1963), 445–468.

7. ———, *Some Hilbert spaces of analytic functions*. II, J. Math. Anal. Appl. **11** (1965), 44–72.

8. ———, *Some Hilbert spaces of analytic functions*. III, J. Math. Anal. Appl. **12** (1965) 149–186.

9. L. de Branges and J. Rovnyak, *Square summable power series*, Holt, Rinehart and Winston, New York, 1966.

10. D. B. Lowdenslager, *On factoring matrix valued functions*, Annals of Math. **78** (1963), 450–454.

11. B. Sz.-Nagy and C. Foiaş, *Sur les contractions de l'espace de Hilbert*. IX. *Factorisations de la fonction caractéristique. Sous-espaces invariants*, Acta Sci. Math. Szeged **25** (1964), 283–316.

PURDUE UNIVERSITY
LAFAYETTE, INDIANA

A Lindelöf Theorem and Analytic Continuation for Functions of Several Variables, with an Application to the Feynman Integral[1]

R. H. CAMERON
D. A. STORVICK

1. **The n-dimensional Lindelöf Theorem.** The first theorem that we shall prove is an n-dimensional analogue of a classical theorem of Lindelöf [**1**, p. 13] on asymptotic values of analytic functions. Its proof, like Lindelöf's original proof, is based on the maximum modulus principle.

THEOREM 1. *Let $w = f(z_1, \cdots, z_m)$ be analytic and bounded in the generalized polycylinder $T = \Delta_1 \times \Delta_2 \times \cdots \times \Delta_m$, $\Delta_j = \{z_j \,|\, |\arg z_j| < \Theta_j, 0 < |z_j| < R_j \leqq \infty\}$. Let $f(z_1, \cdots, z_m)$ tend to a unique limit α as $(z_1, \cdots, z_m) \to (0, \cdots, 0)$ along $\mathscr{L}_1 \times \mathscr{L}_2 \times \cdots \times \mathscr{L}_m$, where $\mathscr{L}_j$ is a continuous path $z_j = z_j(t_j)$ lying in Δ_j and terminating at $z_j = 0$. Then $f(z_1, \cdots, z_m)$ tends uniformly to α as $(z_1, \cdots, z_m) \to (0, \cdots, 0)$ inside any angular subdomain $V_1 \times \cdots \times V_m$, $V_j = \{z_j \,|\, |\arg z_j| < \Theta_j - \delta_j, 0 < \delta_j < \Theta_j, 0 < |z_j| < R_j\}$.*

PROOF. It is clear that elementary conformal mappings enable us to restrict ourselves to the case that $\Theta_j = \pi/3$, $j = 1, \cdots, m$. We prove first that $f(z_1, \cdots, z_m)$ tends to α as $(z_1, \cdots, z_m) \to (0, \cdots, 0)$ along the "angle bisector" $\arg z_j = 0$, $j = 1, \cdots, m$. To this end, since $f(z_1, \cdots, z_m) \to \alpha$ along $\mathscr{L}_1 \times \cdots \times \mathscr{L}_m$ it is possible to find a number R_ε such that $|f(z_1, \cdots, z_m) - \alpha| < \varepsilon$ on that part of $\mathscr{L}_1 \times \cdots \times \mathscr{L}_m$ contained in $D'_1 \times \cdots \times D'_m$, $D'_j = \{z_j \,|\, |z_j| < 2R_\varepsilon < R_0\}$. Let $(z_1^0, \cdots, z_m^0)$ be an arbitrary point of $\arg z_1 = \arg z_2 = \cdots = \arg z_m = 0$ such that $|z_j^0| < R_\varepsilon$ for $j = 1, \cdots, m$. If $(z_1^0, \cdots, z_m^0)$ lies on $\mathscr{L}_1 \times \cdots \times \mathscr{L}_m$, then $|f(z_1^0, \cdots, z_m^0) - \alpha| < \varepsilon$; otherwise we form regular hexagons $\{H_j\}$ with sides of length $|z_j^0|$, one on each of the rays $\arg z_j = \pi/3$ and $\arg z_j = -\pi/3$ bounding the domain Δ_j. We form the product $H_1 \times \cdots \times H_m$. Each curve $\mathscr{L}_j$ divides its respective hexagon H_j into open components, and that component containing z_j^0 (or one of those containing z_j^0 in its closure) we shall call Ω_j^0 for $j = 1, \cdots, m$. We now define the domains Ω_{jk}^0 to be the rotations of Ω_j^0 about z_j^0 through the angles $k\pi/6$, $k = 1, \cdots, 5$. Setting $f(z_1, \cdots, z_m) - \alpha = F(\zeta_1, \cdots, \zeta_m)$ where $\zeta_j = z_j - z_j^0$, we have

[1] Research sponsored by the Air Force Office of Scientific Research, Office of Aerospace Research, United States Air Force, under AFOSR Grants No. AF-AFOSR-381-65 and 66.

that $|F(\zeta_1,\cdots,\zeta_m)|$ is bounded in the transform of T, say $|F(\zeta_1,\cdots,\zeta_m)| < M$, since $|f(z_1,\cdots,z_m)|$ is bounded in T. Furthermore, we define the function

$$\phi(\zeta_1,\cdots,\zeta_m) = \prod_{k_1,\ldots,k_m=0}^{5} F(\omega^{k_1}\zeta_1,\omega^{k_2}\zeta_2,\cdots,\omega^{k_m}\zeta_m)$$

where ω is a complex primitive sixth root of unity. Letting G_j denote the component of $\bigcap_{k=0}^{5}\Omega_{jk}^0$ containing z_j^0, and letting $\Omega = G_1 \times G_2 \times \cdots \times G_m$, we observe that $\phi(\zeta_1,\cdots,\zeta_m)$ is analytic in Ω. At each point of the distinguished boundary of Ω, $Fr(G_1) \times Fr(G_2) \times \cdots \times Fr(G_m)$, we have

$$|\phi(\zeta_1,\cdots,\zeta_m)| = \prod_{k_1,\ldots,k_m=0}^{5} |F(\omega^{k_1}\zeta_1,\omega^{k_2}\zeta_2,\cdots,\omega^{k_m}\zeta_m)| \leqq M^{6^m-1}\varepsilon.$$

Hence it follows by the maximum modulus principle that

$$|f(z_1^0,\cdots,z_m^0)-\alpha| = |\phi(0,\cdots,0)|^{6^{-m}} \leqq M^{(6^m-1)/6^m}\varepsilon^{1/6^m}$$

Thus we have shown that $f(z_1,\cdots z_m)\to\alpha$ as $(z_1,\cdots,z_m)\to(0,\cdots,0)$ along the "angle bisector" $\arg z_j = 0$, $j = 1,\cdots,m$.

It is clear that by this process of showing that $f(z_1,\cdots,z_m)\to\alpha$ as $(z_1,\cdots,z_m)\to(0,0,\cdots,0)$ along the bisector of the angular domain, we may show that $f(z_1,\cdots,z_m)\to\alpha$ on each ray in $T = \Delta_1\times\cdots\times\Delta_m$. The lemma now follows from the observation that, if $f(z_1,\cdots,z_m)$ is analytic in the region $T = \Delta_1\times\Delta_2\times\cdots\times\Delta_m$, $\Delta_j = \{z_j \,|\, |\arg z_j| \leqq \Theta_j', \ 0 < |z_j| \leqq R_j\}$ and if $f(z_1,\cdots,z_m)\to\alpha$ as $(z_1,\cdots,z_m)\to(0,\cdots,0)$ along the 2^n rays ($\arg z_1 = \pm\Theta_1'$, $\arg z_2 = \pm\Theta_2',\cdots,\arg z_m = \pm\Theta_m'$) for independent $\pm$, then $f(z_1,\cdots,z_m)\to\alpha$ uniformly as $(z_1,\cdots,z_m)\to(0,\cdots,0)$ in T.

To prove this assertion, we may assume, as before, that $\Theta_j' = +\pi/3$, $j = 1,\cdots,m$, and that $|f(z_1,\cdots,z_m)-\alpha| < \varepsilon$ on that part of $\arg z_j = \pm\pi/3$, $j = 1,\cdots,m$ contained in $|z_j| < 2R_\varepsilon$, $j = 1,\cdots,m$. For an arbitrary point $(z_1^0,\cdots,z_m^0)$ contained in $D_1'\times\cdots\times D_m'$, $D_j' = \{z_j \,|\, |z_j| < R_\varepsilon\}$, we form equilateral triangles τ_j in D_j with z_j^0 as centroid and with one side lying on the nearest of the rays $\arg z_j = \pm\pi/3$ and we form the generalized polycylinder $\tau_1\times\cdots\times\tau_m$. If $\zeta_j = z_j - z_j^0$ then $|F(\zeta_1,\cdots,\zeta_m)| = |f(z_1,\cdots,z_m)-\alpha|$ will be bounded, say by M; and the function

$$g(\zeta_1,\cdots,\zeta_m) = \prod_{k_1,\ldots,k_m=0}^{2} F(\omega^{k_1}\zeta_1,\cdots,\omega^{k_m}\zeta_m),$$

where ω is a complex third root of unity, will be analytic and bounded

$$|g(\zeta_1,\cdots,\zeta_m)| < \varepsilon M^{3^m-1} \quad \text{in} \quad \tau_1\times\cdots\times\tau_m.$$

Setting $\zeta_1 = \cdots = \zeta_m = 0$, we obtain

$$|\,(z_1^0,\cdots,z_m^0)-\alpha| < \varepsilon^{1/3^m}M^{(3^m-1)/3^m},$$

from which our assertion follows.

2. **An n-dimensional analytic continuation theorem.** In this section we continue our work on analytic continuation for functions of several complex variables begun in [**2**].

LEMMA 1. *Let $f(z,w)$ be defined for $|z| < R_1$, $|w| < R_2$ and let $f(z,w)$ be analytic in z over $|z| < R_1$ for each real w, $|w| < R_2$ and let $f(z,w)$ be analytic in w over $|w| < R_2$ for each z in $|z| < R_1$. Let $f(z,w)$ be bounded in each compact subset of $\{|z| < R_1\} \times \{|w| < R_2\}$. Then $f(z,w)$ is analytic in (z,w) over $\{|z| < R_1\} \times \{|w| < R_2\}$.*

PROOF. In our previous work [**2**] we established the following lemma on analytic continuation.

LEMMA. *Let $f(z,w)$ be defined and bounded on all compact subsets of the three dimensional region obtained by intersecting the product of two circles*

$$\mathscr{R} = \{(z,w) \,\big|\, |z| < R_1 \leqq +\infty,\ |w| < R_2 \leqq +\infty\}$$

with the union of two hyperplanes $\mathscr{P} = \{(z,w)\ (\operatorname{Im} z)(\operatorname{Im} w) = 0\}$. For each real x, $-R_1 < x < R_1$, let $f(x,w)$ be analytic in w throughout $|w| < R_2$, and for each real u, $-R_2 < u < R_2$, let (z,u) be analytic throughout $|z| < R_1$. Then there exists a function $g(z\ w)$ and positive numbers $\delta_1 \geqq R_1/4$, $\delta_2 \geqq R_2/4$ such that $g(z,w)$ is analytic in $\mathscr{R}^* = \{(z,w) \,\big|\, |z| < \delta_1, |w| < \delta_2\}$ and $f(z,w) \equiv g(z,w)$ for $(z,w) \in \mathscr{R}^* \cap \mathscr{P}$.

Take any disk $D(z,a,\rho) = \{z \,\big|\, |z-a| < \rho\}$ with $0 < \rho < R_1 - |a|$ and apply that lemma to the function $f(z,w)$ in the product of this disk $D(z,a,\rho)$ with $D(w,0,R_2)$. By that lemma therefore we may conclude that $f(z,w)$ is analytic in (z,w) throughout $D(z,a,\frac{1}{4}\rho) \times D(w,0,\frac{1}{4}R_2)$. Thus we see that $f(z,w)$ is analytic in the product of any small disk $D(z,a,\frac{1}{4}\rho)$ with $D(w,0,\frac{1}{4}R_2)$ whenever the disk $D(z,a,\rho)$ is contained in $D(z,0,R_1)$.

By the monodromy theorem we may conclude that $f(z,w)$ is analytic in the product of any interior disk $D(z,a,\rho)$ with $D(w,0,\frac{1}{4}R_2)$. Thus $f(z,w)$ is analytic in the product of $D(z,0,R_1)$ with $D(w,0,\frac{1}{4}R_2)$.

We now apply the technique used in the proof of Lemma 2 of the paper cited above to the disk $D(w,a,R_2 - |a|)$ for a real, $|a| < R_2$ to obtain the analyticity of $f(z,w)$ in the product of $D(z,0,R_1)$ with $D(w,a,\frac{1}{4}(R_2 - |a|))$. By repeated applications a diamond shaped region Δ with vertices at $w = R_2, (i/4)R_2, -R_2$, and $-(i/4)R_2$ is obtained in $D(w,0,R_2)$ and $f(z,w)$ is analytic in $D(z,0,R_1) \times \Delta$. The same argument involving diamond shaped regions, all possible star-shaped regions, convexity, rectifiability of the boundary curve enables us to conclude that $f(z,w)$ is analytic in $D(z,0,R_1) \times D(w,0,R_2)$.

By using this lemma we may easily establish the following theorem on analytic continuation for functions of several complex variables.

THEOREM 2. *Let D_1, D_2 be simply connected domains in the z and w-planes respectively. Let l_2 be a line segment contained in D_2. Let $f(z,w)$ be defined in $D_1 \times D_2$, let $f(z,w)$ be analytic in $z \in D_1$ for $w \in l_2 \subset D_2$, let $f(z,w)$ be analytic in $w \in D_2$ for $z \in D_1$. Let $f(z,w)$ be bounded on every compact subset of $D_1 \times D_2$. Then $f(z,w)$ is analytic in $D_1 \times D_2$.*

PROOF. Take any disk $\Delta_2 \subset D_2$ which has a diameter $l \subset l_2$. An application of the preceding lemma yields the conclusion that $f(z,w)$ is analytic in $\Delta_1 \times \Delta_2$ for any disk $\Delta_1 \subset D_1$. An application of the monodromy theorem yields the conclusion that $f(z,w)$ is analytic in $D_1 \times \Delta_2$. By reapplying the technique (using diamond shaped regions, star-shaped regions, etc.) and the monodromy theorem in the w-plane this time we establish the analyticity of $f(z,w)$ in $D_1 \times D_2$.

3. **An application to the Feynman integral.** The above results enable us to establish another translation theorem for the analytic Feynman integral.[1] This is based on the following translation theorem for Wiener integrals [**4**].

THEOREM *Let $x_0(t)$ be a real absolutely continuous function on $[a,b]$ which vanishes at $t=a$, and let $x_0'(t)$ be essentially of bounded variation on $[a,b]$. Let $C[a,b]$ be the space of real continuous functions $x(t)$ on $[a,b]$ that vanish at $t=a$, and let $G(x)$ be a Wiener measurable functional on $C[a,b]$. Then if either member of the following equation exists, both exist and they are equal:*

$$(1)\quad \int_{C[a,b]} G(x+x_0)\,dx = \exp\left(-\frac{1}{2}\int_a^b [x_0'(t)]^2 dt\right) \int_{C[a,b]} G(x)\exp\left(\int_a^b x_0'(t)dx(t)\right)dx.$$

The analytic Wiener integral and analytic Feynman integral have been defined by analytic extension from the Wiener integral as follows.

DEFINITION. Let the complex number λ_0 satisfy $\operatorname{Re}\lambda_0 \geqq 0$ and $\lambda_0 \neq 0$, so that $\lambda_0 = |\lambda_0|\exp(i\theta)$ for some θ on the interval $[-\pi/2, \pi/2]$. Let $F(x)$ be a functional defined on $C[a,b]$ and such that the Wiener integral

$$(2)\qquad J(\lambda) \equiv \int_{C[a,b]} F(\lambda^{-1/2}x)\,dx$$

exists for all real λ in the interval $|\lambda_0| < \lambda < |\lambda_0| + \delta$ for some $\delta > 0$. Then if $J(\lambda)$ can be extended so that it is defined and continuous on the closed region

$$(3)\quad S = \{\lambda : \lambda = \rho e^{i\gamma}, |\lambda_0| \leqq \rho \leqq |\lambda_0| + (1-\gamma\theta^{-1})\delta,\ \gamma \in [0,\theta],\ \text{or}\ \gamma \in [\theta, 0]\}$$

and analytic in its interior, we define

$$(4)\qquad \int_{C[a,b]}^{\text{anw}_{\lambda_0}} F(x)\,dx = J(\lambda_0),$$

[1] See [**3**] for our earlier theorem on this subject.

and we call the left member of (4) the "analytic Wiener integral of $F(x)$ with revar parameter λ_0." If $\theta = 0$, interpret S to be the real interval $[\lambda_0, \lambda_0 + \delta]$, omit the analyticity requirement since the interior of S is empty, and define the analytic Wiener integral to be $J(\lambda_0 +)$.

In particular, if $\lambda_0 = -iq$ where q is real, the integral (4) will be called the "analytic Feynman integral with parameter q," and we write

$$(5) \qquad \int_{C[a,b]}^{\text{anf}_q} F(x)dx = \int_{C[a,b]}^{\text{anw}_{-iq}} F(x)\,dx.$$

If $q = 1$, we shall write simply anf instead of anf_1.

Our new translation theorem for Feynman integrals is the following.

THEOREM 3. *Let $x_0(t)$ be a real absolutely continuous function on $[a,b]$ which vanishes at $t = a$, and let $x_0'(t)$ be essentially of bounded variation on $[a,b]$. Let Λ be the following horn-shaped domain for some $\delta > 0$,*

$$\Lambda = \{\lambda \mid \lambda = \rho e^{i\gamma},\ |\rho - 1| < \delta\,(1 - 2/\pi\,|\gamma|),\ -\pi/2 < \gamma < \delta\}$$

and let M be the union of a horn-shaped domain with some neighborhood of $-i$,

$$M = \{\mu \mid \mu = \rho e^{i\gamma},\ |\rho - 1| < \delta(1 - 2/\pi\,|\gamma|),\ -\pi/2 < \gamma < \delta\} \cup \{\mu \mid |\mu + i| < \delta\}.$$

Let $F(x)$ be a functional defined on Wiener space $C[a,b]$ such that the integral

$$\Phi(\lambda, \mu) \equiv \int_{C[a,b]}^{\text{anw}_\lambda} F(x)\exp\left\{\mu \int_a^b x_0'(t)\,dx(t)\right\} dx$$

exists and is bounded for all $\lambda \in \Lambda$ and all $u \in M$. Let

$$\lim_{(\theta,\phi)\to(-\pi^+/2,\,-\pi^+/2)} \Phi(\exp i\theta, \exp i\phi) = L$$

exist. Then the following integrals exist and equality holds:

$$\int_{C[a,b]}^{\text{anf}} F(x + x_0)\,dx = \exp\left(\frac{i}{2}\int_a^b [x_0'(t)]^2\,dt\right) \int_{C[a,b]}^{\text{anf}} F(x)\exp\left(-i\int_a^b x_0'(t)\,dx(t)\right) dx.$$

COROLLARY. *The hypotheses of the theorem can be weakened by replacing the domain M by the the domain*

$$M^* = \{\mu \mid \mu = \rho e^{i\gamma},\ |\rho - 1| < \delta(1 - 2/\pi\,|\gamma|),\ -\pi/2 < \gamma < \delta,\ \delta > 0\}$$

if in addition it is assumed that for each fixed θ, $-\pi/2 < \theta < -\pi/2 + \delta < 0$ the limit $\lim_{\phi\to-\pi^+/2}\Phi(e^{i\theta}, e^{i\phi})$ exists.

PROOF. By an application of the monodromy theorem $\Phi(\lambda, \mu)$ is analytic in $\lambda \in \Lambda$ for each $\mu \in M$. Moreover for each real $\lambda \in \Lambda$

$$\Phi(\lambda,\mu) = \int_{C[a,b]} F(\lambda^{-1/2}x)\exp\left\{\mu\lambda^{-1/2}\int_a^b x_0'(t)\,dx(t)\right\} dx.$$

We next prove the analyticity of $\Phi(\lambda,\mu)$ in (λ,μ). Let us define ϕ to be the integrand

$$\phi(x,\lambda,\mu) \equiv F(\lambda^{-1/2}x)\exp\left\{\mu\lambda^{-1/2}\int_a^b x_0'(t)\,dx(t)\right\}.$$

We shall next show that if $\mu\in M$ and if λ is real, $1-\delta<\lambda<1+\delta$, the integral below is finite:

$$\int_{C[a,b]} |\phi(x,\lambda,\mu)|\,dx = \int_{C[a,b]} |F(\lambda^{-1/2}x)|\exp\left\{\lambda^{-1/2}\operatorname{Re}\mu\int_a^b x_0'(t)\,dx(t)\right\}dx < \infty.$$

In fact if $-\delta<\operatorname{Re}\mu<1+\delta$, and if λ is real, $1-\delta<\lambda<1+\delta$, the above integral is finite.

For, if $c(x)=\int_a^b x_0'(t)\,dx(t)\geqq 0$ and $-\delta_1<\operatorname{Re}\mu<1+\delta_1$, we would then have $|\exp\{\mu\lambda^{-1/2}c(x)\}| = \exp\{\operatorname{Re}(\mu)\lambda^{-1/2}c(x)\}\leqq\exp\{(1+\delta_1)\lambda^{-1/2}c(x)\}$ and if $c(x)\leqq 0$ $|\exp\{\mu\lambda^{-1/2}c(x)\}|\leqq\exp\{-\delta_1\lambda^{-1/2}c(x)\}$. We may then conclude that:

$$|\phi(x,\lambda,\mu)| \leqq |\phi(x,\lambda,-\delta_1)| + |\phi(x,\lambda,1+\delta_1)|.$$

But the right member is Wiener integrable and hence so is the left member and the above assertion is justified.

Now let Γ be any closed curve contained in M

$$\int \Phi(\lambda,\mu)d\mu = \int_\Gamma\left(\int_{C[a,b]}\phi(x,\lambda,\mu)dx\right)d\mu = \int_{C[a,b]}\left(\int_\Gamma\phi(x,\lambda,\mu)\right)d\mu = 0$$

where the use of the Fubini theorem is justified by the above inequality and the integrability of the right-hand member. Thus by Morera's theorem we conclude that $\Phi(\lambda,\mu)$ is analytic in μ throughout M for every λ, $1-\delta_1<\lambda<1+\delta_1$.

By an application of Theorem 2 above we may conclude that $\Phi(\lambda,\mu)$ is analytic in (λ,μ) throughout $\Lambda\times M$.

By the two-variable Lindelöf theorem we then may conclude that

$$\lim_{(\lambda,\mu)\to(-i,-i);(\lambda,\mu)\in S} \Phi(\lambda,\mu) = L \tag{6}$$

exists uniformly in δ where δ is the product of two angular domains $S_1,S_2;S_1\subset\Lambda,S_2\subset M$. We shall choose the S_j so that they contain horizontal line segments with $-i$ as the left-hand end points.

The functions $\Phi(\lambda,\mu)$ is analytic at $(\lambda,-i)$ for each $\lambda\in\Lambda$ and therefore for each fixed $\lambda_0\in\Lambda$ the limit

$$\lim_{\mu\to -i}\Phi(\lambda_0,\mu) = \Phi(\lambda_0,-i) = \int_{C[a,b]}^{\text{anw}_{\lambda_0}} F(x)\exp\{-i\int_a^b x_0'(t)\,dx(t)\}dx \quad \text{exists.} \tag{7}$$

Since the double limit in equation (6) exists and the simple limit in equation (7) exists we may conclude that the iterated limit

$$\lim_{\lambda_0 \to -i} \lim_{\mu \to -i} \Phi(\lambda_0, \mu) = \lim_{\lambda_0 \to -i} \Phi(\lambda_0, -i)$$

$$= \lim_{\lambda_0 \to -i} \int_{C[a,b]}^{\text{anw}\,\lambda_0} F(x) \exp\left\{ -i \int_a^b x_0'(t)\, dx(t) \right\} dx \tag{8}$$

$$= \int_{C[a,b]}^{\text{anf}} F(x) \exp\left\{ -i \int_a^b x_0'(t)\, dx(t) \right\} dx = L.$$

From equation (6) we may conclude that

$$\lim_{t \to -\pi/2} \Phi(e^{it}, e^{it}) = L.$$

By applying the translation theorem for Wiener integrals [**4**],

$$\int_{C[a,b]} G(x + x_0)\, dx = \exp\left(-\frac{1}{2} \int_a^b [x_0'(t)]^2\, dt \right) \int_{C[a,b]} G(x) \exp\left(\int_a^b x_0'(t)\, dx(t) \right) dx$$

to the functional $G(x) = F(\lambda^{-1/2} x)$ and replacing x_0 by $\lambda^{1/2} x_0$ we obtain

$$\int_{C[a,b]} F(\lambda^{-1/2} x + x_0)\, dx$$

$$= \exp\left(-\frac{\lambda}{2} \int_a^b \{x_0'(t)\}^2 dt \right) \int_{C[a,b]} F(\lambda^{-1/2} x) \exp\left(\lambda^{1/2} \int_a^b x_0'(t)\, dx(t) \right) dx$$

for real λ on $|\lambda - 1| < \delta$ since the right member exists and this implies the measurability of $F(\lambda^{-1/2} x)$. From the definition of $\Phi(\lambda, \mu)$ for real λ and μ we have

$$\Phi(\lambda, \lambda) = \int_{C[a,b]}^{\text{anw}_\lambda} F(x) \exp\left\{ \lambda \int_a^b x_0'(t)\, dx(t) \right\} dx$$

$$= \int_{C[a,b]} F(\lambda^{-1/2} x) \exp\left\{ \lambda^{+1/2} \int_a^b x_0'(t)\, dx(t) \right\} dx.$$

Consequently we obtain

$$\int_{C[a,b]} F(\lambda^{-1/2} x + x_0)\, dx = \exp\left(-\frac{\lambda}{2} \int_a^b [x_0'(t)]^2 dt \right) \Phi(\lambda, \lambda).$$

The right-hand member is analytic for $\lambda \in \Lambda$ so the left-hand member has an analytic extension throughout Λ, namely the analytic Wiener integral

$$\int_{C[a,b]}^{\text{anw}_\lambda} F(x + x_0)\, dx = \exp\left(-\frac{\lambda}{2} \int_a^b [(x_0'(t))]^2 dt \right) \Phi(\lambda, \lambda).$$

We now pass to the limit as $\lambda \to -i$ in Λ within an angular domain $S_1 \subset \Lambda$ to obtain

$$\int_{C[a,b]}^{\text{anf}} F(x+x_0)\,dx = \exp\left(\frac{i}{2}\int_a^b (x_0'(t))^2\,dt\right)\int_{C[a,b]}^{\text{anf}} F(x)\exp\left(-i\int_a^b x_0'(t)\,dx(t)\right)dx.$$

Thus the theorem is established.

From the hypothesis of the corollary and an application of the one variable Lindelöf theorem we may conclude that for $\lambda_0 = e^{i\theta}$,

$$\lim_{\mu\to -i} \Phi(\lambda_0,\mu) = \int_{C[a,b]}^{\text{an}\,w_{\lambda_0}} F(x)\exp\left\{-i\int_a^b x_0'(t)\,dx(t)\right\}dx.$$

The corollary follows from equation (8) (bypassing the second member) with λ_0 restricted to the circular arc, and another application of the 2 variable Lindelöf theorem.

Bibliography

1. E. Lindelöf, *Sur un principe général de l'analyse et ses applications à la theorie de la représentation conforme*, Acta Soc. Sci. Fenn. Nova Ser. A, no. 4 **46** (1915), 1–35.

2. R. H. Cameron and D. A. Storvick, *Analytic continuation for functions of several complex variables*, Trans. Amer. Math. Soc. **125** (1966), 7–12.

3. ———, *A translation theorem for analytic Feynman integrals*, Trans. Amer. Math. Soc. **125** (1966), 1–6.

4. R. H. Cameron and W. T. Martin, *Transformations of Wiener integrals under translation*, Ann. of Math. (2) **45** (1944), 386–396.

University of Minnesota
Minneapolis, Minnesota

On Holomorphic Mappings of Hermitian Manifolds of the Same Dimension*

SHIING-SHEN CHERN

1. **Introduction.** In order to study the holomorphic mappings of one complex manifold into another, it is desirable, and in many cases even necessary, to introduce hermitian metrics in these manifolds. If the two manifolds are of the same dimension, the simplest metrical invariant that arises is the ratio of their volume elements. This has an important bearing on the properties of the mapping itself, and the present paper is devoted to a study of such relations.

More precisely, let $f: M \to N$ be a holomorphic mapping, where M and N are complex manifolds of the same dimension n. Relative to the local coordinates z^i let

$$ds_M^2 = \sum_{i,k} h_{ik} dz^i d\bar{z}^k \tag{1}$$

be a positive definite hermitian metric on M. (*Throughout this paper, all the small Latin indices will run, without exception, from* 1 *to* n.) Similarly, let

$$ds_N^2 = \sum_{j,l} \tilde{h}_{jl} dw^j d\bar{w}^l \tag{1a}$$

be a positive definite hermitian metric on N, with the local coordinates w^j. Under the mapping f, w^j are holomorphic functions of z^i, The function

$$u = \frac{\det(\tilde{h}_{jl})}{\det(h_{ik})} \left| \det\left(\frac{\partial w^r}{\partial z^s}\right) \right|^2 \quad (\geqq 0) \tag{2}$$

is independent of the choice of the local coordinates, and is geometrically the ratio of the volume elements. Our main result is a local formula for Δu where Δ is the Laplacian in M (see formula (60)). The formula is particularly simple when $u > 0$; then we have

$$\tfrac{1}{2}\Delta \log u = R - \mathrm{Tr}(f^*(\mathrm{Ric})). \tag{3}$$

---

* Work done under partial support form NSF grant GP-3990 and ONR contract 3656 (14).

At the right-hand side of (3), R denotes the scalar curvature of M and $\mathrm{Tr}(f^*(\mathrm{Ric}))$ is the trace of the inverse image of the Ricci form of N (The Ricci form "Ric" is an hermitian form in the tangent spaces of N and is mapped by the induced linear map into an hermitian form in the tangent spaces of M.)

To draw geometrical consequences it is necessary to impose some conditions on the domain manifold M and the image manifold N. The first property is:

(DO_K). M is exhausted by a sequence of open submanifolds

$$M_1 \subset M_2 \subset M_3 \subset \cdots \subset M \tag{4}$$

whose closures $\bar{M}_\alpha$ are compact, such that: (1) to each $\alpha = 1, 2, \cdots$ there is a smooth function $v_\alpha \geqq 0$ defined in M_α, which satisfies the inequality

$$\tfrac{1}{2}\Delta v_\alpha \leqq R + K \exp(v_\alpha / n), \tag{5}$$

where K is a given positive constant; (2) $v_\alpha(p_\beta) \to \infty$, if p_β is a divergent sequence of points in M_α. (An infinite sequence of points $p_\beta \in M_\alpha$ is called divergent, if every compact set of M_α contains only a finite number of them.)

For example, we will see that the unit ball D_1 defined by

$$z_1 \bar{z}_1 + \cdots + z_n \bar{z}_n < 1 \tag{6}$$

in the n-dimensional number space C_n with coordinates $(z_1, \cdots, z_n)$ has the property (DO_K), with $K = 2n(n+1)$. It is very likely that all the bounded symmetric domains have this property.

The condition on the image manifold N also involves a positive constant K. To state it, we denote by $H(\xi, \eta)$, where ξ, η are tangent vectors with the same origin in N, the hermitian inner product in N, and by $\mathrm{Ric}(\xi, \eta)$ the Ricci form of the hermitian metric. Then we have

(IM_K). N is said to have the property (IM_K), if

$$\mathrm{Ric}(\xi, \xi) \leqq -(K/n) H(\xi, \xi). \tag{7}$$

In particular, N has the above property, if its hermitian metric is einsteinian and has its scalar curvature bounded above by a negative constant. In general we say therefore that N is *almost einsteinian* if it has the property (IM_K).

If $f: M \to N$ is a holomorphic mapping and M and N have the properties (DO_K) and (IM_K) respectively, then we have an upper bound for the scalar function u (cf. Theorems 2 and 3 below). This includes an n-dimensional version of the Schwarz lemma, which says that if M is the unit n-ball and N is almost einsteinian, the mapping f does not increase volume.

The basic idea of this study is the realization that "negative curvature of the image manifold restricts a holomorphic mapping". Many classical theorems in function theory in one complex variable (such as Schottky's theorem) and on mappings of Riemann surfaces fall under this general theme. However, even if the metric on N does not have the property (IM_K), but is conformally equivalent to one with this property, important geometrical consequences will also follow. This leads in particular to a generalization of the Bloch theorem to n dimensions. (Cf. Theorem 4 below.) In this respect it may be worthwhile to point out that our results are developed for hermitian manifolds, and not merely for kählerian manifolds, because a conformal change of the metric usually destroys the kählerian property. It is not without surprise that the possible lack of the kählerian property does not complicate the problems studied in this paper.

2. **Hermitian geometry.** We will review in this section the basic facts on hermitian manifolds. For details compare [2], [5], [7].

Let M be a complex manifold of dimension n. An hermitian metric in M is given in terms of the local coordinates z^i by the hermitian differential form (1) where h_{ij} are complex valued C^∞ functions satisfying the condition of "hermitian symmetry";

$$\bar{h}_{ij} = h_{ji}, \tag{8}$$

and the further condition that the matrix (h_{ij}) is positive definite.

To the hermitian form (1) is associated the real-valued exterior differential form of degree two and rank $2n$:

$$\Theta = \frac{i}{2} \sum_{j,k} h_{jk} dz^j \wedge d\bar{z}^k, \tag{9}$$

called the associated two-form of ds_M^2. It is so normalized that when $n = 1$ and $z^1 = z = x + iy$ the associated two-form of $dz d\bar{z} = dx^2 + dy^2$ is $dx \wedge dy$.

The hermitian metric (1) is called kählerian if Θ is closed, i.e., if

$$d\Theta = 0. \tag{10}$$

Kähler proved that a necessary and sufficient condition for (10) is the existence of a local "potential", i.e., a real-valued function F, defined locally, such that

$$\Theta = id'd''F/2. \tag{11}$$

The hermitian metric (1) defines an inner product in the tangent spaces of M. A unitary frame is an ordered set of n tangent vectors e_i with the same origin, such that

$$(e_i, e_j) = \delta_{ij}, \tag{12}$$

where the expression at the left-hand side stands for the inner product of e_i, e_j. The dual basis in the cotangent space dual to a unitary frame is called a unitary coframe. A unitary coframe consists therefore of n complex-valued linear differential forms θ_i of type (1,0) such that

(13) $$ds_M^2 = \sum_i \theta_i \bar{\theta}_i.$$

Let B be the bundle of all unitary frames of M. The forms θ_i are forms in B. It is a fundamental theorem in local hermitian geometry that there exists in B a set of complex-valued linear differential forms θ_{ij}, uniquely determined, which satisfy the conditions

(14) $$d\theta = \sum_j \theta_j \wedge \theta_{ji} + \Theta_i,$$

(15) $$\theta_{ij} + \bar{\theta}_{ji} = 0,$$

where Θ_i are of type (2,0). The θ_{ij} are the "connection forms" of the hermitian metric (13).

By the consideration of the exterior derivative of (14), it can be shown that we can put

(16) $$d\theta_{ij} - \sum_k \theta_{ik} \wedge \theta_{kj} = \Theta_{ij},$$

where Θ_{ij} are of the form

(17) $$\Theta_{ij} = \tfrac{1}{2} \sum_{k,l} R_{ijkl} \theta_k \wedge \bar{\theta}_l,$$

and satisfy the conditions

(18) $$\Theta_{ij} + \bar{\Theta}_{ji} = 0.$$

In terms of the coefficients R_{ijkl} in (17), the conditions (18) are expressed by the symmetry properties

(19) $$R_{ijkl} = \bar{R}_{jilk}.$$

The quantities R_{ijkl}, which in our formulation are functions in B, constitute essentially the curvature tensor of the hermitian metric. From them we form the "Ricci tensor"

(20) $$R_{ij} = \sum_k R_{kkij} = \bar{R}_{ji},$$

and the scalar curvature

(21) $$R = \sum_i R_{ii}.$$

The latter is a real-valued function in M.

To establish the relationship between our "Ricci tensor" R_{ij} and the Ricci form mentioned in the Introduction, let ξ and η be tangent vectors with the same origin, which are given in terms of a unitary frame e_i by

(22) $$\xi = \sum_i \xi_i e_i, \qquad \eta = \sum_k \eta_k e_k.$$

Then we have

$$\mathrm{Ric}(\xi,\eta) = \sum_{i,k} R_{ik}\xi_i\bar{\eta}_k. \tag{23}$$

In this respect it may be mentioned that the frame e_i being unitary the hermitian inner product of our metric can be expressed by

$$H(\xi,\eta) = \sum \xi_i\bar{\eta}_i. \tag{24}$$

We set

$$\theta = \sum_k \theta_{kk}. \tag{25}$$

By (16) and (17) we have then

$$d\theta = \tfrac{1}{2} \sum_{i,j} R_{ij}\theta_i \wedge \bar{\theta}_j. \tag{26}$$

The form θ, from which R_{ij} is determined by (26), is completely characterized by the conditions

$$d(\theta_1 \wedge \cdots \wedge \theta_n) = -\theta \wedge \theta_1 \wedge \cdots \wedge \theta_n, \qquad \theta + \bar{\theta} = 0. \tag{27}$$

The hermitian metric is called einsteinian, if $d\theta$ is a multiple of Θ i.e., if

$$d\theta = -(i/n)R\Theta. \tag{28}$$

We now consider differential operators on real-valued C^∞ functions on M. Let u be such a function. We will either lift u into a function in B or choose a local frame field and restrict ourselves to a neighborhood of M; the formalisms are exactly the same. In fact, we put

$$du = \sum_i (u_i\theta_i + \bar{u}_i\bar{\theta}_i). \tag{29}$$

Taking the exterior derivative of this equation, we get

$$\sum_i \Big(du_i - \sum_k u_k\theta_{ik}\Big) \wedge \theta_i + \sum_i \Big(d\bar{u}_i - \sum_k \bar{u}_k\bar{\theta}_{ik}\Big) \wedge \bar{\theta}_i + \sum_i u_i\Theta_i + \sum_i \bar{u}_i\bar{\Theta}_i = 0.$$

This allows us to put

$$du_i - \sum_k u_k\theta_{ik} = \sum_k u'_{ik}\theta_k + \sum_k u_{ik}\bar{\theta}_k, \tag{30}$$

where

$$\sum_{i,k} u'_{ik}\theta_k \wedge \theta_i + \sum_i u_i\Theta_i = 0. \tag{31}$$

From (29) we get

$$d'u = \sum_i u_i\theta. \tag{32}$$

It follows that

(33) $$d'd''u = -dd'u = \sum_{i,j} u_{i\bar{j}}\theta_i \wedge \bar{\theta}_j.$$

We define the Laplacian of u to be

(34) $$\Delta u = 4 \sum_i u_{i\bar{i}}.$$

The factor in (34) is so chosen that for $n = 1$ and $ds_M^2 = dzd\bar{z}$, we have

$$\Delta u = 4\partial^2 u/\partial z\partial\bar{z} = \partial^2 u/\partial x^2 + \partial^2 u/\partial y^2.$$

If $u > 0$, we find

(35) $$\Delta \log u = \frac{1}{u}\Delta u - \frac{4}{u^2} \sum_i u_i \bar{u}_i,$$

a formula which will be useful later on.

It is worth noting that although kählerian manifolds are the ones with which we will be concerned in applications, the kählerian property is not needed in the establishment of our general formulas.

Before proceeding, we wish to consider the example of the unit ball. Let $z_1, \cdots, z_n$ be the coordinates in C_n, and let $r \geqq 0$ be defined by

(36) $$r^2 = \sum_k z_k \bar{z}_k.$$

The unit ball is the domain D_1 in C_n defined by $r < 1$. In it we introduce the kählerian metric whose associated two-form is the form (11) with

(37) $$F = -\log(1 - r^2).$$

It follows that

(38) $$\Theta_\cdot = \frac{i}{2}\left\{\frac{1}{1 - r^2} \sum_k dz_k \wedge d\bar{z}_k + \frac{4r^2}{(1 - r^2)^2} d'r \wedge d''r\right\}$$

and

(39) $$ds_M^2 = \frac{1}{1 - r^2} \sum_k dz_k d\bar{z}_k + \frac{4r^2}{(1 - r^2)^2} d'r\, d''r.$$

We wish to calculate the Ricci curvature of this metric and to show in particular that it is einsteinian. For this purpose we use the definition of the Ricci curvature by (26). It suffices to make the calculation relative to a unitary frame field or, what is the same, a unitary coframe field. This amounts to writing

(40) $$ds_M^2 = \sum_k \theta_k \bar{\theta}_k,$$

where

(41) $$\theta_k = (1-r^2)^{-1/2}(dz_k + z_k L(r) d'r),$$

with $L(r)$ given by

(42) $$rL(r) + 2 = 2(1-r^2)^{-1/2}.$$

It follows that

$$\bigwedge_k \theta_k = (1-r^2)^{-n/2} \bigwedge_k (dz_k + z_k L d'r) = (1-r^2)^{-(n+1)/2} dz_1 \wedge \cdots \wedge dz_n.$$

Differentiating, we get

$$d\left(\bigwedge_k \theta_k\right) = -(n+1)r(1-r^2)^{-(n+3)/2}(d'-d'')r \wedge dz_1 \wedge \cdots \wedge dz_n.$$

Comparison with (27) gives

(43) $$\theta = (n+1)r(1-r^2)^{-1}(d'-d'')r.$$

Its exterior derivative is found to be

(44) $$d\theta = 2i(n+1)\Theta.$$

By comparing with (28) we see that the metric (39) in the unit ball is einsteinian with its scalar curvature equal to $-2n(n+1)$. For $n=1$ we have

(45) $$ds_M^2 = (1-r^2)^{-2} dz d\bar{z},$$

and the scalar curvarture is equal to -4 as is well known.

3. **Holomorphic mapping of hermitian manifolds.** Our purpose is to study the holomorphic mappings of M into an hermetian manifold N of the same dimension n. The treatment of the last section applies also to N. It is necessary, however, to denote the quantities and differential forms pertaining to N by different symbols, and we list the corresponding ones in the following table:

(46) $$\begin{pmatrix} M & \theta_i & \theta_{ij} & \Theta_i & \Theta_{ij} & R_{ijkl} & R_{ij} & R \\ N & \omega_i & \omega_{ij} & \Omega_i & \Omega_{ij} & S_{ijkl} & S_{ij} & S \end{pmatrix}.$$

Thus all the general formulas in §1 remain valid, when the quantities are replaced by the corresponding ones in the second line of (46).

Let $f: M \to N$ be a holomorphic mapping. The most important scalar invariant which arises from such a situation is the ratio of the volume elements, i.e., the volume element of N by that of M. We denote it by $u \geqq 0$ and our basic analytic tool is a formula for Δu.

The derivation of this formula is local, and we will make use of unitary frame fields in both M and N; our final result will of course be independent of these choices. Under the mapping f let us set

$$\omega_i = \sum_j a_{ij}\theta_j. \tag{47}$$

Then we have

$$u = \left|\det(a_{ij})\right|^2. \tag{48}$$

By taking the exterior derivative of (47) and making use of (14) and its corresponding formula for $d\omega_i$, we get

$$\sum_j \left(da_{ij} - \sum_k a_{ik}\theta_{jk} + \sum_k a_{kj}\omega_{ki} \right) \wedge \theta_j - \Omega_i + \sum_j a_{ij}\Theta_j = 0.$$

This allows us to put

$$da_{ij} - \sum_k a_{ik}\theta_{jk} + \sum_k a_{kj}\omega_{ki} = \sum_k a_{ijk}\theta_k \tag{49}$$

where a_{ijk} satisfy the relations

$$\sum_{j,k} a_{ijk}\theta_k \wedge \theta_j - \Omega_i + \sum_j a_{ij}\Theta_j = 0. \tag{50}$$

Similarly, we take the exterior derivative of (49) and make use of the equations (14), (16), (17), and the corresponding equations for the image manifold N. It gives

$$da_{ijk} - \sum_l a_{ilk}\theta_{jl} - \sum_l a_{ijl}\theta_{kl} + \sum_l a_{ljk}\omega_{li} = \sum_l a_{ijkl}\,\theta_l + \sum_l b_{ijkl}\bar{\theta}_l, \tag{51}$$

where b_{ijkl} satisfy the relation

$$\sum_{k,l} b_{ijkl}\bar{\theta}_l \wedge \theta_k = -\sum_h a_{ih}\Theta_{jh} + \sum_h a_{hj}\,\Omega_{hi}. \tag{52}$$

The last equation gives, more explicitly,

$$b_{ijkl} = \tfrac{1}{2}\sum_h a_{ih}R_{jhkl} - \tfrac{1}{2}\sum_{h,r,s} a_{hj}a_{rk}\bar{a}_{sl}S_{hirs}. \tag{53}$$

To calculate Δu, with u given by (48), we put

$$D = \det(a_{ij}) \tag{54}$$

so that

$$u = D\bar{D}. \tag{55}$$

It follows from (49) that dD is of the form

$$dD = D(\theta - \omega) + \sum_i D_i\theta_i, \tag{56}$$

where θ is defined by (25) and similarly

$$\omega = \sum_i \omega_{ii}. \tag{57}$$

The u_i defined in (29) are therefore given by

$$u_i = \bar{D}D_i. \tag{58}$$

The D_i are polynomials in a_{ij} and a_{ijk}. In order to evaluate Δu it is necessary (and sufficient) to find u_{ik} as defined in (30). This can be done by the use of (51) and (53). However, as we are only interested in the quantities u_{ik}, i.e., the coefficients of $\bar{\theta}_k$ in du_i, a simpler way is to take directly the exterior derivative of (56). This gives

$$\sum_i \left\{ dD_i - \sum_k D_k \theta_{ik} - D_i(\theta - \omega) \right\} \wedge \theta_i + D(d\theta - d\omega) + \sum_i D_i \Theta_i = 0,$$

from which it follows that

$$u_{ik} = D_i \bar{D}_k + \frac{u}{2} \left(R_{ik} - \sum_{h,j} S_{hj} a_{hi} \bar{a}_{jk} \right). \tag{59}$$

Therefore we have our basic formula

$$\frac{1}{4} \Delta u = \sum_i D_i \bar{D}_i + \frac{u}{2} \left(R - \sum_{h,j,i} S_{hj} a_{hi} \bar{a}_{ji} \right). \tag{60}$$

If $u > 0$, it follows from (35), (58), and (60) that

$$\frac{1}{2} \Delta \log u = R - \sum_{h,j,i} S_{hj} a_{hi} \bar{a}_{ji}. \tag{61}$$

This is precisely the formula (3) given in the Introduction.

To draw geometrical conclusions we start with some definitions: f is said to be *degenerate* at $p \in M$, if u vanishes at p. Geometrically this means that the induced linear map on the tangent spaces is not univalent at p. f is called *totally degenerate* if u vanishes identically. If the common dimension of M and N is 1 and M is connected, a totally degenerate mapping is a constant mapping. From the definition of u we see that f is *volume decreasing* or *volume increasing* according as $u \leqq 1$ or $u \geqq 1$.

Then we have the following theorem:

THEOREM 1. *Let $f: M \to N$ be a holomorphic mapping, where M, N are hermitian manifolds of the same dimension, with M compact and N einsteinian. Let R and S be their scalar curvatures respectively. Then we have:*

(1) *If $R > 0$, $S \leqq 0$, f is totally degenerate.*

(2) *If $R < 0$, $S \geqq 0$, then there is a point of M at which f is degenerate.*

In fact, the condition for N to be einsteinian is

$$S_{ik} = (S/n)\delta_{ik},$$

in which case (60) reduces to

$$\frac{1}{4}\Delta u = \sum_i D_i\bar{D}_i + \frac{u}{2}\Big(R - \frac{S}{n}\sum_{i,k}|a_{ik}|^2\Big).$$

Since M is compact, u attains its maximum at a point $p_0 \in M$. If u is not identically zero, $u(p_0) > 0$. At p_0 we have

$$(D_i)_{p_0} = 0, \quad (\Delta u)_{p_0} \leqq 0.$$

This is not possible under the conditions in (1). Thus (1) is proved.

Similarly (2) is proved by the consideration of the minimum of u.

4. **A generalization of Schwarz lemma.** The condition (IM_K) for the manifold N to be almost einsteinian can be expsessed analytically by the relation

$$\sum_{i,k} S_{ik}\xi_i\bar{\xi}_k \leqq -\frac{K}{n}\sum_i \xi_i\bar{\xi}_i, \quad \textit{all } \xi_i. \tag{62}$$

By Hadamard's well-known determinant inequality we have

$$\frac{1}{n}\sum_{i,k}|a_{ik}|^2 \geqq |\det(a_{ik})|^{2/n} = u^{1/n}. \tag{63}$$

We now put

$$v = \log u. \tag{64}$$

It follows from (61) that if N satisfies the condition (62) and $u > 0$ we have

$$\tfrac{1}{2}\Delta v \geqq R + K\exp(v/n). \tag{65}$$

THEOREM 2. *Let $f: M \to N$ be a holomorphic mapping, where M and N are hermitian manifolds of the same dimension having the properties* (DO_K) *and* (IM_K) *respectively, with the same positive constant K. Then $v \leqq v_\alpha$.*

To prove this theorem consider the open subset E of M defined by $v > v_\alpha$. In E we have $u > 0$. By (5) and (65) we have

$$\tfrac{1}{2}\Delta(v - v_\alpha) \geqq K(\exp(v/u) - \exp(v_\alpha/u)).$$

Thus $\Delta(v - v_\infty) > 0$ in E, and the function $v - v$ cannot have a maximum in E. Hence $v - v_\alpha$ must approach its least upper bound on a sequence tending to the boundary of E. This sequence cannot have a limit point p_0 in M_α, for at p_0, $v - v_\alpha > 0$, and p_0 would belong to E and be a maximum for $v - v_\alpha$. It also cannot be divergent, for then $v_\alpha \to \infty$ while v is bounded. It follows that the only possibility is that E is vacuous, so that $v \leqq v_\alpha$.

In applications it is useful to replace a metric by a conformal one. Let ds_N^2 denote the hermitian metric on N. A conformal metric defined in a neighborhood

$V \subset N$ of a point $q_0 \in N$ and given by $ds_N^{*2} = \lambda(q)ds_N^2$, $q \in V$, $\lambda(q) > 0$, is said to be a *supporting metric* at q_0 if: (1) $\lambda(q_0) = 1$; (2) $\lambda(q) \leqq 1$, $q \in V$. Then we have:

THEOREM 2a. *The conclusion in Theorem* 2 *remains valid if the property* (IM_K) *of* N *is replaced by the following weaker property*: *To every point* $q_0 \in N$ *there is an hermitian metric which is a supporting metric at* q_0, *and has the property*.((IM_K).

To prove this we consider as above the open subset $E \subset M_\alpha$ defined by $v > v_\alpha$. Suppose that $v - v_\alpha$ has a maximum at $p_0 \in E$. Let $q_0 = f(p_0)$ and let $ds_N^{*2} = \lambda ds_N^2$ be a supporting metric of ds_N^2 at q_0. Let u' be the quotient of the volume element of N with the metric ds_N^{*2} by that of M. Then $u' = \lambda^n u$ and if $v' = \log u'$, $v' - v_\alpha = v - v_\alpha + d \log \lambda$. Since $\log \lambda(q_0) = 0$ and $\log \lambda(q) \leqq 0$ in a neighborhood of q_0, it follows that p_0 is a maximum of the function $v' - v_\alpha$. As the metric ds_N^{*2} has the property (IM_K), this is not possible according to the proof above. The theorem then follows by the rest of the above argument.

To apply Theorem 2 to a more concrete case, we consider the unit ball $r < 1$ discussed at the end of §2. In it let D_ρ be defined by $r < \rho$, where ρ is a constant satisfying $0 < \rho < 1$. We wish to construct in D_ρ the function $v_\rho \geqq 0$ with the property (DO_K), with $K = 2n(n+1)$, i.e., (1) v_ρ satisfies the inequality

$$\tfrac{1}{2}\Delta v_\rho \leqq -2n(n+1)(1 - \exp(v_\rho/n)); \tag{66}$$

(2) $v_\rho(p)$ tends to $+\infty$ as p approaches the boundary of D_ρ.

This function v_ρ will be intimately related to $\log(\rho^2 - r^2)$ and we wish to find the expression for $\Delta \log(\rho^2 - r^2)$. By (33) and (34), this can be read off from the form $d'd'' \log(\rho^2 - r^2)$. We have

$$d'' \log(\rho^2 - r^2) = -(\rho^2 - r^2)^{-1} \sum_k z_k d\bar{z}_k,$$

and

$$-d'd'' \log(\rho^2 - r^2) = -dd'' \log(\rho^2 - r^2) = (\rho^2 - r^2)^{-1} \sum_k dz_k \wedge d\bar{z}_k$$
$$+ 4r^2(\rho^2 - r^2)^{-2} d'r \wedge d''r.$$

From the expressions (39) and (40) for ds_M^2, we see that their associated two-forms are also equal. This gives

$$-d'd'' \log(\rho^2 - r^2) = \frac{1 - r^2}{\rho^2 - r^2} \sum_k \theta_k \wedge \bar{\theta}_k + \frac{4(1 - \rho^2)r^2}{(\rho^2 - r^2)^2(1 - r^2)} d'r \wedge d''r. \tag{67}$$

From (41), (42), and the relation $2rd'r = \sum_k \bar{z}_k dz_k$, we get

$$2rd'r = (1 - r^2) \sum_k \bar{z}_k \theta_k. \tag{68}$$

It follows that

$$(69)\qquad \tfrac{1}{4}\Delta\log(\rho^2 - r^2) = -n\,\frac{1-r^2}{\rho^2 - r^2} - \frac{(1-\rho^2)r^2(1-r^2)}{(\rho^2-r^2)^2}.$$

In particular, this formula is valid for $\rho = 1$. Subtracting them, we get

$$(70)\qquad \frac{1}{4(n+1)}\Delta\log\left(\frac{1-r^2}{\rho^2-r^2}\right)^{n+1} = -n+n\,\frac{1-r^2}{\rho^2-r^2}\left\{1+\frac{1}{n}\frac{(1-\rho^2)r^2}{\rho^2-r^2}\right\}$$
$$\leqq -n+n\left(\frac{1-r^2}{\rho^2-r^2}\right)^{1+1/n}.$$

Hence the function

$$v_\rho = \log\left(\frac{1-r^2}{\rho^2-r^2}\right)^{n+1}$$

has the property (DO_K) with $K = 2n(n+1)$.

By letting $\rho \to 1$, we derive from Theorem 2 an n-dimensional version of the Schwarz lemma:

THEOREM 3. *Let $f, D_1 \to N$ be a holomorphic mapping where D_1 is the unit n-ball with the standard kähler metric and where N is an n-dimensional hermitian einsteinian manifold with scalar curvature $\leqq -2n(n+1)$. Then f is volume-decreasing.*

For the case that N is kählerian a similar theorem has been given by Dinghas [3].

5. **Bloch's problem.** The above considerations lead naturally to a problem which was in a sense initiated by A. Bloch: Let f: $M \to N$ be a holomorphic mapping, where M and N are complex hermitian manifolds of the same dimension. Suppose furthermore that N be simply connected and that the riemannian sectional curvature of N be everywhere $\leqq 0$. An open ball of radius s in N is called *univalent* if there is an open subset $U \subset M$ such that the restriction $f|U$ is a one-one mapping of U onto it, with a functional determinant everywhere nonzero. Let $b(f)$ be the least upper bound of the radii of univalent balls under f. Given a family $\mathscr{F}$ of mappings f, the Bloch problem is to determine the value of $\inf_{f\in\mathscr{F}} b(f)$.

The classical Bloch theorem concerns the case when M is the unit disk, N is the complex line C, and $\mathscr{F}$ is the family of holomorphic mappings f such that $|f'(0)| = 1$. Bloch's theorem states that $\beta = \inf_{f\in\mathscr{F}} b(f) > 0$ and β is called Bloch's constant. Subsequent works of Ahlfors, Grunsky and R. Robinson have established that $0.433 < \beta < 0.472$ (cf., for instance [1]).

This is a case when N does not have an hermitian metric whose Ricci curvature is negative definite and the above results do not apply. However we have the following theorem:

THEOREM 4. *Let N be a simply connected hermitian manifold of dimension n whose metric ds_N^2 has the following properties:*

(1) *Its riemannian sectional curvature is everywhere $\leqq 0$.*

(2) *There exists a smooth nonincreasing function $\lambda(t) > 1$, $0 < t < c$ (c = a fixed constant), such that for every fixed point $q_0 \in N$, the metric $\lambda(\delta(q_0, q))ds_N^2$, $q \in N$, has the property (IM_K) with $K = 2n(n+1)$, where $\delta(q_0, q)$ is the distance between q_0, q.*

Let D be the unit n-ball with the standard kähler metric and let $\mathcal{F}$ be the family of all holomorphic mappings $f: D \to N$, whose Jacobians are nowhere zero and which have the property that the ratio u of volume elements is equal to 1 at a given point a M. Then $\inf_{f \in \mathcal{F}} b(f) \geqq c$.

To prove this theorem consider a fixed mapping $f \in \mathcal{F}$. We shall show that the hypothesis $b(f) < c$ leads to a contradiction.

To a point $q \in N$ let $\rho(q)$ be the radius of the largest univalent ball about q. Then $0 < \rho(q) \leqq b(f) < c$ and $\rho(q)$ is a continuous function in N. Consider the metric $ds_N^{*2} = \lambda(\rho(q))ds_N^2$ in N. We wish to show that at any point $q_0 \in N$ there is a supporting metric with the property (IM_K). In fact, the open ball U of radius $\rho(q_0)$ must have a singular point m on its boundary. Consider the metric

$$d\sigma_N^2 = \lambda(\delta(m,q))ds_N^2 = \lambda(\delta(m,q))ds_N{}^{*2}/\lambda(\rho(q)) ,$$

which is defined for all q such that $\delta(m,q) < c$, i.e. in some neighborhood V of q_0. By the hypothesis (2), $d\sigma_N^2$ has the property (IM_K). Since $\rho(q) \leqq \delta(m,q)$, $q \in V$, and $\lambda(t)$ is nonincreasing, we have

$$\lambda(\delta(m,q))/\lambda(\rho(q)) \leqq 1,$$

and the left-hand side is equal to 1 for $q = q_0$. Therefore $d\sigma_N^2$ is a supporting metric of ds_N^{*2} at q_0. It follows from Theorem 2a that $\lambda^n(\rho(f(p)))u(p) \leqq 1$ for all $p \in D$. This gives the inequality $\lambda(\rho(f(a))) \leqq 1$, when $p = a$, contradicting the assumption $\lambda(t) > 1$. Thus Theorem 4 is proved.

A special case of Theorem 4 is when $n = 1$ and N is the complex line with the coordinate z and the metric $ds_N^2 = dzd\bar{z}$. The condition that the gaussian curvature of the metric $\lambda dzd\bar{z}$ is $\leqq -4$ is

$$\Delta \log \lambda \geqq 4\lambda^2. \tag{72}$$

If $r \geqq 0$ is defined by $r^2 = z\bar{z}$, a solution of (72) is given by

$$\lambda(r) = \tfrac{1}{2} pAr^{p/2-1}(A^2 - r^p)^{-1}, \tag{73}$$

where p, A are positive constants; in fact, for this function $\lambda(r)$, the equality sign holds in (72). $\lambda(r)$ is a nonincreasing function of r, if

$$r^p \leqq (2-p)A^2/(2+p), \tag{74}$$

and is strictly decreasing, when the inequality sign holds in (74). We choose

(75) $$A = 2^{-p}(2+p)^{(2+p)/4}(2-p)^{-(2-p)/4}.$$

If $p < 2$ and r_0 is defined by

(76) $$r_0^p = \frac{2-p}{2+p} A^2,$$

then $\lambda(r_0) = 1$. It follows that $\lambda(r) > 1$, $0 < r < r_0$. By Theorem 4 we have

$$\inf_{f \in \mathcal{F}} b(f) \geqq r_0 = \tfrac{1}{4}(4 - p^2)^{1/2}.$$

Letting $p \to 0$, we get $\inf_{f \in \mathcal{F}} b(f) \geqq \frac{1}{2}$. Thus we have the following corollary:

COROLLARY. *Let $f: D \to C$ be a holomorphic mapping of the unit disk D into the complex line C such that $f'(\zeta)$, $\zeta \in D$, is nowhere zero and $|f'(0)| = 1$. Then there is a univalent disk in C of radius $\geqq \frac{1}{2}$.*

Our constant $\frac{1}{2}$ is larger than Bloch's constant. This can be explained by the fact that the assumption that $f'(\zeta)$ is nowhere zero imposes an additional restriction on the family of mappings than Bloch's case. If this assumption is dropped, we have to choose $p \geqq 1$, and the choice $p = 1$ gives the lower bound $3^{1/2}/4$ for Bloch's constant as established by Ahlfors.

Even in the general case it should be possible to drop in Theorem 4 the condition that the Jacobian is nowhere zero. This will require the consideration of hermitian metrics which are allowed to be positive semidefinite on certain divisors, as is customary in the one-dimensional case. We hope to return to this problem on a later occasion.

BIBLIOGRAPHY

1. L. V. Ahlfors, *An extension on Schwarz's lemma*, Trans. Amer. Math. Soc. **43** (1938), 359–364.

2. S. S. Chern, *Characteristic classes of hermitian manifolds*, Ann. of Math. (2) **47** (1946), 85–121.

3. A. Dinghas, "Ein *n*-dimensionales Analogon des Schwarz-Pickschen Flächensatzes für holomorphe Abbildungen der komplexen Einheitskugel in eine Kähler-Mannigfaltigkeit," in *Festschrift zur Gedächtnisfeier für Karl Weierstrass*, 1815–1965, Westdeutscher Verlag, Cologne, 1966, pp. 477–494.

4. H. Grauert and H. Reckziegel, *Hermitesche Metriken und normale Familien holomorpher Abbildungen*, Math. Z. **89** (1965), 108–125.

5. P. Griffiths, *The extension problem in complex analysis*. II; *embedding with positive normal bundle*, Amer. J. Math. **88** (1966), 366–446.

6. A. Koranyi, *A Schwarz lemma for bounded symmetric domains*, Proc. Amer. Math. Soc. **17** (1966), 210–213.

7. A. Weil, *Introduction à l'étude des variétés kählériennes*, Actualités Sci. Ind. no. 1267, Hermann, Paris, 1958.

UNIVERSITY OF CALIFORNIA
BERKELEY, CALIFORNIA

On the Coefficients of a Class of Univalent Functions

J. CLUNIE

1. At the recent conference on entire functions and related topics held at La Jolla the following problem was raised by Dr. Pommerenke.

Let $f(z) = \sum_{n=1}^{\infty} a_n z^n$ ($|z| < 1$), with $a_1 = 1$, be univalent and bounded in $|z| < 1$. Suppose also that $\arg f'(z)$ is bounded in $|z| < 1$ where, in order to be precise we assume that $\arg f'(0) = 0$ and that $\arg f'(z)$ is defined elsewhere in $|z| < 1$ by continuity. What can be said about the coefficients a_n; and in particular is it true that $a_n = O(1/n)$ $(n \to \infty)$?

The latter result is certainly true in a number of cases where the behaviour of $\arg f'(z)$ is further restricted. Perhaps the most general result of this kind so far published is that of Pommerenke and myself which states that $a_n = O(1/n)$ $(n \to \infty)$ whenever $f(z)$ is close-to-convex and bounded in $|z| < 1$. Because of this result the above problem is a natural one since, very roughly speaking, the boundary of the image of $|z| < 1$ by a function $f(z)$ satisfying the conditions of the problem cannot be much "worse" than the boundary of a bounded close-to-convex domain. It is well known that the images of $|z| < 1$ by two univalent functions may have somewhat similar geometric properties although the coefficients of the two functions have very different behaviour. However, in the particular case of a function $f(z)$ as in the problem one might feel that the relevant geometric properties are so similar to those for a bounded close-to-convex function that the coefficient estimates for the latter class of functions could well be true for the former class of functions. The main result of this paper shows that in fact this is not the case.

At the end of the paper I shall state a number of problems that suggest themselves as a consequence of Dr. Pommerenke's problem and of the examples I am about to give.

2. Before stating our results it is convenient to introduce certain kinds of functions with which we shall be concerned. The sequence $\{n_k\}_{k=1}^{\infty}$ always consists of positive integers with $n_k > 2\sum_{\nu=1}^{k-1} n_\nu$ for $k = 2\ 3, \cdots$. Associated with the sequence $\{n_k\}$ and a number $K > 1$ are the cosine polynomials

$$g_k(\theta) = \prod_{\nu=1}^{k}\left(1 + \frac{\cos n_\nu \theta}{K}\right) = \sum_{\nu=0}^{m_k} a_\nu \cos \nu\theta \quad (k = 1, 2, \cdots),$$

where $m_k = \sum_{v=1}^{k} n_v$. At first sight it might appear that the coefficients a_v should depend on k. However, the condition $n_k > 2 \sum_{v=1}^{k-1} n_v$ ensures that when $g_{k+1}(\theta)$ is obtained from $g_k(\theta)$ by introducing another factor the further terms arising do not overlap those of $g_k(\theta)$. Hence one sees that the a_v depend only on K and the sequence $\{n_k\}$.

Let

$$h_k(z) = \sum_{v=0}^{m_k} a_v z^v, \quad h(z) = \lim_{k\to\infty} h_k(z), \qquad (|z| < 1).$$

From the above definitions it follows that

$$g_{k+1}(\theta) = \left(\sum_{v=0}^{m_k} a_v \cos v\theta\right)\left(1 + \frac{\cos n_{k+1}\theta}{K}\right)$$

$$= g_k(\theta) + \frac{1}{2K} \sum_{=0}^{m_k} a_v \{\cos(n_{k+1} - v)\theta + \cos(n_{k+1} + v)\theta\}$$

and so

(1) $$h_{k+1}(z) = h_k(z) + (z^{n_{k+1}}/2K)\{h_k(1/z) + h_k(z)\}.$$

THEOREM. *In the above notation, if $K > 1$ and $\{n_k\}$ are suitably chosen and $f(z) = \sum_{n=1}^{\infty} b_n z^n$ $(|z| < 1)$, with $b_1 = 1$, is defined by*

$$f'(z) = \{(1 + h(z))/2\}^2 \qquad (|z| < 1),$$

then $f(z)$ is bounded and univalent in $|z| < 1$ and $|\arg f'(z)| < \pi$ $(|z| < 1)$ but $b_n \neq O(1/n)$ $(n \to \infty)$.

COROLLARY 1. *There is a function $f(z) = \sum_{n=1}^{\infty} b_n z^n$ $(|z| < 1)$ which is bounded and univalent in $|z| < 1$ with $|\arg f'(z)| < \pi$ $(|z| < 1)$, but such that $b_n \neq o(\log\log n/n)$ $(n \to \infty)$.*

COROLLARY 2. *There is a function $f(z) = \sum_{n=1}^{\infty} b_n z^n (|z| < 1)$ which is bounded and univalent in $|z| < 1$ with $|\arg f'(z)| < k\pi/2$ $(|z| < 1)$ which is bounded and univalent in $|z| < 1$ with $|\arg f')z)| < k\pi/2$ $(|z| < 1)$, where k is any given positive integer, but such that $b_n \neq o\{(\log\log n)^{k-1}/n\}$ $(n \to \infty)$.*

Corollaries 1 and 2 are obtained by taking a special sequence for $\{n_k\}$. One could in fact prove the theorem for such a special sequence directly. However, I feel that by dealing in the proof of the theorem with an unspecified sequence and considering only its general behaviour the arguments involved will be clearer.

A number of lemmas are required. We always assume that $\{n_k\}$ is a sequence as above, i.e. it consists of positive integers with $n_k > 2 \sum_{v=1}^{k-1} n_v$ for $k = 2, \cdots$. For convenience we define $\rho_k = \exp[-1/(n_k n_{k+1})^{1/2}]$ $(k = 1, 2, \cdots)$; and we set $n_0 = 0$, $\rho_0 = 0$; and we put $h_0(z) \equiv 1$.

LEMMA 1. *Let $\eta > 0$ be given. If $\{n_k\}$ increases sufficiently rapidly then*

$$||h_k(1/z)| - |h_k(z)|| < \eta \qquad (\rho_k \leqq |z| \leqq \rho_{k+1};\ k = 0,1,\cdots),$$

provided $(2/n_1)\exp[-1/n_1^{1/2}] < \eta$.

PROOF. The result is trivial for $k = 0$ so we assume that $k \geqq 1$. Since $h_k(z)$ has real coefficients we see that $|h_k(1/z)| = |h_k(1/\bar{z})|$ and so, with $|z| = r < 1$,

$$\begin{aligned} ||h_k(1/z)| - |h_k(z)|| &= ||h_k(1/\bar{z})| - |h_k(z)|| \\ &\leqq |h_k(1/\bar{z}) - h_k(\bar{z})| = \left| \int_z^{1/\bar{z}} h_k'(\zeta)\, d\zeta \right| \\ &\leqq \frac{1 - r^2}{r} \max_{\zeta \in [z, 1/\bar{z}]} |h_k'(\zeta)|, \end{aligned}$$

where $[z, 1/\bar{z}]$ denotes the segment joining z to $1/\bar{z}$. Now $h_k'(\zeta) = \sum_{\nu=1}^{m_k} \nu a_\nu \zeta^{\nu-1}$ and so, since $a_\nu \geqq 0$, if $|z| \geqq r$ $(0 < r < 1)$,

$$\max_{\zeta \in [z, 1/\bar{z}]} |h_k'(\zeta)| \leqq \sum_{\nu=1}^{m_k} \nu a_\nu r^{-\nu+1}.$$

Hence with $|z| = r$, since also $a_\nu \leqq 1$,

$$||h_k(1/z)| - |h_k(z)|| \leqq (1 - r^2) \sum_{\nu=1}^{m_k} \nu a_\nu r^{-\nu} \leqq (1 - r^2) m_k^2 r^{-m_k}.$$

When r lies in the range of Lemma 1 we obtain from the above that

$$\begin{aligned} ||h_k(1/z)| - |h_k(z)|| &\leqq (1 - \rho_k^2) m_k^2 \rho_k^{-m_k} \\ &\leqq (2m_k^2/(n_k n_{k+1})^{1/2}) \cdot \exp[m_k/(n_k n_{k+1})^{1/2}] \end{aligned}$$

Suppose that $n_{k+1} > m_k^4$ and it follows that when $|z| = r$ satisfies the conditions of Lemma 1 then

$$||h_k(1/z)| - |h_k(z)|| \leqq (2/n_k^{1/2}) \exp[1/n_k^{1/2}].$$

It is not difficult to verify that if $(2/n_1)\exp[1/n_1^{1/2}] < \eta$ then $(2/n_k)\exp[1/n_k^{1/2}] < \eta\ (k \geqq 1)$. Hence the result of Lemma 1 has been proved.

As a consequence of this lemma one sees that if $k \geqq 0$ and $\rho_k \leqq |z| \leqq \rho_{k+1}$ then

$$|h_{k+1}(z)| \geqq |h_k(z)|(1 - 1/K) - \eta/2K. \tag{2}$$

LEMMA 2. *Given $\eta > 0$ if $\{n_k\}$ increases sufficiently rapidly then*

$$\lambda_{k+1}'(z) < \eta/(1 - r) \qquad (\rho_k \leqq |z| \leqq \rho_{k+1};\ k = 0,1,2,\cdots),$$

where $h(z) = \lim_{k\to\infty} h_k(z)$ $(|z| < 1)$ *and* $h(z) = h_{k+1}(z) + \lambda_{k+1}(z)$.

PROOF. By definition we see that $\lambda_{k+1}(z) = \sum_{\nu=l_{k+2}}^{\infty} a_\nu z^\nu$ where $l_{k+2} = n_{k+2} - \sum_{\nu=1}^{k+1} n_\nu$. Hence, with $|z| = r < 1$,

$$\begin{aligned}
|\lambda'_{k+1}(z)| &\leqq \sum_{\nu=l_{k+2}}^{\infty} \nu a_\nu r^{\nu-1} \qquad (\text{since } a_\nu \geqq 0)\\
&\leqq \sum_{\nu=l_{k+2}}^{\infty} \nu r^{\nu-1} \qquad (\text{since } a_\nu \leqq 1)\\
&= \frac{d}{dr}\left(\sum_{\nu=l_{k+2}}^{\infty} r^\nu\right)\\
&= \frac{d}{dr}\left(\frac{r^{l_{k+2}}}{1-r}\right)\\
&= \frac{l_{k+2} r^{l_{k+2}-1}}{1-r} + \frac{r^{l_{k+2}}}{(1-r)^2}.
\end{aligned}$$

Consequently when $|z| = r$ and $\rho_k \leqq r \leqq \rho_{k+1}$ $(k = 0,1,2,\cdots)$, then

$$|\lambda'_{k+1}(z)| \leqq \frac{1}{1-r}\left(l_{k+2}\exp[-(l_{k+2}-1)/(n_{k+1}n_{k+2})^{1/2}] + \frac{\exp[-l_{k+2}/(n_{k+1}n_{k+2})^{\frac{1}{2}}]}{1-\exp[-1/(n_{k+1}n_{k+2})^{1/2}]}\right).$$

For $0 \leqq x \leqq \log 2$ we have $1 - e^{-x} > x/2$ and so, if $1/n_1 < \log 2$ and $l_{k+2} > (n_{k+1}/2) + 1$, we obtain

$$|\lambda'_{k+1}(z)| \leqq (1/(1-r))(n_{k+2}\exp[-(n_{k+2}/2n_{k+1})^{1/2}] + 2(n_{k+1}n_{k+2})^{1/2} \cdot \exp[-(n_{k+2}/2n_{k+1})^{1/2}]).$$

From this inequality the lemma follows provided the sequence $\{n_k\}$ increases sufficiently rapidly.

LEMMA 3. *Given $\eta > 0$, then provided the sequence $\{n_k\}$ increases sufficiently rapidly we have*

$$|\lambda_{k+1}(z)| < \eta \qquad (\rho_k \leqq |z| \leqq \rho_{k+1};\ k = 0,1,2,\cdots).$$

PROOF. With $|z| = r < 1$ we find that

$$\begin{aligned}
|\lambda_{k+1}(z)| &\leqq \sum_{\nu=l_{k+2}}^{\infty} r^\nu = \frac{r^{l_{k+2}}}{1-r}\\
&\leqq 2\exp[-(n_{k+2}/2n_{k+1})]^{1/2}(n_{k+1}n_{k+2})^{1/2}\\
&\qquad\qquad (\rho_k \leqq r \leqq \rho_{k+1};\ k = 0,1,2,\cdots),
\end{aligned}$$

provided $l_{k+2} > n_{k+2}/2$. Hence if the sequence $\{n_k\}$ increases sufficiently rapidly the lemma follows.

LEMMA 4. *Given* $\eta > 0$ *if* $\{n_k\}$ *increases sufficiently rapidly then, provided* K *is suitably chosen, with* $|z| = r$.

$$|h'_{k+1}(z)|/|1 + h(z)| = \eta/(1-r) \qquad (\rho_k \leqq r \leqq \rho_{k+1};\ k = 0, 1, 2, \cdots).$$

PROOF. We have

$$\begin{aligned} h'_{k+1}(z) &= h'_k(z) + (n_{k+1} z^{n_{k+1}-1}/2K)\{h_k(1/z) + h_k(z)\} \\ &\qquad + (z^{n_{k+1}}/2K)\{-(1/z^2)h'_k(1/z) + h'_k(z)\}. \end{aligned} \tag{3}$$

First of all we estimate $|h'_k(\zeta)|$ for $|\zeta| \leqq 1/\rho_k$ and $k > 0$. For such ζ,

$$\begin{aligned} |h'_k(\zeta)| &\leqq \sum_{\nu=1}^{m_k} \nu a_\nu \rho_k^{-(\nu-1)} \\ &\leqq \frac{m_k(m_k+1)}{2} \exp[(1/n_k)^{1/2}] \quad (\text{provided } n_{k+1} > m_k^2) \\ &\leqq m_k^2 \quad (\text{if } n_1 \text{ is large enough}). \end{aligned}$$

In the range of the lemma we have $1 - 1/(n_k n_{k+1})^{1/2} \leqq r$ and so, with ζ as above,

$$|h'_k(\zeta)| \leqq \frac{m_k{}^2}{(n_k n_{k+1})^{1/2}} \cdot \frac{1}{1-r} \leqq \frac{\eta}{1-r},$$

provided the sequence $\{n_k\}$ increases sufficiently rapidly.

Now $\operatorname{Re} h(z) > 0$ $(|z| < 1)$ as can be seen from the construction and hence $|1 + h(z)| > 1$ $(|z| < 1)$. From the above estimates for $h'_k(\zeta)$ we see, from (3), that if $\{n_k\}$ increases fast enough it is sufficient to prove Lemma 4 for $|n_{k+1} z^{n_{k+1}-1}\{h_k(1/z) + h_k(z)\}/2K\{1 + h(z)\}|$. Also from Lemma 3 we see that if $\{n_k\}$ increases fast enough then $\lambda_{k+1}(z)$ is "negligible" in the range of Lemma 4 and so it is clear that if $\{n_k\}$ increases fast enough we need prove the lemma only for $|n_{k+1} z^{n_{k+1}-1}\{h_k(1/z) + h_k(z)\}/2K\{1 + h_{k+1}(z)\}|$. This we now proceed to do.

If $\operatorname{Re}\zeta > 0$ then

$$\begin{aligned} |1+\zeta|^2 &= 1 + 2\xi + |\zeta|^2 \qquad (\operatorname{Re}\zeta = \xi) \\ &\geqq 1 + |\zeta|^2 \geqq \tfrac{1}{2}(1 + |\zeta|)^2. \end{aligned}$$

Hence $1/|1+\zeta| < 2^{1/2}/(1 + |\zeta|)$ and so, using the fact that $\operatorname{Re} h_{k+1}(z) > 0$ $(|z| < 1)$,

$$\left| \frac{n_{k+1} z^{n_{k+1}-1} \left\{ h_k\left(\frac{1}{z}\right) + h_k(z) \right\}}{2K\{1 + h_{k+1}(z)\}} \right| \leqq \frac{n_{k+1} r^{n_{k+1}-1}(2|h_k(z)| + \eta)}{2^{1/2}K\left\{1 - \frac{\eta}{2K} + \left(1 - \frac{1}{K}\right)|h_k(z)|\right\}} \tag{4}$$

for $|z| = r$ in the range considered with $K > \eta$, by Lemma 1 and (2) and the preceding observation. Here and elsewhere when a lemma is quoted it is assumed that our conditions have been adapted to ensure the hypotheses and requirements of the lemma are satisfied.

Suppose that we are given suitable preliminary values η_0, K_0. Then for $\eta < \eta_0$, $K > K_0$ we see that there is a constant $H = H(\eta_0, K_0)$ such that from (4) it follows that

$$\left| \frac{n_{k+1} z^{n_{k+1}-1}\{h_k(1/z) + h_k(z)\}}{2K\{1 + h_{k+1}(z)\}} \right| < \frac{H}{K} n_{k+1} r^{n_{k+1}-1}. \tag{5}$$

If $\xi > 0$ it is easy to show that $\xi r^{\xi} < 1/(1-r)$ $(0 \leqq r < 1)$ and so, from (5), we obtain

$$\begin{aligned} \left| \frac{n_{k+1} z^{n_{k+1}-1}\{h_k(1/z) + h_k(z)\}}{2K\{1 + h_{k+1}(z)\}} \right| &< \frac{H n_{k+1}}{K(n_{k+1} - 1)} (n_{k+1} - 1) r^{n_{k+1}-1} \\ &< \frac{2H}{K} \cdot \frac{1}{1-r} \end{aligned}$$

assuming that $n_1 > 1$. Hence if the sequence $\{n_k\}$ satisfies the conditions which ensure the above is true we see that K can be chosen sufficiently large to give the desired result.

LEMMA 5. *If $\{n_k\}$ increases fast enough then*

$$\int_0^1 h^2(r)\, dr < \infty.$$

PROOF. Considering the preliminary definitions before the statement of the theorem and using $a_\nu \geqq 0$, $K > 1$ we see that if $\{n_k\}$ increases fast enough then

$$h_{k+1}(r) \leqq h_k(r)(1 + r^{n_{k+1} - 2m_k}) \leqq h_k(r)(1 + r^{n_{k+1}/2}).$$

Hence provided $\{n_k\}$ increases sufficientlty rapidly we obtain first that $h(r) \leqq \prod_{k=1}^{\infty}(1 + r^{n_k/2})$ $(0 \leqq r < 1)$ and then $\int_0^1 h^2(r)\, dr < \infty$, by an obvious argument.

We now prove the theorem. If $f(z)$ is defined as in the theorem we find that, for $|z| < 1$,

$$f''(z)/f'(z) = 2h'(z)/(1 + h(z)).$$

Hence, given $\eta > 0$ we can choose K and $\{n_k\}$ so that the preliminary conditions are satisfied and also

$$|f''(z)/f'(z)| \leqq 4\eta/(1-r) \qquad (|z| = r;\ 0 \leqq r < 1), \tag{6}$$

by Lemma 2 and the fact that $\operatorname{Re} h(z) > 0$ $(|z| < 1)$, and Lemma 4. From Cauchy's integral for the derivative we obtain easily from (6) that

(7) $$|(f''(z)/f'(z))'| \leqq 16\eta/(1-r)^2 \qquad (|z| = r;\ 0 \leqq r < 1).$$

From (6) and (7) it follows that $\{f, z\} = (f''/f')' - \frac{1}{2}(f''/f')^2$, the Schwarzian derivative of $f(z)$ with respect to z, satisfies, for $|z| = r$ $(0 \leqq r < 1)$,

$$|\{f, z\}| \leqq (16\eta + 8\eta^2)/(1-r)^2$$

$$\leqq (64\eta + 32\eta^2)/(1-r^2)^2 .$$

If now η is such that $64\eta + 32\eta^2 \leqq 2$ then the corresponding $f(z)$ is univalent in $|z| < 1$ by a well-known criterion of Nehari [2].

We assume that K and $\{n_k\}$ are such that the preliminary conditions are satisfied and such that $f(z)$ is univalent and bounded in $|z| < 1$. That the latter condition can be satisfied follows from Lemma 5 or at once from (6), with a suitably small η. For this function we see that $4nb_n$ is the $(n-1)$th coefficient of $(1 + \sum_{\nu=0}^{\infty} a_\nu z^\nu)^2 = \{1 + h(z)\}^2$. Consider the case when $n = n_k + 1$. From the construction it follows that $1 + \sum_{\nu=0}^{\infty} a_\nu z^\nu$ contains the terms z^{n_k}/K, $z^{n_k - n_1}/2K^2$, $z^{n_k - n_2}/2K^2, \cdots$, $z^{n_k - n_{k-1}}/2K^2/K$ and also the terms 2, $z^{n_1}/K, z^{n_2}/K, \cdots, z^{n_{k-1}}$ Hence, since $a_\nu \geqq 0$, the coefficient of z^{n_k} in $\{1 + h(z)\}^2$ is at least

(8) $$2/K + (k-1)/2K^3 \geqq k/2K^3 \qquad (\text{since } K > 1).$$

Therefore

(9) $$b_{n_k+1} \geqq k/8K^3(n_k + 1) \qquad (k = 1, 2, \cdots)$$

and so the theorem follows.

Corollary 1 is proved as follows. It can be shown that K and a positive integer C can be chosen so that the result of the theorem holds with this K and $n_k^k = C^{C^k}$ $(k = 1, 2 \cdots,)$. In this case

(10) $$k = (\log\log n_k - \log\log C)/\log C$$

and so the result of Corollary 1 follows at once from (9) and (10).

Corollary 2 is proved by defining $f(z)$ by $f'(z) = \{(1 + h(z))/2\}^k$ $(|z| < 1)$ and choosing K and a positive integer C so that with this K and $n_k = C^{C^k}$ $(k = 1, 2, \cdots)$ it follows that $f(z)$ is bounded in $|z| < 1$ and, by Nehari's criterion, that $f(z)$ is univalent in $|z| < 1$. That this is possible follows as above with a few minor modifications. Using the estimates in (8) for the coefficients of z^{n_k} in $\{1 + h(z)\}^2$ together with (10) and arguing inductively we obtain the result of the second corollary.

3. Remarks. (i) If $f(z) = \sum_{n=1}^{\infty} b_n z^n$ $(|z| < 1)$ and $|\arg f'(z)| < \pi/2$ $(|z| < 1)$ then it follows that $f(z)$ is univalent in $|z| < 1$ and that $b_n = O(1/n$ $(n \to \infty)$. For the subclass of bounded univalent functions is this result best possible? By this I mean: if $\gamma > 1$ is given does there exists a bounded univalent

function $f(z) = \sum_{n=1}^{\infty} b_n z^n$ ($|z| < 1$) with $|\arg f'(z)| < (\gamma\pi/2)$ ($|z| < 1$) such that $b_n \neq O(1/n)$ ($n \to \infty$).

(ii) Given $\alpha > 1$ it would be interesting to know what are the correct estimates for the coefficients of a bounded univalent function $f(z) = \sum_{n=1}^{\infty} b_n z^n$ ($|z| < 1$) satisfying $|\arg f'(z)| < (\alpha\pi/2)$ ($|z| < 1$). I have excluded $\alpha < 1$ since it is almost obvious in this case that $b_n = o(1/n)$ ($n \to \infty$) and that this result is best possible. I have also excluded $\alpha < 1$ since in this case it is easily proved that $b_n = O(1/n)$ ($n \to \infty$) and that this is best possible. Hence the problem is only significant when $\alpha > 1$ and it may be true in this case that there is a $\beta = \beta(\alpha) > 0$ such that $b_n = O((\log n)^\beta/n)$ ($n \to \infty$); or perhaps even a stronger result than this can be proved.

References

1. J. Clunie and Ch. Pommerenke, *On the coefficients of close-to-convex univalent functions*, J. London Math. Soc. **41** (1966), 161–165.

2. Z. Nehari, *The Schwarzian derivative and schlicht functions*, Bull. Amer. Math. Soc. **55** (1949), 545–551.

Imperial College of Science and Technology
London, England

Properties of Exceptional Values of Meromorphic Functions

E. F. COLLINGWOOD

1. This brief expository article on one group of problems arising in Nevanlinna Theory contains the substance of a talk at the conference given as a sequel to a more general survey talk on the theory by W. H. J. Fuchs.

I confine myself, for simplicity, to the case of a function f meromorphic and nonrational in the finite z-plane and I begin by recalling the standard notations and definitions. Let a be any complex number, infinity not excluded. For $a \neq \infty$ a zero of $f - a$ will be called an a-point of f. Then $n(r, a) = n(r, 1/(f-a))$ is the number of a-points and $n(r, \infty) = n(r, f)$ the number of poles in the closed disc $|z| \leqq r$, each being counted with its order of multiplicity. The *counting function* for a-points (or poles) is, by definition,

$$N(r, a) = N(r, 1/(f-a)) = \int_0^r \frac{n(t, a) - n(0, a)}{t} dt + n(0, a) \log r,$$

$$N(r, \infty) = N(r, f) = \int_0^r \frac{n(t, \infty) - n(0, \infty)}{t} + n(0, \infty) \log r,$$

where $n(0, a)$ (or $n(0, \infty)$) is the multiplicity of a possible a-point (or pole) at the origin.

The *proximity function* for a value a (or ∞) is

$$m(r, a) = m(r, 1/(f-a)) = \frac{1}{2\pi} \int_0^{2\pi} \log^+ \frac{1}{|f(re^{i\theta}) - a|} d\theta,$$

$$m(r, \infty) = m(r, f) = \frac{1}{2\pi} \int_0^{2\pi} \log^+ |f(re^{i\theta})| \, d\theta.$$

The sum function $T(r) = T(r, f) = m(rf) + N(r, f)$ is the Nevanlinna *characteristic* of f and, for any complex $a \neq \infty$,

$$T(r) = m(r, a) + N(r, a) + O(1).$$

T is an unbounded convex function of $\log r$ and the *deficiency* $\delta(a)$ of the value a is defined by the equation

$$\delta(a) = \liminf_{r\to\infty} \frac{m(r,a)}{T(r)} = 1 - \limsup_{r\to\infty} \frac{N(r,a)}{T(r)}.$$

Plainly, $0 \leqq \delta(a) \leqq 1$. A value a is said to be *deficient* if $\delta(a) > 0$. The *upper deficiency* of the value a is defined by

$$\Delta(a) = \limsup_{r\to\infty} \frac{m(r,a)}{T(r)} = 1 - \liminf_{r\to\infty} \frac{N(r,a)}{T(r)}$$

from which it follows that $0 \leqq \delta(a) \leqq \Delta(a) \leqq 1$.

A classical result is that deficiency of either type is exceptional. In fact, the set of values for which $\delta(a) > 0$ is at most countable and the set for which $\Delta(a) > 0$ is at most of linear measure zero. A value a for which $\Delta(a) = 0$ we shall call *normally distributed.*

The Nevanlinna Theory has been mainly concerned with studying the exceptional values collectively. We may study a whole category of exceptional values, for example, the set of deficient values, or we may fix our attention upon an individual exceptional value and investigate its properties. This is the point of view of this paper.

Results in this direction have been known for a long time. The most extreme exceptional values of a meromorphic function are the Picard values which are not assumed at all near infinity. It was proved independently by Valiron and Iversen in 1913 that a Picard value is an asymptotic value of the function. An obvious example is the function e^z which has Picard values 0 and ∞ which are asymptotic in the left and right half planes respectively.

2. **Picard values are of deficiency** 1. With the discovery of deficiency the question at once arose whether a deficient value of a function is an asymptotic value. We now know that the answer is in the negative, but it took nearly twenty years from the time the question was first raised (in 1924) to find out. Counterexamples were constructed by Teichmüller [**10**b] and Marie-Hélène Schwartz [**7**]. These examples were both meromorphic functions. Subsequently, Hayman [**5**] gave a counterexample for an entire function.

The theory of the deficient value in relation to the structure of the Riemann surface in which the function $f(z)$ maps the finite plane was first based on the work of Iversen whose thesis [**6**] contained a systematic study of the surface in the parabolic case. The most important result established by him was to identify the asymptotic values of $f(z)$ with the winding points of infinite order (the transcendental critical points) of the surface and to classify the critical points as *algebraic points* (corresponding to zeros of $f'(z)$), *direct* (*logarithmic*) and *indirect transcendental points.*

After the Valiron-Iversen theorem on Picard values the next result [Collingwood, **1**] was the following theorem:

I. *A value a such that no branch points of the Riemann surface S_f of an entire function $f(z)$ lie over a disc $|w-a| < \sigma$, where σ is any fixed positive number, is normally distributed, i.e.* $\Delta(a) = 0$.

Alternatively, the theorem may be formulated in this way.

II. *If $\Delta(a) > 0$ then there is either a transcendental singularity of S_f at a or for any $\sigma > 0$ an infinity of algebraic points lies over the disc $|w-a| < \sigma$.*

The original proof of the theorem, in a slightly weaker form for $\delta(a)\delta(a) > 0$, used the observation that if all the pieces of S_f over the disc, where f is an entire function, were schlicht, the Koebe distortion theorem could be applied to show by a very simple argument that if a at the center of the disc has $\delta(a) > 0$ then an arbitrary number of values in the disc have deficiency not less than $\delta(a)$, whereas in fact the number cannot exceed $1/\delta(a)$.

Yet another way of looking at this primitive result is the following:

III. *If there are no branch points over the disc $D_\sigma(a) = \{w:|w-a| < \sigma\}$ so that the pre-images of the sheets of S_f over $D_\sigma(a)$ are domains $G_\sigma(a)$ enclosed by bounded level curves on which $|f(z)-a| = \sigma$ and each containing just one a-point, then the a-points are normally distributed as are all the b-points for every $b \in D_\sigma(a)$.*

3. It was natural to try to replace the condition of univalence of $f(z)$ in the $G_\sigma(a)$ by p-valence for any fixed $p > 1$. This implied that $f^{-1} = z(w)$ would be p-valued on the separate Riemannian domains of S_f over $D_\sigma(a)$. This problem did not yield very quickly.

But before the solution was achieved, Teichmüller [**10**a] was able to reduce the primitive theorem to a more precise and suggestive form, namely

IV. *If over $D_\sigma(a)$ the surface S_f, where f is meromorphic, has only schlicht sheets all of whose boundary points are internal points of S_f, then*

$$m(r,a) = O(1),$$

and hence $\Delta(a) = 0$.

In 1944 H. L. Selberg [**9**] proved the following theorem for a meromorphic function.

V. *If over $D_\sigma(a)$ the surface S_f consists of portions of not more than p sheets all of whose frontier points are internal points of S_f, then*

$$m(r,a) < p\log r + \log^+(1/\sigma) + O(1). \tag{1}$$

Again, we have $\Delta(a) = 0$, and similarly for any $b \in D_\sigma(a)$. But note that Teichmüller's result does not contain the factor $\log r$.

The "direct" form of the theorem is as in III except that the $G_\sigma(a)$ contains not more than p a-points of $f(z)$.

4. The obstacle to making the jump from 1 to p lay in the fact that, in the case of a meromorphic function, for $p > 1$ the domains $G_\sigma(a)$ are not necessarily simply connected. The figure illustrates the situation. Let $G_\sigma^1(a)$ and $G_\sigma^2(a)$ be two neighbouring G's. For $\sigma + \varepsilon$ these may coalesce and for $\sigma + 2\varepsilon$ may form a ring. The conformal mapping methods broke down in this situation which was resolved by Selberg who reduced the problem to one of potential theory which can be solved for multiply connected domains. His method is based upon the following

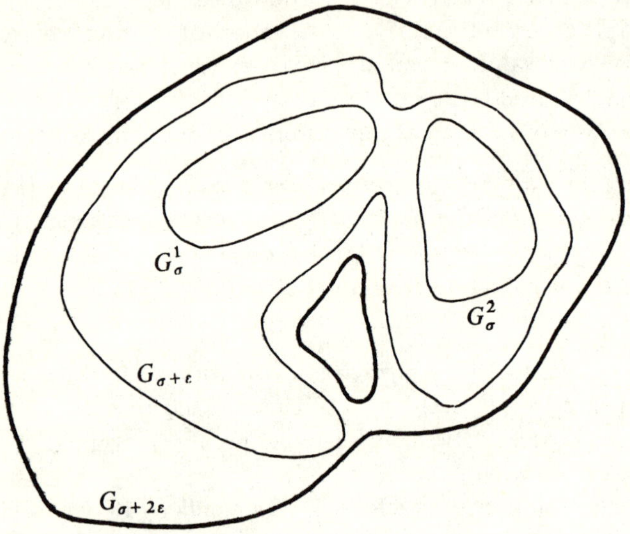

LEMMA. *Let G be a finite domain in the z-plane bounded by a finite number of analytic arcs and let Q be the open set of points of $|z| = r$ intercepted by G. Suppose further that Q consists of a finite number of distinct arcs whose total length is θr and let z_0 be an internal point of G and $g(z, G, z_0)$ be the Green's function for G having its pole at z_0.*

Then

$$\int_Q g(re^{i\phi}, G, z_0)\,d\phi \leqq \pi^2 \tan(\theta/4). \tag{2}$$

We also require an additional inequality. Suppose $z = 0$ is outside G and let r_1 be the distance from $z = 0$ to the frontier Fr G. Then G is contained in the domain outside $|z| = r_1$ so that

$$g(z, G, z_0) \leqq \log\left|(r_1^2 - 2\bar{z}_0)/r_1(z - z_0)\right|.$$

For the right-hand side is the Green's function for the domain outside $|z| = r_1$ and having its pole at z_0. Now

$$\int_0^{2\pi} \log|(r_1^2 - re^{i\phi}\bar{z}_0)/r_1(re^{i\phi} - z_0)|\, d\phi \leqq 2\pi \log^+ (r/r_1)$$

and so

$$(3) \qquad \int_Q g(re^{i\phi}, G, z_0) d\phi \leqq 2\pi \log^+ (r/r_1),$$

which is the inequality required.

5. The proof of (1) follows easily from(2) and (3). There is evidently no restriction in taking $a = \infty$, for we have only to put $w_1 = 1/(w - a)$ and note that

$$m(r, w_1) = m(r, a), \quad N(r, w_1) = N(r, a), \quad T(r, w_1) = T(r, w) + O(1).$$

Our hypothesis is that for every large r the $G_\sigma(\infty)$ intercepted by $|z| = r$ contains not more than p poles and are all bounded. Let $G_\nu(r)$, $\nu \leqq \lambda(r)$ be these domains. On $\mathrm{Fr}\, G_\nu(r)$ we have $|f| = 1/\sigma$. Hence in each domain $G_\nu(r)$, $\log|\sigma f(z)|$ can be represented as the sum of not more than p Green's functions for $G_\nu(r)$ having the poles of $f(z)$ in this domain as the poles of the Green's functions. Now denote by Q_ν the arcs of $|z| = r$ intercepted by $G_\nu(r)$ and let their total length be $\theta_\nu r$. Then we may apply (2) to $G_\nu(r)$ and obtain

$$\int_{Q_\nu} \log|\sigma f(re^{i\phi})|\, d\phi \leqq p\pi^2 \tan(\theta_\nu/4).$$

We may evidently put $\theta_1 \geqq \theta_2 \geqq \cdots \geqq \theta_{\lambda(r)}$ and since the Q_ν do not overlap, it follows that $\theta_\nu \leqq \pi$ for $\nu = 2, 3, \cdots, \lambda(r)$ and $\sum_{\nu=2}^{\lambda(r)} \theta_\nu < 2\pi$. Hence

$$(4) \qquad \sum_{\nu=2}^{\lambda(r)} \log|\sigma f(re^{i\phi})|\, d\phi \leqq p\pi^2 \sum_{2}^{\lambda(r)} \tan \frac{\theta_\nu}{4}$$

$$< p\pi \sum_{\nu=2}^{\lambda(r)} \theta_\nu < 2\pi^2 p.$$

We now have to deal with $G_1(r)$, by means of (3). (We make this distinction because θ_1 could be $> \pi$.) Since, by hypothesis, all the $G_\sigma(\infty)$ are bounded $G_1(r)$ will be outside $|z| = 1$ for sufficiently large r and we have

$$(5) \qquad \int_{Q_1} \log|\sigma f(re^{i\phi})|\, d\phi < 2\pi p \log r.$$

Now, by (4) and (5),

$$\begin{aligned} m(r, \infty) &\leqq \frac{1}{2\pi} \int_0^{2\pi} \log^+|\sigma f(re^{i\phi})|\, d\phi + \log^+ \frac{1}{\sigma} \\ &= \frac{1}{2\pi} \sum_{r=1}^{\lambda(r)} \int_{Q_\nu} \log|\sigma f(re^{i\phi})|\, d\phi + \log^+ \frac{1}{\sigma} \\ &\leqq p(\pi + \log r) + \log^+ \frac{1}{\sigma}. \end{aligned}$$

Transforming back to a general a this proves (1).

The possibilities are not, however, exhausted by Selberg's result. More information can be obtained if we make the constants p and σ functions of r. Suppose, then, that $\sigma(r)\downarrow 0$ or is constant and that $p(r)\uparrow\infty$ or is constant. We can then study the conditions under which an inequality

$$(6)\qquad m(r,a) < (\pi + \log r)p(r) + \log^+(1/\sigma(r)) + O(1)$$

holds for all large values of r.

There is no difficulty in extending Selberg's method to this situation. We need another definition: For a given value of r there will be, for a given $\sigma(r)$, a finite number $\lambda(r)$, which may be zero, of domains $G_\nu(r,a,\sigma(r))$ intersecting $|z| = r$. These G_ν may be bounded or unbounded. Let $p_\nu(r)$ be the valency of $f(z)$ in $G_\nu(r,a,\sigma(r))$ and $P(r,a,\sigma(r)) = \max_{\nu\leqq\lambda(r)} p_\nu(r)$. P may be infinite if one of the G_ν is unbounded, or finite, or zero if $\lambda(r) = 0$, i.e. if $|z| = r$ intersects *no* domain $G_\nu(r,a,\sigma(r))$ so that $|f(z) - a| \geqq \sigma(r)$ on $|z| = r$. With these definitions we have as a generalization of Selberg's theorem V (Collingwood [2a]).

VI. *Suppose that $f(z)$ is meromorphic and that for a given $\sigma(r)\downarrow 0$ or constant and a given $p(r)\uparrow\infty$ or constant the inequality $P(r,a,\sigma(r)) \leqq p(r) < \infty$ is satisfied for all large r, then*

$$(7)\qquad m(r,a) < (\pi + \log r)p(r) + \log^+(1/\sigma(r)) + O(1).$$

6. A more precise formulation can be given if we introduce the sets $E(a,\sigma(r),p(r))$ of values of r for which there are either no domains $G_\nu(r,a,\sigma(r))$, in which the valency $p_\nu(r) \leqq p(r)$, or they are all bounded. We denote complements by $\mathscr{C}$. Then we can deduce

VII. *If, for a given a, $\exists\,\sigma(r)\downarrow$ or $=$ constant as $r\to\infty$ and a $p(r), 0 \leqq p(r) < \infty$ for all $r < \infty$ such that ∞ is a limit point of $E(a,\sigma(r),p(r))$ then for all $r\in E$*

$$m(r,a) < (\pi + \log r)p(r) + \log^+ 1/\sigma(r) + O(1).$$

$E(a,\sigma(r),p(r)) \subset V(a,\sigma(r),p(r))$, the set in which $P(r,a,\sigma(r)) \leqq p(r)$ without the condition that $G_\nu(r,a,\sigma(r))$ is necessarily finite.

VIII. *If, for a given $\sigma(r)$ and $p(r)$, ∞ is not a limit of $\mathscr{C}V(a,\sigma(r),p(r))$ (i.e. if $P(r,a,\sigma(r)) \leqq p(r)$ for all large r), then*

$$\text{(i)}\qquad \Delta(a) \leqq \limsup_{r\to\infty}\frac{p(r)\log r}{T(r)} + \limsup_{r\to\infty}\frac{\log^+ 1/\sigma(r)}{T(r)},$$

$$\text{(ii)}\qquad \delta(a) \leqq \limsup_{r\to\infty}\frac{p(r)\log r}{T(r)} + \liminf_{r\to\infty}\frac{\log^+ 1/\sigma(r)}{T(r)},$$

$$\text{(iii)}\qquad \delta(a) \leqq \liminf_{r\to\infty}\frac{p(r)\log r}{T(r)} + \limsup_{r\to\infty}\frac{\log^+ 1/\sigma(r)}{T(r)}.$$

Many corollaries can be obtained. I content myself with the following.

IX. *If, for a given* $\sigma(r)$ *and* $p(r)$, either

$$\Delta(a) > \limsup_{r\to\infty} \frac{p(r)\log}{T(r)} + \limsup_{r\to\infty} \frac{\log^+ 1/\sigma(r)}{T(r)},$$

$$\text{or} \qquad \delta(a) > \liminf_{r\to\infty} \frac{p(r)\log r}{T(r)} + \limsup_{r\to\infty} \frac{\log^+ 1/\sigma(r)}{T(r)},$$

$$\text{or} \qquad \delta(a) > \limsup_{r\to\infty} \frac{p(r)\log r}{T(r)} + \liminf_{r\to\infty} \frac{\log^+ 1/\sigma(r)}{T(r)},$$

then $P(r,a,\sigma(r)) > p(r)$ *for some large values of* r; *while if*

$$\delta(a) > \limsup_{r\to\infty} \frac{p(r)\log r}{T(r)} + \limsup_{r\to\infty} \frac{\log^+ 1/\sigma(r)}{T(r)}$$

then $P(r,a,\sigma(r)) > p(r)$ *for all large* r.

X. *For a given* $\sigma(r)$ *the valency* $P(r,a,\sigma(r))$ *satisfies the inequalities*

$$\text{(i)} \qquad \Delta(a) \leqq \limsup_{r\to\infty} \frac{P(r,a,\sigma(r))\log r}{T(r)} + \limsup_{r\to\infty} \frac{\log^+ 1/\sigma(r)}{T(r)},$$

$$\text{(ii)} \qquad \delta(a) \leqq \limsup_{r\to\infty} \frac{P(r,a,\sigma(r))\log r}{T(r)} + \liminf_{r\to\infty} \frac{\log^+ 1/\sigma(r)}{T(r)},$$

$$\text{(iii)} \qquad \delta(a) \leqq \liminf_{r\to\infty} \frac{P(r,a,\sigma(r))\log r}{T(r)} + \limsup_{r\to\infty} \frac{\log^+ 1/\sigma(r)}{T(r)}.$$

XI. *If* $\delta(a) > 0$ *and we choose* $\sigma(r)$ *such that*

$$\limsup_{r\to\infty} \frac{\log^+ 1/\sigma(r)}{T(r)} \leqq \theta\delta(a)$$

then, by X. (iii)

$$\liminf_{r\to\infty} \frac{P(r,a,\sigma(r))\log r}{T(r)} \geqq (1-\theta)\delta(a) > 0. \tag{8}$$

For a constant $\sigma(r) = \sigma$ the lower order of $P(r,a,\sigma(r))$ as given by the above is best possible within a constant factor as shown by the example of Marie-Hélène Schwartz. The function in question is

$$f(z) = \prod_1^\infty \left(\frac{1+z/4^n}{1-z/4^n}\right)^{(-2)^n}$$

$f(z)$ is meromorphic of order 1/2 and has $\delta(0) = \delta(\infty) > 0$ while neither 0 nor ∞ is an asymptotic value. It can be shown that

$$P(r,0,\sigma)/T(r) = P(r,\infty,\sigma)/T(r) < k < \infty.$$

The method we have used is equally applicable to the hyperbolic case of a function meromorphic in the unit disc, subject to the condition $T(r)$ unbounded. A case of special interest is that where the set of asymptotic values of the function is of capacity zero (see Collingwood and Lohwater [**3**]).

7. These methods also enable us to establish conditions sufficient for reversal of the second fundamental theorem, namely

$$N_1(r) + \sum_1^q m(r,a_\nu) \leqq 2T(r) + S(r) \tag{*}$$

when $N_1(r) = N(r,1/f') + 2N(r,f) - N(r,f')$ counts the multiple points of f'.

We can prove the following reversal of (*) (Collingwood [**2b**]).

XII. *If for some $\sigma(r)$ such that $\log^+ 1/\sigma(r) = o(T(r))$ and some $p(r) = o(T(r)/\log r)$ there is a finite B such that $\sup \mathscr{C} E(a,\sigma(a,r),p(r)) \leqq B$ for all $a \in \mathscr{C}A$, where $A = \{a_i\}$ $(1 < i < q)$ is a finite set, then*

$$2T(r) \leqq N_1(r) + \sum_1^q m(r,a_i) + S(r) \tag{9}$$

when $S(r) = o(T(r))$.

Here $\sigma(a,r)$ is a modification of the $\sigma(r)$ of the previous section involving the distance $[a,A]$ to ensure that no A-point of f is contained in any $G_\nu(r,a,\sigma(a,r))$.

XIII. *If $\sigma(r)$ and $p(r)$ satisfy the conditions of* XII *and $\sup \mathscr{C} E(a,\sigma(r),p(r)) < B$ for all a, then $\Delta(a) = 0$ for all a and*

$$\lim_{r\to\infty} \frac{N_1(r)}{T(r)} = 2.$$

XIV. *If the winding points of S_f are all algebraic points of uniformly bounded order and if their traces are finite in number then $\Delta(a) = 0$ for all a and*

$$\lim_{r\to\infty} \frac{N_1(r)}{T(r)} = 2.$$

EXAMPLE. Weierstrass' $\wp$ function which has 4 completely multiple points of order 2.

We observe that the conditions of Theorem XII ensure that no points outside the finite set A can have a positive deficiency. In the inequality (9) account must be taken of the term $N_1(r)$. As a corollary of XIV we have

XV. *If the traces of the winding points of S_f form a finite set $a_1, a_2, \cdots, a_\nu$ and if*

$$\liminf_{r\to\infty} \frac{N_1(r)}{T(r)} < 2$$

then at least one of the points a_ν, $\nu = 1, 2, \cdots, q$, is the trace either of a logarithmic point or of an infinite set of algebraic points whose orders are unbounded.

It follows that if

$$\liminf_{r\to\infty} \frac{N_1(r)}{T(r)} < 2$$

then either the traces of the winding points form an infinite set or, if they form a finite set, at least one of them is the trace either of a logarithmic point or of a set of algebraic points whose orders are unbounded. Whether we can deduce the existence of two such points if in the above inequality 2 is replaced by 1 is an open question.

8. In conclusion we may note that theorems analogous to those of §§2, 3 and 5 above for a simply connected domain D of the Riemann sphere whose coverage by the Riemann surface of a meromorphic function $f(z)$ is deficient in the sense of Ahlfors have been proved by Dinghas [**4**] and Parreau [**8**]. Let $D(r)$ be a domain containing D whose boundary is at a distance $\sigma(r)$ from the boundary of D and suppose that none of the Riemannian domains of the surface $S_f(r)$ on which f maps $|z| < r$ lying over $D(r)$ has more than $P(r)$ sheets then, by Parreau's theorem, the Ahlfors deficiency $\delta_A(D)$ of the domain D satisfies the inequality

$$\delta_A(D) \leqq \limsup_{r\to\infty} \frac{P(r)L(r)}{2\sigma(r)\Sigma(r)},$$

where $\pi\Sigma(r)$ is the spherical area of $S_f(r)$ and $L(r)$ is the length of its boundary. The Ahlfors deficiency of D is defined by the equation

$$\delta_A(D) = 1 - \limsup_{r\to\infty} \frac{n(r, D)}{\Sigma(r)}$$

where $n(r, D)$ is the total number of sheets of the islands, i.e. connected portions of $S_f(r)$, covering D whose pre-images and their boundaries lie in $|z| < r$.

REFERENCES

1. E. F. Collingwood, *Sur les valeurs exceptionnelles des fonctions entières d'ordre fini*, C. R. Acad. Sci. Paris **79** (1924), 1125-1127.

2. ———, a. *Exceptional values of meromorphic functions.* Trans. Amer. Math. Soc. **66** (1949), 308–346, b. *Sufficient conditions for reversal of the second fundamental inequality for meromorphic functions*, J. Analyse Math. **2** (1952), 29–50.

3. E. F. Collingwood and A. J. Lohwater, *Application of the theory of cluster sets to a class of meromorphic functions*, Proc. Cambridge Philos. Soc. **53** (1957), 93–105.

4. A. Dinghas, *Über Ausnahmegebiete meromorpher Funktionen*, Math. Z. **45** (1939), 20–24.

5. W. K. Hayman, *An integral function with a defective value that is neither asymptotic nor invariant under change of origin*, J. London Math. Soc. **28** (1953), 369–376.

6. F. Iverson, *Recherches sur les fonctions inverses des fonctions méromorphes*, Thèse Helsingfors, 1914.

7. M. H. Schwartz, *Example d'une fonction méromorphe ayant des valeurs déficientes non-asymptotiques*, C. R. Acad. Sci. Paris **212** (1941), 382–384.

8. M. Parreau, *Sur le théorème de Collingwood-Cartan*, C. R. Acad. Sci. Paris **227** (1948), 1323–1325.

9. H. L. Selberg, *Eine Ungleichung der Potentialtheorie und ihre Anwendung in der Theorie der meromorphen Funktionen*, Comment. Math. Helv. **18** (1946), 309–326.

10. O. Teichmüller, a. *Eine Umkehrung des zweiten Hauptsatzes der Wertverteilungslehre*, Deutsche Math. **2** (1937), 96–107 b. *Einfache Beispiele zur Wertverteilungslehre*, Deutsche Math. **7** (1942–1944), 360–368.

LILBURN TOWER
ALNWICK, NORTHUMBERLAND,
ENGLAND

An Introduction to Complex Fourier Analysis and Lacunary Series[1]

LEON EHRENPREIS

The first example of the type of result we wish to discuss is the classical result that the Taylor series about x_0 of an analytic function converges uniformly on compact sets of the largest disc centered at x_0 in which f is analytic. To express this another way, if we know (x is a complex variable)

(1) the series $\sum c_j x^j$ converges uniformly on compact sets of $|x| < 1$, say to $f(x)$, and

(2) $f(x)$ is analytic in $|x| < 1 + \delta$ for some $\delta > 0$,

then we can conclude

(4) the series converges uniformly on compact sets of $|x| < 1 + \delta$.

(The reason there is no (3) will appear presently.)

There have been many sharpenings of this result in the direction of weakening Condition (2), which is only possible if we place conditions on the c_j. An important result of the type we are interested in is the celebrated Fabry gap theorem:

Suppose (1) holds as above and also

(2′) $f(x)$ is analytic in some neighborhood of the point 1, say in $|x - 1| < \delta$.

(3′) $c_j = 0$ except for j belonging to a subsequence $\{a_j\}$ with the property that the number of a_j such that $a_j \leqq N$ is $o(N)$ as $N \to \infty$.

Then the conclusion (4) holds as before.

Before studying the structure of these theorems more carefully, let us make an exponential change of variables, i.e., replace x by $\exp(ix)$. Then the results are:

I. Suppose

(a) $\sum c_j \exp(ijx)$ converges uniformly on compact sets of $Ix > 0$ to $f(x)$ say. (Here Ix denotes the imaginary part of x.)

(b) $f(x)$ is analytic in $Ix > -\delta$ for some $\delta > 0$.

Then

(d) the series converges uniformly on compact sets of $Ix > -\delta$.

The Fabry gap theorem becomes

II. Suppose (a) as in I and

[1] More details will be found in my book *Fourier analysis in several complex variables*, (to appear soon).

(b′) $f(x)$ is analytic in $|x| < \delta$,

(c′) $c_j = 0$ except for j in a subsequence $\{a_j\}$ as in (3′). Then (d) follows.

Now (a) and (d) can easily be formulated in terms of growth conditions on the c_j, namely, (a) is equivalent to

$$c_j = o(\exp(\varepsilon j)) \qquad \text{for any } \varepsilon > 0,$$

while (d) is equivalent to

$$c_j = o(\exp(-\delta + \varepsilon)j) \qquad \text{for any } \varepsilon > 0.$$

On the other hand, (b) (or (b′)) is not, a priori, a condition on the growth of the c_j. *However*, there is a growth condition on the c_j which is "naturally" associated to (b) (or (b′)), namely that growth condition on the c_j which would force (b) (or (b′)). This is just

$$c_j = o(\exp(-\delta + \varepsilon)j) \qquad \text{for any } \varepsilon > 0.$$

Thus the growth condition for the conclusion (d) is the same as the growth condition related to (b) or (b′).

In order to put the above in a general setting, we must decide on what we shall mean by "growth condition". For many purposes, a natural type of growth condition is that given in the theory of analytically uniform (au) spaces (cf. Ehrenpreis, *Analytically uniform spaces and some applications*, Trans. Amer. Math. Soc. **101** (1961), 52–79, and the author's forthcoming book *Fourier analysis in several complex variables*). An au space is a linear topological vector space W of functions or distributions (of x) which contains all exponentials $\exp(ixz)$ for all comlex z such that.

(1) W is reflexive, that is, W is the dual of its dual W'.

(2) $\exp(ixz)$ depends analytically on z, that is, for any $S \in W'$ the *Fourier transform*

$$\hat{S}(z) = S \cdot \exp(ixz) \tag{1}$$

is an entire function of z.

(The reason for defining the Fourier transform in this way is seen from the following example: If W is the space of C^∞ functions then any continuous function h of compact support defines the linear function S_h: $g \to \int g(x)h(x)\,dx$ on W. If we calculate $\hat{S}_h$ we have

$$\hat{S}_h(z) = S_h \cdot \exp(ixz) = \int h(x)\exp(ixz)\,dx$$

which is just the ordinary Fourier transform of h.)

(3) The linear combinations of $\exp(ixz)$ for varying z are dense in W. (Thus $\hat{S}$ determines S.)

(4) Denote by $\hat{W}'$ the space of Fourier transforms of $S \in W'$ with the topology to make the Fourier transform a topological isomorphism. We can find a family $K = \{k(z)\}$ of continuous positive functions such that for each $S \in W'$, $k \in K$ we have $\hat{S}(z)/k(z) \to 0$ as $|z| \to \infty$ and the seminorms

$$\|\hat{S}\|_k = \max |\hat{S}(z)|/k(z) \tag{2}$$

define the topology of $\hat{W}'$. The set K is called an au structure for W.

Using the Hahn-Banach theorem we deduce easily that every $g \in W$ has a Fourier representation

$$g(x) = \int \exp(ixz)\, d\mu(z)/k(z) \tag{3}$$

for some $k \in K$ and some bounded measure (finite total variation). The integral converges in the topology of W.

EXAMPLE A. Let $W = H(|x| < \delta)$ be the space of functions analytic in $|x| < \delta$. Then $\hat{W}'$ consists of all entire functions $G(x)$ such that for some $\varepsilon > 0$ there is an $A > 0$ so that

$$G(z) \leqq A \exp[(\delta - \varepsilon)|z|]. \tag{4}$$

An au structure consists of all continuous positive $k(z)$ such that the functions on the right side of (4) are $o(k(z))$.

EXAMPLE B. Let $W = H(Ix > b)$ be the space of functions analytic in $Ix > b$. Then $\hat{W}'$ consists of all entire functions $G(z)$ for which there exist A, B, ε so that

$$\begin{aligned} G(z) &\leqq A\exp(B|Iz| + B|Rz|) && \text{for} \quad Rz \leqq 0, \\ &\leqq A\exp(B|Iz| + (b-\varepsilon)|Rz|) && \text{for} \quad Rz \geqq 0. \end{aligned} \tag{5}$$

Again, an au structure consists of all continuous positive $k(z)$ such that the functions on the right side of (5) are $o(k(z))$.

Let us show how to cast Theorems I and II in the language of au spaces. First observe that a series $\sum c_j \exp(ijx)$ is of the form (3) where μ has its support on the positive integers if the c_j are smaller than $[j^2 k(j)]^{-1}$ for some k. In particular our description of Example B shows that Condition (a) is just the statement that the series $\sum c_j \exp(ijx)$ is of the form (3) for some k in an au structure for $H(Ix > 0)$. Conclusion (d) is the statement that the series is of the form (3) for some k in $H(Ix > -\delta)$.

We now observe from Examples A and B that the au structure for $H(Ix > -\delta)$ is obtained by taking functions of the form

$$k(z) = \max(k_1(z), k_2(z)) \tag{6}$$

where k_1 is in an au structure for $H(Ix > 0)$ and k_2 is in an au structure for $|Ix| < \delta$.

This leads us to define, in general, the au intersection $W_1 \cap^a W_2 = W$ of two au spaces W_1, W_2 as the au space whose au structure is given as in (6) for $k_1(k_2)$ in an au structure for $W_1(W_2)$.

We can now formulate the *general problem for lacunary series*: Let W_1, W_2 be two au spaces, and let $\{a_j\}$ be a discrete set of complex numbers. Suppose

(a″) The series $\sum c_j \exp(ia_j x)$ converges in the topology of W_1 to a function $f(x)$ and that $c_j = O(1/k_1(a_j))$ for some k_1 in an au structure for W_1.

(b″) $f(x) \in W_2$.

Find the correct "lacunary" condition on $\{a_j\}$ so as to force

(d″) $f(x) \in W_1 \cap^a W_2$ and $c_j = O(1/k_2(a_j))$ for some k_2 in an au structure for W_2.

RESULT A. Let W_1 be the space of functions analytic in $Ix > 0$ and having distribution boundary values, and W_2 the space of C^∞ functions on $Ix = 0$. Then a "correct" lacunary condition is: $|a_j - a_{j'}| \geq \eta > 0$ for all j,j', and for any real x_0 (large) the number of a_j in $|x - x_0| < \log|x_0|$ is $o(\log|x_0|)$. (Note that this condition is finer than the one in the Fabry theorem.)

RESULT B. (We make a slight change of notation for obvious reasons.) Let $f(s) = \sum c_n n^{-s}$ where the series converges in some right half plane. Suppose that $f(s)$ is an entire function sagtisfying

$$|f(s)| = O(\exp[|s|\log|s| - (1 + \log\pi)|Rs| - A\log|s|]) \tag{7}$$

for any $A > 0$. Then $c_n = O[\exp(-B\exp n)]$ for some $B > 0$.

REMARK. The function $(1 - 2^{1-s})\,\zeta(s)$ where $\zeta(s)$ is the Riemann ζ function shows (using the functional equaton and Stirling's formula) that (7) cannot be improved very much.

COURANT INSTITUTE OF MATHEMATICAL SCIENCES
NEW YORK UNIVERSITY
NEW YORK, NEW YORK

Asymptotic Properties of Entire Functions of Order Less than One

MATTS ESSÉN
TORD GANELIUS

1. Let f be an entire function. We denote $\max |f(z)|$ and $\min |f(z)|$ on $|z| = r$ by $M(r)$ and $m(r)$, respectively. We assume that $f(0) = 1$. Since we are going to consider the behavior of f at infinity, this is certainly not a severe restriction. If f is of order less than one we have the representation

$$f(z) = \prod_{1}^{\infty} (1 - z/z_n).$$

We also define

$$f_1(z) = \prod_{1}^{\infty} (1 + z/|z_n|)$$

and

$$I(R) = \int_0^R r^{-1-\lambda}(\log m(r) - \cos \pi\lambda \log M(r))\, dr.$$

The notations $m_1(r)$, $M_1(r)$ and $I_1(R)$ are used for the similar expressions calculated for f_1 instead of f. Let us also denote by $n(r)$ the number of zeros of f with moduli less than or equal to r.

The following formulas will be of importance in the sequel. They can for example be found in Essén [6].

$$m_1(r) \leqq m(r) \leqq M(r) \leqq M_1(r), \qquad m_1(r)M_1(r) \leqq m(r)M(r), \tag{1.1}$$

$$\log m_1(r) - \cos \pi\lambda \log M_1(r) \leqq \log m(r) - \cos \pi\lambda \log M(r). \tag{1.2}$$

In 1963, B. Kjellberg [10] proved the following theorem.

THEOREM 1. *Let f be an entire function and λ any number satisfying $0<\lambda<1$. If, for all large values of r, $\log m(r) - \cos \pi\lambda \log M(r) \leqq 0$, then $\lim_{r\to\infty} r^{-\lambda} \log M(r)$ exists. The limit is positive or infinite.*

The case $\lambda = 1/2$ is a well-known result given by Heins [8, 1948]. Several authors have been studying the same problem and they have shown that the condition on the minimum modulus can be weakened (Anderson [**1**]; [**2**], Essén [**6**]). We quote the following theorem by Essén [**6**].

THEOREM 2. *Let f be an entire function and λ any number satisfying $0 < \lambda < 1$. If the finite limit $\lim_{r\to\infty} r^{-\lambda}\log M(r) = \alpha$ exists (α may be zero), then $\lim_{r\to\infty} r^{-\lambda}\log M_1(r) = \alpha$, $\lim_{R\to\infty} I_1(R) = -\pi\alpha\sin\pi\lambda$ and $\lim_{R\to\infty} I(R)$ exists, being possibly infinite. Conversely, the existence of the finite limit $\lim_{\to\infty} I(R)$ implies the existence of the possibly infinite limit $\lim_{r\to\infty} r^{-\lambda}\log M(r)$.*

It has been suggested by Kjellberg that something might be said about the rapidity with which $r^{-\lambda}\log M(r)$ approaches its limit. We shall assume that something is known about the rate of convergence of $\lim_{R\to\infty} I_1(R)$ or $\lim_{R\to\infty} I(R)$ and derive estimates for $r^{-\lambda}\log M(r)$.

It should be remarked that the theorems quoted above could have been given as theorems for subharmonic functions and that the same methods of proof apply in this more general situation. This is also true for our results in the present paper (cf. Heins [**8**], Kjellberg [**9**]).

The lemmas we use are collected together in §2. The theorems in §3 are all simple consequences of these lemmas. In §4, we study in greater detail the case when the zeros of the function are not on a half line.

All our lemmas can be obtained by Fourier methods without application of contour integration and similar devices. However, we have chosen a method of presentation better showing the connection with previous work in the field.

2. We introduce the following notation which makes it possible for us to write all the formulas we use in convolution form:

$$\phi(x) = e^{-\lambda x}\log M(e^x),$$

$$v(x) = e^{-\lambda x} n(e^x),$$

$$\theta(x) = \int_x^\infty e^{-\lambda y}(\log m(e^y) - \cos\pi\lambda\log M(e^y))\,dy,$$

$$\psi(x) = \int_x^\infty e^{-\lambda y}(\sin\pi\lambda\log m(e^y) - \pi\cos\pi\lambda\, n(e^y))\,dy.$$

We also let ϕ_1, v_1, θ_1 and ψ_1 denote the similar functions obtained by using f_1 instead of f.

In all the lemmas λ is a number in the interval $(0,1)$. For $g \in L^1(-\infty,\infty)$, the Fourier transform is defined by $\hat{g}(t) = \int_{-\infty}^{\infty} g(x)e^{-itx}dx$.

LEMMA 1. *Assume that the limit $\lim_{x\to\infty}\phi(x) = \alpha$ exists. Then*

$$\phi_1 - \alpha = L * \theta_1 \tag{2.1}$$

where

$$L(x) = (2/\pi \sin \pi\lambda)[(1-\lambda)e^{-2\lambda x} + \lambda e^{2(1-\lambda)x} - 1](e^x - e^{-x})^{-2}$$

is nonnegative and

$$\hat{L}(t) = it/(\cos \pi\lambda - \cos \pi(\lambda + it)).$$

The lemma is closely related to formula (5.4) in Anderson [**1**], but we give a different proof.

PROOF. Let $R > 0$. We integrate

$$\frac{1}{z}\frac{(z/R)^{\lambda} - (z/R)^{-\lambda}}{z^2 - R^2}\log f_1(-z)$$

over a half circle in the upper half plane with its centre at $z = 0$ and a diameter on the real line. We define z^{λ} and $\log f_1(-z)$ to be real on the positive part of the real axis and the negative part of the real axis, respectively. Take real parts and let the radius of the half circle tend to infinity. The singularity at $z = -R$ throws out $\log|f_1(R)|$ and we obtain

$$\begin{aligned} &\pi \sin \pi\lambda\ R^{-\lambda}\log M_1(R) \\ &= -\int_0^{\infty} \frac{1-(r/R)^{2\lambda}}{1-(r/R)^2}(\log m_1(r) - \cos \pi\lambda \log M_1(r))\gamma^{-1-\lambda}dr. \end{aligned} \tag{2.2}$$

This is the formula we need. By introducing $R = e^x$, $r = e^y$, we obtain

$$\pi \sin \pi\lambda\, \phi_1(x) = \int_{-\infty}^{\infty} \frac{1 - e^{-2\lambda(x-y)}}{1 - e^{-2(x-y)}}\ \theta_1'(y)\, dy.$$

It follows from Theorem 2 that θ_1 is defined and that $\theta_1(-\infty) = -\pi\alpha \sin \pi\lambda$. We obviously have $\theta_1(\infty) = 0$, and a partial integration gives (2.1). Since

$$\pi \sin \pi\lambda\, L(x) = \frac{d}{dx}\left(\frac{1 - e^{-2\lambda x}}{1 - e^{-2x}}\right),$$

L is the derivative of an increasing function and hence nonnegative. A residue argument gives $\hat{L}$.

LEMMA 2.

$$\phi_1 = H * v \tag{2.3}$$

where $H(x) = e^{(1-\lambda)x}/(1 + e^x)$ *is nonnegative and* $\hat{H}(t) = \pi/\sin \pi(\lambda + it)$.

PROOF. The lemma follows by introducing $R = e^x$, $r = e^y$ in the well-known formula:

$$\log M_1(R) = R \int_0^\infty \frac{n(r)}{r(r+R)}\,dr.$$

Another residue argument gives $\hat{H}$.

LEMMA 3. *Assume that the limit* $\lim_{x\to\infty}\phi(x) = \alpha$ *exists. Then*

$$\phi_1 - \alpha = G * \psi_1 \tag{2.4}$$

where $G(x) = (1/\pi)(e^x/(e^x+1)^2)$ *is nonnegative and* $\hat{G}(t) = it/\sin \pi it$.

PROOF. This lemma is a consequence of formula (2.3) in Boas [**5**]. The argument is similar to the one used in the proof of Lemma 1. It follows from Theorem 2 that $\lim_{x\to\infty}\phi_1(x) = \alpha$ and we can use (1.9) and (1.10) in Anderson [**1**] to prove that ψ_1 is defined and that $\psi_1(-\infty) = -\alpha\pi$.

LEMMA 4. *Let the limit* $\lim_{x\to\infty}\phi(x) = \alpha$ *exist. Then*

$$\theta_1 = J * (v - \alpha \sin \pi\lambda/\pi) \tag{2.5}$$

where

$$J \in L^1(-\infty, \infty)$$

and

$$\hat{J}(t) = \hat{H}(t)/\hat{L}(t) = \pi(\cos \pi\lambda - \cos \pi(\lambda + it))/it \sin \pi(\lambda + it).$$

PROOF. Obviously $\hat{J} \in L^2(-\infty,\infty)$ and hence $J \in L^2(-\infty,\infty)$. We have the exact form of $\hat{J}$, and it follows by a standard argument that $J \in L^1(-\infty,\infty)$.

According to (2.1) and (2.3), $H * v - \alpha = L * \theta_1$ and since $J * L = H$, we have

$$L*(\theta_1 - J * v + \alpha\pi \sin \pi\lambda) = 0. \tag{2.6}$$

The fact that $\hat{L}(t) \neq 0$ for t real now implies that the translates of L are dense in L^1[**11**, p. 98]. In particular it is possible to approximate the characteristic function of any interval $(0, x)$ and hence it follows from (2.6) and the boundedness of the integrand that

$$\int_0^x (\theta_1(y) - J * v(y) + \alpha\pi \sin \pi\lambda)\,dy = 0.$$

Differentiating we obtain

$$\theta_1 - J * v + \alpha\pi \sin \pi\lambda = 0$$

and the lemma is proved.

REMARK. In exactly the same way, starting from (2.4), we can prove that there exists a function $R \in L^1(-\infty,\infty)$ such that $R * G = H$ and such that

$$\psi_1 = R * (v - \alpha \sin \pi\lambda/\pi).$$

It is of course also possible to obtain Lemma 4 by starting with the formula

$$\log m_1(R) - \cos \pi\lambda \log M_1(R) = \int_0^{\infty} \{\log|1 - R/r| - \cos \pi\lambda \log|1 + R/r|\}\, dn(r)$$

and then using complex integration.

3. Our first result is

THEOREM 3. *Let λ and γ be numbers such that $0 < \lambda < 1$ and $0 < \gamma < 2\lambda$. We assume that the limit $\lim_{R\to\infty} R^{-\lambda}\log M(R) = \alpha$ exists and that*

$$\int_R^{\infty} r^{-1-\lambda}(\log m(r) - \cos \pi\lambda \log M(r))\, dr \leqq O(R^{-\gamma}), \qquad R \to \infty.$$

Then

$$\alpha + o(1) \leqq R^{-\lambda} \log M(R) \leqq R^{-\lambda}\log M_1(R) \leqq \alpha + O(R^{-\gamma}), \qquad R \to \infty. \tag{3.1}$$

Theorem 3 is a consequence of the slightly more general Theorem 3′, which we give in convolution form.

THEOREM 3′. *Let λ satisfy $0 < \lambda < 1$, let the limit $\lim_{x\to\infty}\phi(x) = \alpha$ exist and assume that p is a subadditive, nonnegative, increasing function such that*

$$\int_0^{\infty} e^{p(y)-2\lambda y}\, dy < \infty. \tag{3.2}$$

Then $\theta(x) \leqq O(e^{-p(x)})$, $x \to \infty$, implies that

$$\alpha + o(1) \leqq \phi(x) \leqq \phi_1(x) \leqq \alpha + O(e^{-p(x)}), \qquad x \to \infty. \tag{3.3}$$

If, in Theorem 3′, we choose $p(x) = \gamma x$, $x \geqq 0$, and make our usual change of variables, we obtain Theorem 3.

PROOF OF THEOREM 3′. We may assume that $p(x) = p(0)$ for $x \leqq 0$. According to (1.2), $\theta_1 \leqq \theta$. We now use (2.1):

$$\phi_1 - \alpha = L * \theta_1 \leqq L * \theta \leqq O\left(e^{-p(x)} \int_{-\infty}^{\infty} L(y) e^{p(y)} dy\right).$$

The integral in the last expression is convergent due to condition (3.2) and the fact that p is bounded on the negative half of the real line. Since $\phi \leqq \phi_1$ (cf. (1.1)), the proof is complete.

Other similar results may be obtained by assuming that $\theta(x) \leqq -C/S(x)$ for some positive constant C and some increasing function S. It is also easy to give a similar theorem with θ replaced by ψ (cf. (2.4)).

We also give the following more precise form of a result of Anderson [2, Theorem 5].

THEOREM 4. *If* $\lim_{R\to\infty} R^{-\lambda}\log M(R) = 0$ *and if for some positive* δ,

$$\int_R^\infty r^{-1-\lambda}(\sin \pi\lambda \log m(r) - \pi \cos \pi\lambda\, n(r))dr \leqq o(R^{-\delta}), \qquad R\to\infty, \tag{3.4}$$

then $\lim_{R\to\infty} R^{-(\lambda-\delta)}\log M(R) = 0$ *so that* f *is at most of order* $\lambda - \delta$, *minimal type.*

We prove the equivalent theorem in convolution form.

THEOREM 4′. *If* $\lim_{x\to\infty}\phi(x) = 0$ *and if, for some positive* δ *less than one,* $\psi(x) \leqq o(e^{-\delta x})$, $x\to\infty$, *then* $\lim_{x\to\infty} e^{\delta x}\phi(x) = 0$.

A similar theorem with ψ replaced by θ is true.

It should perhaps be pointed out that in these theorems we only assume a one-sided bound for θ and ψ. Thus we obtain a known estimate also in the case $\lambda = 1/2$ under the assumption of Heins, i.e. $m(r) \leqq C$ implies $R^{-1/2}\log M(R) \leqq \alpha + O(R^{-1/2})$.

PROOF OF THEOREM 4′. According to (1.1), $\psi_1 \leqq \psi$. We now use (2.4) $\phi \leqq \phi_1 \leqq G * \psi_1 \leqq G * \psi \leqq o(e^{-\delta x})$, $x\to\infty$, and the proof is complete.

4. It seems desirable to have a better lower bound than the one given in (3.1). We believe that such a bound can only be obtained if special assumptions are made about the zeros. We shall give one result of this type. Let the zeros of f be $\{\rho_n e^{i\alpha_n}\}_1^\infty$, $-\pi < \alpha_n \leqq \pi$, and observe that

$$\begin{aligned} 0 &\leqq \int_R^\infty r^{-1-\lambda}(\log m(r) - \log m_1(r))\,dr \\ &\leqq \int_R^\infty r^{-1-\lambda}\{\log|f(-r)| - \log|f_1(-r)|\}\,dr \\ &\leqq \sum_1^\infty \int_0^\infty r^{-1-\lambda}\{\log|1 - r/(\rho_n e^{i\alpha_n})| - \log|1 - r/\rho_n|\}\,dr \\ &= \pi\sum_1^\infty \frac{\cos\lambda|\alpha_n| - \cos\lambda\pi}{\lambda\sin\lambda\pi\,\rho_n^\lambda} \leqq C^\lambda \sum_1^\infty \frac{\sin\lambda(\pi - |\alpha_n|)}{\rho_n^\lambda} \end{aligned} \tag{4.1}$$

where we have used the formula $\int_0^\infty r^{-1-\lambda}\log|1 + re^{i\alpha}|\,dr = \pi\cos\lambda\alpha/\lambda\sin\lambda\pi$. There is also an inversion of the order of integration and summation which is justified by the fact that all terms in the sum are positive.

If we assume that the last series in (4.1) converges and add something about the rate of convergence, we can improve the lower bound in (3.1).

THEOREM 5. *Let* λ *and* γ *be numbers in the interval* $(0,1)$ *and assume that* $\lim_{R\to\infty} R^{-\lambda}\log M(R) = \alpha$. *If*

$$\int_R^\infty r^{-1-\lambda}(\log m(r) - \cos\pi\lambda \log M(r))\,dr = O(R^{-\gamma}), \qquad R\to\infty, \tag{4.2}$$

and if

(4.3) $$\sum_{\rho_n \geqq y} \rho_n^{-\lambda} \sin \lambda(\pi - |\alpha_n|) = O(y^{-\sigma}), \qquad y \to \infty,$$

where $\sigma = \gamma(1+\lambda)/(1+\lambda-\gamma)$, *then*

(4.4) $$O(R^{-\gamma/2}) \leqq R^{-\lambda} \log M(R) - \alpha \leqq O(R^{-\gamma}), \qquad R \to \infty.$$

The equivalent convolution theorem is

THEOREM 5′. *Let* λ *and* γ *be numbers in the interval* $(0,1)$ *and assume that* $\lim_{x\to\infty}\phi(x) = \alpha$. *If*

(4.2′) $$\theta(x) = O(e^{-\gamma x}), \qquad x \to \infty,$$

and if (4.3) *is true, then*

(4.4′) $$O(e^{-\gamma x/2}) \leqq \phi(x) - \alpha \leqq O(e^{-\gamma x}), \qquad x \to \infty.$$

Finally we use Theorem 5 and a well-known Tauberian theorem of Beurling to prove the following result.

THEOREM 6. *If* $\lim_{R\to\infty} R^{-\lambda} \log M(R) = \alpha$, *and if* (4.2) *and* (4.3) *are true with* $\gamma < 2\lambda$, *then*

(4.5) $$r^{-\lambda} n(r) = \alpha \sin \pi\lambda_{/}\pi + O(r^{-\gamma/2}), \qquad r \to \infty.$$

The case of real roots and with $\lambda = 1/2$ is given by Beurling [3, p. 22]. Using the remark after the proof of Lemma 4, it is easy to state and prove a similar theorem with θ replaced by ψ.

PROOF OF THEOREM 5. We need two estimates. The first is obtained in the following way:

$$0 \leqq \int_R^\infty r^{-1-\lambda}(\log m(r) - \log m_1(r))\,dr$$

$$\leqq \sum_{\rho_n < y} \int_R^\infty r^{-1-\lambda}(\log|1 + r/(\rho_n e^{i\alpha_n})| - \log|1 - r/\rho_n|)\,dr$$

$$+ C_\lambda \sum_{\rho_n \geqq y} \frac{\sin \lambda(\pi - |\alpha_n|)}{\rho_n^\lambda} \leqq \int_R^\infty r^{-1-\lambda} \int_0^y \log \left|\frac{s+r}{s-r}\right| dn(s)\,dr + O(y^{-\sigma}).$$

In the estimate of the second term, (4.3) has been used. Suppose $y < R/2$ and introduce $n(s) = O(s^\lambda)$ (cf. Boas [4, 2.9.5]) after a partial integration. We obtain

$$0 \leqq \int_R^\infty r^{-1-\lambda}(\log m(r) - \log m_1(r))\,dr$$

$$\leqq \int_R^\infty r^{-1-\lambda}\left[n(y)\log\frac{r+y}{r-y} - \int_0^y n(s)\left(\frac{1}{r+s} + \frac{1}{r-s}\right)ds\right]dr + O(y^{-\sigma})$$

$$\leqq O(y^{1+\lambda}) \int_R^\infty r^{-2-\lambda}\,dr + O(y^{-\sigma}) = O((y/R)^{1+\lambda} + y^{-\sigma}).$$

By choosing $y = R^{(1+\lambda)/(1+\lambda+\sigma)}$, which is certainly less than $R/2$ for large values of R, we obtain

$$0 \leqq \int_R^\infty r^{-1-\lambda}(\log m(r) - \log m_1(r))dr = O(R^{-\gamma}). \tag{4.6}$$

This is our first estimate. Our second estimate is obtained from the second inequality in (1.1):

$$\begin{aligned} 0 &\leqq \int^\infty r^{-1-\lambda}(\log M_1(r) - \log M(r))\,dr \\ &\leqq \int_R^\infty r^{-1-\lambda}(\log m(r) - \log m_1(r))\,dr \leqq O(R^{-\gamma}). \end{aligned} \tag{4.7}$$

For the proof of the theorem we start by using the second inequality in (1.1) to obtain

$$\begin{aligned} 0 &\leqq \log m(r) - \cos \pi\lambda \log M(r) - (\log m_1(r) - \cos \pi\lambda \log M_1(r)) \\ &\leqq (1 + \cos \pi\lambda)\,(\log m(r) - \log m_1(r)). \end{aligned}$$

Hence (4.2) combined with (4.6) gives

$$\theta_1(x) = O(e^{-\gamma x}), \qquad x \to \infty, \tag{4.8}$$

and it follows from (2.1) that

$$\phi_1(x) - \alpha = O(e^{-\gamma x}), \qquad x \to \infty. \tag{4.9}$$

The formulas (1.1) and (4.7) imply that for every $y > x$ we have

$$\int_x^y (\phi(u) - \alpha)\,du \geqq \int_x^y (\phi_1(u) - \alpha)\,du + O(e^{-\gamma x}) \geqq O(e^{-\gamma x}), \qquad x \to \infty.$$

It is a well-known property of ϕ that there is a constant C such that $\phi(x) + Cx$ is nondecreasing. Thus

$$\int_x^y (\phi(y) - \alpha + C(y-u))\,du \geqq \int_x^y (\phi(u) - \alpha)\,du \geqq O(e^{-\gamma x}).$$

Choosing $y - x = e^{-\gamma x/2}$, we obtain

$$\phi(y) - \alpha \geqq O(e^{-\gamma y/2}). \tag{4.10}$$

The formulas (4.9) and (4.10) constitute Theorem 5.

PROOF OF THEOREM 6. Under the assumptions of the theorem, (4.8) is true. We now apply Beurling's theorem [3, p. 22] (for a proof see Ganelius [7]) to formula (2.5). We can in a natural way extend the range of the function $\hat{J}$ for every strip $0 < \operatorname{Im} z < 2\lambda'$, $0 < \lambda' < \lambda$, in such a way that $\{\hat{J}(z)\}^{-1}$ will be holomorphic and $d/dz\{\hat{J}(z)\}^{-1}$ will be bounded in the strip. Since there exists a constant C such that $v(x) + Cx$ is nondecreasing, the Tauberian condition (3.1) in [7] is satisfied, and we can apply Theorem 1 in [7]. Hence formula (2.5), that is $\theta_1 = J * (v - \alpha \sin \pi\lambda/\pi) = O(e^{-\gamma x})$, $x \to \infty$, implies that

$$v(x) = \alpha \sin \pi\lambda/\pi + O(e^{-\gamma x/2}),$$

which is equivalent to (4.5), and our theorem is proved.

References

1. J. M. Anderson, *Asymptotic properties of integral functions of genus zero*, Quart. J. Math. Oxford Ser. (2) **16** (1965), 151–164.

2. ———, *Growth properties of integral and subharmonic functions*, J. Analyse Math. **13** (1964), 355–389.

3. A. Beurling, *Sur les intégrales de Fourier absolument convergentes et leur application à une transformation fonctionnelle*, Congr. Math. Scand., Helsinki, 1938.

4. R. P. Boas, *Entire functions*, Academic Press, New York, 1954.

5. ———, *Integral functions with negative zeros*, Canad. J. Math. **5** (1953), 179–184.

6. M. Essén, *A theorem on the maximum modulus of entire functions*, Math. Scand. **17** (1965), 161–168.

7. T. Ganelius, *The remainder in Wiener's tauberian theorem*, Mathematica Gothoburgensia, **1**. Acta Universitatis Gothoburgensis, Göteborg, 1962.

8. M. Heins, *Entire functions with bounded minimum modulus; subharmonic functions analogues*, Ann. of Math. (2) **49** (1948), 200–213.

9. B. Kjellberg, *On the minimum modulus of entire functions of order less than one*, Math. Scand. **8** (1960), 189–197.

10. ———, *A theorem on the minimum modulus of entire functions*, Math. Scand. **12** (1963), 5–11.

11. N. Wiener, *The Fourier integral and certain of its applications*, Cambridge Univ. Press, New York, 1933.

Royal Institute of Technology,
Stockholm, Sweden

University of Göteborg,
Göteborg, Sweden

The Behavior of an Entire Function as Determined by its Maximum Term

ALFRED GRAY
S. M. SHAH[1]

1. **Introduction.** Let $f(z) = \sum_{n=0}^{\infty} a_n z^n$ be a transcendental entire function and let $M(r)$ denote the maximum modulus, $\mu(r)$ the maximum term and $\nu(r)$ the central index of $f(z)$. Hence we have

$$\mu(r) = \max_n |a_n| r^n = |a_{\nu(r)}| r^{\nu(r)}.$$

By the Wiman-Valiron theory **[21]** we have

$$M(r) < \mu(r)\{\log \mu(r)\}^{1/2+\varepsilon}, \quad (\varepsilon > 0), \tag{1}$$

valid outside an exceptional set E of finite logarithmic measure. Rosenbloom **[18]** gave a proof of (1) based on probabilistic considerations; in fact, he proved the following sharper result:

THEOREM 1. *Let $\psi_1(y)$ and $\psi_2(y)$ be increasing functions of y and suppose that*

$$\int_a^{\infty} \frac{dy}{\psi_j(y)} < \infty \qquad (j = 1, 2).$$

Then if $\psi(y) = \psi_2(\psi_1(y))$ we have

$$M(r)/(\psi(\log M(r)))^{1/2} \leqq K\mu(r) \tag{2}$$

for some constant K outside a set of finite logarithmic measure.

T. Kővári **[16]** proved analogous results for functions regular for $|z| < 1$.

Equation (1) shows that the maximum term determines to some extent the growth rate of an entire function. If more information about a particular entire function is known, the maximum term determines the growth rate to a greater extent, and furthermore something can be said about the value distribution of the entire function. In this paper we discuss some results of this type.

[1] The work of the first author was supported by NSF grant GP-3990 and an NSF Post-doctoral Fellowship and that of the second by NSF grant GP-4311.

2. **Functions of a given order.** Denote by ρ and λ the order and lower order of $f(z)$ and by $f'(z)$ the derivative of $f(z)$. If $\rho < \infty$ it is known that

(3) $$\lim_{r\to\infty} \frac{\log \mu(r)}{\log M(r)} = 1.$$

On the other hand for functions of finite or infinite order we have

(4) $$\liminf_{r\to\infty} \frac{\log \mu(r)}{\nu(r)} \leqq \frac{1}{\rho} \leqq \frac{1}{\lambda} \leqq \limsup_{r\to\infty} \frac{\log \mu(r)}{\nu(r)},$$

and outside a set of finite logarithmic measure

(5) $$\log M(r,f) \sim \log \mu(r,f) \sim \log(rM(r,f')).$$

If $f(z)$ satisfies a regular growth condition we have **[20]**

THEOREM 2. *Let $\Phi(x)$ and $L(x)$ be positive and increasing for $x > a > 0$ and assume $L(2x) \sim L(x)$ as $x \to \infty$. Write $\mu = \mu(r,f)$ and $R = \exp\{M(r,f)^\alpha\}$ and suppose that $f(z)$ satisfies the condition*

$$\limsup_{r\to\infty} \frac{1}{L(r)} \int_a^\mu \frac{dt}{t\Phi(t)} < \liminf_{r\to\infty} \frac{1}{L(r)} \int_a^R \frac{dt}{t\Phi(t)}$$

for all $\alpha > 0$ and for some $L(r)$ and $\Phi(t)$. Then

$$\lim_{r\to\infty} \frac{\log(rM(r,f'))}{\log \mu(r,f)} = 1.$$

In particular (5) holds without any restriction on r if for some integer $k \geqq 2$ and for some $L(r)$, $l_k(M(r,f))/L(r)$ tends to a finite nonzero limit as $r \to \infty$. Here l_k denotes the kth iterate of the logarithm.

3. **Gap conditions.** Suppose $f(0) \neq 0$ and let $\{n_k\}$ denote the range of $\nu(r)$ and $\{\rho(n)\}$ the points of discontinuity of $\nu(r)$ so that $\rho(1) > 0$ and

$$\rho(n_k) < \rho(n_k + 1) = \cdots = \rho(n_{k+1}),$$

for $k = 1, 2, \cdots$. Write

$$\left.\begin{matrix} B \\ b \end{matrix}\right\} = \lim_{k\ \infty} \begin{matrix} \sup \\ \inf \end{matrix} \frac{n_{k+1}}{n_k},$$

$$\left.\begin{matrix} L \\ l \end{matrix}\right\} = \lim_{n\to\infty} \begin{matrix} \sup \\ \inf \end{matrix} \frac{\rho(n+1)}{\rho(n)},$$

$$l^* = \liminf_{k\to\infty} \frac{\rho(n_{k+1})}{\rho(n_k)}.$$

THEOREM 3. *For any entire function we have* **[11]**

$$\text{(7)} \qquad \frac{\log l^*}{B-1} \leqq \liminf_{r\to\infty} \frac{\log \mu(r)}{\nu(r)} \leqq \frac{\log b}{\rho(b-1)} \leqq \frac{1}{\rho}$$

with the conventions that if $B = \infty$, then the left extreme is equal to 0 and the third term is equal to ∞ when $\rho = 0$, and $(\log b)/(b-1)$ is taken equal to 1 when $b = 1$. Also,

$$\text{(8)} \qquad \frac{1}{\lambda} \leqq \frac{b \log b}{\lambda(b-1)} \leqq \limsup_{r\to\infty} \frac{\log \mu(r)}{\nu(r)} \leqq \frac{b \log L}{b-1}$$

with the conventions that the right extreme is equal to ∞ if either $b = 1$ or $L = \infty$ and the second term is equal to 0 when $\lambda = \infty$.

We now consider the limiting values of $\mu(r)/M(r)$ as $r \to \infty$. Let us write

$$\left.\begin{matrix} U \\ u \end{matrix}\right\} = \lim_{r\to\infty} \begin{matrix} \sup \\ \inf \end{matrix} \frac{\mu(r)}{M(r)}.$$

For example, for e^z, $\mu(r) \sim e^r/(2\pi r)^{1/2}$ and so $U = 0$ in this case. Erdös **[3]** conjectured that if $\lim_{r\to\infty}(\mu(r)/M(r))$ exists it must be 0. Clunie and Hayman **[1]** proved this conjecture false by constructing entire functions for which

$$\text{(9)} \qquad \lim_{r\to\infty} \frac{\mu(r)}{M(r)} = \alpha$$

for any α with $0 \leqq \alpha \leqq 1/2$. They also proved that $u < 2/\pi$. It is an open question whether (9) holds for some α with $1/2 < \alpha < 2/\pi$. Another result of Clunie and Hayman is that the Erdös conjecture does hold for functions with positive coefficients. It also holds **[5]** when $L > 1$.

The question arises: What can be said about the value distribution of $f(z)$ when $U > 0$ Erdös and Macintyre **[4]** proved that if

$$\sum \frac{1}{\lambda_{k+1} - \lambda_k} < \infty,$$

then for $f(z) = \sum a_k z^{\lambda_k}$ there exists a sequence $\{r_k\}$ tending to ∞ such that

$$f(z) = a_k z^{\lambda_k}(1 + o(1))$$

for $|z| = r_k$ and $\mu(r_k) = |a_k| r_k^{\lambda_k}$. For this function $U = 1$ and there are no finite asymptotic values or defect values.

A different sort of example for which $U > 0$ is given by the function

$$\sum_{n=0}^{\infty} z^n \lambda^{-n(n+1)/2} \qquad (\lambda > 1)\,,$$

which has order zero. This function also has no asymptotic values or defect values. Thus there seems to be a correlation between the nonexistence of asymptotic values or defect values and the condition $U > 0$. However, nothing very precise is known. Clunie and Hayman [1] have shown that if $f(z)$ never vanishes, then $u = 0$. Another result along this line is the following [12]:

THEOREM 4. *Let* $f(z) = \sum_{n=0}^{\infty} a_n z^n$ *be a transcendental entire function satisfying condition* (i) *and either condition* (ii) *or* (iii):

(i) $a_n \geqq 0$ *for all but a finite number of* n,

(ii) $f(z)$ *has a Borel exceptional value.*

(iii) $f(z)$ *is of mean type integer order* ρ *and there exists a complex number* a *such that the canonical product formed with the zeros of* $f(z) - a$ *is at most zero type order* ρ.

Then $U \leqq 1/2$. *If* $f(z)$ *satisfies* (i) *and* $U > 1/2$ *then* $f(z)$ *has no finite asymptotic value.*

The results obtained by Kövári on Borel exceptional values and by Pólya (1929), Macintyre (1952), Edrei (1957), and Kövári on asymptotic values are of a different nature. Thus Kövári [14] has shown that if $f(z) = \sum a_n z^{\lambda_n}$ is an entire function and if $\lambda_n > n(\log n)^{1+\varepsilon}$, $\varepsilon > 0$, then $f(z)$ can have no Borel exceptional values or finite asymptotic values [15].

4. **Asymptotic values of an entire function with respect to its maximum term.** We shall mainly be concerned with the relation of the growth of $\mu(r)$ to the existence or nonexistence of asymptotic values of $f(z)$. Many of the theorems given here are also true for functions with a finite radius of convergence.

There are two natural ways that the maximum term $\mu(r)$ can be extended to a complex function. We set

$$\mu(z) = |a_{\nu(r)}| z^{\nu(r)} = \mu(r) e^{i\nu(r)\theta}$$

and

$$m(z) = a_{\nu(r)} z^{\nu(r)} \qquad \text{for } |z| = r > 0.$$

DEFINITION. Let $\gamma(t)$ be a (continuous) curve such that $|\gamma(t)| \to \infty$ as $t \to \infty$. If

$$f(\gamma(t))/\mu(\gamma(t)) \quad \text{or} \quad f(\gamma(t))/m(\gamma(t))$$

tends to a limit ω ($0 \leqq |\omega| \leqq \infty$) as $t \to \infty$ we say that ω is a μ or m *asymptotic value*, respectively, and $\gamma(t)$ is a μ or m *asymptotic path*, respectively.

The importance of the above notion is evidenced in the following lemma, whose proof is immediate:

LEMMA. *If* f *has an ordinary finite asymptotic value along a path* γ, *then* f *has* μ *and* m *asymptotic value* 0 *along* γ.

Thus in order to show that f has no finite asymptotic values it suffices to prove that f has no μ or m asymptotic values. The determination of the μ and m asymptotic values of a function f is similar but more difficult than the problem of determining the limiting values of $\mu(r)/M(r)$. The latter is not a special case of the former since the curve on which the maximum modulus is assumed need not be continuous μ.

Although it is possible to construct examples with [cf. 21, p. 27] asymptotic values which are not m asymptotic values and vice versa, the question of existence of the two kinds is quite similar, and so we usually restrict ourselves to μ asymptotic values. Of course the two notions coincide for positive coefficients.

We discuss two classes of functions; the first class always has μ asymptotic values of a certain type and the second has none.

Let $f(z)$ be real for real z and $f(r) > 0$ for large enough r. Write

$$a(r) = rf'(r)/f(r) \quad \text{and} \quad b(r) = ra'(r).$$

These are the first and second logarithmic derivatives of $M(r)$. Hayman **[13]** has defined f to be *admissible* provided $b(r) \to \infty$ as $r \to \infty$ and there exists a function $\delta(r)$ defined for $R_0 < r < \infty$ satisfying $0 < \delta(r) < \pi$ such that

$$f(re^{i\theta}) \sim f(r)\exp[i\theta a(r) - \tfrac{1}{2}\theta^2 b(r)] \tag{10}$$

as $r \to \infty$ uniformly for $|\theta| \leqq \delta(r)$, while uniformly for $\delta(r) \leqq |\theta| \leqq \pi$,

$$f(re^{i\theta}) = o(f(r)/b(r)^{1/2}). \tag{11}$$

Examples of admissible functions are $e^{P(z)}$ where $P(z)$ is a real polynomial such that $e^{P(z)}$ attains its maximum modulus on the real axis and the Mittag-Leffler function

$$E_\alpha(z) = \sum_{n=0}^{\infty} \frac{z^n}{\Gamma(1+\alpha n)} \qquad \alpha > 0.$$

If $f(z)$ is admissible so is $e^{f(z)}$.

THEOREM 5. *If f is admissible, the positive real axis is a μ asymptotic path with μ asymptotic value ∞. Any path in an angle outside the positive real axis is a μ asymptotic path with μ asymptotic value* 0. *If c is a real number with $0 < c < \infty$ then c is a μ asymptotic value of f provided*

$$a(r) - v(r) = o((b(r)/\log b(r))^{1/2}).$$

PROOF. Hayman **[13]** has shown that for admissible functions

$$\mu(r) \sim f(r)/(2\pi b(r))^{1/2}.$$

It is immediate that $f(r)/\mu(r) \to \infty$. Furthermore

$$|f(re^{i\theta})/\mu(re^{i\theta})| \sim (2\pi b(r))^{1/2}\exp[-\tfrac{1}{2}\theta^2 b(r)] \qquad (|\theta| \leqq \delta(r)),$$

$$|f(re^{i\theta})/\mu(re^{i\theta})| = o(1/(b(r)^{1/2}) \qquad (\delta(r) \leqq |\theta| \leqq \pi).$$

This proves the second statement. The last statement can be proved by choosing a path suitably near the positive real axis.

Next we give conditions that an entire function f have no μ asymptotic values. Define

$$S = \limsup_{k\to\infty} (n_{k+1} - n_k),$$

$$\left.\begin{matrix}\Phi \\ \phi\end{matrix}\right\} = \lim_{k\to\infty} \begin{matrix}\sup \\ \inf\end{matrix} (n_{k+1} - n_k) \log \left\{\frac{\rho(n_{k+1})}{\rho(n_k)}\right\},$$

$$\left.\begin{matrix}\Xi \\ \xi\end{matrix}\right\} = \lim_{k\to\infty} \begin{matrix}\sup \\ \inf\end{matrix} (n_k - n_{k-1}) \log \left\{\frac{\rho(n_{k+1})}{\rho(n_k)}\right\}.$$

We have $\rho(n_k) = |a_{n_{k-1}}/a_{n_k}|^{1/(n_k - n_{k-1})}$.

THEOREM 6. *If $L > 1$ and $S < \infty$, then $f(z)$ has no asymptotic values.*

THEOREM 7. *Assume $f(z)$ has the form*

$$f(z) = \sum_{k=0}^{\infty} a_{n_k} z^{n_k}$$

where $\{n_k\}$ is the range of $\nu(r)$. If either

$$\phi = \Xi = \infty, \qquad \xi > 0,$$

or

$$\xi = \Phi = \infty, \qquad \phi > 0,$$

then $f(z)$ has no μ or m asymptotic values.

THEOREM 8. *Suppose $f(z)$ has the form*

$$f(z) = \varepsilon(0) + \sum_{k=1}^{\infty} \frac{\varepsilon(k) z^{n_k}}{\rho(1) \cdots \rho(n_k)}$$

where $|\varepsilon(k)| = 1$ and $\varepsilon(k)$ has period h, a positive integer. If

$$0 < \phi = \Phi < \infty \quad \text{and} \quad 0 < \xi = \Xi < \infty$$

then $f(z)$ has no μ asymptotic values.

We give an outline of the proof of Theorem 6. The following lemma will be necessary.

LEMMA. *Let $1 \leqq r \leqq \rho(n_{k+1})/\rho(n_k)$. Then*

$$M(\rho(n_k)r)/\mu(\rho(n_k)r) \leqq \pi(1 + r^{n_{k-1} - n_k})/4.$$

The proof is a trick using the Cauchy integral formula.

Now for the proof of the theorem. Without loss of generality we may assume $f(0) = 1$. Let $1 < \alpha < \beta < L_1 < L$ and

$$\alpha < (\pi/(4-\pi))^{1/S}.$$

There exists a sequence $\{k(p)\}$ of integers such that

$$\rho(k(p+1))/\rho(k(p)) > L_1.$$

Write $\phi(w) = f(\rho(k(p+1)w)/\mu(\rho(k(p))w)$

for $w \in \Omega_1 = \{w \mid 1 < |w| < L_1\}$. It is straightforward to show that the estimate

$$|\phi_p(w)| < 1 + \sum_{j=1}^{\infty} \left(\frac{|w|}{L_1}\right)^j + \sum_{j=1}^{\infty} |w|^{-j}$$

holds.

Therefore, $\{\phi_p(w)\}$ is uniformly bounded on compact subsets of Ω_1 and so it is a normal family. Thus there is a subsequence of $\{\phi_p(w)\}$ which converges uniformly on compact subsets of Ω_1. We may therefore assume that $\{k(p)\}$ has been so chosen that $\{\phi_p(w)\}$ itself converges uniformly on compact subsets to a holomorphic function $G(w)$.

We shall show that $G(w)$ is nonconstant, for suppose $G(w) \equiv C$ on Ω_1. The constant term in the Laurent expansion of $\phi_p(w)$ about the origin has absolute value 1, and so for $1 < r < L_1$ we have

$$|C| = \left|\frac{1}{2\pi i}\int_{|w|=r} \frac{G(w)}{w} dw\right| = \lim_{p\to\infty}\left|\frac{1}{2\pi i}\int_{|W|=r} \frac{\phi_p(w)}{w} dw\right| = 1.$$

Thus $|G(w)| \equiv 1$ on Ω_1. But by the lemma

$$M(r, G) = \lim_{p\to\infty} M(r, \phi_p) \geqq \frac{\pi}{4}(1 + r^{-S})$$

for $1 < r < L_1$. In particular for $r = \alpha$ we have $M(\alpha, G) > 1$, a contradiction. Hence G is nonconstant.

Let $\Omega = \{w \mid \alpha \leqq |w| \leqq \beta\}$ and suppose that $f(z)$ has μ asymptotic value ω. Then there exists a curve $\gamma(t)$ with $|\gamma(t)| \to \infty$ as $t \to \infty$ such that $f(\gamma(t))/\mu(\gamma(t)) \to \omega$.

There exists an unbounded set I with the following property: for each $t \in I$ there is a unique integer p such that

$$\rho(n_{k(p)}) \leqq |\gamma(t)| \leqq L_1 \rho(n_{k(p)}).$$

Write $\gamma(t) = \rho(k(p))\gamma_p(t)$. Then

$$1 \leqq |\gamma_p(t)| \leqq L_1 \rho(n_{k(p)})$$

so $\{\gamma_p(t)\}$ is bounded. We now consider the set T of limit points of $\{\gamma_p(t)\}$ as $t \to \infty$, $t \in I$, which lie in Ω and prove that they are an uncountable set on which $G(w)$ is constant. In fact, let Σ be the intersection of Ω with the positive real axis, and define $\chi: \Sigma \to T$ follows. For each p and each $x \in \Sigma$ there exists $t_p \in I$ such that $|\gamma(t_p)| = \rho(n_{k(p)})x$; then $|\gamma_p(t_p)| = x$. Choose a limit point v of $\{\gamma_p(t_p)\}$ and define $\chi(x) = v$. Then χ is one-one since $|\chi(x)| = x$. Thus T is uncountable since χ is.

Furthermore, $G(w)$ is constant on T, for suppose $\gamma_p(t_s) \to b \in T$ for a sequence $\{t_s\}$ with $t_s \in I$. By virtue of uniform convergence $\phi_p(\gamma_p(t_s)) \to G(b)$. But we are assuming that ω is a μ asymptotic value and so $G(b) = \omega$. Hence G is constant on T. This is a contradiction; therefore G has no μ asymptotic values.

REFERENCES

1. J. Clunie and W. K. Hayman, *The maximum term of a power series.* I, J. Analyse Math. **12** (1964), 143–186.

2. ———, *The maximum term of a power series.* II, J. Analyse Math. **14** (1965), 15–65.

3. P. Erdös, *Some unsolved problems*, Michigan Math. J. **4** (1957), 291–300.

4. P. Erdös and A. J. Macintyre, *Integral functions with gap power series*, Proc. Edinburgh Math. Soc. (2) **10** (1954), 62–70.

5. A. Gray and S. M. Shah, *A note on entire functions and a conjecture of Erdös*, Bull. Amer. Math. Soc. **69** (1963), 573–577.

6. ———, *A note on entire functions and a conjecture of Erdös.* II, J. Analyse Math. **12** (1964), 83–104.

7. ———, *Holomorphic functions with gap power series*, Math. Z. **86** (1965), 375–394.

8. ———, *Asymptotic values of a holomorphic function with respect to its maximum term*, Pacific J. Math. **18** (1966), 111–120.

9. ———, *Asymptotic values of holomorphic functions of irregular growth*, Bull. Amer. Math. Soc. **71** (1965), 747–749.

10. ———, *Holomorphic functions with gap power series.* II, J. Math. Anal. Appl. **13** (1966), 370–382.

11. ———, *Holomorphic functions with gap power series.* III, J. Math. Mech. **16** (1966), 297–310.

12. ———, *Borel exceptional values of entire functions*, Math. Ann. (to appear).

13. W. K. Hayman, *A generalization of Stirling's formula*, J. Reine Angew. Math. **196** (1956), 67–95.

14. T. Kövári, *On the Borel exceptional values of lacunary integral functions*, J. Analyse Math. **9** (1961), 71–109.

15. ———, *On the asymptotic paths of entire functions with gap power series*, J. Analyse Math. **15** (1965), 281–286.

16. ———, *On the maximum modulus and maximum term of functions analytic in the unit disc*, J. London Math. Soc. **41** (1966), 129–137.

17. G. Pólya and G. Szegö, *Aufgaben und Lehrsätze aus der Analysis*, vol. II, Springer, Berlin, 1925.

18. P. C. Rosenbloom, "Probability and entire functions" in *Studies in mathematical analysis and related topics*, Stanford Univ. Press, Stanford, Calif., 1962, pp. 325–332.

19. S. M. Shah, *The maximum term of an entire series*, Math. Student **10** (1942), 80–82.

20. ———, *On entire functions of infinite order*, Arch. Math. **14** (1963), 323–327.

21. G. Valiron, *Lectures on the General Theory of Integral Functions*, Chelsea, New York, 1949.

UNIVERSITY OF CALIFORNIA,
BERKELEY, CALIFORNIA

UNIVERSITY OF KENTUCKY,
LEXINGTON, KENTUCKY

Note on Hadamard's Convexity Theorem

W. K. HAYMAN

1. **Introduction and statement of results.** Suppose that $f(z)$ is regular in an annulus $r_1 < |z| < r_2$, and

$$M(r) = \sup_{|z|=r} |f(z)|, \qquad r_1 < r < r_2,$$

$$b(r) = \left(r \frac{d}{dr}\right)^2 \log M(r), \qquad r_1 < r < r_2.$$

The function $b(r)$ exists and is continuous except at isolated points and Hadamard's Theorem asserts that $b(r) \geqq 0$. In certain situations we can say a little more. We have

THEOREM 1. *If $f(z)$ is regular for $r_0 \exp[-\pi^2/4] < |z| < r_0 \exp[\pi^2/4]$, and $f(z_0) = 0$ where $|z_0| = r_0 \neq 0$, then there exists r in the annulus*

$$r_0 \exp[-\pi^2/4] < r < r_0 \exp[\pi^2/4],$$

such that $b(r) \geqq A_0$, where A_0 is an absolute constant, such that $A_0 > .18$.

We deduce immediately

THEOREM 2. *If $f(z)$ is a polynomial other than cz^n, then*

$$\sup_{r>0} b(r) \geqq A_0.$$

THEOREM 3. *If $f(z)$ is a transcendental integral function, then*

$$\limsup_{r\to\infty} b(r) \geqq A_0.$$

In the opposite direction we have

THEOREM 4. *The conclusions of Theorems 1 to 3 are false if $A_0 > \frac{1}{4}$.*

It seems quite probable that $A_0 = \frac{1}{4}$ is the correct value at least for Theorems 2 and 3.

2. **Proof of Theorems 1 to 3.** It is evident that the conclusion of Theorem 1 holds if $M(r)$ and hence $b(r)$ is unbounded in the annulus. Thus we assume that $\log M(r)$ is bounded.

Set

$$\sigma = \log r - \log |z_0|,$$

and let $L(\sigma) = B\sigma + C$, where the real constants B, C are so chosen that

$$\lim_{\sigma \to \mp \pi^2/4} \{\log M(r_0 e^\sigma) - L(\sigma)\} = 0.$$

Since $\log M(r)$ is a bounded convex function of $\log r$, the limits exist and determine $L(\sigma)$ uniquely. Also by Hadamard's Theorem

$$|f(z_0 e^{\sigma + i\tau})| \leqq e^{L(\sigma)} \tag{1}$$

in the strip S: $-\pi^2/4 < \sigma < \pi^2/4$, $-\infty < \tau < +\infty$.

We now note that $f(z_0 e^\zeta)$ has zeros at all the points $\zeta = 2k\pi i$, where k is an integer. Let N be a positive integer and consider in S the function

$$\phi_N(\zeta) = f(z_0 e^\zeta) e^{-L(\zeta)} \prod_{n=-N}^{N+1} \cot\left(\frac{\zeta - 2n\pi i}{\pi}\right).$$

Then $\phi_N(\zeta)$ is regular in S and in view of (1) and the fact that $|\cot(\mp\pi/4 + i\tau)| = 1$ for any real τ, we deduce that

$$\limsup |\phi_N(\zeta)| \leqq 1, \tag{2}$$

as ζ approaches any finite boundary point of S from inside S. Similarly (2) holds as $\zeta \to \infty$ in S. Thus $|\phi_N(\zeta)| \leqq 1$ in S by the maximum principle. In particular we deduce for $\sigma = 0$, $0 \leqq \tau \leqq 2\pi$,

$$\begin{aligned} |f(z_0 e^{i\tau})| &\leqq e^C \prod_{n=-N}^{N+1} \left|\tanh \frac{\tau - 2n\pi}{\pi}\right| \\ &= e^C \prod_{n=0}^{N} \left\{\tanh \frac{2(n+1)\pi - \tau}{\pi} \tanh \frac{2n\pi + \tau}{\pi}\right\} \\ &= e^C \prod_{n=0}^{N} \frac{\cosh(4n+2) - \cosh[2(1-\tau/\pi)]}{\cosh(4n+2) + \cosh[2(1-\tau/\pi)]} \\ &\leqq e^C \prod_{n=0}^{N} \frac{\cosh(4n+2) - 1}{\cosh(4n+2) + 1}. \end{aligned}$$

On letting n tend to ∞ we deduce that

$$\log M(r_0) \leqq L(0) - A_1, \tag{3}$$

where

$$A_1 = 2\log \prod_{n=0}^{\infty} \left\{\frac{1 + e^{-(4n+2)}}{1 - e^{-(4n+2)}}\right\}.$$

Suppose now that, with the hypotheses of Theorem 1,

$$b(r) < A, \qquad r_0\exp[-\pi^2/4] < r < r_0\exp[\pi^2/4].$$

Then

$$\psi(\sigma) = L(\sigma) - \log M(r_0 e^{\sigma}) + A(\sigma^2 - \pi^4/16)/2$$

is strictly convex in σ for $|\sigma| < \pi^2/4$ and, in view of this and (1), $\psi(\sigma) < 0$ in this range. Setting $\sigma = 0$ we deduce that

$$\log M(r_0) > L(0) - A\pi^4/32.$$

This contradicts (3) if

$$A \leqq (32/\pi^4)A_1,$$

and so Theorem 1 is proved with

$$A_0 = (32/\pi^4)A_1 = .180\cdots.$$

If $f(z)$ is a polynomial other than cz^n, $f(z)$ must have a zero z_0, where $z_0 \neq 0$. Now Theorem 2 follows at once from Theorem 1. We note that if $f(z) = cz^n$, then

$$\log M(r) = n\log r + c, \quad b(r) = 0, \qquad 0 < r < \infty.$$

To prove Theorem 3 we may suppse without loss in generality that

$$\limsup_{r\to\infty} b(r) < +\infty,$$

since otherwise the result is trivial. In this case

$$\log M(r) = O(\log r)^2, \qquad \text{as} \quad r \to \infty,$$

so that $f(z)$ has infinitely many zeros. Thus we may apply Theorem 1 with r_0 as large as we please and Theorem 3 follows. In the opposite direction we note that if $f(z)$ is a polynomial

$$f(z) = c_0 z^n + c_1 z^{n-1} + \cdots,$$

then $M(r,f)$ possesses near infinity a power series expansion

$$M(r,f) = |c_0|r^n + |c_1|r^{n-1} + \cdots$$

so that

$$b(r) = O(r^{-1}), \qquad \text{as} \quad r \to \infty.$$

Thus Theorem 3 is false for polynomials.

3. **Proof of Theorem** 4. Consider $f(z) = 1 + z$. Here we have

$$\log M(r) = 1 + r,$$

$$b(r) = r/(1+r)^2 \leqq \tfrac{1}{4}, \quad 0 < r < \infty.$$

Thus Theorems 1 and 2 are false with $A_0 > \frac{1}{4}$. To construct a counterexample for Theorem 3 let r_n be an increasing sequence of real positive numbers such that

$$r_{n+1}/r_n \to \infty, \qquad \text{as } n \to \infty \tag{4}$$

and set

$$f(z) = \prod_{n=1}^{\infty} \left(1 + \frac{z}{r_n}\right).$$

Then clearly

$$M(r,f) = f(r),$$

$$b(r) = \sum_{n=1}^{\infty} \frac{r\, r_n}{(r + r_n)^2}.$$

It is easy to see that, for $(r_{n-1} r_n)^{1/2} < r < (r_n r_{n+1})^{1/2}$, we have in view of (4)

$$b(r) = r\, r_n/(r + r_n)^2 + o(1) \leqq \tfrac{1}{4} + o(1),$$

as $n \to \infty$. Thus for this function

$$\limsup_{r \to \infty} b(r) = \tfrac{1}{4}.$$

4. **Two further problems suggest themselves.** Firstly can we have a function $f(z)$ other than cz^n, which is regular on $|z| = r$ and for which $b(r+0) = b(r-0) = 0$? If we set

$$f(z) = (1 - z)e^z,$$

then we easily see that

$$M(r,f) = \exp[r^2/2], \qquad r \leqq 2,$$

$$M(r,f) = (r-1)e^r, \qquad r \geqq 2.$$

Thus

$$b(r) = 2\, r^2, \qquad r < 2,$$

$$b(r) = r(r-2)/(r-1)^2, \qquad r > 2,$$

so that $b(2+0) = 0$, $b(2-0) = 8$. This example is due to J. Clunie.

Secondly can $b(r)$ approach a finite limit α as $r \to \infty$, when $f(z)$ is a transcendental integral function? Theorem 3 shows that α must be at least A_0 in this case. We note that if $f(z) = e^z$, then $b(r) = r$, which tends to infinity.

IMPERIAL COLLEGE OF SCIENCE AND TECHNOLOGY
LONDON, ENGLAND

Bounds for the Deficiencies of Meromorphic Functions of Finite Order

SIMON HELLERSTEIN
DANIEL F. SHEA[1]

Introduction. The properties of entire functions with negative zeros have been investigated extensively throughout the development of the theory of entire functions. In this note we propose to study one aspect of the growth and value-distribution properties of meromorphic functions with negative zeros and positive poles. Such a study is of some interest since this class of functions has recently been found to display certain striking and useful extremal properties.

We begin with a brief discussion of a few classical results in the theory of entire and meromorphic functions. Consider, for example, the entire functions f of genus zero:

$$f(z) = az^k \prod (1 - z/a_\nu) \qquad (\sum (1/|a_\nu|) < +\infty,\ 0 \leqq k = \text{integer}).$$

It is a well known, and elementary, observation that each such function has a maximum modulus on $|z| = r$ which is dominated by that of the associated function

$$\hat{f}(z) = |a| z^k \prod (1 + z/|a_\nu|),$$

i.e. $M(r,f) \leqq M(r,\hat{f}) = \hat{f}(r)$ for $r > 0$. Thus, in the class of entire functions of genus zero, the functions with negative zeros (or zeros on any one ray) have an extermal growth behavior.

Typically, essential use is made of this observation in the following way: Consider the classical inequality of Pólya [**17**],

$$\limsup_{r\to\infty} \frac{n(r, 1/f)}{\log M(r,f)} \geqq \frac{\sin \pi\lambda}{\pi}, \tag{1}$$

valid for all entire functions of genus zero. Here, as usual, $n(r, 1/f)$ denotes the counting function for the number of zeros in the disk $|z| \leqq r$, and $\lambda = \lambda(f)$ is the *order* of f. (We shall assume familiarity with the terminology, standard notations and basic results of Nevanlinna's theory as given for example in [**9**], [**14**], [**15**].)

[1] This research was supported by NSF grant GP–5728.

Since $n(r,1/\hat{f}) = n(r,1/f)$ and $\lambda(\hat{f}) = \lambda(f)$, it is clear that the validity of (1) is a consequence of its validity for the "extremal" functions $\hat{f}$. Further, a proof of (1) for the functions $\hat{f}$ follows readily from a well-known representation for $\log \hat{f}(r)/r$ as a Stieltjes transform of $n(r,1/\hat{f})/r$.

A recent result of A. A. Gol'dberg [7] shows that in the class of meromorphic functions f of genus zero, the functions $\hat{f}$ with negative zeros and positive poles have the largest growth. More precisely, if the counting functions for the zeros and poles of f and $\hat{f}$ are identical, then

$$T(r,f) \leqq T(r,\hat{f}) \qquad (r > 0). \tag{2}$$

To see how Gol'dberg's inequality has led to important new results in the value-distribution theory of meromorphic functions, we first recall a classical result due to Nevanlinna [**14**, p. 51] [**9**, p. 101].

For any meromorphic function f of finite nonintegral order λ and any two values $\boldsymbol{a}, \boldsymbol{b}$, Nevanlinna showed that

$$\limsup_{r\to\infty} \frac{N(r,a) + N(r,b)}{T(r,f)} \geqq \kappa(\lambda) > 0. \tag{3}$$

Here κ is a function of λ alone, and a problem of some forty years' standing is that of finding the exact value of $\kappa(\lambda)$. Nevanlinna himself conjectured that the best possible choice of $\kappa(\lambda)$ is

$$\begin{aligned} \kappa(\lambda) &= |\sin \pi\lambda|/(q + |\sin \pi\lambda|) \qquad && (q \leqq \lambda < q + \tfrac{1}{2}), \\ &= |\sin \pi\lambda|/(q + 1) && (q + \tfrac{1}{2} \leqq \lambda \leqq q + 1), \end{aligned} \tag{4}$$

where q is a nonnegative integer, and he gave examples to show that equality could hold in (3) for all λ with this choice of κ.

It is clear, by the well-known quasi-invariance of Nevanlinna's characteristic $T(r,f)$ under bilinear transformations of f, that there is no loss of generality in setting $\boldsymbol{a} = 0$ and $\boldsymbol{b} = \infty$ in (3). Using this remark, and Gol'dberg's inequality (2), Edrei and Fuchs [**3**] were able to verify Nevanlinna's conjecture (4) for orders $\lambda < 1$. An important feature of their proof is a representation for $T(r,\hat{f})$ in terms of $N(r,0)$ and $N(r,\infty)$. This representation is analogous to—but much more complicated than—the one for $\log \hat{f}(r)$ in terms of $n(r,0)$ mentioned above in connection with Pólya's inequality (1).

For the best known bounds on $\kappa(\lambda)$ when $\lambda > 1$, see the results of Edrei and Fuchs in [**2**].

For convenience, we shall denote by $\mathscr{M}_\lambda$ the class of meromorphic functions of order λ whose zeros lie on the negative real axis and whose poles lie on the positive real axis.

The extremal growth of the functions in $\mathcal{M}_\lambda$, when $\lambda < 1$, also enabled Edrei and Fuchs to completely characterize the possible values of the pairs $(\delta(\boldsymbol{a}), \delta(\boldsymbol{b}))$ for any two distinct values $\boldsymbol{a}$ and $\boldsymbol{b}$ of a meromorphic function f of order $\lambda < 1$.

If we set

$$(5) \quad u = \limsup_{r\to\infty} \frac{N(r,0)}{T(r,f)} = 1 - \delta(0), \qquad v = \limsup_{r\to\infty} \frac{N(r\ \infty)}{T(r,f)} = 1 - \delta(\infty),$$

then their result [3] is

THEOREM A. *Let f be a meromorphic function of order λ $(0 < \lambda < 1)$. Then, in addition to the trivial inequalities, $0 \leqq u \leqq 1$ and $0 \leqq v \leqq 1$, u and v satisfy*

$$(6) \qquad u^2 + v^2 - 2uv \cos \pi\lambda \geqq \sin^2 \pi\lambda.$$

If $u < \cos \pi\lambda$, then $v = 1$; and if $v < \cos \pi\lambda$, then $u = 1$. Further, all values u and v compatible with these restrictions are actually possible.

Recently, one of us [19] found the following complement of this result.

THEOREM B. *Let $f \in \mathcal{M}$. If $0 < \lambda < 1$, then the quantities*

$$(7) \quad X = \liminf_{r\to\infty} \frac{N(r,0)}{T(r,f)} = 1 - \Delta(0), \qquad Y = \liminf_{r\to\infty} \frac{N(r,\infty)}{T(r,f)} = 1 - \Delta(\infty)$$

satisfy

$$(8) \qquad X^2 + Y^2 - 2XY \cos \pi\lambda \leqq \sin^2 \pi\lambda,$$

provided that[2]

$$(9) \qquad X \geqq Y \cos \pi\lambda \quad \text{and} \quad Y \geqq X \cos \pi\lambda.$$

When $\lambda = 1$, then

$$(10) \qquad X = Y = 0.$$

Theorem B is sharp in the sense that equality is possible in (8) for all orders λ $(0 < \lambda < 1)$ when (9) is satisfied. Also, when X and Y fail to satisfy (9), then these quantities are subject only to the trivial restrictions

$$X < Y \cos \pi\lambda \quad \text{or} \quad Y < X \cos \pi\lambda$$

and

$$0 \leqq X \leqq 1, \quad 0 \leqq Y \leqq 1.$$

The reader will notice that condition (9) is automatically satisfied when $\frac{1}{2} \leqq \lambda < 1$.

Theorem B, and the proof of Theorem A, reveal certain singular aspects of the character of the functions in $\mathcal{M}_\lambda$. For other aspects of this behavior, see the classical results of Laguerre (cf. e.g. [21, p. 268]), Pólya [17], Valiron [22],

[2] The conditions (9) are easily seen to be equivalent, for orders $\lambda < 1$, to the inequalities (2.4) in [19].

Teichmüller [20] and others, as well as more recent results of MacLane [13] and Aissen, Edrei, Schoenberg and Whitney [1], for example.

This paper concerns the bounds which restrict the deficiencies of the functions in $\mathscr{M}_\lambda$ for *arbitrary* finite λ.

In particular, we shall be able to extend Theorem B to higher orders. In view of the obvious symmetry between the inequalities (6) and (8), it is hoped that this extension might suggest the correct generalization of Theorem A as well.

1. **Statement of results.** Our first theorem generalizes Theorem B to arbitrary nonintegral orders λ.

THEOREM 1. *Let $f \in \mathscr{M}_\lambda$ for some finite nonintegral value of λ, and let q be the integer determined by*

$$q < \lambda < q + 1. \tag{1.1}$$

Then the quantities X and Y defined in (7) *must satisfy*

$$0 \leqq X \leqq 1/(q+1), \quad 0 \leqq Y \leqq 1/(q+1). \tag{1.2}$$

Further, if X and Y satisfy

$$X \geqq Y \cos \pi\lambda \quad \text{and} \quad Y \geqq X \cos \pi\lambda, \tag{1.3}$$

then

$$\begin{aligned} X^2 + Y^2 - 2XY\cos\pi\lambda &\leqq \sin^2\pi\lambda/(q+1)^2 \qquad (\textit{if } q \textit{ is even}), \\ &\leqq (1 - X - Y)^2 \sin^2\pi\lambda/q^2 \qquad (\textit{if } q \textit{ is odd}). \end{aligned} \tag{1.4}$$

The inequalities (1.2) and (1.4) are sharp, as we show in §7.

Obviously, the conditions (1.3) are always satisfied for orders λ such that

$$2k + 1/2 \leqq \lambda \leqq 2k + 3/2, \qquad k = 0, 1, 2, \cdots. \tag{1.5}$$

An inspection of the inequality (1.4) suggests that when λ is an odd integer, the functions in $\mathscr{M}_\lambda$ would be such that $X = Y = 0$. This is true, as we show in §6. We also consider there the possible restrictions on X and Y for functions of even integral order, and show that in this case

$$X + Y < 4/(\lambda + 1) \qquad (\lambda = 2, 4, 6, \cdots). \tag{1.6}$$

Since there exist such functions with

$$X = Y = 1/(\lambda + 1) \qquad (\lambda = 2, 4, 6, \cdots),$$

it is clear that the bound in (1.6) has the correct order of magnitude for large λ.

One sees from (1.2) and (1.4) of Theorem 1 that the functions in $\mathscr{M}_\lambda$ display strong growth properties even for orders λ greater than one. These strong growth properties are also in evidence in the following result of Edrei, Fuchs and Hellerstein [5]:

THEOREM C. *Let* $f \in \mathcal{M}_\lambda$, *where* $1 < \lambda < +\infty$. *Then*

$$\limsup_{r\to\infty} \frac{N(r,0)+N(r,\infty)}{T(r,f)} \leqq A_0, \tag{1.7}$$

where A_0 (<1) *is an absolute constant.*

It follows from (1.7) that the quantities u, v defined in (5) satisfy

$$u \leqq A_0, \qquad v \leqq A_0, \tag{1.8}$$

and thus 0 and ∞ are deficient values in the sense of Nevanlinna for the functions of Theorem C.

Since (8) implies that $\Delta(0) \geqq 1-\sin\pi\lambda$ and $\Delta(\infty) \geqq 1-\sin\pi\lambda$ when $1/2 < \lambda \leqq 1$, 0 and ∞ are always deficient values in the sense of Valiron for the functions in $\mathcal{M}_\lambda$ $(1/2 < \lambda \leqq 1)$, and thus Theorem B may be regarded as the extension of Theorem C to these orders λ.

It should be noted that the simple inequalities (1.2) already show that the quantities X and Y must satisfy considerably more stringent conditions than (1.8) for the functions of Theorem 1. Also, (1.4) represents a "best possible" relation between X and Y and the order of f.

Theorem 1 is a consequence of the more general

THEOREM 2. *Let* $f \in \mathcal{M}_\lambda$, *where* λ *satisfies* (1.1). *Then necessarily*

$$\liminf_{r\to\infty} \frac{AN(r,0)+BN(r,\infty)}{T(r,f)} \leqq |\sin\pi\lambda|, \tag{1.9}$$

where A *and* B *are defined in the following way:*

Set

$$D = (X^2 + Y^2 - 2XY\cos\pi\lambda)^{-1/2}$$

where X *and* Y *are defined by* (7).

Assume, with no loss of generality,[3] *that* $X \geqq Y$.

I. *If* q *is an even integer, then*

$$\begin{aligned} A &= (q+1)(X - Y\cos\pi\lambda)D, \\ B &= (q+1)(Y - X\cos\pi\lambda)D \end{aligned} \tag{1.10}$$

when $Y \geqq X\cos\pi\lambda$, *and*

$$A = \sin\pi\lambda + q(X - Y\cos\pi\lambda)D, \qquad B = q(Y - X\cos\pi\lambda)D \tag{1.11}$$

---

[3] If $Y > X$, then consideration of $\tilde{f}(z) = 1/f(-z)$ shows at once that A and B should be defined in this case by interchanging the formulas for A and B given in (1.10)–(1.13).

when $Y < X\cos\pi\lambda$.

II. *If* q *is odd, then*

$$A = |\sin\pi\lambda| + q(X - Y\cos\pi\lambda)D, \tag{1.12}$$
$$B = |\sin\pi\lambda| + q(Y - X\cos\pi\lambda)D$$

when $Y \geqq X\cos\pi\lambda$, *and*

$$A = (q+1)(X - Y\cos\pi\lambda)D, \tag{1.13}$$
$$B = |\sin\pi\lambda| + (q+1)(Y - X\cos\pi\lambda)D$$

when $Y < X\cos\pi\lambda$.

It is easy to see from (1.9), (1.11) and (1.13) that inequalities analogous to (1.4) still hold when (1.3) is replaced by

$$Y < X\cos\pi\lambda, \tag{1.14}$$

at least when $\lambda > 1$ and the limit

$$Y = \lim_{r\ \infty} \frac{N(r,\infty)}{T(r,f)} \tag{1.15}$$

exists. In this case we have the inequalities

$$X^2 + Y^2 - 2XY\cos\pi\lambda \leqq ((1-X)^2/q)^2\sin^2\pi\lambda \qquad (q \text{ even}), \tag{1.16}$$
$$\leqq ((1-Y)/(q+1))^2\sin^2\pi\lambda \qquad (q \text{ odd}).$$

The remaining case, characterized by the inequality

$$X < Y\cos\pi\lambda, \tag{1.17}$$

leads to the corresponding inequalities obtained when one interchanges the roles of X and Y in (1.15) and (1.16), as considerations of symmetry show.

The limit in (1.15) surely exists when f is entire, and in this case the inequalities (1.4) and (1.16) reduce to the following simple

COROLLARY 1. *Let* f *be an entire function of finite nonintegral order* λ, *all of whose zeros lie on the negative real axis.*

Then

$$X \leqq \kappa(\lambda), \tag{1.18}$$

where κ *is the function defined in* (4).

The bound (1.18) remains valid for integer values of λ, as one of us has shown **[19]**.

Corollary 1 was first established by I. V. Ostrovskii **[16]**, for canonical products f of nonintegral order.

Corollary 1 reveals that the following surprising phenomenon holds for entire functions of order $\lambda \leqq 1$, and, if Nevanlinna's conjecture (4) is valid, for entire functions of all finite orders-

Let f, g be any two entire functions of order λ, f having only negative zeros. Then necessarily

$$\liminf_{r\to\infty}\frac{N(r,1/f)}{T(r,f)} \leqq \kappa(\lambda) \leqq \limsup_{r\to\infty}\frac{N(r,1/g)}{T(r,g)}, \tag{1.19}$$

for all such pairs f, g. In particular, (1.19) holds when the zeros $\{z_n\}$ of g are related to the zeros $\{a_n\}$ of f by

$$a_n = -|z_n| \qquad (n = 1, 2, \cdots).$$

Suppose now that f, instead of having no poles at all, has its zeros and poles symmetric with respect to the origin. Such a class of functions was studied extensively by O. Teichmüller [**20**, pp. 175–188], who proved

THEOREM D. *Let $f \in \mathscr{M}_\lambda$ for some $\lambda < 1$, and assume that the zeros $\{a_n\}$ and the poles $\{b_n\}$ of f satisfy*

$$a_n = -b_n \qquad (n = 1, 2, \cdots). \tag{1.20}$$

Then

$$u = v \geqq \cos(\pi\lambda/2), \tag{1.21}$$

where u and v are defined by (8).

The very restrictive hypothesis (1.20) was later relaxed somewhat by A. A. Gol'dberg ([7]; cf. also [**9**, p. 115]), who showed that (1.21) remains valid for any meromorphic function of order less than one whose zeros and poles satisfy

$$|a_n| = |b_n| \qquad (n = 1, 2, \cdots).$$

More recently, Edrei and Fuchs established Theorem A quoted above; it is clear that their result contains the theorem of Teichmüller-Gol'dberg as a very special case.

In view of the importance of Theorem D as a guide to the possible relations that exist between the order of a meromorphic function and the distribution of two of its values, the following complement of this result—valid for all nonintegral orders—may be worth pointing out.

COROLLARY 2. *Let $f \in \mathscr{M}_\lambda$, where $q < \lambda < q + 1$ for some integer q. If* (1.20) *holds, then*

$$\begin{aligned} X = Y &\leqq |\cos(\pi\lambda/2)|/(q+1) \qquad (q \text{ even}), \\ &\leqq |\cos(\pi\lambda/2)|/(q + 2|\cos(\pi\lambda/2)|) \qquad (q \text{ odd}), \end{aligned} \tag{1.22}$$

where X and Y are defined by (7). *These bounds are sharp.*

Our proofs depend upon a new way of using the following theorem of Pólya **[18]** on the singularities of certain integral transforms. This result has a tauberian character which allows us to deduce information on the relations between

$$N(r,0), \quad N(r,\infty), \quad T(r,f)$$

from integral means involving these quantities.

THEOREM E. *Suppose that the functions*

$$\phi_1(s), \quad \phi_2(s), \cdots, \quad \phi_n(s)$$

of the real variable s *are positive (i.e.* $\geqq 0$*) and measurable for* $s \geqq b$ *and that the integrals*

$$\int_b^\infty \phi_1(s)e^{-s\alpha}ds, \quad \int_b^\infty \phi_2(s)e^{-s\alpha}ds, \cdots, \quad \int_b^\infty \phi_n(s)e^{-s\alpha}ds$$

are all convergent for $\operatorname{Re}\alpha > \lambda$, *but not all convergent for any values of* α *for which* $\operatorname{Re}\alpha < \lambda$. *Suppose further that the analytic functions*

$$g_1(\alpha), \quad g_2(\alpha), \cdots, \quad g_n(\alpha)$$

of the complex variable α *are all regular in a certain neighborhood of the point* $\alpha = \lambda$ *and take there real values for real values of* α; *and especially suppose that*

$$g_1(\lambda) > 0, \quad g_2(\lambda) > 0, \cdots, \quad g_n(\lambda) > 0.$$

Then the point $\alpha = \lambda$ *is a singular point for the function*

$$H(\alpha) = \int_b^\infty [\phi_1(s)g_1(\alpha) + \cdots + \phi_n(s)g^n(\alpha)]\, e^{-s\alpha}\, ds.$$

2. **Some preliminary lemmas.** Let $E(z,q)$ denote the Weierstrass primary factor of genus q:

$$E(z,q) = (1-z)\exp(z + \tfrac{1}{2}z^2 + \cdots + (1/q)z^q) \qquad (q > 0),$$

$$E(z,0) = 1 - z.$$

We shall require the following elementary lemma.

LEMMA 1. *Let* q $(\geqq 0)$ *be an arbitrary integer, and let* α *be a constant such that*

$$q < \alpha < q + 1.$$

Then

$$\int_0^\infty \log|E(-te^{i\theta},q)|\frac{dt}{t^{1+\alpha}}=\frac{\pi\cos\theta\alpha}{\alpha\sin\pi\alpha} \tag{2.1}$$

is valid for all θ in $-\pi\leqq\theta\leqq\pi$.

Further, this integral is absolutely convergent for each value of θ.

PROOF. We start from Valiron's formula [**22**]

$$\log E(-te^{i\theta},q)=(-1)^q\int_1^\infty\frac{(te^{i\theta})^{q+1}}{x^{q+1}(x+te^{i\theta})}dx. \tag{2.2}$$

Dividing both sides of (2.2) by $t^{1+\alpha}$, integrating from 0 to $+\infty$, and interchanging the order of the integrations, we deduce

$$\int_0^\infty \log E(-te^{i\theta},q)\frac{dt}{t^{1+\alpha}}=(-1)^q e^{i(q+1)\theta}\int_1^\infty\frac{dx}{x^{q+1}}\int_0^\infty\frac{t^{q-\alpha}}{x+te^{i\theta}}dt$$

$$=\alpha^{-1}(-1)^q e^{i(q+1)\theta}\int_0^\infty\frac{s^{q-\alpha}}{1+se^{i\theta}}ds \qquad (|\theta|<\pi),$$

by an obvious change of variable.

Using the formula

$$\int_0^\infty\frac{s^{q-\alpha}}{1+se^{i\theta}}ds=(-1)^q e^{-i(q+1)\theta}e^{i\theta\alpha}\frac{\pi}{\sin\pi\alpha},$$

easily verified by a standard contour integration, we find that

$$\int_0^\infty \log E(-te^{i\theta},q)\frac{dt}{t^{1+\alpha}}=\frac{\pi}{\alpha\sin\pi\alpha}e^{i\theta\alpha} \qquad (|\theta|<\pi), \tag{2.3}$$

and (2.1) follows for $-\pi<\theta<\pi$ on taking real parts.

By the definition of the primary factor $E(z,q)$,

$$\log|E(-te^{i\theta},q)|=\log|1+te^{i\theta}|+\sum_{k=1}^{q}\frac{(-1)^k}{k}t^k\cos k\theta,$$

the sum on the right side being omitted when $q=0$.

Thus

$$\big|\log|E(-te^{i\theta},q)|\big|=\left|\sum_{k=q+1}^{\infty}\frac{(-1)^k}{k}t^k\cos k\theta\right|<2t^{q+1}$$

for all t such that $0<t<1/2$, and

$$\left|\log\left|E(-te^{i\theta},q)\right|\right| < \log(1+t)+2t^q$$

when $t>2$. Also

$$\left|\log\left|E(-te^{i\theta},q)\right|\right| < \log 3 + q2^q - \log|1-t|$$

for $1/2 \leqq t \leqq 2$, and since $q<\alpha<q+1$ it follows that the integral in (2.1) converges absolutely and uniformly for $-\pi \leqq \theta \leqq \pi$. In particular, the formula (2.1) is valid for all θ in the closed interval $-\pi \leqq \theta \leqq \pi$, and this observation completes the proof of the lemma.

LEMMA 2. *Let $\{a_n\}$ denote an arbitrary sequence of negative numbers with finite exponent of convergence ξ, and let α be an arbitrary positive number, not an integer, such that*

$$\sum (1/|a_n|^\alpha) < +\infty. \tag{2.4}$$

Define

$$g(z) = \prod E(z/a_n, q), \tag{2.5}$$

where q is the greatest integer in α. If α satisfies

$$\max(\xi, q) < \alpha < q+1,$$

then

$$\int_0^\infty \log|g(te^{i\theta})|\frac{dt}{t^{1+\alpha}} = \frac{\pi\cos\theta\alpha}{\alpha\sin\pi\alpha}\sum\frac{1}{|a_n|^\alpha} \tag{2.6}$$

is valid for each θ in $-\pi \leqq \theta \leqq \pi$. If the series converges for $\alpha=\xi$, and ξ is not an integer, then the integral in (2.6) *also is convergent for $\alpha=\xi$.*

Clearly, g is an entire function of order $\lambda=\xi$ if the product (2.5) is canonical; otherwise, g has integral order $\lambda = q$ $(\geqq \xi)$.

PROOF. Since by (2.4) the series

$$\sum_{n=1}^\infty \int_0^\infty \left|\log\left|E\left(\frac{te^{i\theta}}{a_n},q\right)\right|\right|\frac{dt}{t^{1+\alpha}} = \int_0^\infty \left|\log\left|E(-xe^{i\theta},q)\right|\right|\frac{dx}{x^{1+\alpha}}\sum_{n=1}^\infty\frac{1}{|a_n|^\alpha}$$

converges, it follows that

$$\sum_{n=1}^\infty \int_0^\infty \log\left|E\left(\frac{te^{i\theta}}{a_n},q\right)\right|\frac{dt}{t^{1+\alpha}} = \int_0^\infty \log|g(te^{i\theta})|\frac{dt}{t^{1+\alpha}}. \tag{2.7}$$

Applying Lemma 1 to each term on the left side of (2.7) we deduce (2.6).

It will be convenient to restate (2.6) in a slightly different way. Using the well-known identity

$$\sum_{n=1}^{\infty} \frac{1}{|a_n|^{\alpha}} = \alpha^2 \int_0^{\infty} \frac{N(t,0)}{t^{1+\alpha}} dt,$$

we have from (2.6)

$$\int_0^{\infty} \log|g(te^{i\theta})| \frac{dt}{t^{1+\alpha}} = \frac{\pi\alpha}{\sin \pi\alpha} \cos \theta\alpha \int_0^{\infty} \frac{N(t,0)}{t^{1+\alpha}} dt. \tag{2.8}$$

We remark that Lemmas 1 and 2 generalize some observations of Pólya [**18**], I. V. Ostrovskii [**16**], and others. Pólya used the integrals

$$\int_0^{\infty} \frac{\log(1+t)}{t^{1+\alpha}} dt = \frac{\pi}{\alpha \sin \pi\alpha}, \quad \int_0^{\infty} \frac{\log|1-t|}{t^{1+\alpha}} dt = \frac{\pi}{\alpha \tan \pi\alpha} \tag{2.9}$$

to deduce results on the growth of entire functions of order less than one, and a pair of formulas somewhat analogous to these was used by Littlewood in his famous paper [**12**]. It is clear that the identities (2.9) are contained in (2.1) (take $q = 0$ and $\theta = 0, \pi$).

3. **Proof of Theorem** 2. Since the order λ of f is not an integer, we can write

$$f(z) = e^{P(z)} \prod E(z/a_n, q) \Big/ \prod E(z/b_n, q) = e^{P(z)} g(z), \tag{3.1}$$

where P is a polynomial of degree $d \leqq q$ and at least one of the two products in (3.1) is canonical. Thus Lemma 2 and (2.8) imply that the identity

$$\int_0^{\infty} \left\{ \log|g(te^{i\theta})| - \frac{\pi\alpha}{\sin \pi\alpha} [\cos \theta\alpha\, N(t,0) - \cos(\pi - \theta)\alpha\, N(t,\infty)] \right\} \frac{dt}{t^{1+\alpha}} = 0 \tag{3.2}$$

is valid for all α in $\lambda < \alpha < q+1$, and for θ such that $0 \leqq \theta \leqq \pi$.

In fact, since both sides of (3.2) are holomorphic functions of α in the strip

$$\lambda < \operatorname{Re} \alpha < q+1, \tag{3.3}$$

the identity (3.2) persists throughout this region.

Let E be any measurable subset of $0 \leqq \theta \leqq \pi$, and define the integral mean

$$\tilde{m}(t) = \tilde{m}(t,g,E) = \frac{1}{\pi} \int_E \log|g(te^{i\theta})|\, d\theta.$$

Clearly,

$$m(t,g) \geqq \tilde{m}(t), \tag{3.4}$$

since $\log|g(te^{i\theta})|$ is an even function of θ.

Integrating both sides of (3.2) over the set E, we obtain

$$\int_0^{\infty} \{\sin \pi(\alpha - q)\, T(t,g) - [U_\alpha N(t,0) + V_\alpha N(t,\infty) + \sin \pi(\alpha - q)\, p(t)]\} \frac{dt}{t^{1+\alpha}} = 0, \tag{3.5}$$

where

$$U_\alpha = \alpha(-1)^q \int_E \cos\theta\alpha \, d\theta, \tag{3.6}$$

$$V_\alpha = \alpha(-1)^{q+1} \int_E \cos(\pi-\theta)\alpha \, d\theta + \sin\pi(\alpha - q), \tag{3.7}$$

$$p(t) = m(t,g) - \tilde{m}(t).$$

It is important to notice that, in view of (3.4), $p(t) \geqq 0$ for all $t > 0$.

By using Pólya's Theorem E quoted above, we shall deduce from (3.5) the following inequality:

$$\liminf_{r\to\infty} \frac{U_\lambda N(r,0) + V_\lambda N(r,\infty) + |\sin\pi\lambda|\, p(r)}{T(r,g)} \leqq |\sin\pi\lambda|. \tag{3.8}$$

In order to replace $T(r,g)$ by $T(r,f)$ in (3.8), it is important that we obtain some information on the relative rates of growth of $g(z)$ and $f(z)$. We shall show that there exists an increasing, unbounded sequence $\{r_n\}$ such that (3.8) remains valid when r tends to infinity restricted to the values of this sequence, and such that

$$\lim_{n\to\infty} \frac{\log T(r_n, g)}{\log r_n} = \lambda. \tag{3.9}$$

In view of the obvious inequalities

$$T(r,g) = T(r, f\cdot e^{-P}) \leqq T(r,f) + O(r^d) \leqq T(r,g) + O(r^d),$$

(3.9) shows that

$$T(r_n, g) \sim T(r_n, f) \qquad (n \to \infty)$$

since $d \leqq q < \lambda$.

Using this relation in (3.8) and taking into account the positivity of $p(r)$ we find that

$$\liminf_{r\to\infty} \frac{U_\lambda N(r,0) + V_\lambda N(r,\infty)}{T(r,f)} \leqq |\sin\pi\lambda|. \tag{3.10}$$

Theorem 2 will follow from this inequality, after we have chosen the set E so that U_λ and V_λ give the strongest possible bounds in (3.10).

We proceed to establish the inequality

$$\liminf_{r\to\infty} \frac{U_\lambda N(r\ 0) + V_\lambda N(r,\infty) + |\sin\pi\lambda|\, p(r) + r^\sigma}{T(r,g)} \leqq |\sin\pi\lambda| \tag{3.11}$$

for any $\sigma < \lambda$. In fact, if it were false then there would exist numbers $a > 1$, $\varepsilon > 0$, $\sigma < \lambda$ such that

$$(U_\lambda - \varepsilon)\, N(t,0) + (V_\lambda - \varepsilon) N(t,\infty) + (|\sin \pi\lambda| - \varepsilon) p(t) + t^\sigma > (|\sin \pi\lambda| + \varepsilon)\, T(t,g) \qquad (t \geqq a). \tag{3.12}$$

Rewriting (3.5), we obtain

$$- \int_0^a \{\sin \pi(\alpha - q)\, T(t,g) - U_\alpha N(t,0) - V_\alpha N(t,\infty) - \sin \pi(\alpha - q) p(t)\} \frac{dt}{t^{1+\alpha}}$$

$$= \int_a^\infty \{\sin \pi(\alpha - q)\, T(t,g) - U_\alpha N(t,0) - V_\alpha N(t,\infty) - \sin \pi(\alpha - q) p(t)\} \frac{dt}{t^{1+\alpha}}$$

$$\begin{aligned} = \int_a^\infty &\{[(U^\lambda - \varepsilon) N(t,0) + (V_\lambda - \varepsilon) N(t,\infty) + (|\sin \pi\lambda| - \varepsilon) p(t) + t^\sigma \\ &- (|\sin \pi\lambda| + \varepsilon) T(t\ g)] + [U_\alpha - U\ + \varepsilon] N(t,0) + [V_\alpha - V_\lambda + \varepsilon] N(t,\infty) \\ &+ [\sin \pi(\alpha - q) - |\sin \pi\lambda| + \varepsilon] p(t) \\ &+ [|\sin \pi\lambda| - \sin \pi(\alpha - q) + \varepsilon] T(t,g)\} \frac{dt}{t^{1+\alpha}} - \int_a^\infty t^{\sigma - \alpha - 1} dt. \end{aligned} \tag{3.13}$$

Denoting the first integral in (3.13) by $G(\alpha)$, it is clear that $G(\alpha)$ is holomorphic everywhere in the strip

$$q < \operatorname{Re} \alpha < q + 1.$$

The right-hand side of (3.13) has the form

$$H(\alpha) - (\alpha - \sigma)^{-1} a^{\sigma - \alpha},$$

where

$$H(\alpha) = \int_a^\infty \left\{ \sum_{k=1}^{5} g_k(\alpha) \phi_k(t) \right\} \frac{dt}{t^{1+\alpha}},$$

$g_1(\alpha) = 1, \quad g_2(\alpha) = U_\alpha - U_\lambda + \varepsilon, \quad g_3(\alpha) = V_\alpha - V_\lambda + \varepsilon,$
$g_4(\alpha) = \sin \pi(\alpha - q) - |\sin \pi\lambda| + \varepsilon, \quad g_5(\alpha) = |\sin \pi\lambda| - \sin \pi(\alpha - q) + \varepsilon,$
$\phi_1(t) = (U_\lambda - \varepsilon) N(t,0) + (V_\lambda - \varepsilon) N(t,\infty) + (|\sin \pi\lambda| - \varepsilon) p(t) + t^\sigma - (|\sin \pi\lambda| + \varepsilon) T(t,g),$
$\phi_2(t) = N(t,0), \quad \phi_3(t) = N(t,\infty), \quad \phi_4(t) = p(t), \quad \phi_5(t) = T(t,g).$

Hence the functions $g_k(\alpha)$ are each entire functions of α, real for real α, and such that

$$g_k(\lambda) > 0 \qquad (k = 1 - 5).$$

Further the functions $\phi_k(t)$ are measurable, positive for $t \geqq a$ (in view of the assumption (3.12)), and such that the integrals

(3.14) $$\int_a^\infty \phi_k(t)\frac{dt}{t^{1+\alpha}} \qquad (k = 1-5)$$

converge for all α in the strip (3.3) while at least one of the integrals (3.14) is divergent when $\operatorname{Re}\alpha < \lambda$ (cf. [**14**, p. 49]).

Using Pólya's Theorem E (with $s = \log t$ and $n = 5$) we find that the above conditions imply that the function $H(\alpha)$ has a singularity at the point $\alpha = \lambda$. But this contradicts the regularity of

$$G(\alpha) + (\alpha - \sigma)^{-1}a^{\dot\sigma - \alpha}$$

at $\alpha = \lambda$, since this function coincides with $H(\alpha)$ throughout the region (3.3).

This contradiction proves (3.11), and thus shows that there exist positive sequences $\{r_n\}, \{\sigma_n\}$ such that

$$\lim_{n\to\infty} r_n = \infty, \qquad \lim_{n\to\infty} \sigma_n = \lambda$$

and

$$\liminf_{n\to\infty} \frac{U_\lambda N(r_n, 0) + V_\lambda N(r_n, \infty) + |\sin \pi\lambda|\, p(r_n) + r_n^{\dot\sigma n}}{T(r_n, g)} \leqq |\sin \pi\lambda|.$$

This relation together with the trivial inequalities

$$N(t,0) \leqq T(t,g), \qquad N(t,\infty) \leqq T(t,g)$$

show that (3.9) is true. Thus (3.10) holds, and the proof of Theorem 2 will be complete if we can find a subset E of $0 \leqq \theta \leqq \pi$ such that the definitions (3.6), (3.7) yield

(3.15) $$U_\lambda = A, \qquad V_\lambda = B$$

where $A = A(X, Y, \lambda)$ and $B = B(X, Y, \lambda)$ are defined in the statement of Theorem 2.

In fact it is not difficult to verify that the set

(3.16) $$E = \{\theta; (-1)^q X \cos\theta\lambda > (-1)^q Y \cos(\pi - \theta)\lambda,\ 0 < \theta < \pi\}$$

has just the desired property.

This is most easily seen by observing that E is the union of a finite number of disjoint open intervals whose end points are simply those zeros of the function

$$k(\theta) = (-1)^q X \cos\theta\lambda - (-1)^q Y \cos(\pi - \theta)\lambda$$

which lie in $0 < \theta < \pi$ possibly augmented by one or both of the points $\theta = 0$, $\theta = \pi$.

Assume as we did in the statement of Theorem 2 that $X \geqq Y$.

If we denote by γ the smallest positive root of $k(\theta) = 0$ then

(3.17) $$X \cos\gamma\lambda = Y \cos(\pi - \gamma)\lambda,$$

$$(3.18) \qquad (X - Y\cos \pi\lambda)\cos \gamma\lambda = Y\sin \pi\lambda \sin \gamma\lambda$$

and hence

$$0 < \gamma \leqq \pi/2\lambda \quad \text{or} \quad \pi/2\lambda \leqq \gamma < \pi/\lambda,$$

the first inequality holding if q is even and the second if q is odd.

Now (3.18) is easily seen to be equivalent to

$$(3.19) \qquad \sin \gamma\lambda = (X - Y\cos \pi\lambda)D,$$

where

$$D = (X^2 + Y^2 - 2XY\cos \pi\lambda)^{-1/2}.$$

On the other hand, (3.17) implies

$$Y^2 \sin^2(\pi - \gamma)\lambda = Y^2 - X^2 + X^2 \sin^2 \gamma\lambda$$

and thus (3.19) yields

$$Y^2D^{-2}\sin^2(\pi - \gamma)\lambda = Y^2(Y^2 - 2XY\cos \pi\lambda + X^2\cos^2\pi\lambda),$$

$$(3.20) \qquad \sin(\pi - \gamma)\lambda = (Y - X\cos \pi\lambda)D.$$

In order to complete the proof of Theorem 2, it is convenient to distinguish four cases.

Case I (i). Assume that q is even and

$$(3.21) \qquad Y \geqq X\cos \pi\lambda.$$

Obviously (3.21) is equivalent to $k(\pi) \leqq 0$, and a moment's thought will make clear the fact that this inequality implies E is comprised of $1 + q/2$ intervals of the form

$$0 < \theta < \gamma,$$

$$\gamma + (2j - 1)\pi/\lambda < \theta < \gamma + 2j\pi/\lambda \qquad (j = 1, 2, \cdots, q/2 \text{ if } q > 0).$$

Using the definitions (3.6) and (3.7), we deduce

$$(3.22) \qquad U_\lambda = \int_0^{\gamma\lambda} \cos\phi d\phi + (q/2)\int_{\gamma\lambda+\pi}^{\gamma\lambda+2\pi} \cos\phi d\phi = (q+1)\sin\gamma\lambda$$

and, similarly,

$$(3.23) \qquad V_\lambda = (q+1)\sin(\pi - \gamma)\lambda.$$

Combining (3.22) with (3.19) and (3.23) with (3.20) yields (3.15), when q is even and (3.21) holds. This completes the proof of Theorem 2 in this case.

Case I (ii). If q is even and

(3.24) $$Y < X \cos \pi\lambda,$$

we have $k(\pi) > 0$.

When $q = 0$, then (3.16) shows that E is the whole interval

$$0 < \theta < \pi.$$

When $q \geqq 2$, then E is the union of the two intervals

$$0 < \theta < \gamma,$$

$$\gamma + (q-1)\pi/\lambda < \theta < \pi,$$

together with the $q/2 - 1$ intervals

$$\gamma + (2j-1)\pi/\lambda < \theta < \gamma + 2j\pi/\lambda \qquad (j = 1, 2, \cdots, q/2 - 1, \quad \text{if} \quad q > 2).$$

Then (3.6) and (3.7) lead to

(3.25) $$U_\lambda = q \sin \gamma\lambda + \sin \pi\lambda, \quad V_\lambda = q \sin(\pi - \gamma)\lambda.$$

Taking into account (3.19) and (3.20), (3.25) yields (3.15)—and hence also Theorem 2—in this case.

The proof of Theorem 2 is completed by considering, in a similar way, the Cases II (i) and II (ii) corresponding to q an odd integer and X, Y satisfying either (3.21) or (3.24).

4. **Proof of Theorem** 1. Let $f(z) \in \mathscr{M}_\lambda$ have nonintegral order λ, and determine q by (1.1).

The definitions (3.6) and (3.7) imply that

(4.1) $$U_\lambda = (-1)^q \int_{\mathscr{E}} \cos \phi d\phi,$$

(4.2) $$V_\lambda = |\sin \pi\lambda| + (-1)^{q+1}\{\cos \pi\lambda \int_{\mathscr{E}} \cos \phi \, d\phi + \sin \pi\lambda \int_{\mathscr{E}} \sin \phi \, d\phi\},$$

where $\mathscr{E}$ denotes an arbitrary subset of $0 \leqq \phi \leqq \pi\lambda$.

If q is even and positive set

$$\mathscr{E} = [0, (\lambda - q)\pi] \cup \bigcup_{k=0}^{q/2-1} [(\lambda - 2k - 1)\pi, (\lambda - 2k)\pi],$$

and observe that

$$\int_{\mathscr{E}} \cos \phi \, d\phi = (q+1) \sin \pi\lambda, \qquad \int_{\mathscr{E}} \sin \phi \, d\phi = 1 - (q+1) \cos \pi\lambda,$$

so that by (4.1) and (4.2)

$$U_\lambda = (q+1)\sin \pi\lambda, \qquad V = 0. \tag{4.3}$$

For this choice of U_λ and V, (3.10) yields

$$X = \liminf_{r\to\infty} \frac{N(r,0)}{T(r,f)} \leqq \frac{1}{q+1}. \tag{4.4}$$

Since $1/f(-z) \in \mathscr{M}_\lambda$, (4.4) shows that $Y \leqq 1/(q+1)$; this inequality together with (4.4) proves the assertion (1.2) of Theorem 1, in case q is even.

If q is odd, we can choose

$$\mathscr{E} = [q\pi, \lambda\pi] \cup \bigcup_{k=1}^{(q-1)/2} [(2k-1)\pi, 2k\pi]$$

if $q > 1$, and take

$$\mathscr{E} = [\pi, \lambda\pi]$$

when $q = 1$. Then

$$\int_{\mathscr{E}} \cos\phi \, d\phi = \sin \pi\lambda, \qquad \int_{\mathscr{E}} \sin\phi \, d\phi = -q - \cos\pi\lambda,$$

so that (4.1) and (4.2) yield

$$U_\lambda = |\sin\pi\lambda|, \qquad V_\lambda = (q+1)\,|\sin\pi\lambda|. \tag{4.5}$$

Using (3.10) with (4.5) leads to

$$X + (q+1)Y \leqq 1,$$

and by symmetry, also to

$$(q+1)X + Y \leqq 1.$$

Since these inequalities are stronger than (1.2), the first assertion of Theorem 1 is true for all integers q.

To show that (1.4) follows from Theorem 2, assume first that X and Y satisfy

$$X \geqq Y \geqq X\cos\pi\lambda.$$

Then the quantities A, B defined by (1.10), (1.12) are nonnegative, and hence (1.9) implies

$$AX + BY \leqq |\sin\pi\lambda|.$$

This leads to

$$\begin{aligned}[][X(X - Y\cos\pi\lambda) + Y(Y - X\cos\pi\lambda)]D &\leqq \sin\pi\lambda/(q+1) \quad (q \text{ even}),\\ &\leqq ((1 - X - Y)/q)|\sin\pi\lambda| \quad (q \text{ odd}),\end{aligned}$$

which is equivalent to (1.4) in view of the definition of D.

If on the other hand

$$Y > X \geqq Y\cos\pi\lambda,$$

then (1.4) remains valid, as consideration of $\tilde{f}(z) = 1/f(-z)$ shows, since (1.4) is symmetric in X and Y.

This completes the proof of Theorem 1.

5. **Proof of Corollaries** 1 **and** 2. Corollary 1 is a trivial consequence of (1.4), (1.16) and the observation that for entire functions $N(r, \infty) \equiv 0$ and $Y = 0$.

To prove Corollary 2, it is only necessary to notice that (1.20) implies

$$N(r,0) \equiv N(r,\infty)$$

and hence, in particular, $X = Y$. Using this information in (1.9), the definitions (1.10) and (1.12) yield

$$X \leqq \sin\pi\lambda/(A+B) \leqq |\sin\pi\lambda|/(q+1)(2(1-\cos\pi\lambda))^{1/2} \quad (q \text{ even});$$

$$\leqq |\sin\pi\lambda|/(q(2(1-\cos\pi\lambda))^{1/2} + 2|\sin\pi\lambda|) \quad (q \text{ odd}).$$

Corollary 2 now follows from the identity

$$|\sin\pi\lambda| = |\cos(\pi\lambda/2)|(2(1-\cos\pi\lambda))^{1/2}.$$

6. **Case of integral orders** λ. To complement the preceding results on the functions in $\mathscr{M}_\lambda$ for nonintegral values of λ, we prove

THEOREM 3. *Let $f \in \mathscr{M}$ for some positive integral order λ. Then*

$$X = Y = 0 \qquad (\lambda = 1, 3, 5, \cdots) \tag{6.1}$$

and

$$X + Y < 4/(\lambda+1) \qquad (\lambda = 2, 4, 6, \cdots). \tag{6.2}$$

PROOF. Let q denote the smallest nonnegative integer such that

$$\Sigma\,(1/|a_n|^{q+1}) + \Sigma\,(1/b_n^{q+1}) < +\infty,$$

where the a's and b's are the zeros and poles of f, and let d denote the degree of the polynomial P defined by

$$f(z) = e^{P(z)}\prod E(z/a_n, q)/\prod E(z/b_n, q).$$

We shall assume that $d \leqq q$ since otherwise $f(z)$ is dominated by the exponential factor $\exp(P(z))$ and the study of its asymptotic behavior is trivial (cf. [**19**, p. 214]).

Now let p denote a positive integer, set

$$w = \exp(2\pi i/p), \tag{6.3}$$

and define

$$F(z^p) = \prod_{k=0}^{p-1} f(w^k z). \tag{6.4}$$

Clearly, $F(z)$ is meromorphic, and $F \in \mathscr{M}_{\lambda/p}$ if p is odd.

If λ is odd, set $p = \lambda$ and notice that F has order $\lambda_0 = 1$, so that by Theorem B quoted above

$$X_0 = Y_0 = 0, \tag{6.5}$$

where

$$X_0 = \liminf_{r\to\infty} \frac{N(r,1/F)}{T(r,F)}, \qquad Y_0 = \liminf_{r\to\infty} \frac{N(r,F)}{T(r,F)}. \tag{6.6}$$

It is easy to see from (6.4), (6.6) and the definitions of $N(r,c)$ and $T(r,F)$ that

$$X \leqq X_0, \qquad Y \leqq Y_0 \tag{6.7}$$

(cf. [**19**, p. 224]). Then (6.1) follows from (6.5) and (6.7).

If λ is even, choose $p = \lambda + 1$ and notice that F has order $\lambda_0 = \lambda/(\lambda + 1)$. Hence Theorem B (or Theorem 1) implies that

$$X_0^2 + Y_0^2 - 2X_0 Y_0 \cos \pi\lambda_0 \leqq \sin^2 \pi\lambda_0. \tag{6.8}$$

In view of (6.7) and (6.8), X and Y must satisfy

$$X^2 + Y^2 + 2XY\cos(\pi/(\lambda+1)) \leqq \sin^2(\pi/(\lambda+1))$$

and obvious estimates lead from this inequality to the simpler (6.2).

7. **The "best possible" character of Theorem 1.** To show that equality is possible in (1.2), (1.4), (1.16), (1.18) and (1.22), we consider quotients of suitably modified "Lindelöf functions" of order λ,

$$f_\lambda(z) = \prod_{n=1}^{\infty} E(z/a_n, q) \qquad (q < \lambda < q + 1), \tag{7.1}$$

whose zeros $\{a_n\}$ are "regularly" distributed along the negative real axis.

We begin by considering a notion of "regular variation" which is particularly natural for our purposes. Following Karamata [**10**], we say that a positive function $L(r)$ is *slowly varying* if, for each positive number σ,

$$\lim_{r\to\infty} \frac{L(\sigma r)}{L(r)} = 1. \tag{7.2}$$

Karamata showed [**10**], [**6**] that the condition (7.2) is equivalent to writing

(7.3) $$L(r) = c(r)\exp\left(\int_a^r \frac{\varepsilon(t)}{t}dt\right) \qquad (a > 0),$$

where the limit

(7.4) $$c = \lim_{r\to\infty} c(r)$$

exists and is positive, and where

(7.5) $$\lim_{t\to\infty} \varepsilon(t) = 0.$$

To avoid meaningless difficulties, let us agree to identify asymptotically equivalent functions $L(r)$; thus, in view of (7.3), there will be no loss of generality in assuming that *all our slowly varying functions are of the form*

(7.6) $$L(r) = c\exp\left(\int_1^r \frac{\varepsilon(t)}{t}dt\right),$$

where c is a positive constant and $\varepsilon(t)$ is continuous and satisfies (7.5).

It follows from the representation (7.6) that

(7.7) $$\lim_{r\to\infty} \frac{\log L(r)}{\log r} = 0,$$

and also that $L(r)$ satisfies (7.2) *uniformly* for σ in any interval $\sigma_0 \leqq \sigma \leqq \sigma_1$ $(\sigma_0 > 0)$, as Karamata showed [**10**]. (For an independent proof of this latter fact, see Besicovitch's lemma in [**8**].)

Levin [**11**, p. 32], and others, have observed that if $\lambda(r)$ is a *proximate order* (in the modern sense of [**11**, p. 32]), then the function $L(r)$ defined by

(7.8) $$r^{\lambda(r)} = r^{\lambda}L(r)$$

is slowly varying. We point out here that *the converse is also true*; *If* $L(r)$ *varies slowly, then the function* $\lambda(r)$ *defined by the equation* (7.8) *is a proximate order.*

Indeed, it is obvious from (7.7) and (7.8) that

(7.9) $$\lim_{r\to\infty} \lambda(r) = \lambda,$$

while (7.6) shows that

(7.10) $$\lim_{r\to\infty} \lambda'(r)\, r \log r = 0;$$

but (7.9) and (7.10) are just the conditions that characterize proximate orders.

We shall make essential use of this observation in the following way: Let us assume that the zeros $\{a_n\}$ of the product (7.1) satisfy the regularity conditions

(7.11) $$a_n < 0 \qquad (n = 1, 2, \cdots)$$

and

(7.12) $$n(r, 1/f_\lambda) \sim r^\lambda L(r) \qquad (r \to \infty;\ q < \lambda < q+1),$$

where q is a nonnegative integer and $L(r)$ varies slowly. In view of the above-remarked connection between functions of the form (7.12) and proximate orders, it follows from a classical [22] [11, p. 64] asymptotic expansion for $\log f_\lambda(z)$ that

(7.13) $$\log f_\lambda(re^{i\theta}) = (\pi/\sin \pi\lambda)[e^{i\theta\lambda} + o(1)]\, r^\lambda L(r) \qquad (r \to \infty),$$

uniformly for $|\theta| \leqq \pi - \eta$ $(\eta > 0)$.

Now let $F(z) \in \mathscr{M}_\lambda$ be such that

(7.14) $$n(r, 1/F) = (1 + o(1))\, r^\lambda L(r),$$

(7.15) $$n(r, F) = (\beta + o(1))\, r^\lambda L(r)$$

as $r \to \infty$, where $q < \lambda < q + 1$ and β satisfies

(7.16) $$0 \leqq \beta \leqq 1.$$

Then

(7.17) $$F(z) = e^{P(z)} f_\lambda(z)/g\,(-z),$$

where f_λ and g_λ are Weierstrass products of genus q having only negative zeros (g_λ may have only finitely many zeros, or be $\equiv 1$), and P is a polynomial of degree $d \leqq q$. Hence

(7.18) $$\begin{aligned}\log F(re^{i\theta}) &= \log f_\lambda(re^{i\theta}) - \log g_\lambda(-re^{i\theta}) + O(r^d)\\ &= (\pi/\sin \pi\lambda)[1 - \beta e^{-i\pi\lambda}]e^{i\theta\lambda} r^\lambda L(r) + o(r^\lambda L(r))\end{aligned}$$

as $r \to \infty$, uniformly for θ in $\eta \leqq \theta \leqq \pi - \eta$ $(\eta > 0)$, at least if $\beta > 0$.

To see that (7.18) holds even if $\beta = 0$, suppose that the product g_λ has negative zeros satisfying

(7.19) $$n(r, 1/g_\lambda) = o(r^\lambda L(r)) \qquad (r \to \infty).$$

Hence, if $\varepsilon > 0$ we may choose r_0 so that

$$n(r, 1/g\ ) < \varepsilon r^\lambda L(r) \qquad (r \geqq r_0)$$

and notice that, by the well known [22] formula

(7.20) $$\log g_\lambda(z) = (-1)^q z^{q+1} \int_0^\infty \frac{n(t, 1/g_\lambda)}{t^{q+1}(t+z)}\,dt \qquad (|\arg z| < \pi),$$

we have

$$|\log g_\lambda(re^{i\theta})| \leqq r^{q+1}\int_0^\infty \frac{n(t,1/g_\lambda)}{t^{q+1}|t+re^{i\theta}|}dt$$

$$< \frac{\varepsilon}{\sin(\eta/2)}r^{q+1}\int_{r_0}^\infty t^{\lambda-q-1}L(t)\frac{dt}{t+r}+Kr^q \qquad (r\geqq 1, |\theta|\leqq \pi-\eta)$$

$$< \frac{2\varepsilon}{\sin(\eta/2)}\frac{\pi}{|\sin \pi\lambda|}r^\lambda L(r) \qquad (r\geqq r_0, |\theta|\leqq \pi-\eta).$$

The last inequality above follows from (7.20) and (7.13), with $\theta=0$.

Hence (7.19) implies

$$\log g_\lambda(re^{i\theta}) = o(r^\lambda L(r)) \qquad (r\to\infty, |\theta|\leqq\pi-\eta),$$

so that (7.18) holds even if $\beta=0$.

By well known properties of slowly varying functions [**10**], (7.14) and (7.15) imply

(7.21) $$N(r,1/F)\sim(1/\lambda)r^\lambda L(r), \quad N(r,F)=(\beta/\lambda+o(1))r^\lambda L(r) \qquad (r\to\infty).$$

In order to use (7.18) to estimate $m(r,F)$, we notice that (7.13) implies

(7.22) $$|\log|f(re^{i\theta})||<Kr^\lambda L(r)$$

for $|\theta|\leqq\pi-\eta$ and $r>r_0(\eta)$, where η (>0) is arbitrarily small.

On the other hand, Edrei and Fuchs [**3**, p. 244] have shown that there exists a constant A, independent of ϕ and η, such that

(7.23) $$\int_\phi^{\phi+\eta}|\log|f_\lambda(re^{i\theta})||\,d\theta < A\eta\left(1+\overset{+}{\log}\frac{1}{\eta}\right)T(2r,f),$$

provided r is sufficiently large and $-\pi\leqq\phi<\phi+\eta\leqq\pi$. By choosing η small enough, it follows from (7.22) and (7.23) that

(7.24) $$\int_{-\pi}^{\pi}|\log|f_\lambda(re^{i\theta})||\,d\theta<Kr^\lambda L(r)+2^{-\lambda-2}T(2r,f_\lambda) \qquad (r>r_0),$$

and hence, using (7.21), that

(7.25) $$G(r)<K+\tfrac{1}{2}G(2r) \qquad (r>r_0),$$

where

(7.26) $$G(r)=T(r,f_\lambda)/r^\lambda L(r).$$

Iterating (7.25) leads to

$$G(r) < 2K + 2^{-n}G(2^n r) \qquad (r > r_0, n \geqq 1)$$

and thus to

$$G(r) < 2K \qquad (r > r_0)$$

and

$$T(r, f_\lambda) < Kr^\lambda L(r) \qquad (r > r_0). \tag{7.27}$$

The same arguments show that (7.23) and (7.27) also hold with g_λ instead of f_λ. Thus (7.17), (7.23) and (7.27) imply

$$\int_\phi^{\phi+\eta} \left|\log |F(re^{i\theta})|\right| d\theta < A\eta \left(1 + \overset{+}{\log} \frac{1}{\eta}\right) r^\lambda L(r) \qquad (r > r_0), \tag{7.28}$$

where ϕ and η satisfy $0 \leqq \phi < \phi + \eta \leqq \pi$.

Combining (7.18) and (7.28) leads to

$$m(r, F) \sim \frac{1}{\pi} \int_0^\pi \overset{+}{\log} |F(re^{i\theta})| \, d\theta \sim \frac{H(\beta)}{|\sin \pi\lambda|} r^\lambda L(r) \qquad (r \to \infty), \tag{7.29}$$

where

$$H(\beta) = \int_0^\pi \{(-1)^q [\cos \theta\lambda - \beta \cos(\theta - \pi)\lambda]\}^+ \, d\theta. \tag{7.30}$$

Thus (7.21) implies

$$T(r, F) \sim (H(\beta)/|\sin \pi\lambda| + \beta/\lambda) r^\lambda L(r) \qquad (r \to \infty). \tag{7.31}$$

It follows from (7.21) and (7.31) that the limits

$$X = \lim_{r\to\infty} \frac{N(r, 1/F)}{T(r, F)}, \qquad Y = \lim_{r\to\infty} \frac{N(r, F)}{T(r, F)} \tag{7.32}$$

exist and are connected by

$$Y/X = \beta. \tag{7.33}$$

Further, from (7.21), (7.31) and (7.32) we get

$$\sin^2 \pi\lambda = [\lambda H(\beta) + \beta |\sin \pi\lambda|]^2 X^2. \tag{7.34}$$

To compute $H(\beta)$, we first select the smallest number $\gamma (> 0)$ such that

$$\cos \gamma\lambda = \beta \cos(\pi - \gamma)\lambda, \tag{7.35}$$

i.e. such that

$$(1 - \beta \cos \pi\lambda) \cos \gamma\lambda = \beta \sin \pi\lambda \sin \gamma\lambda, \tag{7.36}$$

and then distinguish four cases, as in the proof of Theorem 2.

If q is even and $\beta \geqq \cos \pi\lambda$, then (7.30) and (7.35) show that

$$\lambda H(\beta) = \int_0^{\pi\lambda} [\cos\phi - \beta\cos(\phi - \pi\lambda)]^+ d\phi$$

$$= \left\{ \int_0^{\gamma\lambda} + \frac{q}{2} \int_{\gamma\lambda+\pi}^{\gamma\lambda+2\pi} \right\} [\cos\phi - \beta\cos(\phi - \pi\lambda)]\, d\phi$$

$$= (q+1)[\sin\gamma\lambda + \beta\sin(\pi - \gamma)\lambda] - \beta\sin\pi\lambda.$$

Thus

$$[\lambda H(\beta) + \beta\sin\pi\lambda]^2(q+1)^{-2} = [(1 - \beta\cos\pi\lambda)\sin\gamma\lambda + \beta\sin\pi\lambda\cos\gamma\lambda]^2,$$

and (7.36) is equivalent to

$$0 = [(1 - \beta\cos\pi\lambda)\cos\gamma\lambda - \beta\sin\pi\lambda\sin\gamma\lambda]^2.$$

Adding the last two equations leads to

$$[\lambda H(\beta) + \beta\sin\pi\lambda]^2(q+1)^{-2} = (1 - \beta\cos\pi\lambda)^2 + \beta^2\sin^2\pi\lambda = 1 + \beta^2 - 2\beta\cos\pi\lambda.$$

This relation, together with (7.34) and (7.33), yields

(7.37) $$X^2 + Y^2 - 2XY\cos\pi\lambda = \sin^2\pi\lambda/(q+1)^2,$$

valid when q is an even integer and $Y \geqq X\cos\pi\lambda$.

When q is even and $Y < X\cos\pi\lambda$, then a similar analysis leads to

$$\lambda H(\beta) = q[\sin\gamma\lambda + \beta\sin(\pi - \gamma)\lambda] + \sin\pi\lambda \qquad (q \geqq 2),$$

$$= (1 - \beta)\sin\pi\lambda \qquad (q = 0),$$

and hence to

(7.38) $$X^2 + Y^2 - 2XY\cos\pi\lambda = ((1 - X)^2/q^2)\sin^2\pi\lambda \qquad (q \geqq 2),$$

$$X = 1, \quad Y = \beta \qquad (q = 0),$$

in this case.

The two cases to be considered for odd integers q are treated in a similar way; we omit the computations involved, since they involve no new ideas, and state the results of our analysis in the following

THEOREM 4. *Let $f(z) \in \mathscr{M}_\lambda$ for some value of λ $(q < \lambda < q+1)$, and assume that*

(7.39) $$n(r,0) = (U + o(1))r^\lambda L(r),$$

(7.40) $$n(r,\infty) = (V + o(1))r^\lambda L(r)$$

as $r \to \infty$, *where* $L(r)$ *is a slowly-varying function and at least one of* U, V *is positive.*

Then

$$\log f(re^{i\theta}) = (\pi/\sin \pi\lambda)(U - Ve^{-i\pi\lambda})e^{i\theta\lambda}r^{\lambda}L(r) + o(r^{\lambda}L(r)) \qquad (r \to \infty),$$

uniformly in any interval $\eta \leqq \theta \leqq \pi - \eta$, $(\eta > 0)$. *Further,*

$$T(r,f) \sim Cr^{\lambda}L(r) \qquad (r \to \infty),$$

so that the limits

$$X = \lim_{r\to\infty} \frac{N(r,0)}{T(r,f)}, \qquad Y = \lim_{r\to\infty} \frac{N(r,\infty)}{T(r,f)}$$

exist and are connected by the relations

$$U = \lambda CX, \qquad V = \lambda CY,$$

$$X^2 + Y^2 - 2XY\cos\pi\lambda = \sin^2\pi\lambda/(q+1)^2 \qquad (q \text{ even}),$$

$$= ((1 - X - Y)^2/q^2)\sin^2\pi\lambda \qquad (q \text{ odd})$$

if $X \geqq Y\cos\pi\lambda$ *and* $Y \geqq X\cos\pi\lambda$, *and by*

$$X^2 + Y^2 - 2XY\cos\pi\lambda = ((1 - M)^2/q^2)\sin^2\pi\lambda \qquad (q \text{ even}, \ q \geqq 2),$$

$$= ((1 - m)^2/(q+1)^2)\sin^2\pi\lambda \qquad (q \text{ odd})$$

otherwise. Here $M = \max(X, Y)$, $m = \min(X, Y)$.

If $q = 0$ *and* $m < M\cos\pi\lambda$, *then* $M = 1$ *and* $m = V/U$ *if* $V < U$, *with* $m = U/V$ *in case* $U < V$.

To see that Theorem 4 follows from the above arguments, we first assume that $U \geqq V$, and replace $L(r)$ in (7.39), (7.40) by $L_1(r) = U^{-1}L(r)$. Then $L_1(r)$ varies slowly, and (7.14), (7.15) are satisfied with $L_1(r)$ in place of $L(r)$ (and with $\beta = V/U$). Thus all the assertions of Theorem 4 follow from (7.31), (7.32), (7.33), (7.37) and (7.38) (and the corresponding relations for odd integers q). When $U < V$, consideration of $1/F(-z)$ shows that Theorem 4 is true in this case as well.

A comparison of Theorem 4 with the bounds (1.4), (1.16), (1.18), and (1.22) shows that equality may hold in any of these relations, for each nonintegral value of λ.

We remark that Theorem 4 extends a result of Edrei and Fuchs [**4**, Theorem 2] to arbitrary nonintegral orders. This observation suggests that the main (tauberian) result of [**4**] may also remain valid for higher orders.

REFERENCES

1. M. Aissen, A. Edrei, I. J. Schoenberg and A. Whitney, *On the generating functions of totally positive sequences*, Proc. Nat. Acad. Sci. U.S.A. **37** (1951), 303–307.

2. A. Edrei and W. H. J. Fuchs, *On the growth of meromorphic functions with several deficient values*, Trans. Amer. Math. Soc. **93** (1959), 292–328.

3. ———, *The deficiencies of meromorphic functions of order less than one*, Duke Math. J. **27** (1960), 233–250.

4. ———, *Tauberian theorems for a class of meromorphic functions with negative zeros and positive poles*, Proc. Int. Conf. in Function Theory, Erevan, Armenian S.S.R.

5. A. Edrei, W. H. J. Fuchs and S. Hellerstein, *Radial distribution and deficiencies of the values of a meromorphic function*, Pacific J. Math. **11** (1961), 135–151.

6. W. Feller, *Introduction to probability theory and its applications*. Vol. II, Wiley, New York, 1966.

7. A. A. Gol'dberg, *On the deficiencies of meromorphic functions*, Dokl. Akad. Nauk SSSR **98** (1954), 893–895 (Russian).

8. G. H. Hardy and W. W. Rogosinski, *Notes on Fourier series*. III, Quart. J. Math. **16** (1945), 49–58.

9. W. K. Hayman, *Meromorphic functions*, Oxford Univ. Press, New York, 1964.

10. J. Karamata, *Sur un mode de croissance régulière des fonctions*, Mathematica (Cluj) **4** (1930), 38–53.

11. B. Ja. Levin, *Distribution of zeros of entire functions*, Transl. Math. Monograph, vol. 5, Amer. Math. Soc., Providence, R. I., 1964.

12. J. E. Littlewood, *A general theorem on integral functions of finite order*, Proc. London Math. Soc. (2) **6** (1908), 189–204.

13. G. R. MacLane, *Riemann surfaces and asymptotic values associated with real entire functions*, Rice Institute Pamphlet, Houston, 1952.

14. R. Nevanlinna, *Le théorème de Picard-Borel et la théorie des fonctions méromorphes*, Paris, 1930.

15. ———, *Eindeutige analytische Funktionen*, Springer-Verlag, Berlin, 1953.

16. I. V. Ostrovskii, *Some asymptotic properties of entire functions with negative zeros*, Zap. Mat. Otd. Fiz. — Mat. i Khar'kov Mat. Obsc. (4) **28** (1961), 23–32 (Russian).

17. G. Pólya, *Bemerkungen uber unendlichen Folge und ganzen Funktionen*, Math. Ann. **88** (1923), 169–183.

18. ———, *On the minimum modulus of integral functions of order less than unity*, J. London Math. Soc. **1** (1926), 78–86.

19. D. F. Shea, *On the Valiron deficiencies of meromorphic functions of finite order*, Trans. Amer. Math. Soc. **124** (1966), 201–227.

20. O. Teichmüller, *Vermutungen und Satze uber die Werteilung gebrochener Funktionen endlicher Ordnung*, Deutsche Math. **4** (1939), 163–190.

21. E. C. Titchmarsh, *The theory of functions*, Oxford, 1939.

22. G. Valiron, *Sur les fonctions entières d'ordre fini et d'ordre nul, et en particulier les fonctions à correspondence régulière*, Ann. Fac. Sci. Univ. Toulouse (3) **5** (1913), 117–257.

THE UNIVERSITY OF WISCONSIN
MADISON, WISCONSIN

The Counting Function for Points of Maximum Modulus[1]

F. HERZOG
G. PIRANIAN

Corresponding to an entire function f (or a function f holomorphic in the unit disk D), let $\nu(r) = \nu(r,f)$ denote the number of points on the circle $|z| = r$ where $|f(z)| = M(r) = M(r,f)$. In a private communication, P. Erdös asked whether $\nu(r)$ can be unbounded if f is not of the form $f(z) = cz^n$ (see W. K. Hayman [**1**, Problem 2.16]). We shall show that the answer is affirmative.

THEOREM 1. *There exists a nontrivial univalent function in D for which* $\nu(r)$ *is unbounded.*

To prove this theorem, we construct a univalent function $w = f(z)$ mapping the unit disk onto a domain G that consists of a neighborhood of the origin (in the w-plane) together with certain approximately radial spikes. The spikes of G naturally fall into countably many sets. The jth set $(j = 1, 2, \cdots)$ consists of j spikes, and the tips of these spikes lie slightly beyond a circle $|w| = \rho_j$, where $\{\rho_j\}$ is an increasing sequence. For each j, the spikes of the $(j+1)$st set are much sharper than those of the jth set, so that (with the appropriate fine adjustment of the lengths of the individual spikes) the image of a certain circle $|z| = r_j$ meets the circle $|w| = \rho_j$ in j points but does not penetrate the domain $|w| > \rho_j$.

To come to grips with the details, we first define the sequence $\{\rho_j\}$ by one of the two equations

$$\rho_j = 1 - 2^{-j}, \qquad \rho_j = j,$$

depending on whether we want a bounded or an unbounded function f. We also choose a set

$$\{z_{11}, z_{21}, z_{22}, z_{31}, z_{32}, z_{33}, z_{41}, \cdots\}$$

of points on the unit circle, in such a way that no limit point of the set belongs to the set, and we construct a family $\{S_{jm}\}$ of disjoint open sectors of D such that each S_{jm} contains the radius vector of the corresponding point z_{jm}.

[1] This research was supported by the National Science Foundation.

Our building blocks are the functions

$$g_{jm}(z) = z_{jm}[1 - (1 - z/z_{jm})^{k_j}],$$

where $0 < k_j < 1$, and where the expression in brackets represents the branch of the corresponding function that takes the value 0 at the point $z = 0$. Since the expansion

$$1 - (1 - z)^k = k[z + ((1 - k)/2!)z^2 + ((1 - k)(2 - k)/3!)z^3 + \cdots]$$

has decreasing positive coefficients beginning with k, we see that the maximum modulus of $g_{jm}(z)$ in the closure of D is 1 and that we can make $g_{jm}(z)$ as small as we like, in each closed disk $|z| \leqq r < 1$ (as well as in the set $D \backslash S_{jm}$) by choosing k_j sufficiently small.

It will be convenient to choose the radii r_j in such a way that they are automatically determined by the values of the exponents k_j. Since we shall require that the maximum of $|g_{jm}(z)|$ on $|z| = r_j$ is nearly 1, we can for example choose

$$r_j = 1 - \exp(-k_j^{-2});$$

then

$$g_{jm}(r_j z_{jm}) = z_{jm}[1 - \exp(-k_j^{-1})] = z_{jm}[1 - o(1)].$$

Without specifying the values of the exponents k_j, we now write a formal series

$$\sum_{j,m} \rho_j g_{jm}(z).$$

Clearly, this series converges to a holomorphic function f_0 in D, provided $k_j \downarrow 0$ fast enough. Moreover, the function f_0 is univalent, since the derivative of each of its terms has a positive real part throughout D.

Let C_{jm} denote the intersection of the circle $|z| = r_j$ and the sector S_{jm}. Obviously, $|f_0(z)|$ has a local maximum on C_{jm}; however, the value of this local maximum differs slightly from ρ_j, not only because the maximum of $|g_{jm}(z)|$ is slightly less than 1, on C_{jm}, but also because all other terms of the series defining f_0 contribute to the value of $f(z)$ at the point of the local maximum.

To achieve a partial correction of this defect, we replace each coefficient ρ_j in our series by some appropriate number L_{jm1}. Each of the numbers L_{jm1} is determined by the requirement that it be positive and that the maximum value on C_{jm} of the function

$$|f_0(z) + (L_{jm1} - \rho_j)g_{jm}(z)|$$

be ρ_j. *In other words, each term in our series undergoes a correction that would bring the maximum modulus of f_0 on C_{jm} to the desired value, if all the other terms remained unchanged.*

On each arc C_{jm}, the function $f_1(z) = \sum L_{jm1} g_{jm}(z)$ has a local maximum modulus. Again, we determine positive numbers L_{jm2} by the requirement that the maximum value on C_{jm} of

$$|f_1(z) + (L_{jm2} - L_{jm1})g_{jm}(z)|$$

be ρ_j, and we write $f_2(z) = \sum L_{jm2}g_{jm}(z)$. The continuation of the process is obvious. It remains only to show that if $k_j \downarrow 0$ fast enough, then the sequences $\{L_{jmq}\}_{q=1}^{\infty}$ converge uniformly to limits L_{jm}.

The sum of the absolute values of the first corrections satisfies the inequality

$$\sum |L_{jm1} - \rho_j| < 1/2,$$

provided k_1 is small enough and $k_j \downarrow 0$ fast enough. Also, the absolute value of each of the corrections at the qth stage is bounded by a linear function of the absolute values of the corrections at the $(q-1)$st stage, the coefficients in this linear function being independent of q, and their sum being less than $1/2$ if the k_j are sufficiently small. It follows that for an appropriate sequence $\{k_j\}$ (that is, for sufficiently sharp spikes on the image of D under the mapping f_0) the sequence $\{f_q\}$ converges to some function

$$f(z) = \sum L_{jm}g_{jm}(z).$$

Univalence of f is established by the same argument as univalence of f_0. Finally, if $k_j \downarrow 0$ fast enough, then the quantity $M(r_j)$ is equal to the j local maxima ρ_j of $|f(z)|$ on the j arcs C_{jm} $(m = 1, 2, \cdots, j)$, and our theorem is proved.

THEOREM 2. *There exists a nontrivial entire function for which $\nu(r)$ is unbounded.*

The proof is similar to the proof of Theorem 1, except for certain rather obvious modifications in the early phases.

First, we can not permit the point set $\{z_{jm}\}$ to have a finite limit point. Therefore we choose the j points z_{jm} $(m = 1, \cdots, j)$ on the circle $|z| = j$. (The number j will also serve as the radius r_j of the circle on which $M(r) = \rho_j$.)

Second, because the function $1 - (1 - z/z_{jm})^{k_j}$ is not an entire function, we replace it with a section P_{jm} of its Taylor series about the origin. Since the Taylor series converges absolutely on the circle $|z| = j$, we can choose P_{jm} so that $P_{jm}(z) - g_{jm}(z)$ is as small as we like on this circle.

Third, because the functions P_{jm} have large maximum modulus on the circles $|z| = r_h$ $(h > j)$, it now becomes necessary to make an inductive choice of the maximum moduli ρ_j. For example, we can begin with $\rho_1 = 1$ and $k_1 = 1$, and we can then take $P_{11}(z) = g_{11}(z) = z$. Thereafter, we must choose ρ_2 large compared with 2 (since $|P_{11}(z)| = 2$ on $|z| = 2$), take k_2 small enough so that $2\rho_2[1-(1 - z/2)^{k_2}]$ is small at $z = 1$ (and so that at each point of the circle $|z| = 2$, one of the two functions g_{21} and g_{22} is small), and replace the functions g_{21} and

g_{22} with appropriate polynomials P_{2m}. At the third stage, we choose ρ_3 large compared with the sum of the maximum moduli of $\rho_1 P_{11}(z)$, $\rho_2 P_{21}(z)$, and $\rho_2 P_{22}(z)$ on $|z| = 3$, take k_3 small enough so that $3\rho_3[1 - (1 - z/3)^{k_3}]$ is small at $z = 2$ (and so that at each point of $|z| = 3$, two of the three functions g_{3m} are small), and replace the three functions g_{3m} with appropriate polynomials P_{3m}.

When the polynomials P_{jm} ($j = 1, 2, \cdots$; $m = 1, 2, \cdots, j$) have been defined, we form the function

$$f_0(z) = \sum_{j,m} L_{jm0} P_{jm}(z),$$

with $L_{jm0} = \rho_j$. We then replace the coefficients L_{jm0} with coefficients L_{jm1} in such a way that if only one of the coefficients (with indices j_0 and m_0, say) were replaced, the corresponding local maximum modulus on $|z| = j$ of the modified function would be ρ_j. The successive approximations $f_1, f_2, \cdots$ are defined inductively, as in the proof of Theorem 1. The proof that the process converges (provided at each stage we choose ρ_j large enough and k_j small enough) can be carried out much as before. We omit the details.

REFERENCE

1. W. K. Hayman, *Research problems in classical function theory*, London Univ. Press, London, 1967.

MICHIGAN STATE UNIVERSITY
EAST LANSING, MICHIGAN
UNIVERSITY OF MICHIGAN
ANN ARBOR, MICHIGAN

Quasi-Analytic Functions

JERRY KAZDAN[1]

Introduction. One of the most exotic properties of an analytic function is that it is uniquely determined by its values on very thin sets. The most elementary aspect of this is

THEOREM A. *Let f be holomorphic in some domain $\mathscr{D}$ containing the origin. If $f^{(j)}(0) = 0$, $j = 0, 1, 2, \cdots$, then $f \equiv 0$.*

This result has many natural generalizations if we restate the condition that f is holomorphic as: f is a solution of the Cauchy-Riemann equations $\partial f/\partial \bar{z} = 0$, that is, a solution of $Lf = 0$ where L is an elliptic partial differential operator. In fact, if L is any linear (or even nonlinear) elliptic operator with analytic coefficients and $Lf = 0$, then Theorem A is true as stated for the simple reason that a solution of $Lf = 0$ is ipso facto an analytic function. Moreover, the result has been extended by many people to the case where the coefficients of L are not necessarily analytic, beginning with Carleman (see [**6a**]) and most recently by Calderón [**5**]. Since for nonanalytic operators L the solutions of $Lf=0$ may not be infinitely differentiable, the vanishing condition in Theorem A is replaced by either a zero of infinite order at the origin ($\lim_{|x|\to 0} |x|^{-n} f(x) = 0$ for $n = 0, 1, 2, \cdots$) or possibly vanishing in a whole neighborhood of the origin.

Theorem A was extended by Holmgren to the case where L is not necessarily elliptic, however with an important modification of the condition that f vanish on some set. A simplified version is

THEOREM B (HOLMGREN). *Let f be a solution of $Lf = 0$, where L is any linear differential operator of order m with analytic coefficients. If the hyperplane $x_1 = 0$ is noncharacteristic for L and if*

$$f = \partial f/\partial x_1 = \cdots = \partial^{m-1} f/\partial x_1^{m-1} = 0 \quad on \quad x_1 = 0,$$

then $f \equiv 0$.[2]

[1] This work was supported by the Army Research Office (Durham).
[2] See [**8**] for a proof.

The point here is that f is not assumed to be analytic, but merely sufficiently differentiable to satisfy the equation. For example, if L is the wave operator, $L = \partial^2/\partial x_1^2 - \partial^2/\partial x_2^2$, then the condition on $x_1 = 0$ states that both the initial position $f(0, x_2)$ and initial velocity $(\partial f/\partial x_1)(0, x_2)$ are both zero, so the conclusion that $f \equiv 0$ is certainly natural from a physical point of view. This theorem has also been extended, again beginning with Carleman [6a] and most recently by Calderón [5] to operators whose coefficients are not necessarily analytic.

With these two theorems as background, we are led to state the following *general quasi-analytic problem*: Let $A(\mathscr{D})$ be some class of functions defined on a domain $\mathscr{D}$. To find a domain $\mathscr{D}_1 \subset \mathscr{D}$ (better yet, all domains $\mathscr{D}_1$) such that if "f vanishes" in $\mathscr{D}_1$, then $f \equiv 0$ in the larger set $\mathscr{D}$. The condition that "f vanishes" in $\mathscr{D}_1$ may involve conditions on the derivatives of f in $\mathscr{D}_1$ (see Theorem 1 below). We shall then refer to the triple $(A; \mathscr{D}, \mathscr{D}_1)$ as being quasi-analytic, or perhaps to $\mathscr{D}_1$ as being quasi-analytic for A.

This general quasi-analytic problem includes uniqueness problems for differential operators. Thus, for example, if $A(\mathscr{D})$ is the set of harmonic functions in a compact set $\mathscr{D}$, then one could take $\mathscr{D}_1 = \partial\mathscr{D}$—or one could take $\mathscr{D}_1$ to be any open set in $\mathscr{D}$—while the corresponding condition for the wave operator is contained in Holmgren's Theorem. The problem of finding sets of uniqueness where $A(\mathscr{D})$ is the set of solutions of a differential equation $Lf = 0$ is an important part of the issue of finding well-posed problems (in the sense of Hadamard) for differential operators.

Before going on, we would like to draw attention to a more refined result of the above nature for analytic functions. Assume $f \in H_\infty$, the Hardy space, in the unit disc and $f(\xi_j) = 0$. If the ξ_j either tend to an interior point, or do not tend to the boundary too rapidly, so that $\sum (1 - |\xi_j|) = \infty$, then $f \equiv 0$. This theorem is the key to many proofs (e.g. one proof of Muntz's theorem on the completeness of $\{x^{n_d}\}$) but has yet to find any suitable generalization even to functions of two complex variables. The standard proof of the above follows from Jensen's theorem, i.e. essentially from the First Main Theorem of Nevanlinna's theory of meromorphic functions. There is still no suitable generalization to holomorphic functions of two or more complex variables. For example $f(z, \zeta) = z\zeta$ vanishes on all of $z = 0$ and $\zeta = 0$ but $f \not\equiv 0$. Although many people are currently working in this area (see related articles in this volume), the needed results are apparently still not known.

We have mentioned several generalizations of Theorem A to differential operators. There is another, more standard, type of generalization. The first of these is the beautiful result found by Denjoy and Carleman in 1921 signaling the beginning of the classical theory of quasi-analytic functions. Observe that if the holomorphic hypothesis of Theorem A is relaxed to merely C^∞, then the conclusion is not valid, illustrating the extreme flexibility of C^∞ functions. (An even more striking indication of the flexibility of C^∞ functions is a theorem due to Borel:

given *any* sequence $\{a_j\}$ of real numbers, there is a function $f \in C^\infty$ such that $f^{(j)}(0) = a_j$.) Denjoy and Carleman noticed that by restricting one's attention to subset of C^∞, the conclusion of Theorem A is valid. Their idea was to impose conditions on the successive derivatives $f^{(j)}$ similar to the Cauchy inequalities for analytic functions.

Thus, let $B = \{b_j\}$ be a sequence of real numbers normalized with $b_0 = 1$, and let

$$C_B = \{F \in C^\infty(\mathbf{R}) \mid \| f^{(j)} \|_\infty \leqq c r^j b_j\},$$

where c and r will depend on the function f.

THEOREM 1 (DENJOY-CARLEMAN). *Let $f \in C_B$. Then C_B is quasi-analytic (that is, $f^{(j)}(0) = 0$, $j = 0, 1, 2, \cdots$ implies $f \equiv 0$) if and only if*

$$\int_{-\infty}^{\infty} \frac{\log \lambda(x)}{1 + x^2} \, dx = \infty,$$

where

$$\lambda(x) = \sum_{j=0}^{\infty} \frac{|x|^j}{b_j}.$$

The condition means that the b_j's do not get big too rapidly. For example, if $b_j = j!$ or j^j, then C_B is quasi-analytic (even analytic) while if $b_j = (j!)^2$ or $(kj)!$, $k > 1$, then C_B is not quasi-analytic. A proof of the theorem can be found in [**6**b] or [**14**].

Generalizations. We shall consider only one way in which the Denjoy-Carleman Theorem can be extended for other classes of functions of one variable and then pass immediately to the more subtle situations which arise for functions of several variables. In the next theorem, we replace the inequality on $(d/dx)^j f$ by inequalities on only the successive mth derivatives, $(d^m/dx^m)^j f$.

THEOREM 2. *Let $L = d^m/dx^m$ and define*

$$C_B(L) = \{f \in C^\infty(\mathbf{R}) \mid \| L^j f \|_\infty \leqq c r^j b_j\},$$

$$C_{B'}(d/dx) = \{f \in C^\infty(\mathbf{R}) \mid \| (d/dx)^j f \|_\infty \leqq c r^j b_{[j/m]}\}, \qquad j = 0, 1, 2, \cdots,$$

where $[\alpha]$ is the greatest integer in α. Then $C_B(L)$ is quasi-analytic if and only if $C_{B'}(d/dx)$ is quasi-analytic (where in both cases we mean $f^{(j)}(0) = 0$ for $j = 0, 1, 2, \cdots$ implies $f \equiv 0$).

PROOF. Since one can consider the inequalities defining $C_B(L)$ as inequalities on all derivatives by using infinity for the missing derivative inequalities, the Denjoy-Carleman function for $C_B(L)$ is $\lambda(x) = \sum(|x|^{mj}/b_j)$ while for $C_{B'}(d/dx)$ we have $\lambda'(x) = \sum |x|^j / b_{[j/m]}$. Since

$$\lambda'(x) = (1 + |x| + \cdots + |x|^{m-1})\lambda(x),$$

the corresponding integrals in the Denjoy-Carleman Theorem converge or diverge together, so the proof is completed.

REMARK. The analytic function $f(x) = x^k$ shows that the theorem is false if we delete $f^{(k)}(0) = 0$ for even one value of k.

We turn now to quasi-analytic questions for functions of several independent variables. Here we find a lovely blending of the various issues mentioned above for differential operators with those of Denjoy and Carleman.

First, we state the simplest generalization. It was apparently first mentioned by Bochner and Taylor in [3]. As usual, we write

$$\| D^j f \|_\infty = \sup_{x \in R^n} \left| \sum_{j_1 + \cdots + j_n = j} \frac{\partial^j f(x)}{\partial x_1^{j_1} \cdots \partial x_n^{j_n}} \right| .$$

THEOREM 3. *Let $C_B = \{ f \in C^\infty(R^n) \mid \| D^j f \|_\infty \leqq cr^j b_j \}$, $\lambda(x) = \sum_j \| x \|^j / b_j$ and let $\lambda_l(x) =$ the restriction of $\lambda(x)$ to a straight line l through the origin. Then C_B is quasi-analytic (again in the sense that if f and all of its derivatives vanish at the origin, then $f \equiv 0$) if and only if*

$$\int_{-\infty}^{\infty} \frac{\log \lambda_l(x)}{1 + x^2} \, dx = \infty$$

for all lines l through the origin.

In the same direction as the Holmgren Theorem and Theorem 2, it is natural to replace the above inequalities $\| D^j f \|_\infty \leqq cr^j b_j$ on all the powers of all partial derivatives by inequalities only on the powers of a given differential operator. We now give two such applications.

THEOREM 4 (BOCHNER-TAYLOR [3]). *Let*

$$C_B(\Delta) = \{ f \in C^\infty(R^n) \mid \| \Delta^j f \|_\infty \leqq cr^{2j} b_{2j},\ j = 0, 1, 2, \cdots \},$$

and let $\mathscr{D}_1 \subset R^n$ be a set with the property that if a real analytic function vanishes on $\mathscr{D}_1$, then it is identically zero. Then $C_B(\Delta)$ is quasi-analytic with respect to $\mathscr{D}_1$ (in the sense that $(\Delta^j f)(\xi) = 0$, $j = 0, 1, 2, \cdots$, for all $\xi \in \mathscr{D}_1$ implies $f \equiv 0$) if and only if

$$\int_{-\infty}^{\infty} \frac{\log \lambda(t)}{1 + t^2} \, dt = \infty,$$

where

$$\lambda(t) = \sum_{j=0}^{\infty} \frac{|t|^j}{b_{2j}}.$$

Outline of Proof. The idea is to introduce a new parameter t, transfer the properties of the successive powers of Δ to the powers of $\partial/\partial t$, and then apply the one variable Denjoy-Carleman Theorem. This is most naturally done by using the heat equation. Thus, let $F(x,t)$ be defined by $\partial F/\partial t = \Delta F$, $F(x,0) = f(x)$, so

$$F(x,t) = \frac{1}{(\pi t)^{n/2}} \int_{\mathbf{R}^n} f(y) \exp[-|x-y|^2/t]\, dy.$$

Then F is a real analytic function of x for $t > 0$, and

$$\sup_x |F(x,t)| \leqq \|f\|_\infty \qquad \text{for all} \quad t > 0.$$

Moreover, $(\partial^j/\partial t^j)F = \Delta^j F$. The remainder of the proof is a straightforward matter of observing that as a function of t, F satisfies the conditions of the Denjoy-Carleman Theorem for $x \in \mathscr{D}_1$, and is analytic for $t > 0$. (If f grows too quickly at infinity for the integral defining $F(x,t)$ to make sense, just approximate f by a C^∞ function with compact support.)

Remark. The penalty for requiring that $(\Delta^j f)(\xi) = 0$, i.e., only *some* of the derivatives vanish at a point, and not all of the derivatives, is that the set of points ξ for which this holds must be thicker that the one point set which sufficed earlier. Of course, sets of analytic determination are still not very thick (any open set will do). As mentioned above, very little is known about sets of analytic determination in more than one variable.

Since the Laplacian is an elliptic differential operator, we were still able to get away with using a set of "analytic determination" for $\mathscr{D}_1$ in the above theorem. This rather thin set cannot be made much thinner as one can see even in the one variable case, taking f to be an odd analytic function, as x^{odd} or $\sin x$. For both functions all of the even derivatives vanish at the origin, while for $\sin x$ this vanishing takes place at an infinite number of points (but not a set of analytic determination). It is possible to prove similar results for a wide class of elliptic operators, although more machinery is needed. The guiding idea is that bounds on an elliptic operator of order m yield bounds not only on the derivative of lower order, but also on *all* the derivatives of order m too. To illustrate the spirit of this, consider the Laplacian for example. By Ploncherel's theorem if $\Delta f \in L_2(\mathbf{R}^2)$, then (modulo factors of π)

$$\|f_{xy}\|^2 = \int |f_{xy}|^2 = \int |\xi\eta\hat{f}|^2$$

$$\leqq \tfrac{1}{4}\int |(\xi^2+\eta^2)\hat{f}|^2 = \tfrac{1}{4}\int |\Delta f|^2 = \tfrac{1}{4}\|\Delta f\|^2,$$

so

$$\left\| \sum_{i+j=2} \frac{\partial^2 f}{\partial x^i \partial y^j} \right\| \leqq \text{const}\, \|\Delta f\|.$$

Such inequalities using various topologies are now standard. See Kazdan **[10]** for statements and proofs of some quasi-analytic results for elliptic operators.

If one leaves the realm of elliptic operators and requires inequalities of the form $\| L^j f \| \leqq cr^j b_j$, where L is an arbitrary differential operator (necessarily with C^∞ coefficients for the powers L^j, $j = 0, 1, \cdots$ to make sense), then the problem of finding quasi-analytic classes of functions is much more difficult. Here, we are not only generalizing the Denjoy-Carleman Theorem, but we are also directly generalizing the Holmgren Theorem.

For example, if L is the wave operator, $L = (\partial^2/\partial t^2) - (\partial^2/\partial x^2)$ and $b_1 = 0$, then we are considering honest solutions of the wave equation

$$\partial^2 f/\partial t^2 - \partial^2 f/\partial x^2 = 0.$$

Holmgren's Theorem tells us that if $f(0,x) \equiv 0$ and $f_t(0,x) = 0$ for all $x \in \boldsymbol{R}$, then $f \equiv 0$. Thus, we are naturally led to expect that for general differential operators L, a somewhat thicker set $\mathscr{D}_1$ of "vanishing" will be required than in the elliptic case where rather thin sets of "analytic determination" sufficed.

I am aware of only one result in this direction. It is due to Ehrenpreis [7]. He restricts himself to a differential operator with constant coefficients because the proof critically utilizes his generalized Fourier analysis (i.e. analytically uniform spaces). We shall write $(x,t) = (x_1, \cdots x_n, t)$ for points in $\boldsymbol{R}^{n+1}$. Think of t as the time coordinate for the motion $f(x,t)$ of a medium in $\boldsymbol{R}^n$, and $f(x,0)$ as the initial position.

THEOREM 5 (EHRENPREIS). *Let L be a constant coefficient differential operator of order m. Define*

$$C_B(L) = \{ f \in C^\infty(\boldsymbol{R}^{n+1}) \mid \| L^j f \|_\infty \leqq cr^j b_j \},$$

and

$$C_B(d^m/dt^m) = \{ \zeta \in C^\infty(\boldsymbol{R}') \mid \| (d^m/dt^m)^j \zeta \|_\infty \leqq cr^j b_j \}, \qquad j = 0, 1, 2, \cdots.$$

Assume the n-dimensional hyperplane $t = 0$ is noncharacteristic for L. Then $C_B(L)$ is quasi-analytic for the set $t = 0$ (in the sense that $(\partial^j/\partial t^j) f)(0,x) = 0$ for all $x \in \boldsymbol{R}^n$, $j = 0, 1, \cdots$, implies $f \equiv 0$) if and only if $C_B(d^m/dt^m)$ is quasi-analytic in the Denjoy-Carleman sense (*Theorems* 1–2).

OUTLINE OF PROOF. If $C_B(L)$ is quasi-analytic, it is immediate that $C_B(d^m/dt^m)$ is quasi-analytic by merely considering functions $f(t) \in C^\infty(\boldsymbol{R}^{n+1})$ which depend only on the single variable t. Then, since the hyperplane $t = 0$ is not characteristic for L, we have $Lf = d^m f/dt^m +$ lower order t derivatives.

But the proof of the converse lies much deeper. Because L has constant coefficients, we have $L = P((1/i)(\partial/\partial t), 1/i)(\partial/dx))$, where $P(\tau, \xi)$ is a constant coefficient polynomial. Now Ehrenpreis obtains a generalized Fourier representation

$$f(x,t) = \int_{C^{n+1}} e^{i(t\tau + x\xi)} \frac{d\mu(\tau,\xi)}{k(\tau,\xi)}$$

for functions $f \in C_B(L)$. Here μ is a measure of finite total mass and k is a weight factor which incorporates the fact that $f \in C_B(L)$. It is easy to guess the properties k must have to represent a function in $C_B(L)$. In order that the integral make sense for all real (t,x), we need $k(\tau,\xi)$ to dominate the linear exponential when τ and ξ are pure imaginary. To insure that $f \in C^\infty$, we need k to dominate all polynomials. Moreover, because

$$(L^j f)(x,t) = \int_{C^{n+1}} P^j(\tau,\xi) e^{i(t\tau + x\xi)} \frac{d\mu(\tau,\xi)}{k(\tau,\xi)},$$

and because we want $\| L^j f \|_\infty \leqq c r^j b_j$, we must further require that p dominate all functions $\lambda(\tau,\xi)$ of the form

$$\lambda(\tau,\xi) = \sum \frac{|P(\tau,\xi)|^j}{b_j} r^{-j}.$$

That such a representation is possible is a consequence of Ehrenpreis' Representation Theorem for his analytically uniform spaces.

We proceed naively as follows. Let $q(t,x)$ be any polynomial and consider the integral

$$I(\sigma) = \int_{x^2 + t^2/\sigma^2 = 1;\, t \geqq 0} fq = \int f q_\sigma,$$

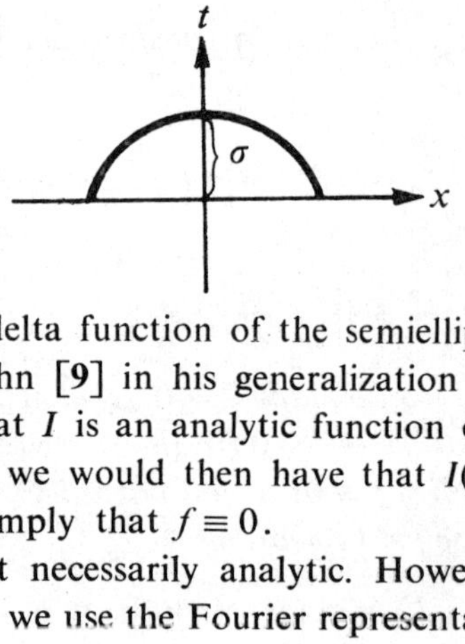

where q_σ is q times the delta function of the semiellipse $t \geqq 0$, $x^2 + t^2/\sigma^2 = 1$. Using an idea due to John [9] in his generalization of Holmgren's Theorem, one attempts to prove that I is an analytic function of σ. Since by hypothesis $I(0) = 0$ (for $f(0,x) = 0$), we would then have that $I(\sigma) \equiv 0$. Because q is any polynomial, this would imply that $f \equiv 0$.

Unfortunately, I is not necessarily analytic. However, a modification does go through. Here is where we use the Fourier representation for f. By Plancherel's theorem,

$$\begin{aligned} I(\sigma) &= \int \hat{f} \hat{q}_\sigma = \int \hat{q}_\sigma \frac{d\mu}{k} \\ &= \int_{\text{inside}} \hat{q}_\sigma \frac{d\mu}{k} + \int_{\text{outside}} \hat{q}_\sigma \frac{d\mu}{k} \equiv I_\xi(\sigma) + I_\tau(\sigma), \end{aligned}$$

where by inside and outside, we mean domains in C^{n+1} inside and outside an angle defined by $|\tau| \geqq a(1+|\xi|)$, with a some constant. We want to contain the algebraic variety where $P(\tau, \xi)=0$ inside the angle. For hyperbolic equations, the consant a is essentially the signal speed and the angle is essentially that which separates the space-like from the time-like domains.

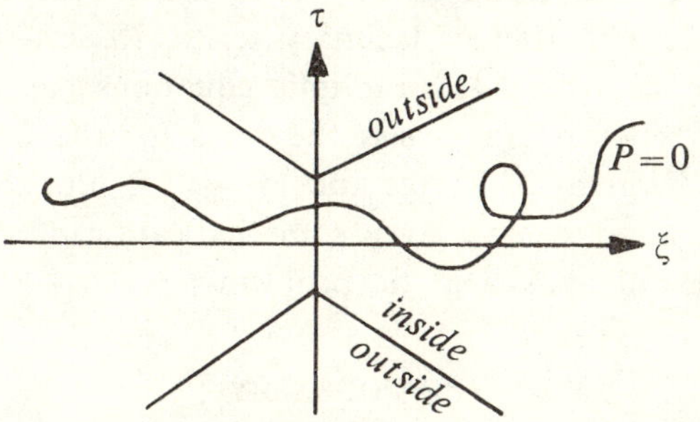

It is now possible to prove that $I_\xi(\sigma)$ is an analytic function and, if $C_B(d^m/dt^m)$ is quasi-analytic, that $I_\tau(\sigma)$ is quasi-analytic. Therefore $I(\sigma)$ is the sum of an analytic function and a quasi-analytic function, so it is quasi-analytic. Again, since q is any polynomial, we can conclude that the quasi-analyticity of $I(\sigma)$ implies that of the class of functions $f \in C_B(L)$.

Supplementary Remarks and Problems. We shall close this report by mentioning a few open problems and some related work on quasi-analytic functions.

First of all one should generalize Ehrenpreis' Theorem to linear differential operators with analytic coefficients. Perhaps a first step in this direction would be to treat a constant coefficient operator but replace the hyperplane $t=0$ by any hypersurface which is nowhere tangent to a characteristic.

In another direction, it would be interesting (and difficult) to determine minimal sets $\mathscr{D}_1$ on which the vanishing of a function and its derivatives implies its vanishing everywhere. There is a vast gap between the results for elliptic operators and those for an arbitrary operator.

One should also consider the case where the hyperplane $t=0$ is a characteristic for the operator, as it is for the heat operator $L=\partial/\partial t-\partial^2/\partial x^2$. Here one will need to impose additional growth conditions for the functions at infinity. This is true for the classical uniqueness proofs for the heat equation. Ehrenpreis [7] has treated the case of some constant coefficient operators.

An entirely different type of problem is to generalize Borel's Theorem mentioned earlier. Let $\mathscr{S}_B$ be the set of sequences $\{\beta_j\}$ such that $\beta_j=O(r^j b_j)$ as $j\to\infty$, where the constant $r>0$ may depend on the particular sequence $\{\beta_j\}$, and let C_B be as in Theorem 1. There is a natural map $\Phi\colon C_B\to\mathscr{S}_B$ defined by $\Phi(f)=\{f^{(j)}(0)\}$. The Denjoy-Carleman Theorem answers the question, "Is Φ one-to-one?", but not "Is Φ onto?", and "If Φ is not onto, what is its range?". Bang [1] has shown that if C_A is a quasi-analytic class but not analytic, then Φ cannot be onto. Ehrenpreis [7] has recently found necessary and sufficient conditions for Φ to be onto.

Other quasi-analytic problems have been considered by many people. Bernstein, Beurling [2], and Ehrenpreis [7] have treated the case where the inequalities $\| D^j f \| \leqq \cdots$ are replaced by conditions on approximating f by analytic functions. And Mandelbrojt [13] has taken an approach which reveals the intimate relationship of our topic with that of lacunary series. Recent work by Malgrange [12], Lax [11], and Browder [4] on elliptic equations has shown that there is a close connection between uniqueness theorems for the Cauchy Problem and the approximation theorems of Runge and Mergelian. But this part of the story must wait. We only hope to have given some indication of the scope and recent progress on the general quasi-analytic problem.

BIBLIOGRAPHY

1. T. Bang, *Om quasi-analytiske Funktioner*, Thesis, University of Copenhagen, 1946. [See also C. R. Dixième Congrès Math. Scand. 1946, pp. 249–254, Jul. Gjellerups Forlag, Copenhagen, 1947.]

2. A. Beurling, *On quasi-analyticity and general distributions*, Mimeographed notes, Stanford University, Stanford, Calif., 1960.

3. S. Bochner, and A. Taylor, *Some theorems on quasi-analyticity for functions of several variables*, Amer. J. Math. **61** (1939), 303–329.

4. F. Browder, *Functional analysis and partial differential equations* II, Math. Ann. **145** (1962), 81–226.

5. A. Calderón, *Uniqueness in the Cauchy problem for partial differential equations*, Amer. J. Math. **80** (1958), 16–36.

6. (a). T. Carleman, *Sur un problème d'unicité pour les systèmes d'équations aux dérivees partielles à deux variables independents*, Arkiv. Mat., Astr. 'Fys. **26**B (1939), 17.

(b). ———, *Les functions quasi-analytiqués*, Paris, 1926.

7. L. Ehrenpreis, *Fourier analysis in several complex variables*, Harper and Row, New York, 1968.

8. P. Garabedian, *Partial differential equations*, Wiley, New York, 1964.

9. F. John, *On linear partial differential equations with analytic coefficients: Unique continuation of Data*, Comm. Pure Appl. Math. **2** (1949), 209–253.

10. J. Kazdan, *Quasi-analyticity for elliptic differential operators* (to appear).

11. P. Lax, *A stability theory of abstract differential equations and its applications to the study of local behaviour of solutions of elliptic equations*, Comm. Pure Appl. Math. **8** (1956), 747–766.

12. B. Malgrange, *Existence et approximation des solutions des equations aux derivées partielles et des equations de convolution*, Ann. Inst. Fourier Grenoble **6** (1956), 271–355.

13. S. Mandelbrojt, *Quasi-analyticity and analytic continuation — a general principle*, Trans. Amer. Math. Soc. **55** (1944), 96–131.

14. W. Rudin, *Real and complex analysis*, McGraw-Hill, New York, 1966.

HARVARD UNIVERSITY
CAMBRIDGE, MASSACHUSETTS

UNIVERSITY OF PENNSYLVANIA
PHILADELPHIA, PENNSYLVANIA

Volume Elements, Holomorphic Mappings and Schwarz's Lemma

SHOSHICHI KOBAYASHI[1]

1. **Introduction.** The classical Schwarz's lemma states that if f is a holomorphic mapping of the open unit disk $D = \{z; |z| < 1\}$ into the closed unit disk $\bar{D} = \{z; |z| \leqq 1\}$ such that $f(0) = 0$, then $f'(0) < 1$ and $|f(z)| < |z|$ for $0 < |z| < 1$ unless f is a linear function of the form $f(z) = e^{ia}z$. An invariant formulation of Schwarz's lemma says that if f is a holomorphic mapping of D into itself, then $f^*(ds^2) \leqq ds^2$ where ds^2 denotes the Poincaré metric in D and the inequality holds everywhere unless f is a holomorphic diffeomorphism.[2]

For a holomorphic mapping of a classical bounded symmetric domain into itself, a generalization of Schwarz's lemma was given by Look [**10**]. Recently Korányi [**8**] obtained the following generalization whose proof is independent of classification theory:

Let M be a hermitian symmetric space of noncompact type of rank l and $f: M \to M$ a holomorphic mapping. Then, for the Bergman metric ds^2,

$$f^*(ds^2) \leqq l \cdot ds^2.$$

If there exists a point $p \in M$ such that $f^*(ds^2)_p = ds^2{}_p$, then f is a holomorphic automorphism of M.

In the case of complex dimension 1, a holomorphic mapping f such that $f^*(ds^2) \leqq ds^2$ may be viewed as a volume-decreasing map as well as a length-decreasing map. In this sense, the following theorem of Chern [**4**] is also a generalization of Schwarz's lemma:[3]

Let $f: D_1 \to N$ be a holomorphic mapping where D_1 is the open unit n-ball with the standard Kähler metric and where N is an n-dimensional hermitian Einstein manifold with scalar curvature $\leqq -2n(n+1)$. Then f is volume-decreasing.

A similar theorem has been obtained for the case N is kählerian by Dinghas [**5**].

---

[1] Supported partially by NSF Grant GP-5798.

[2] This intrinsic formulation of Schwarz's lemma is due to Pick [**12**]. See also Nevanlinna [**11**].

[3] For the case of complex dimension 1, see Ahlfors [**1**].

We shall generalize the theorem above of Chern to a more general domain space M well as a more general image space N. Our result includes in particular the case where M is a symmetric bounded domain. Our method is similar to that of Chern but is a little more elementary. Essential tools are volume elements, i.e., positive $2n$-forms and the associated hermitian forms which have been extensively used by Koszul [9] and by Hano-Kobayashi [6].

2. **The hermitian form associated with a volume element.** Let M be a complex manifold of complex dimension n and v_M a volume element of M. By definition v_M is a $2n$-form which is positive in the following sense. In terms of a local coordinate system $z^1, \cdots, z^n$ of M we may write

$$v_M = i^n K dz^1 \wedge d\bar{z}^1 \wedge \cdots \wedge dz^n \wedge d\bar{z}^n,$$

where K is a positive function. To this volume element v_M we associate the 2-form ϕ_M defined by

$$\phi_M = i \sum \frac{\partial^2 \log K}{\partial z^\alpha \partial \bar{z}^\beta} dz^\alpha \wedge d\bar{z}^\beta$$

together with the hermitian form h_M defined by

$$h_M = \sum \frac{\partial^2 \log K}{\partial z^\alpha \partial \bar{z}^\beta} dz^\alpha d\bar{z}^\beta .$$

It is well known that if v_M is the volume element defined by a hermitian metric on M, then $-h_M$ is the Ricci tensor.[4]

Similarly, let N be a complex manifold of complex dimension n and v_N a volume element on N which may be expressed as

$$v_N = i^n L dw^1 \wedge d\bar{w}^1 \wedge \cdots \wedge dw^n \wedge d\bar{w}^n$$

in terms of a local coordinate system $w^1, \cdots, w^n$ of N. To v_N we associate a 2-form ϕ_N and a hermitian form h_N in the same manner as above.

Let $f: M \to N$ be a holomorphic map. We are interested in the ratio

$$u = f^*(v_N)/v_M.$$

Clearly u is a nonnegative function on M. On the open subset of M where u is positive, we have

$$i \sum \frac{\partial^2 \log u}{\partial z^\alpha \partial \bar{z}^\beta} dz^\alpha \wedge d\bar{z}^\beta = i \sum \frac{\partial^2 (f^*(\log L) - \log K)}{\partial z^\alpha \partial \bar{z}^\beta} dz^\alpha \wedge d\bar{z}^\beta$$

$$= f^*\phi_N - \phi_M.$$

4 The Ricci tensor with respect to the canonical hermitian connection in the sense of Chern [4], not with respect to the Riemannian connection. In this paper the reader needs no knowledge of hermitian differential geometry if he takes $-h_M$ as a definition of the Ricci tensor of a hermitian manifold with volume element v_M .

Similarly,

$$\sum \frac{\partial^2 \log u}{\partial z^\alpha \partial \bar{z}^\beta} dz^\alpha d\bar{z}^\beta = f^* h_N - h_M.$$

3. **The case where the domain space M is compact.** We consider here a holomorphic mapping f of a *compact* complex manifold M into a (not necessarily compact) complex manifold N. The reasoning in this section will help the understanding of the subsequent sections.

In general, for a hermitian form h on a complex vector space V we denote by $\dim^+ h$ (resp. $\dim^- h$) the dimension of a maximal dimensional complex subspace of V on which h is positive (resp. negative) definite. Hence h is positive (resp. negative) definite if and only if $\dim^+ h = \dim V$ (resp. $\dim^- h = \dim V$).

The following theorem generalizes Theorem 1 in Chern [4].

THEOREM 1. *Let M and N be complex manifolds of complex dimension n with volume elements v_M and v_N and the associated hermitian forms h_M and h_N respectively. Let $f: M \to N$ be a holomorphic mapping. Assume that M is compact. Then*

(i) *if $\dim^+ h_N > \dim^+ h_M$ for all points of M and N, then*

$$f^*(v_N) = 0,$$

i.e., f is degenerate everywhere on M;

(ii) *if $\dim^- h_N > \dim^- h_M$ for all points of M and N, then $f^*(v_N)$ vanishes at least at one point of M, i.e., f is degenerate at least at one point of M.*

PROOF. (i) Assuming the contrary, let p be a point of M where $u = f^*(v_N)/v_M$ attains its (positive) maximum. Then the complex Hessian form of the function $\log u$ is defined and negative semidefinite at p. By the fundamental formula in §2 this complex Hessian form is equal to $f^*(h_N) - h_M$. Since f is assumed to be non-degenerate at p, $\dim^+ h_N$ at $f(p)$ coincides with $\dim^+ f^*(h_N)$ at p. Hence $f^*(h_N) - h_M$ cannot be negative semidefinite at p under the assumption in (i). This is a contradiction.

(ii) Assuming the contrary, let p be a point of M where u attains its (positive) minimum. The rest of the proof is the same as above; interchange "plus or positive" with "minus or negative".

COROLLARY. *Let $f: M \to N$ be as in Theorem 1. Then*

(i) *if h_N is everywhere positive definite on N and if h_M is nowhere positive definite on M, then f is degenerate everywhere on M;*

(ii) *if h_N is everywhere negative definite on N and h_M is nowhere negative definite on M, then f is degenerate somewhere on M.*

REMARK 1. As we remarked earlier, if M and N are hermitian manifolds in Theorem 1 or its Corollary, then $-h_M$ and $-h_N$ are the Ricci tensors of M and N

respectively. Hence Corollary contains Theorem 1 of Chern [**4**] in which he assumes that the Ricci tensor of N is negative (or positive) definite and is proportional to the metric tensor and that the scalar curvature of M (i.e., the trace of $-h_M$) is positive (or negative).

REMARK 2. As an example of manifold N with volume element v_N such that h_N is positive definite, we mention a manifold admitting the Bergman metric; v_N is the kernel form of Bergman and h_N is the Bergman metric of N (cf. Kobayashi [7]).

4. **A generalized Schwarz lemma.** We shall now consider the case where both h_M and h_N are positive definite. Starting with the case where M is compact, we shall gradually weaken the assumption on M.

THEOREM 2. *Let M and N be complex manifolds of complex dimension n with volume elements v_M and v_N respectively. Let ϕ_M and h_M (resp. ϕ_N and h_N) be the 2-form and the hermitian form associated with v_M (resp. v_N). Let $f\colon M \to N$ be a holomorphic mapping. Assume*

(a) *both h_M and h_N are positive definite;*

(b) *for the function r_M on M and the function r_N on N defined by*

$$\phi_M^n = r_M v_M \quad and \quad \phi_N^n = r_N v_N,$$

we have

$$r_N(q)/r_M(p) \geqq 1 \quad for \quad p \in M \quad and \quad q \in N;$$

(c) *M is compact.*

Then f is volume-decreasing in the sense that $f^* v_N / v_M \leqq 1$.

PROOF. Let p be a point of M where $u = f^*v_N/v_M$ attains its maximum which may be assumed to be positive; otherwise $u = 0$. Then $f^*h_N - h_M$ is negative semidefinite at p. This fact combined with (a) implies $f^*(\phi_N^n)/\phi_M^n \leqq 1$ at p. From (b) we obtain $f^*v_N/v_M \leqq 1$ at p. Since u attains its maximum at p, $u \leqq 1$ everywhere on M.

In Theorem 2 as well as in (i) of Theorem 1, the compactness of M was used only to insure the existence of a point where u attains its maximum. Hence

THEOREM 2′. *In Theorem 2, f is volume-decreasing if assumption* (c) *is replaced by*

(c′) *$u = f^*v_N/v_M$ tends to zero at the boundary in the sense that for every positive number a the set $\{p \in M; u(p) \geqq a\}$ is compact.*

A further generalization of Theorem 2 is given by

THEOREM 2″. *In Theorem 2, f is volume-decreasing if assumption* (c) *is replaced by* (c″) *there exists a sequence* $M_1 \subset M_2 \subset \cdots \subset M$ *of open submanifolds such that*

(c″.1) $\bigcup_k M_k = M$;

(c″.2) *each* M_k *carries a volume element* v_k *for which the map* $f: M_k \to N$ *satisfies* (a), (b) *and* (c′) *of Theorem* 2′ *and moreover*

$$\lim_{k\to\infty} v_k = v_M \quad (\textit{pointwise on } M).$$

PROOF. Applying Theorem 2′ to $f: M_k \to N$ we obtain

$$f^*v_N/v_k \leqq 1 \quad \text{on} \quad M_k.$$

Hence

$$f^*v_N/v_M = \lim_{k \quad \infty} f^*v_N/v_k \leqq 1.$$

We shall now apply Theorem 2″ to the case where M is a bounded domain in C^n. With respect to the natural coordinate system $z^1, \cdots, z^n$ in C^n, the Bergman kernel form b_M can be written as

$$b_M = i^n K(z, \bar{z}) dz^1 \wedge d\bar{z}^1 \wedge \cdots \wedge dz^n \wedge d\bar{z}^n,$$

where $K(z, \bar{z})$ is the Bergman kernel function of the domain M.[5]

LEMMA. *Let M be a bounded in c with the Bergman kernel form* b_M. *Assume that*

(i) *the Bergman kernel function K of M tends to infinity at the boundary of M;*

(ii) *M is star-like in the sense that, for each* $0 \leqq a < 1$, *the set*

$$M_a = \{az \in C^n; z \in M\} \quad (z \text{ being considered as a vector})$$

is contained in M.

Denote by b_a *the Bergman kernel form of* M_a *for* $0 < a < 1$. *Then*

$$\lim_{a\to 1} b_a = b_M$$

and for any complex manifold N of complex dimension n with volume element v_N *and for any holomorphic map* $f: M \to N$, *the function* f^*v_N/b_a *on* M_a *tends to zero at the boundary of* M_a.

PROOF. For each $0 < a < 1$, the map $g_a: M \to M_a$ defined by $g_a(z) = az$ is a holomorphic isomorphism. Hence $g_a^*(b_a) = b_M$. If we write

$$b_M = i^n K(z, \bar{z}) dz^1 \wedge d\bar{z}^1 \wedge \cdots \wedge dz^n \wedge d\bar{z}^n$$

and

---

[5] For Bergman metrics and Bergman kernel functions, see Bergman [2] and for Bergman kernel forms, see Kobayashi [7].

$$b_a = i^n K_a(z, \bar{z}) dz^1 \wedge d\bar{z}^1 \wedge \cdots \wedge dz^n \wedge d\bar{z}^n,$$

then the invariance $g_a^*(b_a) = b_M$ means

$$K(z, \bar{z}) = a^{2n} K_a(az, a\bar{z}) \quad \text{for} \quad z \in M.$$

The first assertion is now clear. If we write

$$f^* v_N = i^n L(z, \bar{z}) dz^1 \wedge d\bar{z}^1 \wedge \cdots \wedge dz^n \wedge d\bar{z}^n,$$

then the function L is continuous on M and hence is bounded on the closure $\bar{M}_a$ for each a. Since K goes to infinity at the boundary of M, so does K_a at the boundary of M_a. Hence L/K_a tends to zero at the boundary of M_a. This completes the proof of the Lemma.

In general, let M be a bounded domain in C^n with the Bergman kernel form b_M and the Bergman metric g_M; g_M is the hermitian form associated with b_M. Let v_M be the volume element defined by the metric tensor g_M. Consider now the special case where the ratio v_M/b_M is a constant function; such is the case if M is homogeneous for both b_M and v_M are invariant by every holomorphic automorphism of M. Then the hermitian form h_M associated with v_M coincides with the metric tensor g_M. (Differential geometrically, this says that —(Ricci tensor) $= h_M = g_M$ metric tensor.) In the notations of Theorem 2 we then have $r_M = \phi_M{}^n/v_M = 1$. If v_M/b_M is constant, then b_M and b_a in the Lemma above may be replaced by the volume element v_M and v_a defined by the Bergman metrics of M and M_a. From Theorem 2″ and Lemma we obtain

THEOREM 3. *Let M be a bounded domain in C^n with the Bergman kernel form b_M and the volume element v_M defined by the Bergman metric. Assume*

(i) *the Bergman kernel function K of M goes to infinity at the boundary of M;*

(ii) *M is star-like;*

(iii) *the ratio v_M/b_M is a constant function on M.*

Let N be a complex n-dimensional manifold with volume element v_N. Assume

(iv) *the hermitian form h_N associated with v_N is positive semidefinite and the 2-form ϕ_N associated with v_N satisfies*

$$\phi_N^n / v_N \geqq 1.$$

Then every holomorphic map $f: M \to N$ is volume-decreasing, i.e.,

$$f^* v_N / v_M \leqq 1.$$

We shall mention two cases to which Theorem 3 can be applied.

Case 1. M is a hermitian symmetric space of noncompact type and v_M is the volume element defined by the Bergman metric. N is a hermitian manifold with metric g_N and its volume element v_N such that the hermitian form h_N associated with v_N satisfies

$$0 \leqq h_N - g_N ,$$

i.e., $h_N - g_N$ is positive semidefinite.

Since $-h_N$ is called the Ricci tensor of the hermitian manifold N, the condition about N can be stated in terms of its Ricci tensor.

The following is a special case of Case 1.

Case 2. M is as above. N is a homogeneous bounded domain (or more generally, a homogeneous complex manifold admitting the Bergman metric) and v_N is the volume element defined by the Bergman metric of N.

As we have seen already, in Case 2 the Bergman metric g_N of N and the hermitian form h_N associated with v_N coincide.

In view of recent works by Pyatetski-Shapiro and others, it is most likely that Theorem 3 can be applied to all homogeneous bounded domains M (with respect to suitable realizations).

5 **Concluding remarks.** In the classical Schwarz's lemma a holomorphic map f of the open unit disk D into itself is strictly distance-decreasing everywhere unless f is a holomorphic automorphism. As in Korányi **[8]** this can be also generalized to some extent.

According to H. Cartan and Carathéodory (cf. Bochner-Martin **[3]**) a holomorphic map f of a bounded domain M into itself is a holomorphic automorphism if it has a fixed point and the Jacobian of f at this fixed point has determinant $= 1$.

Let M be a homogeneous bounded domain with an invariant volume element v_M. Let f be a holomorphic map of M into itself. If f^*v_M and v_M coincides at one point of M, then f is a holomorphic automorphism. In fact, if $f^*v_M = v_M$ at $p \in M$, then choose a holomorphic automorphism h of M such that $h(f(p)) = p$. Then $h \circ f$ has a fixed point p and $(h \circ f)^*v_M = v_M$. Our assertion now follows from the result of Cartan and Carathéodory.

For a holomorphic map f of one manifold M into another, a stronger condition is needed to conclude that f is a holomorphic isomorphism. Let M and N be complex manifolds admitting the Bergman metric. Let b_M and b_N be the Bergman kernel forms of M and N respectively. If $f: M \to N$ satisfies $f^*b_N = b_M$ everywhere on M, then f is a local isometry. If moreover M is complete with respect to the Bergman metric (in particular, if M is homogeneous) and if N is simply connected, then f is a global isomorphism.

ADDED IN PROOF. In recent papers (*Generalization of Schwarz-Pick lemma to invariant volume in Kahler manifold*, Trans. Amer. Math. Soc. **128** (1967), 221–231 and the same title II, to appear), K. T. Hahn and J. Mitchell consider *biholomorphic* maps from a bounded homogeneous domain D in $\mathbf{C}^n$ into a Kähler manifold N_n whose Ricci tensor satisfies certain boundedness conditions. Lemma 1 of their first paper shows that a bounded homogeneous domain M in $\mathbf{C}^n$ satisfies condition (vi) of

Theorem 3 in our paper. The reasoning in the proof of Theorem 1 in their second paper shows that Theorem 3 in our paper holds for any bounded homogeneous domain M in $\mathbf{C}^n$ and a complex manifold N satisfying condition (iv) thanks to the result of Vinberg-Gindikin-Pyatetski-Shapiro (*Classification and canonical realization of complex homogeneous domains*, Trudy Moscow, Mat. Obsc. **12** (1963), 359–388 = Trans. Moscow Math. Soc. **12** (1963), 404–437).

Bibliography

1. L. V. Ahlfors, *An extension of Schwarz's lemma*, Trans. Amer. Math. Soc. **43** (1938), 359–364.

2. S. Bergman, *Über die Kernfunktion eines Bereiches und ihr Verhalten am Rande*, J. Reine Angew. Math. **169** (1933), 1–42.

3. S. Bochner and W. T. Martin, *Several complex variables*, Princeton Univ. Press, Princeton, N. J., 1948.

4. S. S. Chern, *On holomorphic mappings of hermitian manifolds of the same dimension*, these Proceedings.

5. A. Dinghas, "Ein *n*-dimensionales Analogon des Schwarz-Pickschen Flächensatzes für holomorph Abbildungen der komplexen Einheitskugel in eine Kähler-Mannigfaltigkeit" in *Fest schrift zur Gedächtnesfeir für Karl Weierstrass*, 1815–1965, Westdeutscher Verlag, Cologne, 1966, pp. 477–494.

6. J. I. Hano and S. Kobayashi, *A fibering of a class of homogeneous complex manifolds*, Trans. Amer. Math. Soc. **94** (1960), 233–243.

7. S. Kobayashi, *Geometry of bounded domains*, Trans. Amer. Math. Soc. **92** (1959), 261–291.

8. A. Korányi, *A Schwarz lemma for bounded symmetric domains*, Proc. Amer. Math. Soc. **17** (1966), 210–213.

9. J. L. Koszul, *Sur la forme hermitienne canonique des espaces homogènes complexes*, Canad. J. Math. **7** (1955), 562–576.

10. K. H. Look, *Schwarz lemma and analytic invariants*, Sci. Sinica **7** (1958), 453–504.

11. R. Nevanlinna, *Eindeutige analytische Funktionen*, Springer, Berlin, 1936.

12. G. Pick, *Über eine Eigenschaft der konformen Abbildung kreisförmiger Bereiche*, Math. Ann. (2) **77** (1916).

University of California
Berkeley, California

Limits of Polynomials Whose Zeros Lie in a Given Set

JACOB KOREVAAR[1]

Survey of recent work, published and unpublished, by the author and his former students Judith M. Elkins G. W. Hedstrom, J. E. Lange, T L. McCoy and M. D. Thompson, as well as recent work of others, notably T. Ganelius.

1. Polynomial approximation sets.

1.1. **Introduction.** Let D be a simply connected region (connected open set) in the (extended) complex plane which does not contain the point at infinity, and let J be a simple arc with the end points removed. We wish to represent holomorphic functions in D (or continuous functions on J) as limits of polynomials; the convergence must be uniform on every compact subset of D (or of J). By Runge's theorem [cf. **34**] (of a theorem of Walsh [**37, 38**]) such representation is possible for all functions of the kind indicated. However, we impose one additional condition: *the zeros of the approximating polyncmials must lie in a prescribed set* R. It is no restriction to assume that R is closed: the approximable functions for the closure of R are the same as those for R. Polynomials whose zeros lie in R will be called *R-polynomials*.

Definition. *A set R will be called a polynomial axproximation set relative to D (or J) if every zero free holomorphic function in D (or every continuous function on J) is a limit of R-polynomials.*

It is not difficult to show that R is a polynomial approximation set if and only if all exponentials e^P, with P a polynomial, are approximable. If R is a polynomial approximation set relative to D, then all holomorphic functions in D whose zeros lie in $D \cap R$ are approximable. A polynomial approximation set relative to D is also a polynomial approximation set relative to every simply connected subregion, and relative to every simple arc J in D.

The main problem is to determine the polynomial approximation sets R for a given simply connected region D. There is another way of comparing this problem with Runge's theorem. Let z_0 be some finite point outside D (assuming that there is one). By one form of Runge's theorem [cf. **34**], every holomorphic function $f(z)$ in D is a limit of rational functions of the form

[1] Paper prepared with support from NSF grant GP 4106.

$$R_m(z) = \sum_j \frac{b_{mj}}{(z_0 - z)^j}.$$

$R_m(z)$ has only one pole, but (in general) the order of the pole will tend to infinity as $m \to \infty$. One can replace this single pole of high order by a large number of first order poles at points close to z_0. It is not hard to show that if $\{z_k\}$ is any sequence of points (distinct from z_0) that tends to z_0, then $f(z)$ can be represented as a limit of sums

$$S_n(z) = \sum_k \frac{c_{nk}}{z_k - z}.$$

Thus if R is any set with a finite limit point outside D, one can approximate every $f(z)$ by sums $S_n(z)$, with the z_k in R. This result holds also if R has infinity as its only limit point.

In comparison, *approximation of zero free holomorphic* $F(z)$ *in* D *by* R*-polynomials* $A\prod_{k=1}^{n}(z - z_k)$ *is equivalent to approximation of holomorphic* $f(z) = -F'(z)/F(z)$ *by sums of the form*

$$T_n(z) = \sum_{k=1}^{n} \frac{1}{z_k - z},$$

with the z_k *in* R *and coefficients equal to* 1. In trying to approximate by sums $T_n(z)$, one will in general have to let the z_k depend on n, and allow multiple occurrence of certain z_k.

In physical terms, the problem is the following. For which sets R can one represent an arbitrary (regular) gravitational or electrostatic field $\mathscr{F}$ in D as a limit of fields due to distributions of unit point masses or point charges ("electrons") on R? Indeed, every regular field $\mathscr{F}$ in D can be represented as the gradient of a harmonic function Φ. In complex form,

$$\mathscr{F}(z) = \frac{\partial\Phi}{dx} + i\,\frac{\partial\Phi}{\partial y} = \frac{\partial\Phi}{\partial x} - i\,\frac{\partial\Psi}{\partial x} = \left\{\frac{d}{dz}(\Phi + i\Psi)\right\}^*,$$

where Ψ is a harmonic function conjugate to Φ, and * denotes the complex conjugate. On the other hand, assuming attractive forces inversely proportional to the distance, and using appropriate normalization, the expression

$$T_n^*(z) = \left(\sum_{k=1}^{n} \frac{1}{z_k - z}\right)^*$$

represents the field at the point z due to unit masses at the points z_k.

The problem of approximating gravitational fields by fields due to distributions of unit point masses can be formulated in any number of dimensions.

1.2. *Results.* We first consider the case of bounded sets R, and take R disjoint from D and J.

THEOREM 1. *A bounded closed set R disjoint from D (or J) is a polynomial approximation set for D (or J) if and only if it separates D (or J) from the point at infinity.*

In other words, R is a polynomial approximation set if and only if it divides the plane, and D (or J) belongs to a bounded component of its complement R^c.

COROLLARY 1. *The boundary ∂D of a bounded simply connected region D is a polynomial approximation set for the interior.*

A necessary and sufficient condition is known also for the case of sets R that lie in a half-plane:

THEOREM 2. *A set R disjoint from D and J that lies in a half-plane H is a polynomial approximation set for D (or J) if and only if the following pair of conditions is satisfied:*
(i) *R separates D (or J) from the complement H^c of H;*
(ii) *the component of R^c which contains D (or J) does not contain a half-plane.*

COROLLARY 2. *If the simply connected region D is contained in a half-plane but does not contain one, the boundary ∂D is a polynomial approximation set for D.*

For example, the pair of rays $\arg z = \pm\alpha$, where $0 < \alpha < \frac{1}{2}\pi$, is a polynomial approximation set for the angle $|\arg z| < \alpha$. The boundary of a strip is a polynomial approximation set for the interior. The imaginary axis by itself is not a polynomial approximation set for the half-plane $\operatorname{Re} z > 0$, but the union of the imaginary axis and an arbitrary sequence of points $\{w_k\}$ such that $\operatorname{Re} w_k \to \infty$ is.

For the case of arbitrary unbounded sets R, no geometric conditions are known that are both necessary and sufficient. However, one can state simple sufficient conditions for R to be a polynomial approximation set for the plane (and hence for every D and J). These conditions require that R extend to infinity in sufficiently many directions. For the precise result, see Theorem 3 (§2.2).

History of Theorems 1 *and* 2. G. R. MacLane **[24]** proved about 1949 that if ∂D is a rectifiable Jordan curve, then ∂D is a polynomial approximation set for the interior D. This result was reproved in a different way, and extended to a somewhat larger class of Jordan curves by M. D. Thompson **[35, 36]** in 1962. The general form of Corollary 1 and Theorem 1 is due to the author **[15]**.

M. D. Thompson **[35, 36]** also obtained special cases of Corollary 2, including the examples mentioned above. The general form of Corollary 2 is due to Judith M. Elkins **[4]**; Theorem 2 follows from her work.

It is not known if Corollary 1 is valid for arbitrary bounded open sets D with connected complement. However, the author has shown [unpublished work] that in this case, every holomorphic function in D is the limit of sums $\sum 1/(z_k - z)$ where the z_k lie on ∂D.

The author has also done some work [17] on the higher dimensional problem indicated at the end of §1.1. If D is a bounded region with connected complement and sufficiently smooth boundary, it is possible to represent every regular gravitational field $\mathscr{F}$ in D ($\mathscr{F} = \operatorname{grad} \Phi$, with $\Delta\Phi = 0$) as the limit of fields generated by sets of unit point masses on the boundary ∂D.

§§1.3, 1.4 contain an outline of the proof of the sufficiency of the condition in Theorem 1. It is easy to prove the necessity of the condition that R separate D or J from the point at infinity. Indeed, suppose the sums $T_n(z)$ (§1.1) converge in D or on J. Since the sequence $\{(1/n)\, T_n(z)\}$ is uniformly bounded on the set of all z whose distance to R exceeds $\delta > 0$, it follows that $(1/n)\, T_n(z) \to 0$ throughout the component D^+ of R^c which contains D. Hence if this component were unbounded, $T_n(z)$ would be $o(n)$ at points arbitrarily far from R. However, far from R all n terms in the sum for $T_n(z)$ have approximately the same argument.

1.3 *Asymptotically neutral distributions of electrons.* It is assumed from here on that the component D^+ of the complement of R which contains D is bounded. Let D^∞ be the unbounded component of the complement of the closure of D^+ (relative to the extended plane). Then D^∞ is a simply connected region which contains the point at infinity. The boundary of D^∞ will be denoted by R^-.

The first step in the proof of Theorem 1 consists in showing that *for each point ζ in R^-, the function $-1/(\zeta - z)$ in D is a limit of sums $\sum 1/(z_k - z)$, with the z_k in R^-.* This result follows from the existence of *asymptotically neutral families of points* ("*electrons*") on R^-, that is, families of finite sequences of points

$$z_{n1}, z_{n2}, \cdots, z_{nn} \quad \text{on} \quad R^-, \qquad \text{with} \quad n = n_j \to \infty,$$

such that

$$\sum_{k=1}^{n} \frac{1}{z_{nk} - z} \to 0 \quad \text{as} \quad n = n_j \to \infty,$$

uniformly on every compact subset of D. In view of the remarks preceding §1.3, the points z_{nk} of such a family must be dense on R^-.

If $R = R^-$ is the unit circumference $|z| = 1$, and $D = D^+$ the unit disc $|z| < 1$, an asymptotically neutral family is obtained by taking the nth roots of unity, $n = 1, 2, \cdots$. In this case

$$\sum_{k=1}^{n} \frac{1}{z_{nk} - z} = -\frac{d}{dz}\log(1 - z^n),$$

which tends to 0 uniformly for $|z| \leqq \rho < 1$ as $n \to \infty$.

In the general case one may try to use conformal images of roots of unity. Let $z = \Phi(w)$ denote the one to one conformal mapping of the exterior $|w| > 1$ of the unit circle in the (extended) w-plane onto the region D^∞ in the z-plane, normalized by the conditions $\Phi(\infty) = \infty$, $\Phi'(\infty) > 0$. One can extend the mapping to the boundary by setting $\Phi(e^{i\theta}) = \lim \Phi(re^{i\theta})$ as $r \searrow 1$ wherever the limit exists; for the exceptional θ (which form a set of measure zero), one may take $\Phi(e^{i\theta})$ equal to one of the radial limit values. Observe that all points $z = \Phi(e^{i\theta})$ will belong to R^-.

Images of roots of unity have been used as interpolation points by Fejér [5] and others. That they should be useful in the present problem is made plausible by the following consideration. Writing $\Phi(e^{i\theta}) = \phi(\theta)$ one has $\Phi(re^{i\theta}) \to \phi(\theta)$ as $r \searrow 1$ for almost all θ, and boundedly on $(0, 2\pi)$; it follows that

$$\int_0^{2\pi} \frac{1}{\Phi(\theta) - z}\, d\theta = \int_{|w| = r > 1} \frac{1}{\Phi(w) - z} \frac{dw}{iw} = 2\pi \frac{1}{\Phi(\infty) - z} = 0$$

for all z in D. Thus if the boundary function $\phi(\theta)$ is sufficiently well behaved, the special Riemann sums

$$\frac{2\pi}{n} \sum_{k=1}^{n} \frac{1}{\phi(2\pi k/n) - z}$$

will be good approximations, for large n, to the value 0 of the integral. Of course, one needs a very good approximation, one of order $o(1/n)$, in order to conclude that

$$(*) \qquad \sum_{k=1}^{n} \frac{1}{\phi(2\pi k/n) - z} \to 0$$

as $n \to \infty$!

For the special case where R^- is a rectifiable Jordan curve, M. D. Thompson [35, 36] has shown that relation (*) holds uniformly on every compact subset of D. In the general case one may combine the area formula of conformal mapping with Parseval's formula to show that there exist asymptotically neutral families on R^- consisting of images of suitably shifted roots of unity [15].

The existence of asymptotically neutral families has the following physical interpretation. Think of R as a (single) closed hollow conductor, with simply connected interior D. By the preceding it is possible to put a large number n of electrons on the outer boundary R^- in such a way that the resulting field is small throughout D (uniformly small on compact subsets). In the context of this physical picture it would be natural to expect that the field is in some sense minimal in D when the n electrons are placed at equilibrium or Fekete points, that is, points z_k on R^- for which $\prod_{j<k} |z_j - z_k|$ is a maximum (the potential energy is a minimum for such a configuration). *It is an open problem if for Fekete*

points z_k, *all or certain sums* $\sum_{k=1}^{n} 1/(z_k - z)$ *are indeed small in* D. The answer is unknown even in the case of rectifiable or analytic Jordan curves R^-.

1.4. *Completion of the proof of Theorem* 1. The approximability of $-1/(\zeta - z)$ in D by sums $\sum 1/(z_k - z)$, with ζ and the z_k on R^-, implies the approximability of $-\log(\zeta - z)$ by sums $a + \sum \log(z - z_k)$, and hence the approximability of the real functions $-\log|\zeta - z|$ by real sums

$$(*) \qquad \alpha + \sum \log|z - z_k|, \qquad z_k \text{ on } R^-.$$

We will indicate how the last result can be used to prove the approximability of arbitrary real harmonic functions $\log|F(z)|$ in D by sums (*), and hence (via the Borel-Carathéodory inequality) the approximability of zero free holomorphic functions $F(z)$ in D by R-polynomials. By the maximum principle, it is sufficient to prove the approximability of $\log|F(z)|$ on all rectifiable Jordan curves γ in D.

For given γ, let $C(\gamma)$ be the Banach space of the real continuous functions on γ under the maximum norm, let $H(\gamma)$ be the subspace of the restrictions to γ of harmonic functions in D, and let $A(\gamma)$ denote the subclass of functions approximable by sums (*). It is not clear a priori that $A(\gamma)$ is a vector space. In order to apply the usual method of testing by continuous linear functionals, one therefore begins by looking for a large subset of $A(\gamma)$ which is a subspace of $C(\gamma)$.

Since the functions $\pm \log|\zeta - z|$ are in $A(\gamma)$ for every point ζ on R^-, the same is true for the functions

$$v\{\log|\zeta'' - z| - \log|\zeta' - z|\},$$

where v is an arbitrary integer, and ζ' and ζ'' lie on R^-. At each point ζ of the connected set R^-, there is at least one "tangential" direction s with the following property: there are sequences $\{\zeta_n'\}, \{\zeta_n''\}$ of points on R^- such that $\zeta_n' \to \zeta$, $\zeta_n'' \to \zeta$, and the direction of the vector from ζ_n to ζ_n'' tends to s. For such sequences, and arbitrary real λ, letting [ ] denote the greatest integer function,

$$[\lambda/|\zeta_n'' - \zeta_n'|]\{\log|\zeta_n'' - z| - \log|\zeta_n' - z|\} \to \lambda(\partial/\partial s)\log|\zeta - z|$$

uniformly for z on γ. One concludes that $\lambda(\partial/\partial s)\log|\zeta - z|$ *is in* $A(\gamma)$ *for every tangential direction* s *at* ζ *of* R^- *and every real number* λ.

We now let $A_0(\gamma)$ be the closed subspace of $C(\gamma)$ generated by the constant 1 and the functions $(\partial/\partial s)\log|\zeta - z|$, where ζ runs over R^-, s over the tangential directions at ζ. We will show that $A_0(\gamma)$ contains $H(\gamma)$.

Let $d\mu$ be any nonzero real measure on γ orthogonal to $A_0(\gamma)$:

$$\int_\gamma d\mu(z) = 0, \qquad \int_\gamma \frac{\partial}{\partial s} \log|\zeta - z|\, d\mu(z) = 0$$

for all ζ on R^- and the corresponding tangential directions s. Consider the logarithmic potential

$$u(\zeta) = \int_\gamma \log|\zeta - z|\,d\mu(z) = \int_\gamma \log|1 - z/\zeta|\,d\mu(z).$$

The function $u(\zeta)$ is harmonic outside γ and vanishes at infinity. On R^-, its tangential derivatives $\partial u/\partial s$ are equal to 0. As a consequence (and this is not difficult to prove rigorously, cf. [**15**]), $u(\zeta) = \text{constant}$ on R^-, hence $u(\zeta) = \text{constant}$ outside γ. Thus $u(\zeta) = 0$ outside γ, that is, $A_0(\gamma)$ contains all limits of sums $\alpha + \sum \beta_k \log|\zeta_k - z|$, with ζ_k outside γ. It now follows easily from Runge's theorem that $A_0(\gamma)$ contains the restrictions of all harmonic functions $\log|F(z)|$ in D, hence $A(\gamma) \supset A_0(\gamma) \supset H(\gamma)$.

The ideas in the proof of Theorem 2 are similar to those used for Theorem 1, but the details are much more complicated [**4**].

2. Approximable Functions in Terms of Asymptotic Directions

2.1 **Introduction.** In this part, R will always be an unbounded closed set; as before, R-polynomials will be those polynomials whose zeros lie in R. For given R. we ask which entire functions can bė represented as limits of R-polynomials; the convergence is required to be uniform on every compact set. The class of approximable entire functions (different from the zero function) will be denoted by $E(R)$; the zeros of a function in $E(R)$ most lie in R.

Definition. *If $E(R)$ consists of all entire functions (not identically zero) whose zeros lie in R we call R regular; otherwise, R is called singular.*

A set R is regular if and only if all exponentials

$$\exp\{cz^k\},\ k = 1,2,\cdots, c \text{ arbitrary complex}$$

are approximable; in the terminology of §1, a regular set is nothing else than a polynomial approximation set relative to the whole plane.

The important questions are:

(i) Which sets R are regular?

(ii) If R is singular, what can one say about $E(R)$?

A number of answers can be given in terms of the directions in which points of R tend to infinity, *the asymptotic directions of R*.

Example. We will determine $E(R)$ for the case where R has just one asymptotic direction θ. To begin with, let $e^{g(z)}$ be a zero free function in $E(R)$. That is, $-g'(z)$ is a limit of sums

$$\sum_k \frac{1}{z_{nk} - z}, \quad \text{where } z_{nk} \in R, \quad \min_k |z_{nk}| \to \infty.$$

We saw in §1.1 that every entire function $h(z)$ is a limit of sums $\sum_k c_{nk}/(z_{nk} - z)$ with z_{nk} in R. However, the story is different when the c_{nk} are all real and $\geqq 0$. In that case,

$$\arg \sum_k \frac{c_{nk}}{z_{nk} - z} \to -\theta = \arg h(z).$$

Thus by the open mapping theorem,

$$h(z) = \text{constant} = \lambda e^{-i\theta}, \qquad \lambda \geqq 0.$$

In particular, $-g'(z) = \lambda e^{-i\theta}$. Taking sums $\sum_k 1/(z_{nk} - z)$ in which all z_{nk} are equal to some far-out number ζ_n in R, one shows that every limit of the form $\lambda e^{-i\theta}$ is obtainable. Hence the zero free approximable functions are precisely the functions

$$\exp\{-\lambda e^{-i\theta} z\}.$$

Now let $f(z)$ be an arbitrary function in this class $E(R)$. It may be assumed by translation and normalization that $f(0) = 1$, and that the zeros $z_{n1}, z_{n2}, \cdots$ of the approximating R-polynomials all have argument close to θ. Then

$$\sum_k \frac{1}{|z_{nk}|} \approx \left| \sum_k \frac{1}{z_{nk}} \right| \to |f'(0)|.$$

It follows that the zeros $z_1, z_2, \cdots$ of $f(z)$ are such that the series $\sum_k 1/|z_k|$ converges. A little additional argument shows that the functions $f(z)$ in $E(R)$ with $f(0) = 1$ are the functions of the form

$$e^{az} \prod_k (1 - z/z_k),$$

with z_k in R, $\sum_k 1/|z_k| < \infty$, and $\arg - a^* = \theta$.

We return now to the case of arbitrary R. *Let $A(R)$ denote the set of all rays $\arg z = \theta$, where θ runs through the asymptotic directions of R, and let $HA(R)$ be the convex hull of $A(R)$*. The set $HA(R)$ is either (i) the plane, or (ii) an angle (infinite sector) of opening $< \pi$ or a ray, or (iii) a half-plane or a line. Observing that the functions $\exp\{-\lambda e^{-i\theta} z\}$ are approximable for every asymptotic direction θ of R, and that products of approximable functions are approximable, one concludes that e^{az} is approximable for all a such that $-a^*$ belongs to $HA(R)$. It can be shown that in cases (i) and (ii) the last relation gives all approximable functions e^{az}.

It will be convenient to introduce certain powers of the set $A(R)$. *For $j = 1, 2, \cdots$ we define $A^j(R)$ as the set of all rays $\arg z = j\theta \pmod{2\pi}$, where θ runs through the asymptotic directions of R. The convex hull of $A^j(R)$ will be denoted by $HA^j(R)$.*

Assuming that e^{az} is in $E(R)$ for every complex number a, which functions $\exp\{bz^2\}$ will be approximable? It will be necessary and sufficient for approximability that $-2b$ be a limit of sums $\sum_k 1/(z_{nk} - z)^2$, with z_{nk} in R. Thus $\exp[bz^2]$ is certainly approximable if $-b^*$ belongs to $HA^2(R)$.

Considerations such as the foregoing provide the proofs of Theorems 3 and 4 below.

2.2. **Results.** We begin by stating a simple sufficient condition for regularity.

THEOREM 3. *A set R such that $HA^j(R)$ is the plane for every positive integer j is regular.*

Suppose now that $HA^j(R)$ is not always the plane. Let m be the *smallest* positive integer such that $HA^m(R)$ is contained in a half-plane.

THEOREM 4. *Suppose that $HA^m(R)$ is an angle $< \pi$ or a ray. Then $E(R)$ consists of the entire functions which can be represented in the form*

$$A \exp(c_1 z + \cdots + c_m z^m) z^r \prod_k (1 - z/z_k) \exp(z/z_k + \cdots + z^{m-1}/(m-1)z_k^{m-1}),$$

with z_k in R, $\sum_k |z_k|^{-m} < \infty$, $A \neq 0$, and $-c_m^*$ in $HA^m(R)$; r can be > 0 only if the origin is in R.

What can one say if $HA^m(R)$ is a half plane or a line? If there is a positive integer p such that $A^p(R)$ belongs to an angle $< \pi$ or a ray, then R is singular and all functions in $E(R)$ are of order $\leqq p$. If $A^j(R)$ never belongs to an angle $< \pi$ the asymptotic directions of R are not sufficient to determine whether R is regular or singular.

EXAMPLES. An angle of opening $> \pi$ is regular, one of opening $< \pi$ is singular. A half-line, a line and an infinite strip are singular. A half-plane is singular; however, the union of the half-plane $\operatorname{Re} z \geqq 0$ and a sequence of points $\{w_k\}$ such that $\operatorname{Re} w_k \to -\infty$ is regular. The sequence $\{w_k\}$ may be chosen in such a way that $\arg w_k \to \frac{1}{2}\pi$; in this case, $A(R)$ is a half-plane.

There exist a number of results independent of the geometry of R. For a singular set, the class $E(R)$ is in a sense quite small:

THEOREM 5. *For a singular set R, there is a finite least upper bound $\rho(R)$ to the orders of the functions in $E(R)$.*

Let q be the largest integer less than $\frac{1}{2}\rho(R)$. Then $E(R)$ contains all entire functions (not identically zero) of order less than $q+1$ whose zeros lie in R (and this bound is best possible). If e^g is any zero free entire function in $E(R)$, then $E(R)$ contains all functions of the form $e^{\lambda g}$, with $\lambda \geqq 0$.

History. For the special cases of a half-line, a line and an angle of opening $\leqq \pi$ the approximable functions have been known for a long time, mainly through the work of É. Laguerre [**20**] and G. Pólya [**28, 29, 30, 31**] (cf. the survey by N. Obrechkoff [**27**]). The singularity of a set for which $A^p(R)$ belongs to an angle $< \pi$ may be derived from the work of E. Lindwart and G. Pólya [**23**]. Theorems 3 and 4 were proved by the author in 1948 [**10, 11**], Theorem 5 and the associated results in 1953 [**12**]. Recently, J. E. Lange and the author [**21, 22**] obtained a geometrical characterization of the class $E(R)$ for all radial sets R (that is, sets of rays starting at the origin). It is easy to see that a radial set R is regular if and only if no power $A^j(R) = R^j$ belongs to a half-plane.

2.3. **From local to global convergence.** We will say that a set R has the *local* *to global property* if uniform convergence of R-polynomials on the unit disc, to a limit function different from the zero function, implies uniform convergence on every bounded set. It was shown by E. Lindwart and G. Pólya **[23]** that if R (or a power of R) lies in a half-plane with boundary through 0, then R has the local to global property. In 1953, T. Ganelius **[6]** extended this result to the case where one component of the complement R^c of R contains a half plane, as well as part of the unit disc. Some years ago, the author conjectured that the following strong alternative is true:

CONJECTURED ALTERNATIVE. *Suppose that every component of R^c meets the unit disc. Then either R is regular, or R has the local to global property.*

As a consequence, either all zero free holomorphic functions in the unit disc would be approximable by R-polynomials, or only those that are the restriction of rather special entire functions (the functions in $E(R)$).

It follows from the preceding that the alternative holds for all radial sets R. Quite recently, T. L. McCoy **[26]** has developed a potential-theoretic method which establishes the alternative for a large class of sets R with well-behaved boundaries.

There are also cases where uniform convergence on an arc (to a limit different from the zero function) implies uniform convergence on every bounded set. Thus it was shown by C. Loewner and the author **[18]** that if R lies in the half-plane $\operatorname{Im} z \geqq 0$, then R has the local to global property relative to every arc in the closed half plane $\operatorname{Im} z \leqq 0$. In particular, the only continuous functions (not identically zero) on the interval $[-1, 1]$ that can be uniformly approximated by polynomials with real zeros, or zeros with imaginary part $\geqq 0$, are the restrictions of very special entire functions. Judith M. Elkins **[4]** has shown that if one component Q of R^c contains a half-plane, then R has the local to global property relative to every Jordan arc in Q, and every free Jordan arc in the closure of Q. The above conjecture can be extended to the case of convergence on an arc.

3. THE CASE OF PARTIAL SUMS OF POWER SERIES.

Let $\sum_0^\infty a_n z^n$ be a formal power series different from a polynomial,

$$s_n(z) = \sum_{k=0}^{n} a_k z^k$$

its partial sum of order n. *By $v_n(\delta)$ we denote the minimum number of zeros of s_n in any angle of fixed opening δ and vertex O.*

If for some $\delta > 0$,

$$v_n(\delta) = o(n) \qquad \text{as} \quad n \to \infty,$$

then it follows from the work of Jentzsch **[9]**, Carlson **[1, 2]**, Rosenbloom **[32, 33]** and the author **[11]** that the series $\sum a_n z^n$ represents an entire function of order zero. That is,

$$a_n = O\{\exp(-Cn\log n)\}$$

for arbitrarily large C.

Until very recently it was not known what the implications are of restrictions stronger than $v_n(\delta) = o(n)$. In particular, what can one say if $v_n(\delta) = 0$ for some $\delta > 0$ and all n? More generally, suppose that for some $\delta > 0$ and an α with $0 \leqq \alpha < 1$,

$$(*) \qquad v_n(\delta) \leqq An^{\alpha}, \qquad n = 1, 2, \cdots.$$

In order to get information on the possible implications of (*), G. W. Hedstrom and the author [8] made a very detailed study of the zeros of the partial sums of the special series $\sum_0^\infty \exp(-n^\beta)z^n$. It had long been known that for $\beta \geqq 2$ all zeros of all s_n are negative real. For $1 < \beta < 2$ they found that all but $O(n^{2-\beta})$ zeros of s_n are negative real. The exceptional zeros are distributed over all angles, and $v_n(\delta) \sim cn^{2-\beta}$, with $c = c(\delta, \beta) > 0$. They conjectured that this is about as small as the $v_n(\delta)$ can be for power series with coefficients a_n of comparable size. That is, if the $v_n(\delta)$ are smaller, the $|a_n|$ must be smaller than $\exp(-n^\beta)$.

However, it took the ingenuity and the powerful techniques of T. Ganelius [7] to establish the conjecture. During a 1962 analysis conference at the University of Wisconsin he proved that (*) implies an estimate of the form

$$(**) \qquad a_n = O\{\exp(-\varepsilon n^{2-\alpha})\}, \qquad \varepsilon > 0.$$

In particular, if every partial sum s_n has a zero free angle of fixed opening $\delta > 0$ and vertex O, then

$$a_n = O\{\exp(-\varepsilon n^2)\} \quad \text{or} \quad \log M(r) = O\ \{(\log r)^2\},$$

a result previously known only in the simpler case $\delta \geqq \pi$ [3] and other special cases [25].

An estimate of the size of ε in (**) was obtained by T. L. McCoy and the author [19]. They also showed that if the partial sums have a zero free angle of fixed opening and vertex O, the power series cannot have unbounded gaps.

References

1. F. Carlson, *Sur les fonctions entières*, Comptes Rendus Acad. Sci. Paris **179** (1924), 1583–1585.

2. ———, *Sur les fonctions entières*, Ark. Mat. Astr. Fys. **35A** (1948) No. 14, 1–18.

3. A. Edrei, *Power series having partial sums with zeros in a half-plane*, Proc. Amer. Math. Soc. **9** (1958), 320–324.

4. Judith M. Elkins, *Approximation by polynomials with restricted zeros*, Ph.D. thesis, University of Wisconsin, 1966.

5. L. Fejér, *Interpolation und konforme Abbildung*, Nachr. Ges. Wiss. Göttingen, 1918, 319–331.

6. T. Ganelius, *Sequences of analytic functions and their zeros*, Ark. Mat. **3** (1953) No. 1, 1–50.

7. ———, *The zeros of the partial sums of power series*, Duke Math. J. **30** (1963), 533–540.

8. G. W. Hedstrom and J. Korevaar, *The zeros of the partial sums of certain small entire functions*, Duke Math. J. **30** (1963) 519–532.

9. R. Jentzsch, *Untersuchungen zur Theorie der Folgen analytischer Funktionen*, Acta Math. **41** (1918), 219–251.

10. J. Korevaar, *Approximation and interpolation applied to entire functions*, Ph.D. thesis, University of Leiden, Amsterdam, 1949.

11. ———, *The zeros of approximating polynomials and the canonical representation of an entire function*, Duke Math. J. **18** (1951) 573–592.

12. ———, *Entire functions as limits of polynomials*, Duke Math. J. **21** (1954), 533–548.

13. ———, *Approximation by polynomials whose zeros lie on a circle*, Nieuw Arch. Wisk. (3) **10** (1962), 11–16.

14. ———, *Limits of polynomials with restricted zeros*, Studies in Mathematical Analysis and Related Topics (Essays in honor of G. Pólya), Stanford, Calif., pp. 183–190, 1962.

15. ———, *Asymptotically neutral distributions of electrons and polynomial approximation*, Ann. of Math. **80** (1964), 403–410.

16. ———, *Approximation by polynomials whose zeros lie in a given set*, ISNM vol. 5, On approximation theory, Basel, 1964, pp. 161–171.

17. ———, *Distributions of unit masses and approximation of harmonic functions* [unpublished].

18. J. Korevaar and C. Loewner, *Approximation on an arc by polynomials with restricted zeros*, Nederl. Akad. Wetensch. Proc. Ser. **A 67** (1964), 121–128.

19. J. Korevaar and T. L. McCoy, *Power series whose partial sums have few zeros in an angle*, J. Math. Anal. Appl. **8** (1964), 461–473.

20. É. Laguerre, *Sur les fonctions du genre zéro et du genre un*, Comptes Rendus Acad. Sci. Paris **98** (1882); Oeuvres **1** (1898), 174–177.

21. J. E. Lange, *Entire functions as limits of zero-restricted polynomials*, Ph.D. thesis, University of Wisconsin, 1961.

22. J. E. Lange and J. Korevaar, *Limits of polynomials whose zeros lie in a radial set*, Trans. Amer. Math. Soc. **114** (1965), 65–79.

23. E. Lindwart and G. Pólya, *Über einen Zusammenhang zwischen der Konvergenz von Polynomfolgen und der Verteilung ihrer Wurzeln*, Rend. Circ. Mat. Palermo **37** (1914), 297–304.

24. G. R. MacLane, *Polynomials with zeros on a rectifiable Jordan curve*, Duke Math. J. **16** (1949), 461–477.

25. T. L. McCoy, *Entire functions with restraints on the zeros of the partial sums*, Ph. D. thesis, University of Wisconsin, 1961.

26. ———, *Equilibrium distributions of charge on curves and locally convergent sequences of polynomials*, these Proceedings.

27. N. Obrechkoff, *Quelques classes de fonctions entières limites de polynomes et de fonctions méromorphes limites de fractions rationnelles*, Actualités Sci. Ind. No. 891, Hermann, Paris, 1941.

28. G. Pólya, *Über Annäherung durch Polynome mit lauter reellen Wurzeln*, Rend. Circ. Mat. Palermo **36** (1913), 279–295.

29. ———, *Über Annäherung durch Polynome, deren sämtlichen Wurzeln in einen Winkelraum fallen*, Nachr. Ges. Wiss. Göttingen, 1913, 326–330.

30. ———, *Sur les opérations fonctionnelles linéaires, échangeables avec la dérivation et sur les zéros des polynomes*, Comptes Rendus Acad. Sci. Paris **183** (1926), 413–414.

31. G. Pólya I. and Schur, *Über zwei Arten von Faktoren-folgen in der Theorie der algebraischen Gleichungen*, J. Reine Angew. Math. **144** (1914), 89–113.

32. P. Rosenbloom, *On sequences of polynomials, especially sections of power series*, Ph.D. thesis, Stanford University, 1944.

33. ———, *Distribution of zeros of polynomials*, Lectures on functions of a complex variable, Ann Arbor, 1955, pp. 265–285.

34. S. Saks and A. Zygmund, Monografie Matematyczne, *Analytic functions*, Warsaw, 1965.

35. M. D. Thompson, *Approximation by polynomials whose zeros lie on a given curve*, Ph. D. thesis, University of Wisconsin, 1962.

36. ———, *Approximation by polynomials whose zeros lie on a curve*, Duke Math. J. **31** (1964), 255–265.

37. J. L. Walsh, *Über die Entwicklung einer Funktion einer komplexen Veränderlichen nach Polynomen*, Math. Ann. **96** (1926), 437–450.

38. ———, *Interpolation and approximation*, Amer. Math. Soc. Colloq. Publ. Vol. 20, New York, 1935.

University of California, San Diego
La Jolla

Discrete Sets of Uniqueness For Bounded Holomorphic Functions $f(z, w)$

JACOB KOREVAAR

SIMON HELLERSTEIN [1]

1 Introduction.

1.1. PROBLEM AND RESULTS. In the space C^2 of two complex variables $z = x + iy$, $w = u + iv$, let H^2 denote the Cartesian product of the right half-planes $x > 0$, $u > 0$. We consider the class $B(H^2)$ of all bounded holomorphic functions $f(z,w)$ on H^2. It is not difficult to show (cf. §1.2) that a function f in $B(H^2)$ is uniquely determined by its values on the lattice points (m,n) (points with integral coordinates) in the first quadrant Q of the real (x,u)-plane, or by its values on the lattice points in an angle $\lambda x < u < \mu x$ in Q. We are interested in obtaining non-trivial subsets Ω of the lattice points in Q that are sets of uniqueness for the class $B(H^2)$. These are, of course, sets Ω with the property that if f in $B(H^2)$ vanishes everywhere on Ω, then f must be the zero function.

For sets Ω consisting of lattice points in Q, one may define an upper density δ by setting

$$\delta = \limsup_{r \to \infty} \frac{1}{\frac{1}{4}\pi r^2} N(r),$$

where $N(r)$ is the number of points (m,n) in Ω with $m^2 + n^2 < r^2$. We will show (§3) that *every set of lattice points in Q with positive upper density is a set of uniqueness for* $B(H^2)$. Positive upper density is by no means necessary for a set of uniqueness; Ω is a set of uniqueness whenever, in some sense, a sufficient number of points of Ω are concentrated near a ray $u = \lambda x$ (cf. §§1.2, 3.5).

The above result can be extended to the case of more than two complex variables. However, we have not been able to prove an analogous result for the case where the lattice point condition on a discrete set Ω in Q is replaced merely by a separation condition. Of course, there are approximation and moment problems where it is natural to consider sets of lattice points (cf. §1.3).

The method of proof is perhaps more interesting than our specific results It is based on a precise description of the growth of functions $f(z,w)$ in $B(H^2)$. For such functions, we introduce an *angular type*

[1] Work of the first author supported, in part, by NSF grant GP 4106, that of the second author by NSF grant GP 5728.

$$\tau(\lambda,\mu) = \limsup_{x\to\infty} \frac{1}{x} \log \operatorname*{l.u.b.}_{\lambda x<u<\mu x} |f(x,u)|,$$

defined whenever $0 \leqq \lambda < \mu \leqq \infty$. Naturally, $\tau(\lambda,\mu) \leqq 0$. We prove the following results (§2):

(i) $\tau(\lambda,\mu) = \tau(\lambda,\infty)$ *for all* $\mu > \lambda$;

we set $\tau(\lambda,\infty) = \tau(\lambda)$, $0 \leqq \lambda < \infty$.

(ii) *If* f *is bounded by* K, *then actually*

$$|f(z,w)| \leqq K e^{\tau(\lambda)x}$$

for all (z,w) *with* $x > 0$, $\lambda x < u < \infty$. *In particular, if* $\tau(\lambda) = -\infty$ *for some* λ, *then* $f = 0$.

(iii) *If* f *is not the zero function*, $\tau(\lambda)$ *is continuous for* $0 \leqq \lambda < \infty$, *and if* $\lambda' > \lambda$,

$$0 \leqq \tau(\lambda) - \tau(\lambda') \leqq (1/\lambda)|\tau(\lambda)|\,(\lambda'-\lambda).$$

Suppose now that a function f in $B(H^2)$ vanishes on a set of lattice points Ω which, in an appropriate sense, has a great many points near the ray $u = \lambda x$. If $f \neq 0$, the angular type $\tau(\lambda)$ of f has a finite value τ_0. However, we show that because f vanishes at many points near the ray $u = \lambda x$, actually $\tau(\lambda) < \tau_0$; this contradiction will prove that $f = 0$. Our argument makes essential use of two estimates on the growth of holomorphic functions of one complex variable, one a simple lemma (§1.2), the other a far from simple theorem due to N. Levinson (§3.1).

1.2. SOME SIMPLE SETS OF UNIQUENESS. We begin with a review of the 1-dimensional situation. Let $B(H)$ denote the class of the bounded holomorphic functions $g(z)$ in the half plane $H = \{z = x + iy \mid x > 0\}$. It has been known for a long time that an increasing sequence of positive integers $\{m_k\}$ is a set of uniqueness for $B(H)$ if and only if

$$\sum_{k=1}^{\infty} \frac{1}{m_k} = \infty.$$

This result follows from a more general uniqueness theorem for bounded holomorphic functions in the unit disc obtained by W. Blaschke in 1915 [**1**]. It can also be obtained from J. L. Jensen's formula which dates back to 1899 [**5**]. A simple proof can be derived from the following (well-known) lemma which will be used later.

LEMMA 1.2. *Suppose the function* $g(z)$ *in* $B(H)$ *is equal to* 0 *at positive integers* $m_1 < m_2 < \cdots$. *Let* K *denote an upper bound for* $|g(z)|$. *Then for* $0 < x \leqq m_1$, *and for any* N *not exceeding the number of points* m_k,

$$|g(x)| \leqq K \prod_{k=1}^{N} \frac{1 - x/m_k}{1 + x/m_k} \leqq K \exp\left(-2x \sum_{k=1}^{N} \frac{1}{m_k}\right).$$

The proof of the lemma follows from the fact that the product

$$g(z) \prod_{k=1}^{N} \frac{m_k + z}{m_k - z}$$

is holomorphic in the right half-plane H and bounded by K (it is bounded by $K + \varepsilon$ on lines $x = \delta > 0$ close to the imaginary axis, and on large semicircles with center 0).

To prove the uniqueness theorem stated above, observe that if $\sum 1/m_k = \infty$, one can let N tend to ∞ in the lemma to show that $g(x) = 0$ for $0 < x \leqq m_1$, hence $g = 0$. However, if $\sum 1/m_k < \infty$, the product $\prod_k (m_k - z)/(m_k + z)$ defines a function in $B(H)$ different from the zero function which vanishes at the points m_k.

The 1-dimensional result enables one to construct simple sets of uniqueness for $B(H^2)$.

EXAMPLE 1. Let $0 < n_1 < n_2 < \cdots$ be an increasing sequence of positive integers such that $\sum 1/n_k = \infty$. Let Ω be a set of lattice points $(m_{kj}, n_k), j = 1,2 \cdots$, $k = 1,2,\cdots$ such that $0 < m_{k1} < m_{k2} < \cdots$ and $\sum_j 1/m_{kj} = \infty$ for each k. Then Ω is a set of uniqueness. Indeed, suppose that f in $B(H^2)$ vanishes on Ω. Then for fixed k, the function $f(z, n_k)$ is in $B(H)$ and equal to 0 on the sequence $\{m_{kj}\}$, hence by the preceding, $f(z, n_k) = 0$. This holds for $k = 1,2,\cdots$; applying the above result to $f(z,w)$ with fixed z, one concludes that $f = 0$.

This example shows that sets of uniqueness Ω need not have positive upper density. In fact, $N(r)$ may be $o(r)$; more precisely, there are sets of uniqueness of this type for which $\int_1^R N(r) r^{-2} dr$ tends to ∞ as slowly as one pleases.

In Example 1 the lines $u = n_k$ carry "many" points of Ω. However, the lines that carry "many" points of Ω need not be horizontal.

EXAMPLE 2. Let $\{n_k\}$ be a sequence as in Example 1, and let Ω consist of those lattice points on the lines $u = n_k$ that belong to a (fixed) angle $\lambda x < u < \mu x$, where $0 \leqq \lambda < \mu \leqq \infty$. Then if p/q is a rational number between λ and μ, an f in $B(H^2)$ which vanishes on Ω will vanish on all half-lines in Q with slope p/q that contain "many" points of Ω. Roughly speaking, at least one out of every q "consecutive" half-lines with slope p/q that pass through a lattice point contains "many" points of Ω. It follows that Ω is a set of uniqueness.

It is interesting to compare these examples with a special case of a result obtainable by the method of this paper. Let $\{n_k\}$ be a sequence as in Example 1, and let Ω consist of those lattice points on the lines $u = n_k$ that lie between two parabolas $u = \lambda x^{1/2}$ and $u = \mu x^{1/2}$, where $0 \leqq \lambda < \mu < \infty$. Then Ω is a set of uniqueness for $B(H^2)$ (cf. §3.5).

1.3. APPROXIMATION BY LINEAR COMBINATIONS OF MONOMIALS $s^m t^n$. Let I be the unit interval $0 \leqq t \leqq 1$, and let $L^p(I)$ $(1 \leqq p < \infty)$ and $C(I)$ be the usual Banach spaces of functions; $C_0(I)$ will denote the subspace of $C(I)$ consisting of the functions that vanish at $t = 0$. Finally, let $\{m_k\}$ be an increasing sequence of positive integers. By the well-known theorem of C. H. Müntz [**8**] and O. Szász [**10**; cf. **9**], the linear combinations of powers t^{m_k} are dense in $L^p(I)$, and also in $C_0(I)$, if and only if the series $\sum 1/m_k$ diverges. In other words, the set of powers $\{t^{m_k}\}$ spans the spaces $L^p(I)$ and $C_0(I)$ if and only if the integers m_k form a set of uniqueness for the class $B(H)$ (§1.2). If the set $\{t^{m_k}\}$ does not span, the only approximable functions are the restrictions of power series $\sum a_k t^{m_k}$ that converge in the unit disc [**3,9,6**].

Very little is known about the corresponding approximation problems in several variables. Let I^2 be the unit square $0 \leqq s \leqq 1$, $0 \leqq t \leqq 1$, and $C_0(I^2)$ the subspace of $C(I^2)$ consisting of the functions that vanish where st does. We consider sets of monomials

$$A = \{s^m t^n \,|\, (m, n) \text{ in } \Omega\},$$

where Ω is a set of points with positive integral coordinates. The problems are for which sets Ω the set A spans spaces such as $L^p(I^2)$ or $C_0(I^2)$, and what the closed span of A is when it is not the whole space. The first question is closely related to the problem of sets of uniqueness for $B(H^2)$.

To take the case of C_0, a continuous function ϕ on I^2 will belong to the closed span $\bar{S}(A)$ if and only if for every complex Borel measure $d\mu$ on I^2 that is orthogonal to A, also $d\mu \perp \phi$. With a given $d\mu$ we associate the function

$$f_\mu(z, w) = \iint_{I^2} s^z t^w d\mu(s, t), \quad x > 0, \quad u > 0.$$

It is clear that f_u is in $B(H^2)$. *Assume now that* Ω *is a set of uniqueness and take* $d\mu \perp A$. Then we will have $f_\mu = 0$, hence in particular $f_\mu(m, n) = 0$ for all lattice points (m, n), m, $n = 1, 2, \cdots$. It follows that *all* monomials $s^m t^n$ are in $\bar{S}(A)$, and since the set of all monomials $s^m t^n$ spans $C_0(I^2)$, *one concludes that* $\bar{S}(A) = C_0(I^2)$. In the other direction one can prove the following. If $\bar{S}(A) = C_0(I^2)$, then the set Ω' of all points $(m + 1, n + 1)$, with (m, n) in Ω, is a set of uniqueness for $B(H^2)$. This raises the question if for lattice points, the property of being a set of uniqueness is preserved under translation (unit translation to the left, say).

2. The growth of functions of the class $B(H^2)$.

It is assumed throughout §2 that $f(z, w)$ is holomorphic in H^2 and (in absolute value) bounded by 1. We will begin by studying the growth of f in the first quadrant Q of the real (x, u)-plane.

2.1. GROWTH ALONG A RAY $u = \lambda x$ IN Q. For $0 < \lambda < \infty$ we define a *radial type* $\tau_r(\lambda)$ by setting

$$\tau_r(\lambda) = \limsup_{x \to \infty} \frac{1}{x} \log |f(x, \lambda x)|.$$

Clearly $\tau_r(\lambda) \leqq 0$. Observe that $f(z, \lambda z)$ is holomorphic for $x > 0$, bounded by 1, and of type $\tau_r(\lambda)$ along the positive real axis. Thus by a well-known inequality of G. Pólya and G. Szegö [cf. **2**, §6.2],

$$|f(z,\lambda z)| \leqq \exp[\tau_r(\lambda)x], \quad x > 0.$$

In particular, if $\tau_r(\lambda) = -\infty$, then $f(x, \lambda x) = 0$.

It turns out that the radial type is not too satisfactory, mainly because it is not a nice function of λ. In fact, $\tau_r(\lambda)$ can be equal to $-\infty$ for one or several values of λ without being $-\infty$ everywhere: f may vanish on several rays $u = \lambda x$ without being equal to the zero function. Example:

$$f(z,w) = \prod_\lambda \frac{w - \lambda z}{(w+1)(\lambda z + 1)}.$$

However, $\tau_r(\lambda)$ cannot be equal to $-\infty$ throughout an interval unless $f = 0$: if f vanishes throughout an angle in Q, then f is the zero function. We will next investigate the implications of a bound on the growth of f in an angle.

2.2. GROWTH IN AN ANGLE $\lambda x < u < \mu x$ IN Q. For $0 \leqq \lambda < \mu \leqq \infty$ we define an *angular type* $\tau(\lambda, \mu)$ of f by

$$\tau(\lambda, \mu) = \limsup_{x \to \infty} \frac{1}{x} \log \operatorname*{l.u.b.}_{x < u < \mu x} |f(x,u)|.$$

Note that $\tau(\lambda, \mu) \leqq 0$, and that for all ν between λ and μ, one has $\tau_r(\nu) \leqq \tau(\lambda, \mu)$. Thus by §2.1,

$$|f(x,u)| \leqq e^{\tau(\lambda,\mu)x} \tag{2.2.1}$$

for $x > 0$, $\lambda x < u < \mu x$. If $\tau(\lambda, \mu) = -\infty$, then $f = 0$.

LEMMA 2.2. *One has*

$$|f(x,u)| \leqq e^{\tau(\lambda,\mu)x} \quad \textit{for} \quad x > 0, \quad \lambda x < u < \infty.$$

PROOF. Let (x', u') be any point with $x' > 0$, $\lambda x' < u' < \infty$. We set $h = u' - \lambda x'$, and consider the function

$$g(z) = f(z, \lambda z + h).$$

This function is holomorphic for $x > 0$, bounded by 1, and of type $\leqq \tau(\lambda, \mu)$ along the positive real axis (for all sufficiently large x, the points of the line $u = \lambda x + h$ belong to the angle $\lambda x < u < \mu x$). Thus by the inequality of Pólya and Szegö used in §2.1,

$$|g(z)| \leqq e^{\tau(\lambda,\mu)x}, \quad x > 0.$$

In particular $g(x') = f(x', u')$ is bounded by $\exp\{\tau(\lambda, \mu)x'\}$.

COROLLARY. *For all* $\mu > \lambda$

$$\tau(\lambda, \mu) = \tau(\lambda, \infty).$$

Indeed, by the definition of angular type, $\tau(\lambda, \mu) \leqq \tau(\lambda, \infty)$, and by the lemma, $\tau(\lambda, \infty) \leqq \tau(\lambda, \mu)$.

2.3. THE ANGULAR TYPE $\tau(\lambda)$. We will set

$$\tau(\lambda,\infty)=\tau(\lambda),\qquad 0\leqq\lambda<\infty,$$

and call $\tau(\lambda)$ the angular type associated with the ray $u=\lambda x$. It is clear that $\tau(\lambda)$ *is a nonincreasing function of* λ.

It follows from (2.2.1) and the nonpositivity of $\tau(\lambda,\mu)=\tau(\lambda)$ that

$$|f(x,u)|\leqq e^{\tau(\lambda)u/\mu}\qquad\text{for}\quad u>0,\quad u/\mu<x<u/\lambda.$$

Thus by the argument used to prove Lemma 2.2, interchanging the roles of x and u,

$$|f(x,u)|\leqq e^{\tau(\lambda)u/\mu}\qquad\text{for}\quad u>0,\quad u/\mu<x<\infty.$$

This is true for every $\mu>\lambda$. Assuming that $\lambda>0$, we can let μ tend to λ to obtain

$$|f(x,u)|\leqq e^{\tau(\lambda)u/\lambda}\qquad\text{for}\quad u>0,\quad u/\lambda\leqq x<\infty.$$

Taking $\nu<\lambda$ and $u/\lambda<x<u/\nu$, and remembering that $\tau(\lambda)\leqq 0$, we conclude that

$$|f(x,u)|\leqq e^{\tau(\lambda)\nu x/\lambda}\qquad\text{for}\quad x>0,\quad \nu x<u<\lambda x.$$

Hence by the definition of the angular type $\tau(\nu,\lambda)$,

$$\tau(\nu)=\tau(\nu,\infty)=\tau(\nu,\lambda)\leqq\tau(\lambda)\nu/\lambda.$$

In other words, $\tau(\lambda)/\lambda$ *is a nondecreasing function of* λ. It follows that $\tau(\lambda)$ *satisfies a Lipschitz condition*:

LEMMA 2.3. *For* $0<\lambda<\lambda'<\infty$,

$$0\leqq\tau(\lambda)-\tau(\lambda')\leqq(1/\lambda)|\tau(\lambda)|(\lambda'-\lambda);$$

the function $\tau(\lambda)$ *is continuous for* $0\leqq\lambda<\infty$.

PROOF. The inequalities express the monotonicity of $\tau(\lambda)$ and $\tau(\lambda)/\lambda$. They imply that $\tau(\lambda)$ is continuous for $\lambda>0$. To prove the continuity at $\lambda=0$, we observe first that $\tau(0+)\geqq\tau(\lambda)$ for all $\lambda>0$, hence

$$|f(x,u)|\leqq e^{\tau(0+)x}$$

for all (x,u) in Q. Thus $\tau(0)\leqq\tau(0+)$. On the other hand $\tau(0+)\leqq\tau(0)$, hence $\tau(0)=\tau(0+)$.

2.4. AN INEQUALITY FOR $f(z,w)$. We can now prove

LEMMA 2.4. *For all* (z,w) *with* $x>0$, $\lambda x<u<\infty$,

$$|f(z,w)|\leqq e^{\tau(\lambda)x}.$$

If $\tau(\lambda)=-\infty$, *then* $f=0$.

PROOF. For $x>0$, $u>0$, set

$$g(z,w) = f(z, \lambda z + w).$$

We then have to show that wherever g is defined,

$$|g(z,w)| \leqq e^{\tau x}, \quad \text{with} \quad \tau = \tau(\lambda)$$

It follows from Lemma 2.2 that $g(x,u)$ is bounded by $e^{\tau x}$. Applying the inequality of Pólya and Szegö (§2.1) to $g(z,u)$ with fixed u, one finds that $|g(z,u)| \leqq e^{\tau x}$, $x > 0$, $u > 0$. Thus $g(x + ih, \rho x)$ is bounded by $e^{\tau x}$ for all $x > 0$, all real h and all $\rho > 0$. Application of the inequality of Pólya and Szegö to $g(z + ih, \rho z)$ with fixed h and ρ next shows that

$$|g(z + ih, \rho z)| \leqq e^{\tau x}, \qquad x > 0, \quad h \text{ real}, \quad \rho > 0.$$

Since all points (z', w') with $x' > 0, u' > 0$ can be represented in the form $(z + ih, \rho z)$. the last inequality contains the desired result.

2.5. REMARKS. The preceding results were originally obtained with the aid of a Laplace transform representation. Setting

$$\tilde{f}(z,w) = f(z,w)/(1 + z)^2(1 + w)^2,$$

one can write

$$\tilde{f}(z,w) = \iint_{s,t \leqq 0} e^{zs + wt} \phi(s,t)\, d(s,t).$$

Here ϕ, the inverse Laplace transform of $\tilde{f}$, is continuous and bounded on the (s,t)-plane, and equal to 0 outside the third quadrant. It can be shown that *the angular type $\tau(\lambda, \mu)$ of f is equal to the smallest number γ such that $\phi(s,t) = 0$ at all points (s,t) with $s + \lambda t \geqq \gamma$.*

One obtains an interesting diagram by superimposing the (s,t)-plane on the (x,u)-plane (of the (x,u)-plane, only the first quadrant is used, and of the (s,t)-plane, only the third quadrant). In such a diagram, the angular type $\tau(\lambda, \mu)$ is given by the s-intercept of that supporting line to the convex hull of the support of ϕ which is perpendicular to the direction λ and separates the support of ϕ from the first quadrant. The t-intercept of the same line gives $\tau(\lambda, \mu)/\lambda$. The above characterization implies that $\tau(\lambda, \mu)$ is independent of μ, hence $\tau(\lambda, \mu) = \tau(\lambda, \infty)$. Setting $\tau(\lambda, \infty) = \tau(\lambda)$, it is clear geometrically that $\tau(\lambda)$ is a continuous nonincreasing function of λ, and that $\tau(\lambda)/\lambda$ is a nondecreasing function of λ.

In our discussion of angular type, we have favored the variable x in the (x,u)-plane because this is convenient for the present application. Obviously, there are parallel results with the roles of x and u interchanged. It is also possible to base the whole treatment on the use of the symmetric variable $r = (x^2 + u^2)^{1/2}$, and to introduce a *directional type* $\tau_d(\theta)$ as the limit of the (new) angular type for angles that contract onto the ray $u = x \tan\theta$. This directional type is equal to the "supporting function" of the convex hull of the support of ϕ.

3 Discrete sets of uniqueness for the class $B(H^2)$.

In the following Ω will always denote a set of points (m,n) with positive integral coordinates. For $0 \leqq \alpha < \beta \leqq \frac{1}{2}\pi$, $N(r,\alpha,\beta)$ will be the number of points of Ω with $m^2 + n^2 < r^2$, $\tan \alpha < n/m < \tan \beta$. The formula

$$\delta = \limsup_{r \to \infty} N(r, 0, \tfrac{1}{2}\pi) / \tfrac{1}{4}\pi r^2$$

gives the upper density of Ω. Our principal result is

THEOREM 3. *Let $f(z,w)$ be a bounded holomorphic function in* Re $z > 0$, Re $w > 0$ *which vanishes on a set of lattice points Ω with positive upper density δ. Then $f = 0$.*

3.1. OUTLINE OF THE PROOF. Ω is located in the first quadrant of the real $(x,)$-u plane, that is, in the angle $0 \leqq \theta = \arctan u/x \leqq \frac{1}{2}\pi$. By the hypothesis on Ω, we can use repeated bisection to obtain a nested sequence of angles $\alpha \leqq \theta \leqq \beta$, with $\beta - \alpha$ tending to 0 through the sequence $\{\pi/2^k\}$, such that

$$\limsup_{r \quad \infty} N(r,\alpha,\beta) / \tfrac{1}{2}(\beta - \alpha) r^2 \geqq \delta \tag{3.1.1}$$

for each of the angles. The angles $\alpha \leqq \theta \leqq \beta$ will contract onto a common ray $\theta = \theta_0$; every ray that can be obtained in this manner will be called a *ray of density* for Ω. Interchanging z and w, if necessary, we may assume that our ray $\theta = \theta_0$ is not the positive u-axis.

If our ray of density $\theta = \theta_0$ is the positive x-axis, we have $\alpha = \theta_0 = 0$, and

$$\limsup_{r \to \infty} N(r, \theta_0, \theta_0 + \omega) / \tfrac{1}{2}\omega r^2 \geqq \delta \tag{3.1.2}$$

for a sequence of positive ω tending to 0. Suppose next that for our ray of density $\theta = \theta_0$ we have $0 < \theta_0 < \frac{1}{2}\pi$. Then for an angle $\alpha \leqq \theta \leqq \beta$ containing θ_0 for which (3.1.1) holds, we will either have (3.1.1) with α replaced by θ_0 and $\beta > \theta_0$, or (3.1.1) with β replaced by θ_0 and $\alpha < \theta_0$ (or both). Taking a subsequence of the angles $\alpha \leqq \theta \leqq \beta$ and interchanging z and w if necessary, we may thus again assume that (3.1.2) is satisfied for a sequence of ω tending to 0.

Now let f be a function in $B(H^2)$ that vanishes on Ω. Setting $\tan \theta_0 = \lambda_0$, it will be shown that the angular type $\tau(\lambda_0)$ of f cannot be finite. We base the proof of this fact on §2, Lemma 1.2 and the following special case of a theorem of N. Levinson for holomorphic functions of one variable [7, p. 107; cf. 2, p. 200].

THEOREM OF LEVINSON. *Let $h(z)$ be holomorphic and bounded in the half plane $x > 0$. Let $\{x_k\}$ be a sequence of positive real numbers such that*

$$x_{k+1} - x_k \geqq d > 0, \; k = 1, 2, \cdots \quad \textit{and} \quad \sum 1/x_k = \infty.$$

Then

$$\limsup_{x\to\infty} \frac{1}{x} \log|h(x)| = \limsup_{k\to\infty} \frac{1}{x_k} \log|h(x_k)|.$$

In his result, Levinson allows more general sequences, but he imposes the additional condition that k/x_k tend to a limit $D \geqq 0$. It is easy to show that a sequence $\{x_k\}$ as in the theorem must have a subsequence that satisfies this condition with $D = 0$, in addition to the conditions stated in the theorem. Thus by Levinson's original result, the type of h on such a subsequence is equal to the type of h on the positive x-axis. Since the type of h on the original sequence $\{x_k\}$ is not smaller than the type on the subsequence, and not larger than the type on the positive x-axis, the present theorem follows.

The proof of Theorem 3 will be carried out in full detail for the case $\lambda_0 = 0$ (§3.2). In §3.3 we indicate the modifications required in the case where λ_0 is a positive rational number. In §3.4, finally, we deal with the more difficult case where λ_0 is irrational; here we need the fact that $\tau(\lambda)$ satisfies a Lipschitz condition (Lemma 2.3). Once it has been shown that $\tau(\lambda_0) = -\infty$, it follows from Lemma 2.4 that $f = 0$.

3.2. Proof of Theorem 3 in the case $\lambda_0 = 0$. Given that $\theta = 0$ is a ray of density for Ω, there are (arbitrarily small) numbers $\omega > 0$ and corresponding sequences of r tending to ∞ such that

(3.2.1) $$N(r,0,\omega)/\tfrac{1}{2}\omega r^2 > \cdot 9\delta$$

(cf. 3.1.2). We choose one such $\omega < \frac{1}{2}\pi$. For the corresponding r's, let S_r denote the sector $x^2 + u^2 < r^2$, $0 < \theta < \omega$; for the time being, we will also keep r fixed.

We will consider the horizontal segments which the lines $u = n$, $n = 1, 2, \cdots$ have in common with the sector S_r. There are $\{r \sin \omega\}$ such segments, where $\{r \sin \omega\}$ denotes the largest integer less than r s n ω. Let the number of these segments that contain fewer than $\frac{1}{4}\delta r$ points of Ω be equal to v_r. Since the other segments contain not more than r points of Ω apiece, we have

$$.25\,\delta r\, v_r + (\{r\sin\omega\} - v_r)r \geqq N(r,0,\omega) > .45\,\delta r\{r\sin\omega\}$$

(cf.3.2.1). It follows that $v_r < (1 - \cdot 2\delta)\,\{r\sin\omega\}$.

We will focus on the set U_r of those values of n for which the segment of the line $u = n$ in S_r contains at least $\frac{1}{4}\delta r$ points (m,n) of Ω. Since for these points one has $m \leqq r$, *the sum* $\sum 1/m$ *over the points* (m,n) *in* Ω *belonging to a fixed* n *in* U_r *is at least equal to* $\frac{1}{4}\delta$. By the above, *the number of integers in* U_r *is not less than* $.\ 2\delta\{r \sin \omega\}$.

It is no restriction to assume that our f in $B(H^2)$ which vanishes on Ω is bounded by 1. We suppose for the time being that f is not the zero function. Then Lemma 2.4 shows that $\tau(0)$ has a finite value τ, and that

$$|f(z,u)| \leqq e^{\tau x}, \quad x > 0, \quad u > 0.$$

We will apply Lemma 1.2 to the functions

$$g(z) = e^{-\tau z} f(z,u), \qquad \text{with } u = n \text{ in } U_r.$$

Each g is holomorphic in the half-plane $x > 0$ and bounded by 1, and it vanishes at distinct integers $m > n \cot \omega$ such that $\sum 1/m \geqq \frac{1}{4}\delta$. Thus by Lemma 1.2, $|g(x)| \leqq e^{-\delta x/2}$ for $0 < x \leqq n \cot \omega$, *hence*

(3.2.2) $$|f(x,u)| \leqq e^{(\tau - \delta/2)x}$$

on each of the segments

(3.2.3) $$u = n \quad in \quad U_r, \qquad 0 < x \leqq n \cot \omega.$$

Let U denote the union of the sets U_r corresponding to a sequence of r tending to ∞ for which (3.2.1) is satisfied. We arrange the integers in U in an increasing sequence $n_1 < n_2 < \cdots$. Since for each r in an unbounded sequence, U contains at least $.2\delta\{r \sin \omega\}$ positive integers $\leqq \{r \sin \omega\}$, it is easy to see that $\sum 1/n_k = \infty$. By (3.2.2), (3.2.3)

(3.2.4) $$|f(x,n_k)| \leqq e^{(\tau-\delta/2)x}, \qquad 0 < x \leqq n_k \cot \omega, \quad k = 1,2,\cdots.$$

We are now in a position to apply Levinson's theorem (§3.1) to the bounded holomorphic functions

$$h(z) = f(z, \lambda z), \qquad \lambda \geqq \tan \omega$$

in the half-plane $x > 0$. For fixed λ, let us set $x_k = n_k/\lambda$, $k = 1,2,\cdots$. Then $0 < x_k \leqq n_k \cot \omega$, $x_{k+1} - x_k \geqq 1/\lambda$, $k = 1,2,\cdots$, and $\sum 1/x_k = \infty$. By (3.2.4), the type of h on the sequence $\{x_k\}$ does not exceed $\tau - \frac{1}{2}\delta$. Thus by Levinson's theorem, the same is true for the type of h on the positive x-axis. Since h is bounded by 1, *we conclude that* (cf. §2.1)

(3.2.5) $$|h(x)| = |f(x, \lambda x)| \leqq e^{(\tau-\delta/2)x}, \qquad x > 0, \quad \lambda \geqq \tan \omega.$$

Inequality (3.2.5) shows that *the angular type* $\tau(\tan \omega)$ *of* f *is bounded by* $\tau - \frac{1}{2}\delta$. We finally let ω tend to 0 through a sequence for which (3.1.2) is satisfied, with $\theta_0 = 0$. The continuity of $\tau(\lambda)$ (Lemma 2.3) then shows that

$$\tau = \tau(0) = \lim \tau(\tan \omega) \leqq \tau - \tfrac{1}{2}\delta.$$

This contradiction proves that $f = 0$.

From here on it is assumed that every ray of density for Ω has positive slope.

3.2.3. PROOF OF THEOREM 3 IN THE CASE $\lambda_0 = p/q > 0$. We assume in this section that Ω has a ray of density $\theta = \theta_0 > 0$ with $\lambda_0 = \tan \theta_0$ rational, and such that (3.1.2) is satisfied for a sequence of positive ω tending to 0. (The latter is no restriction; if $\tan \theta_0$ is rational so is $\tan(\frac{1}{2}\pi - \theta_0)$; cf. §3.1.) We will write $\lambda_0 = p/q$, where p and q are positive integers with greatest common divisor 1. For a fixed permissible $\omega < \frac{1}{2}\pi - \theta_0$, we consider a sequence of r's tending to ∞ such that

(3.3.1) $$N(r, \theta_0, \theta_0 + \omega)/\tfrac{1}{2}\omega r^2 \geqq .9\delta.$$

For each of these r's, let S_r denote the sector $x^2 + u^2 < r^2$, $\theta_0 < \theta < \theta_0 + \omega$. This time we consider, instead of horizontal segments, the segments which the lines

$u = (p/q)x + v/q$, $v = 1, 2, \cdots$ have in common with S_r. On these lines the lattice points are a distance q sec θ_0 apart, but the lines themselves are only a distance $(\cos\theta_0)/q$ apart. A calculation similar to the one in §3.2 shows that the number of segments in S_r with at least $\frac{1}{4}\delta r\,(\cos\theta_0)/q$ points (m, n) of Ω is given by a fraction not less than $.2\delta$ of the total number of segments in S_r. Since for the points (m,n) in S_r, $m \leqq r\cos\theta_0$, the sum $\sum 1/m$ over the points of Ω on each of the "good" segments in S_r is at least equal to $\frac{1}{4}\delta/q$.

Suppose now that f in $B(H^2)$ vanishes on Ω but not everywhere in H^2. Then $\tau(\lambda_0)$ has a finite value τ. Proceeding as in §3.2, one arrives at the estimate

$$\tau(\lambda') \leqq \tau - \tfrac{1}{2}\delta/q, \qquad \text{where} \quad \lambda' = \tan(\theta_0 + \omega). \tag{3.3.2}$$

(We note for later use that this estimate is also valid if p/q is a reducible fraction representing λ_0.) Letting ω tend to 0 through a permissible sequence, we again arrive at the contradiction $\tau < \tau$, hence $f = 0$.

3.4. Proof of Theorem 3 in the case of irrational λ_0. We assume in this section that every ray of density for Ω has irrational slope. Let $\theta = \theta_0$ be one of these rays, and let $\lambda_0 = \tan\theta_0$. Our proof depends on approximation of λ_0 by rational numbers. We consider a fraction p/q with a large denominator q such that

$$|\lambda_0 - p/q| < 1/q^2$$

[cf. **4**, p. 140]. Next, we determine an angle $\alpha \leqq \theta \leqq \beta$ containing the ray $\theta = \theta_0$, and with opening $\beta - \alpha = \omega$ comparable to $q^{-3/2}$, such that (3.1.1) is satisfied (one may take $1 \leqq q^{3/2}\omega < 2$; cf. §3.1). Then either

$$\limsup_{r\to\infty} N(r, \theta_0, \theta_0 + \omega)/\tfrac{1}{2}\omega r^2 \geqq \tfrac{1}{2}\delta,$$

or the analogous relation is true for $N(r, \theta_0 - \omega, \theta_0)$. It is no restriction to assume the former. Indeed, if only the latter is true, we can interchange z and w. This does not spoil anything; the approximation of $1/\lambda_0$ by q/p is $O(1/p^2)$, and ω is comparable to $p^{-3/2}$.

Assuming (3.4.1), we set $\arctan p/q = \theta_1$. Since $\theta_1 - \theta_0 = O(q^{-2})$ while ω is comparable to $q^{-3/2}$, we derive from (3.4.1) that for large q,

$$N(r, \theta_1, \theta_1 + \omega)/\tfrac{1}{2}\omega r^2 \geqq .45\delta \tag{3.4.2}$$

for a sequence of r tending to ∞. Thus by §3.3 (cf. 3.3.1, 3.3.2), supposing that our f in $B(H^2)$ which vanishes on Ω is not the zero function,

$$\tau(\lambda') \leqq \tau(p/q) - \tfrac{1}{4}\delta/q, \qquad \text{where} \quad \lambda' = \tan(\theta_1 + \omega). \tag{3.4.3}$$

The proof is completed with the aid of the Lipschitz condition for $\tau(\lambda)$. By Lemma 2.3,

$$\begin{aligned} \tau(p/q) - \tau(\lambda') &\leqq (q/p)|\tau(p/q)|(\lambda' - p/q) \\ &\sim (1/\lambda_0)|\tau(\lambda^2)|\,(1 + \lambda_0)\omega = O(q^{-3/2}). \end{aligned} \tag{3.4.4}$$

For sufficiently large q, (3.4.4) contradicts (3.4.3). Thus again $f = 0$.

3.5. FINAL REMARKS. The above method can be used to obtain results not covered by Theorem 3. When it is known that a set of lattice points Ω has many points near a certain ray $u = \lambda x$, much weaker conditions than positive upper density suffice to make Ω a set of uniqueness for $B(H^2)$. We will restrict ourselves here to an example where the ray is the positive x-axis.

Let $n_1 < n_2 < \cdots$ be an increasing sequence of positive integers such that $\sum 1/n_k = \infty$. Suppose Ω is a set of lattice points (m_{kj}, n_k), $j = 1.2, \cdots$, $k = 1,2,\cdots$, with $0 < m_{k1} < m_{k2} < \cdots$ for each k, such that the following condition is satisfied. There is a constant $\varepsilon > 0$ such that for the sums

$$s_k(\lambda) = \textstyle\sum_j^* 1/m_{kj},$$

extended over those j for which $\lambda m_{kj} > n_k$,

$$\liminf_{k\to\infty} s_k(\lambda) > \varepsilon \tag{3.5.1}$$

for every $\lambda > 0$. Then Ω is a set of uniqueness for $B(H^2)$. Indeed, the method of §3.2 shows that for a function f in $B(H^2)$ which vanishes on Ω, $\tau(\lambda) \leqq \tau(0) - 2\varepsilon$ for all $\lambda > 0$. It follows that $\tau(0) = -\infty$, hence $f = 0$.

Condition (3.5) is satisfied by the set Ω consisting of those lattice points on the lines $u = n_k$, with the n_k as above, that lie between two parabolas $u = \alpha x^{1/2}$, $u = \beta x^{1/2}$, $0 < \alpha < \beta$.

REFERENCES

1. W. Blaschke, *Eine Erweiterung des Satzes von Vitali über Folgen analytischer Funktionen*, Ber. Kön. Sächs. Ges., Akad. Wiss., Leipzig, Math.-Phys. Kl. **67** (1915), 194–200.
2. R. P. Boas, *Entire functions*, Academic Press, New York, 1954.
3. J. A. Clarkson and P. Erdös, *Approximation by polynomials*, Duke Math. J. **10** (1943), 5–11.
4. G. H. Hardy and E. M. Wright, *An introduction to the theory of numbers*, Oxford Univ. Press, Oxford, 1954.
5. J. L. Jensen, *Sur un nouvel et important théorème de la théorie des fonctions*, Acta Math. **22** (1899), 359–364.
6. J. Korevaar, *A characterization of the submanifold of C[a, b] spanned by the sequence* $\{x^{n_k}\}$, Nederl. Akad. Wetensch. Proc. **50** (1947), 750–758.
7. N. Levinson, *Gap and density theorems*, Amer. Math. Soc. Colloq. Publ. **26**, New York, 1940.
8. C. H. Müntz, *Über den Approximationssatz von Weierstrass*, H. A. Schwarz Festschrift, Berlin, 1914, 303–312.
9. L. Schwartz, *Étude des sommes d'exponentielles réelles*. Actualités Sci. Ind. no. 959, Hermann, Paris, 1943.
10. O. Szász, *Über die Approximation stetiger Funktionen durch lineare Aggregate von Potenzen*, Math. Ann. **77** (1915), 482–496.

UNIVERSITY OF CALIFORNIA, SAN DIEGO
LA JOLLA, CALIFORNIA
UNIVERSITY OF WISCONSIN
MADISON, WISCONSIN

Noncontinuous Indicators for Entire Functions of $n \geqq 2$ Variables and of Finite Order

P. LELONG

1. **Introduction.** For an entire function $F(z)$, $z = (z_1, \cdots, z_n)$ of finite order $\rho > 0$ and of finite type, the radial indicator of center ζ is defined by

$$L_r(\zeta; z) = \limsup t^{-\rho} \log |F(\zeta_k + tz_k)|, \qquad t \to +\infty. \tag{1.1}$$

Let us consider first the case $n = 1$. The indicator is independent of ζ. If $\rho = 1$, F is of exponential type and $L(z) = L(\zeta; z)$ reduces to a convex function $g(x, y)$, $z = x + iy$, in the complex plane. The indicator is therefore a continuous function. Such a result remains true for $\rho > 0$ and $n = 1$.

For $n \geqq 2$ we consider the following problem: is it possible to find entire functions of order $\rho > 0$, with discontinuous indicators? The answer is yes, for $n \geqq 2$, and each order. The aim of this paper is to construct such entire functions F: we give examples of entire functions having discontinuous functions for both the circular and the radial main indicator (the radial main indicator is the regularization with respect to z of (1.1); it does not depend on the center ζ; it is by definition an upper semicontinuous function of z).

This paper is closely connected[1] with my paper [**4**, e] and contains extensions for the order $\rho > 0$ of results given in [**4**, e] only for the exponential type (see Theorems 3.1 and 3.2).

2. **Definition of the indicators.** Recall some definitions given in [**4**, d] and [4, e]. We use the norm $\|z\|$ in C^n; $\|z\|^2 = \sum_1^n z_q \bar{z}_q$. An entire function $F(z) = F(z_1, \cdots, z_n)$ is said to be of order ρ if

$$\limsup (\log_2 |F(z)| / \log \|z\|) = \rho, \qquad \|z\| \to +\infty, \quad z \in C^n. \tag{2.1}$$

It is of type γ (of the order ρ) if

$$\limsup \|z\|^{-\rho} \log |F(z)| = \gamma, \qquad \|z\| \to +\infty, \quad z \in C^n. \tag{2.2}$$

It is of exponential type γ if it is of order 1 and of type γ. E_n denotes the ring

---

[1] I gave in a lecture at the Summer institute on entire functions, La Jolla, July 1966, a survey of some new results on functions of finite order obtained recently by C. O. Kiselman [3], A. Martineau [5, a, b], and myself [4, d, e].

of the entire functions of exponential type in C^n, and E_n^ρ denotes the ring of the entire functions of order ρ and of finite type; $E_n = E_n^1$.

If f is a real valued function,

$$f^*(z) = \text{Reg.}\, f(z) = \limsup_{z' \to z} f(z'), \qquad z' \in C^n$$

denotes the smallest upper semicontinuous majorant of f and is called *the regularization* of f.

We recall the following definitions (cf. [**4**, e]):

DEFINITION 2.1. *Given $F(z) = F(z_1, \cdots, z_n) \in E_n^\rho$, we define the circular indicator of F of center $\zeta \in C^n$ and of variable $z \in C^n$ as follows:*

$$L_c(\zeta; z) = \limsup_{|u| \to +\infty} |u|^{-\rho} \log |F(\zeta + uz)|, \qquad u \text{ complex}. \tag{2.3}$$

If $n = 1$, L_c reduces to a constant value which is the type of the order ρ. For $\rho = 1$, L_c is the exponential type.

DEFINITION 2.2. *Given $F(z) \in E_n^\rho$, the radial indicator of center $\zeta \in C^n$ and of variable $z \in C^n$ is defined by*

$$L_r(\zeta; z) = \limsup_{t \to +\infty} t^{-\rho} \log |F(\zeta + tz)|, \qquad t \text{ real}, \quad t > 0. \tag{2.4}$$

We consider first the case $\rho = 1$: as a classical result, for $n = 1$, the radial indicator $L_r(\zeta; z)$ is a convex, bounded and therefore continuous function of $z \in C^1$ and does not depend on the center ζ. For $\rho \neq 1$, $n = 1$, the radial indicator is convex with respect to the functions H which are harmonic and positively homogeneous of order ρ; such functions are given by

$$H(z) = Ar^\rho \cos \rho(\phi - \phi_0), \qquad z = re^{i\phi}$$

and depend on two constants A, ϕ_0. It is easy to prove that such a condition of convexity implies the continuity of the radial indicator.

As a consequence the two indicators for $n = 1$ are constants or continuous functions. Such results are no longer true for $n > 1$. Instead of convex functions, we need plurisubharmonic functions. For the convenience of the reader, I recall the definition and some properties of plurisubharmonic functions, a class I introduced in 1942 (cf. [**4**, a, b]) and which is now a classical tool of the complex analysis for $n > 1$ variables (for $n = 1$, the plurisubharmonic class reduces to the subharmonic class in the complex plane). We denote by $P(\Omega)$ the set of the plurisubharmonic functions which are defined in an open set Ω of a topological complex vector space.

DEFINITION 2.3. *A function f belongs to $P(\Omega)$ if and only if*

(I) *f is real valued, $-\infty \leqq f < +\infty$, $f \not\equiv -\infty$,*

(II) *f is upper semicontinuous, i.e. the set $[z; f(z) < c]$ is an open set for every real c,*

(III) *the function* $\phi(u) = f(z^{(1)} + uz^{(2)})$, $u \in C^1$, $z^{(1)} \in C^n$, $z^{(2)} \in G^n$, *is subharmonic or the constant* $-\infty$ *for all* $z^{(1)}$, $z^{(2)} \in C^n$ *in the open subsets of* $C^1(u)$ *where it is defined* (*if the open subset has many components, it is allowed for f to be in some components the constant* $-\infty$ *and to be a subharmonic function in the others*).

In the following we suppose that Ω is a domain of C^n. Given a family $f \in P(\Omega)$, $s \in S$, locally bounded above in Ω, let us consider the function

$$W_1 = \sup_s f_s, \qquad s \in S.$$

If S is a directed set, we consider also

$$W_2 = \limsup_s f_s, \qquad s \in S.$$

The functions $W_1(z)$, $W_2(z)$ are not in $P(\Omega)$. But the regularizations $W_1^* = \text{Reg.} W_1$ $W_2^* = \text{Reg.} W_2$ are in $P(\Omega)$. *For the proof see* [**4**, b, c].

Now we recall some properties of the indicators which were given in [**4**, e]:

(a) Let us consider the circular indicator

$$L_c(\zeta; z) = \limsup |u|^{-1} \log |F(\zeta + uz)|, \qquad |u| \to +\infty.$$

The regularization L_c^* of L_c with respect to $z = (z_1, \cdots, z_n)$ is a plurisubharmonic function. Further it was proved in [**4**, e] that it does not depend on the center ζ.

(b) The same properties are true for the regularization L_r^* of L_r; the functions L_c^* and L_r^* depend only on the given function F and have special interest; therefore we state the following definitions:

DEFINITION 2.4. *The regularizations* $L_c^*(z)$ *and* $L_r^*(z)$, *with respect to* z, *of the indicators* $L_c(\zeta; z)$, $L_r(\zeta; z)$—*see* (2. 3) *and* (2. 4)—*are called the main indicators* (L_c^* *is the circular main indicator and* L_r^* *the radical main indicator*) *of the entire function F of finite order* ρ.

As direct consequences of the Definitions 2.1 and 2.2 we have

$$L_c(\zeta; uz_k) = |u|^\rho L_c(\zeta; z_k) \quad \text{and} \quad L_c^*(uz_k) = |u|^\rho L_c(z_k), \tag{2.5}$$

$$L_r(\zeta; tz_k) = t^\rho L_r(\zeta; z_k) \quad \text{and} \quad L_r^*(tz_k) = t^\rho L_r(z_k) \quad \text{for } t \text{ real} > 0. \tag{2.6}$$

(c) There exists an obvious relation between the main indicators. Given a point $z = (z_1, \cdots, z_n) \in C^n$, $z \neq 0$, we consider the mapping $z \to i(z)$ defined by

$$i(z) = [z_k'; z_k' = z_k e^{i\theta}, \quad 0 \leqq \theta \leqq 2\pi] \tag{2.7}$$

which associates to each point $z \neq 0$ a circle $i(z)$; given a point $z \neq 0$, we consider $r(z)$ determined by z and given by

$$r(z) = [z_k'; z_k' = tz_k, \quad 0 < t < \infty, \ t \text{ real}] \tag{2.8}$$

and the complex line $l(z)$ determined from a given point, $z = (z_k)$ by

(2.9) $$l(z) = [z'_k; z'_k = uz_k,\ u \text{ complex}].$$

Obviously, $l(z) = i[r(z)]$, and we obtain

THEOREM 2.1. (a) $L_c(\zeta_k; z_k) = \sup_\theta L_r(\zeta_k; z_k e^{i\theta})$,
(b) $L_c^*(z_k) = \sup_\theta L_r^*(z_k e^{i\theta})$.

PROOF. We prove (b): by Definition 2.1, $L_r(\zeta_k; z_k) \leqq L_c(\zeta_k; z_k)$, which has as consequence

$$L_r^*(z_k) \leqq L_c^*(z_k).$$

Let us suppose

(2.10) $$\sup_\theta L_r^*(z_k^0 e^{i\theta}) = b < a = L_c^*(z_k^0),\ z^0 \neq 0$$

and prove that this leads to a contradiction. The set of plurisubharmonic functions

(2.11) $$V_t(z) = t^{-1} \log |F(tz)|, \qquad t \geqq 1$$

is bounded above on every compact of C^n. We consider the circle $i(z^0)$: $z_k = z_k^0 e^{i\theta}$, $0 \leqq \theta < 2\pi$; the equality $\|z\| = \|z_0\|$ holds on $i(z^0)$. We take $\varepsilon > 0$ sufficiently small to have $b + 2\varepsilon < a$. The inequality $L_r^*(z) < (b+\varepsilon)\|z\|\|z^0\|^{-1}$ defines an open neighborhood ω of the compact set $i(z^0)$. We consider a compact neighborhood $\bar{\omega}_1 \subset \omega$ of $i(z^0)$ and $\bar{\omega}_2 = \omega_1 \cap [\|z\| = \|z^0\|]$. If $z \in \bar{\omega}_2$, we obtain

$$\limsup V_t(z) \leqq L_r^*(z) \leqq (b+\varepsilon)\|z\|\ \|z^0\|^{-1}$$

Now by the Hartog's theorem (applied to plurisubharmonic functions $V_t(z)$ and to the compact $\bar{\omega}_2$), there exists t_0, $(t_0 > 1)$, such that

$$V_t(z) \leqq (b+2\varepsilon) \quad \text{for} \quad z \in \bar{\omega}_2 \quad \text{and} \quad t > t_0,$$

$$\log |F(z)| \leqq (b+2\varepsilon)\ \|z\|\ \|z^0\|^{-1}$$

for $\|z\| \geqq t_0 \|z^0\|$ and $z \in \Gamma(\bar{\omega}_2)$, where $\Gamma(\bar{\omega}_2)$ denotes the cone with vertex at the origin and basis $\bar{\omega}_2$. Obviously $\Gamma(\bar{\omega}_2)$ is a neighbourhood of the complex line $l(z^0)$ and (2.10) has as a consequence

$$L_c(0, z) \leqq (b+2\varepsilon)\|z\|\ \ \|z^0\|^{-1} \quad \text{if } z \in \Gamma(\bar{\omega}_2).$$

Hence we deduce $L_c^*(z^0) \leqq b + 2\varepsilon < a$; clearly this violates the supposition (2.10).

Now to prove (a), it is sufficient to apply (b) to the function of one complex variable which is the restriction of F to the complex line $l(z^0)$.

3. **Properties of the main circular indicator.** We gave a complete characterization of the main circular indicators $L_c^*(z)$ in [**4**, d, e] for $\rho = 1$, (exponential type). Now we denote by Λ_c^ρ the class of the main circular indicators of the entire functions of order ρ of finite type [i.e. $\|z\|^{-\rho} \log |F(z)|$ is bounded above as $\|z\| \to +\infty$, see (2.2)]. Then we prove the following lemma.

LEMMA 3.1. *Let us consider a pseudo-convex domain* $\Omega < C^n$ *and suppose that the origin* 0 *belongs to* Ω; *for a given* $z \in C^n$, *we define*

$$R(z) = [\sup |u|,\ i(uz) \subset \Omega] \tag{3.1}$$

where $i(z)$ *is the circle associated to* z *by* (2.9). *Then* $R(uz) = |u|^{-1}R(z)$ *and* $-\log R(z)$ *is a plurisubharmonic function defined in* C^n.

We give an alternate proof.

FIRST PROOF. By the Oka-Norguet theorem, Ω is a domain of holomorphy. We consider some function $F(z_1, \cdots, z_n)$ of which Ω is the domain of holomorphy. Then the series

$$F(z_1, \cdots, z_n) = \sum_0^{\infty} P_q(z_1, \cdots, z_n). \tag{3.2}$$

P_q homogeneous polynomial, degree $P_q = q$, converges in the domain

$$\Omega' = [z; S(z) < 1],$$

where

$$\log S(z) = \text{Reg}_z [\limsup_q q^{-1} \log | P_q(z_1, \cdots, z_n)|]. \tag{3.3}$$

$S(z)$ is a plurisubharmonic function and has the property

$$S(uz) = |u| S(z) \text{ for every } u \text{ complex.} \tag{3.4}$$

By a classical result in the Hartogs series (3.2) Ω' is the domain defined in C^n by

$$\Omega' = [(z'_k) \in C^n;\ z'_k = uz_k,\ u \text{ complex};\ |u| < R(z)]$$

i.e. the largest domain in Ω which is invariant by the transformation $z \to i(z)$. As consequence let us consider, for each $z \in C^n$, a point $z' = zR(z)$ located on the boundary of Ω'. We obtain for $z \in C^n$

$$S(z') = S[zR(z)] = R(z)S(z) = 1,$$

$$\log S(z) = -\log R(z).$$

This proves by (3.3) that $-\log R(z)$ is plurisubharmonic in C^n.

SECOND PROOF. On the other hand it is easy to prove Lemma 3.1 using only more elementary properties of the pseudo-convex domains in C^n.

First let us suppose $\bar{\Omega}$ compact and the boundary $b\Omega$ smooth and strictly pseudo-convex. We assume for each $M \in b\Omega$ the existence of a neighborhood U_M such that $\Omega \cap U_M = [z; W_M(z) < 0]$, with differentiable $W_M(z)$ and a strictly positive Levi's form $L(W) = d_z d_{\bar{z}} W$ in U_M. Then there exists a polynomial Q_M of degree 2 and a neighborhood U'_M, $\bar{U}'_M \subset U_M$ such that the set

$$\sigma_M = [z \in U'_M;\, Q_M(z) = 0]$$

has the property: each point $P \in \sigma_m$, $P \neq M$, is an exterior point of Ω. Now we put

$$\Phi = \sum_M \bar{\sigma}_M.$$

We obtain $b\Omega \subset \Phi$, and $\Phi \cap \Omega = \varnothing$. Moreover Ω is an open component of the set $C^n - \Phi$. For each M contained in the compact set $b\Omega$, the construction can be made in such a way that U'_M is a ball $B(M, a)$ the radius of which $a > 0$, is independent of M. Then we calculate $R(z)$ using the equations $Q_M = 0$, by a familiar method (see [**4**, h, p. 195]). Given $z^0 \in C^n$ and $\zeta^0 \in b\Omega$, with $\zeta_k^0 = u_0 z_k^0$, $1 \leqq k \leqq n$, we consider the complex line $l(z)$ for z in a neighborhood of z^0, and the intersection

$$l(z) \cap \sigma_{\zeta^0}$$

given by $\zeta_k = u z_k$, $z = (z_k)$

$$Q_{\zeta^0}(z_1 u, \cdots, z_n u) = 0.$$

This equation has solutions $u_j(z_1, \cdots, z_n)$ and we consider in a neighborhood of z^0 the function:

$$\psi[\zeta^0; z] = \sup_j\, [-\log |u_j(z_1, \cdots, z_n)|];$$

it is a plurisubharmonic function of z in a neighborhood of z^0.

Now we consider the ball $B = [\zeta; \|\zeta - \zeta_0\| < a/4]$. The smoothness of $b\Omega$ implies that there exists a neighborhood ω of z^0 such that $l(z)$ contains a boundary point ζ of Ω in B for each $z \in \omega$.

Then we consider the family of the functions of z obtained for $\zeta \in B$, $z \to \psi(\zeta; z)$, $\zeta \in B$, $z \in \omega$ which are all defined and plurisubharmonic in ω. We obtain

$$-\log R(z) = \sup_\zeta \psi(\zeta; z), \quad \zeta \in B. \tag{3.5}$$

Now $-\log R(z)$ is obviously an upper semicontinuous function of z; as consequence $-\log R(z)$ is a plurisubharmonic function in ω. The identity $R(uz) = |u|^{-1} R(z)$ is an obvious consequence of the definition (3.1). Further the property remains valid for $\Omega = \lim \Omega_n$, if it is true for Ω_n and if Ω_n is an increasing sequence of domains, and the proof is fulfilled.

THEOREM 3.2. *For a function V defined in C^n and real valued, the following conditions are equivalent* ($\rho > 0$, $\rho' > 0$):

(i) $V \in \Lambda_c^\rho$.

(ii) $V(z)$ *is plurisubharmonic and* $V(uz) = u^\rho V(z)$ *for each complex* u.

(iii) $V(z) = \exp U(z)$; $U(z)$ *is plurisubharmonic and* $U(uz) = U(z) + \rho \log |u|$.

PROOF. (a) Using the Definition 2.1 and equality (2.3), we obtain $L_c(\zeta; uz) = |u^\rho| \, L_c(\zeta; z)$ and by regularization

$$L_c^*(uz) = |u|^\rho L_c^*(z).$$

Further L_c^* is plurisubharmonic; (i) → (ii) is so proved.

(b) Now we prove (ii) → (iii). To begin we shall prove $V(z) \geqq 0$ for each z. If $V \not\equiv -\infty$, there exists $z^0 \in C^n$ where $V(z^0) > -\infty$; then $V(uz^0) = u^\rho V(z_o)$, $\rho > 0$, implies $V(0) = 0$. Then if there exists z^1 with $V(z^1) < 0$, we consider the restriction $\phi(u) = V(uz^1)$ of V to the complex line $l(z^\circ)$; $\phi(u)$ is not the constant $-\infty$, because $\phi(0) = V(0) = 0$. Hence $\phi(u)$ is subharmonic; but $\phi(u) < |u|^\rho V(z^1)$ tends to $-\infty$ when $|u| \to -\infty$. We obtain a contradiction, which proves $V(z) \geqq 0$ for each $z \in C^n$.

Now we consider the open set

$$\Omega_V = [z; V(z) < 1]. \tag{3.6}$$

The identity $V(uz) = |u|^\rho . V(z)$ has as a consequence that Ω_V is the union of circles

$$\Omega_V = \Sigma_\lambda \gamma_\lambda, \qquad \|\lambda\| = 1, \qquad \lambda = (\lambda_j) \in C^n,$$
$$\gamma_\lambda = [z_k; z_k = \lambda_k u, \ \|\lambda\| = 1, \ |u| < |R(\lambda)],$$

with $R(\lambda) > 0$. Then each point $z \in \Omega_V$ can be joined to the origin by means of a curve in the complex line $l(z)$ and Ω_V is connected: Ω_V is a pseudo-convex domain by (3.6) and the assumption that V is plurisubharmonic.

Now we define for each $z \in C^n$

$$R(z) = [\sup |u|, \ i(uz) \subset \Omega_V] \tag{3.7}$$

where $i(z)$ is the circle associated to z by (2.9). Then $R(uz) = |u|^{-1} R(z)$ for each $z \in C^n$. Moreover if z belongs to $b\Omega_V$, the following relations hold.

$$z = (z_k = u\lambda_k), \qquad \|\lambda\| = 1, \qquad |u| = R(\lambda),$$
$$V(z) = V(|u|\lambda) = |u^\rho| V(\lambda) = [R(\lambda)]^\rho V(\lambda) = 1,$$
$$\log V(\lambda) + \rho \log R(\lambda) = 0.$$

Hence we have $\log V(z) = -\rho \log R(z)$ if $\|z\| = 1$; but this relation remains true for all z, as consequence of the identities $V(uz) = |u|^\rho V(z)$ and $R(uz) = |u|^{-1} R(z)$.

Then the relation

$$U(z) = \log V(z) = -\rho \log R(z) \tag{3.8}$$

and the Lemma 3.1 prove that $U(z)$ is plurisubharmonic for $z \in C^n$, and (ii) → (iii) is proved.

(c) The converse (iii) $\to$ (ii) is immediate: if U is plurisubharmonic $V = e^U$ is also plurisubharmonic.

(d) Now we shall prove (ii) $\to$ (i): given a plurisubharmonic function V having the properties (ii) and (iii), we shall construct an entire function $F(z)$ of order $\rho > 0$ such that V is the main circular indicator L_c^* of F. The argument is quite similar to the one used for $\rho = 1$ (exponential type)—see [**4**, e].

Let us consider the domain

$$\Omega_V = [z; V(z) < 1].$$

It is the holomorphy domain of some function

$$G(z_1, \cdots, z_n) = \sum_0^\infty P_q(z_1, \cdots, z_n)$$

where P_q is a homogeneous polynomial of degree q; now we consider

$$G(uz_1, \cdots, uz_n) = \sum_0^\infty u^q P_q(z_1, \cdots, z_n)$$

and by a classical result on the Hartogs series we obtain

$$S(z) = -\log R(z) = \mathrm{Reg}_z[\limsup_q 1/q \log |P_q(z_1, \cdots, z_n)|]. \tag{3.9}$$

Now we consider a sequence A_q of complex numbers and define the function

$$F(z_1, \cdots, z_n) = \sum_0^\infty A_q P_q(z_1, \cdots, z_n),$$

and we determine the constants A_q so that F is of order ρ and has V as main circular indicator. We take

$$A_q = (e\rho/q)^{q/\rho}. \tag{3.10}$$

Then F is entire and a direct calculation of the order $\bar{\rho}(z)$ of the function

$$g_z(u) = F(uz_1, \cdots, uz_n)$$

gives

$$-\frac{1}{\bar{\rho}(z)} = \limsup_q \frac{\log |A_q| + \log |P_q(z)|}{q \log q}.$$

By (3.10) we obtain $\lim_q (\log |A_q| / q \log q) = -1/\rho$; moreover by (3.9), we obtain $\limsup (\log |P_q(z)| / q \log q) = 0$ almost everywhere.

Then $\bar{\rho}(z) = \rho$ almost everywhere, and $\bar{\rho}(z) \leqq \rho$ everywhere. As consequence F is of order ρ. Now the type of $g_z(u)$ is $L_c(0, z)$ and we obtain by a classical calculation:

$$\begin{aligned} \log L_c(0, z) &= \limsup_q [\log q(e\rho)^{-1} + \rho q^{-1} \log |A_q P_q(z)|] \\ &= \rho \limsup_q q^{-1} \log |P_q(z)|. \end{aligned}$$

Then F is an entire function of order ρ and its main circular indicator $L_c^*(z)$ is given by

$$\log L_c^*(z) = \rho \operatorname{Reg}_z[\limsup_q q^{-1} \log |P_q(z)|] = \rho S(z).$$

The comparison with (3.8) and (3.9) leads to the conclusion

$$\log L_c^*(z) = -\rho \log R(z) = \log V(z)$$

and proves that $V \in \Lambda_\rho$; then (ii) $\to$ (i), and Theorem (3.2) is proved.

COROLLARY 3.3. (I) *Let us denote by* Δ_c *the class of the holomorphy domains in* C^n *which are invariant by* $z_k \to z_k e^{i\theta}$, $1 \leqq k \leqq n$, *and defined by* $[z = (z_k);\ z_k = \lambda_k u;\ |u| < R(\lambda),\ \|\lambda\| = 1]$, *and* $R(\lambda) > 0$.

We put $R(z) = \|z\|^{-1} R(z\|z\|^{-1})$; *then* $U(z) = -\log R(z)$ *is a plurisubharmonic function in* C^n.

There exists a bijective mapping

$$\Lambda_\rho \to \Lambda_c \quad \text{and} \quad \Delta_c \to \Lambda_\rho$$

from the set Λ_ρ *of the main circular indicators of entire functions of order* $\rho > 0$ *and of finite type, onto* Δ_c. *The mappings are given by*

$$L_c^*(z) = -\rho \log R(z).$$

(II) *If we consider* $\rho > 0$, $\rho' > 0$, $\rho \neq \rho'$, *there exists a bijective mapping*

$$\Lambda_\rho \rightleftarrows \Lambda_{\rho'}$$

given by

$$-\log R(z) = \rho L_c^*(z) = \rho' L_c^{*\prime}(z)$$

$$L_c^* \in \Lambda_\rho;\ L_c^{*\prime} \in \Lambda_{\rho'}.$$

4. Plurisubharmonic functions of minimal growth. Let us put

$$M(r) = \sup V(z),\ \|z\| = r;$$

the growth

$$M(r) = O(\log r)$$

is the minimal one for nonconstant plurisubharmonic functions V. If $V = \log|F|$, F entire, we obtain $F = P$, P polynomial.

The minimal class has special properties (see [**4**, e]). We give the following definition.

DEFINITION 4.1. (a) We denote by S_n the class of the functions U which are plurisubharmonic in C^n and satisfy

(4.1) $$\limsup (\log\|z\|)^{-1} U(z) = a < \infty.$$

(b) We denote by $S_{n,\rho}$ the subclass of S_n which has the property

(4.2) $$U(uz) = U(z) + \rho \log|u|, \quad \rho > 0.$$

5. **The mapping $S_n \to S_{n+1,1}$ associated to the mapping $C^n \to P^n$.** We consider the canonical mapping

(5.1) $$z' \in C^n\colon z' = (z_1', \cdots, z_n') \to z = (z_0, z_1, \cdots, z_n) \in P^n,$$

where P^n is the complex projective space and $z_k' = z_k z_0^{-1}$.

Given a plurisubharmonic function $U(z) = U(z_1, \cdots, z_n) \in S_n$, we define first that

(5.2) $$U_1(z_0, z_1, \cdots, z_n) = a^{-1} U(z_1 z_0^{-1}, \cdots, z_n z_0^{-1}) + \log|z_0|$$

(5.3) $$a = \limsup (\log \|z\|)^{-1}, \qquad U(z) < \infty.$$

We assume that $a > 0$, (if $a \leqq 0$, U reduces to a constant). Then U_1 is a plurisubharmonic function in the domain

$$G = [z = (z_0, \cdots, z_n);\ z_0 \neq 0].$$

THEOREM 5.1. *Given $U(z) \in S_n$, the function U_1 defined by* (5.2) *has a unique extension $\bar{U}(z_0, \cdots, z_n)$ which is plurisubharmonic in C^{n+1}, and belongs to $S_{n+1,1}$.*

We recall the following theorem (cf [**4**, g]). Given $V \in P(D - E)$, where D is a domain in C^n and E a closed set in D, which is of vanishing R^{2n}-capacity, if V is bounded above in $D - E$ in the neighbourhood of each point of E, then V has a unique plurisubharmonic extension $V \in P(D)$, and the values of $V(z)$ for $z \in E$ are given by

$$V(z) = \limsup_{z' \to z} V(z'), \qquad z' \in D - E.$$

The set $[z = (z_0, \cdots, z_n) = (z_0, z');\ z_0 \neq 0]$ is an analytic set in C^{n+1} and therefore a C^{n+1}-polar set. So it is a set of vanishing capacity. Moreover $U_1(z_0, \cdots, z_n)$ defined by (5.2) is upper bounded in $G = [z = (z_0, z'); z_0 \neq 0]$, $z' = (z_1, \cdots, z_n)$, on each set

$$K_r = [|z_0| \leqq 1,\ z' = (z_1, \cdots, z_n)|;\ \|z'\| \leqq r,\ z_0 \neq 0].$$

Now we consider

$$M(r) = \sup U(z_1, \cdots, z_n) \qquad \text{for } \|z\| \leqq r;$$

$M(r)$ is a convex function of $\log r$, and from (5.3) we deduce

$$M(r) \leqq M(1) + a \log r.$$

For $(z_0, z_1, \cdots, z_n) \in K_r$ we have as consequence the following majorization of U_1:

(5.4) $U_1(z_0, z_1, \cdots, z_n) \leqq a^{-1}[M(1) + a \log r |z_0|^{-1}] + \log |z_0| = \log r + a^{-1}M(1).$

Then U_1 is bounded above in K_r in the neighborhood of each point of the set $z_0 = 0$ and has a plurisubharmonic extension $\bar{U}(z_0, z_1, \cdots, z_n)$. By (5.2), we obtain

$$\bar{U}(uz_0, \cdots, uz_n) = \bar{U}(z_0, \cdots, z_n) + \log |u|$$

for the extension $\bar{U} \in P(C^{n+1})$ and $\bar{U} \in S_{n+1,1}$. The previous construction defines a mapping

$$\psi \colon U \in S_n \to \bar{U} \in S_{n+1\ 1}.$$

We have a converse mapping

$$\psi^{-1} \colon V(z_0, z_1, \cdots, z_n) \in S_{n+1,1} \to V(1, z_1, \cdots, z_n)$$

and ψ is a bijective mapping of S_n onto $S_{n+1,1}$.

6. Proof of the main result.

THEOREM 6.1. *There exist entire functions of finite order $\rho > 0$ having as main circular indicator L_c^* and as main radial indicator L_r^* plurisubharmonic functions which are not continuous.*

For the proof, let us consider a potential in C^1.

$$S^\mu(z) = \int d\mu(a) \log |a - z|,$$

μ is a positive measure with total norm $\|\mu\| = \rho > 0$; the support of μ is the unit disk $|a| \leqq 1$. Obviously

$$S^\mu(z) \leqq \|\mu\| \log |z| + \|\mu\| \log 2$$

and $S^\mu(z)$ belongs to S_1. We choose μ such that $S^\mu(z)$ is discontinuous at the origin $z = 0$ (see for example [2]).

We define for $z_0 \neq 0$

(6.1) $$U(z_0, z_1) = \|\mu\|^{-1} S^\mu(z_0, z_1^{-1}).$$

By the Theorem 5.1, U has a plurisubharmonic extension $\bar{U}(z_0, z_1) \in S_{2,1}$. We consider $V(z_0, z_1) = \exp \rho \bar{U}(z_0, z_1) = \exp S^\mu(z_1 z_0^{-1})$.

Then by the Theorem 3.2, there exists an entire function $F_0(z_0, z_1)$, of order ρ, having $V(z_0, z_1)$ as main circular indicator.

Now we prove that $V(z_0, z_1)$ is not continuous at the points of the set $A = [z = (z_0, z_1);\ z_0 \neq 0,\ z_1 = 0]$. If V were continuous at the point $(z_0, 0)$, the same property would hold for $\bar{U}(z_0, z_1)$ and for $U(z_0, z_1)$. By (6.1) and $\|\mu\| = \rho > 0$, the continuity of U at a point $(z_0, 0)$ of A implies the continuity of the potential S^μ at the origin; the continuity of U gives

$$|U(z_0', z_1') - U(z_0, 0)| < \varepsilon$$

if $(z_0', z_1') \in Q_\eta$, where Q_η $[|z_0' - z_0| < \eta, |z_1'| < \eta]$, $\eta < |z_0|/2$. By the mapping $P^2 \to C^1$ defined by $z = z_1 z_0^{-1}$, the image of Q_η contains a neighbourhood of the origin $z = 0$ given by $|z| < \eta |z_0|^{-1}$. Therefore, we have

$$|S^\mu(z) - S^\mu(0)| < \varepsilon$$

if $|z| < \eta |z_0|^{-1}$, and the potential S^μ would be continuous at the origin. As a consequence a contradiction is obtained, which proves that the main circular indicator is not continuous at the point $(z_0, 0) \in A$.

Now, given $n \geqq 2$, $F_0(z_0, z_1)$ can be considered as an entire function of n complex variables $(z_0, z_1, \cdots, z_n)$, and is of order ρ in C^n. The example is then valid for all n.

Now we consider the main radial indicator L_r^* of $D_0(z_1, z_0)$. If L_r^* were continuous, the same property would be true [by Theorem 2.1, (b)] for L_c^*, main circular indicator of F_0. We obtain a contradiction: therefore L_r^* is not continuous and the second part of Theorem 6.1 is proved.

REMARK. The Corollary 3.3 of §3 gives a general method to construct main circular indicators L_c. Let us consider a pseudo-convex domain D with the origin as interior point, invariant by the transformations

$$g_\theta : z_k \to z_k e^{i\theta}, \qquad 0 \leqq \theta < 2\pi$$

and given by

$$D = [\, z = (z_k);\ z_k = \lambda_k u; \qquad |u| < R(\lambda),\ \|\lambda\| = 1]. \tag{6.2}$$

The domain D defines (see §3) a function $R(z)$, and $-\log R(z)$ is a plurisubharmonic function in C^n. By the Theorem 3.2, there exists at least one entire function of order ρ, of finite type, of which the main circular indicator is given by

$$L_c^*(z) = \exp - \rho \log R(z) = R^{-\rho}(z), \qquad \rho > 0.$$

Then in order to prove the existence of entire functions the main circular indicator of which has a given property, it is sufficient to construct domains D in C^n, given by (6.2) with corresponding properties for the function $R(z)$.

BIBLIOGRAPHY

1. R. P. Boas, *Entire functions*, Academic Press, New York, 1954.

2. M. Brelot, *Etude des fonctions sousharmoniques au voisinage d'un point*, Hermann, Paris, 1934.

3. C. O. Kiselman, *On entire functions of exponential type and indicators of analytic functionals* (mimeographed).

4. P. Lelong.(a) *Définition des fonctions plurisousharmoniques.* C. R. Acad. Sci. Paris, **215** (1942) 398; see also: **215** (1942), 454 and **216** (1942), 107.

(b) *Les fonctions plurisousharmoniques*, Ann. Sci. Ecole Norm. Sup., **62** (1945), 301–338.

(c) *Fonctions plurisousharmoniques et formes différentielles positives*, Colloque du C.I.M.E. Varenne, 1963. Editions Cremonese, Rome.

(d) *Fonctions entières (n-variables) et fonctions plurisousharmoniques de type exponentiel.* C. R. Acad. Sci. Paris **260** (1965), 1663.

(e) *Fonctions entières de type exponentiel dans C^n*, Ann. Inst. Fourier Grenoble, no. 2 **16** (1966).

(f) *Plurisubharmonic functions and entire functions of exponential type*, Summer Institute, La Jolla, 1966.

(g) *Ensembles singuliers impropres des fonctions plurisousharmoniques*, Journal de Math. **36** (1957), 263–303.

(h) *Domaines convexes par rapport aux fonctions plurisousharmoniques*, J. Analyse Math. **2**, (1952), 178–208.

5. A. Martineau, (a) *Sur les fonctionnelles analytiques et la transformation de Fourier-Borel* J. Analyse Math. **11** (1963), 1–162.

(b) *Fonctions entières de type exponentiel*, Séminaire P. Lelong, exposé 8, June 1966, Paris, Institut Henri Poincaré, 1965–1966.

INSTITUT H. POINCARÉ
PARIS, FRANCE

Differential Equations of Infinite Order Hyperdirichlet Series and Entire Functions of Bounded Index

BENJAMIN LEPSON

Dedicated to the memory of Joseph Fels Ritt

1. **Ordinary linear differential operators of infinite order.** Consider the "entire differential operator of genus zero",

(1) $$A = (1 - D/a_1)(1 - D/a_2)\cdots(1 - D/a_n)\cdots,$$

where D denotes differentiation and where the complex constants a_n (not necessarily distinct) satisfy the condition

(2) $$\sum_{n=1}^{\infty} \frac{1}{|a_n|} < \infty.$$

This operator was first studied by Ritt **[11]**. Its most notable property is that its domain consists of all analytic functions. Specifically, Ritt showed the following, in which the operator A can be applied either as an infinite product, as shown above, or as the corresponding expansion in formal power series.

THEOREM 1. *If $f(z)$ is holomorphic in a domain d, $Af(z)$ converges in d, the convergence being uniform in every compact subset of d.*

It is easy to see that condition (2) has a vital part in establishing the truth of this theorem. It can also be shown that the order of the factors of A is immaterial, and that A is commutative with any power of D.

THEOREM 2. *If A and B are two such operators, their product BA (defined by applying A first and then B) is also an operator of the same class, and its factors will be the combined factors of A and B. Therefore, multiplication of operators is commutative.*

Theorem 2 can also be extended to an infinite (denumerable) sequence of operators.

2. **Linear homogeneous differential equations of infinite order with constant coefficients.** Let A be an entire differential operator of genus zero and consider the differential equation

$$Af(z) = 0 \tag{3}$$

which we call the "homogeneous equation of genus zero". We are interested in analytic solutions of this equation, and as in the preceding section, we summarize the pertinent results due to Ritt [**11**]. It is convenient to change our notation concerning the zeros of A so that unless otherwise specified, it will be assumed that the a_n are distinct and that the multiplicity of each a_n will be designated by the positive integer p_n. Note that condition (2) now becomes

$$\sum_{n=1}^{\infty} \frac{p_n}{|a_n|} < \infty. \tag{4}$$

We also note that, as a direct consequence of Theorem 1, if $f(z)$ satisfies (3) in an arbitrarily small neighborhood, it will satisfy (3) in its entire domain of existence. We then have the fundamental result on solutions of (3).

THEOREM 3. *If the series*

$$\sum_{n=1}^{\infty} Q_n(z)\, e^{a_n z}, \tag{5}$$

where each $Q_n(z)$ *is an arbitrary polynomial of degree* $p_n - 1$ *with complex coefficients, is uniformly convergent in some domain, its sum satisfies* (3) *in that domain.*

It is easy, on the basis of Theorem 3, to give an example of an analytic solution of (3) which is not an entire function (and which therefore satisfies no finite order linear differential equation with constant coefficients). Such an example is given by the Dirichlet series

$$\sum_{n=0}^{\infty} e^{-n^2 z}$$

which converges only for $\mathscr{R}(z) > 0$, uniformly in any compact subset of the right halfplane. The sum of this series is analytic and satisfies (3) in the right halfplane (with $a_n = -n^2$ and $p_n = 1$ for all n), but is singular at the origin, since it is real on the real axis and approaches $+\infty$ as z tends to zero along the positive real axis.

However, it is possible to establish the following properties of all solutions of (3), whether or not they can be represented by series of the form (5).

THEOREM 4. *The solutions of* (3), *and the successive integrals of such solutions, are all single-valued functions.*

THEOREM 5. *If a solution of* (3) *is analytic on the entire circumference of a circle, it is analytic throughout the interior of the circle; therefore its domain of existence is simply connected and it has no isolated singularities.*

3. **Hyperdirichlet series.** We will now investigate in some detail the convergence behavior of series (5) and the function-theoretic properties of their sums. At the same time, we will relax condition (4) so as to obtain results of more general interest.

The first investigation of this type was conducted by Valiron [**13**]. He replaced condition (4) by the two conditions:

(6) $$\lim_{n\to\infty}\frac{p_n-1}{a_n}=0 \quad\text{and}\quad \lim_{n\to\infty}\frac{\log n}{a_n}=0.$$

The first of these conditions is, of course, an immediate consequence of (4), while the second, which is the familiar condition that the domains of convergence and absolute convergence coincide for a Dirichlet series with real or complex exponents (see Hardy and Riesz [**4**], Valiron [**14**], and Hille [**5**]), also follows from (4) if the a_n are so arranged (which can always be done if $|a_n|\to\infty$) that their moduli form a nondecreasing sequence. His results can be stated as follows:

THEOREM 6. *Let C_n be the maximum of the moduli of the coefficients of $Q_n(z)$. Then*

(7) $$\sum_{n=1}^{\infty} C_n e^{a_n z}$$

is a Dirichlet series with complex exponents, whose domain of absolute convergence is convex and which diverges outside of the closure of that domain. The interiors of the sets of convergence and absolute convergence of (5) *both coincide with the interior of this domain, and the convergence of* (5) (*as well as* (7)) *is uniform in any compact subset of this interior.*

Unlike (7), though, (5) may have points of convergence or absolute convergence outside of the closure of the above domain (due primarily to the influence of the zeros of the $Q_n(z)$). These "exceptional sets" are shown to have (Carathéodory or Hausdorff) linear measure zero under the stronger assumption

(8) $$\lim_{n\to\infty}\frac{p_n\log n}{a_n}=0,$$

which clearly implies (6).

It is natural to ask whether it is possible to obtain results of this type if no condition on the rate of growth of the exponents a_n is assumed. This can be done (Lepson [**7**]) if the exponents satisfy the usual conditions of (real) general Dirichlet series: $0<\lambda_0<\lambda_1<\cdots\to+\infty$. (We write λ_n instead of a_n to emphasize the analogy with Dirichlet series, but we assume $\lambda_0>0$ instead of $\lambda_0\geqq 0$ for convenience in stating the results. As a further change in notation, we write $P_n(z)$ instead of $Q_n(z)$ and designate its degree by μ_n.) It is easy to see, though, that we must keep the first condition of (6), in order that the exponentials, rather than the polynomials, should dominate the asymptotic behavior of the series. We note that the results on absolute convergence are true for arbitrary positive λ_n and nonnegative integers μ_n such that (10) holds (which does not imply that $\lambda_n \to \infty$), and that this theory includes that for (real) general Dirichlet series as a special case.

DEFINITION 1. A *hyperdirichlet series* is a series of the form

$$\sum_{n=0}^{\infty} P_n(z)e^{-\lambda_n z} \tag{9}$$

where the $P_n(z)$ are polynomials of degrees μ_n respectively and the λ_n are positive numbers tending monotonically to infinity.

DEFINITION 2. A hyperdirichlet series is called *regular* if the condition

$$\lim_{n\to\infty} \frac{\mu_n}{\lambda_n} = 0 \tag{10}$$

holds.

DEFINITION 3. A series of functions $\sum_{n=0}^{\infty} F_n(z)$ is *M-convergent* on a set S if $\sum_{n=0}^{\infty} M_n < \infty$, where $M_n = \text{l.u.b.}_S |F_n(z)|$.

DEFINITION 4. The *set* (*region*) *of convergence* of a series or sequence is the set of points (interior points) for which it converges. Similar phrases are used for absolute convergence and boundedness of partial sums (for series only) and termwise boundedness.

DEFINITION 5. C_n is the maximum of the moduli of the coefficients of $P_n(z)$.

DEFINITION 6. A is the abscissa of absolute convergence of the Dirichlet series

$$\sum_{n=0}^{\infty} C_n e^{-\lambda_n z}. \tag{11}$$

DEFINITION 7. E is the set of limit points of the roots of the $P_n(z)$, including those which appear infinitely often. F is the intersection of the sets analogous to E for all subseries of (9) whose corresponding subseries of (11) have abscissa of absolute convergence A.

We need the following results of Carleson (unpublished) on capacity and Fekete on transfinite diameter (Szegö [12]):

LEMMA 1. *Let $\{F_n(x)\}$ be a sequence of real Borel measurable functions which converges on a bounded plane point set of capacity $k > 0$. Then, given any $\varepsilon > 0$, there exists a closed subset F of E of capacity greater than $k - \varepsilon$ on which the convergence is uniform.* (*Valid for logarithmic capacity and for any capacity whose potential function is nonnegative.*)

LEMMA 2. *Let $P(z)$ be a polynomial of degree n with leading coefficient unity, and let E be a compact set of transfinite diameter $d > 0$. Then*

$$\max_E |P(z)| \geqq d^n.$$

Certain preliminary results which may be of interest in themselves are the following:

THEOREM 7. (An analog of Osgood's theorem **[10]** on uniform convergence in subdomains of convergent sequences of analytic functions.)

Let a series of functions, each analytic in a domain D, converge absolutely at every point of D. Then there is a domain contained in D on which the series is M-convergent.

(For a discussion of the relation of these two results, see Lepson **[8]**.)

THEOREM 8. *Let E be a closed set contained in the circle* $|z| \leqq R \geqq 1$ *of transfinite diameter* $d > 0$. *Let* $P(z)$ *be a polynomial of degree n and let M be the maximum of the moduli of its coefficients. Then*

$$\max_E |P(z)| \geqq M\left(\frac{d}{3R}\right)^n$$

with equality only for $n = 0$.

THEOREM 9. *Let* $\{P_n(z)\}$ *be a sequence of polynomials of degrees* $r_n > 0$ *with maximum coefficients unity. Let S be the set of points for which*

$$\lim_{n\to\infty} [P_n(z)]^{1/r_n} = 0.$$

Then S is a set of logarithmic capacity zero.

THEOREM 10. *Let* $\{P_n(z)\}$ *be a sequence of polynomials of degrees* $r_n \geqq 0$ *with maximum coefficients unity. Let* $\{\lambda_n\}$ *be a sequence of positive numbers such that* $\lim_{n\to\infty} r_n/\lambda_n = 0$. *Let S' be the set of points for which*

$$\limsup_{n\to\infty} \; [P_n(z)]^{1/\lambda_n} < 1.$$

Then S' is an F_σ *set of outer logarithmic capacity zero.*

We now state the principal results of the convergence behavior of regular hyperdirichlet series.

THEOREM 11. *If* (9) *converges absolutely at a point z' not in E, then it is M-convergent in any compact subset of the halfplane* $\mathscr{R}(z) > \mathscr{R}(z')$ *and more generally in every sector* $S: |z - z'| \leqq K\mathscr{R}(z - z')$ *with* $K > 0$.

Since E, as well as F, may be the entire plane, this theorem gives no information as to the region of absolute convergence in general. Such information is given by the following global result.

THEOREM 12. *The region of absolute convergence of* (9) *is the halfplane* $\mathscr{R}(z) > A$. *Also,* (9) *is M-convergent in every sector* $S: |z - z'| \leqq K\mathscr{R}(z - z')$, *where* $\mathscr{R}(z') > A$, *and therefore represents a function analytic in the halfplane* $\mathscr{R}(z) > A$. *Let R be that portion of the set of absolute convergence of* (9) *in the halfplane* $\mathscr{R}(z) < A$. *Then R is an* F_σ *set of outer logarithmic capacity zero and a subset of F of first category in the plane.*

The fact that R is a subset of F might lead one to suspect that E could be replaced by F in Theorem 11. This, however, is false. In fact, the example

$$S(z) = \sum_{n=1}^{\infty} P_n(z) e^{-n^3 z}$$

where $P_n(z) = n^{4n} z^n$ if n is odd and $P_n(z) = n^{-2}$ if n is even (which even satisfies Ritt's condition $\sum_{n=1}^{\infty} (\mu_n + 1)/\lambda_n < \infty$), for which E consists of the single point 0 and F is empty, shows that the Abel limit theorem may fail for a point of absolute convergence, which lies on the boundary of the region of absolute convergence, and which lies in E but not in F.

In order to discuss convergence and boundedness of partial sums, we study the series

$$\sum_{n=0}^{\infty} P_n(z) e^{-\lambda_n w} \tag{12}$$

in the two complex variables z and w. For fixed z, (12) is a Dirichlet series in w. Let its abscissa of convergence be $C(z)$, and let $C = \text{l.u.b.}\, C(z)$ over all z. Then it is possible to state the main results on convergence and boundedness of partial sums as follows:

THEOREM 13. *The regions of convergence and boundedness of partial sums of* (9) *are both the halfplane* $\mathscr{R}(z) > C$. *Let S be that portion of the set of boundedness of partial sums of* (9) *in the halfplane* $\mathscr{R}(z) < C$. *Then S is an* F_σ *set of outer logarithmic capacity zero of first category in the plane.*

However, these methods do not seem able to give results concerning uniformity of convergence. Such results have only been obtained by the use of a direct, but quite involved, "infinite series" type of argument. One possible reason for this difficulty is the fact, which follows from the example considered earlier, that boundary points of convergence are not generally included in sets of uniform convergence. (Although Theorem 11 gives a sufficient condition that a boundary point of absolute convergence be included in such sets, no analogous condition is known for boundary points of convergence.) The principal result on uniform convergence is

THEOREM 14. *The series* (9) *is uniformly convergent in every sector* S:

$|z - z_0| < K\mathscr{R}(z - z_0)$ for which $\mathscr{R}(z_0) > C$, and therefore represents a function analytic in the halfplane $R(z) > C$.

It is now possible to obtain a uniqueness theorem which includes that for Dirichlet series [**14**, p. 8].

LEMMA 3. *A regular hyperdirichlet series for which not all coefficients are zero has at most a finite number of zeros in every sector $S: |z - z_0| \leqq K\mathscr{R}(z - z_0)$ for which $\mathscr{R}(z_0) > C$.*

PROOF. Follows from the decomposition

$$\sum_{n=0}^{\infty} P_n(z)e^{-\lambda_n z} = e^{-\lambda_0 z}\left[P_0(z) + \sum_{n=1}^{\infty} P_n(z)e^{-(\lambda_n - \lambda_0)z}\right]$$

and uniformity of convergence in sectors.

THEOREM 15. *If two regular hyperdirichlet series converge to the same sum in a domain D, then the corresponding terms are identical.*

PROOF. Immediate consequence of Lemma 3.

4. **Generalizations of hyperdirichlet series.** We now ask whether it is possible to replace the polynomials in (9) by a larger class of functions and still obtain results similar to those given in the previous section. This was done successfully by John J. MacDonnell, S. J., who replaced the polynomials by certain "entire functions of bounded index" (see below, also MacDonnell [**9**]), Lawrence E. Isenecker, S. J., who used a class of rational functions (Isenecker [**6**]), and Sister Margaret Grace Elsey, S.L., who combined both of these approaches by considering quotients of entire functions of bounded index, to produce a class of series known as "Ritt series" (Elsey, [**2**]). Naturally, in the two latter cases, the zeros of the denominators, which produce poles in the individual terms, also play a significant part in determining the convergence behavior of the series. We cannot enter here into the detailed results of these investigations but, with respect to the first approach, let us look in some detail at the class of entire functions of bounded index.

5. **Entire functions of bounded index.** Let $\mathscr{B}$ be a class of entire functions which includes all polynomials. If the members of this class are to replace the polynomials in (9), then since condition (10) plays a central role in the preceding theory, it would appear necessary that, to each member of $\mathscr{B}$, we could associate a real number directly analogous to the degree of a polynomial, and that the sequence of these numbers would satisfy (10). This is the basis for the following definition.

DEFINITION 8. Let $\{a_n(w)\}$ be the sequence of coefficients in the Taylor series expansion of the entire function $f(z) \not\equiv 0$ about the point w. Let $C(w)$ be the maximum of their moduli, and let $I(w)$ be the largest value of n for which

$|a_n(w)| = C(w)$. If there is an integer N such that $I(w) \leqq N$ for all w, then $f(z)$ is said to be of *bounded index*, and the smallest such integer, which we denote by $I(f)$, is called the index of $f(z)$. If $I(w)$ is unbounded, then $f(z)$ is said to be of *unbounded index*, and we write $I(f) = \infty$.

EXAMPLES:

(a) If $P(z)$ is a polynomial of degree n, then $I(P) \leqq n$.

(b) If $f(z) = \exp(cz)$, then $I(f) = [|c|]$.

(c) If $f(z) = \sin(cz)$ or $\cos(cz)$, then $I(f)$ is one of the three integers $[|c|] - 1$, $[|c|]$, $[|c|] + 1$.

(d) If $f(z) = \exp(z^2)$, then $I(f) = \infty$.

We show that $C(w)$ is continuous for all w and that $I(w)$ is bounded on bounded sets as follows:

LEMMA 4. *Let $f(z)$ be analytic in $|z - z_0| \leqq 1 + \delta$. Then $\{f^{(n)}(z)/n!\}$ tends to zero uniformly in z for $|z - z_0| \leqq \delta$.*

PROOF. It is clear that the sequence converges pointwise to zero for $|z - z_0| \leqq \delta$. But also, for $|z - z_0| \leqq \delta$, if $M = \max |f(z)|$ on the circle $|z - z_0| = 1 + \delta$, we have

$$\left| \frac{f^n(z)}{n!} \right| = \left| \frac{1}{2\pi i} \int_{|z - z_0| = 1 + \delta} \frac{f(\xi)}{(\xi - z_0)^{n+1}} d\xi \right|$$

$$\leqq \frac{2\pi(1 + \delta)}{2\pi} \cdot \frac{M}{1} = M(1 + \delta).$$

Therefore, by Vitali's theorem, the sequence converges uniformly in any compact subset of $|z - z_0| < \delta$. Hence, if we replace δ in the hypothesis by a slightly larger value, the conclusion follows.

COROLLARY. *Let $f(z)$ be an entire function and let E be any bounded set. Then $\{f^{(n)}(z)/n!\}$ tends to zero uniformly for z in E.*

THEOREM 16. *Let $f(z) \not\equiv 0$ be an entire function, and let E be any bounded set. Then $I(w)$ is bounded on E.*

PROOF. Let K be a closed disk containing E. Suppse $f(z) \neq 0$ in K. Then $|f(z)| > \delta > 0$ in K. But, by the above corollary, $|f^{(n)}(z)/n!| < \delta$ in K for sufficiently large n, which completes the proof in this case. In general, let $z_1, z_2, \cdots, z_p$ be the zeros of $f(z)$ in K. We may assume that $f(z) \neq 0$ on the boundary of K. For each point z_r, there is at least one derivative $f^{(n_r)}(z)$ which does not vanish at z_r. Let $C_1, C_2, \cdots, C_p$ be circles about $z_1, z_2, \cdots, z_p$ respectively such that no two of them intersect, each is contained in the interior of K, and $f^{(n_r)}(z)$ does not vanish in or on the circle C_r. Then there are positive numbers δ_r such that $|f^{(n_r)}(z)| > \delta_r$ in and on the circles C_r. Also $|f(z)| > \delta$ for all z in K but external to the circles C_r.

Let $M = \max(n_1, n_2, \cdots, n_p)$. Then, for all z in K,

$$\max_{n=0,\dots,M}\left[\frac{|f^{(n)}(z)|}{n!}\right] > \min\left(\delta, \frac{\delta_1}{1!}, \dots, \frac{\delta_p}{p!}\right) = \varepsilon > 0.$$

But, by the above corollary, we can find an integer $N > M$ so large that $|f^{(n)}(z)/n!| < \varepsilon$ for all $n > N$ and all z in K. Hence $I(w) \leqq N$ in K.

COROLLARY. *$C(w)$ is continuous for all w.*

PROOF. Let K be the closed disk in the w plane. Let N be an upper bound for $I(w)$ for all w in K. Then, in K, $C(w) = \max_{n=0,\dots,N}[|f^{(n)}w|/n!]$. It follows $C(w)$ is continuous in K, and therefore in the entire plane.

It is clear that the question as to whether a given entire function is of bounded index is closely related to its order and the distribution of its zeros. No characterization in these terms is known, although it can be shown that an entire function of bounded index is necessarily of exponential type. (The proof of this result will appear elsewhere.) However, there exist entire functions of unbounded index of all orders, including order zero, and in fact of arbitrarily slow rate of growth (but which must, of course, grow faster than polynomials). We will close this discussion by giving an example of the latter.

THEOREM 17. *Let $F(r)$ be a positive, real, continuous function of r for $0 \leqq r < \infty$ which grows faster than any power of r; i.e., for every positive integer k, we have*

$$\lim_{r\to\infty} \frac{r^k}{F(r)} = 0.$$

Then there exists an entire function $f(z)$ of unbounded index such that

$$|f(z)| < F(r)$$

for all complex z, where $r = |z|$.

PROOF. It is clear that $F(r)$ has a positive minimum, which we may assume to be greater than unity, therefore $F(r) \geqq 1 + \varepsilon$ for all r. Then there is an integer n_1 such that $F(r) > 1 + r/n_1$ for all r. Let $F_1(r) = F(r)/(1 + r/n_1)$. Then $F_1(r)$ is also a positive, real, continuous function of r which grows faster than any power of r. But $F_1(r) > 1$ for all r, hence $F_1(r) \geqq 1 + \varepsilon_1$ for some ε_1 and all r. Thus there is an integer $n_2 > n_1$ such that $F_1(r) > 1 + r/n_2$ for all r. Let $F_2(r) = F_1(r)/(1 + r/n_2)$. Then $F_2(r)$ has the same properties as $F_1(r)$, and $F(r) = (1 + r/n_1)F_1(r) = (1 + r/n_1)(1 + r/n_2)F_2(r) > (1 + r/n_1)(1 + r/n_2)$. For each integer k, there is a function $F_k(r)$ with the above properties and an integer $n_k > n_{k-1}$ such that $F_k(r) = F_{k-1}(r)/(1 + r/n_k)$ and $F(r) = \prod_{s=1}^{k}(1 + r/n_s)\, F_k(r) > \prod_{s=1}^{k}(1 + r/n_s)$.

Therefore $\prod_{s=1}^{\infty}(1 + z/n_s)$ converges absolutely for all z, uniformly in every compact set, and therefore defines an entire function $f_1(z)$ satisfying the relation $|f_1(z)| \leqq F(r)$. We now construct $f(z)$ by modifying $f_1(z)$ as follows:

$$f(z) = \prod_{s=1}^{\infty} \left(1 + \frac{z}{m_s}\right)$$

where $m_1 = n_1$, $m_2 = m_3 = n_3$, $m_4 = m_5 = m_6 = n_6$, etc. Then, except for $z = 0$ (for which $f(z) = 1$, $F(r) > 1$),

$$|f(z)| \leqq \prod_{s=1}^{\infty} \left(1 + \frac{r}{m_s}\right) < \prod_{s=1}^{\infty} \left(1 + \frac{r}{n_s}\right) \leqq F(r).$$

Also, $f(z)$ has a simple zero at $-n_1$, a double zero at $-n_3$, etc. Therefore, $f(z)$ is of unbounded index, since it is clear that an entire function having zeros of arbitrarily large order is of unbounded index, as the Taylor expansion about a zero of order k has its first k coefficients equal to zero.

Bibliography

1. L. Carleson, *On a class of meromorphic functions and its associated exceptional sets*, Thesis, University of Uppsala, 1950.

2. Sister M. G. Elsey, S. L., *Some convergence theorems for normal Ritt series*, Catholic Univ. Press, Washington, D. C., 1963.

3. M. Fekete, *Über den transfiniten Durchmesser ebener Punktmengen. Erste Mitteilung*, Math. Z. **32** (1930), 108–114.

4. G. H. Hardy and M. Riesz, *The general theory of Dirichlet's series*, Cambridge Univ. Press, reprinted by Stechert-Hafner, New York, N. Y., 1964.

5. E. Hille, *Note on Dirichlet's series with complex exponents*, Ann. of Math. (2) **25** (1924), 261–278.

6. L. E. Isenecker, S. J., *On the absolute and uniform convergence of Dirichlet-type series whose coefficients are rational functions of a complex variable and on meromorphic functions represented by such series*, Catholic Univ. Press, Washington, D.C., 1962.

7. B. Lepson, *On hyperdirichlet series and on related questions of the general theory of functions*, Trans. Amer. Math. Soc. **72** (1952), 18–45.

8. ———, *On irregular points of normal convergence and M-convergence for series of analytic functions*, Proc. Int. Cong. Math. **1** (1950), 416–417.

9. J. J. MacDonnell, S. J., *Some convergence theorems for Dirichlet-type series whose coefficients are entire functions of bounded index*, Catholic Univ. Press, Washington, D.C., 1957.

10. W. F. Osgood, *Note on the functions defined by infinite series whose terms are analytic functions of a complex variable ...*, Ann. of Math. (2) **3** (1902), 25–34.

11. J. F. Ritt, *On a general class of linear homogeneous differential equations of infinite order with constant coefficients*, Trans. Amer. Math. Soc. **18** (1917), 27–49.

12. G. Szegö, *Bemerkung zu einer Arbeit von Herrn M. Fekete: Über die Verteilung der Wurzeln bei gewissen algebraischen Gleichungen mit ganzzahliger Koeffizienten*, Math. Z. **21** (1924), 203–208.

13. G. Valiron, *Sur les solutions des équations différentielles linéaires d'ordre infiné et a coefficients constants*, Ann. École Norm. (3) **46** (1929), 25–53.

14. ———, *Théorie générale des séries de Dirichlet*, Gauthier-Villars, Paris, 1926.

Naval Research Laboratory
Washington, D.C.
Catholic University of America
Washington, D.C.

Equilibrium Distributions of Charge on Curves and Locally Convergent Sequences of Polynomials

T. L. McCOY

1. **Introduction.** Let $\mathfrak{R}$ be a closed set in the complex plane. By an $\mathfrak{R}$-*polynomial* we mean a polynomial, all of whose zeros lie in $\mathfrak{R}$. An $\mathfrak{R}$-function is an entire function with all zeros in $\mathfrak{R}$. *Local convergence* means uniform convergence on the compact subsets of the open unit disc, and *global convergence* means uniform convergence on the compact subsets of the plane. We always require that the limit function be different from the zero function.

The region $\mathfrak{R}$ is said to be *regular* if any $\mathfrak{R}$-function is the global limit of a sequence of $\mathfrak{R}$-polynomials. The question of regularity of regions has been studied by J. Korevaar **[5]** and **[6]**. It turns out that, in a sense, most unbounded regions are regular. Another interesting property of regions is the *local-to-global property*. We say $\mathfrak{R}$ has the local-to-global property if any locally convergent sequence of $\mathfrak{R}$-polynomials is globally convergent. It has been known for some time that the half plane $\operatorname{Re} e^{i\alpha}z \geqq 0$ has the local-to-global property (**[13]**, cf. **[7]** and **[8a]** for additional background).

In this paper we concern ourselves with a conjecture of Korevaar (see **[7]**, §5) which we state rather loosely as follows.

KOREVAAR'S CONJECTURE. Under mild geometric restrictions on $\mathfrak{R}$, either $\mathfrak{R}$ is regular or $\mathfrak{R}$ has the local-to-global property.

One restriction we must place on $\mathfrak{R}$ is that $\mathfrak{R}$ does not contain a closed curve surrounding the unit disc. If γ were such a closed curve, z_0 exterior to γ, then we could approximate locally to $(z - z_0)^{-1}$ by γ-polynomials **[8]**, so we would have a locally convergent sequence of γ-polynomials which clearly does not converge globally. If, for some r, $\mathfrak{R} \cap (|z| \geqq r)$ lies in an angle of opening less than π, $\mathfrak{R}$ is not regular **[5]**. The requirement that $\mathfrak{R}$ does not contain a closed curve surrounding the unit disc may well be the only restriction necessary for Korevaar's dichotomy to hold. If $\mathfrak{R}$ is a *radial set* ($z_0 \in \mathfrak{R}$ implies $\lambda z_0 \in \mathfrak{R}$ for all $\lambda > 0$) then the dichotomy follows from the characterization of the global limits obtained by J. E. Lange and J. Korevaar **[9]**, **[10]**.

We will be able to prove Korevaar's conjecture for a rather large class of regions with smooth boundaries, our basic tool being the physical idea of replacing point masses in $\mathfrak{R}$ by a continuous spread of mass along the boundary of a domain containing $\mathfrak{R}$. Theorem 9.1 is the formal statement of our principal result. Because our proof is rather complicated in technical detail, it seems worthwhile to provide a brief outline. This we do in the next section, in which we also collect some preliminary definitions and lemmas.

Let us note next that a regular set does not have the local-to-global property. To see this, we notice that for $|z| < 1$, $1/(z-1)$ has an expansion

$$1/(z-1) = \prod_{j=1}^{\infty} \exp\{A_j z^j\}.$$

Assuming that $\mathfrak{R}$ is regular, there is a sequence $p_{n,m}$ of $\mathfrak{R}$-polynomials converging globally to

$$\prod_{j=1}^{m} \exp\{A_j z^j\}.$$

An appropriate sequence $p_{n_m, m}$ converges locally to $1/(z-1)$. Thus Korevaar's alternative, in one direction, is easily established.

I wish to express my gratitude to Professor J. Korevaar, who has given much valuable advice both at early stages of this work and during the final preparation of the manuscript. Thanks are also due Professor S. E. Warschawski for a helpful discussion.

2. **Outline of proof. Preparatory lemmas and definitions.** If $\mathfrak{S}$ is any set, $\mathfrak{S}^k$ (k an integer) has the obvious meaning: the set filled out by z^k as z ranges throughout $\mathfrak{S}$. $\mathfrak{C}\mathfrak{S}$ and $\partial\mathfrak{S}$ denote, respectively, the complement and the boundary of $\mathfrak{S}$. $\rho(\mathfrak{S}_1, \mathfrak{S}_2)$ denotes the distance between the sets $\mathfrak{S}_1$ and $\mathfrak{S}_2$.

The ray $\operatorname{Arg} z = \alpha$ is called an *asymptotic direction of* $\mathfrak{R}$ if there is a sequence $\{z_n\} \subset \mathfrak{R}$ with $|z_n| \to \infty$ and $\operatorname{Arg} z_n \to \alpha$. $\mathfrak{A}(\mathfrak{R})$ denotes the set of all asymptotic directions of $\mathfrak{R}$.

The angular index, $I(\mathfrak{S})$, of a set $\mathfrak{S}$ is the least positive integer k such that $\mathfrak{A}(\mathfrak{S}^k)$ lies in an angle with vertex at the origin and opening strictly less than π. Of course, $I(\mathfrak{S})$ may not exist.

We say $\mathfrak{R}$ is k-ample if, for any real α, $\varepsilon > 0$, $\lambda > 0$, there exists a sequence $p_n(z) = \prod_j (1 - z/z_{j,n}) = \exp\{\sum_m (-(1/m)\sum_j z_{j,n}^{-m}) z^m$ of $\mathfrak{R}$-polynomials such that

(2.1) $$\sum_j z_{j,n}^{-k} \to \lambda e^{i\alpha},$$

(2.1′) $$\left|\sum_j z_{j,n}^{-(k+\nu)}\right| < \varepsilon, \qquad \nu \geqq 1,$$

(2.1″) $$\left|\sum_j z_{j,n}^{-(h+\nu)}\right| < \varepsilon^\nu, \qquad \nu \geqq 1,$$

for some h depending only on $\mathfrak{R}$.

To motivate the definition of k-ample, let us notice that $\mathfrak{R}$ is regular if we can approximate globally to $\exp\{A_k z^k\}$ for all positive integers k and any corresponding choice of A_k [5]. If in our definition we take $\lambda e^{i\alpha} = -kA_k$, we see that the sequence p_n in that definition nearly forms an approximating sequence for $\exp\{A_k z^k\}$ the difficulty being that we may have no control over the numbers $\sum_j z_{j,n}^{-m}$ for m less than k. If $\mathfrak{R}$ is also m-ample for $m = 1, 2, \cdots, (k-1)$ we can systematically subtract off the logarithmic coefficients $\sum_j z_{j,n}^{-m}$ for m less than k, hence approximate globally to $\exp\{A_k z^k\}$. This proves the first clause in the following lemma, which summarizes some auxiliary results in work of J. Korevaar [5] and [6].

LEMMA 2.1. (i) *$\mathfrak{R}$ is regular if $\mathfrak{R}$ is k-ample for every positive integer k.*

(ii) *$\mathfrak{R}$ is k-ample if* (2.1), (2.1′) *and* (2.1″) *hold for arbitrary $\varepsilon > 0$, $\lambda > 0$ and three or four values of $e^{i\alpha}$ not contained in an arc of the unit circle of length π.*

(iii) *If $\beta \in \mathfrak{A}(\mathfrak{R}^k)$, then* (2.1), (2.1′) *and* (2.1″) *hold for arbitrary $\varepsilon > 0$, $\lambda > 0$ with $\alpha = -\beta$.*

We introduce the usual counting function $N_n(r)$ for a sequence $p_n(z)$ of polynomials:

$$N_n(r) = \{\text{number of zeros of } p_n(z) \text{ in } |z| \leqq r\}.$$

A sequence $p_n(z)$ is called *thin* if $N_n(r)$ is, for any fixed r, bounded independently of n. (In the sequel, there will never be any ambiguity about which sequence of polynomials $N_n(r)$ refers to.) A set $\mathfrak{S}$ will be called *thin* if any locally convergent sequence of $\mathfrak{S}$-polynomials is thin. As an example, it is rather easy to see by a geometric argument that the half plane $\operatorname{Re} e^{i\alpha} z \geqq 0$ is thin.

In §3, we shall see that a region $\mathfrak{R}$ has the local-to-global property if and only if $\mathfrak{R}$ is thin. The central problem then reduces to showing that, given the positive integer k, if $\mathfrak{R}$ is not thin then $\mathfrak{R}$ is k-ample.

Briefly, our method is as follows. We embed $\mathfrak{R}$ in a simply-connected domain $\mathfrak{D}^-$ of a special type, (a so-called sector-like domain, defined below) with $\mathfrak{D}^+$ the complementary open set. Each component of $\mathfrak{D}^+$ is to intersect the unit disc, one (unbounded) component containing the origin. If $p_n(z)$ is a sequence of $\mathfrak{R}$-polynomials, we have representations

$$-p_n'(z)/p_n(z) = \int_{\partial\mathfrak{D}^-} \frac{U_n(\tilde{z})\,d\tilde{z}}{\tilde{z}-z}, \qquad z \in \mathfrak{D}^+$$

where $U_n(\tilde{z})$ is constructed from Green's functions, and $U_n(\tilde{z})\,d\tilde{z}$ is positive real along $\partial\mathfrak{D}^-$. Studying certain normalized distributions $V_n(\tilde{z})$ of charge, we shall be led to a limiting distribution $V(\tilde{z})$ which is an equilibrium distribution in the sense that for some positive integer J,

$$\int_{\partial\mathfrak{D}^-} \frac{V(\tilde{z})\,d\tilde{z}}{(\tilde{z}-z)^J} = 0, \qquad z \in \mathfrak{D}^+.$$

To determine the possible $V(\tilde{z})$, we first exhibit $V(\tilde{z})$ as the boundary value of $\Lambda'(z)$, where $\Lambda(z)$ is regular in $\mathfrak{D}^-$ and maps each component of $\partial\mathfrak{D}^-$ onto a subset of a line parallel to the real axis. To further determine the possible $V(\tilde{z})$, we need to go through some calculations with Cauchy integrals and with conformal mappings. These calculations are the bulk of §§5 and 6. Lemma 8.2 is the key lemma which sums up our conclusions concerning the behavior of $V(\tilde{z})$.

To prove that $\mathfrak{R}$ is k-ample, we will show that (2.1), (2.1′) and (2.1″) hold for four values of $e^{i\alpha}$ not contained in an arc of length π. In the nontrivial cases, we shall be assured the existence of three distinct points $e^{i\alpha}$ which may, however, belong to an arc of length π. We then make an appropriate choice of sector-like domain $\mathfrak{D}^-$ in which to embed $\mathfrak{R}$, and using the properties of $V(\tilde{z})$ together with the fact that

$$\int_{\partial\mathfrak{D}^-} \frac{V_n(\tilde{z})}{\tilde{z}^k} d\tilde{z} \to 0,$$

we produce distributions of charge on $\partial\mathfrak{D}^-$ which satisfy integral analogues of (2.1), (2.1′) and (2.1″) with an appropriate fourth value of $e^{i\alpha}$. Assuming $\mathfrak{D}^-$ fits $\mathfrak{R}$ rather closely we can replace the continuous spreads of charge by distributions of point charges along $\partial\mathfrak{D}^-$, then move these point charges onto $\partial\mathfrak{R}$. The details vary according to the nature of $\mathfrak{A}(\mathfrak{R}^k)$, so that our proof of the main theorem, carried out in §8, has to be split into several cases. Rather strong geometric conditions must be placed on $\mathfrak{R}$ to ensure the possibility of properly embedding $\mathfrak{R}$ in an appropriate sector-like domain.

§7 contains an unrelated result obtained incidentally by our methods. Theorem 8.1 we believe is also of independent interest; we obtain here a sufficient condition for thinness which generalizes a theorem of Ganelius **[2]**. §9 contains the formal statement of the main theorem, as well as some concluding comments and open questions.

It is convenient now to describe more fully the idea of a sector-like domain. The definition involves a certain type of curve, occurring so frequently in the sequel that it is convenient to have a special name for such curves. Let us say that the curve

$$\gamma: z = z(s), \qquad 0 \leqq s < \infty$$

where s is arc-length, is a *standard arc* if

(i) $z(s)$ is twice continuously differentiable,

(ii) $z(s)$ is analytic except for finitely many values of s,

(iii) as s tends to infinity, Arg z tends to a limit,

(iv) as z tends to infinity along γ, the curvature $K(z)$ satisfies the condition

$$K(z) = O(|z|^{-(1+\varepsilon)})$$

for some positive ε.

We shall say $\mathfrak{D}^-$ is a *sector-like domain* it $\mathfrak{D}^-$ is a simply-connected domain bounded by finitely many curves, and each semi-infinite arc of $\partial\mathfrak{D}^-$ is a standard arc.

The sector-like domain $\mathfrak{D}^-$ is a *standard* sector-like domain if each component of $\mathfrak{C}\mathfrak{D}^-$ intersects the open unit disc, one component having the origin as interior point, and if $\operatorname{Arg} z$ tends to distinct limits (mod 2π) as z tends to infinity along distinct standard arcs of $\partial\mathfrak{D}^-$

The intersection of a sector-like domain with the exterior of a large circle consists of finitely many disjoint sets, which we call the *sectors of* $\mathfrak{D}^-$. In particular, a phrase such as "z tends to infinity in the sector $\mathfrak{S}_j$ of $\mathfrak{D}^-$" has an unambiguous meaning. We regard the sectors $\mathfrak{S}_j$ of $\mathfrak{D}^-$ as being numbered in counterclockwise order. When we speak of $\mathfrak{D}^-$, we will understand $\mathfrak{D}^+$ to be the complementary open set. We shall from now on use Γ to denote the boundary of $\mathfrak{D}^-$, Γ understood to have the positive orientation with respect to $\mathfrak{D}^-$. We shall commonly use $\tilde{z}$ rather than z for points on Γ.

We close this section with the following simple lemma:

LEMMA 2.2. *If Γ is the boundary of a sector-like domain, Γ has a finite angular index.*

PROOF. The intersection of $\mathfrak{A}(\Gamma)$ with the unit circumference is a finite set. The lemma follows immediately from a well-known result of number theory ([**3**], Theorem 201).

3. **Thin regions.** We start with two definitions.

DEFINITION 3.1. $\mathfrak{R}$ is *locally regular* if every function regular and zero-free in the open unit disc is the local limit of a sequence of $\mathfrak{R}$-polynomials. (In the terminology of other writers, $\mathfrak{R}$ is a polynomial approximation set for the unit disc.)

DEFINITION 3.2. The *order* ρ of a sequence $\{p_n(z)\}$ of polynomials is

$$\rho = \text{g.l.b.}\left\{s \,\middle|\, \limsup_{n\to\infty} \sum_j |z_{j,n}|^{-s} < \infty\right\},$$

$z_{j,n}$ being the zeros of $p_n(z)$.

The first result of this section is the analogue for local convergence of a theorem obtained by Korevaar [**6**] in the context of global convergence. We state it as a lemma rather than a theorem because, as we shall see shortly, it is true in a somewhat vacuous way.

LEMMA 3.1. *Let $\mathfrak{R}$ be thin. Then if there are locally convergent sequences of $\mathfrak{R}$-polynomials which are of infinite order, or of arbitrarily large finite order, $\mathfrak{R}$ is locally regular.*

PROOF. The analogue of our lemma is the case of arbitrarily large τ in ([**6**], Theorem 5.1). The proof of our lemma consists of inspection of Korevaar's proof, and making appropriate modifications to fit the context of local convergence. We therefore just briefly indicate the small changes needed. The reader may wish to consult [6] in order to follow our proof.

We form an analogue of ([6], Lemma 3.1) simply by reinterpreting convergence to mean local convergence. We find that we have a correct proof of our analogous lemma if, throughout the proof of ([6], Lemma 3.1) we replace convergence by local convergence, global limits by local limits. Lemmas 3.2, 3.3, and 3.5 of [6] concern relations between the order of polynomials sequences and properties of their sets of directions of s-abundance. The analogous lemmas, formed by reinterpreting the meaning of convergence, are easily proved by noticing exactly where the convergence is used in [6], and convincing ourselves that local convergence plus thinness permits us to draw the same conclusions.

Lemmas 4.1 and 4.2 of [6] are geometric, not involving convergence. We form an analogue of ([6], Lemma 4.3) in the now familiar way. Because of thinness and local convergence the statements in the first paragraph in Korevaar's argument remain correct, and the rest of the proof goes through without change.

We obtain the appropriate analogues of Corollaries 3.4 and 4.4 of ([6]) in precisely the same manner as in ([6]). Now, with our new interpretation of convergence in mind, a word by word repetition of the argument of ([6], Theorem 5.1) proves our lemma.

We remarked above that Lemma 3.1 holds in a rather vacuous way. This is true because of the next result, which in essence was communicated to the author by J. Korevaar.

LEMMA 3.2. *If $\mathfrak{R}$ is locally regular, $\mathfrak{R}$ is not thin.*

PROOF. If $\mathfrak{R}$ is a subset of the closed unit disc, $\mathfrak{R}$ must contain the unit circumference; for if, e.g., the set $|z-1| < \varepsilon$ lies outside $\mathfrak{R}$ there are constants C and $\delta > 0$ such that for any $\mathfrak{R}$-polynomial $p(z)$,

$$|p'(z)/p(z)| > C \qquad (\text{degree of } p) \tag{3.1}$$

for $|z-1| < \delta$, as is easily seen by considering the force-field which is the physical representation of the left-hand side of (3.1). The unit circumference is certainly not a thin set as the example

$$p_n(z) = 1 - z^n \to 1$$

shows.

Thus we suppose $\mathfrak{R} \supset z_0$ with $|z_0| > 1$. The functions $(1 - z/z_0)^{-n}$ are regular and zero-free in $|z| < 1$, so there is for each n a sequence $p_{j,n}(z)$ of $\mathfrak{R}$-polynomials such that locally

$$p_{j,n}(z)\,(1 - z/z_0)^n \to 1.$$

We choose a sequence

$$p_n^*(z) = p_{j_n,n}(z)\,(1 - z/z_0)^n$$

of $\mathfrak{R}$-polynomials converging locally to 1. The p_n^* are clearly not thin.

THEOREM 3.1. *Let $\mathfrak{R}$ be thin. Then $\mathfrak{R}$ has the local-to-global property.*

PROOF. By Lemmas 3.1 and 3.2, any locally convergent sequence of $\mathfrak{R}$-polynomials has finite order. The global convergence follows immediately from Vitali's Theorem ([**14**], p. 168) once we establish boundedness in any fixed circle.

Let $p_n(z)$ be the locally convergent sequence, $z_{j,n}$ the zeros of p_n, and let ρ be the order of the sequence. We write $p_n(z) = q_n(z)\,r_n(z)$, where

$$q_n(z) = p_n(0) \prod_{|z_{j,n}| \leqq 2R} (1 - z/z_{j,n}).$$

The $q_n(z)$ are clearly bounded in $|z| \leqq R$ since they are of bounded degree, and $p_n(0)$ is a convergent sequence. We see also that the $r_n(z)$ are of order $\leqq \rho$.

It remains to prove that the $r_n(z)$ are bounded in $|z| \leqq R$. We have

$$\log r_n(z) = \sum_{\nu=1} A_{\nu,n} z^\nu, \qquad |z| < 2R,$$

where

$$A_{\nu,n} = (-1)^\nu/\nu \sum_{|z_{j,n}| > 2R} z_{j,n}^{-\nu}.$$

If t is the first integer exceeding ρ,

$$\begin{aligned} |A_{\nu,n}| &< 1/(2R)^{\nu-t} \sum |z_{j,n}|^{-t} \\ &< C/(2R)^{\nu-t} \end{aligned} \tag{3.2}$$

for some constant C. (3.2) gives a satisfactory bound for the $A_{\nu,n}$ with $\nu > t$. The $A_{\nu,n}$ for $\nu \leqq t$ are uniformly bounded because of the uniform boundedness of the $r_n(z)$ in any compact subset of the unit disc.

Since a globally convergent sequence of polynomials is obviously thin, we have

COROLLARY 3.1. *$\mathfrak{R}$ has the local-to-global property if and only if $\mathfrak{R}$ is thin.*

More important to us, however, is the fact that we can rephrase the central problem as follows: show that if $\mathfrak{R}$ is not thin, then $\mathfrak{R}$ is k-ample for every positive integer k.

We also have

COROLLARY 3.2. *If $\mathfrak{R}$ is thin, there is a finite upper bound for the orders of locally convergent sequences of $\mathfrak{R}$-polynomials.*

4. Representation of polynomials. Our principal tool is the idea of representing the logarithmic derivative of a polynomial in terms of the force field due to a continuous spread of charge on a curve. Let $\mathfrak{D}^-$ be a standard sector-like domain (defined in §2) with boundary Γ, $\mathfrak{D}^+$ the complementary open set. Suppose that $z_0 \in \mathfrak{D}^-$, and let z be in $\mathfrak{D}^+$. If we fix z then $(\tilde{z} - z)^{-1}$ is a regular function of $\tilde{z}$ in $\mathfrak{D}^-$, hence

$$(z_0 - z)^{-1} = -\frac{1}{2\pi}\int_\Gamma (\tilde{z} - z)^{-1}\frac{\partial G(\tilde{z}, z_0)}{\partial n_{\tilde{z}}} ds_{\tilde{z}},$$

G being Green's function for $\mathfrak{D}^-$. It is well known that $-(1/2\pi)\partial G(\tilde{z}, z_0)/\partial n_{\tilde{z}}$ is a positive kernel, total mass 1.

We define

$$U(\tilde{z}, w)d\tilde{z} = -\frac{1}{2\pi}\frac{\partial G(\tilde{z}, w)}{\partial n_{\tilde{z}}} ds_{\tilde{z}}$$

and we have then, for a $\mathfrak{D}^-$-polynomial $p_n(z)$ with zeros $z_{j,n}$,

$$\sum_j (z_{j,n} - z)^{-1} = \int_\Gamma \sum_j U(\tilde{z}, z_{j,n})\, d\tilde{z}/(\tilde{z} - z) = \int_\Gamma \frac{U_n(\tilde{z})}{\tilde{z} - z} d\tilde{z}, \quad z \in \mathfrak{D}^+.$$

This relation defines U_n, hence $U_n(\tilde{z})d\tilde{z}$ is a positive distribution of charge representing the negative of the logarithmic derivative of p_n. We shall be interested in the normalized distributions $V_n(\tilde{z})$, defined by

$$V_n(\tilde{z}) = U_n(\tilde{z}) \Big/ \int_{\Gamma_0} U_n(t)\, dt$$

where Γ_0 is a finite subarc of Γ, which we regard as chosen once and for all.

We introduce the function $w = f(z)$, which maps $\mathfrak{D}^-$ conformally onto $|w| < 1$, with Γ_0 corresponding to the right half of the unit circumference. It is convenient to let $\theta = \theta(\tilde{z})$ stand for the boundary correspondence. By our assumption on $\mathfrak{D}^-$, $\theta(\tilde{z})$ is infinitely differentiable at all but a finite number of points of Γ, and, by **15**, continuously differentiable everywhere on Γ. (Throughout this paper, $f(z)$ will have the meaning of this paragraph. We suppose that some choice of the pre-image of $w = 1$ is made so that $f(z)$ is uniquely determined.) We also introduce the conformal distance $D(\tilde{z}, \mathfrak{R})$ by $D(\tilde{z}, \mathfrak{R}) = \rho(f(\tilde{z}), f(\mathfrak{R}))$. We have

LEMMA 4.1. *Let V_n be the normalized distribution corresponding to an $\mathfrak{R}$-polynomial, $\mathfrak{R} \subset \mathfrak{D}^-$. Then except for finitely many points where Γ is not analytic we have*

$$\left|[d^{(j-1)}/d\tilde{z}^{(j-1)}]V_n(\tilde{z})\right| \leq A_j \operatorname*{Max}_{1 \leq h \leq j} \left|\theta^{(h)}(\tilde{z})\right|^j D^{-2j}(\tilde{z}, \mathfrak{R}).$$

For $j = 1,2$ the inequality holds everywhere on Γ.

PROOF. It is enough to prove

$$\left|[d^{(j-1)}/d\tilde{z}^{(j-1)}]\, U(\tilde{z}, z_0)\right| \leqq A_j \operatorname*{Max}_{1 \leqq h \leqq j} \left|\theta^{(h)}(\tilde{z})\right|^j D^{-2j}(\tilde{z}, \mathfrak{R}) \int_{\Gamma_0} U(t, z_0)\, dt$$

for then we just replace z_0 by the zeros z_ν of the polynomial, and sum on ν. Take $w_0 = f(z_0)$ in the following calculation.

Because harmonic measure is a conformal invariant, we have

$$U(\tilde{z}, z_0)\, d\tilde{z} = P(e^{i\theta}, w_0)\, d\theta,$$

P denoting Poisson's kernel, and

$$\text{(4.1)} \quad \begin{aligned} U(\tilde{z}, z_0) \Big/ \int_{\Gamma_0} U(t, z_0)\, dt &= P(e^{i\theta(\tilde{z})}, w_0)\, \theta'(\tilde{z}) \Big/ \int_{-\pi/2}^{\pi/2} P(e^{i\lambda}, w_0)\, d\lambda \\ &= \frac{\theta'(\tilde{z})}{\left|e^{i\theta(\tilde{z})} - w_0\right|^2} \Big/ \int_{-\pi/2}^{\pi/2} \frac{d\lambda}{\left|e^{i\lambda} - w_0\right|^2}. \end{aligned}$$

We note that

$$\int_{-\pi/2}^{\pi/2} \frac{d\lambda}{\left|e^{i\lambda} - w_0\right|^2} > \pi/4.$$

Now simply differentiate (4.1), which we can do off a finite set of points, and the lemma is proved.

From here on, let $p_n(z)$ be a locally convergent sequence of $\mathfrak{R}$-polynomials $\prod_j (1 - z/z_{j,n})$. The corresponding $V_n(\tilde{z})$ and their derived sequences are all locally uniformly bounded, hence all equicontinuous, and any sequence therefore contains a subsequence which converges uniformly, on compact subsets of Γ which avoid the set where Γ is not analytic, to a limit distribution $V(\tilde{z})$. We can by diagonalization arrange that all derived sequences converge. Notice that $V(\tilde{z})$ is locally integrable on all of Γ. $V(\tilde{z})$ is not the trivial function, since its integral over Γ_0 is unity.

The possible limiting distributions are of particular interest.

Corresponding to $p_n(z)$, we define

$$\text{(4.2)} \quad Q_n(z) = \int_\Gamma (V_n(\tilde{z})/(\tilde{z} - z))\, d\tilde{z} = \frac{-p_n'(z)}{p_n(z)} \Big/ \int_{\Gamma_0} U_n(t)\, dt, \qquad z \in \mathfrak{D}^+$$

and let

$$Q_{n,m}(z) = \int_\Gamma (V_n(\tilde{z})/(\tilde{z} - z)^m)\, d\tilde{z}.$$

Suppose now that for some r_0, $N_n(r_0)$ is unbounded. By deleting subscripts, we may as well suppose $N_n(r_0) \to \infty$. It follows from the relation between U and P that

$$\int_{\Gamma_0} U_n(t)\, dt > A(r_0)\, N_n(r_0)$$

for some nonvanishing function $A(r)$, so that the $Q_n(z)$ converge uniformly to zero in any compact subset of $\mathfrak{D}^+ \cap (|z| < 1)$.

Now let I be the angular index of Γ. We take Γ_1 to be a subset of Γ, consisting of all but a bounded part of Γ, and such that $\Gamma_1{}^I$ lies in an angle with vertex 0 and opening less than π. Without loss of generality we suppose this angle symmetric with respect to the positive real axis. Then for some positive A,

$$\left| \operatorname{Re} \int_{\Gamma_1} (V_n(\tilde{z})/\tilde{z}^I)\, d\tilde{z} \right| > A \int_{\Gamma_1} \left| (V_n(\tilde{z})/(\tilde{z}^I)\, d\tilde{z} \right| \tag{4.3}$$

since $V_n(\tilde{z})\, d\tilde{z}$ is positive real along Γ. By (4.2) and the fact that $\mathfrak{D}^+$ contains the origin, the numbers

$$Q_{n,I}(0) = \int_\Gamma (V_n(\tilde{z})/\tilde{z}^I)\, d\tilde{z}$$

are a bounded sequence, so we see by (4.3) that the integrals over Γ_1 of the functions $|V_n(\tilde{z})/\tilde{z}^I|$ are a bounded sequence of numbers. If $\Gamma_{1,r} = \Gamma_1 \cap (|z| \leqq r)$ we have

$$\int_{\Gamma_{1,r}} |(V(\tilde{z})/\tilde{z}^I) d\tilde{z}| = \lim_{n\to\infty} \int_{\Gamma_{1,r}} |(V_n(\tilde{z})/\tilde{z}^I)\, d\tilde{z}| \leqq \limsup_{n\to\infty} \int_{\Gamma_1} |(V_n(\tilde{z})/\tilde{z}^I)\, d\tilde{z}|$$

so that $V(\tilde{z})/\tilde{z}^I$ is absolutely integrable over Γ_1, and hence over Γ.

We can now define, for $m \geqq I$,

$$Q_m(z) = \int_\Gamma (V(\tilde{z})/(\tilde{z} - z)^m)\, d\tilde{z},$$

a function sectionally holomorphic in $\mathfrak{D}^-$ and in $\mathfrak{D}^+$.

LEMMA 4.2. *$Q_{I+1}(z)$ vanishes throughout $\mathfrak{D}^+$.*

PROOF. Fix z in $\mathfrak{D}^+$, choose $r > 2|z|$, and let A be an upper bound for the integrals over Γ of $|V_n(\tilde{z})/\tilde{z}^I|$, $|V(\tilde{z})/\tilde{z}^I|$, with respect to $|dz|$. Let Γ_r be that part of Γ lying inside $|z| = r$. We have

$$\begin{aligned} |Q_{I+1}(z) - Q_{n,I+1}(z)| &\leqq \left| \int_{\Gamma_r} \frac{V(\tilde{z}) - V_n(\tilde{z})}{(\tilde{z} - z)^{I+1}} d\tilde{z} \right| + \left| \int_{\Gamma - \Gamma_r} \frac{V(\tilde{z})}{(\tilde{z} - z)^{I+1}} d\tilde{z} \right| \\ &\quad + \left| \int_{\Gamma - \Gamma_r} \frac{V_n(\tilde{z})}{(\tilde{z} - z)^{I+1}} d\tilde{z} \right| \\ &\leqq \left| \int_\Gamma \frac{V(\tilde{z}) - V_n(\tilde{z})}{(\tilde{z} - z)^{I+1}} d\tilde{z} \right| + 2^{I+2} A/r, \end{aligned}$$

and we choose first r, then n, to make this as small as desired (a subsequence of the V_n converges to V on Γ).

Since $Q_{n,I+1}(z)$ converges to zero in $\mathfrak{D}^+ \cap (|z| < 1)$ by (4.2), $Q_{I+1}(z)$ vanishes in this set, hence throughout $\mathfrak{D}^+$ (every component of $\mathfrak{D}^+$ meets the unit disc).

5. **Analytic properties of** $V(\tilde{z})$. In this section, we show that $V(\tilde{z})$ is the boundary function of a function $V(z)$ regular in $\mathfrak{D}_-$, and obtain a growth inequality. We take Γ_1 to be the intersection of Γ with the annulus $r/2 < |z| < 2r$, r chosen so large that Γ_1 is analytic. Then

$$(5.1) \qquad Q_I(z) = \int_{\Gamma-\Gamma_1} \frac{V(\tilde{z})\,d\tilde{z}}{(\tilde{z}-z)^I} + \int_{\Gamma_1} \frac{V(\tilde{z})\,d\tilde{z}}{(\tilde{z}-z)^I} = Q_I^*(z) + Q_I^{**}(z),$$

say. We have

$$|Q_I^*(re^{i\alpha})| \leqq 2^I \int_\Gamma \frac{V(\tilde{z})\,d\tilde{z}}{|\tilde{z}^I|},$$

and in the formula for Q_I^{**} we may integrate by parts to get

$$(5.2) \quad Q_I^{**}(z) = -\sum_{j=0}^{I-2} \frac{(I-j-2)!}{(I-1)!} \frac{V^{(j)}(\tilde{z})}{(\tilde{z}-z)^{I-j-1}} \Bigg|_{\Gamma_1} + \frac{1}{(I-1)!} \int_{\Gamma_1} \frac{V^{(I-1)}(\tilde{z})\,d\tilde{z}}{\tilde{z}-z}.$$

If $|z| = r$, the sum of integrated terms has modulus not exceeding

$$(5.3) \quad \begin{aligned} &I \cdot 2^I \operatorname*{Max}_{\tilde{z} \in \Gamma_1} \operatorname*{Max}_{0 \leqq j \leqq I} |V^{(j)}(\tilde{z})| \\ &< A \operatorname*{Max}_{r/2 < |\tilde{z}| < 2r} \operatorname*{Max}_{1 \leqq j \leqq I+1} |f^{(j)}(\tilde{z})|^{I+1} \cdot D^{-2I-2}(\tilde{z}, R) = AE(I, r), \end{aligned}$$

defining the quantity $E(I,r)$. We have used Lemma 4.1 in this calculation.

It remains to estimate the integral

$$(5.4) \qquad Q_I^{***}(z) = \int_{\Gamma_1} \frac{V^{(I-1)}(\tilde{z})}{\tilde{z}-z}\,d\tilde{z}.$$

For $\tilde{z}$, $\tilde{z}_0$ on an arc of Γ_1, we integrate along the arc of Γ_1 joining them to find that

$$(5.5) \quad |V^{(I-1)}(\tilde{z}) - V^{(I-1)}(\tilde{z}_0)| = \left| \int_{\tilde{z}_0 \, \Gamma_1}^{\tilde{z}} V^{(I)}(t)\,dt \right| \leqq 2|\tilde{z}-\tilde{z}_0|\, AE(I,r),$$

at least if we assume r large enough so that

$$\left| \int_{\tilde{z}_0 \, \Gamma_1}^{\tilde{z}} dt \right| \leqq 2|\tilde{z}-\tilde{z}_0|.$$

We use the standard device of writing

$$Q_I^{***}(z) = \int_{\Gamma_1} \frac{V^{(I-1)}(\tilde{z}_0)}{\tilde{z}-z}\,d\tilde{z} + \int_{\Gamma_1} \frac{V^{(I-1)}(\tilde{z}) - V^{(I-1)}(\tilde{z}_0)}{\tilde{z}-z}\,d\tilde{z},$$

so that by (5.5)

$$(5.6)\quad |Q_I^{***}(z)| \leqq |V^{(I-1)}(\tilde{z}_0)|\,\left|\int_{\Gamma_1} d\tilde{z}/(\tilde{z}-z)\right| + 2AE(I,r)\int_{\Gamma_1} |(\tilde{z}-z_0)/(\tilde{z}-z)|\,|d\tilde{z}|.$$

Now set $|z| = |\tilde{z}_0| = r$ in (5.6). The first integral appearing in (5.6) is uniformly bounded, while the integrand of the second integral is uniformly bounded, even as z tends to $\tilde{z}_0$, and we have obtained earlier the bound $AE(I,r)$ for $V^{(I-1)}(\tilde{z}_0)$. Collecting the bounds for Q_I^*, Q_I^{**}, and Q_I^{***} we have proved the inequality

$$(5.7)\qquad Q_I(re^{i\alpha}) = O(rE(I,r)).$$

We now give a key lemma.

LEMMA 5.1. *$V(z) = \Lambda'(z)$, where $\Lambda(z)$ is regular in $\mathfrak{D}^-$ and continuous in the closure. $\Lambda(z)$ maps each component of Γ onto a (possibly unbounded) line segment parallel to the real axis.*

We have $\Lambda(z) = AR(z)$, A a constant, $R(z)$ an $(I+1)$-fold antiderivative of $Q_{I+1}(z)$. It follows from (5.7) that

$$(5.8)\qquad \Lambda^{(j)}(z) = O(|z|^{I+1-j}E(I,|z|)) \qquad \text{for } 0 \leqq j \leqq I.$$

We point out that $V(z)$ thus far was defined only on Γ, and that (5.8) gives information about the growth of $V(z)$ throughout $\mathfrak{D}^-$.

PROOF OF THE LEMMA. Let $Q_I(\tilde{z}_0{}^-)$ and $Q_I(\tilde{z}_0{}^+)$ represent the limits of $Q_I(z)$ as z tends to $\tilde{z}_0 \in \Gamma$ from within $\mathfrak{D}^-$ and from within $\mathfrak{D}^+$ respectively. We suppose Γ is analytic at $\tilde{z}_0$. By (5.1) and (5.2)

$$(5.9)\qquad Q_I(\tilde{z}_0{}^-) - Q_I(\tilde{z}_0{}^+) = (1/(I-1)!)\{Q_I^{***}(\tilde{z}_0^-) - Q_I^{***}(\tilde{z}_0{}^+)\},$$

Q_I^{***} defined by (5.4), provided these latter limits exist. By a theorem of Plemelj ([**11**], p. 42) the right-hand side of (5.9) exists, and has the value $\pi i\, V^{(I-1)}(\tilde{z}_0)/(I-1)!$.

By Lemma 4.2 $Q_I(z)$ is constant in each component of $\mathfrak{D}^+$, hence on each component of Γ,

$$V^{(I)}(z) = ((I-1)!/\pi i)\,Q_I(\tilde{z}^-) + \text{Constant}.$$

We can use this identity to define $V^{(I-1)}(z)$ as a function regular in $\mathfrak{D}^-$, and integrating we have $V(z) = \Lambda'(z)$.

Since $V(\tilde{z})$ is locally integrable, $\Lambda(z)$ is continuous in the closure of $\mathfrak{D}^-$.

Recall that $V(\tilde{z})\,d\tilde{z}$ is a positive real differential along Γ. Thus if $\tilde{z}_1$, $\tilde{z}_2$ belong to a single component of Γ, we have

$$\Lambda(\tilde{z}_2) - \Lambda(\tilde{z}_1) = \int_{\tilde{z}_1 \atop \Gamma}^{\tilde{z}_2} V(\tilde{z})\, d\tilde{z},$$

proving that the components of Γ are mapped as described. This completes the proof of Lemma 5.1.

Now let $\tau = g(z)$ be a conformal mapping of $\mathfrak{D}^-$ onto the upper half plane, such that as z tends to infinity in any sector $\mathfrak{S}_j$ of $\mathfrak{D}^-$, $g(z)$ tends to a finite real point τ_j. (This meaning of $g(z)$ and τ_j will be permanent.) Define

$$\Omega(\tau) = \Lambda(g^{-1}(\tau)).$$

LEMMA 5.2. *$\Omega'(\tau)$ is a single-valued function, bounded at infinity, having singularities only at the points τ_j.*

PROOF. We employ the technique used, for instance, in [**12**] to study Schwarz-Christoffel transformations. Note that the counterclockwise ordering of the $\mathfrak{S}_j$ induces the cyclic ordering of the τ_j on the closed real axis.

$\Omega(\tau)$ maps each segment (τ_{j-1}, τ_j) onto a subset of a line parallel to the real axis, hence can be continued across by the Reflection Principle, so the τ_j are the only singularities.

To obtain the other possible determinations of $\Omega(\tau)$ in the upper half plane we reflect back through (τ_{k-1}, τ_k), getting a new value $\Omega_1(\tau)$, say. We see that

$$\Omega_1(\tau) - \Omega(\tau) = iC, \qquad C \text{ real},$$

hence the derivative is single-valued.

We shall need another property of $\Omega(\tau)$ which requires no proof, but is crucial enough to state as

LEMMA 5.3. *$\Omega(\tau)$ is direction-preserving. That is, as τ describes (τ_{j-1}, τ_j) from left to right, $\Omega(\tau)$ describes the image from left to right.*

The importance of $\Omega(\tau)$ follows from the obvious relation

$$V(z) = \Omega'(\tau) g'(z). \tag{5.10}$$

We shall need precise asymptotic information about $V(z)$, and this requires knowledge about the conformal mapping of $\mathfrak{D}^-$ onto the unit disc. This is the object of the next section.

6. **Conformal mapping of $\mathfrak{D}^-$.** We shall need a theorem of O. D. Kellogg [**4**] which we state as Lemma 6.1. (See [**15**] for a short proof of a more general result.)

LEMMA 6.1. *Let $\mathfrak{U}$ be a domain bounded by a closed Jordan curve μ in the w-plane. Let $\phi(s)$ be the tangent angle to μ, s being arc length. We suppose $\phi(s)$ exists and satisfies*

$$|\phi(s) - \phi(s_0)| \leqq K|s - s_0|^\alpha, \qquad \alpha > 0$$

with K and α independent of s and s_0.

Then if $w = h(z)$ is the conformal mapping from $|z| < 1$ onto $\mathfrak{U}$,

(i) $h'(z_0) = \lim_{z \to z_0} (h(z) - h(z_0))/(z - z_0)$

exists for unrestricted approach in $|z| \leqq 1$, $|z_0| \leqq 1$,

(ii) *$h'(z)$ is continuous in $|z| \leqq 1$, and*

(iii) *$h'(z)$ does not vanish in $|z| \leqq 1$.*

If $z = H(w)$ is the inverse mapping, we see easily that $H'(w)$ has the same properties in the closure of $\mathfrak{U}$, ($H'(w)$ defined in the obvious way on μ).

LEMMA 6.2. *Suppose that, in addition to satisfying the requirements of Lemma* 6.1., *μ is piecewise analytic. Then*

(i) $$A_1|H(w) - H(w_0)| \leqq |w - w_0| \leqq A_2|H(w) - H(w_0)|$$

for all w, w_0 belonging to $\mathfrak{U}$, A_1 and A_2 being positive constants, and

(ii) *if w_0 is a point on μ where μ is analytic, the derivatives of H satisfy*

$$|H^{(j)}(w_0)| \leqq A[L(w_0)]^{-j},$$

A a constant determined by μ, $L(w_0)$ the distance from w_0 to the nearest point of nonanalyticity of μ.

PROOF. μ is a fortiori a rectifiable curve with a continuously turning tangent line, and can be continuously parametrized in terms of arc length, thus by a routine continuity argument we prove that there is a positive number η_0 such that, if C is any disc with center on μ, and radius less than η_0, $\mu \cap C$ is a connected subarc of μ.

Next, let $d(w_1, w_2)$ be the length of the shortest arc in $\mathfrak{U}$ which connects w_1 to w_2. We shall show the existence of an upper bound η_1 for the quotient $d(w_1, w_2)/|w_1 - w_2|$. This is also basically a topological result, apparently more difficult than the existence of η_0. We give a simple direct verification using the mapping function $h(z)$ of Lemma 6.1. We have

$$d(w_1, w_2) = \int_\Delta |h'(z)|\,|dz|,$$

where Δ is some curve connecting the pre-images z_1 and z_2 in $|z| \leqq 1$. Since z_1 and z_2 can be connected by a line segment we get

$$d(w_1, w_2) \leqq M|z_1 - z_2|, \tag{6.1}$$

where M is the maximum of $|h'(z)|$ in the closed unit disc. Now let m be the minimum of $|h'(z)|$. By uniform continuity we can find a $\delta > 0$ such that

$|h'(z) - h'(\zeta)| < m/2$ if z, ζ belong to the closed unit disc and $|z - \zeta| < \delta$. If $|H(w_2) - H(w_1)| < \delta$ we have (understanding the path of integration to be a line segment)

$$\begin{aligned} |w_2 - w_1| &= \left| \int_{z_1}^{z_2} h'(z)\,dz \right| \\ &= \left| \int_{z_1}^{z_2} h'(z_1)\,dz + \int_{z_1}^{z_2} (h'(z) - h'(z_1))\,dz \right| \qquad (6.2) \\ &\geqq |z_2 - z_1| \left(|h'(z_1)| - m/2 \right) \geqq (m/2)|z_2 - z_1|. \end{aligned}$$

By uniform continuity of $H(w)$, then, we compare (6.1) and (6.2) to see that there is a $\sigma > 0$ such that

$$d(w_1, w_2)/|w_2 - w_1'| < 2M/m,$$

if $|w_2 - w_1| < \sigma$. If $|w_2 - w_1| \geqq \sigma$, then

$$d(w_1, w_2)/|w_2 - w_1| \leqq K/\sigma,$$

where

$$K = \underset{w,\omega}{\text{l.u.b.}}\; d(w, \omega).$$

Now we prove (i). Letting γ_z be the line segment connecting z to z_0 in the closed unit disc we have

$$\begin{aligned} \frac{|w - w_0|}{|H(w) - H(w_0)|} &= \frac{|h(z) - h(z_0)|}{|z - z_0|} \\ &= \frac{1}{|z - z_0|} \left| \int_{\gamma_z} h'(z)\,dz \right| < M, \end{aligned}$$

proving the second part of the inequality (i). Letting γ_w be a path of minimal length connecting w_0 to w in U, we have

$$\frac{|H(w) - H(w_0)|}{|w - w_0|} = \frac{1}{|w - w_0|} \left| \int_{\gamma_w} H'(w)\,dw \right| < \eta_1/m,$$

completing the proof of (i).

To prove (ii) we shall use an argument based on analytic continuation. We take $R(w_0)$ to be the smaller of the numbers η_0, $L(w_0)$, so that the disc with center at w_0 and radius $R(w_0)$ intersects μ in a single arc χ. Let β be that part of the circumference lying inside $\mathfrak{U}$, and let δ be the image of β under the mapping $z = H(w)$, and γ the image of χ. E_w and E_z, as indicated on the figure, are the regions bounded by χ, β and by γ, δ respectively. We assume in what follows that δ is a proper subset of the unit circumference.

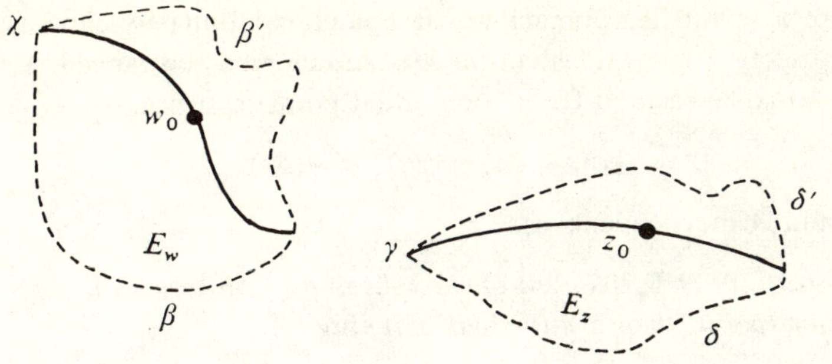

If we take θ to be the argument of z, then χ is parametrized analytically by

$$(6.3) \qquad w - w_0 = h(z) - z_0 = h^*(\theta),$$

say, where we extend θ by the definition

$$(6.4) \qquad \theta = -i \log z.$$

(6.3) then gives a univalent mapping of E_w onto a domain E_θ, χ corresponding to a line segment. E_θ is also made to correspond univalently to E_z. Now if δ' is formed by inversion of δ through the arc γ, we see by the Inversion Principle ([**12**], p. 184) that the mapping from E_z to E_θ extends to a function analytic and univalent in the region bounded by δ and δ'. Since θ, as defined by (6.4), is also univalent in this region, we see that $h(z)$ is univalent there.

We let β' be the image of δ' under $h(z)$. By the second part of inequality (i), $\rho(z_0, \delta) \geqq R(w_0)/A_2$, and the inverse arc must therefore satisfy $\rho(z_0, \delta') \geqq R(w_0)/A_2$, for when we invert points inside the unit disc with respect to the unit circumference we increase the distance from any point on the circumference ([**14**], p. 191). $h(z)$ therefore is univalent in the disc with center z_0 and radius $R(w_0)/A_2 = r(z_0)$, say. We may write

$$h(z) = w_0 + (r(z_0)h'(z_0))\, k(u),$$

in this disc, with $u = (z - z_0)/r(z_0)$. $k(u)$ is a normalized univalent mapping of the unit disc. By the One-Quarter Theorem ([**14**], p. 209), $|k(u)| > 1/4$ on $|u| = 1$. Therefore $H(w)$ is analytic in a circle with center w_0 and radius $\frac{1}{4}r(z_0)\,h'(z_0) \geqq mR(w_0)/4\,A_2$.

Describing a circle of radius $mR(w_0)/4A_2$ about each point w_0 of analyticity of μ, we obtain a domain $\mathfrak{U}'$ in which $H(w)$ is analytic. $H(w)$ is uniformly bounded away from zero in any subdomain $|w - h(0)| > C$. If we suppose, without loss of generality, that $R(w_0) < C/2$, we have a uniform bound B for $H(w)$ in $\mathfrak{U}'$. Now we complete the proof by using Cauchy's integral representation of the derivatives. ($R(w_0)$ and $L(w_0)$ coincide when $L(w_0)$ is small.)

Our idea for mapping $\mathfrak{D}^-$ onto the unit disc is of course to compare with the mapping of a sector-like domain whose boundary ultimately consists of straight lines. Let us say a curve Γ is *uniformly Lipschitz* if the tangent angle $\phi(s)$, (s being arc length measured from some intial point) satisfies

$$|\phi(s+\Delta s)-\phi(s)| < A(\Delta s)^{\alpha},$$

$\alpha > 0$, α and A independent of s, Δs.

LEMMA 6.3. (i) *If Γ_z is a uniformly Lipschitz finite arc in the z-plane, the origin an endpoint, then a univalent mapping*

$$w = h(z) = z^p + a_1 z^{2p} + \cdots, \qquad p > 1,$$

sends Γ_z onto a uniformly Lipschitz arc Γ_w.

(ii) *Let Γ_z be a finite arc issuing from the origin, t the arc-length measured from the origin. Let Γ_z have a radius of curvature $K(t)$ satisfying*

$$|K(t)| < A/t^{\lambda}, \qquad t \neq 0, \qquad \lambda < 1,$$

λ, A independent of t. Then Γ_z is uniformly Lipschitz.

(iii) *Let Γ_1 be a semi-infinite arc in the ζ-plane, with curvature K_1 satisfying*

$$K_1 = O(|\zeta|^{-(1+\varepsilon)}), \qquad \varepsilon \to 0.$$

Then the mapping $z = \zeta^{-1/2}$ sends Γ_1 onto a curve Γ with curvature K satisfying

$$|K(s)| < A/s^{\lambda}, \qquad s \neq 0$$

for some $\lambda < 1$, A and λ independent of s.

PROOF. (i) Rotate Γ_z so that it is tangent to the real axis at the origin, and let $z(t) = x(t) + iy(t)$, t being arc length measured from the origin, $\phi(t)$ the tangent angle. Let s be arc length of Γ_w measured from the origin, $\psi(t)$ its tangent angle. We have

$$\psi = \operatorname{Arg} dw/dt = \operatorname{Arg} dw/dz + \operatorname{Arg} dz/dt.$$

We find that an increment Δt produces an increment

$$\Delta\psi = (p-1)\Delta \operatorname{Arg} z + \Delta\phi(t) + \Delta \operatorname{Im}\log(1 + a_{n+1}z + \cdots)$$

and if A, α, are the Lipschitz numbers for Γ_z we have

$$|\Delta\psi| \leqq (p-1)|\Delta \operatorname{Arg} z| + A(\Delta t)^{\alpha} + A_1\Delta t.$$

To estimate the increment in $\operatorname{Arg} z$ we use the positioning of Γ_z to write

$$|d/dt(\operatorname{Arg} z)| = |\operatorname{Im}(z'(t)/z(t))| \sim |y'(t)/x(t)| \sim |y'(t)/t| < At^{\alpha-1}$$

hence

$$|\Delta \operatorname{Arg} z| < (A/\alpha)[(t_0 + \Delta t)^\alpha - t_0^\alpha] \leqq A/[\alpha \Delta t]^\alpha.$$

Thus $\Delta\psi/(\Delta t)^\alpha$ is uniformly bounded, and the increment in arc length s along Γ_w is

$$\Delta s = \int_{t_0}^{t_1} |h'(z)| dt \sim (t_0 + \Delta t)^p - t_0^p \geqq (\Delta t)^p$$

so that $\Delta\psi/(\Delta s)^{\alpha/p}$ is uniformly bounded.

(ii) follows by integrating the inequality for $K(t)$.

For (iii) let Γ be the image of Γ_1 in the form $z = z(t)$. The formula for curvature may be written

$$K = \operatorname{Im} \overline{z'(t)}\, z''(t)/|z'(t)|^3.$$

Writing $z'(t) = (dz/d\zeta)(d\zeta/dt)$, etc., and letting the parameter t be arc-length along Γ_1, we find after a short calculation that

$$\begin{aligned} K &= |\zeta|^{3/2}\{2K_1 - 3\operatorname{Im}(\zeta'(t)/\zeta)\} \\ &= 2|\zeta|^{3/2}K_1 - 2|\zeta|^{1/2} \sin\Theta(\zeta'(t),\zeta), \end{aligned} \tag{6.5}$$

$\Theta(\zeta',\zeta)$ being the angle between ζ' and ζ.

Without loss of generality, suppose the arguments of ζ and ζ' tend to zero as $|\zeta|$ tends to infinity, and let $\zeta = x + iy$. Then

$$|1 - \zeta'(t_0)| = \left| -\int_{t_0}^{\infty} \zeta''(t)\, dt \right| = O(t_0^{-\varepsilon})$$

and

$$\operatorname{Arg} \zeta(t_0) \sim y(t_0)/x(t_0) = O(t_0^{1-\varepsilon})/t_0$$

so (6.5) now becomes

$$K = (|\zeta|^{1/2-\varepsilon}) = O(s^{-(1-2\varepsilon)})$$

as arc length s tends to zero along Γ. This completes the proof of Lemma 6.3.

Now we can describe the mapping $f(z)$ of a sector-like domain $\mathfrak{D}^-$ onto the unit disc. ($f(z)$ is the function of §4.) We approximate $\mathfrak{D}^-$ by a sector-like domain $\mathfrak{D}^*$ whose boundary consists for large z of rays. Let $w = f^*(z)$ map $\mathfrak{D}^*$ conformally onto the unit disc. Let $\mathfrak{S}_j^*$ be a sector of $\mathfrak{D}^*$ of opening π/α_j, with $f^*(z) \to w_j^*$ as z tends to infinity in $\mathfrak{S}_j^*$. By applying the Inversion Principle ([**12**], p. 184) we find that the function $F^*(\zeta)$ defined by

$$F^*(\zeta) = f^*(\zeta^{1/\alpha_j}), \qquad z = \zeta^{1/\alpha_j}$$

is regular at infinity, hence

$$f^* = w_j^* + a_1 z^{-\alpha_j} + a_2 z^{-2\alpha_j} + \cdots$$

as z tends to infinity in $\mathfrak{S}_j^*$. The element of linear magnification at $\tilde{z}$ is $|f^{*\prime}(\tilde{z})|$, and since $f^{*\prime}$ changes very little in a circle of radius $\rho(\tilde{z}, \mathfrak{R})$ about $\tilde{z}$, we have

$$\rho(f^*(\tilde{z}), f^*(\mathfrak{R})) \geqq A|z|^{-(\alpha_j+1)}\rho(\tilde{z}, \mathfrak{R}).$$

Now $\Gamma = \partial\mathfrak{D}^-$ itself maps into a curve which, as we see by Lemma 6.3 and part (iii) of the definition of a sector-like domain, satisfies the hypotheses of Lemma 6.2. Thus

$$D(\tilde{z}, \mathfrak{R}) \geqq A|\tilde{z}|^{-(\alpha_j+1)}\rho(\tilde{z}, \mathfrak{R}) \tag{6.6}$$

as z tends to infinity in $\mathfrak{S}_j$, and

$$f'(z) \sim B\, z^{-(\alpha_j+1)} \tag{6.7}$$

for some constant B_j, $B_j \neq 0$. From the series for f^* we see that

$$f^*(z) - w_j^* \sim a_1 z^{-\alpha_j} \tag{6.8}$$

so for the quantity $L(w)$ referred to in Lemma 6.2 we have

$$L(f^*(z)) \sim a_1|z|^{-\alpha_j},$$

hence by the second part of Lemma 6.2

$$|f^{(j)}(z)| \leqq B_j'|z|^{j\alpha_j}. \tag{6.9}$$

Now the mapping $\tau = g(z)$ of $\mathfrak{D}^-$ onto the upper half plane is effected by following $f(z)$ by a bilinear transformation, which must have its singularity away from the $H(w_j^*)$, since the points $H(w_j^*)$ are carried onto the points τ_j. From (6.7) and (6.8) we read off the asymptotic relations

$$g'(z) \sim A_j z^{-(\alpha j+1)}, \qquad A_j \neq 0, \qquad z \in \mathfrak{S}_j \tag{6.10}$$

$$\tau - \tau_j \sim A_j' z^{-\alpha_j}, \qquad A_j' \neq 0, \qquad z \in \mathfrak{S}_j. \tag{6.11}$$

Now we compute bounds for $\Omega(\tau)$, (Ω being the function introduced in §5). Combining (6.10) and (6.11) with (5.8) we have, taking $j = 0$ in (5.8), that

$$\Omega(\tau) = O(|\tau - \tau_j|^{-(I+1)\alpha j} E(I, |z|)). \tag{6.12}$$

Now recall the definition (5.3) of $E(I, |z|)$ and use (6.6) and (6.11) to conclude that

$$E(I, |\tilde{z}|) = O(|\tilde{z}|^N \rho(\tilde{z}, \mathfrak{R})^{-2I-2}),$$

N some integer depending on I and α_j. Using (6.11) again, we get

$$\Omega(\tau) = O(|\tau - \tau_j|^{-M} \rho(\tilde{z}, \mathfrak{R})^{-2I-2}) \tag{6.13}$$

for some M, as z tends to infinity in $\mathfrak{S}_j$.

7. **$\mathfrak{CR}$ Containing an exponential strip.** The theorem of this section is a side issue of our investigation, but serves as a simple application of our method.

THEOREM 7.1. *Let the complement of $\mathfrak{R}$ be connected, containing a neighborhood of the origin, and a strip*

$$|y| \leqq Ae^{-x^{\lambda}}, \qquad A, \lambda, x > 0$$

about the positive real axis. Let $\lambda < 1/4$.

Then any sequence of $\mathfrak{R}$-polynomials $p_n(z)$ (with p_n of degree n) having the property that

$$\log^+ |p_n(z)| = o(n)$$

in a neighborhood of the origin also satisfies

$$N_n(r) = o(n)$$

for any fixed r.

PROOF. Enclose $\mathfrak{R}$ in a simply connected domain $\mathfrak{D}^-$ with boundary Γ having bounded curvature at all points, and coinciding for large z with the curve

$$y = \pm A/2e^{-x^{\lambda}}.$$

We suppose the complement of $\mathfrak{D}^-$ is simply connected and contains a neighborhood of the origin. We see easily that $\mathfrak{D}^-$ is a sector-like domain (with opening 2π).

Suppose for some r_0 and some subsequence n_h we have $N_{n_h}(r_0) > Bn_h$. By deleting subscripts we may suppose $N_n(r_0) > Bn$, from which it follows that

$$\int_{\Gamma_0} U_n(\tilde{z})\, d\tilde{z} > Cn,$$

C a constant depending on the geometry, and hence

$$\int_{\Gamma} V(\tilde{z})\, d\tilde{z} < 1/C. \tag{7.1}$$

Now consider the function $\Omega(\tau)$. $\Omega(\tau)$ cannot be an entire function, since otherwise being bounded at infinity it would reduce to a constant, and by (5.10) $V(\tilde{z})$ would vanish identically, which we have noted is impossible. There is then one singularity of $\Omega(\tau)$, which we may take to be at $\tau = 0$. The argument of Lemma 5.2 shows that in this case $\Omega(\tau)$ itself is single-valued. The angular index of Γ is 1, so by (6.13) we have

$$\Omega(\tau) = O(|\tau|^{-M} e^{4|z|^{\lambda}}) = O(|\tau|^{-M} \exp\{4|\tau|^{-2\lambda}\}$$

as τ tends to zero. (We use (6.11) in the last step.) By (7.1) we find that $\Omega(\tau)$ maps the real axis onto a segment of length $< 1/C$.

The function $\Omega^*(\sigma) = \Omega(1/\sigma)$ is then an entire function of order at most 2λ, mapping the real axis onto a bounded set. But if $\Omega^*(\sigma)$ is of order $< 1/2$, its minimum modulus is unbounded ([**1**], p. 39).

8. **The alternative for a class of sets.** For our work in this section it is convenient to make some preliminary definitions.

We define $\mathfrak{B}(\mathfrak{R})$ to be the set $\mathfrak{A}(\mathfrak{R}) \cap (|z| = 1)$.

DEFINITION 8.1. The closed set $\mathfrak{R}$ is said to be *correctly bounded* with respect to the interval $(\alpha, \beta) \subset (0, 2\pi)$ if there is a component $\mathfrak{S}$ of $\mathfrak{C}\mathfrak{R}$, containing the origin, and such that for sufficiently large r, $\mathfrak{S} \cap (|z| > r)$ is a simply connected domain bounded by an arc of $|z| = r$ and by two standard arcs, γ_α and γ_β, which $\operatorname{Arg} z$ tends respectively to α or β as z tends to infinity.

DEFINITION 8.2. The closed set $\mathfrak{R}$ is called *admissible* if $\mathfrak{B}(\mathfrak{R})$ is the nonempty union of finitely many disjoint closed intervals, and $\mathfrak{R}$ is correctly bounded with respect to each open subinterval in the complement of $\mathfrak{B}(\mathfrak{R})$.

To clarify the concept of an admissible set, we show how to construct an admissible set $\mathfrak{R}$ for which $\mathfrak{B}(\mathfrak{R})$ is a given set $\mathfrak{E}$, $\mathfrak{E}$ the union of disjoint closed intervals. Let $\mathfrak{R}^*$ be a radial set (defined in §1) with $\mathfrak{B}(\mathfrak{R}^*) = \mathfrak{E}$. Form a new set $\mathfrak{R}^{**}$, with $\mathfrak{A}(\mathfrak{R}^{**}) = \mathfrak{E}$ by distorting the boundary of $\mathfrak{R}^*$ slightly, so that each semi-infinite arc of $\partial\mathfrak{R}^{**}$ is still a standard arc, and removing $\mathfrak{R}^* \cap (|z| < 1)$. Now remove any arbitrary subset $\mathfrak{S}$ of $\mathfrak{R}^{**}$, such that $\mathfrak{S} \cap \partial\mathfrak{R}^{**}$ is empty, and such that removal of $\mathfrak{S}$ does not alter the set $\mathfrak{A}(\mathfrak{R}^{**})$. Note that we can remove practically all of $\mathfrak{R}^{**}$, just leaving the boundary and a sequence of points tending to infinity in any direction of $\mathfrak{E}$.

DEFINITION 8.3. Let M be any nonnegative integer. We say $\mathfrak{R}$ is *correctly embedded* in the sector-like domain $\mathfrak{D}^-$, *with tightness* M, if $\mathfrak{R} \subset \mathfrak{D}^-$, and if as z tends to infinity along any semi-infinite arc of Γ $(= \partial\mathfrak{D}^-)$ we have an inequality

$$A_1 |\tilde{z}|^{-M} \leqq \rho(\tilde{z}, \mathfrak{R}) \leqq A_2 |\tilde{z}|^{-M} \tag{8.1}$$

with A_1, A_2 nonzero constants.

In situations where the value of M in (8.1) is immaterial, we shall simply say that $\mathfrak{R}$ is *correctly embedded* in $\mathfrak{D}^-$.

Throughout this section, as in §4, Γ is to be the oriented boundary of $\mathfrak{D}^-$. We shall take Γ_r to be the intersection of Γ with $|z| \leqq r$.

LEMMA 8.1. *Let $\mathfrak{R}$ be any admissible region, for which $\mathfrak{A}(\mathfrak{R})$ is not the entire plane, and let M be any nonnegative integer. Then there is a standard sector-like domain $\mathfrak{D}^-$ such that $\mathfrak{R}$ is correctly embedded in $\mathfrak{D}^-$ with tightness M.*

Furthermore, if $\mathfrak{B}_1, \mathfrak{B}_2, \cdots, \mathfrak{B}_N$ is any finite subcollection of the open intervals in $\mathfrak{C}\mathfrak{B}(\mathfrak{R})$, we can arrange that $\mathfrak{C}\mathfrak{B}(\mathfrak{D}^-) = \mathfrak{B}_1 \cup \mathfrak{B}_2 \cup \cdots \cup \mathfrak{B}_N$.

PROOF. $\mathfrak{R}$ is correctly bounded with respect to the open interval $\mathfrak{B}_1 = (\alpha, \beta)$, thus there is a component $\mathfrak{T}$ of $\complement\mathfrak{R}$, containing the origin, such that for sufficiently large r, $\mathfrak{T} \cap (|z| > r)$ is bounded by an arc of $|z| = r$ and two standard arcs γ_α and γ_β.

After appropriate rotation, γ_α has an equation $y = Y(x)$, where $Y(x)$ tends to zero as z tends to infinity. Now let $Y^*(x) = Y(x) \pm x^{-M}$, the sign chosen according to whether we want the graph of $Y^*(x)$ to lie above or below the graph of $Y(x)$. The standard Cartesian formula for curvature shows curvature is subadditive. Thus $y = Y^*(x)$ is a standard arc. Now we rotate the figure back to obtain the curve γ_α^* approximating γ_α (with tightness M).

We operate on γ_β the same way, obtaining two semi-infinite arcs γ_α^* and γ_β^*. Now we can certainly extend γ_α^* and γ_β^* so that together they bound a simply connected domain $\mathfrak{T}^* \subset \mathfrak{T}$, with $\mathfrak{T}^*$ intersecting the open unit disc. Note that the plane with $\mathfrak{T}^*$ removed is a sector-like domain.

Form $\mathfrak{D}^-$ by carrying out this construction for each $\mathfrak{B}_m$, $m = 1, 2, \cdots, N$. We can arrange that one component of $\complement\mathfrak{D}^-$ contains the origin, and that $\mathfrak{D}^-$ is simply connected, by modifying the bounding curves if necessary in the open set $\complement\mathfrak{R} \cap (|z| < 1)$.

The next lemma is of fundamental importance in the considerations of this section.

LEMMA 8.2. *Let $\mathfrak{R}$ be admissible and correctly embedded in the standard sector-like domain $\mathfrak{D}^-$. Then as z tends to infinity in the sector $\mathfrak{S}_j$ of opening $\pi/\alpha_j (> 0)$ we have one of the following asymptotic behaviors for $V(z)$:*

(i) $V(z) = O(|z|^{-(\alpha_j + 1)})$

(ii) $V(z) \sim C_j z^{B_j \alpha_j - 1}$,

B_j a positive integer, $C_j \neq 0$.

Alternative (ii) *holds in at least one sector of $\mathfrak{D}^-$.*

PROOF. As before we use (6.13) with (6.11) to estimate $\Omega(\tau)$ near τ_j, finding this time that, because of the first part of (8.1), $\Omega(\tau)$ is bounded by a power of $(\tau - \tau_j)^{-1}$ near τ_j. It follows that the singularities of $\Omega'(\tau)$ are all poles. Thus combining (6.10), (6.11) with (5.10),

$$V(z) \sim C_j (\tau - \tau_j)^{-N_j} z^{-(\alpha_j + 1)} \sim C^j z^{(N_j - 1)\alpha_j - 1}$$

if Ω' has a pole of order N_j at τ_j, while if τ_j is not a singularity of $\Omega(\tau)$ we have $V(z) \sim A_j g'(z)$ which is (i).

We need to rule out the possibility that $\Omega'(\tau)$ has a simple pole at τ_j. If τ_j is a simple pole we have

$$\Omega(\tau) = C_j \log (\tau - \tau_j) + \Omega_1(\tau),$$

where $\Omega_1(\tau)$ is regular at τ_j. Such a form for $\Omega(\tau)$ contradicts Lemma 5.3, since any determination of $\Omega(\tau)$ which is continuous in the closed upper half plane with τ_j omitted will not preserve direction near τ_j.

If alternative (ii) never holds our proof shows that $\Omega(\tau)$ has no singularities. But then $\Omega(\tau)$, being bounded at infinity, reduces to a constant. By (5.10) then $V(\tilde{z})$ vanishes identically, which we have noted is impossible.

LEMMA 8.3. *Let $\mathfrak{R}$ be admissible, correctly embedded in the standard sector-like domain $\mathfrak{D}^-$.*

Then for any $\varepsilon > 0$ there exists r_0 such that

$$\int_{\Gamma-\Gamma_r} |V_n(\tilde{z})/\tilde{z}^I|\,|d\tilde{z}| < \varepsilon$$

for $r > r_0$.

PROOF. By Lemma 4.2 $Q_I(z)$ is constant in each component of $\mathfrak{D}^+$, and letting z tend to infinity in $\mathfrak{D}^+$ in such a way that $\rho(z,\Gamma) > A|z|$, $A > 0$, we find that

$$|Q_I(z)| \leqq \left| \int_{\Gamma_s} \frac{V(\tilde{z})}{(\tilde{z}-z)^I} d\tilde{z} \right| + \left| \int_{\Gamma-\Gamma_s} \frac{V(\tilde{z})}{(\tilde{z}-z)^I} d\tilde{z} \right|$$

which is made arbitrarily small by choosing s first, then z. Thus $Q_I(z) \equiv 0$ in $\mathfrak{D}^+$.

We have, (with notation established in §4)

$$(8.2)\quad Q_I(0) - Q_{n,I}(0) = \int_{\Gamma_r} \frac{V(\tilde{z}) - V_n(\tilde{z})}{\tilde{z}^I} d\tilde{z} + \int_{\Gamma-\Gamma_r} \frac{V(\tilde{z}) d\tilde{z}}{\tilde{z}^I} d\tilde{z} - \int_{\Gamma-\Gamma_r} \frac{V_n(\tilde{z})}{\tilde{z}^I} d\tilde{z}.$$

First choose r so large that the second integral on the right-hand side of (8.2) is of modulus less than $\varepsilon/3$, then choose n to make the integral over Γ_r less than $\varepsilon/3$ in modulus. By the preceding paragraph and (4.2) we can also suppose, by taking n sufficiently large, that the left-hand side of (8.2) has modulus less than $\varepsilon/3$ Then we have

$$\left| \int_{\Gamma-\Gamma_r} (V_n(\tilde{z})/\tilde{z}^I)\, d\tilde{z} \right| < \varepsilon.$$

Now $V_n(\tilde{z})\,d\tilde{z}$ is a positive real differential along Γ, and we may suppose r was taken so large that the argument of $\tilde{z}^I$ is nearly constant on $\Gamma - \Gamma_r$.

THEOREM 8.1. *Let $\mathfrak{R}$ be an admissible set, and suppose that, for some positive integer k, $\mathfrak{R}^k$ lies in a half plane. Then $\mathfrak{R}$ is thin.*

We emphasize that the half plane in the statement of the theorem is arbitrary; it is not assumed that the origin is a boundary point.

PROOF. We suppose that $\mathfrak{R}^k$ lies in the half plane $\operatorname{Im} z > -B$, where $B > 0$. $\mathfrak{R}$ can be correctly embedded in a sector-like domain $\mathfrak{D}^-$, for which Γ^k coincides with the line $\operatorname{Im} z = -2B$ when z is large. We arrange that $\rho(\Gamma, \mathfrak{R})$ be bounded away from zero.

Now suppose $\mathfrak{R}$ is not thin. Then we are led to a limiting distribution $V(z)$, obtained from $\Omega(\tau)$ in the now familiar manner. Alternative (ii) of Lemma 8.2 holds in some of the sectors of $\mathfrak{D}^-$, with $\alpha_j = k$. Since Γ has angular index $\leqq 2k$, and $V(\tilde{z})/\tilde{z}^{2k}$ therefore must be absolutely integrable over Γ by the argument of §4, we have $B_j = 1$. Now in the sectors in which alternative (ii) holds we have

$$V(z)\,dz \sim C_j z^{k-1} dz = (1/k)\,C_j d(z^k), \tag{8.3}$$

and since $V(\tilde{z})\,d\tilde{z}$ is a positive real differential along Γ, as is $d(\tilde{z}^k)$ for z large, we conclude that the C_j must be positive real numbers.

Now

$$\lim_{n\to\infty} \int_{\Gamma_r} (V_n(\tilde{z})/\tilde{z}^k)\,d\tilde{z} = \int_{\Gamma_r} (V(\tilde{z})/\tilde{z}^k)\,d\tilde{z}.$$

Since $V(z)/z^k$ is analytic in $\mathfrak{D}^-$ we have

$$\int_{\Gamma_r} (V(\tilde{z})/\tilde{z}^k)\,d\tilde{z} = -\int_{\gamma} (V(z)/z^k)\,dz,$$

where $\gamma_r = \mathfrak{D}^- \cap (|z| = r)$. Take r so large that each arc of γ_r subtends at the origin an angle larger than $\pi/2k$ and use (8.3) to conclude that, for sufficiently large r,

$$\lim_{n\to\infty} \operatorname{Im} \int_{\Gamma_r} (V_n(\tilde{z})/\tilde{z}^k)\,d\tilde{z} = -\operatorname{Im} \int_{\gamma_r} (V(\tilde{z})/\tilde{z}^k)\,d\tilde{z} < -C_0\pi/2k,$$

C_0 being the maximal C_j. Since $Q_{n,k}(0) \to 0$ this implies that

$$\limsup_{n\to\infty} \operatorname{Im} \int_{\Gamma-\Gamma_r} (V_n(\tilde{z})/\tilde{z}^k)\,d\tilde{z} > C_0\pi/2k. \tag{8.4}$$

But since $V_n(\tilde{z})\,d\tilde{z}$ is real along Γ,

$$\begin{aligned}\operatorname{Im} \int_{\Gamma-\Gamma_r} (V_n(\tilde{z})/\tilde{z}^k)d\tilde{z} &= \int_{\Gamma-\Gamma_r} \operatorname{Im}[\overline{\tilde{z}^k}(V_n(\tilde{z})/|\tilde{z}|^{2k})\,d\tilde{z}] \\ &= 2B \int_{\Gamma-\Gamma_r} (V_n(\tilde{z})/|\tilde{z}|^{2k})\,d\tilde{z} < 2B\varepsilon\end{aligned}$$

by Lemma 8.3, for proper choice of r depending on ε. We have supposed in this calculation that r is so large that $(\Gamma - \Gamma_r)^k$ consists of rays.

For a suitable choice of ε our last inequality contradicts (8.4).

We remark that Theorem 8.1, with $k = 1$, is essentially due to T. Ganelius [2] who showed directly that $\mathfrak{R}$ has the local-to-global property.

The next lemma will permit us to shift zeros from Γ into $\mathfrak{R}$.

LEMMA 8.4. *Let $W_m(\tilde{z})$ be a sequence of functions defined on Γ, with support $\Gamma - \Gamma_{r_m}$, where r_m tends to infinity. We suppose that $W_m(\tilde{z})\, d\tilde{z}$ is a positive real differential along Γ.*

Suppose that for some nonnegative integer M, $\mathfrak{R}$ is correctly embedded in $\mathfrak{D}^-$ with tightness M, and that there is a uniform bound for the integrals over Γ of the functions $W_m(\tilde{z})/|\tilde{z}|^M$.

Then for each W_m there is an $\mathfrak{R}$-polynomial p_m, zeros $z_{j,m}$, satisfying

$$\left| \int_\Gamma \frac{W_m(\tilde{z})}{\tilde{z}^{k+\nu}} d\tilde{z} - \sum_j z_{j,m}^{-(k+\nu)} \right| < C r_m^{-(k+\nu)}, \qquad \nu \geqq 0,$$

C a constant.

PROOF. Along any component Γ_0 of Γ, we may write $W_m(\tilde{z})\, d\tilde{z} = d\tilde{W}_m(s)$, s being arc length. Replace $\tilde{W}_m(s)$ by its integral part, forming a distribution of point masses with the same total mass. We have

$$\begin{aligned} \left| \int_{\Gamma_0} \frac{W_m(\tilde{z})}{\tilde{z}^k} d\tilde{z} - \int_{\Gamma_0} \frac{d[\tilde{W}_m(s)]}{\tilde{z}(s)^k} \right| &\leqq \left| \frac{\tilde{W}_m(s) - [\tilde{W}_m(s)]}{\tilde{z}(s)^{k+\nu}} \right| \Gamma_0 \Big| \\ &\quad + (k+\nu) \left| \int_{\Gamma_0} \frac{\tilde{W}_m(s) - [\tilde{W}_m(s)]}{\tilde{z}^{k+v}} d\tilde{z} \right| \\ &< A r_m^{-(k+\nu)} \end{aligned}$$

for some constant A depending only on Γ. Performing this operation on all components of Γ we get a Γ-polynomial p_m^*, with zeros $z_{j,m}^*$ such that

$$\left| \sum_j (z_{j,m}^*)^{-(k+\nu)} - \int_\Gamma \frac{W_m(\tilde{z})}{\tilde{z}^{k+\nu}} d\tilde{z} \right| < A r_m^{-(k+\nu)}.$$

Now we derive $z_{j,m}$ from $z_{j,m}^*$ by moving $z_{j,m}^*$ about the shortest arc of the circle $|z| = |z_{j,m}^*|$ connecting $z_{j,m}^*$ to $\mathfrak{R}$. Now

$$\begin{aligned} \left| z_{j,m}^{-(k+\nu)} - (z_{j,m}^*)^{-(k+\nu)} \right| &= (k+\nu) \int_{z_{j,m}}^{z_{j,m}} dz/z^{k+\nu+1} \\ &< (k+\nu) |z_{j,m}^* - z_{j,m}^*| / |z_{j,m}|^{k+\nu+1} \end{aligned}$$

thus

$$\left| \sum_j \left\{ z_{j,m}^{-(k+v)} - (z_{j,m}^*)^{-(k+v)} \right\} \right| \leqq (k+v) \sum_j \frac{|z_{j,m} - z_{j,m}^*|}{|z_{j,m}|^{k+v+1}}$$

(8.5)

$$< (k+v)\pi \int_r^\infty \frac{L(r)}{r^{k+v+1}} dN_m(r),$$

$N_m(r)$ the number of zeros of p_m^* of modulus not exceeding r, $L(r)$ the maximum of $\rho(z, \mathfrak{R})$ for $z \in \Gamma$, $|z| = r$. The sum on the left-hand side of (8.5) therefore does not exceed

$$(k+v)\pi r_m^{-(k+v+1)} \int_0^\infty dN_m(r)/r^M.$$

The last integral is bounded independently of m by hypothesis and because of the relation between N_m and $\tilde{W}_m$; if γ is the intersection of Γ with the annulus $r_1 < |z| < r_2$ then

$$\int_{r_1}^{r_2} dN_m(r) = \int_\gamma d\tilde{W}_m(s).$$

The proof of Lemma 8.4 is concluded.

We now proceed to show that, given the positive integer k, if the admissible region $\mathfrak{R}$ is not thin then $\mathfrak{R}$ is k-ample. Then proof requires consideration of several cases separately. If $\mathfrak{A}(\mathfrak{R}^k)$ is empty (which is the same as saying that $\mathfrak{R}$ is bounded) or if $\mathfrak{A}(\mathfrak{R}^k)$ lies in an angular sector with opening less than π (which means $\mathfrak{R}^k$ lies in a half plane) then by Theorem 8.1 $\mathfrak{R}$ is thin. If $\mathfrak{A}(\mathfrak{R}^k)$ does not lie in a half plane then by Lemma 2.1 $\mathfrak{R}$ is k-ample. So we may restrict ourselves (after a preliminary rotation) to the case where $\mathfrak{A}(\mathfrak{R}^k)$ lies in the upper half plane but not in any angular sector of opening less than π.

THEOREM 8.2. *Let $\mathfrak{R}$ be admissible, and let $\mathfrak{A}(\mathfrak{R})$ be the set* Im $z^k \geqq 0$. *Then if $\mathfrak{R}$ is not thin, $\mathfrak{R}$ is k-ample.*

PROOF. We first make a suitable choice of sector-like domain $\mathfrak{D}^-$ in which to embed $\mathfrak{R}$. By Lemma 8.1 $\mathfrak{R}$ may be correctly embedded, with tightness $2k$, in a sector-like domain $\mathfrak{D}^-$ having k "sectors" of opening π/k. If $\mathfrak{R}$ is not thin, then there exists an equilibrium distribution $V(\tilde{z})$ along Γ.

Now refer to Lemma 8.2, where now we take $\alpha_j = k$. The angular index I of Γ is just $2k$, and as noted in §4, $V(\tilde{z})/\tilde{z}^I$ is absolutely integrable over Γ. Thus in the sectors where alternative (ii) of Lemma 8.2 holds we must have $B_j = 1$. As in the proof of Theorem 8.1 we conclude that the corresponding C_j are positive real.

Let $\delta_r = \mathfrak{D}^- \cap (|z| = r)$. Since $V(z)/z^{k+v}$ is analytic in $\mathfrak{D}^-$ we have

$$\int_{\Gamma_r} (V(z)/z^{k+\nu})\,dz = -\int_{\delta_r} (V(z)/z^{k+\nu})\,dz\,.$$

Thus as r tends to infinity

$$\int_{\Gamma_r} (V(z)/z^k)\,dz \to -C_0\pi i$$

where C_0 is the sum of the C_j over the sectors in which alternative (ii) of Lemma 8.2 holds. Because of (4.2) we have $\int_\Gamma (V_n(\tilde{z})/\tilde{z}^k)\,d\tilde{z} \to 0$.
Since

$$\int_{\Gamma_r} (V(\tilde{z})/\tilde{z}^k)\,d\tilde{z} = \lim_{j\to\infty} \int_{\Gamma_r} (V_{n_j}(\tilde{z})\tilde{z}^{\,k}d\tilde{z}$$

for some subsequence n_j, there exists a sequence $r_m \to \infty$ and corresponding n_m such that

$$\text{(8.6)} \qquad \int_{\Gamma-\Gamma r_m} (V_{n_m}(\tilde{z})/\tilde{z}^k)\,d\tilde{z} \to C_0\pi i\,.$$

By the same argument

$$\text{(8.6')} \qquad \int_{\Gamma-\Gamma r_m} (V_{n_m}(\tilde{z})/\tilde{z}^{k+\nu})\,d\tilde{z} \to 0, \qquad \nu \geqq 1$$

and since the integrals of the $|V_n(\tilde{z})/\tilde{z}^{2k}|$ over Γ are uniformly bounded we have

$$\text{(8.6'')} \qquad \int_{\Gamma-\Gamma r_m} (V_{n_m}(\tilde{z})/\tilde{z}^{2k+\nu})\,d\tilde{z} < Br_m^{-\nu}\,, \qquad \nu \geqq 1,$$

B some constant.

We notice that (8.6), (8.6′) and (8.6″) are integral analogues of the relations occurring in the definition of a k-ample region. Let us form

$$W_m(\tilde{z}) = (\lambda/C_0\pi)V_{n_m}(\tilde{z}), \quad \tilde{z}\in\Gamma-\Gamma_{r_m}$$

$$= 0, \quad \tilde{z}\in\Gamma_{r_m}$$

Then apply Lemma 8.4 to $W_m(\tilde{z})$ with $\mathrm{M} = 2k$. We obtain a polynomial sequence $p_m(z)$, zeros $z_{j,m}$, satisfying, for any preassigned positive ε and m sufficiently large depending on ε, the relations

(8.7) $$\sum_j z_{j,m}^{-k} \to \lambda i$$

(8.7′) $$\left| \sum_j z_{j,m}^{-(k+\nu)} \right| < \varepsilon^\nu, \quad \nu \geqq 1, \quad m > m_0.$$

(8.7″) $$\left| \sum_j z_{j,m}^{-(2k+\nu)} \right| < \varepsilon^\nu, \quad \nu \geqq 1, \quad m > m_0.$$

Since $\mathfrak{A}(\mathfrak{R}^k)$ certainly contains three distinct rays in the upper half plane, (8.7), (8.7′) and (8.7″) together with Lemma 2.1 show that $\mathfrak{R}$ is indeed k-ample, and Theorem 8.2 is proved.

We now consider the case where $\mathfrak{A}(\mathfrak{R})$ does not completely fill out the set $\operatorname{Im} z^k \geqq 0$. This case we handle in a different manner from the case of Theorem 8.2. First let us rule out two trivial situations. If $\mathfrak{A}(\mathfrak{R}^k)$ reduces to the real axis, then certainly $\mathfrak{R}^{2k}$ lies in a half plane, and $\mathfrak{R}$ is thin by Theorem 8.1. Another special case is that in which a sector of $\mathfrak{A}(\mathfrak{R})$ intersects a sector of $\operatorname{Im} z^k \geqq 0$ either in that full sector or the empty set. Here, we embed $\mathfrak{R}$ in a sector-like domain $\mathfrak{D}^-$, each sector of which has opening π/k, but $\mathfrak{D}^-$ has fewer than k sectors. The argument of Theorem 8.2 goes through without change. With the understanding throughout the proof that these trivial cases are ruled out, we give our next theorem.

THEOREM 8.3. *Let $\mathfrak{A}(\mathfrak{R})$ be a proper subset of $\operatorname{Im} z^k \geqq 0$. Then if $\mathfrak{R}$ is admissible and not thin, $\mathfrak{R}$ is k-ample.*

PROOF. We first must make a suitable choice for $\mathfrak{D}^-$. Let us say that the sector $\mathfrak{S}_j$ of $\mathfrak{D}^-$ is *k-narrow* if $\mathfrak{A}(\mathfrak{S}_j^k)$ is an angular sector of positive opening less than π. Because of the trivial situations ruled out above, $\mathfrak{R}$ can be correctly embedded in a sector-like domain $\mathfrak{D}^-$ having at least one k-narrow sector, by Lemma 8.1. Having chosen $\mathfrak{D}^-$, let I be the angular index of Γ. Now if necessary we adjust Γ slightly so that $\mathfrak{R}$ is correctly embedded in $\mathfrak{D}^-$ with tightness I. Assuming $\mathfrak{R}$ is not thin, we have an equilibrium distribution $V(\tilde{z})$ along Γ.

Case (i); *in some k-narrow sector, say $\mathfrak{S}_1$, of $\mathfrak{D}^-$, alternative* (ii) *of Lemma* 8.2 *holds*. We choose a semi-infinite arc Γ' of Γ, $\Gamma' \subset \partial\mathfrak{S}_1$, and such that as z tends to infinity on Γ',

$$\operatorname{Arg} z^k \to \beta, \qquad 0 < \beta < \pi.$$

For some sequence r_m tending to infinity let

(8.8) $$\delta_m = \Gamma' \cap \{r_m < |z| < r_m^3\}.$$

For sufficiently large m we have

$$\int_{\delta_m} |(V(\tilde{z})/\tilde{z}^k)\, d\tilde{z}| > (1/2) \int_{r_m}^{r_m^3} dt/t = \log r_m.$$

Because of the convergence of the $\{V_n\}$ to V, we can find a sequence n_m such that

$$\int_{\delta_m} |(V_{n_m}(\tilde{z})/\tilde{z}^k)\,d\tilde{z}| = C_m \to \infty \tag{8.9}$$

while at the same time

$$\left| \int_{\delta_m} (V_{n_m}(\tilde{z})/\tilde{z}^{k+\nu})\,d\tilde{z} \right| < 2C_m/r_m^{\nu}, \qquad \nu \geqq 1. \tag{8.9'}$$

Notice that since $V_{n_m}(\tilde{z})\,d\tilde{z}$ is positive real along Γ,

$$\operatorname{Arg} \int_{\delta_m} (V_{n_m}(\tilde{z})/\tilde{z}^k)\,d\tilde{z} \to \beta + \pi. \tag{8.9''}$$

Now we form some new distributions of charge along Γ. Choose λ any positive number. Form

$$\begin{aligned} Y_m(\tilde{z}) &= (\lambda/C_m)\, V_{n_m}(\tilde{z}), \quad \tilde{z} \notin \delta_m \\ &= 0, \quad \tilde{z} \in \delta_m . \end{aligned}$$

Then

$$\int_{\Gamma} (Y_m(\tilde{z})/\tilde{z}^k)\,d\tilde{z} \to \lambda e^{i\beta}, \tag{8.10}$$

$$\int_{\Gamma} (Y_m(\tilde{z})/\tilde{z}^{k+\nu})\,d\tilde{z} \to 0, \qquad \nu \geqq 1. \tag{8.10'}$$

Next we modify the $\{Y_m\}$ to obtain distributions having support off compact sets. Since the $\{V_{n_m}\}$ are uniformly bounded on compact sets, the $\{Y_m\}$ tend uniformly to zero on compact sets. It follows that by defining

$$\begin{aligned} W_m(\tilde{z}) &= Y_m(\tilde{z}), \quad |\tilde{z}| > s_m \\ &= 0, \quad |\tilde{z}| \leqq s_m \end{aligned}$$

for some appropriate sequence $\{s_m\}$ tending to infinity sufficiently slowly we satisfy (8.10) and (8.10′) with Y_m replaced by W_m.

Now apply Lemma 8.4 to conclude the existence of a sequence $p_m(z)$ of $\mathfrak{R}$-polynomials, zeros $z_{j,m}$, satisfying

$$\sum_j z_{j,m}^{-m} \to \lambda e^{i\beta} \tag{8.11}$$

$$\sum_j z_{j,m}^{-(k+\nu)} \to 0, \qquad \nu \geqq 1, \tag{8.11'}$$

and notice that our argument proves

(8.11″) $$\sum_j z_{j,m}^{-(l+\nu)} = O(r_m^{-(l+\nu)}), \qquad \nu \geqq 1.$$

Now $\mathfrak{A}(\mathfrak{R}^k)$ contains the real axis and the ray $\operatorname{Arg} z = \beta$. Thus (8.11), (8.11′) and (8.11″) together with Lemma 2.1 show that $\mathfrak{R}$ is k-ample.

Case (ii). *Alternative* (i) *of Lemma* 8.2 *holds in all k-narrow sectors.* We have to consider two subcases.

Subcase (a). If Γ' is any semi-infinite subarc of Γ such that $\mathfrak{A}(\Gamma'^k)$ lies in the open upper half plane, the numbers

$$\int_{\Gamma'} |(V_n(\tilde{z})/\tilde{z}^{2k})\, d\tilde{z}|$$

are a bounded sequence.

Subcase (b). Along some such semi-infinite subarc Γ', the numbers

$$\int_{\Gamma'} |(V_n(\tilde{z})/\tilde{z}^{2k})\, d\tilde{z}|$$

are unbounded.

Subcase (b) is slightly easier, and we dispose of it first. Let $\operatorname{Arg} z^k$ tend to β as z tends to infinity along Γ', and suppose that

$$\int_{\Gamma'} |(V_{n_m}(\tilde{z})/\tilde{z}^{2k})\, d\tilde{z}| \to \infty, \qquad m \to \infty.$$

Then a fortiori the numbers

$$\int_{\Gamma'} |(V_{n_m}(\tilde{z})/\tilde{z}^{k})\, d\tilde{z}|$$

tend to infinity. Now for some sequence $\{r_m\}$ tending to infinity sufficiently slowly, we define δ_m by (8.8) in such a way that (8.9), (8.9′) and (8.9″) hold. Now complete the argument precisely as in Case (i). The difference is that, here, we arrive at (8.9), (8.9′) and (8.9″) without the intermediate use of $V(\tilde{z})$.

Now we consider subcase (a). By Lemma 8.2 there is a sector of $\mathfrak{D}^-$ with opening π/k in which alternative (ii) holds. The argument used in proving Theorem 8.2 goes through without change, once we convince ourselves that $V(\tilde{z})/\tilde{z}^{2k}$ is absolutely integrable over Γ.

We may write $\Gamma = \Gamma_1 \cup \Gamma_2$, where as $\tilde{z}$ tends to infinity along any semi-infinite arc of Γ_1, $\operatorname{Arg} \tilde{z}^k$ tends to 0 or π, while as $\tilde{z}$ tends to infinity along a semi-infinite arc of Γ_2, $\operatorname{Arg} \tilde{z}^k$ tends to a number different from 0 or π. By the hypothesis defining subcase (a),

$$\int_{\Gamma_2} |(V_n(\tilde{z})/\tilde{z}^{2k})\, d\tilde{z}| = O(1), \qquad n \to \infty.$$

If $\Gamma_1{}^*$ is any subset of Γ_1 consisting of all but a bounded portion of Γ_1 and such that $(\Gamma_1{}^*)^{2k}$ is contained in an angle of opening less that π, we have

$$\int_{\Gamma_1^*} |(V_n(\bar{z})/\bar{z}^{2k})\, d\bar{z}| < A \left| \operatorname{Re} e^{i\alpha} \int_{\Gamma_1^*} (V_n(\bar{z})/\bar{z}^{2k})\, d\bar{z} \right| \tag{8.12}$$

for some real constants A and α.

The numbers $Q_{n,2k}(0)$ are bounded, and the numbers

$$\operatorname{Re}\left\{e^{i\alpha} \int_{\Gamma - \Gamma_1^*} (V_n(\bar{z})/\bar{z}^{2k})\, d\bar{z}\right\}$$

are bounded, since the path of integration includes all of Γ except for the union of Γ_2 with a bounded set, thus the numbers

$$\operatorname{Re}\left\{e^{i\alpha} \int_{\Gamma_1^*} (V_n(\bar{z})/\bar{z}^{2k})\, d\bar{z}\right\}$$

are bounded.

Now, using (8.12) as starting point, imitate the argument used in §4 to prove the absolute integrability over Γ of $V(\bar{z})/\bar{z}^I$. We find that $V(\bar{z})/\bar{z}^{2k}$ is absolutely integrable over Γ_1. because of the inequality defining alternative (i) of Lemma 8.2, $V(\bar{z})/\bar{z}^{2k}$ is also absolutely integrable over $\Gamma - \Gamma_1 = \Gamma_2$.

9. **The main theorem and concluding remarks.** Our principal result is that Korevaar's conjectured alternative holds for sets which are admissible, according to Definition 8.2. That is,

THEOREM 9.1. *Let $\mathfrak{R}$ be an admissible set. Then either $\mathfrak{R}$ is regular, or $\mathfrak{R}$ has the local-to-global property.*

Since our definition of admissibility is rather complicated, we wish to give, as the following corollary, a simplified version of our main result, which involves only the terminology of our Introduction.

COROLLARY 9.1. *Let $\mathfrak{R}$ be a closed region whose complement is connected, and contains the origin. Let $\mathfrak{R}$ be bounded by finitely many piece wise analytic curves which are everywhere twice continuously differentiable with respect to arc length. Suppose further that as z tends to infinity along any semi-infinite arc of $\partial\mathfrak{R}$, the curvature $K(z)$ satisfies the condition*

$$K(z) = O(|z|^{-(1+\varepsilon)})$$

for some positive ε.

Then either $\mathfrak{R}$ is regular or $\mathfrak{R}$ has the local-to-global property.

It would of course be very interesting to prove Korevaar's alternative (that is, either $\mathfrak{R}$ is regular or has the local-to-global property) with no restrictions on $\mathfrak{R}$. As pointed out, in §1, $\mathfrak{C}\mathfrak{R}$ must have an unbounded component intersecting the unit disc, and as far as is presently known this may well be the only restriction we need to place on $\mathfrak{R}$. Our theorem is not the most general possible, for we can relax somewhat the smoothness conditions imposed in our definition of standard arc; we have not used the full power of Warschawski's result [**15**]. Also, we note that at no point in our development did the specific shape of the unit disc play a role, so that we may redefine local convergence in terms of any domain $\mathfrak{E}$ with compact closure, and our theorems remain valid. One generalization that seems particularly interesting to attempt is the case where $\mathfrak{R}$ is, perhaps, smoothly bounded, but $\mathfrak{A}(\mathfrak{R})$ has isolated points.

We remark that the hypotheses of Theorem 7.1 can be weakened in the following manner; instead of

$$\log^+ |p_n(z)| = o(n) \tag{9.1}$$

in a full neighborhood of the origin, it is enough if (9.1) holds uniformly on an infinite sequence of points having a limit point in the open unit disc. For then $Q_{l+1}(z)$ would be regular in $\mathfrak{D}^+$, vanishing on a set with limit point in $\mathfrak{D}^+$, and the conclusion of Lemma 4.2 would still hold. The determination of $V(\tilde{z})$ is based on this lemma. In the same way we may weaken the hypotheses of the theorems of §8, and therefore of Theorem 9.1, by replacing convergence on the compact subsets of $\mathfrak{C}\mathfrak{R} \cap (|z| < 1)$ by uniform boundedness on a set of points having limit points in the set $\mathfrak{C}\mathfrak{R} \cap (|z| < 1)$.

The problem of §7 raises questions that appear to be of some interest. Is the result best possible as regards the narrowness of the strip $\mathfrak{R}$ leaves out, and can one improve the bound for $N_n(r)$ when $\mathfrak{C}\mathfrak{R}$ is larger, say bounded by algebraic curves? Finally, we note that our method is insensitive to the rate at which $(1/n)\log^+ |p_n(z)|$ tends to zero in $\mathfrak{E}$. It would be interesting to connect this rate more closely to the size of $N_n(r)$.

REFERENCES

1. R. P. Boas, Jr., *Entire functions*, Academic Press, New York, 1954.

2. T. Ganelius, *Sequences of analytic functions and their zeros*, Ark. Mat. **3**(1953), 1–50.

3. G. H. Hardy and E. M. Wright, *An introduction to the theory of numbers*. 3rd ed., Oxford, at the Clarendon Press, 1954.

4. O. D. Kellogg, *Harmonic functions and Green's integral*, Trans. Amer. Math. Soc. **13** (1912), 109–132.

5. J. Korevaar, *The zeros of approximating polynomials and the canonical representation of an entire function*, Duke Math. J. **18** (1951), 573–592.

6. ———, *Entire functions as limits of polynomials*, Duke Math. J. **21** (1954), 533–548.

7. ———, *Limits of polynomials with restricted zeros*. Studies in Mathematical Analysis and Related Topics (Essays in honor of G. Pólya) pp. 183–190. Stanford Univ. Stanford, Press, California, 1962.

8. ———, *Asymptotically neutral distributions of electrons and polynomial approximation* Ann. of Math. **80** (1964), 403–410.

8a. ———, *Limits of polynomials whose zeros lie in a given set*, these Proceedings pp. 261–272.

9. J. E. Lange. *Entire functions as limits of zero-restricted polynomials.* Ph. D. thesis, University of Wisconsin, 1961.

10. J. E. Lange and J. Korevaar, *Limits of polynomials whose zeros lie in a radial set*, Trans. Amer. Math. Soc. **114** (1965), 65–79.

11. N. I. Muskhelishveli, *Singular integral equations.* Second ed., OGIZ Moscow, 1946.

12. Z. Nehari, *Conformal mapping*, McGraw-Hill, New York, 1952.

13. G. Pólya and E. Lindwart, *Ueber einen Zusammenhang zwischen der Konvergenz von Polynomfolgen und der Verteilung ihrer Wurzeln*, Rend. Circ. Mat. Palermo **37** (1914), 297–304.

14. E. C. Titchmarsh, *The theory of functions*, Second ed., Oxford Univ. Press, London, 1939.

15. S. W. Warschawski, *On differentiability at the boundary in conformal mapping*, Proc. Amer. Math. Soc. **12** (1961), 614–620.

MICHIGAN STATE UNIVERSITY
EAST LANSING, MICHIGAN

A Parabolic Riemannian Ball[1]

MITSURU NAKAI
LEO SARIO

We wish to report on some joint work **[6]** we have carried out this year in Los Angeles on the classification theory of Riemannian spaces. Such an enterprise is always a bit risky, as it may end up in a more or less trivial extension of what is known for Riemann surfaces. Fortunately this did not turn out to be the case, but rather surprising new phenomena have cropped up.

The first one concerns the type problem. We recall that a Riemann surface is, by definition, parabolic or hyperbolic according as the harmonic measure of the boundary is zero or positive. It is well known **[1]** that the existence of a proper continuum on the boundary guarantees hyperbolicity. We shall now show that this is no longer true for Riemannian spaces of higher dimensions. Even a ball in n-space can, despite its very strong sphere-boundary, be endowed with a Riemannian metric that makes it parabolic.

1. **Definitions.** But let us first give precise definitions to our concepts. We consider a Riemannian space V, i.e., a connected, orientable, C^∞ n-manifold with a strictly positive metric tensor g_{ij}, whose matrix (g_{ij}) has the inverse matrix (g^{ij}) and the determinant g. We shall use the standard notations d for the exterior derivative, δ for the coderivative, and Δ for the Laplace-Beltrami operator $\Delta = d\delta + \delta d$.

A function is called harmonic if it satisfies the equation $\Delta u = 0$. In terms of local coordinates this takes the form of the selfadjoint elliptic partial differential equation

$$\Delta u = -\frac{1}{g^{1/2}} \sum_{i=1}^{n} \frac{\partial}{\partial x^i} \left(g^{1/2} g^{ij} \frac{\partial u}{\partial x^j} \right), \tag{1}$$

where we have used the Einstein summation convention with respect to j.

---

[1] An invited hour address delivered by Leo Sario at the American Mathematical Society 1966 Summer Institute in La Jolla on July 12, and at the meeting of the Mathematical Society of Japan in Tokyo on July 25, 1966.

The work was sponsored by the U. S. Army Research Office–Durham, Grant DA-AROD-31–124–G742, University of California, Los Angeles.

The length of the gradient is given by

$$|\operatorname{grad} u|^2 = g^{ij} \frac{\partial u}{\partial x^i} \frac{\partial u}{\partial x^j}. \tag{2}$$

For the Euclidean metric $g^{ij} = \delta_{ij}$ we recognize the conventional square sum of the partials. The volume element dV, that is $*1$, is $g^{1/2}dx^1 \cdots dx^n$, and the Dirichlet integral is, by definition,

$$D(u) = \int_V |\operatorname{grad} u|^2 dV = \int_V du \wedge *du, \tag{3}$$

where in $du \wedge *du$ we again see the square sum of the partials in the case of the Euclidean metric.

The harmonic measure of the ideal boundary is defined as follows. Given a Riemannian space V we exhaust it by regular regions $\{V_m\}_0^\infty$ with boundaries $\beta_0, \cdots, \beta_m, \cdots$. In the region between β_0 and β_m we take the harmonic function u_m with boundary values $u_m | \beta_0 = 0$, $u_m | \beta_m = 1$. The sequence $\{u_m\}$ is monotone decreasing and converges to a harmonic limiting function u. If $u \equiv 0$, we call V *parabolic*. The class of such spaces is denoted by O_G, in reference to the fact, easily established, that there are no Green's functions of (1) on V. In other cases V is called *hyperbolic*.

2. **Parabolic ball.** Now to our parabolic Riemannian ball. In Euclidean n-space with $n \geqq 3$ we use the notation $|x|^2 = \sum_1^n (x^i)^2$ and consider the unit ball B: $|x| < 1$. We exhaust it by balls with boundaries

$$\beta_m : |x| = 1 - 1/(2+m). \tag{4}$$

Thus β_0 is the sphere with radius 1/2, and the β_m tend to the unit sphere. Between β_0 and β_m we take

$$u_m = 1/(m(1-|x|)) - 2/m. \tag{5}$$

One verifies at once that the boundary conditions are satisfied: $u_m | \beta_0 = 0$ and $u_m | \beta_m = 1$. But clearly u_m is not harmonic in the Euclidean metric, and we have to find a metric where it is.

We choose g_{ij} to be C^∞ in B with

$$g_{ij}(x) = \lambda(x)\delta_{ij} \tag{6}$$

in $1/2 \leqq |x| < 1$. Then (g_{ij}) is diagonal with elements λ and (g^{ij})has elements λ^{-1}, g is λ^n, and (1) takes the form

$$\Delta u = -\lambda^{-n/2} \sum_{i=1}^n \frac{\partial}{\partial x^i} \left(\lambda^{(n-2)/2} \frac{\partial u}{\partial x^i} \right). \tag{7}$$

The problem is to find a λ which makes this $\equiv 0$ for u_m.

We set

(8) $$\lambda(x) = |x|^{(2-2n)/(n-2)}(1-|x|)^{4/(n-2)}.$$

Then in $\lambda^{(n-2)/2}$ the exponent of $(1-|x|)$ becomes 2, but this cancels against -2 in $\partial u/\partial x^i$. The exponent of $|x|$ becomes $1-n$, but here the 1 cancels against the -1 in $\partial u/\partial x^i$. All that remains is, up to a constant factor, $|x|^{-n}$ multiplied by the innermost derivative x^i in $\partial u/\partial x^i$, and the new differentiation gives

$$\sum \frac{\partial}{\partial x^i}(|x|^{-n}x^i) = \sum |x|^{-n-2}(-n(x^i)^2+|x|^2) \equiv 0.$$

We conclude that the u_m are harmonic in our new metric, and the limiting function u is the harmonic measure. But this is $\equiv 0$, and we have a parabolic case. We have proved:

THEOREM 1. *A Riemannian ball can be parabolic.*

More precisely, this can be so in a metric (5) conformally equivalent to the Euclidean one. We know that this cannot take place for $n = 2$, a fact suggested by the impossibility of this value in (8).

One might remark that since λ goes rapidly to zero as $|x| \to 1$, we are actually shrinking the unit sphere to a point. This is indeed so, but the phenomenon goes deeper: even for $n=2$ we can take $\lambda(x)=1-|x|$ to make the unit circle reduce to a point, yet the disk remains hyperbolic.

3. **Hyperbolic punctured torus.** What about the other extreme? For Riemann surfaces it is known that a point boundary implies parabolicity. We shall now show that this is not true for higher dimensional Riemannian spaces.

Consider a flat torus, that is, an n-dimensional cube with the opposite faces identified by pairs. Remove from it a point, the center of a unit ball B: $|x| < 1$. Exhaust the resulting punctured torus V by subregions with boundaries

(9) $$\beta_m: |x| = 1/(1+m).$$

Thus β_0 is the unit sphere, and the β_m close in on the point boundary. Between β_0 and β_m take

(10) $$u_m \equiv \frac{1-|x|}{1-1/(1+m)}.$$

Again the boundary conditions $u_m|\beta_0 = 0$, $u_m|\beta_m = 1$ are clearly satisfied, and again the function is not harmonic in the Euclidean metric. To find a metric where it is we let g_{ij} be C^∞ in V and $g_{ij}(x) = \lambda(x)\delta_{ij}$ in B, but this time we choose

(11) $$\lambda(x) = |x|^{(2-2n)/(n-2)}.$$

Then in $\lambda^{(n-2)/2}$ the exponent of $|x|$ is $1-n$ which again is reduced by $\partial u/\partial x$ to $-n$, and we end up with the same expression $\partial(x^{-n}x^i)/\partial x^i$ as before. Thus u_m is again harmonic, but now the limiting function

$$u = 1-|x| \not\equiv 0, \tag{12}$$

and we have a hyperbolic space.

THEOREM 2. *A punctured flat torus can be hyperbolic.*

Again, the new metric is conformally equivalent to the Euclidean one, and this is impossible for $n = 2$.

4. **Conformal invariance.** But the surprises do not end here. For Riemann surfaces Pfluger [7] has shown that the class O_G is invariant under quasi-conformal mappings. We shall now prove that this is not so for higher dimensional Riemannian spaces. For these the quasi-conformal mappings are defined as follows:

Let T be a homeomorphism between two Riemannian spaces V and $\tilde{V}$. The distance between two points p, q in V is, by definition,

$$d(p,q) = \inf \int_\alpha (g_{ij}dx^i dx^j)^{1/2},$$

the infimum taken among all rectifiable arcs α connecting p and q. Distances $\tilde{d}$ in $\tilde{V}$ are defined similarly. A geodesic ball with center p and radius r is mapped into a distorted ellipsoid. We denote by L and l the longest and shortest distances of its boundary from $T(p)$. Explicitly,

$$\begin{aligned} L(p,r) &= \sup_{d(p,q)=r} \tilde{d}(T(p),T(q)), \\ l(p,r) &= \inf_{d(p,q)=r} \tilde{d}(T(p),T(q)). \end{aligned} \tag{13}$$

The ratio L/l gives in the limit $r \to 0$ the dilation of the infinitesimal ellipsoid. The mapping is called *quasi-conformal* if this dilation is bounded,

$$\limsup_{r\to 0} \frac{L(p,r)}{l(p,r)} \leqq K. \tag{14}$$

For $K = 1$ we have a *conformal* mapping.

In our first example we take for V the unit ball B with the Euclidean metric, and for $\tilde{V}$ the same ball with our new metric (6). For T we choose the mapping induced by the identity mapping of the underlying space B. Because of the special choice (6) of g_{ij}, the change of linear scale is independent of the direction from a given point, i.e., $\lim_{r\to 0} L/l = 1$, and the mapping is conformal.

Clearly V is hyperbolic, the harmonic measure being

$$u = \frac{|x|^{2-n} - 2^{n-2}}{1 - 2^{n-2}} \not\equiv 0.$$

Since $\tilde{V}$ was shown to be parabolic, we have proved:

THEOREM 3. *O_G is not preserved under quasi-conformal and not even conformal mappings.*

This is in striking contrast with the situation on Riemann surfaces, where conformality is in the very definition of equivalence.

5. **Quasi-isometries.** In view of this behavior, or one might say misbehavior, of higher dimensional Riemannian spaces one might wonder if there are any significant mappings under which O_G is invariant. Since the restrictions we imposed on L and l turned out to be insufficient, we have to tighten them. In Riemannian spaces one considers distances in the large, and we are led, in a natural manner, to require that *the relative change of distances is bounded.* Explicitly,

(15) $$K^{-1}r \leqq l(p,r) \leqq L(p,r) \leqq Kr$$

for some K. If these conditions are satisfied we call the homeomorphism T a *quasi-isometry.* In the special case $K = 1$, T is an *isometry.*

Clearly quasi-isometries are quasi-conformal mappings, for $L/l \leqq K^2$.

We are going to show that O_G is quasi-isometrically invariant. But we have a long way to go to see that.

We start by characterizing O_G in terms of an equivalence relation between function spaces which we shall now introduce.

6. **Function spaces.** For functions f on V we consider the norm

(16) $$[f]_\infty = \sum_0^\infty \frac{1}{2^m} \sup_{Vm} \frac{|f|}{1 + |f|}.$$

When it goes to zero, so does each term, and for any compact set we have $\sup |f| \to 0$ on it, uniformly. In view of this compact convergence we shall refer to (16) as the C-norm. By the D-norm we understand the norm

(17) $$\|f\| = (D(f))^{1/2},$$

which gives Dirichlet convergence, and by the CD-norm the combination

(18) $$\langle f \rangle = [f]_\infty + \|f\|$$

providing both compact and Dirichlet convergence.

For preciseness we note that (16) and (18) are actually quasi-norms and do not give rise to inner products, whereas (17) is a semi-norm and furnishes an inner product in terms of the mixed Dirichlet integral.

For a parametric ball B we denote by $ACL(B)$ the class of functions which are absolutely continuous on almost all lines parallel to the coordinate axes. Similarly we set

$$ACL(V) = \{f \mid f \mid B \in ACL(B)\}, \tag{19}$$

where the property $f \mid B \in ACL(B)$ is to hold for all parametric balls B and all coordinate systems in them.

Of fundamental importance in our study is the space

$$\mathscr{D} = \{f \mid f \in C(V) \cap ACL(V), D(f) < \infty\}. \tag{20}$$

It is not complete with respect to the D-norm. But it is so with respect to the CD-norm. For the subspace

$$\mathscr{D}_0 = \{f \mid f \in \mathscr{D}, \operatorname{supp} f \text{ compact}\}$$

we take the closure $\overline{\mathscr{D}}_0$ in $\mathscr{D}$ under the CD-norm.

7. **Characterization of O_G.** The alternative of $\overline{\mathscr{D}}_0 \subset \mathscr{D}$ comprising or not comprising all of $\mathscr{D}$ gives us the desired characterization of O_G:

LEMMA 1. *$V \in O_G$ if and only if $\mathscr{D} = \overline{\mathscr{D}}_0$.*

This statement is an immediate extension to Riemannian spaces of a result by Royden [8], [9] and Nakai [4], [5] and we quote it here without proof. The intuitive meaning is that a hyperbolic boundary is strong enough to hold $\mathscr{D}$-functions that are not within reach of functions with compact supports, while a parabolic boundary is too weak to do it. It is something like a needle, not strong enough to support a soap film which slides right through it.

8. **Dirichlet mappings.** To make use of Lemma 1 we introduce *Dirichlet mappings*, i.e., homeomorphisms T with the following properties. First, $\mathscr{D}$ is preserved:

$$f \circ T \in \mathscr{D}(V) \quad \text{if and only if } f \in \mathscr{D}(\tilde{V}). \tag{21}$$

Second, the relative change of $D(f)$ is bounded:

$$K^{-1}D(f) \leqq D(f \circ T) \leqq KD(f) \tag{22}$$

for some constant K.

These conditions are tailor-made to give the following intermediate result.

LEMMA 2. *O_G is preserved under Dirichlet mappings.*

In fact, we only have to show that $\mathscr{D} = \overline{\mathscr{D}}_0$ is preserved. But $\mathscr{D}$ is so by (21), $\mathscr{D}_0$ is so since compacts go into compacts, and for the same reason the compact closing of $\mathscr{D}_0$ is preserved. The same is true of the Dirichlet closing since by (22) the Dirichlet integrals tend to zero simultaneously.

The significance of Dirichlet mappings is, first of all, in that for $n = 2$ quasi-conformal mappings can be shown [**6**] to be Dirichlet mappings, and we obtain a new proof of Pfluger's theorem. But for higher dimensions quasi-conformal mappings cannot be Dirichlet mappings as we have just seen by Theorem 3. In contrast, we shall show:

LEMMA 3. *Quasi-isometries are Dirichlet mappings.*

As to condition (21), quasi-isometries are quasi-conformal mappings, and these have been shown by Gehring [**3**] and Väisälä [**11**] to preserve the ACL-property, to be totally differentiable a.e., and to have a positive Jacobian a. e. All our formulas in the sequel shall be understood to hold a.e.

The finiteness of $D(f)$ in (21) will be a consequence of (22) which we proceed to establish. This is the crucial step in our reasoning, and it will occupy the rest of our paper.

We are to compare the Dirichlet integrals. In view of (3) we shall first compare volume elements, then gradients.

9. **Volume elements under quasi-isometries.** We start with the quasi-isometric condition (15) applied to arc elements:

$$K^{-1} ds \leqq d\tilde{s} \leqq K ds.$$

On squaring we obtain

$$K^{-2} g_{ij} dx^i dx^j \leqq \tilde{g}_{ij} \frac{\partial \tilde{x}^i}{dx^k} \frac{\partial \tilde{x}^j}{\partial x^l} dx^k dx^l \leqq K^2 g_{ij}\, dx^i dx^j. \tag{23}$$

In terms of the matrix $M = (\partial \tilde{x}^i / \partial x^k)$ this is equivalent to the matrix inequality

$$K^{-2}(g_{ij}) \leqq M^t(\tilde{g}_{ij})M \leqq K^2(g_{ij}), \tag{24}$$

which is to be understood in the following sense: for any two $n \times n$ matrices A and B the relation $A \leqq B$ means that $B - A$ is a positive matrix, i.e., the corresponding quadratic form is positive semidefinite. (For a strict inequality $A < B$, positive definite.) In the present case, we have in (23) nothing but quadratic forms corresponding to the three members of (24).

But what kind of matrices do we have here? The matrix $(\tilde{g}_{ij})$ is strictly positive by the definition of Riemannian space, and it remains so after multiplication by matrices M and M^t, nonsingular a.e. Thus each member of (24) is a strictly positive matrix.

For such matrices an inequality $A \leqq B$ can be replaced by the corresponding inequality $|A| \leqq |B|$ of the determinants. To see this we multiply $A \leqq B$ through by the following sequence of matrices on the right and by their transposes on the left. First by an orthogonal matrix we rotate so as to diagonalize B. Then by a diagonal matrix we shrink to make B a unit matrix. Next we rotate again by a orthogonal matrix to diagonalize A. But B remains diagonal since it already was a unit matrix. Now, for diagonal matrices an inequality is equivalent to the inequality of the elements by pairs, hence of their products, i.e., the determinants. Obviously the conclusion carries back to the original determinants: $|A| \leqq |B|$.

In the present case of (24) we obtain

$$K^{-2n}g \leqq \tilde{g}|M|^2 \leqq K^{2n}g. \tag{25}$$

We take square roots and multiply through by $dx^1 \cdots dx^n$. Since

$$|M|\, dx^1 \cdots dx^n = d\tilde{x}^1 \cdots d\tilde{x}^n,$$

we obtain

$$K^{-n}dV \leqq d\tilde{V} \leqq K^n\, dV. \tag{26}$$

This is the relation for volume elements we set out to prove.

10. **Gradients under quasi-isometries.** We proceed to compare gradients. Again by the device of simultaneous diagonalization we conclude that, for strictly positive matrices, $A \leqq B$ implies $B^{-1} \leqq A^{-1}$. On taking inverses in (24) we obtain

$$K^{-2}(g^{ij}) \leqq M^{-1}(\tilde{g}^{ij})(M^t)^{-1} \leqq K^2(g^{ij}). \tag{27}$$

On left and right multiplying the second inequality by M and M^t respectively, we arrive at the first inequality below, and the second one is obtained in an analogous manner from the first inequality (27):

$$K^{-2}\tilde{g}^{ij} \leqq M(g^{ij})M^t \leqq K^2\tilde{g}^{ij}. \tag{28}$$

In terms of $(\partial f/d\tilde{x}^i)$ the corresponding quadratic forms are

$$K^{-2}\tilde{g}^{ij}\frac{\partial f}{\partial \tilde{x}^i}\frac{\partial f}{\partial \tilde{x}^j} \leqq g^{ij}\frac{\partial \tilde{x}^k}{\partial x^i}\frac{\partial \tilde{x}^l}{\partial x^j}\frac{\partial f}{\partial \tilde{x}^k}\frac{\partial f}{\partial \tilde{x}^l} \leqq K^2\tilde{g}^{ij}\frac{\partial f}{\partial \tilde{x}^i}\frac{\partial f}{\partial \tilde{x}^j}. \tag{29}$$

In view of (2), we have here the gradients:

$$K^{-2}|\mathrm{grad} f|^2 \leqq |\mathrm{grad} f \circ T|^2 \leqq K^2|\mathrm{grad}\, f|^2. \tag{30}$$

It remains only to multiply (26) and (30) to obtain (22). The proof of Lemma 3 is herewith complete.

11. **Quasi-isometric invariance.** On combining Lemmas 2 and 3 we have immediately:

THEOREM 4. *O_G is preserved under quasi-isometries.*

We might add that an analogous reasoning can be carried out for two other null classes of Riemannian spaces. Let HD be the class of harmonic functions with finite Dirichlet integrals, and KD the subclass characterized by a vanishing flux $\int_\gamma * du$ across all dividing $(n-1)$-cycles γ. Denote by O_{HD} and O_{KD} the classes of Riemannian spaces on which there are no nonconstant functions HD and KD respectively. For Riemann surfaces it was proved by Royden [**8**], [**9**], and Nakai [**4**] that O_{HD} and O_{KD} are quasi-conformally invariant. Again, it can be shown [**5**] that for higher dimensions this ceases to be true.

For quasi-isometries no such distinction is needed but quasi-isometric invariance is a universal property of classes O_G, O_{HD}, O_{KD}, true for all dimensions.

In view of such unexpected phenomena it seems that there is a new, promising field of research in the classification theory of Riemannian spaces.

References

1. L. Ahlfors and L. Sario, *Riemann surfaces*, Princeton Univ. Press, Princeton, N. J., 1960.

2. F. Gehring, *Rings and quasi-conformal mappings in space*, Trans. Amer. Math. Soc. **103** (1962), 353–393.

3. ———, *Quasi-conformal mappings in space*, Bull. Amer. Math. Soc. **69** (1963), 146–164.

4. M. Nakai, *On a ring isomorphism induced by quasi-contormal mappings*, Nagoya Math. J. **14** (1959), 201–221.

5. ———, *A measure on the harmonic boundary of a Riemann surface*, Nagoya Math J. **17** (1960), 181–218.

6. M. Nakai and L. Sario, *Classification and deformation of Riemannian spaces*, Math. Scand. **20** (1967), 193–208.

7. A. Pfluger, *Sur une propriété de l'application quasiconforme d'une surface de Riemann ouverte*, C. R. Acad. Sci. Paris **227** (1948), 25–26.

8. H. Royden, *Harmonic functions on open Riemann surfaces*, Trans. Amer. Math. Soc. **73** (1952), 40–94.

9. ———, *On a class of null-bounded Riemann surfaces*, Comment. Math. Helv. **34** (1960), 52–66.

10. L. Sario, M. Schiffer and M. Glasner, *The span and principal functions in Riemannian spaces*, J. Analyse Math. **15** (1965), 115–134.

11. J. Väisälä, *On quasiconformal mappings in space*, Ann. Acad. Sci. Fenn. Ser. AI, No. 298 (1961), 36 pp.

NAGOYA UNIVERSITY
NAGOYA, JAPAN
UNIVERSITY OF CALIFORNIA
LOS ANGELES, CALIFORNIA

Principal Fields, Semifields and Forms on Riemannian Spaces[1]

MITSURU NAKAI
LEO SARIO[2]

1. Background.

1. We would like to discuss some joint work **[4]** we have carried out this year on constructing harmonic fields on Riemannian spaces. For background we refer to the special problem, long solved, of constructing harmonic functions with a given singularity on a Riemann surface R. Explicitly, given a point ζ in R and a regular neighborhood $U \subset R$ of ζ, let σ be in the class $H(\bar{U} - \zeta)$ of harmonic functions on $\bar{U} - \zeta$. The problem is to find a $\rho \in H(R - \zeta)$ that imitates the behavior of σ in $U - \zeta$. (In the present case this imitation means that $\sigma - \rho$ has a harmonic continuation to all of U.)

We consider the Riemann surface V obtained by puncturing R at ζ. Then $U - \zeta$ becomes a neighborhood of an ideal boundary component of V corresponding to ζ. This neighborhood is a component of a neighborhood V_1 of the entire ideal boundary of V, so chosen that $V_0 = V - \bar{V}_1$ is a regular open set. Our problem becomes a special case of the following so-called principal function problem. Given $\sigma \in H(\bar{V}_1)$, find a principal function $\rho \in H(V)$ that imitates the behavior of σ. Depending on the choice of σ and the mode of this imitation one obtains as special cases various domain functionals, Abelian integrals, conformal mappings, extremal functions and, as a consequence, inclusion relations between classes of Riemann surfaces.

In all cases the solvability of the problem is completely characterized by the behavior of σ on the relative boundary Γ of V_1. Explicitly, the flux $\int_\Gamma * d\sigma$ must vanish.

2. New problem.

2. Thus the problem has been solved for harmonic functions, hence for exact harmonic differentials. In contrast, it has remained open for arbitrary harmonic differentials. But these are nothing but harmonic fields, and we are led, more generally, to a study of harmonic p-fields on arbitrary Riemannian spaces of

[1] An invited hour lecture delivered by Leo Sario at the American Mathematical Society 1966 Summer Institute in La Jolla on July 13, and at Nagoya University, Japan, on July 29, 1966.

[2] The work was sponsored by the U. S. Army Research Office-Durham, Grant DA-AROD-31-124-G742, University of California, Los Angeles.

dimension n. We continue depicting the spaces as surfaces in 3-space, and we retain our notations.

Our first problem, that of constructing harmonic fields with given singularities, was solved in 1949 by Kodaira in his celebrated paper [**2**]. The corresponding *principal field problem*, as we shall call it, seems to have received no attention. We shall present a solution to this problem in terms of necessary and sufficient conditions on σ. Again the behavior of σ on Γ alone will be of significance.

The problem, then, is this. Let σ be in the space $F(\bar{V}_1)$ of harmonic fields, characterized by $d\sigma = \delta\sigma = 0$, in $\bar{V}_1$. Such fields are C^∞. We extend $\sigma | \bar{V}_1$ as C^∞ to all of V, and continue denoting this extension by σ. We write briefly $\sigma \in F(\bar{V}_1) \cap C^\infty(V)$. The problem is to find a $\rho \in F(V)$ that imitates the behavior of σ. We shall call ρ the *principal field* corresponding to σ.

3. We start by describing our choice for the imitation. Let E^{loc} be the space of locally square integrable differential forms on V. We shall actually use this symbol for the subspace of p-forms, but the degree of the form will always be clear from the context. For this reason we refrain here and later from using an index p in our notations of function spaces.

For $\phi, \psi \in E^{\text{loc}}$ the inner product is defined by the integral, if it exists,

$$(\phi, \psi) = \int_V \phi \wedge *\psi.$$

The subspace $E \subset E^{\text{loc}}$ is charecterized by the additional condition $\| \phi \| < \infty$, where $\| \phi \|^2 = (\phi, \phi)$. We are also interested in the subspace $D \subset E$ defined by

$$D = \{\phi | \phi \in E \cap C^\infty, \text{ supp } \phi \text{ compact}\}. \tag{1}$$

For $\phi \in D$, we form the exterior derivative $d\phi$, the class dD of these derivatives, and the closure $E_d = \overline{dD}$ in E. The coderivative δ is, by definition, the adjoint operator:

$$(d\phi, \psi) = (\phi, \delta\psi)$$

for all $\psi \in D$. We set $E_\delta = \overline{\delta D}$. Since $(d\phi, \delta\psi) = 0$, we have by continuity $E_d \perp E_\delta$. Therefore the direct sum can be written

$$E_{d\delta} = E_d + E_\delta = \overline{dD + \delta D}. \tag{2}$$

It is the membership in this space that we take for the "imitation" of ρ to σ, i.e., we require that

$$\sigma - \rho = \tau \in E_{d\delta}. \tag{3}$$

This means two things. First, $\| \rho - \sigma \| < \infty$. Second, in view of the elements of $E_{d\delta}$ being limits of forms with compact supports, $\rho - \sigma$ has, in some sense, zero data on the ideal boundary β of V: $\rho - \sigma \big| \beta = 0$. In a manner of speaking, ρ solves the Dirichlet problem for harmonic fields with boundary values σ. Thus the imitation is close indeed.

3. **Geometric interpretation.**

4. Our choice of imitation is further illustrated by a simple geometric interpretation which we give in connection with establishing the following intermediate result:

THEOREM 1. *If* $\|\sigma\| < \infty$, *then the principal field* ρ *always exists.*

To see this, take $\alpha \in E^{\mathrm{loc}}$. If there exists a $\beta \in E^{\mathrm{loc}}$ such that

$$(\alpha, \delta\phi) = (\beta, \phi) \tag{4}$$

for all $\phi \in D$, we call β the (weak) exterior derivative $d\alpha$ of α, and say that β is exact. If $d\alpha = 0$, then α is, by definition, closed. We denote by E_c the space of such differentials in E. By the very definition (4),

$$E_c = E_\delta^{\perp}. \tag{5}$$

Similarly, if there exists a $\gamma \in E^{\mathrm{loc}}$ with

$$(\alpha, d\phi) = (\gamma, \phi) \tag{6}$$

for all $\phi \in D$, we call γ the coderivative $\delta\alpha$ of α, and say that γ is coexact. If $\delta\alpha = 0$, then α is coclosed, and we denote by E_{cc} the space of such differentials in E. Again, definition (6) gives

$$E_{cc} = E_d^{\perp}. \tag{7}$$

5. Geometrically, these spaces can be depicted as follows. The 3-space shall represent the space E of square integrable forms. We choose as the x-axis the space E_d of exact forms, and as the y-axis the space E_δ of coexact forms. The xy-plane is the space $E_{d\delta}$ that shall characterize the imitation. The xz-plane is the space E_c of closed forms, and the yz-plane the space E_{cc} of coclosed forms. The z-axis $E_c \cap E_{cc}$ is characterized by $d\phi = \delta\phi_\zeta = 0$, and is therefore, the space F of harmonic fields.

It is a deep result, known as Weyl's lemma, or Kodaira's or de Rham's decomposition theorem, that $F \subset C^\infty$. Of importance is that this is a local property, resulting from

$$(\alpha, d\phi + \delta\psi) = 0$$

for all $\phi, \psi \in D$.

6. Now to our principal field problem. If $\|\sigma\| < \infty$, then σ is a point in our

3-space. Let $\rho \in F$ be its z-coordinate. Since the projection τ onto the xy-plane is in $E_{d\delta}$, ρ is the desired principal field. We have obtained the solution of our problem simply by reading it from the orthogonal decomposition of the space E.

7. We insert here the remark that ρ can also be characterized in a form familiar from the theory of principal functions. In fact, denote by L the projection operator to the $E_{d\delta}$-space: $\tau = L\sigma$. Since $L\rho = 0$, we can write

$$\rho - \sigma = L(\rho - \sigma), \tag{8}$$

in complete analogy with the equation characterizing principal functions. If $\| \sigma \| = \infty$, the projection L is not defined, but the corresponding operator can be determined by the properties $L\sigma \in E_{d\delta}$,

$$(\sigma, d\phi + \delta\psi) = (L\sigma, d\phi + \delta\psi)$$

for all $\phi, \psi \in D$.

Finding the principal field ρ again amounts to solving equation (8) for ρ.

4. Main theorem.

8. Resuming our original notations, we turn to a systematic study of the general case where $\| \sigma \|$ is not assumed to be finite. Then σ cannot be presented as a point in our geometric model. This is where the interest of the problem lies. The solution is as follows.

THEOREM 2 (MAIN THEOREM). *Let $\sigma \in F(\bar{V}_1) \cap C^\infty(V)$. A necessary and sufficient condition for the existence of the principal field $\rho \in F(V)$ with $\sigma - \rho \in E_{d\delta}$ is that*

$$\int_\Gamma \sigma \wedge *\phi = O(\| \delta\phi \|) \tag{9}$$

and

$$\int_\Gamma *\sigma \wedge \phi = O(\| d\phi \|) \tag{10}$$

for all $\phi \in D$.

The conditions may appear somewhat implicit, but we shall see that they are met by forms with surprisingly simple explicit properties.

In the proof we shall use the convenient notation $\varepsilon = \pm 1$, in the sense that for any quantities a, b, the equation $a = \varepsilon b$ means $a = b$ or $a = -b$. For instance, $**\phi = \varepsilon\phi$, $*d*\phi = \varepsilon\delta\phi$, $d*\phi = \varepsilon*\delta\phi$, and $d(\phi \wedge \psi) = d\phi \wedge \psi + \varepsilon\phi \wedge d\psi$. This notation eliminates the sometimes cluttered exponents of -1 occurring in exterior calculus, and is useful here since we are only interested in magnitudes.

9. To prove the necessity, suppose there exists a $\rho \in F(V)$ with $\sigma - \rho = \tau \in E_{d\delta}$. Then

$$\int_{\Gamma} \sigma \wedge *\phi = \int_{V_0} d\sigma \wedge *\phi + \varepsilon \int_{V_0} \sigma \wedge *\delta\phi .$$

Since $d\sigma \mid V_1 = 0$ and $d\rho = 0$ on V, the first term on the right becomes

$$(d\sigma, \phi) = (d\tau, \phi) = (\tau, \delta\phi) = O(\| \tau \| \cdot \| \delta\phi \|) = O(\| \delta\phi \|).$$

The second term gives

$$O(\| \sigma \|_{V_0} \cdot \| \delta\phi \|_{V_0}) = O(\| \delta\phi \|_{V_0}) = O(\| \delta\phi \|).$$

By a similar reasoning one sees that the assumptions $\delta\sigma \mid V_1 = 0$ and $\delta\rho = 0$ on V imply $\int_{V_0} *\sigma \wedge \phi = O(\| d\phi \|)$.

10. For the sufficiency we first show that $(\sigma, \delta\phi)$ is continuous in $\delta\phi$. We have, again by $d\sigma \mid V_1 = 0$,

$$(\sigma, \delta\phi) = (d\sigma, \phi) = \int_{V_0} d\sigma \wedge *\phi.$$

We recognize in the integral the first term of the derivative of $\sigma \wedge *\phi$ and obtain by Stokes' formula

$$\int_{\Gamma} \sigma \wedge *\phi + \varepsilon \int_{V_0} \sigma \wedge *\delta\phi .$$

By assumption (9) and by Schwarz' inequality we obtain

$$(\sigma, \delta\phi) = O(\| \delta\phi \|). \tag{11}$$

The continuity of $(\sigma, \delta\phi)$ thus established, we can extend this inner product continuously to the closure E_δ of δD. We conclude that there exists an element $\tau_\delta \in E_\delta$ such that

$$(\sigma, \delta\phi) = (\tau_\delta, \delta\phi) \tag{12}$$

for all $\phi \in D$.

We next consider $(\sigma, d\psi)$. In a manner similar to that above one shows that the assumptions $\delta\sigma \mid V_1 = 0$ and (10) imply

$$(\sigma, d\psi) = O(\| d\psi \|). \tag{13}$$

By virtue of this continuity, $(\sigma, d\psi)$ can be continuously extended to E_d, and there exists a $\tau_d \in E_d$ such that

$$(\sigma, d\psi) = (\tau_d, d\psi) \tag{14}$$

for all $\psi \in D$.

We set $\tau = \tau_d + \tau_\delta \in E_{d\delta}$ and obtain on adding (12) and (14)

$$(\sigma, \delta\phi + d\psi) = (\tau, \delta\phi + d\psi). \tag{15}$$

In fact, $\tau_d \perp \delta\phi$ and $\tau_\delta \perp d\psi$. We infer that the difference $\rho = \sigma - \tau$ is $\perp E_{d\delta}$, hence in F. The proof of our main theorem is herewith complete.

11. We note in passing that both in the necessity and the sufficiency proofs, $d\sigma$ and $\delta\sigma$ were treated separately. This corresponds to the fact that τ is the sum of its x- and y-components, i.e., projections τ_d and τ_δ of σ on the x- and y-axes. These projections can be taken from our geometric model, or analytically from (12) and (14). The differences $\rho_c = \sigma - \tau_\delta \in E_c$ and $\rho_{cc} = \sigma - \tau_d \in E_{cc}$ are closed and coclosed, respectively, and we have the following by-product:

For a closed σ in $\bar{V}_1$ there exists a closed principal form ρ_c with a coexact $\sigma - \rho_c$ if and only if condition (9) *is satisfied.*

The same is true of the coclosed σ, coclosed ρ_{cc}, and exact $\sigma - \rho_{cc}$ if and only if (10) is met.

5. **Applications.**

12. We return to principal fields. The conditions (9) and (10) of our main theorem are in terms of $\sigma|\Gamma$ alone, and they can be satisfied by forms with strikingly simple properties. For instance:

THEOREM 3. *Let $\sigma \in F(\bar{V}_1) \cap C^\infty(V)$. If*

$$\sigma|\Gamma = d\alpha, \quad *\sigma|\Gamma = d\beta, \tag{16}$$

for some forms α, β defined in a neighborhood U of Γ, then the principal field ρ exists.

We depict U as an interval on the x-axis in a 2-dimensional scheme, V_0 between the origin and the point Γ, and α appearing as an arc above U. Multiply α by an $f \in C^\infty(U)$ which is $\equiv 1$ in a smaller neighborhood U_0 and then comes down to 0 towards the origin. The resulting form, extended as 0 to the rest of V_0, shall be denoted by α_0. Similarly we form β_0. Then

$$\sigma|\Gamma = d\alpha_0|\Gamma, \quad *\sigma|\Gamma = d\beta_0|\Gamma.$$

The integrals in (9) and (10) become

$$\int_\Gamma d\alpha_0 \wedge *\phi = \varepsilon \int_{V_0} d\alpha_0 \wedge *\delta\phi = O(\|\delta\phi\|),$$

$$\int_\Gamma d\beta_0 \wedge \phi = \varepsilon \int_{V_0} d\beta_0 \wedge d\phi = O(\|d\phi\|),$$

and the principal field ρ exists.

13. Perhaps somewhat surprisingly, no conditions whatever are needed on σ if the components Γ_i are *sphere-like.* By this we mean that they have disjoint neighborhoods, each homeomorphic to a spherical shell A (annulus if $n = 2$)

such that Γ_i goes to the sphere $S^{n-1} \subset A$. We know from advanced calculus that a closed form γ on S^{n-1} can be extended to an exact form on all of A, if the degree of γ is different from the dimension $n-1$ of S^{n-1}. If it is the same, then $\int_{S^{n-1}} \gamma$ must vanish to assure this extendability.

We thus have the following result:

THEOREM 4. *Let the Γ_i be sphere-like. Then the principal field ρ exists for*
(a) *all σ if $1 < p < n-1$,*
(b) *σ with $\int_{\Gamma_i} *\sigma = 0$ if $p = 1$,*
(c) *σ with $\int_{\Gamma_i} \sigma = 0$ if $p = n-1$.*

In fact, in case (a), deg $*\sigma$ is also in $(1, n-1)$; in case (b), only $*\sigma$ has the critical degree $n-1$; in case (c), σ alone needs a condition.

14. It is clear that the Γ_i are not always sphere-like, as is exemplified by the solid torus. But for $n = 2$ this condition is automatically fulfilled, and we have:

THEOREM 5. *On Riemannian 2-spaces (hence on Riemann surfaces) the principal 1-field ρ exists if*

$$\int_{\Gamma_i} \sigma = \int_{\Gamma_i} *\sigma = 0. \tag{17}$$

Thus the problem for harmonic differentials is solved. For exact differentials, i.e., for functions, the first part of (17) is satisfied by definition, and in the second part we recognize the flux condition.

15. We can now also solve the problem we referred to in the beginning of our presentation, that of constructing harmonic fields with given singularities. In a Riemannian space V_0, let K be a compact discrete set contained in a finite set B of disjoint balls B_i. Let σ be a harmonic field in $\bar{B} - K$ with singularities at K. The problem is to find a harmonic field ρ in $V_0 - K$ with behavior σ in $B - K$. We also require that $\| \rho \|_{V_0 - B} < \infty$.

Let B' be a set of slightly larger disjoint balls $B_i' \supset B_i$. For the space V of our main theorem we take $V_0 - K$ and set $V_1 = (B - K) \cup (V - \bar{B}')$. Then the Γ_i are sphere-like. For σ we take the given σ in $\bar{B} - K$ and set $\sigma = 0$ in $V - B'$. We have the following

THEOREM 6. *The conditions* (a)–(c) *of Theorem* 4, *with the integrals extended over the ∂B_i, guarantee the existence of a principal field ρ with the given singularity σ, such that $\| \sigma - \rho \|_{B-K} < \infty$ and $\| \rho \|_{V_0 - B} < \infty$.*

6. **A generalization.**

16. Thus far we have assumed that $\sigma | V_1$ is a harmonic field. Can we relax this condition? Yes, we can, and even allow $\sigma | V_1$ to be only locally square integrable, if we tighten conditions (9) and (10):

THEOREM 7. *If $\sigma \in E^{loc}(V)$, then the principal field $\rho \in F(V)$ with $\sigma - \rho \in E_{d\delta}$ exists if and only if*

$$\int_\Gamma \sigma \wedge *\phi + \int_{V_1} d\sigma \wedge *\phi = O(\|\delta\phi\|), \tag{18}$$

and

$$\int_\Gamma *\sigma \wedge \phi + \int_{V_1} *d\sigma \wedge \phi = O(\|d\phi\|). \tag{19}$$

The proof is much the same as before, with the second integral now being added since $d\sigma$ and $*d\sigma$ do not vanish. The conditions are not as explicit as the earlier ones, but there is perhaps some beauty in the fact that no differentiability properties whatever are required of σ, yet we can find a ρ in $F(V)$ with its C^∞ smoothness.

7. Harmonic semifields.

17. Thus we have solved the principal field problem. But harmonic fields are special cases of harmonic forms. Can we solve the problem of principal harmonic forms?

We first consider forms that are intermediate between harmonic forms and harmonic fields. We shall call them *harmonic semifields* and denote their class by S. The definitions are as follows. Let

$$\Delta' = \delta d, \quad \Delta'' = d\delta, \quad \Delta = \Delta' + \Delta'' \tag{20}$$

Then

$$\begin{aligned} F &= \{\alpha \mid d\alpha = \delta\alpha = 0\}, \\ S &= \{\alpha \mid \Delta'\alpha = \Delta''\alpha = 0\}, \\ H &= \{\alpha \mid \Delta\alpha = 0\}. \end{aligned} \tag{21}$$

Clearly $F \subset S \subset H$.

We form the spaces

$$E_{\Delta'} = \overline{\Delta' D}, \quad E_{\Delta''} = \overline{\Delta'' D}, \quad E_{\Delta'\Delta''} = E_{\Delta'} + E_{\Delta''}, \tag{22}$$

where the closures are in E, and the spaces

$$O_{\Delta'} = \{\alpha \mid \Delta'\alpha = 0\}, \quad O_{\Delta}'' = \{\alpha \mid \Delta''\alpha = 0\}. \tag{23}$$

In contrast with δ and d, the operator Δ' is self-adjoint,

$$(\delta d\alpha, \beta) = (\alpha, \delta d\beta),$$

and the same is true of Δ''. As a consequence,

(24) $$E_{\Delta'} \perp E_{\Delta''}, \quad O_{\Delta'} = E_{\Delta'}^{\perp}, \quad O_{\Delta''} = E_{\Delta''}^{\perp},$$

and our geometric scheme takes the following form. The 3-space shall again stand for the space E of square integrable forms. For the x-axis we take $E_{\Delta'}$, and for the y-axis $E_{\Delta''}$, so that the xy-plane is $E_{\Delta'\Delta''}$. The xz-plane becomes $O_{\Delta''}$ and the yz-plane $O_{\Delta'}$. The z-axis is the space S of harmonic semifields: $\Delta'\alpha = \Delta''\alpha = 0$. Again $S \subset C^{\infty}$.

18. If $\|\sigma\| < \infty$, then we take the projection τ of σ on $E_{\Delta'\Delta''}$, and the desired principal semifield $\rho = \sigma - \tau$ can again be directly taken from our model. In the general case the solution is as follows (Larsen, Nakai and Sario [3]):

THEOREM 8. *If $\sigma \in S(\bar{V}_1) \cap C^{\infty}(V)$, then the principal semifield $\rho \in S(V)$ with $\sigma - \rho \in E_{\Delta'\Delta''}$ exists if and only if*

(25) $$\int_{\Gamma} \sigma \wedge *d\phi - \phi \wedge *d\sigma = O(\|\Delta'\phi\|)$$

and

(26) $$\int_{\Gamma} \delta\sigma \wedge *\phi - \delta\phi \wedge *\sigma = O(\|\Delta''\phi\|).$$

The proof follows along the same lines as that for harmonic fields. In particular, the Δ' and Δ'' parts can be separated, corresponding to the fact that τ is the sum of the projections $\tau_{\Delta'}$ and $\tau_{\Delta''}$ of σ on the $E_{\Delta'}$- and $E_{\Delta''}$-axes. As a by-product we obtain the existence of principal forms $\rho_{\Delta'}$ in $O_{\Delta'}$ and $\rho_{\Delta''}$ in $O_{\Delta''}$.

8. **Harmonic forms.**

19. We turn to harmonic forms. Again Δ is self-adjoint,

$$(\Delta\alpha, \beta) = (\alpha, \Delta\beta),$$

and the space H is the orthogonal complement of $E_{\Delta} = \overline{\Delta D}$. The space E can now be represented by the 2-space, with the x-axis E_{Δ} and the y-axis H. If $\|\sigma\| < \infty$, the y-coordinate of σ is the desired principal harmonic form ρ. In the general case we have (Larsen, Nakai and Sario [3]):

THEOREM 9. *Let $\sigma \in H(\bar{V}_1) \cap C^{\infty}(V)$. Then the principal harmonic form $\rho \in H(V)$ with $\sigma - \rho \in E_{\Delta}$ exists if and only if*

(27) $$\int_{\Gamma} \sigma \wedge *d\phi + \delta\sigma \wedge *\phi - \phi \wedge *d\sigma - \delta\phi \wedge *\sigma = O(\|\Delta\phi\|).$$

The proof is again based on expressing the integral over Γ by one over V_0, where by Green's formula we have two terms, this time $\Delta\sigma \wedge *\phi$ and $\sigma \wedge *\Delta\phi$. Each can be estimated using a similar technique as before.

Condition (27) is in terms of $\sigma | \Gamma$ alone and can be met by harmonic forms with quite simple properties. E.g., if σ agrees on Γ with an exact and a coexact form, then the existence of the principal harmonic form is assured.

Bibliography

1. L. Ahlfors and L. Sario, *Riemann surfaces*, Princeton Univ. Press, Princeton, N. J., 1960.

2. K. Kodaira, *Harmonic fields in Riemannian manifolds (generalized potential theory)*, Ann. of Math. **50** (1949), 587–665.

3. K. Larsen, M. Nakai and L. Sario, *Principal harmonic forms on Riemannian spaces*, Bull. Sci. Math. **18** (1967), 245–257.

4. M. Nakai and L. Sario, *Harmonic fields with given boundary behavior in Riemannian spaces*, J. Analyse Math. (to appear).

5. G. de Rham, *Variétés différentiables. Formes, courants, formes harmoniques*. Actualités Sci. Ind., no. 1222, Publ. Inst. Math. Univ. Nancago. III. Hermann, Paris, 1960.

NAGOYA UNIVERSITY
NAGOYA, JAPAN
UNIVERSITY OF CALIFORNIA
LOS ANGELES, CALIFORNIA

Fischer Spaces of Entire Functions[1]

D. J. NEWMAN
H. S. SHAPIRO

Introduction. We begin by setting ourselves the following problem: to devise an inner product among polynomials such that the operations of multiplication by x and differentiation become adjoint.

It suffices to define $(x^m, x^n) = a_{m,n}$ and the adjointness condition requires that $(x \cdot x^m, x^n) = (x^m, dx^n/dx)$ or $a_{m+1,n} = na_{m,n-1}$. Thus $a_{m,n} = n!f(m-n)$ and by symmetry $a_{m,n} = a_{n,m}$ so that $f(m-n)$ must equal **0** for $m \neq n$. Normalizing by choosing $f(0) = 1$, then, gives

$$\begin{aligned} a_{m,n} &= n! \quad \text{for } m = n \\ &= 0 \quad \text{for } m \neq n. \end{aligned}$$

This choice forces the definition $(\sum a_n x^n, \sum b_n x^n) = \sum n!\, \bar{a}_n b_n$. Various other expressions for the inner product are $(P(x), Q(x)) = \boldsymbol{P}(d/dx)Q(x)|_0$ (where $\boldsymbol{P}$ means the polynomial defined from P by conjugating its coefficients and $f(x)|_0$ means $f(0)$), and $(P(x), Q(x)) = 1/\pi \int_{\text{plane}} \bar{P}(z)Q(z)\exp[-|z|^2]\, dA$ (dA being the ordinary Lebesgue measure on the plane).

The inner product is positive definite and so leads to a Hilbert space completion which we call *Fischer space*. This Fischer space is clearly the same as the space of entire functions $f(z)$ normed by

$$\|f(z)\|^2 = \frac{1}{\pi}\int |f(z)|^2 \exp[-|z|^2]\, dA.$$

(Crudely these are functions whose growth is "well dominated" by $\exp|z|^2/2$.)

This completes the definitions of the Fischer space in one variable. For several variables the situation is identical. We seek an inner product on polynomials which renders x_i and $\partial/\partial x_i$ adjoint for all $i = 1, 2, \cdots, k$ and we are led to the definition $\|f(Z)\|^2 = (1/\pi^k)\int_{C^k} |f(Z)|^2 \exp[-|Z|^2]dA_Z$ (here $Z = (z_1, z_2, \cdots, z_k)$, $|Z|^2 = |z_1|^2 + |z_2|^2 + \cdots + |z_k|^2$, and dA_Z is Lebesgue measure in C^k).

---

[1] This paper is essentially an account of a talk by the first-named author, and the presentation is somewhat informal. Some further references and background may be found in the author's paper, *Certain Hilbert spaces of entire functions*, Bull. Amer. Math. Soc. **72** (1966), 971–977.

We briefly denote $(1/\pi^k)\exp[-|Z|^2]dA_Z$ by $d\sigma_Z$, and the Fischer space by $\mathscr{F}_Z$
It is worth mentioning, at this point, that these spaces clearly have a strong connection with quantum theory and as such have received some attention from physicists. [See e.g. Bargmann, Comm. Pure Appl. Math. **14** (1961).] A somewhat different approach from Bargmann's but still using Fischer spaces for the field of quantum theory is to be found in the work of I. E. Segal. At any rate we find it worthwhile to investigate problems in this Fischer space norm. (These problems may perhaps be looked at in other norms as well, but the Fischer norm has a built-in special importance.)

The present account will be a survey of some recent work of ours and will have little overlap with the work of the aforementioned authors.

Historical remarks. The first real application of these spaces was made by Ernst Fischer. He proved, for example, that nth degree homogeneous polynomials in the k variables

$$x_1, x_2, \cdots, x_k$$

have a unique representation as a sum of an nth degree *harmonic* polynomial and a multiple of $x_1^2 + x_2^2 + \cdots + x_k^2$.

The proof of Fischer's theorem is given by "soft" algebra. Namely, the operators $A = \partial^2/\partial x_1^2 + \partial^2/\partial x_2^2 + \cdots + \partial^2/\partial x_k^2$ on nth degree forms and $B = (x_1^2 + x_2^2 + \cdots + x_k^2)$ on $(n-2)$th degree forms are adjoints under the Fischer inner product and we are simply asserting that the whole space (nth degree forms) splits into the null space of A and the range of B. Indeed the null space of an operator is simply the orthogonal complement of the range of its adjoint.

This lovely method of Fischer's was easily generalized by him to forms other than $x_1^2 + \cdots + x_k^2$ because the underlying spaces are always finite dimensional. The situation changes radically, however, when we consider nonhomogeneous polynomials or even nonpolynomials. Here "soft" algebra gives way to "hard" analysis and the dimension becomes infinite.

Differential operators. Let $\phi(Z)$ be an entire function in the variables $Z = (z_1, z_2, \cdots, z_k)$ and suppose that

$$\phi(Z) = O(\exp[|Z|^2/2] - A|Z|]) \qquad \text{for all } A > 0.$$

Surely such ϕ will belong to Fischer space $\mathscr{F}_Z$, but even more is true since every exponential times ϕ also lies in $\mathscr{F}_Z$. We call the class of such ϕ, K, the *operator class*. For each $\phi \in K$ we define the operator $\boldsymbol{\phi}(D)$ $[D = (\partial/\partial z_1, \partial/\partial z_2, \cdots, \partial/\partial z_k)]$, by setting[2] $\boldsymbol{\phi}(D)f(Z) = \int \overline{\phi(\zeta)}\, f(\zeta)\, e^{\bar{\zeta} Z} d\sigma_\zeta$, for all $f(Z) \in \mathscr{F}_Z$.

For $\phi(Z)$ a polynomial or even an exponential polynomial this agrees with the

[2] Here ζ denotes $(\zeta_1, \ldots, \zeta_k)$ and $\bar{\zeta} Z = \Sigma_1^k \bar{\zeta}_i z_1$. The cofactor of Z in exponentials in this paper is always understood to be a complex k-tuple; $a\, b$ denotes $\Sigma_{i=1}^k a_i b_i$.

usual definition. It also represents a formal adjoint to multiplication by $\phi(Z)$ ($e^{\bar{\zeta}Z}$ being the reproducing kernel) and so we adopt it as the "correct" generalization.

Statements of problems.

I. Existence problems. For which $\phi \in K$ can we assert that $\phi(D)f(Z) = g(Z)$ has a solution $f \in \mathscr{F}_Z$ for every $g \in \mathscr{F}_Z$?

II. For which $\phi \in K$ can we obtain a "Fischer theorem"? More precisely: call M_ϕ the subset of $\mathscr{F}_Z$ which is annihilated by $\phi(D)$ and call N_ϕ the set of $f \in \mathscr{F}_Z$ for which $f(Z)/\phi(Z)$ is entire.

When are M_ϕ and N_ϕ orthogonal complements?

We will prove the affirmative result for I and II when ϕ is an exponential polynomial. The question remains as to how far the class of exponential polynomials can be extended. For the case of problem I, this extension could only be slight, as certain counterexamples show. The situation is completely different for problem II. No counterexamples at all are known! It may even be affirmative for all $\phi \in K$, but, on the other hand, we cannot even prove it for functions of exponential type.

Our principal tool will be an apparently new identity in operational calculus which we will call our *isometry theorem*. As a prelude to this theorem we now introduce (following Bargmann, loc. cit.)

The translation isometries: The translation map $f(Z) \to f(Z + a)$ does *not* take $\mathscr{F}_Z$ into itself, nor does the multiplication map $f(Z) \to e^{aZ} f(Z)$. By properly composing these two maps, however, we can obtain in fact an isometry map of $\mathscr{F}_Z$ onto $\mathscr{F}_Z$.

THEOREM 1. *The maps* $T_a f(Z) = \exp[-|a|^2/2]\, e^{-\bar{a}Z} f(Z + a)$ *are isometries of* $\mathscr{F}_Z$ *onto* $\mathscr{F}_Z$.

The proof can be given directly by change of variables in the norm formula. We do not give details since this theorem will be a corollary of the isometry theorem. There is a nice conclusion we can draw from Theorem 1 and that is

THEOREM 2. *Let $f(Z)$ be a given entire function (not necessarily in $\mathscr{F}_Z$). The set, A, of a for which $f(Z + a) \in \mathscr{F}_Z$ is convex.*

PROOF. By Theorem 1, $f(Z + a) \in \mathscr{F}_Z$ is equivalent to $e^{\bar{a}Z} f(Z) \in \mathscr{F}_Z$. Let $a, b \in A$ and $c = ta + (1 - t)b$, $0 \leqq t \leqq 1$. Thus $e^{\bar{a}Z} f(Z) \in \mathscr{F}_Z$, $e^{\bar{b}Z} f(Z) \in \mathscr{F}_Z$ and we wish to conclude that $e^{\bar{c}Z} f(Z) \in \mathscr{F}_Z$. This follows, however, from Hölder's inequality since

$$\begin{aligned} \int |e^{\bar{c}Z} f|^2 d\sigma &= \int |e^{\bar{a}Z} f|^{2t} |e^{\bar{b}Z} f|^{2(1-t)} d\sigma \\ &\leqq \left(\int |e^{\bar{a}Z} f|^2 d\sigma \right)^t \left(\int |e^{\bar{b}Z} f|^2 d\sigma \right)^{1-t}. \end{aligned}$$

Theorems 1 and 2 will be useful later on in our investigations of problem I.

The main isometry theorem. We now introduce an isometry from a subspace of one Fischer space into another Fischer space in twice the number of variables. This isometry may be symbolically thought of as $\phi(Z)f(Z) \to \phi(W+D)f(Z)$. For example with $\phi(z) = z$, $k = 1$ we obtain the map $zf(z) \to wf(z) + f'(z)$. To say that this is an isometry is to say that $\| zf(z) \|^2 = \| f(z) \|^2 + \| f'(z) \|^2$. Again specialize $\phi(Z) = e^{\bar{a}Z}$ and observe that the isometry takes the form

$$e^{\bar{a}Z} f(Z) \to e^{aW} e^{aD} f(Z) = e^{aW} f(Z+a)$$

which means that

$$\begin{aligned} \| e^{\bar{a}Z} f(Z) \|^2 &= \| e^{aW} \|_W^2 \, \| f(Z+a) \|^2 \\ &= e^{|a|^2} \| f(Z+a) \|^2 \end{aligned}$$

and this last identity is simply Theorem 1.

We proceed to the rigorous statement of our isometry theorem.

THEOREM 3. *Let* $\phi(Z)f(Z) \in \mathscr{F}_Z$. *Define* $G(Z,W) = e^{-ZW} \int \overline{\phi(\zeta)}\, e^{\bar{\zeta}Z} e^{\zeta W} f(\zeta) d\sigma_\zeta$. *Then* $G(Z,W) \in \mathscr{F}_{Z,W}$ *and* $\| G \| = \| \phi f \|$.

PROOF. Let $f(\zeta)\overline{\phi(\zeta)} \exp[-|\zeta|^2/2] = F(\zeta)$ and $G(Z,W)\exp[-|Z|^2/2 - |W|^2/2] = B(Z,W)$.

If we change variables by setting $Z = (u+v)/2$, $W = (\bar{u} - \bar{v})/2$, in the definition of G we obtain

$$B(Z,W) = \frac{\exp[i \operatorname{Im} u \bar{v}]}{\pi^k} \int F(\zeta) \exp[-|\zeta - u|^2/2]\, e^{i \operatorname{Im}(\bar{\zeta} v)} dA_\zeta$$

so that $|B| = \pi^{-k} |\mathscr{T}(F(\zeta)\exp[-|\zeta - u|^2/2])|$ where $\mathscr{T}$ is the fourier transform! By the fact that the Fourier transform is an isometry on $L^2(dA_\zeta)$ we conclude that

$$\int |B|^2 dA_v = c_1 \int |F(\zeta)|^2 \exp[-|\zeta - u|^2]\, dA_\zeta$$

(c_1 the appropriate constant). Integrating this identity dA_u gives

$$\begin{aligned} \iint |B|^2 dA_Z dA_W &= c_2 \int |B|^2 dA_v dA_u, \\ &= c_2 c_1 \int |F(\zeta)|^2 \int \exp[-|\zeta - u|^2]\, dA_u dA_\zeta, \\ &= c_3 \int |F(\zeta)|^2 dA_\zeta \end{aligned}$$

and this is the desired result once we show that $c_3 = 1$. That $c_3 = 1$ can be shown in a variety of ways. We could of course just be careful all along and calculate the various constants which arise. Easier still we may simply set $f(Z) \equiv 1$ at the very end. This leads to $\| \phi(Z) \|_Z^2 = c_3 \| \phi(W) \|_W^2$ and so, indeed, completes the proof.

Analysis of problem I. The first step is to exploit the formal adjointness of $\phi(Z)\cdot$ and $\phi(D)$. Thus to say that $\phi(D)$ is *onto* is the same as saying that $\phi(Z)$ has a bounded inverse. Our differential equation problem, then, simply becomes the question of the existence of a $\delta > 0$ for which

$$\| \phi(Z) f(Z) \| \geqq \delta \| f(Z) \| . \tag{1}$$

To see the mechanism by which (1) is obtained for exponential polynomials via the isometry theorem consider the special case $\phi(z) = z$, $k = 1$.

Here we have $\| z f(z) \|^2 = \| f(z) \|^2 + \| f'(z) \|^2 \geqq \| f(z) \|^2$ so that (1) follows with $\delta = 1$. We conclude that the D.E. $f'(z) = g(z)$ has a solution $f \in \mathscr{F}_Z$ for every $g \in \mathscr{F}_Z$.

This example is quite trivial; a more recondite example is given by $\phi(z) = e^z - 1$, $k = 1$. We wish to prove (1) and we do so by employing Theorem 3. This tells us that

$$\begin{aligned} \| (e^z - 1) f(z) \|^2 &= \| e^w f(z+1) - f(z) \|^2_{w,z} \\ &= \| f(z+1) - f(z) \|^2 + (1 + 1/2! + 1/3! + \cdots) \| f(z+1) \|^2 \\ &= \| f(z+1) - f(z) \|^2 + (e-1) \| f(z+1) \|^2 . \end{aligned}$$

The latter is $\geqq (1 - 1/e) \| f(z) \|^2$ by the elementary inequality $(y - x)^2 + (e-1)y^2 \geqq (1 - 1/e)x^2$ and we conclude (1) with $\delta = (1 - 1/e)^{1/2}$.

Thus the difference equation $f(z+1) - f(z) = g(z)$, for $g \in \mathscr{F}_Z$ has solution $f \in \mathscr{F}_Z$ (in fact with $\| f \| \leqq (1 - 1/e)^{-1/2} \| g \|$, $(1 - 1/e)^{-1/2}$ being unimprovable).

As a final example consider the Laplace equation, i.e. $\phi(Z) = Z \cdot Z$. Theorem 3 gives

$$\| Z Z f(Z) \|^2 = \| \Delta f \|^2 + 4 \sum_{i=1}^{k} \| \frac{\partial f}{\partial z_i} \|^2 + 2k \| f \|^2 \geqq 2k \| f \|^2 .$$

Thus $\Delta f = g$ has solution with $\| f \| \leqq (2k)^{-1/2} \| g \|$ with $(2k)^{-1/2}$ the correct bound.

Rather than examining further examples let us draw one simple conclusion: When $\phi(Z)$ is a polynomial the isometry theorem can be written as

$$\| \phi(Z) f(Z) \|^2 = \sum_n \frac{\| \phi^{(n)}(D) f(Z) \|^2}{n!}$$

where the sum goes over all vectors $n = (n_1, n_2, \cdots, n_k)$ of nonnegative integers, and $n!$ means $n_1!\, n_2! \cdots n_k!$. This formula is not important but what counts is the fact that for at least one of the terms this sum $\phi^{(n)}(D)$ is a nonzero constant. This implies (1) and so tells us that any partial differential equation has solutions, i.e., problem I is solved affirmatively for polynomials.

For the promised case of exponential polynomials we must use some care. So simple an equation as $f(z+1) = g(z)$ $(\phi(z) = e^z)$ obviously does not have a solution. To formulate the correct theorem we introduce the notion of a "balanced" exponential polynomial.

An exponential polynomial, ϕ, will be called *unbalanced* if all its exponents lie in a half space (so that ϕ goes to 0 throughout a half space), otherwise it is *balanced*. Clearly this condition is equivalent to requiring that 0 lie in the convex hull of the exponents. Thus $e^z - 1$, $ze^z - e^{-z}$, etc. are balanced while e^z, $5e^z + e^{iz}$ etc. are unbalanced.

THEOREM 4. *Let $\phi(Z)$ be a balanced exponential polynomial. The differential-difference equation $\phi(D)f(Z) = g(Z)$ has a solution $f \in \mathscr{F}_Z$ for every $g \in \mathscr{F}_Z$.*

PROOF. We must verify (1). Write

$$\phi(Z) = \sum_{\lambda} P_\lambda(Z) e^{\lambda Z} ,$$

P_λ being polynomials. Theorem 3 gives

$$\| \phi(Z) f(Z) \|^2 = \| \sum_{\lambda} P_\lambda(W + D) e^{\bar{\lambda} W} f(Z + \bar{\lambda}) \|^2$$

$$= \| \sum_{n,\lambda} \frac{P_\lambda^{(n)}(W)}{n!} e^{\bar{\lambda} W} f^{(n)}(Z + \bar{\lambda}) \|_{Z,W}^2$$

But consider the expression

$$\| \sum_{n,\lambda} \frac{P_\lambda^{(n)}(W)}{n!} e^{\bar{\lambda} W} C_{n,\lambda} \|_W^2 ;$$

this is clearly a positive definite quadratic form in the $C_{n,\lambda}$. These $C_{n,\lambda}$ are finite in number and we conclude that

$$\| \sum_{n,\lambda} \frac{P_\lambda^{(n)}(W)}{n!} e^{\bar{\lambda} W} C_{n,\lambda} \|_W^2 > \varepsilon \sum_{n,\lambda} | C_{n,\lambda} |^2 \qquad (\varepsilon > 0).$$

This means that

$$\| \sum_{n,\lambda} \frac{P_\lambda^{(n)}(W)}{n!} e^{\bar{\lambda} W} f^{(n)}(Z + \bar{\lambda}) \|_{W,Z}^2 > \varepsilon \sum_{n,\lambda} \| f^{(n)}(Z + \bar{\lambda}) \|_Z^2 \geqq \varepsilon \sum_{\lambda} \| f(Z + \bar{\lambda}) \|^2 .$$

But the $\bar{\lambda}$ contain 0 in the convex hull and so Theorem 2 tells us that this last sum is $\geqq \varepsilon' \|f(Z)\|^2$ and (1) is verified.

It might be thought that this last result could be extended to functions of exponential type but it cannot. Indeed for $\phi(z) = (e^z - 1)/z$ we obtain the equation

$$\int_0^1 f(z+t)\,dt = g(z) \quad \text{or} \quad f(z+1) - f(z) = g'(z)$$

and this does not necessarily have solution in $\mathscr{F}_Z$ for $g \in \mathscr{F}_Z$.

Analysis of problem II. We begin by pointing out some interesting reformulations of the Fischer decomposition. Many of these have analogues outside of our present framework and lead to interesting side problems. We will mention these as they arise.

Recall the present formulation:

$$M_\phi = \text{space of all } f(Z) \in \mathscr{F}_Z \text{ for which } \phi(D)f(Z) = 0,$$

$$N_\phi = \text{space of all } f(Z) \in \mathscr{F}_Z \text{ for which } f(Z)/\phi(Z) \text{ is entire.}$$

1. M_ϕ and N_ϕ are orthogonal complements in $\mathscr{F}$.

Consider now the space $N'_\phi \subseteq N_\phi$ consisting of all functions $f(Z)$ which are multiples of $\phi(Z)$ by exponential polynomials. Let us compute the orthogonal complement of $\overline{N'_\phi}$ ($\overline{}$ means closure). The condition for $f(Z)$ to lie in this complement means that $(\phi(Z)e^{aZ}, f(Z)) = 0$ for all a. This in turn means that $\int e^{\bar{a}z}\phi(Z)f(Z)\,d\sigma_Z = 0$ for all a, i.e., that $\phi(D)f(Z) = 0$.

The orthocomplement of $\overline{N'_\phi}$ is M_ϕ and so we have our second formulation.

2. $N_\phi = \overline{N'_\phi}$.

Equivalently we may express 2 as

2′. In the Hilbert space of entire functions square integrable with respect to the measure $|\phi(Z)|^2 d\sigma_Z$ the exponential polynomials are dense.

It is easy to replace the exponential polynomials by ordinary polynomials and obtain the equivalent formulation:

2″. In the Hilbert space of entire functions square integrable with respect to the measure $|\phi(Z)|^2 d\sigma_Z$ the polynomials are dense.

Again since $N_\phi \supseteq N'_\phi$ the statement 1 is equivalent to simply

3. $M_\phi \cap N_\phi = 0$, i.e., there is no nontrivial solution to $\phi(D)f(Z) = 0$ which vanishes (with correct multiplicity) on $\phi(Z) = 0$.

Now let us consider the space M'_ϕ consisting of all exponential polynomials in M_ϕ. These are (aside from multiplicity) spanned by the $e^{\bar{a}Z}$ for which $\phi(a) = 0$. Since, however, $e^{\zeta Z}$ is the reproducing kernel, $f(Z)$ orthogonal to $e^{\bar{a}Z}$ means $f(a) = 0$. Thus (when multiplicities are included) $f(Z) \perp \overline{M'_\phi}$ means $f \in N_\phi$.

Our next formulation, then, is

4. $M_\phi = \overline{M'_\phi}$; i.e., the exponential solutions[3] to $\boldsymbol{\phi}(D)f = 0$ are dense in the set of all solutions.

Finally we state

5. $\boldsymbol{\phi}(D)$ is *truly* adjoint to $\phi(Z)$ according to the standard definition of adjointness for unbounded operators.

Note that $5 \Rightarrow 3$ for if $f(Z) \in N_\phi$ it can be written as $\phi(Z)\cdot F(Z)$. We obtain

$$\| \phi F \|^2 = (F, \boldsymbol{\phi}(D)f) = (F, 0) = 0$$

since $f \in M_\phi$. Hence $\|f\| = 0$ and $f = 0$.

Actually, 5 is *equivalent* to the other formulations but, since we never use this fact, we omit the proof.

We will prove the Fischer decomposition theorem (1–5) only for $\phi(Z)$ an exponential polynomial but we remind the reader that we have no counterexamples at all and it may even hold for all $\phi \in K$.

Before turning to the proofs, however, let us elaborate some of the side problems suggested by the above discussion.

In particular formulation 3 has an extension outside Fischer space in case ϕ is a polynomial. We may well ask whether any entire $f(Z)$ can exist which vanishes with multiplicities on $\phi(Z) = 0$ and which satisfies $\boldsymbol{\phi}(D)f(Z) = 0$. The negative answer is easily obtained when $k = 1$ or when the polynomial ϕ is homogeneous. We also have obtained the result when ϕ is the sum of two homogeneous polynomials. No counterexamples are known!!

We also wish to urge questions like 2″ on the reader. These need not be tied to Fischer space. We may ask generally "When are the analytic polynomials in $L^2(dm)$ dense in the analytic functions in $L^2(dm)$?" The problem seems deeper than at first sight. Examples show that the mere fact that *all* polynomials are dense in *all* of $L^2(dm)$ does not imply the previous statement.[4]

We now turn to the proof of 5 for ϕ an exponential polynomial. The trick is to establish the property † below for ϕ an exponential polynomial and then to establish that † ⇒ 5.

† For any entire $F(Z)$ such that $\phi(Z)F(Z) \in \mathscr{F}_Z$, the function $\phi(Z+W)F(Z) \in \mathscr{F}_Z$ for almost all W and $\int \| \phi(Z+W)F(Z) \|_Z \exp[-|W|^2/2]\, dA_W < \infty$.

Let us first establish that † ⇒ 5. So let $\phi(Z)F(Z) \in \mathscr{F}_Z$ and $f(Z) \in \mathscr{F}_Z$. The formula for $(F(Z), \boldsymbol{\phi}(D)f(Z))$ is, by definition,

$$\int\int \overline{F(Z)}\,\overline{\phi(\zeta)}\, e^{\bar{\zeta} Z} f(\zeta)\, d\sigma_\zeta\, d\sigma_Z$$

while $(\phi(Z)F(Z), f(Z))$ is equal to

---

[3] More precisely, the finite linear combinations of the *exponential-monomial* solutions $z_1^{n_1} \cdots z_k^{n_k} e^{aZ}$

[4] Recent work of L. I. Hedberg, F. Lisin, and others has begun to clarify the situation in one variable.

$$\int \overline{\phi(\zeta)}\,\overline{F(\zeta)}\,f(\zeta)d\sigma_\zeta = \int\int \overline{F(Z)}\,\overline{\phi(\zeta)}\,e^{\bar{\zeta}Z}f(\zeta)\,d\sigma_Z d\sigma_\zeta .$$

Thus 5 is really just an interchange of double integrals. To validate this change, of course, it suffices to show that

$$f(Z)\,\phi(\zeta)\,e^{\bar{\zeta}Z}f(\zeta) \in L^1(d\sigma_Z\, d\sigma_\zeta)$$

or, setting $\zeta = Z + W$, this requirement becomes

$$\int\int |F(Z)\,\phi(Z+W)f(Z+W)e^{-\bar{W}Z}|\, d\sigma_W d\sigma_W < \infty .$$

Applying Schwarz' inequality to the inner integral shows the sufficiency of the finiteness of

$$\int \| F(Z)\,\phi(Z+W)\,\|_Z \, \| f(Z+W)\,e^{-Z\bar{W}}\|_Z \, d\sigma_W .$$

But by Theorem 1, $\|f(Z+W)\,e^{-Z\bar{W}}\|_Z = \exp[|W|^2/2]\,\|f\|$, and so the above integral is simply equal to

$$\|f\| \cdot \int \| F(Z)\,\phi(Z+W)\,\|Z \exp[-|W|^2/2]\,dA_W .$$

That this is finite, however, is exactly the condition †.

The final burden remains, then, to prove † when $\phi(Z)$ is an exponential polynomial. To give the idea of the proof let us first do it for $\phi(Z)$ a polynomial. We apply our isometry theorem to obtain

$$\|\phi(Z)\,F(Z)\|^2 = \sum_n \frac{\|\phi^{(n)}(D)\,F(Z)\|^2}{n!} .$$

Applying this to $\phi^{(m)}(Z)\,F(Z)$ gives

$$\|\phi^{(m)}(Z)\,F(Z)\|^2 = \sum_n \frac{\|\phi^{(n+m)}(D)\,F(Z)\|^2}{n!} .$$

Since all the terms on the right are contained among those of the first sum we conclude that

$$\|\phi^{(m)}(Z)\,F(Z)\| < \infty .$$

Knowing this we conclude that

$$\phi(Z+W)\,F(Z) = \sum_m \frac{\phi^{(m)}(Z)\,F(Z)}{m!} W^m$$

has $\mathscr{F}_Z$ norm which is finite for each W and grows only as fast as a polynomial in the W variables. † follows easily.

Finally we give the argument for general exponential polynomials. Again we apply the isometry theorem. This gives, with $\phi(Z) = \sum_\lambda P_\lambda(Z)e^{\lambda Z}$

$$\|\phi F\| = \|\sum_\lambda P_\lambda(Z)e^{\lambda Z}\cdot F(Z)\| = \|\sum_\lambda P_\lambda(W+D)e^{\bar\lambda W}F(Z+\bar\lambda)\|_{Z,W}$$

$$= \|\sum_{n,\lambda}\frac{P_\lambda^{(n)}(W)}{n!}e^{\bar\lambda\cdot W}F^{(n)}(Z+\bar\lambda)\|_{Z,W}.$$

It follows, then, that each $F^{(n)}(Z+\bar\lambda)\in\mathscr{F}_Z$ for all n,λ for which $P_\lambda^{(n)}\neq 0$. This being the case, what can we say about $\|\phi(Z+a)F(Z)\|$? Applying the previous argument we obtain

$$\|\phi(Z+a)F(Z)\|_Z = \|\sum_{n,\lambda}\frac{P_\lambda^{(n)}(W+a)}{n!}\exp[\bar\lambda W+\bar\lambda a]F^{(n)}(Z+\bar\lambda)\|_{Z,W}$$

and since each $F^{(n)}(Z+\bar\lambda)$ has finite $\mathscr{F}_Z$ norm and the dependence on a is only exponential, we conclude that $\|\phi(Z+a)F(Z)\|_Z$ grows at worst exponentially in the variables a. Once again this is sufficient for † and the proof is complete.

YESHIVA UNIVERSITY
NEW YORK, NEW YORK
UNIVERSITY OF MICHIGAN
ANN ARBOR, MICHIGAN

Extremal Length and Geometric Inequalities[1]

BURTON RODIN

The theme of this paper is the systematic use of potential-theoretic methods for obtaining differential-geometric inequalities. An elegant illustration is provided by C. Loewner's proof of the length-area inequality for quadrilaterals (see A. S. Besicovitch [2] for a purely geometric proof).

Let R be a Riemannian 2-space, $\mathscr{F}_i$ $(i = 1,2)$ a family of curves on R, and let l_i be the infimum of the lengths of the curves in $\mathscr{F}_i$. Trivially, an inequality of the form

$$l_1^2 \leqq (\text{const.})\,(\text{area of } R)$$

amounts to an estimate for the extremal length $\lambda(\mathscr{F}_1)$ of $\mathscr{F}_1$ on the underlying Riemann surface. It should be observed that an inequality of the form

$$l_1 l_2 \leqq (\text{area of } R)$$

can also be cast into the standard extremal length framework; namely, it follows from a conjugate extremal length statement such as $\lambda(\mathscr{F}_1)\cdot\lambda(\mathscr{F}_2) = 1$. We shall derive several inequalities of the above forms from this extremal length point of view. In particular, if $\mathscr{F}_1$ is the homology class of a cycle we find a family $\mathscr{F}_2$ which is conjugate to $\mathscr{F}_1$ in the above sense.

Quadrilaterals. As a first illustration we present a generalization of Loewner's theorem mentioned above. Although the results are stated for plane regions thye apply as well to bordered Riemann surfaces.

Let Γ be a closed Jordan curve in the plane and let A, B, C, D be points on Γ in cyclic order. Let β be a closed subset of the plane contained entirely inside of Γ, and such that the complement of β is connected. Let R be the region obtained by removing β from the inside of Γ. Consider the family $\mathscr{F}$ of all open arcs in R which in one direction tend to the subarc AB of Γ, and tend in the other direction to the subarc CD. Let $\mathscr{F}^*$ be the family of curves in R which separate the arcs AB and CD. That is, each element $c \in \mathscr{F}^*$ is a finite or countable union

[1] This work was supported in part by the N.S.F. under grant GP 4106 at the University of California, San Diego.

of open arcs in R such that every arc in $\mathscr{F}$ intersects c. If $\lambda(\mathscr{F})$, $\lambda(\mathscr{F}^*)$ denote the extremal length in R of the families $\mathscr{F}$, $\mathscr{F}^*$ then one has

$$\lambda(\mathscr{F}) \cdot \lambda(\mathscr{F}^*) = 1. \tag{1}$$

We have been unable to locate a reference for this result in its present form. However, it follows from the results of [6]; in fact, the present situation is much simpler due to the free border Γ. At any rate, we shall regard (1) as a known result in extremal length and consider its implications.

Given R, let l denote the infimum of the Euclidean lengths of the arcs in $\mathscr{F}$. Let l^* be the corresponding quantity for $\mathscr{F}^*$. Let a be the area of R. Then $\lambda(\mathscr{F})$ and $\lambda(\mathscr{F}^*)$ are bounded from below by l^2/a and $(l^*)^2/a$ respectively. Hence (1) yields $ll^* \leqq a$. This gives the corollary:

Let l be the infimum of the lengths of all arcs in R which join AB to CD. Let l^* be the corresponding quantity for all curves in R which separate AB and CD. Then

$$ll^* \leqq (\text{area of } R). \tag{2}$$

If $\beta = \varnothing$ this inequality is precisely Loewner's Theorem. The lack of symmetry for arbitrary β can be remedied as follows.

Let β be the disjoint union of closed sets β_1, β_2. We say that a point set $X \subset R$ separates two subarcs of Γ modulo β_2 if every arc in $R \cup \beta_1$ which joins these subarcs intersects X. Now redefine $\mathscr{F}$ as the family of all curves in R which separate AD and BC modulo β_2. Let $\mathscr{F}^*$ now consist of all curves in R which separate AB and CD modulo β_1. (The earlier definitions correspond to the special case $\beta_2 = \varnothing$.) (1) and the corresponding inequality (2) remain valid in this more general case.

Conjugate extremal length. (1) is an example of a conjugate extremal length statement. We have seen that such statements lead automatically to inequalities of the form (2). Several other examples may be obtained from the conjugate extremal length theorem of [**6**, p. 256].

Let R be a connected orientable 2-manifold with Riemannian metric. Let c be a closed Jordan curve on R such that $R - c$ is connected. Let c^+ and c^- be the right and left "sides" of c. Define

$\mathscr{G}$ = all arcs on $R - c$ which join c^+ to c^-,

$\mathscr{G}^*$ = all curves on $R - c$ which separate c^+ from c^-,

$\mathscr{H}^*$ = all arcs on $R - c$ which separate c^+ from c^-,

$\mathscr{H}$ = all curves on $R - c$ which join c^+ to c^- modulo the ideal boundary.

More precisely, a curve d belongs to $\mathscr{H}$ if and only if every arc in $\mathscr{H}^*$ intersects d. The following conjugate extremal length statements follow from the theorem in [6] mentioned above, applied to the underlying Riemann surface structure of $R - c$:

$$\lambda(\mathscr{G}) \cdot \lambda(\mathscr{G}^*) = 1, \qquad \lambda(\mathscr{H}) \cdot \lambda(\mathscr{H}^*) = 1.$$

We therefore obtain the differential-geometric inequalities

$$g \cdot g^* \leqq (\text{area of } R), \qquad h \cdot h^* \leqq (\text{area of } R)$$

where g, g^*, h, h^* are the infima of the lengths of the elements of $\mathscr{G}$, $\mathscr{G}^*$, $\mathscr{H}$, $\mathscr{H}^*$ respectively.

Möbius bands. We shall apply the following elementary extremal length theorem to obtain length-area inequalities on a Möbius band.

In the annulus $\{r_1 < |z| < r_2\}$ *let* $\mathscr{F}$ *be the family of curves which are freely homotopic to* $\{|z| = (r_1 + r_2)/2\}$, *and let* $\mathscr{F}^*$ *consist of all arcs which join the two contours. Then*

$$\lambda(\mathscr{F}) = 1/\lambda(\mathscr{F}^*) = 2\pi/\log(r_2/r_1). \tag{3}$$

Now let M be a Möbius band with Riemannian metric. Let A be the two-sheeted covering surface of M with cyclic fundamental group. A is conformally equivalent to an annulus whose radii are $r_1 < r_2$ say. Let the homotopy class of γ generate the fundamental group of M. A closed curve c in M lifts to a closed curve in $\mathscr{F}$ if and only if c is homotopic to γ^2. Hence, if

$$l = \inf\{\text{length of } c : c \sim \gamma^2\}, \qquad L = \inf\{\text{length of } d : d \in \mathscr{F}\}$$

then $l \leqq L$. Let l^* be the infimum of the lengths of all transversals in M (i.e., all open arcs which have intersection number ± 1 with each cycle homotopic to γ). Let $L^* = \inf\{\text{length of } d : d \in \mathscr{F}^*\}$. Then $l^* \leqq L^*$ since the projection of each $d \in \mathscr{F}^*$ is a transversal in M.

By (3) we have

$$L^2/(\text{area of } A) \leqq \lambda(\mathscr{F}) = 2\pi/\log(r_2/r_1),$$

$$(L^*)^2/(\text{area of } A) \leqq (2\pi)^{-1}\log(r_2/r_1),$$

$$LL^* \leqq (\text{area of } A).$$

Since A has twice the area of M we obtain

$$l^2 \leqq \frac{4\pi}{\log(r_2/r_1)}(\text{area of } M), \tag{4}$$

$$(l^*)^2 \leqq \frac{\log(r_2/r_1)}{\pi}(\text{area of } M), \tag{5}$$

$$ll^* \leqq 2\,(\text{area of } M). \tag{6}$$

The last inequality is similar to that of C. Blatter [**4**], since l may be replaced by the infimum of the lengths of all cycles in M which are homotopically non-trivial. Inequality (4) is of the type studied by P. M. Pu [7].

An extremal length theorem. Let R be a Riemann surface and let Γ_h be the Hilbert space of first order square-integrable harmonic differentials on R. If c is a 1-chain on R then the functional $\omega \to \int_c \omega$ is continuous for $\omega \in \Gamma_h$. (Unless stated otherwise all chains and cycles have integer coefficients.) Hence it is represented by an inner product with a fixed element of Γ_h, say $*\sigma_c$. The differential σ_c is called the reproducing differential for the 1-chain c. By definition we have

$$(7) \qquad \int_c \omega = (\omega, *\sigma_c) = \int_R\int \sigma_c \wedge \omega \qquad (\omega \in \Gamma_h)$$

and if c and d are 1-cycles on R then

$$(8) \qquad \int_d \sigma_c = c \times d$$

where $c \times d$ is the intersection number of c with d. Note that the left hand side of (8) makes sense for any 1-chains c, d. Therefore we use (8) as a definition of $c \times d$, even if c, d are not cycles. It should be noted that this generalized intersection notion need not be integer valued. If c is a 1-cycle on R let $\mathscr{F}_c$ denote the homology class of c. Let $\mathscr{F}_c^*$ denote the family of 1-chains d on R with $c \times d = 1$.

What is the geometric interpretation of this generalized intersection number? We have, as yet, no satisfactory answer to this question. If R is compact then to any 1-chain d there is associated a real 1-cycle z such that $\int_z \omega = \int_d \omega$ for all harmonic 1-forms ω. Indeed, σ_d determines a real cohomology class which is associated to a real 1-cycle z by the Poincaré duality theorem. Or equivalently, the map $c \to \sigma_c$ of 1-cycles to harmonic 1-forms may be extended linearly to all real 1-cycles and the extended map is surjective. Now if the 1-chains d_1, d_2 correspond to real 1-cycles z_1, z_2 then we see that $d_1 \times d_2 = z_1 \times z_2$, and the latter intersection number has the standard interpretation.

Let us consider an example which will also be useful later. Let R be the torus given by a parallelogram in the plane with opposite edges identified in the usual way. Suppose the vertices are $0, 1, 1+\tau, \tau$. Let c be the cycle on R corresponding to the "upper" edge from $1+\tau$ to τ. If d is any arc from the lower edge to the upper edge then $c \times d = 1$. Indeed, $\sigma_c = (\operatorname{Im}\tau)^{-1} dy$ and hence

$$c \times d = \int_d (\operatorname{Im}\tau)^{-1} dy = 1.$$

Furthermore, any $d \in \mathscr{F}_c^*$ has Euclidean length $\geqq \operatorname{Im}\tau$ since

$$\int_d (dx^2 + dy^2)^{1/2} \geqq \int_d dy = (\operatorname{Im}\tau)(x \times d).$$

If c is a cycle on a Riemann surface then the extremal length of the homology class $\mathscr{F}_c$ of c is given by

(9) $$\lambda(\mathscr{F}_c) = \|\sigma_c\|^2,$$

as was proved by C. Blatter [3] and R. Accola [1]. We now extend this result to a conjugate extremal length statement:

Let R be a compact Riemann surface with or without border. For any cycle c on R we have

(10) $$\lambda(\mathscr{F}_c) = \|\sigma_c\|^2 = 1/\lambda(\mathscr{F}_c^*)$$

where

$$\mathscr{F}_c^* = \{\text{1-chains } d : c \times d = 1\}.$$

For the proof, we choose a cycle $c' \in \mathscr{F}_c$ along which σ_c vanishes. We require further that in each component R_i of $R - c'$, $\sigma_c = du_i$ where u_i is a harmonic function in R_i with locally constant boundary values of 0 or 1. That such a c' exists was shown in R. Accola [1]. With suitable orientation, $*\sigma_c$ increases along c' and $\int_{c'} *\sigma_c = \|\sigma_c\|^2$. Thus we can parametrize c' with $t \in [0, \|\sigma_c\|^2]$ by means of $\int *\sigma_c$. From the point on c' corresponding to t there emanates an arc c_t along which $*\sigma_c$ vanishes and σ_c increases. c_t is a level line of some u_i and we continue it through all of R_i. We omit the finitely many values of t for which c_t leads to a critical point of a u_i. Clearly $\int_{c_t} *\sigma_c = 1$ and hence $c_t \in \mathscr{F}_c^*$. Let $\rho|dz|$ be a linear density on R. Using σ_c as a local coordinate we have

$$L^2(\rho, \mathscr{F}_c^*) \leqq \left(\int_{c_t} \rho|dz|\right)^2 \leqq \int_{c_t} \rho^2 du_i$$

where

$$L(\rho, \mathscr{F}_c^*) = \inf_{d \in \mathscr{F}_c^*} \int_d \rho|dz|.$$

Integrating over $0 \leqq t \leqq \|\sigma_c\|^2$ gives

$$\|\sigma_c\|^2 L^2(\rho, \mathscr{F}_c^*) \leqq A(\rho)$$

where $A(\rho) = \iint \rho^2 dxdy$ is the ρ-area of R. Thus we have

(11) $$\lambda(\mathscr{F}_c^*) \leqq \|\sigma_c\|^{-2}.$$

Consider now the linear density $\rho_0|dz| = |\sigma_c + i*\sigma_c|$. If $d \in \mathscr{F}_c^*$ we have $\int_d \rho_0 |dz| \geqq \int_d \sigma_c = c \times d = 1$. Furthermore, $A(\rho_0) = \|\sigma_c\|^2$. Therefore

(12) $$\lambda(\mathscr{F}_c^*) \geqq \|\sigma_c\|^{-2}.$$

The proof of (10) follows from (9), (11), (12). This proves the following inequalities.

Let R be a compact orientable Riemannian 2-space. Let c be a 1-cycle on R

and let l be the length of the shortest cycle homologous to c. Let l^* be the length of the shortest 1-chain d with $c \times d = 1$. Then

$$ll^* \leqq (\text{area of } R), \tag{13}$$

$$l^2 \leqq \|\sigma_c\|^2 (\text{area of } R), \tag{14}$$

$$(l^*)^2 \leqq \|\sigma_c\|^{-2} (\text{area of } R). \tag{15}$$

C. Blatter [3] has shown that $\|\sigma_c\|^2 < (2/\pi)((g+1)!)^{1/g}$ for some $c \nsim 1$ if R has genus g.

As a special case, suppose R is a torus with Riemannian metric. R is conformally equivalent to a flat torus represented by a parallelogram $0, 1, 1+\tau, \tau$. We may always choose τ in a fundamental region of the modular group, so

$$-1/2 \leqq \operatorname{Re}\tau \leqq 1/2, \quad |\tau| \geqq 1, \operatorname{Im}\tau < 0.$$

Hence $\operatorname{Im}\tau \geqq 3^{1/2}/2$. For the cycle c corresponding to the upper edge of the parallelogram we have seen that $\sigma_c = (\operatorname{Im}\tau)^{-1}dy$. Hence $\|\sigma_c\|^2 = (\operatorname{Im}\tau)^{-1}$, so from (14) we have $l^2 \leqq (2/3^{1/2})$ (area of R). In particular, if L is the infimum of the lengths of *all* nontrivial cycles on R then

$$L^2 \leqq (2/3^{1/2}) (\text{area of } R). \tag{16}$$

According to P. M. Pu [7], this result was proved by C. Loewner in 1949 by a different method. It may be of interest to give the above proof of (10) as it applies to this special case. In the parallelogram R we have a linear density $\rho(z)$. Let c_y denote the line $\operatorname{Im} z = y$ intersected with R. Then

$$L^2 \leqq \left(\int_{c_y} \rho |dz|\right)^2 \leqq \int_{c_y} \rho^2 dx.$$

Integrating over $y \in [0, \operatorname{Im}\tau]$ gives

$$(\operatorname{Im}\tau) L^2 \leqq \int\int_R \rho^2 dxdy = (\rho\text{-area of } R).$$

Since $(\operatorname{Im}\tau) \geqq 3^{1/2}/2$, this gives Loewner's result (16).

Flows. Let c be a 1-cycle on a compact orientable Riemannian 2-space. Then, in the sense of Klein [5], there is a steady fluid flow on R corresponding to c; namely, the flow which has the property that along any closed curve d the change in potential is $c \times d$. It is easily seen that $\mathscr{F}_c^*$ can be interpreted as the family of 1-chains along which the potential drops by $+1$. Indeed, the differential of the velocity potential is precisely σ_c. Thus from (10) we have the following result:

Let c be a 1-cycle on a compact orientable Riemannian 2-space R, and consider the corresponding steady flow. Let a be the length of the shortest geodesic in the homology class of c. Let b the length of the shortest geodesic along which the potential drop is 1. Then $a \cdot b \leqq$ (area of R).

References

1. R. D. M. Accola, *Differentials and extremal length on Riemann surfaces*, Proc. Nat. Acad. Sci. U.S.A. **46** (1960), 540–543.

2. A. S. Besicovitch, *On two problems of Loewner*, J. London Math. Soc. **27** (1952), 141–144.

3. C. Blatter, *Über Extremallängen auf geschlossenen Flächen*, Comment Math. Helv.. **35** (1961), 153–168.

4. ———, *Zur Riemannschen Geometrie im Grossen auf dem Möbiusband*, Compositio Math. **15** (1961), 88–107.

5. Felix Klein, *On Riemann's theory of algebraic functions and their integrals*, Dover, New York, 1963.

6. A. Marden and B. Rodin, *Extremal and conjugate extremal distance on open Riemann surfaces with applications to circular-radial slit mappings*, Acta Math. **115** (1966), 237–269.

7. P. M. Pu, *Some inequalities in certain nonorientable Riemannian manifolds*, Pacific J. Math. **2** (1952), 55–71.

University of California, San Diego
La Jolla, California

On the Growth of Entire Functions Along Half Rays

W. J. SCHNEIDER[1]

1. **Introduction.** In Bagemihl and Seidel [2, pp. 187–190] the following rather striking theorem is proved:

THEOREM A. *Let $M(r)$ be an arbitrary continuous positive monotone increasing function on $[0, \infty)$. Then there exists an entire analytic function $f(z)$ [harmonic function $h(z)$] and a subset $S \subset [0, 2\pi)$ of measure 2π such that*

$$\lim_{r\to\infty} \frac{|f(re^{i\theta})|}{M(r)} = \infty \quad \textit{if} \quad \theta \in S,$$

$$\left(\lim_{r\to\infty} \frac{h(re^{i\theta})}{M(r)} = \infty \quad \textit{if} \quad \theta \in S\right).$$

In the light of the above theorem it seems natural to pose the following questions.

(i) Do there exist entire analytic (harmonic) functions[2] whose modulus goes to infinity (which go to plus infinity) along every half ray emanating from the origin?

(ii) Do there exist entire analytic (harmonic) functions whose modulus eventually grows arbitrarily fast (which eventually go to plus infinity arbitrarily fast) along every half ray emanating from the origin?

If the answer to (ii) is no then:

(iii) How fast can an entire analytic (harmonic) function grow on how big a set of half rays?

Theorems 1 and 2 stated below will give a complete answer to the above questions. However in order to state these theorems we shall first need the following two definitions:

DEFINITION 1. We shall say an entire analytic function is of infinite order along a half ray $\arg z = \theta$ if

1 The author is indebted to Karl Barth for his many helpful comments and criticisms in the preparation of this paper.

2 Throughout this paper whenever we refer to an entire analytic or entire harmonic function, it is assumed these functions are neither polynomials nor the real parts of polynomials.

$$\lim_{r\to\infty} \frac{|f(re^{i\theta})|}{\exp[r^\rho]} = \infty$$

for every positive number ρ.

DEFINITION 2. We shall say an entire harmonic function $h(z)$ grows transcendentally along a half ray $\arg z = \theta$ if

$$\lim_{r\to+\infty} \frac{h(re^{i\theta})}{r^n} = \infty$$

for every positive integer n.

THEOREM 1. *If S is a set of second category on $[0, 2\pi)$, then there do not exist any entire analytic functions which are of infinite order along the set of half rays* $\arg z = \theta$ $(\theta \in S)$. *Also given any $\rho < \infty$ there exist entire analytic functions which grow in modulus faster than* $\exp[r^\rho]$ *along each half ray.*

THEOREM 2. *If S is a set of second category on $[0, 2\pi)$, then there do not exist any entire harmonic functions which grow transcendentally along the set of half rays* $\arg z = \theta$ $(\theta \in S)$. *Also entire harmonic functions exist which go to plus infinity along every half ray.*

A number of other related examples will be given and a number of still open questions, relating to the growth rates of entire analytic and harmonic functions, along sets of half rays, will be posed. We shall also spend some time discussing analogous theorems in the disk.

2. **Proof of Theorem** 1.

PROOF OF THE FIRST PART. Let S^* be the closure of S and let $\tilde{S} = \{\theta_1, \theta_2, \cdots\}$ be a dense subset of S. Then, by the "Picard-Schottky theorem in an angle" for every $\varepsilon > 0$ the image of the set $R_m(\varepsilon) = \{z \mid |\arg z - \theta_m| < \varepsilon\}$ under $f(z)$ covers every finite value except at most one. This means there exists a sequence of points z_{mn} going to infinity such that (i) $\arg z_{mn}$ approaches θ_m as n goes to infinity and (ii) $|f(z_{mn})| < 1$. Therefore, about each point z_{mn} there exists a disk of D_{mn} of radius less than one such that $f(D_{mn}) \subset \{w \mid |w| < 1\}$.

Now let us consider the set of half rays emanating from the origin which strike at least one D_{mn}, at a distance greater than one from the origin. The set of arguments of these half rays contains an open dense subset of S^*. Call this set T_1. Now consider the half rays with arguments in T_1 which strike at least one D_{mn}, at a distance greater than two from the origin. The set of arguments of these half rays contains an open dense subset of S^*. Call this set $T_2 \cdots$. But $T = \bigcap_{n=1}^{\infty} T_n$ is a residual set on S^* and if $\theta \in T$ then there exist points z_n, on the ray $\arg z = \theta$ arbitrarily far out from the origin for which $|f(z_n)| < 1$. Therefore $T \cap S = \varnothing$ and we have a contradiction to S being a set of second category on S^*.

PROOF OF THE SECOND PART. To construct an example of the type described in the theorem we first recall some of the properties of the well-known Mittag-Leffler functions [**3**, pp. 272–275] and also some properties of a not quite so well-known function from a Pólya-Szegö exercise [**6**, pp. 116–117]:

(i) Given any α ($< \pi$) there exists an entire function $M_\alpha(z)$ (usually referred to as the Mittag-Leffler function of order α) with the following properties:

(a) $M_\alpha(z)$ is bounded in the sector

$$\{z \mid \alpha \leqq \arg z \leqq 2\pi - \alpha\}.$$

(b) For any two positive numbers ε and σ the function

$$h(r) = \min_{|\theta| < \alpha - \varepsilon} \left| M_\alpha(re^{i\theta}) \right|$$

has the property that

$$\lim_{r \to \infty} \frac{h(r)}{\exp[r^{\alpha - \sigma}]} = \infty.$$

(ii) There exists an entire function $P(z)$ with the following properties:

(a) $|P(z)|$ is bounded outside the semi-infinite strip

$$S = \{z = x + iy \mid x > 0, \ -\pi < y < \pi\}.$$

(b) $P(r)$ grows transcendentally as r goes to plus infinity.

Now we choose the order α of the Mittag-Leffler function M_α so that the following conditions are satisfied:

(i) $M_\alpha(z)$ grows faster than $\exp[r^\rho]$ in the critical sector.

(ii) π/α is an integer ($= n$).

It is now easy to see that

$$f(z) = \sum_{k=0}^{k=n-1} M_\alpha(\exp[2\alpha k i]z) + \sum_{k=1}^{k=n-1} P(\exp[i(\alpha + 2\alpha k)]z)$$

has the desired properties.

3. **Proof of Theorem** 2.

PROOF OF THE FIRST PART. Suppose the theorem false for an entire harmonic function $h(z)$. Let $h^*(z)$ be the harmonic conjugate of $h(z)$. Let $f(z) = \exp[h(z) + ih^*(z)]$ and apply the first part of Theorem 1 to $f(z)$. Contradiction.

Some remarks on the proof of the first part of Theorem 2. One could, of course, have proved Theorem 2 independently of Theorem 1 by making use of the following lemma on the rate of growth of harmonic functions in an angular sector:

LEMMA ON HARMONIC FUNCTIONS: *If* $h(z)$ *is harmonic in* $S^{\varepsilon}=\{z \mid |\arg z| < \varepsilon, \varepsilon>0\}$ *and* $h(z)$ *grows transcendentally on the real line, then* $h(z)$ *must take on all possible negative values in* S.

The lemma follows by means of a conformal mapping and the fact that if $h(z)$ is positive and harmonic in the upper half plane, then $h(re^{i\theta}) = o(r^{1+\varepsilon})$ for every $\varepsilon > 0$ and every θ $(0 < \theta < \pi)$ [**9**, pp. 149–151].

We might note in concluding the proofs of the first parts of Theorems 1 and 2 that infinity does *not* play a special role in either theorem. This follows from the fact that it is only a technical matter to get a Picard type theorem in an angle (and all real values taken on in an angle theorem) if any finite value is approached too fast on a half ray by an analytic (harmonic) function.

PROOF OF THE SECOND PART. We will first need a lemma which follows immediately from the method of proof in Schneider [**8**].

LEMMA. *Let* S *be the Riemann sphere,* $R_0 = \{z = x + iy \mid x > 0, -1 < y < 1\}$ *and* R_0^* *the closure of* R_0 *in* S. *Let* $f(z)$ *be analytic in the interior of* $S - R_0$ *continuous on* $S - R_0$. *Then given any* $\varepsilon > 0$, *we can find an entire function* $E(z)$ *which has the property* $\sup_{z \in S - R_0^*} |E(z) - f(z)| < \varepsilon$.

Let $R_n = \{z = x + iy \mid x > n + 2, -1 < y < 1\}$ and denote by $f_n(z)$ the conformal map of $S - R_n$ onto $\{z = x + iy \mid -\sigma_n < x < n, \ -\sigma_n < y < \sigma_n\}$ with $f_n(0) = 0$. Also let $E_n(z)$ be an entire function, as in the lemma, with $|E_n(z) - f_n(z)| < \varepsilon_n$. Now by taking ε_n and σ_n small enough we can make $E_n(z)$ less than $1/2^n$ in $|z| < n$. Hence, by the Weierstrass M-test, $E(z) = \sum_{n=1}^{\infty} E_n(z)$ will converge to an entire function.

We can see now that $\lim_{z \to \infty; z \in S - R_0} \operatorname{Re} E(z) = \infty$ since:

(i) If ρ_n is large enough, $E_n(z)$ is approximately equal to n in the set $\{|z| > \rho_n\} - R_n$.

(ii) $E_n(z)$ is within $1/2^{n-1}$ of a positive number in $S - R_0$. Hence $\operatorname{Re} E(z)$ satisfies the conditions of the theorem.

4. Some examples.

Comment on the first example. Relative to the second part of Theorem 1 one can ask the more subtle question,

> "Given any $\rho (0 < \rho < \infty)$ does there exist an entire function of order ρ on each half ray?"

Also one might ask,

> "Can a function of order ρ grow *uniformly* at a growth rate of order ρ on a large set of half rays?"

The first example answers both of these questions.

FIRST EXAMPLE. *Given any $\rho (0 < \rho < \infty)$ there exists an entire function $f(z)$ of order ρ such that for any two positive numbers ε and σ we have:*

(i) *For all θ*

$$\lim_{r \to \infty} \frac{|f(re^{i\theta})|}{\exp[r^{\rho-\sigma}]} = \infty.$$

(ii) *There exists a subset S of $[0,2\pi]$ with measure $2\pi - \varepsilon$ such that*

$$\lim_{r \to \infty} \left[\min_{\theta \in S} \frac{|f(re^{i\theta})|}{\exp[r^{\rho-\sigma}]} \right] = \infty.$$

Construction of the first example. This construction follows in essentially the same manner as the construction in the second part of Theorem 1 except that we must make use of the following very precise representation of the Mittag-Leffler functions [**3**, p. 275]:

$$M_\alpha(z) = (1/\alpha)\exp(z^{1/\alpha}) + O(|1/z|).$$

Now all that is required is to form as in Theorem 1 the appropriate linear combination of enough functions of the form $M_\alpha(\exp[i\theta_n]z)$. Since we did not add a function of infinite order to take care of things, as we did previously, there is the possibility that in finitely many of the half rays the growth rates might cancel. However, the case of order one for $f(z) = e^z + e^{-z} + e^{iz} + e^{-iz}$ is now typical and the growth rates can actually be arranged so that they add.

Comment on the second example. The form of the Bagemihl-Seidel theorem stated at the beginning of the paper was not stated in its strongest form. In its strongest form [**2**, pp. 187–190], which we quote as Example 2, it (in conjunction with our first two theorems) answers completely the question,

"On how large a set of half rays can one arbitrarily precribe the growth rate of an entire analytic or harmonic function?"

SECOND EXAMPLE. *Let $M(r)$ be any continuous positive monotone increasing function on $[0, \infty)$, and let S be any subset of $[0, 2\pi)$ of first category. Then there exists an entire function $E(z)$ with the property*

$$\lim_{r \to \infty} \frac{\operatorname{Re} E(re^{i\theta})}{M(r)} = \infty \quad \textit{if} \quad \theta \in S.$$

Comment on third example. In the second example we have no knowledge of the behavior of the function $E(re^{i\theta})$ $(0 \leq r < \infty)$ when $\theta \notin S$. The following example shows that on any countable set of half rays we can prescribe an arbitrary growth rate for an entire analytic function $E(z)$ while on all other half rays $E(z)$ has the property that

$$\lim_{r \to \infty} \inf |E(re^{i\theta})| = 0.$$

THIRD EXAMPLE. *Given any continuous positive monotone increasing function $M(r)$ on $[0,\infty)$ and any countable set $\theta_1, \theta_2, \cdots (\theta_k \in [0, 2\pi))$ there exists an entire analytic function $E(z)$ with the properties*:

(i) $\lim_{r\to\infty} |E(re^{i\theta_k})|/M(r) = \infty$

(ii) $\liminf_{\theta\neq\theta_k} |E(re^{i\theta})| = 0$.

Construction of the third example. The function $P(z)$, which we mentioned previously in our proof of the second part of Theorem 1, also has the following property [**6**, pp. 115–116]:

Given any $\varepsilon > 0$, in the semi-infinite strip $S = \{z = x + iy \mid x > 0, |y| < \pi - \varepsilon\}$ we have

$$P(z) = \exp[\exp z] + O(1/|z|^2).$$

Now we can choose three sequences of positive constants $\{k_n\}$, $\{r_n\}$, $\{t_n\}$ with $k_n \to \infty$, $r_n \to \infty$, $t_n \to 0$ such that the series

$$f(z) = \sum_{n=1}^{\infty} t_n P[k_n(z - r_n)]$$

will converge, by the Weierstrass M-test, to an entire function and $f(z)$ will have the properties:

(i) $M(r) = o(f(r))$,

(ii) $f(z)$ bounded outside of S.

Now we can choose two sequences of positive constants $\{s_n\}$, $\{c_n\}$ such that

$$G(z) = \sum_{n=1}^{\infty} s_n f[\exp i\theta_n(z - c_n)]$$

will converge by the Weierstrass M-test to an entire function which has the properties:

(i) $G(z)$ is uniformly bounded outside of a set of countably many disjoint semi-infinite strips S_n, each straddling the half ray $\arg z = \theta_n$ from some point onward;

(ii) In each S_n, $G(z)$ grows faster than $M(r)$, along the half ray $\arg z = \theta_n$.

Now $E(z) = (G(z) - G(0))/z$ has the desired properties.

Comments on fourth example. By Theorem 1 it is not possible for an entire analytic function to grow arbitrarily fast on every half ray or even on quite a large set of them. However, it is possible to show that given any arbitrary growth rate $M(r)$ we can find an entire analytic function $E(z)$ and a set of paths P_θ each asymptotically approaching the half ray $\arg z = \theta$ and $|E(z)|$ will grow faster than $M(r)$ along P_θ. An analogous theorem holds for harmonic functions just by considering the real part of the function $E(z)$ in this example.

In the fourth example we also have in some sense a uniform growth rate along all the P_θ (i.e. at each point z_r where the curve P_θ intersects $\{|z| = r\}$ we have $|E(z_r)| > M(r)$ (independent of θ)).

We might also note that if in the second example we had wanted

$$\lim_{r\to\infty} \min_{\theta\in S} \frac{\operatorname{Re} E(re^{i\theta})}{M(r)} = \infty,$$

then it is not known whether for an arbitrary set S of first category one can always find a corresponding $E(z)$. The best that is known is that this is possible if S is closed and nowhere dense [**2**, p. 190].

One might also ask whether given any $M^*(r) \to 0$ we could also in addition to the P_θ paths have a set of paths $P^*(\theta)$ such that $P^*(\theta)$ would asymptotically approach $\arg z = \theta$ and along each $P^*(\theta)$, $|E(z)|$ would go to zero faster than $M^*(r)$. By transplanting to the disk in the obvious way and applying Bagemihl's ambiguous point theorem [**1**] we see that the set of θ, for which there exist such $P^*(\theta)$ paths, would have to be countable.

FOURTH EXAMPLE. *Given any continuous positive monotone increasing function $M(r)$ on $[0, \infty)$, there exists an entire analytic function $E(z)$ with the property that given any half ray* $\arg z = \theta$ *there exists a continuous curve*

$$\gamma(t) = te^{i\phi(t)} \qquad (0 \leqq t < \infty)$$

such that

(i) $\lim_{t\to\infty} \phi(t) = \theta$,

(ii) $|E(te^{i\phi(t)})| > M(t)$ *for all* t.

Construction of the fourth example. Divide the arc of the circle $|z| = 1$ in the first quadrant into two equal pieces by one division point. Divide the arc of the circle $|z| = 2$ in the first quadrant into three equal pieces by two division points. $\cdots$ Divide the arc of the circle $|z| = n$ in the first quadrant into $2^{n-1} + 1$ equal pieces by 2^{n-1} division points. $\cdots$ Now join the origin to the division point on $|z| = 1$ and the division point on $|z| = 1$ to the two division points on $|z| = 2$. Now join each division point on $|z| = 2$ to the two closest division points on $|z| = 3$. $\cdots$ Now join each division point of $|z| = n$ to the two closest division points of $|z| = n+1$. $\cdots$ Construct a set in the same manner in each of the three other quadrants and call the union of these four sets S. Let $S_n = S \cap D_n \, (= \{z \mid |z| \leq n\})$.

From now on our proof consists of just a slight modification of a method first introduced in Bagemihl and Seidel [**2**, pp. 187–190]. Since S_1 is compact and does not separate the plane, by Mergelyan's theorem [**5**], there exists a nonconstant polynomial $P_1(z)$ such that

$$|P_1(z) - M(2)| < 1/2^2 \quad \text{on} \quad S_1.$$

Since $S_2 \cup D_1$ is compact and does not separate the plane, by Urysohn's lemma and Mergelyan's theorem there exists a polynomial $P_2(z)$ such that

(i) $|P_2(z)| < 1/2^2$ on D_1,

(ii) $|P_2(z) + P_1(z) - M(|z| + 1)| < 1/2$ on $S_2 - S_1$.

$\cdots$ Continuing inductively, since $S_n \cup D_{n-1}$ is compact and does not separate the plane by Urysohn's lemma and Mergelyan's theorem there exists a polynomial $P_n(z)$ such that

(i) $|P_n(z)| < 1/2^{n+1}$ on D_n,

(ii) $|\sum_{k=1}^{n} P_k(z) - M(|z| + 1)| < 1/2^n$ on $S_{n+1} - S_n$.

Now, by the Weierstrass M-test, $E(z) = \sum_{k=1}^{\infty} P_k(z)$ represents an entire function, and clearly for any half ray R $(= \{z | \arg z = \theta\})$ one can always find a path in S which approaches R, in the manner specified in the example, along which $E(z)$ has the desired growth properties.

5. **Some open questions.** Relative to Example 1, one can pose the following more precise questions:[1]

Question 1. Can an entire analytic function of order ρ and minimal type have a growth rate of order ρ along each half ray?

Question 2. Can an entire analytic function of order ρ and mean type τ have a growth rate of order ρ and mean type τ along each half ray (our first example is a function of mean type which has a growth rate of mean type along each half ray but not always the maximal possible mean type)?

Question 3. Can an entire analytic function of order ρ, and maximal type have a growth rate of order ρ and maximal type along each half ray?

Question 4. Is it possible to prove some sort of analog of the first example for harmonic functions?

6. **Remarks concerning analogous theorems for functions on the unit disk.** Since the following analog of the "Picard-Schottky theorem in an angle" holds for the disk:

THEOREM. *If $f(z)$ is analytic in $|z| < 1$ and if $\lim_{r \to 1} |f(r)| / \exp[1/(1-r)]^n = \infty$ for all positive integers n then for all $\varepsilon > 0$ the image of S_ε $(= \{z | |z| < 1, |\arg(z-1)| < \varepsilon\})$ under $f(z)$ omits at most one finite value.*

It is also possible to prove analogs of the first parts of Theorems 1 and 2 for the disk.

Also the Bagemihl-Seidel results are valid for the disk; hence complete analogs of Theorem A and the second example hold for functions analytic in the disk.

It is also possible to construct an analog of the third example for the disk. Instead of using the function $P(z)$ (of Example 3) as the fundamental building block, we would use $P^*(z) = P(f(z))$ where $f(z)$ is a conformal map from the disk onto the right half plane. Of course one needs an analog of the function $1/z$ to multiply by as we did in the last step of the construction of Example 3.

[1] For the definitions of maximal, mean and minimal type see Boas [4, p. 8].

The existence of such an analog follows from the results of Bagemihl-Seidel [**2**, pp. 187–190] and the analog is the following:

THEOREM B. *Given any continuous paths* $\gamma(t) = te^{i\phi(t)}$ $(0 \leqq t < 1)$, *where* $\lim_{t\to 1} \phi(t) = \infty$ *in* $|z| < 1$, *there exists a nonconstant analytic function* $f(z)$ *in* $|z| < 1$ *such that* $\lim_{t\to 1} f(te^{i\phi(t)}) = 0$.

There can be no analogs for either of the second parts of Theorems 1 and 2 (let alone the first example) due the classical uniqueness theorem of Lusin and Privalow [**7**, pp. 231–232]:

THEOREM OF LUSIN AND PRIVALOW. *An analytic function* $f(z)$ *in* $|z| < 1$ *cannot have* $\lim_{r\to 1} f(re^{i\theta}) = \infty$ *for each* $\theta \in S$ *if* S *is a set of second category and of positive measure* (to see that this theorem also implies a uniqueness theorem for harmonic functions follows from the method of proof of the first part of Theorem 2).

Due to the above uniqueness theorem for analytic functions in the unit disk, the fact that an analog of Example 4 exists might be considered a little surprising. However, our previous proof can be transplanted to the disk almost word for word and such an analog does exist.

REFERENCES

1. F. Bagemihl, *Curvilinear cluster sets of arbitrary functions*, Proc. Nat. Acad. Sci. U.S.A. **41** (1955) 379–382.

2. F. Bagemihl and W. Seidel, *Some boundary properties of analytic functions*, Math. Z. **61** (1954), 186–199.

3. L. Bieberbach, *Lehrbuch der Funktionentheorie*. Bd. II, 2nd, ed., Chelsea, New York, 1945.

4. R. P. Boas, *Entire functions*, Academic Press, New York, 1954.

5. S. N. Mergelyan, *On the representation of functions by series of polynomials on closed sets*, Dokl. Akad. Nauk SSSR **78** (1951), 405–408; English transl., Amer. Math. Soc. Transl. (1) **3** (1962), 287–293.

6. G. Pólya and G. Szegö, *Aufgaben und Lehrsätze aus der Analysis*. Bd. I, 3rd ed., Springer-Verlag, Berlin 1964.

7. I. I. Privalow, *Randeigenschaften analytischer Funktionen*, VEB Deutscher Verlag, Berlin, 1956.

8. W. J. Schneider, *An entire transcendental function whose inverse takes sets of finite measure into sets of finite measure*, Bull. Amer. Math. Soc. **72** (1966), 841–842.

9. M. Tsuji, *Potential theory in modern function theory*, Maruzen, Tokyo, 1959.

SYRACUSE UNIVERSITY
SYRACUSE, NEW YORK

Trigonometric Series with Nonnegative Partial Sums

S. M. SHAH[1]

1. **Introduction.** In §§1–3 of this paper we consider trigonometric series with nonnegative partial sums and, assuming an additional condition, prove that the series is a Fourier series. In §4 we consider the minimum value of a trigonometric sum. Let the series be

(1) $$\frac{a_0}{2} + \sum_1^\infty (a_n \cos nx + b_n \sin nx) \equiv \sum_0^\infty \rho_n(\cos(nx + c_n))$$

where we may assume $\rho_0 \equiv a_0/2 > 0$, $c_0 = 0$. It is known that if the partial sums are nonnegative,

(2) $$S_n(x) = \sum_0^n \rho_k \cos(kx + c_k) \geqq 0, \quad 0 \leqq x \leqq 2\pi; \ n = 0, 1, \cdots,$$

then $\rho_n \to 0$. In fact Helson [**8**] proved that if

(3) $$\int_0^{2\pi} |S_n(x)|\, dx = O(1),$$

then $\rho_n \to 0$. If we assume a condition on the distribution of c_n, then it is possible to prove the following [**3**], [**4**]:

THEOREM 1. *Let* $\{\rho_n\}$ *and* $\{c_n\}$ *be two sequences of real numbers such that*

(4) $$\rho_0 > 0, \quad c_0 = 0, \quad \rho_n \geqq 0, \quad n = 1, 2, \cdots.$$

Suppose $a > 0$ *and*

(5) $$S_n(x) = \sum_0^n \rho_k \cos(kx + c_k) \geqq 0, \quad n = 0, 1, 2, \cdots, \quad 0 \leqq x \leqq a,$$

(6) $$\sup_{n \geqq 1} \int_0^a \frac{\cos(x + c_n)}{x^\beta}\, dx \leqq -d$$

[1] The work of this author was supported by U.S. National Science Foundation grant GP–4311.

where β is a constant such that $0 < \beta < 1$, and $d > 0$ is a constant depending on a and β. Then

$$\sum_{k=1}^{n} \frac{\rho_k}{k^{1-\beta}} \leqq \frac{\rho_0}{d} \frac{a^{1-\beta}}{(1-\beta)} \tag{7}$$

and the series $\sum_1^\infty \rho_k/k^{1-\beta}$ is convergent.

2. **Remarks.**

(i) Theorem 1 is best possible in the following sense. Given a positive number t, there exists a trigonometric series $\sum_0^\infty \rho_n \cos nx$ satisfying the conditions of the theorem and $\sum_1^\infty \rho_k k^t/k^{1-\beta}$ is divergent.

EXAMPLE. Consider the series

$$\sum_1^\infty \frac{\cos kx}{k^\alpha}, \qquad \alpha > 0. \tag{8}$$

It is known that [**16**, pp. 191–192] that there is a number β_0, $0 < \beta_0 < 1$, such that for each $\alpha \geqq \beta_0$, the partial sums of (8) are uniformly bounded below and for each $\alpha < \beta_0$ they are not: β_0 is the unique root of the equation

$$I(\alpha) = \int_0^{3\pi/2} \frac{\cos u}{u^\alpha} du = 0, \qquad 0 < \alpha < 1.$$

The value of β_0 ($= .30844...$) has been calculated to fifteen places of decimal [**6**], [**10**]. Furthermore,

$$I(\alpha) > 0 \quad \text{when} \quad \alpha > \beta_0; \qquad I(\alpha) < 0 \quad \text{when} \quad \alpha < \beta_0.$$

Let $\alpha = \beta_0 + t/2$ in (8), $\beta = \beta_0 - t/2$ in (6). Take $c_n = 0$, $a = 3\pi/2$. Then there exists ρ_0 such that the partial sums of

$$\rho_0 + \sum_1^\infty \frac{\cos kx}{k^\alpha}$$

are positive over $[0, 2\pi]$. Further $I(\beta_0 - t/2) < 0$. But, since

$$\rho_k = 1/k^\alpha, \quad \sum \frac{(\rho_k)k^t}{k^{1-\beta}} = \infty.$$

(ii) We can replace, in Theorem 1, $\cos x$ by any real valued bounded function $F(x)$ integrable over $[-\pi, a+\pi]$ for some $a > 0$, provided we assume, $|c_n| \leqq \pi$.

3. **Conditions for (1) to be a Fourier series.** J. E. Littlewood proposed the following problem (cf: [**9**]). Does (2) imply that (1) is a Fourier series? We noticed earlier that $(2) \Rightarrow \rho_n \to 0$. Mary Weiss (1959) showed that (3) does not imply

that (1) is a Fourier series and Katznelson [9] showed that (2) does not imply that (1) is a Fourier series. We assume below an additional condition which together with (2) will imply that (1) is a Fourier series.

A sequence a_n of nonnegative numbers is said to be quasi-monotone if $a_{n+1} \leqq a_n(1+\alpha/n)$ for some constant $\alpha > 0$ and all $n > n_0(\alpha)$ [13].

(a) Let $C(x) = \sum_0^\infty \rho_n \cos nx$ and suppose that $\{\rho_n\}$ is quasimonotone and $S_n(x) \geqq 0$ over $[0, 2\pi]$. Then we have from Theorem 1 that $\sum \rho_n/n^{1-\beta} < \infty$ where $\beta < \beta_0$ and hence $\rho_n n^\beta = o(1)$, $C(x)$ is a Fourier series (see [2], [14]) and $x^{-\beta}C(x) \in L(0,\pi)$.

This series is convergent for every x save possibly $x \equiv 0 \pmod{2\pi}$.

(b) Let $S(x) = \rho_0 - \sum_1^\infty \rho_n \sin nx$ and suppose that $\{\rho_n\}$ is quasi-monotone and $S(x) \geqq 0$ over $[0, a]$. Then $\sum_1^\infty \rho_n/n^{1-\beta} < \infty$ where β can be any number in $(0,1)$. Further [14] this series is a Fourier series, $x^{-\beta}S(x) \in L(0,\pi)$, and the series is convergent everywhere and uniformly convergent over $0 < \delta \leqq x \leqq 2\pi - \delta$.

(c) Suppose that $\{a_n\}_1^\infty$ and $\{b_n\}_1^\infty$ are of bounded variation and $S_n(x) \geqq 0$ over $[0, 2\pi]$. Then (1) is a Fourier series.

For $\rho_n \to 0$ and so $a_n \to 0$, $b_n \to 0$. Since $\{a_n\}$ is of bounded variation the series $\sum_1^\infty a_n \cos nx$ is convergent except possibly when $x \equiv 0 \pmod{2\pi}$. Similarly the series $\sum_1^\infty b_n \sin nx$ is convergent over $[0, 2\pi]$. Since $S_n(x) \geqq 0$ and $S_n(x) \to S(x)$ except possibly when $x \equiv 0 \pmod{2\pi}$, we have by Fatou's lemma

$$\int_0^{2\pi} S(x)dx \leqq \liminf_{n\to\infty} \int_0^{2\pi} S_n(x)dx = \pi a_0 .$$

Hence $S(x) \in L(0, 2\pi)$ and so by a known theorem [7, pp. 91–92] (1) is the Fourier series of $S(x)$.

(d) More generally, (1) is a Fourier series if $S_n(x) \geqq 0$ over $[0, 2\pi]$ and if

$$\liminf_{r\to 1} \sum_0^\infty r^n \rho_n \cos(nx + c_n)$$

is finite except possibly in a countable set of values of x [16, p. 356].

(e) Suppose that for the series (1), $S_n(x) \geqq 0$ over $[0, 3\pi/2]$. Then (1) is a Fourier series if any one of the following conditions hold.

(i) $0 \leqq b_n \leqq ka_n$ where $k < 1$ is a fixed constant, for $n = 1, 2, \cdots$ and the sequences $\{a_n\}$ and $\{b_n\}$ are quasi-monotone.

(ii) $\{a_n\}$ and $\{-b_n\}$ are quasi-monotone.

(iii) $0 < \delta b_n \leqq |a_n|$ where $\delta > 0$ for $n = 1, 2, \cdots$ and $\{-a_n\}$ and $\{b_n\}$ are quasi-monotone.

(iv) $\{-a_n\}$ and $\{-b_n\}$ are quasi-monotone.

(v) Suppose $S_n(x) \geqq 0$ over $[0, 2\pi]$ and $\{\rho_n\}$ is quasi-monotone and $c_n = c$, $n = 1, 2, \cdots$. Then (1) is a Fourier series and $x^{-\beta}S(x) \in L(-\pi, \pi)$ where β is any fixed number less than β_0.

(f) A sequence $\{a_n\}$ is called δ-quasi-monotone [3] if $a_n \to 0$, $a_n > 0$ ultimately and $\Delta a_n \geqq -\delta_n$ where $\{\delta_n\}$ is a sequence of positive numbers. Two theorems of Boas [3, Theorems 4, 5] help us to obtain the same conclusions in (a), (b), (e) if we replace the hypothesis "quasi-monotone" by "δ-quasi-monotone".

(g) We say a sequence $\{a_n\}_1^\infty$ of real numbers is well partitioned order k, if it can be partitioned into k disjoint sequences a_{rk+i} $(i = 1, 2, \cdots, k)$ each of which ultimately decreases or increases steadily to zero [11]. With the help of Theorems 1 and 2 of Robertson [11] we get the same conclusions in (a), (b), (e) if we replace the hypothesis "quasi-monotone" by "well partitioned order k".

(h) Finally we note that given any interval $[a, b]$ such that $b - a < \pi$ and $a \leqq c_n \leqq b$, $n = 1, 2, \cdots$, we can find a number β, $0 < \beta < 1$ such that condition (6) in Theorem 1 is satisfied. Hence if (5) holds, we can get the convergence of $\sum \rho_n / n^{1-\beta}$.

4. **Applications.** We now consider applications to trigonometric sums. Hardy and Littlewood raised the following question in 1948. Is it true or not that

$$(9) \qquad \int_0^{2\pi} \left| \sum_{n=1}^{N} \cos a_n x \right| dx > A \log N$$

for all sets of positive integers a_n and for some positive A independent of N?

If we take $a_n = n$ then (9) is true. Write

$$F(x) = \sum_{n=1}^{N} \exp(i a_n x), \qquad f(x) = \operatorname{Re} F(x), \quad m(N) = \min_{0 \leqq x \leqq 2\pi} f(x).$$

P. J. Cohen (1960) proved that for some $A>0$ and all $N>3$, the L^1 norm of $F(x)$

$$(10) \qquad I = \int_0^{2\pi} |F(x)|\, dx > A \left(\frac{\log N}{\log \log N} \right)^a$$

where $a = 1/8$. H. Davenport (1960) showed that (10) holds with $a = 1/4$. R. Salem (1955) proved that if $\log a_n = O(\log n)$, then

$$(11) \qquad \limsup_{N \to \infty} \int_0^{2\pi} |f(x)| (\log N)^{-1/2}\, dx > 0.$$

Hence we have with the same hypothesis,

$$\limsup_{N \to \infty} \left\{ \frac{-m(N)}{(\log N)^{1/2}} \right\} > 0.$$

Chowla [5] proved that if $a_n = O(n^{1+\varepsilon})$, then

$$\limsup_{N \to \infty} \{-m(N)/N^{\gamma}\} > 0$$

for any fixed $\gamma < \beta_0$. His conjecture is that if $a_1, a_2, \cdots, a_N$ be any set of distinct integers, then $-m(N) > AN^{1/2}$ for some absolute constant $A > 0$ (see also [**15**, pp. 1189–1890)]. We consider here

$$f(x, N, c) = \cos(a_1 x + c_{a_1}) + \cdots + \cos(a_N x + c_{a_N}).$$

Write

$$m(a, N, c) = \min_{0 \leqq x \leqq a} f(x, N, c), \quad 0 < a \leqq 2\pi.$$

THEOREM 2. [**4**]. *Let*

$$0 < \beta < 1, \quad 1 \leqq b < 1/(1-\beta), \quad 0 < \gamma < 1 - b(1-\beta).$$

Suppose c_{a_n}, $n = 1, 2, \cdots$ *satisfy condition* (6) (*with some* $d > 0$) *and the sequence of positive integers* a_n, $n \geqq 1$ *satisfy*

$$1 \leqq a_1 < a_2 < \cdots; \quad a_n \leqq Kn^b$$

where K *is a constant. Then*

$$\limsup_{N \to \infty} \left\{ \frac{-m(a, N, c)}{N^\gamma} \right\} > 0.$$

THEOREM 3. *Let* $F(x)$ *be strictly increasing over* $[1, \infty)$ *and suppose that*

$$\log n \leqq F(n) \leqq n, \quad n = 1, 2, \cdots$$

and

$$a_p < A \exp(F(p))$$

where $A > 0$ *is an absolute constant, for* $p = 1, 2, \cdots$. *Then it is possible to choose the numbers* c_{a_k}, $k = 1, 2, \cdots, N$ *such that* $-\pi < c_{a_k} \leqq \pi$, $k = 1, 2, \cdots$ *and* $-m(2\pi, N, c) < B(NF(N))^{1/2}$ *where* $B > 0$ *is an absolute constant.*

PROOF. Our proof is similar to that of Salem ([**12**]: cf. [**1**, pp. 332–334]. Write

$$f(x) = f(x, N, c), \quad M = \sup_{0 \leqq x \leqq 2\pi} |f(x)|, \quad c_{a_k} = -\alpha_k,$$

then

$$\int_0^{2\pi} f^2(x)\,dx = \pi N \leqq 2\pi M^2.$$

Consider

$$I_k(\alpha_1, \cdots, \alpha_N) = I_k = \int_0^{2\pi} \{f(x)\}^{2k} dx$$

where k is a positive integer. This function has a minimum and at the point where it is attained

$$\frac{\partial^2 I}{\partial \alpha_p^2} \geqq 0 \qquad \text{for } p = 1, 2, \cdots, N.$$

Hence we get

$$\int_0^{2\pi} \{f(x)\}^{2k} dx \leqq 2\pi(2k-1)^k N^k.$$

Considering the function $|f(x)|$, for which the integral I_k has a minimum value, we get

$$M^2 < A_1 kN \left(\sum_1^N a_p^2/N \right)^{1/k}$$

where A_1 is a constant. Choose $k = [F(N)] + 1$. Then

$$M^2 < 2A_1 \exp(2 \log A + 2) NF(N)$$

and the theorem is proved.

References

1. N. K. Bary, *A treatise on trigonometric series*. I, Macmillan, New York, 1964.

2. R. P. Boas, Jr., *Integrability of trigonometric series*. III, Quart. J. Math. Oxford Ser. **3** (1952) 217–221.

3. ———, *Quasi-positive sequences and trigonometric series*, Proc. London Math. Soc. (3) **14a** (1965), 38–46.

4. J. Chidambaraswamy and S. M. Shah, *Trigonometric series with non-negative partial sums*, J. Reine Angew. Math. (to appear).

5. S. Chowla, *Some applications of a method of A. Selberg*, J. Reine Angew. Math. **217** (1965), 128–132.

6. R. D. Halbgewachs and S. M. Shah, *Trigonometric sums and Fresnel-type integrals*, Proc. Indian Acad. Sci. Sect. A. **65** (1967), 227–232.

7. G. H. Hardy and W. W. Rogosinski, Fourier series, Cambridge Univ. Press, New York, 1962.

8. H. Nelson, *Proof of a conjecture of Steinhaus*, Proc. Nat. Acad. Sci. U.S.A. **40** (1954), 205–206.

9. Y. Katznelson, *Trigonometric series with positive partial sums*, Bull. Amer. Math. Soc. **71**(1965), 718–719.

10. Y. L. Luke, Wyman Fair, Geraldine Coombs and Rosemary Moran, *On a constant in the theory of trigonometric series*, Math. Comp. **19** (1965), 501–502.

11. M. M. Robertson, *Integrability of trigonometric series*, Math. Z. **83**(1964), 119–122.

12. R. Salem. *Essais sur les séries trigonométriques*, Actualités Sci. Indust. no. 862, Paris (1940).

13. S. M. Shah, *A note on quasi-monotone series*, Math. Student **15** (1947), 19–24.

14. ———, *Trigonometric series with quasi-monotone coefficients*, Proc. Amer. Math. Soc. **13** (1962) 266–273.

15. S. Uchiyama, *On the mean modulus of trigonometric polynomials whose coefficients have random signs*, Proc. Amer. Math. Soc. **16** (1965) 1185–1190.

16. A. Zygmund, *Trigonometric series*. I, Cambridge Univ. Press, New York, 1959.

UNIVERSITY OF KENTUCKY
LEXINGTON, KENTUCKY

About Entire and Meromorphic Functions of Exponential Type[1]

WILHELM STOLL

Dedicated to Professor R. Catesby Taliaferro in honor of his sixtieth birthday.

Let f be a meromorphic function on C^p with $p \geqq 1$. If $s \geqq 0$, then the characteristic function $T_f(r, s)$ of f is defined. If $\rho > 0$, the type of order ρ of f is defined by

$$\tau_\rho[f] = \limsup_{r \to \infty} r^{-\rho} T_f(r, s),$$

which is independent of s. The set $\mathfrak{F}_\rho$ of all meromorphic functions f on C^p with $\tau_\rho[f] < \infty$ is a field. Similarly, the set $\mathfrak{F}_\rho^0$ of all meromorphic functions f on C^p with $\tau_\rho[f] = 0$ is a subfield of $\mathfrak{F}_\rho$. Let $\mathfrak{E}_\rho$ be the integral domain of all entire functions in $\mathfrak{F}_\rho$, and $\mathfrak{E}_\rho^0$ the integral domain of all entire functions in $\mathfrak{F}_\rho^0$. It will be proved that *$\mathfrak{F}_\rho$ is the field of quotients of $\mathfrak{E}_\rho$ and $\mathfrak{F}_\rho^0$ is the field of quotients of $\mathfrak{E}_\rho^0$ if $p > 1$*. This affirms a conjecture of Rubel **[6]**.

In the case of one variable, $p = 1$, Rubel and Taylor **[7]** proved that $\mathfrak{F}_\rho$ is the field of quotients of $\mathfrak{E}_\rho$. Actually they consider the more general case of the field $\mathfrak{F}(\lambda)$ of all meromorphic functions of finite λ-type. ADDED IN PROOF: This more general case of functions of finite λ-type and of zero λ-type in several complex variables has been solved by R. O. Kujala **[14]** in 1967. Also, Taylor has obtained some results in this direction. Their methods are different. If $\lambda(r) = r^p$, then $\mathfrak{F}(\lambda) = \mathfrak{F}_p$. They use Fourier analysis to obtain their results. Of importance for their proof is also a generalization of a theorem of Lindelöf.

In the case of $p > 1$ and $\lambda(r) = r^\rho$, which is treated here, it is not necessary to use Fourier analysis. The Lindelöf Theorem can be extended to several variables and together with an application of the Jensen-Poisson formula yields the desired result. The proof uses the approach to functions of finite order in several complex variables as developed by Kneser **[1]** and **[2]**, and Stoll **[8]**, **[9]**, **[12]** and **[13]**. In order to make the proofs of the new results intelligible, it is necessary to review most of the content of these papers.

1. **Differential forms.** Let M be a complex manifold of pure dimension m. A complex manifold N of pure dimension n with $n \leqq m$ is said to be a *smooth*

[1] This research was partially supported by the National Science Foundation under grant NSF GP 3988.

complex submanifold of M, if and only if N is a subset of M carrying the induced topology and if the inclusion map $j_N: N \to M$ is holomorphic and its Jacobian has rank n everywhere on N. An exterior differential form ϕ of bidegree (n,n) on M is said to be *positive* (respectively *nonnegative*) if and only if for every smooth complex submanifold N of pure dimension n the lifted form $j_N^*(\phi)$ defines a positive (nonnegative) density on N.

The exterior derivative $d = \partial + \bar{\partial}$ splits into a complex component ∂ and a conjugate complex component $\bar{\partial}$. Define

$$d^{\perp} = i(\partial - \bar{\partial}).$$

If N is a pure n-dimensional analytic subset of M, the set $\dot{N}$ of simple points of N is a smooth pure n-dimensional complex submanifold of M. Let $j: \dot{N} \to M$ be the inclusion map. Then an exterior differential form ϕ of bidegree (n,n) induces an exterior differential form $j^*(\phi)$ on $\dot{N}$. If $A \subseteq N$, define

$$\int_A \phi = \int_{A \cap \dot{N}} j^*(\phi)$$

if this integral exists. For instance, the integral exists if $A \cap \dot{N}$ is measurable, if $\bar{A}$ is compact and if ϕ is continuous on M.

For $\mathfrak{z} = (z_1, \cdots, z_p) \in \boldsymbol{C}^p$ and $\mathfrak{w} = (w_1, \cdots, w_p) \in \boldsymbol{C}^p$ define

$$(\mathfrak{z}|\mathfrak{w}) = \sum_{\alpha=1}^{p} z_\alpha \bar{w}_\alpha,$$

$$|\mathfrak{z}| = (\mathfrak{z}|\mathfrak{z})^{1/2},$$

$$B(r) = \{\mathfrak{z} \mid |\mathfrak{z}| < r\},$$

$$S(r) = \{\mathfrak{z} \mid |\mathfrak{z}| = r\} = \overline{B(r)} - B(r),$$

$$R(r,s) = B(s) - \overline{B(r)} \quad \text{if} \quad 0 < r < s,$$

$$R(r,\infty) = \boldsymbol{C}^p - \bar{B}(r) = \{\mathfrak{z} \mid |\mathfrak{z}| > r\}.$$

The sphere $S(r)$ is regarded as a $(2p-1)$-dimensional, real manifold of class C^∞, which is oriented to the exterior of $B(r)$. The positive euclidean volume element of $S(r)$ is denoted by σ_r. The volume of $S(r)$ is

$$\Phi(r) = \Phi_p(r) = (2\pi^p/(p-1)!)\, r^{2p-1} .$$

Define

$$W(r) = W_{p-1}(r) = (\pi^{p-1}/(p-1)!)\, r^{2p-2} \quad \text{and} \quad W = W(1).$$

Then

$$\Phi(r) = 2\pi r\, W(r).$$

On $\boldsymbol{C}^p$ define the positive exterior forms υ and υ_q by

$$\upsilon(\mathfrak{z}) = \tfrac{1}{4} d^{\perp} d(\mathfrak{z} | \mathfrak{z}),$$
$$\upsilon_q = (1/q!)\upsilon \wedge \cdots \wedge \upsilon \qquad (q \text{ times}).$$

Let N be a pure q-dimensional analytic subset of the open set G of $\boldsymbol{C}^p$. If $A \subseteq N$ is measurable, then $\int_A \upsilon_q$ is the $2q$-dimensional Hausdorff measure of A.

If $\sigma: \boldsymbol{C}^p \to \boldsymbol{C}^p$ is a unitary map, i.e., $\sigma: \boldsymbol{C}^p \to \boldsymbol{C}^p$ is linear and $|\sigma(\mathfrak{z})| = |\mathfrak{z}|$ for all $\mathfrak{z} \in \boldsymbol{C}^p$, then $\sigma^*(\upsilon_q) = \upsilon_q$. For $\mathfrak{c} \in \boldsymbol{C}^p$, define $\tau_{\mathfrak{c}}: \boldsymbol{C}^p \to \boldsymbol{C}^p$, by $\tau_{\mathfrak{c}}(\mathfrak{z}) = \mathfrak{z} + \mathfrak{c}$ for all $\mathfrak{z} \in \boldsymbol{C}^p$. Then $\tau_{\mathfrak{c}}^*(\upsilon_q) = \upsilon_q$.

On $\boldsymbol{C}^p - \{0\}$ define the nonnegative exterior forms ω and ω_q by

$$\omega(\mathfrak{z}) = \tfrac{1}{4} d^{\perp} d \log(\mathfrak{z} | \mathfrak{z}),$$
$$\omega_q = (1/q!)\omega \wedge \cdots \wedge \omega \qquad (q \text{ times}).$$

If $\sigma: \boldsymbol{C}^p \to \boldsymbol{C}^p$ is a unitary map, then $\sigma^*(\omega_q) = \omega_q$. Let $\boldsymbol{P}^{p-1}$ be the complex-projective space of dimension $p-1$. Denote $\boldsymbol{P} = \boldsymbol{P}^1$. A surjective, holomorphic map $\rho: \boldsymbol{C}^p - \{0\} \to \boldsymbol{P}^{p-1}$ is uniquely defined by

$$\rho^{-1}(\rho(\mathfrak{z})) = \{\lambda \mathfrak{z} | 0 \neq \lambda \in \boldsymbol{C}\}.$$

The map ρ is called the residual map of $\boldsymbol{C}^p - \{0\}$ onto $\boldsymbol{P}^{p-1}$. One and only one real-analytic, exterior form $\ddot{\omega}$ of bidegree (1,1) exists on $\boldsymbol{P}^{p-1}$ such that $\rho^*(\ddot{\omega}) = \omega$. The form $\ddot{\omega}$ is the exterior form of a Kaehler metric on $\boldsymbol{P}^{p-1}$. Especially, $\ddot{\omega}$ is positive. Define

$$\ddot{\omega}_q = (1/q)\, \ddot{\omega} \wedge \cdots \wedge \ddot{\omega} \quad (q \text{ times}).$$

Then

$$\rho^*(\ddot{\omega}_q) = \omega_q.$$

The total volume of $\boldsymbol{P}^{p-1}$ is

$$W = W(1) = W_{p-1}(1) = \int_{\boldsymbol{P}^{p-1}} \ddot{\omega}_{p-1}.$$

If F is a function on $\boldsymbol{C}^p - \{0\}$ such that $F(a\mathfrak{z}) = F(\mathfrak{z})$ for all $a \in \boldsymbol{C} - \{0\}$ and $\mathfrak{z} \in \boldsymbol{C}^p - \{0\}$, then one and only one function $\ddot{F}$ on $\boldsymbol{P}^{p-1}$ exists such that $\ddot{F} \circ \rho = F$. If no confusion is possible, write

$$\int_{\boldsymbol{P}^{p-1}} \ddot{F} \ddot{\omega}_{p-1} = \int_{\boldsymbol{P}^{p-1}} F \ddot{\omega}_{p-1}.$$

LEMMA 1.1.[2] *Let F be defined and measurable on $\boldsymbol{C}^p - \{0\}$. Suppose that $F(a\mathfrak{z}) = F(\mathfrak{z})$ if $a \in \boldsymbol{C} - \{0\}$ and $\mathfrak{z} \in \boldsymbol{C}^p - \{0\}$. If one of the integrals in* (1.1) *exists the other two integrals in* (1.1) *exist and*

---

[2] See Stoll **[13]**, Lemma 1.1 and Stoll **[8]**, Hilfssatz 1, p. 142.

$$(1.1) \qquad \frac{1}{\pi^p}\int_{C^p} F(\mathfrak{z})e^{-|\mathfrak{z}|^2}\upsilon_{2p} = \frac{1}{\Phi(r)}\int_{S(r)} F\sigma_r = \frac{1}{W}\int_{P^{p-1}} \ddot{F}\ddot{\omega}_{p-1}$$

if $r > 0$.

LEMMA 1.2.[3] *Let F be defined and integrable over $S(r)$. Then*

$$J(\mathfrak{a}) = \frac{1}{2\pi}\int_0^{2\pi} F(re^{i\phi}\mathfrak{a})\,d\phi$$

exists for almost all $\mathfrak{a} \in S(1)$. Moreover $J(\mathfrak{a}) = J(e^{i\theta}\mathfrak{a})$ for $0 \leqq \theta \leqq 2\pi$. Define $\ddot{J}$ by $\ddot{J} \circ \rho = J$. Then

$$\frac{1}{\Phi(r)}\int_{S(r)} F\sigma_r = \frac{1}{W}\int_{P^{p-1}} \ddot{J}\ddot{\omega}_{p-1}.$$

2. **Divisors.** Let $f \not\equiv 0$ be a holomorphic function on the open, connected neighborhood G of $\mathfrak{a}$ in C^p. Then

$$f(\mathfrak{z}) = \sum_{m=\nu}^{\infty} P_m(\mathfrak{z} - \mathfrak{a})$$

converges uniformly in an open neighborhood of $\mathfrak{a}$, where P_m is either identically zero or a homogeneous polynomial of degree m and where $P_\nu \not\equiv 0$. The number

$$\nu = \nu_f(\mathfrak{a}; 0)$$

is uniquely defined by f and $\mathfrak{a}$. It is called the *zero-multiplicity* of f at $\mathfrak{a}$.

A function ν on an open subset G of C^p is said to be a *divisor* on G if and only if for every point $\mathfrak{a} \in G$ an open, connected neighborhood U of $\mathfrak{a}$ and on U holomorphic functions $g \not\equiv 0$ and $h \not\equiv 0$ exist such that

$$\nu(\mathfrak{z}) = \nu_g(\mathfrak{z}; 0) - \nu_h(\mathfrak{z}; 0) \quad \text{if} \quad \mathfrak{z} \in U.$$

Under addition of functions, the set $\mathfrak{D}(G)$ of divisors on G form a module. The zero-divisor is defined by the zero-function. The module $\mathfrak{D}(G)$ is partially ordered by the usual partial ordering of real valued functions. A divisor ν is said to be nonnegative if $\nu(\mathfrak{z}) \geqq 0$ for all $\mathfrak{z} \in G$. A nonnegative divisor $\nu \in \mathfrak{D}(G)$ is said to be *positive* if $\nu(\mathfrak{a}) > 0$ for at least one $\mathfrak{a} \in G$. A divisor ν is nonnegative if and only if for every point $\mathfrak{a} \in G$ an open, connected neighborhood U of $\mathfrak{a}$ and on an U holomorphic function $g \not\equiv 0$ exist such that $\nu(\mathfrak{z}) = \nu_g(\mathfrak{z}; 0)$ for all $\mathfrak{z} \in U$.

If ν is a divisor on the open subset G of C^p, then two nonnegative divisors ν^+ and ν^- on G are uniquely associated to ν such that the following property holds.

If $\mathfrak{a} \in G$, then an open, connected neighborhood U of $\mathfrak{a}$ and on U holomorphic functions $g \not\equiv 0$ and $h \not\equiv 0$ exists such that

[3] See Stoll [13], Lemma 1.2 and Stoll [8], Satz 7, Zusatz 1 and Zusatz 2 and p. 142.

$$\nu(\mathfrak{z}) = \nu_g(\mathfrak{z};0) - \nu_h(\mathfrak{z};0) \quad if \quad \mathfrak{z} \in U$$

and such that the germs of g and h at $\mathfrak{a}$ *are coprime in the ring of germs of holomorphic functions at* $\mathfrak{a}$. *Moreover, if any such U, g and h are selected then*

$$\nu^+(\mathfrak{a}) = \nu_g(\mathfrak{a};0), \qquad \nu^-(\mathfrak{a}) = \nu_h(\mathfrak{a};0).$$

If ν_1 and ν_2 are nonnegative divisors such that $\nu = \nu_1 - \nu_2$, then $\nu_1 = \nu^+ + \delta$ and $\nu_2 = \nu^- + \delta$, where δ is a nonnegative divisor on G. The divisor δ is called the *greatest common divisor of* ν_1, ν_2.

Let ν be a divisor on the open subset G of $\mathbf{C}^p$. The *carrier* of ν is defined as

$$\mathfrak{M}(\nu) = \overline{\{\mathfrak{z} \mid \nu(\mathfrak{z}) \neq 0\}}.$$

It is either empty or an analytic set of pure dimension $p-1$ in G. Obviously, $\mathfrak{M}(\nu)$ is empty if and only if $\nu \equiv 0$. If $\nu \not\equiv 0$, then ν is constant on each connectivity component of the set of simple points of $\mathfrak{M}(\nu)$. If $\nu \geqq 0$, then

$$\mathfrak{M}(\nu) = \{\mathfrak{z} \mid \nu(\mathfrak{z}) > 0\}.$$

If ν_1 and ν_2 are divisors then

$$\mathfrak{M}(\nu_1) \cup \mathfrak{M}(\nu_2) \supseteq \mathfrak{M}(\nu_1 + \nu_2),$$

$$\mathfrak{M}(\nu_1) \cup \mathfrak{M}(\nu_2) \supseteq \mathfrak{M}(\nu_1 - \nu_2),$$

$$\mathfrak{M}(\nu^+) \cup \mathfrak{M}(\nu^-) = \mathfrak{M}(\nu),$$

$$\mathfrak{M}(a\nu) = \mathfrak{M}(\nu) \quad \text{if } 0 \neq a \in \mathbf{Z}.$$

If ν_1 and ν_2 are nonnegative divisors, then

$$\mathfrak{M}(\nu_1) \cup \mathfrak{M}(\nu_2) = \mathfrak{M}(\nu_1 + \nu_2).$$

Let H be open in $\mathbf{C}^q$ and let G be open in $\mathbf{C}^p$. Let $\alpha: H \to G$ be a holomorphic map. Let ν be a divisor on G. Suppose that $\alpha(H') \nsubseteq \mathfrak{M}(\nu)$ for every connectivity component of H. Then α lifts ν to a divisor $\alpha^*(\nu)$ on H which is uniquely defined as follows: Take $\mathfrak{a} \in H$. Take an open, connected neighborhood V of $\alpha(\mathfrak{a})$ in G and holomorphic functions $g \not\equiv 0$ and $h \not\equiv 0$ on V such that $\nu(\mathfrak{z}) = \nu_g(\mathfrak{z};0) - \nu_h(\mathfrak{z};0)$ on V. Take an open, connected neighborhood U of $\mathfrak{a}$ with $\alpha(U) \subseteq V$. Then $g \circ \alpha \not\equiv 0$ and $h \circ \alpha \not\equiv 0$ on U. Define $\alpha(\mathfrak{a}) = \mathfrak{b}$. Define

$$\alpha^*(\nu)(\mathfrak{a}) = \nu_{g \circ \alpha}(\mathfrak{b};0) - \nu_{h \circ \alpha}(\mathfrak{b};0).$$

Obviously, if $\alpha^*(\nu_1)$ and $\alpha^*(\nu_2)$ are defined then

$$\alpha^*(\nu_1) + \alpha^*(\nu_2) = \alpha^*(\nu_1 + \nu_2),$$

$$\alpha^*(\nu_1) - \alpha^*(\nu_2) = \alpha^*(\nu_1 - \nu_2).$$

If $\nu \geqq 0$, then $\alpha^*(\nu) \geqq 0$. If $\alpha: H \to G$ is biholomorphic, then $\alpha^*(\nu) = \nu \circ \alpha$.

Let ν be a divisor on the open subset G of C^p. For $\mathfrak{a} \in C^p - \{0\}$ define $G(\mathfrak{a}) = \{z \mid z\mathfrak{a} \in G, z \in C\}$. Define $j_{\mathfrak{a}}: G(\mathfrak{a}) \to G$ by $j_{\mathfrak{a}}(z) = z\mathfrak{a}$. The divisor ν is said to be *restrictable for* $\mathfrak{a} \in C^p - \{0\}$, if and only if $G(\mathfrak{a}) \neq \emptyset$ and $j_{\mathfrak{a}}(U) \nsubseteq \mathfrak{M}(\nu)$ for every connectivity component U of $G(\mathfrak{a})$. If ν is restrictable for $\mathfrak{a} \in C^p - \{0\}$, then $j_{\mathfrak{a}}^*(\nu)$ is defined on $G(\mathfrak{a})$. It is called the restriction of ν to $\mathfrak{a}$ and denoted by

$$\nu[\mathfrak{a}] = j_{\mathfrak{a}}^*(\nu).$$

Denote also

$$\nu[\mathfrak{a}](z) = \nu(z, \mathfrak{a}).$$

Suppose that $R(r,s) \subset G$ with $0 < r < s \leqq \infty$. Let f be a measurable function on $\mathfrak{M}(\nu) \cap \bar{R}(r,s)$. Define

$$J_{f,\nu}(\mathfrak{a}) = \sum_{r<|z|<s} \nu(z,\mathfrak{a}) f(z\mathfrak{a})$$

for $\mathfrak{a} \in S(1)$, if ν is restrictable for $\mathfrak{a}$. If ν is not restrictable for $\mathfrak{a}$ define $J_{f,\nu}(\mathfrak{a}) = 0$. Then

$$J_{f,\nu}(\mathfrak{a}) = J_{f,\nu}(e^{i\theta}\mathfrak{a}) \quad \text{if} \quad 0 \leqq \theta \leqq 2\pi.$$

Hence $\ddot{J}_{f,\nu}$ is well defined by $\ddot{J}_{f,\nu} \circ \rho = J_{f,\nu}$.

LEMMA 2.1.[4] *Suppose that* $0 \leqq r < s \leqq \infty$. *If one of the integrals*

$$\int_{\mathfrak{M}(\nu) \cap N(r,s)} |\nu f| \omega_{p-1}, \quad \int_{S(1)} J_{|f|,|\nu|} \sigma_1, \quad \int_{P^{p-1}} \ddot{J}_{|f|,|\nu|} \ddot{\omega}_{p-1}$$

exist, all three integrals exist. Moreover,

$$\frac{1}{W} \int_{\mathfrak{M}(\nu) \cap R(r,s)} \nu f \omega_{p-1} = \frac{1}{\Phi(1)} \int_{S(1)} J_{f,\nu} \sigma_1 = \frac{1}{W} \int_{P^{p-1}} \ddot{J}_{f\,\nu} \ddot{\omega}_{p-1}.$$

If $r = 0$ *and* $s > 0$, *then*

$$\frac{1}{W} \int_{\mathfrak{M}(\nu) \cap B(s)} \nu f \omega_{p-1} + \nu(0) f(0)$$

$$= \frac{1}{\Phi(1)} \int_{S(1)} \sum_{|z|<s} \nu(z,\mathfrak{a}) f(z\mathfrak{a}) \sigma_1(\mathfrak{a})$$

$$= \frac{1}{W} \int_{P^{p-1}} \sum_{|z|<s} \nu(z,\mathfrak{a}) f(z\mathfrak{a}) \ddot{\omega}_{p-1}.$$

Observe, that $S(r) \cap \mathfrak{M}(\nu)$ is a set of measure zero on $\mathfrak{M}(\nu)$ if $r > 0$. Hence

$$\frac{1}{W} \int_{\mathfrak{M}(\nu) \cap S(r)} \nu f \omega_{p-1} = 0.$$

[4] See Stoll [**8**, Lemma 1.3] and Stoll [**8**], Satz 9, p. 148.

Therefore, in Lemma 2.1, $R(r,s)$ can be replaced by $\bar{R}(r,s)$ or $\bar{R}(r,s) - S(r)$ or $\bar{R}(r,s) - S(s)$. Similarly, $\sum_{r<|z|<s}$ may be replaced by $\sum_{r\leqq|z|\leqq s}$ or $\sum_{r<|z|\leqq s}$ or $\sum_{r\leqq|z|<s}$, if $r>0$. $\sum_{|z|<s}$ may be replaced by $\sum_{|z|\leqq s}$, if $s>0$.

Let ν be a divisor on the open subset G of the complex plane $\mathbf{C}$. Suppose that $B(R) \subseteq G$ with $0 < R \leqq \infty$. Define the *counting function* n_ν by

$$n_\nu(t) = \sum_{|z|\leqq t} \nu(z) \quad \text{if } 0 \leqq t < R.$$

For $0 < s \leqq r < R$ define the *valence of* ν by

$$N_\nu(r,s) = \int_s^r n_\nu(t)\,\frac{dt}{t}.$$

If $\nu(0) = 0$, then $s = 0$ is also admissible. Denote $N_\nu(r,0) = N_\nu(r)$. Define

$$\begin{aligned} \phi(r,s,t) &= 0 \quad \text{for } t \geqq r, \\ &= \log(r/t) \quad \text{for } s \leqq t \leqq r, \\ &= \log(r/s) \quad \text{for } 0 \leqq t \leqq s. \end{aligned}$$

Then

$$N_\nu(r,s) = \sum_{|z|\leqq r} \nu(z)\,\phi(r,s,|z|).$$

If $\nu \geqq 0$, then $n_\nu(r)$ and $N_\nu(r,s)$ are nonnegative, monotonically increasing functions of r.

LEMMA 2.2.[5] *Let ν be a divisor on $\mathbf{C}$. Let f be a function of class C^1 on $\{t \mid a \leqq t \leqq b\}$ where $0 \leqq a < b < \infty$. Then*

$$\sum_{a<|z|\leqq b} \nu(z) f(|z|) = n_\nu(b)f(b) - n_\nu(a)f(a) - \int_a^b n_\nu(t) f'(t)\,dt.$$

LEMMA 2.3.[6] *Let s be a nonnegative, monotonically decreasing function of class C^1 on $\{t \mid t \geqq a\}$ such that the integral*

$$\int_a^\infty f'(t)\,s(t)\,dt$$

exists. Suppose that $f(t) \to 0$ for $t \to \infty$. Then

$$s(r)f(r) \to 0 \quad \textit{for} \quad r \to \infty.$$

LEMMA 2.4.[7] *Let ν be a nonnegative divisor on $\mathbf{C}$. Let f be a nonnegative, monotonically decreasing function of class C^1 on $\{t \mid t \geqq a\}$, where $a \geqq 0$. Suppose that $f(r) \to 0$ for $r \to \infty$. Then*

---

[5] This formula is well known. A proof can be found, for instance, in **[13]**, Lemma 5.4.

[6] See Stoll, [13] Lemma 5.5.

[7] This follows from Lemma 2.2 and Lemma 2.3. See Stoll **[13]**, Lemma 5.6.

$$\sum_{a<|z|} \nu(z)f(|z|) = -n_\nu(a)f(a) - \int_a^\infty n_\nu(t)f'(t)\,dt.$$

Let ν be a divisor on the open subset G of C^p with $p > 1$. Suppose that $B(R) \subseteq G$ with $0 < R \leqq \infty$. Define *the counting function* n_ν *by*

$$n_\nu(r) = \frac{1}{W(r)} \int_{\mathfrak{M}(\nu)\cap B(r)} \nu\upsilon_{p-1}$$

for $0 < r < R$. Then

$$\begin{aligned} n_\nu(r) &= \frac{1}{W} \int_{\mathfrak{M}(\nu)\cap B(r)} \nu\omega_{p-1} + \nu(0) \\ &= \frac{1}{\Phi(1)} \int_{S(1)} n_{\nu[\mathfrak{a}]}\sigma_1(\mathfrak{a}) \\ &= \frac{1}{W} \int_{P^{p-1}} n_{\nu[\mathfrak{a}]}\ddot{\omega}_{p-1}. \end{aligned}$$

For $r = 0$, define $n_\nu(r) = \nu(0)$. Then

$$n_\nu(r) - n_\nu(s) = \frac{1}{W} \int_{\mathfrak{M}(\nu)\cap R(s,r)} \nu\omega_{p-1}$$

for $0 \leqq s \leqq r < R$. Hence n_ν is continuous on $\{r \mid 0 \leqq r < R\}$. If $\nu \geqq 0$, then n_ν is nonnegative and monotonically increasing.[8]

For $0 < s \leqq r < R$ define the *valence of* ν by

$$N_\nu(r,s) = \int_s^r n_\nu(t)\frac{dt}{t}.$$

If $0 \notin \mathfrak{M}(\nu)$, then

$$N_\nu(r) = N_\nu(r,0) = \int_0^r n_\nu(t)\frac{dt}{t}$$

is defined. If $\nu \geqq 0$, then $N_\nu(r,s)$ is a nonnegative, monotonically increasing function of r on $\{r \mid s \leqq r < R\}$. For $0 < s < r < R$ define

$$\begin{aligned} \psi(r,s,t) &= 0 \quad \text{for} \quad t \geqq r, \\ &= (1/(2p-2))[(r/t)^{2p-2} - 1] \quad \text{for } s \leqq t < r, \\ &= (1/(2p-2))[(r/s)^{2p-2} - 1] \quad \text{for } 0 \leqq t < s. \end{aligned}$$

Then[9]

---

[8] See Kneser **[2]**, p. 8, Stoll **[8]**, Satz 10 and Stoll **[13]**, Lemma 1.4.

[9] See Kneser **[2]**, pp. 14 and 15, Stoll **[13]**, Lemma 1.5 and Stoll **[11]**, p. 157

$$N_\nu(r,s) = \frac{1}{W(r)} \int_{\mathfrak{M}(\nu)} \nu(\mathfrak{z})\,\psi(r,s,|\mathfrak{z}|)\,\upsilon_{p-1}$$

$$= \frac{1}{W} \int_{\mathfrak{M}(\nu)} \nu(\mathfrak{z})\,\phi(r,s,|\mathfrak{z}|)\,\omega_{p-1} + \nu(0)\,\phi(r,s,0).$$

If $0 < s < r < R$, or if $0 = s < r < R$ and $0 \notin \mathfrak{M}(\nu)$, then[10]

$$N_\nu(r,s) = \frac{1}{\Phi(1)} \int_{S(1)} N_{\nu[\mathfrak{a}]}(r,s)\sigma_r(\mathfrak{a}) = \frac{1}{W} \int_{P^{p-1}} N_{\nu[\mathfrak{a}]}(r,s)\,\ddot{\omega}_{p-1}.$$

If $p \geqq 1$, then

$$-n_\nu = n_{-\nu}, \qquad -N_\nu = N_{-\nu},$$

$$n_{\nu_1} + n_{\nu_2} = n_{\nu_1+\nu_2},$$

$$N_{\nu_1} + N_{\nu_2} = N_{\nu_1+\nu_2}.$$

LEMMA 2.5. *Let ν be a divisor on C^p with $p > 1$. Let g be a measurable function on $I = \{r \mid a \leqq r \leqq b\}$. Suppose that g is bounded. Then*

$$\int_a^b n_{\nu[\mathfrak{a}]}(t)\,g(t)\,dt$$

exists for all $\mathfrak{a} \in S(1)$ and is integrable over $S(1)$. Moreover,

$$\frac{1}{\Phi(1)} \int_{S(1)} \int_a^b n_{\nu[\mathfrak{a}]}(t)g(t)dt\sigma_1(\mathfrak{a}) = \int_a^b n_\nu(t)g(t)dt.$$

PROOF. At first assume that $\nu \geqq 0$. An entire function f exists on C^p such that $\nu_f(\mathfrak{z};0) = \nu(\mathfrak{z})$ for every $\mathfrak{z} \in C^p$. If ν is restrictable for $\mathfrak{a}$, which is the case for almost all $\mathfrak{a} \in S(1)$, then Jensen's formula implies

$$N_{\nu[\mathfrak{a}]}(r,s) = \frac{1}{2\pi} \int_0^{2\pi} \log|f(re^{i\phi}\mathfrak{a})|\,d\phi - \frac{1}{2\pi} \int_0^{2\pi} \log|f(se^{i\phi}\mathfrak{a})|\,d\phi.$$

Hence $N_{\nu[\mathfrak{a}]}(r,s)$ is a measurable function of $(\mathfrak{a}, r)$ on $S(1) \times I$. The function $n_{\nu[\mathfrak{a}]}(r)$ is continuous from the right. Hence the right-hand derivative of $N_{\nu[\mathfrak{a}]}(r,s)$ exists and is $(1/r)\,n_{\nu[\mathfrak{a}]}(r)$ if $r > 0$. Therefore $n_{\nu[\mathfrak{a}]}(t)$ is measurable on $S(1) \times I$. Because $n_{\nu[\mathfrak{a}]}(t) \geqq 0$ and because g is bounded, $n_{\nu[\mathfrak{a}]}(t)g(t)$ is integrable over $S(1) \times I$. Moreover

$$\infty > \int_a^b n_\nu(t)\,g(t)\,dt = \int_a^b \frac{1}{\Phi(1)} \int_{S(1)} n_{\nu[\mathfrak{a}]}(t)\,\sigma_1(\mathfrak{a})\,g(t)\,dt$$

$$= \frac{1}{\Phi(1)} \int_{S(1)} \int_a^b n_{\nu[\mathfrak{a}]}(t)\,g(t)\,dt\,\sigma_1(\mathfrak{a}).$$

---

10 See Kneser [2], p. 12, Stoll [8], Satz 9 and Stoll [13], Lemma 1.4.

In the general case is $v = v^+ - v^-$ and $v[\mathfrak{a}] = v^+[\mathfrak{a}] - v^-[\mathfrak{a}]$. Hence, simple subtraction proves the lemma. q.e.d.

LEMMA 2.6. *If v is a divisor on $\mathbf{C}^p$ with $p > 1$ and if $0 \notin \mathfrak{M}(v)$, then $v[\mathfrak{a}]$ exists for all $\mathfrak{a} \in S(1)$. Moreover a monotonically increasing function K of r in $\{r \mid r \geqq 0\}$ exists such that*

$$|n_{v[\mathfrak{a}]}(r)| \leqq K(r).$$

PROOF. At first suppose that $v \geqq 0$. Because $0 \notin \mathfrak{M}(v)$ clearly $v[\mathfrak{a}]$ exists. An entire function f exists on $\mathbf{C}^p$, such that $f(0) = 1$ and $v_f(\mathfrak{z}; 0) = v(\mathfrak{z})$ for all $\mathfrak{z} \in \mathbf{C}^p$. Define

$$M_f(r) = \operatorname*{Max}_{\mathfrak{z} \in S(r)} |f(\mathfrak{z})|, \qquad K(r) = \log M_f(er).$$

Then $K(r) \geqq 0$ is monotonically increasing. According to Jensen's formula,

$$n_{v[\mathfrak{a}]}(r) \leqq \int_r^{er} n_{v[\mathfrak{a}]}(t) \frac{dt}{t} \leqq N_{v[\mathfrak{a}]}(er) = \frac{1}{2\pi} \int_0^{2\pi} \log |f(ere^{i\phi}\mathfrak{a})| \, d\phi$$
$$\leqq \log M_f(er) = K(r).$$

In the general case, $v = v^+ - v^-$. Let K^+ be a bound for $n_{v^+[\mathfrak{a}]}$ and K^- a bound for $n_{v^-[\mathfrak{a}]}$. Define $K = K^+ + K^-$. Then

$$|n_{v[\mathfrak{a}]}(r)| \leqq |n_{v^-[\mathfrak{a}]}(r) - n_{v^+[\mathfrak{a}]}(r)| \leqq K^+(r) + K^-(r) = K(r). \qquad \text{q.e.d.}$$

LEMMA 2.7. *Let v be a divisor on $\mathbf{C}^p$ with $p > 1$. Let f be a function of class C^1 on $\{r \mid a \leqq r \leqq b\}$ where $0 \leqq a < b < \infty$. Then*

$$\frac{1}{W} \int_{\mathfrak{M}(v) \cap B(r)} v f(|\mathfrak{z}|) \omega_{p-1}$$
$$= n_v(b) f(b) - n_v(a) f(a) - \int_a^b n_v(t) f'(t) \, dt.$$

PROOF. According to Lemma 2.1, Lemma 2.3 and Lemma 2.5,

$$\frac{1}{W} \int_{\mathfrak{M}(v) \cap R(a,b)} v f(|\mathfrak{z}|) \omega_{p-1} = \frac{1}{\Phi(1)} \int_{S(1)} \sum_{a < |z| \leqq b} v(z, \mathfrak{a}) f(|z|) \sigma_1(\mathfrak{a})$$
$$= \frac{1}{\Phi(1)} \int_{S(1)} \left[n_{v[\mathfrak{a}]}(b) f(b) - n_{v[\mathfrak{a}]}(a) f(a) - \int_a^b n_{v[\mathfrak{a}]}(t) f'(t) \, dt \right] \sigma_1(\mathfrak{a})$$
$$= n_v(b) f(b) - n_v(a) f(a) - \int_a^b n_v(t) f'(t) \, dt. \qquad \text{q.e.d.}$$

LEMMA 2.8. *Let v be a nonnegative divisor on $\mathbf{C}^p$ with $p > 1$. Let f be a nonnegative, monotonically decreasing function of class C^1 on $\{t \mid t \geqq a\}$, where $a \geqq 0$. Suppose that $f(r) \to 0$ for $r \to \infty$. Then*

$$\frac{1}{W}\int_{\mathfrak{M}(\nu)\cap R(a,\infty)} \nu f(|\mathfrak{z}|)\,\omega_{p-1} = -n_\nu(a)f(a) - \int_a^\infty n_\nu(t)f'(t)\,dt$$

if at least one of the integrals exists.

PROOF. Take $r > a$. Then

$$\frac{1}{W}\int_{\mathfrak{M}(\nu)\cap R(a,r)} \nu f(|\mathfrak{z}|)\omega_{p-1}$$

$$= n_\nu(r)f(r) - n_\nu(a)f(a) + \int_a^r n_\nu(t)(-f'(t))\,dt$$

where $-f'(t) \geqq 0$. If $\int_{\mathfrak{M}(\nu)\cap R(a,\infty)} \nu f(|\mathfrak{z}|)\omega_{p-1}$ exists, then

$$\infty > \frac{1}{W}\int_{\mathfrak{M}(\nu)\cap R(a,\infty)} \nu f(|\mathfrak{z}|)\omega_{p-1} \geqq \int_a^r n_\nu(t)(-f'(t))\,dt.$$

Hence $\int_a^\infty n_\nu(t)f'(t)\,dt$ exists. If this latter integral exists, then $n_\nu(r)f(r) \to 0$ for $r \to \infty$ according to Lemma 2.3. Consequently, the limit $r \to \infty$ proves the lemma. q.e.d.

Let f be a meromorphic function on the open subset G of C^p with $p \geqq 1$. Take $\mathfrak{a} \in G$ and $a \in C \cup \{\infty\}$. Suppose that f is not identical to a on the component G' of G which contains $\mathfrak{a}$. If f is identical to zero on G', define $\nu_f^a(\mathfrak{a}) = 0$. If f is not identical to zero on G', a connected, open neighborhood U of $\mathfrak{a}$ with $U \subseteq G'$ and holomorphic functions $g \not\equiv 0$ and $h \not\equiv 0$ on U exist such that $hf = g$ on U and such that the germs of h and g at $\mathfrak{a}$ are coprime in the ring of germs of holomorphic functions at $\mathfrak{a}$. Take $(\alpha, \beta) \in C^2$ such that $a = -\beta/\alpha$ if $a \in C$ and $\alpha = 0$, $\beta \neq 0$ if $a = \infty$. Then $\alpha g + \beta h$ is not identical to zero on U. The number

$$\nu_f^a(\mathfrak{a}) = \nu\alpha_{g+\beta h}(\mathfrak{a}; 0)$$

does not depend on the choices of U, g, h, α and β. The function ν_f^a is a nonnegative divisor on G, if f is not identical to a on any component of G. Let I be the set of points of indetermination of f. Then

$$\mathfrak{M}_f(a) = \mathfrak{M}(\nu_f^a) = \{\mathfrak{z} \mid \mathfrak{z} \in G - I,\ f(\mathfrak{z}) = a\} \cup I.$$

If f is holomorphic on G, then $\nu_f^0(\mathfrak{z}) = \nu_f(\mathfrak{z}; 0)$ for every $\mathfrak{z} \in G$. If f is meromorphic on G, but not identical to zero on any component of G, define

$$\nu_f = \nu_f^0 - \nu_f^\infty$$

as the divisor of f. Then

$$\nu_f^+ = \nu_f^0, \qquad \nu_f^- = \nu_f^\infty,$$

$$\nu_{f.g} = \nu_f + \nu_g, \qquad \nu_{1/f} = -\nu_f.$$

The divisor ν_f is nonnegative on G, if and only if f is holomorphic on G, that is, if and only if $\nu_f^\infty \equiv 0$. If f is meromorphic on G, define $\iota_f(\mathfrak{z}) = \operatorname{Min}(\nu_f^0(\mathfrak{z}), \nu_f^\infty(\mathfrak{z}))$. Here ι_f is not a divisor. Its support is the set of points of indetermination of f:

$$I = \{\mathfrak{z} \mid \iota_f(\mathfrak{z}) > 0\}.$$

Abbreviate $\mu = \nu_f^a$ and denote

$$n_f(r;a) = n_\mu(r), \qquad n_f(r) = n_{\nu_f}(r),$$

$$N_f(r,s;a) = N_\mu(r,s), \qquad N_f(r,s) = N_{\nu_f}(r,s),$$

$$N_f(r;a) = N_\mu(r), \qquad N_f(r) = N_{\nu_f}(r).$$

If $\mathfrak{a} \in \boldsymbol{C}^p - \{0\}$, define $G(\mathfrak{a}) = \{z \mid z\mathfrak{a} \in G\}$. If f is meromorphic on the open set $G \subseteq \boldsymbol{C}^p$ with $p > 1$, and if ν_f^∞ is restrictable for $\mathfrak{a}$, define $f \mid \mathfrak{a}$ as a meromorphic function on $G(\mathfrak{a})$ by $(f \mid \mathfrak{a})(z) = f(z\mathfrak{a})$ for $z \in G(\mathfrak{a})$. If ν_f^a exists and is restrictable for $\mathfrak{a}$, then

$$\nu_f^a[\mathfrak{a}](z) = \nu_{f|\mathfrak{a}}^a(z) \tag{2.1}$$

if $z \in G(\mathfrak{a})$ and $z\mathfrak{a}$ is not a point of indetermination of f; at points of indetermination (2.1) may be wrong. However, in all cases, the equality

$$\nu_f[\mathfrak{a}] = \nu_{f|\mathfrak{a}}$$

holds. Therefore

$$n_f(t) = \frac{1}{\Phi(1)} \int_{S(1)} n_{f|\mathfrak{a}}(t)\,\sigma_1(\mathfrak{a}),$$

$$N_f(r,s) = \frac{1}{\Phi(1)} \int_{S(1)} N_{f|\mathfrak{a}}(r,s)\,\sigma_1(\mathfrak{a}).$$

If $0 \in \boldsymbol{C}^p$ is not a point of indetermination, then

$$n_f(t;a) = \frac{1}{\Phi(1)} \int_{S(1)} n_{f|\mathfrak{a}}(t;a)\,\sigma_1(\mathfrak{a}),$$

$$N_f(r,s;a) = \frac{1}{\Phi(1)} \int_{S(1)} N_{f|\mathfrak{a}}(r,s;a)\,\sigma_1(\mathfrak{a}).$$

3. The Jensen Formulas and the First Main Theorem. For $p \in N$ define

$$L(z) = L_p(z) = \frac{1}{(p-1)!}\,\frac{d^{p-1}}{dz^{p-1}}[z^{p-1}\operatorname{Log}(1-z)]$$

if $|z| < 1$, where Log denotes the principal value.

PROPOSITION 3.1[11] (JENSEN FORMULA). *Let $f \not\equiv 0$ be a meromorphic function on the open, connected neighborhood G of $\overline{B(r)}$ in C^p with $p > 1$. Suppose that $B(s) \cap \mathfrak{M}(v_f) = \emptyset$ with $0 < s < r$, then*

$$N_f(r,s) = \frac{1}{\Phi(r)} \int_{S(r)} \log|f|\,\sigma_r - \frac{1}{\Phi(s)} \int_{S(s)} \log|f|\,\sigma_s.$$

THEOREM 3.2[12] (JENSEN-POISSON FORMULA). *Let $f \not\equiv 0$ be a meromorphic function on the open, connected subset G of C^p. Suppose that $\overline{B(r)} \subseteq G$ and $\mathfrak{M}(v_f) \cap B(s) = \emptyset$ where $0 < s < r < \infty$. Suppose that $f(0) = 1$. On $B(s)$ define* $\log f$ by $\log f(0) = 0$. *If $\mathfrak{z} \in B(s)$, then*

$$\log f(\mathfrak{z}) = \frac{2}{\Phi(r)} \int_{S(r)} \log|f(\mathfrak{y})| \left[\frac{r^{2p}}{(r^2 - (\mathfrak{z}|\mathfrak{y}))^p} - 1\right] \sigma_r(\mathfrak{y})$$

$$+ \frac{1}{W} \int_{\mathfrak{M}(v_f) \cap B(r)} v_f(\mathfrak{y})\, L\left(\frac{(\mathfrak{z}|\mathfrak{y})}{(\mathfrak{y}|\mathfrak{y})}\right) \omega_{p-1}(\mathfrak{y})$$

$$- \frac{1}{W} \int_{\mathfrak{M}(v_f) \cap B(r)} v_f(\mathfrak{y})\, L\left(\frac{(\mathfrak{z}|\mathfrak{y})}{r^2}\right) \omega_{p-1}(\mathfrak{y})$$

where the first and third integral define holomorphic functions in $B(r)$ and where the second integral is holomorphic in $B(s)$.

PROPOSITION 3.3.[13] *The same assumptions as in Theorem 3.2 are made. Then*

$$\log f = \sum_{m=1}^{\infty} F_m$$

in $B(s)$, where F_m is identically zero or a homogeneous polynomial of degree m. Moreover

$$F_m^1(\mathfrak{z}) = \frac{2}{\Phi(r)} \int_{S(r)} \log|f(\mathfrak{y})| \frac{(\mathfrak{z}|\mathfrak{y})^m}{(\mathfrak{y}|\mathfrak{y})^m} \sigma_r(\mathfrak{y}),$$

$$F_m^2(\mathfrak{z}) = \frac{1}{W} \int_{\mathfrak{M}(v_f) \cap B(r)} v_f(\mathfrak{y}) \frac{(\mathfrak{z}|\mathfrak{y})^m}{(\mathfrak{y}|\mathfrak{y})^m} \omega_{p-1}(\mathfrak{y}),$$

$$F_m^3(\mathfrak{z}) = \frac{1}{W} \int_{\mathfrak{M}(v_f) \cap B(r)} v_f(\mathfrak{y}) \frac{(\mathfrak{z}|\mathfrak{y})^m}{r^{2m}} \omega_{p-1}(\mathfrak{y})$$

are homogeneous polynomials of degree m or identically zero, such that

---

[11] See Stoll **[13]**, Proposition 1.6 and Stoll **[11]**, p. 156.

[12] See Stoll **[10]**, (unpublished) and Stoll **[13]**, Theorem 1.7.

[13] See Kneser **[2]**, p. 22 and 23 and Stoll **[10]**, (unpublished).

$$F_m = \binom{m+p-1}{p-1}\left[F_m^1 - \frac{1}{m}F_m^2 + \frac{1}{m}F_m^3\right].$$

PROOF. For $|z| < 1$,

$$\frac{1}{(1-z)^p} = \sum_{m=0}^{\infty}\binom{m+p-1}{p-1} z^m.$$

Hence

$$\frac{r^{2p}}{(r^2 - (\mathfrak{z}\,|\,\mathfrak{y}))^p} - 1 = \sum_{m=1}^{\infty}\binom{m+p-1}{p-1}\left(\frac{(\mathfrak{z}\,|\,\mathfrak{y})}{(\mathfrak{y}\,|\,\mathfrak{y})}\right)^m$$

which converges uniformly in $(\mathfrak{y},\mathfrak{z}) \in S(r) \times K$ if K is a compact subset of $B(r)$. Hence

$$\frac{2}{\Phi(r)}\int_{S(r)} \log|f(\mathfrak{y})| \left[\frac{r^{2p}}{(r^2 - (\mathfrak{z}\,|\,\mathfrak{y}))^p} - 1\right]\sigma_r(\mathfrak{y}) = \sum_{m=1}^{\infty}\binom{m+p-1}{p-1} F_m^1(\mathfrak{z})$$

for $\mathfrak{z} \in B(r)$. If $|z| < 1$ then

$$L(z) = -\frac{1}{(p-1)!}\frac{d^{p-1}}{dz^{p-1}}\sum_{m=1}^{\infty}\frac{1}{m}z^{m+p-1} = -\sum_{m=1}^{\infty}\binom{m+p-1}{p-1}\frac{1}{m}z^m.$$

Hence

$$L\left(\frac{(\mathfrak{z}\,|\mathfrak{y})}{(\mathfrak{y}\,|\,\mathfrak{y})}\right) = -\sum_{m=1}^{\infty}\binom{m+p-1}{p-1}\frac{1}{m}\left(\frac{(\mathfrak{z}\,|\,\mathfrak{y})}{)\mathfrak{y}\,|\,\mathfrak{y})}\right)^m$$

which converges uniformly for $(\mathfrak{y},\mathfrak{z}) \in (\mathfrak{M}(v_f) \cap B(r)) \times K$ if K is a compact subset of $B(s)$. Therefore

$$\frac{1}{W}\int_{\mathfrak{M}(v_f)\cap B(r)} v_f(\mathfrak{y})\, L\left(\frac{(\mathfrak{z}\,|\,\mathfrak{y})}{(\mathfrak{y}\,|\,\mathfrak{y})}\right)\omega_{p-1}(\mathfrak{y}) = -\sum_{m=1}^{\infty}\binom{m+p-1}{p-1}\frac{1}{m}F_m^2(\mathfrak{z})$$

if $\mathfrak{z} \in B(s)$. For $(\mathfrak{y}\,|\,\mathfrak{z}) \in (\mathfrak{M}(v_f) \cap B(r)) \times K$, if K is a compact subset of $B(r)$, the series

$$L\left(\frac{(\mathfrak{z}\,|\,\mathfrak{y})}{r^2}\right) = -\sum_{m=1}^{\infty}\binom{m+p-1}{p-1}\frac{1}{m}\left(\frac{(\mathfrak{z}\,|\,\mathfrak{y})}{r^2}\right)^m$$

converges uniformly. Therefore

$$\frac{1}{W}\int_{\mathfrak{M}(v_f)\cap B(r)} v_f(\mathfrak{y})\, L\left(\frac{(\mathfrak{z}\,|\,\mathfrak{y})}{r^2}\right)\omega_{p-1}(\mathfrak{y}) = -\sum_{m=1}^{\infty}\binom{m+p-1}{p-1}\frac{1}{m}F_m^3(\mathfrak{z})$$

if $\mathfrak{z} \in B(r)$. Consequently, the Jensen-Poisson Formula implies

$$F_m = \binom{m+p-1}{p-1}[F_m^1 - F_m^2 + F_m^3]. \qquad \text{q.e.d.}$$

Suppose that f is a meromorphic function on C^p with $p \geqq 1$. For $r \geqq 0$ the integral[14]

$$A_f(r) = \frac{1}{\pi W(r)} \int_{B(r)} \frac{i}{2} \frac{df \wedge d\bar{f}}{(1+|f|^2)^2} \wedge \upsilon_{p-1}$$

exists and has a nonnegative integrand. If f is not identically zero, define $\iota_f(0) = \mathrm{Min}(\nu_f^0(0), \nu_f^\infty(0))$. Then[14]

$$A_f(r) = \frac{1}{\pi W} \int_{B(r)} \frac{i}{2} \frac{df \wedge d\bar{f}}{(1+|f|^2)^2} \wedge \omega_{p-1} + \iota_f(0).$$

The function A_f is monotonically increasing, nonnegative and continuous. It is[14]

$$A_f(r) = \frac{1}{\Phi(1)} \int_{S(1)} A_{f|\mathfrak{a}}(r)\,\sigma_1(\mathfrak{a}) + \iota_f(0).$$

The *characteristic* of f is defined by[14]

$$T_f(r,s) = \int_s^r A_f(t) \frac{dt}{t}$$

for $0 \leqq s < r < \infty$. The function T_f of r is nonnegative, monotonically increasing and of class C^1 for $r \geqq s$. It is[14]

$$T_f(r,s) = \frac{1}{\Phi(1)} \int_{S(1)} T_{f|\mathfrak{a}}(r,s)\,\sigma_1(\mathfrak{a}) + \iota_f(0) \log \frac{r}{s}.$$

PROPOSITION 3.4. *Let f and g be meromorphic functions on C^p. Take $s \geqq 0$. Then constant numbers c_1 and c_2 exist such that*[15]

$$T_{f+g}(r,s) \leqq T_f(r,s) + T_g(r,s) + c_1,$$

$$T_{fg}(r,s) \leqq T_f(r,s) + T_g(r,s) + c_2,$$

$$T_{1/f}(r,s) = T_f(r,s) \quad \text{if} \quad f \not\equiv 0$$

for all $r \geqq s$.

PROPOSITION 3.5.[16] *Let $f \not\equiv 0$ be a meromorphic function on C^p with $p \geqq 1$. Let $g \not\equiv 0$ and $h \not\equiv 0$ be entire functions on C^p such that $hf = g$. Then*

$$\delta = \nu_g - \nu_f^0 = \nu_h - \nu_f^\infty$$

---

14 See Kneser [2], p. 15, Stoll [9], Stoll [10], §8 and §9 and Stoll [11], §24.

15 See Stoll [9], p. 234.

16 See Stoll [9], p. 233, Stoll [10], Satz 9.1 and Stoll [11], p. 157.

is a nonnegative divisor, which is the zero divisor if and only if g and h are coprime at every point of C^p. *For* $0 < s \leqq r < \infty$,

$$T_f(r,s) = \frac{1}{\Phi(r)}\int_{S(r)} \log(|g|^2+|h|^2)^{1/2}\sigma_r - \frac{1}{\Phi(s)}\int_{S(s)} \log(|g|^2+|h|^2)^{1/2}\sigma_s - N_\delta(rs).$$

PROPOSITION 3.6.[17] *Let f be an entire function on* C^p *with* $p \geqq 1$. *For* $r \geqq 0$ *define* $M_f(r) = \mathrm{Max}_{\mathfrak{z}\in S(r)}|f(\mathfrak{z})|$. *Take* $s > 0$. *Then constant numbers* $K_1 > 0$ *and* $K_2 > 0$ *exist such that*[18]

$$\log^+ M_f(r) \leqq K_1 T(8er, s) + K_2,$$

$$T_f(r,s) \leqq \log^+ M_f(r) + \tfrac{1}{2}\log 2$$

if $r \geqq s$.

Let f be a meromorphic function on C^p with $p \geqq 1$. Take $a \in C \cup \{\infty\}$. Take $(\alpha, \beta) \in C^2$ such that $\alpha \neq 0$ and $a = -\beta/\alpha$ if $a \in C$, and such that $\alpha = 0$ and $\beta \neq 0$ if $a = \infty$. Then the integral[19]

$$m_f(r;a) = \frac{1}{\Phi(r)}\int_{S(r)} \log\frac{((1+|f|^2)(|\alpha|^2+|\beta|^2))^{1/2}}{|\alpha f+\beta|}\sigma_r$$

exists and is nonnegative. It does not depend on the special choice of (α, β). The function m_f is called the *proximity function*. It is[19]

$$m_f(r;a) = \frac{1}{\Phi(1)}\int_{S(1)} m_{f|\mathfrak{a}}(r;a)\,\sigma_1(\mathfrak{a}).$$

THEOREM 3.7[20] (FIRST MAIN THEOREM). *Let f be a nonconstant meromorphic function on* C^p *with* $p \geqq 1$. *Take* $0 < s < r < \infty$. *Then*

$$T_f(r,s) = N_f(r,s;a) + m_f(r;a) - m_f(s;a)$$

for every $a \in C \cup \{\infty\}$.

4. **The canonical function of a divisor.** Let $s > 0$. Define $I_s = \{r \mid r \geqq s\}$. Let g be a monotonically increasing, nonnegative function on I_s. For $0 < \rho < \infty$, define

$$\tau_\rho(g) = \limsup_{r\to\infty} \frac{g(r)}{r^\rho}.$$

Define

---

17 See Stoll [9], Satz 2 and its proof.

18 $\log^+ x = \mathrm{Max}\,(0, \log x)$.

19 See Kneser [2], Stoll [9], p. 233, Stoll [10] and Stoll [11].

20 See Kneser [2], p. 12, Stoll [10], Satz 8.2 and Stoll [11], p. 157.

$$C(g) = \left\{\mu \,\middle|\, \int_s^\infty g(t)\,t^{-\mu-1}dt < \infty;\ \mu > 0\right\},$$

$$D(g) = \left\{\mu \,\middle|\, \int_s^\infty g(t)\,t^{-\mu-1}dt = \infty;\ \mu > 0\right\},$$

$$\operatorname{Ord} g = \limsup_{r\to\infty} \frac{\log g(r)}{\log r}.$$

If $\rho < \operatorname{Ord} g$, then $\tau_\rho(g) = \infty$; if $\rho > \operatorname{Ord} g$, then $\tau_\rho(g) = 0$. If $\mu < \operatorname{Ord} g$, then $\mu \in D(g)$, if $\mu > \operatorname{Ord} g$, then $\mu \in C(g)$. If $0 < \operatorname{Ord} g = \lambda < \infty$, define

$$\begin{aligned} \text{type } g &= \text{maximal type } \textit{iff } \tau_\lambda(g) = \infty, \\ &= \text{normal type } \textit{iff } 0 < \tau_\lambda(g) < \infty, \\ &= \text{minimal type } \textit{iff } \tau_\lambda(g) = 0. \\ \text{class } g &= \text{convergence class } \textit{iff } \lambda \in C(g), \\ &= \text{divergence class } \textit{iff } \lambda \in D(g). \end{aligned}$$

If v is a nonnegative divisor on $\mathbf{C}^p$ with $p \geqq 1$, a number $\mu > 0$ is said to be an *exponent of convergence* if and only if

$$\frac{1}{W} \int_{\mathfrak{N}(v)\cap R(s,\infty)} v(\mathfrak{z})\,|\mathfrak{z}|^{-\mu}\,\omega_{p-1} < \infty \quad \text{for} \quad p > 1,$$

$$\sum_{|z|\geqq s} v(z)\,|z|^{-\mu} < \infty \quad \text{for} \quad p = 1$$

for some (hence every) $s > 0$. Then the *exponent of* v is defined by

$$\begin{aligned} \operatorname{expon} v &= \inf\{\mu \mid \mu > 0 \text{ exponent of convergence}\} \\ &= \sup\{\mu \mid \mu > 0 \text{ exponent of divergence}\} \end{aligned}$$

where at least one set is nonempty.

PROPOSITION 4.1.[21] *Let v be a nonnegative divisor on $\mathbf{C}^p$ with $p \geqq 1$. Then $\mu > 0$ is an exponent of convergence if and only if $\mu \in C(n_v)$, i.e., if and only if $\mu \in C(N_v)$. Moreover*

$$\begin{aligned} \operatorname{Ord} N_v &= \operatorname{Ord} n_v = \operatorname{expon} v = \lambda, \\ \text{type } N_v &= \text{type } n_v \quad \textit{if} \quad 0 < \lambda < \infty, \\ \text{class } N_v &= \text{class } n_v \quad \textit{if} \quad 0 < \lambda < \infty, \\ \tau_\rho(N_v) &\leqq (1/\rho)\tau_\rho(n_v) \quad \textit{if} \quad 0 < \rho < \infty, \\ \tau_\rho(n_v) &\leqq e^\rho \tau_\rho(N_v) \quad \textit{if} \quad 0 < \rho < \infty. \end{aligned}$$

---

21 See Stoll [9], Satz 1.

For $\mathfrak{c} \in \mathbf{C}^p$ define $\gamma: \mathbf{C}^p \to \mathbf{C}^p$ by $\gamma(\mathfrak{z}) = \mathfrak{z} + \mathfrak{c}$. Then $\nu \circ \gamma$ is a nonnegative divisor on $\mathbf{C}^p$ such that

$$n_\nu(r) \leqq (1 + |\mathfrak{c}|/r)^{2p-2} n_{\nu\circ\gamma}(r + |\mathfrak{c}|),$$
$$N_\nu(r,s) \leqq (1 + |\mathfrak{c}|/s)^{2p-1} N_{\nu\circ\gamma}(r + |\mathfrak{c}|, s).$$

Hence, Ord, class and type of N_ν and $N_{\nu\circ\gamma}$, and ord, class and type of n_ν and $n_{\nu\circ\gamma}$ coincide. Moreover

$$\operatorname{expon} \nu = \operatorname{expon} \nu\circ\gamma,$$
$$\tau_\rho(n_\nu) = \tau_\rho(n_{\nu\circ\gamma}) \quad \text{if } 0 < \rho < \infty,$$
$$\tau_\rho(N_\nu) = \tau_\rho(N_{\nu\circ\gamma}) \quad \text{if } 0 < \rho < \infty.$$

For a meromorphic function f on $\mathbf{C}^p$ define

$$\operatorname{Ord} f = \operatorname{Ord} T_f(\cdot, s),$$
$$\operatorname{type} f = \operatorname{type} T_f(\cdot, s) \quad \text{if } 0 \leqq \operatorname{Ord} f < \infty,$$
$$\operatorname{class} f = \operatorname{class} T_f(\cdot, s) \quad \text{if } 0 \leqq \operatorname{Ord} f < \infty,$$
$$\tau_\rho[f] = \tau_\rho(T_f(\cdot, s)) \quad \text{if } 0 < \rho < \infty.$$

These definitions are independent of $s > 0$. It is

$$\tau_\rho[f \cdot g] \leqq \tau_\rho[f] + \tau_\rho[g],$$
$$\tau_\rho[f + g] \leqq \tau_\rho[f] + \tau_\rho[g].$$

If f is an entire function define $M_f(r) = \operatorname{Max}_{\mathfrak{z} \in S(r)} |f(\mathfrak{z})|$. Then[22]

$$\operatorname{Ord} f = \operatorname{Ord} \overset{+}{\log} M_f = \lambda,$$
$$\operatorname{type} f = \operatorname{type} \overset{+}{\log} M_f \quad \text{if } 0 < \lambda < \infty,$$
$$\operatorname{class} f = \operatorname{class} \overset{+}{\log} M_f \quad \text{if } 0 < \lambda < \infty.$$

$$\tau_\rho[f] \leqq \tau_\rho(\overset{+}{\log} M_f) \leqq 8^{3p+1}(8e)^\rho \tau_\rho[f] \tag{4.1}$$

if $0 < \rho < \infty$.

Let f be a nonconstant, meromorphic function on $\mathbf{C}^p$. Take $a \in \mathbf{C} \cup \{\infty\}$. The First Main Theorem implies

$$\begin{aligned} &\operatorname{expon} \nu_f^a \leqq \operatorname{Ord} f, \\ &\tau_\rho(N_f(\cdot, s; a)) \leqq \tau_\rho[f] \quad \text{if } 0 < \rho < \infty. \end{aligned} \tag{4.2}$$

PROPOSITION 4.2. *Let f be a meromorphic function on $\mathbf{C}^p$. Take $\mathfrak{c} \in \mathbf{C}^p$.*

22 See Proposition 3.6, and Stoll [9], Satz 2.

Define $\gamma: \mathbf{C}^p \to \mathbf{C}^p$ *by* $\gamma(\mathfrak{z}) = \mathfrak{z} + \mathfrak{c}$. *Then* $f \circ \gamma$ *is a meromorphic function on* $\mathbf{C}^p$ *such that*

$$A_f(r) \leqq \left(1 + \frac{|\mathfrak{c}|}{r}\right)^{2p-2} A_{f \circ \gamma}(r + |\mathfrak{c}|),$$

$$T_f(r,s) \leqq \left(1 + \frac{|\mathfrak{c}|}{s}\right)^{2p-1} T_{f \circ \gamma}(r + |\mathfrak{c}|, s).$$

Hence f and $f \circ \gamma$ *have the same* Ord, type *and* class. *Moreover,*

$$\tau_\rho[f] = \tau_\rho[f \cdot \gamma]. \tag{4.3}$$

PROOF. For the proof of the estimates, see Stoll [**9**], p. 215 and Kneser [**2**] p. 16. Hence

$$\tau_\rho[f] \leqq \left(1 + \frac{|\mathfrak{c}|}{s}\right)^{2p-1} \tau_\rho[f \circ \gamma].$$

Because $\tau_\rho[f]$ and $\tau_\rho[f \circ \gamma]$ do not depend on s, the limit $s \to \infty$ implies $\tau_\rho[f] \leqq \tau_\rho[f \circ \gamma]$. Moreover $\tau_\rho[f \circ \gamma] \leqq \tau_\rho[f \circ \gamma \circ \gamma^{-1}] = \tau_\rho[f]$. q.e.d.

Let g be a nonnegative, monotonically increasing function on the positive real axis. If $s > 0$ exists such that $g(x) = 0$ for $0 \leqq x < s$ and if $0 \leqq q \in Z$ define

$$\kappa = 3e(q+1)(2 + \log(q+1)),$$

$$K_q^1(r) = K_q^1(r; g) = \kappa r^q \int_0^r g(t)\, t^{-q-1} dt.$$

If $q + 1 \in C(g)$, define

$$K_q^2(r) = K_q^2(r; g) = \kappa r^{q+1} \int_r^\infty g(t)\, t^{-q-2} dt \quad \text{for} \quad r > 0.$$

If both assumptions hold, define

$$K_q(r) = K_q(r; g) = K_q^1(r) + K_q^2(r).$$

LEMMA 4.3. *Let g be an increasing, nonnegative function on the nonnegative real axis such that* $g(x) = 0$ *for* $0 \leqq x < s$. *Let q be a nonnegative integer. Define* κ *as before and* κ_1 *by*

$$\kappa_1 = (\kappa/q)(1 - 1/2^q) \quad \text{if } q > 0,$$
$$= \kappa \log 2 \quad \text{if } q = 0.$$

Take $\rho > q$. *Then*

$$2^{-\rho} \kappa_1 \tau_\rho(g) \leqq \tau_\rho(K_q^1) \leqq (\kappa/(\rho - q))\, \tau_\rho(g).$$

PROOF. Suppose that $\tau_\rho(g) < \infty$. Take $\varepsilon > 0$. Then $r_0 > 0$ exists such that $g(t) \leqq (\tau_\rho(g) + \varepsilon)\, t^\rho$ for $t \geqq r_0$. Hence

$$K_q^1(r)\, r^{-\rho} \leqq \kappa r^{q-\rho} \int_0^{r_0} g(t)\, \frac{dt}{t^{q+1}} + \frac{\kappa}{\rho - q}(\tau_\rho(g) + \varepsilon).$$

Because $\rho > q$, this implies

$$\tau_\rho(K_q^1) \leqq (\kappa/(\rho - q))(\tau_\rho(g) + \varepsilon)$$

for every $\varepsilon > 0$. Hence $\tau_\rho(K_q^1) \leqq (\kappa/(\rho - q))\, \tau_\rho(g)$, which is also true if $\tau_\rho(g) = \infty$.

If $q > 0$, then

$$K_q^1(2r)(2r)^{-\rho} \geqq \kappa(2r)^{q-\rho} \int_r^{2r} g(t)\, t^{-q-1}\, dt \geqq \frac{\kappa}{q} \frac{1}{2^\rho} \frac{g(r)}{r^\rho} \left(1 - \frac{1}{2^q}\right).$$

Hence

$$\tau_\rho(K_q^1) \geqq \kappa_1 2^{-\rho} \tau_\rho(g).$$

If $q = 0$, then

$$\begin{aligned} K_q^1(2r)(2r)^{-\rho} &\geqq \kappa(2r)^{-\rho} \int_r^{2r} g(t) \frac{dt}{t} \\ &\geqq 2^{-\rho} \kappa \log 2 \cdot g(r)/r^\rho . \end{aligned}$$

Hence

$$\tau_\rho(K_q^1) \geqq 2^{-\rho} \kappa_1 \tau_\rho(g). \qquad \text{q.e.d.}$$

LEMMA 4.4. *Let g be an increasing, nonnegative function on the nonnegative real axis. Let q be a nonnegative integer with $q + 1 \in C(g)$. If $q \leqq \rho < q + 1$, then*

$$(\kappa/(q + 1))\, \tau_\rho(g) \leqq \tau_\rho(K_q^2) \leqq (\kappa/(q + 1 - \rho))\, \tau_\rho(g).$$

If $\rho \geqq q + 1$, then $\tau_\rho(g) = \tau_\rho(K_q^2) = 0$.

PROOF. For $r > 0$,

$$K_q^2(r)\, r^{-\rho} = \kappa r^{q+1-\rho} \int_r^\infty \frac{g(t)}{t^{q+2}}\, dt \geqq (\kappa/(q + 1))\, (g(r)/r^\rho).$$

Hence

$$\tau_\rho(K_q^2) \geqq (\kappa/(q + 1))\, \tau_\rho(g).$$

If $\rho \geqq q + 1$ then $K_q^2(r)\, r^{-\rho} \to 0$ for $r \to \infty$. Hence $\tau_\rho(K_q^2) = 0$. Consequently, also $\tau_\rho(g) = 0$. Suppose that $q \leqq \rho < q + 1$. Assume that $\tau_\rho(g) < \infty$. Take $\varepsilon > 0$. Then $r_0 > 0$ exists such that $g(t) \leqq (\tau_\rho(g) + \varepsilon)\, t^\rho$ for $t \geqq r_0$. Take $r > r_0$. Then

$$K_q^2(r)r^{-\rho} \leqq \kappa r^{q+1-\rho}(\tau_\rho(g)+\varepsilon)\int_r^\infty t^{\rho-q-2}\,dt \leqq (\kappa/(1+q-\rho))(\tau_\rho(g)+\varepsilon).$$

Hence

$$\tau_\rho(K_q^2) \leqq (\kappa/(q+1-\rho))(\tau_\rho(g)+\varepsilon)$$

for every $\varepsilon > 0$. Hence $\tau_\rho(K_q^2) \leqq (\kappa/(q+1-\rho))\,\tau_\rho(g)$; q.e.d.

Lemma 4.3 and Lemma 4.4 imply

LEMMA 4.5.[23] *Let g be an increasing nonnegative function on the nonnegative real axis such that $g(x) = 0$ if $0 \leqq x < s$. Suppose that q is a nonnegative integer with $q+1 \in C(g)$. If $q < \rho < q+1$, then*

$$(2^{-\rho}\kappa_1 + \kappa/(q+1))\,\tau_\rho(g) \leqq \tau_\rho(K_q) \leqq \kappa(1/(\rho-q) + 1/(q+1-\rho))\,\tau_\rho(g).$$

If $\rho \geqq q+1$, then $\tau_\rho(g) = \tau_\rho(K_q) = 0$. Moreover, if $q \leqq \operatorname{Ord} g \leqq q+1$, then $\operatorname{Ord} K_q = \operatorname{Ord} g = \lambda$ *and,*

$$\text{type } K_q = \text{type } g \quad \textit{if } q < \lambda \leqq q+1,$$

$$\text{class } K_q = \text{class } g \quad \textit{if } q < \lambda < q+1.$$

Let q be a nonnegative integer. The Weierstrass *prime factor* is defined by

$$E(z,q) = (1-z)\exp\left[\sum_{\mu=1}^{q} \frac{1}{\mu} z^\mu\right]$$

If $|z| < 1$, then $\log E(z,q)$ is defined by $\log E(0,q) = 0$, i.e., by

$$\log E(z,q) = \sum_{\mu=q+1}^{\infty} \frac{z^\mu}{\mu}$$

If $p \in N$, then e_p is defined for $|z| < 1$ by

$$e_p(z,q) = (1/(p-1)!)(d^{p-1}/dz^{p-1})[z^{p-1}\log E(z,q)] = \sum_{\mu=q+1}^{\infty} \binom{\mu+p-1}{\mu}\frac{z^\mu}{\mu}.$$

The function e_p is called the *Kneser Kernel.*

THEOREM 4.6.[24] *Let v be a positive divisor on $\mathbf{C}^p$ with $p > 1$. Suppose that the positive integer $q+1$ is an exponent of convergence of v. Suppose that*

$$\eta = \inf\{r \mid \mathfrak{M}(v) \cap B(r) \neq \varnothing\}$$

is positive. Then the following assertions are true.

1. *The integral*

---

23 See Nevanlinna [5], pp. 227–228.

24 See Stoll [9], Satz 3, Satz 4 and Satz 5. If $v \equiv 0$, define the constant 1 as the canonical function to v and q.

$$H(\mathfrak{z}) = \frac{1}{W} \int_{\mathfrak{M}(\nu)} \nu(\mathfrak{y})\, e_p\left(\frac{(\mathfrak{z} \mid \mathfrak{y})}{(\mathfrak{y} \mid \mathfrak{y})}, q\right) \omega_{p-1}(\mathfrak{y})$$

exists for every $\mathfrak{z} \in B(\eta)$ *and defines a holomorphic function* H *on* $B(\eta)$ *with* $\nu_H(0) \geqq q + 1$. *The derivatives of* H *can be computed by differentiation under the integral.*

2. *One and only one entire function* h *on* C^p *exists such that* $h = e^H$ *on* $B(\eta)$. *The function* h *is called the canonical function to* ν *and* q.

3. *On* C^p *is* $\nu = \nu_h$.

4. *For* $r > 0$ *define* $M_h(r) = \mathrm{Max}_{\mathfrak{z} \in S(r)} |h(\mathfrak{z})|$. *Then*

$$\log M_h(r) \leqq 8^{3p}[K_q(4r, n_\nu) + N_\nu(4r, 0) + n_\nu(4r)].$$

5. *For* $\mathfrak{a} \in C^p - \{0\}$,

$$h(z\mathfrak{a}) = \lim_{r \to \infty} \prod_{|y| \leqq r} E\left(\frac{y}{z}, q\right)^{\nu(z,\mathfrak{a})}.$$

If $\mathrm{expon}\, \nu < q + 1$, *then this product converges absolutely and* $\lim_{r\to\infty} \prod_{|y| \leqq r}$ *can be replaced by* $\prod_{y \in C}$.

6. *If* $q + 1$ *is the smallest, positive integer, which is an exponent of convergence of* ν, *then*

$$q \leqq \mathrm{Ord}\, h = \mathrm{expon}\, \nu = \lambda \leqq q + 1,$$

$$\mathrm{type}\, h = \mathrm{type}\, n_\nu \quad \text{if } q < \lambda \leqq q + 1,$$

$$\mathrm{class}\, h = \mathrm{class}\, n_\nu \quad \text{if } q < \lambda < q + 1.$$

For the proof of these assertions, except 5, see Stoll [**9**], Satz 3, Satz 4 and Satz 5. For $\mathfrak{a} \in C^p - \{0\}$ is $\nu[\mathfrak{a}] = \nu_{h|\mathfrak{a}}$. Because $q + 1$ is an exponent of convergence, it is

$$\tau_{q+1}(n_\nu) = \tau_{q+1}(N_\nu) = \tau_{q+1}(K_q(\cdot, n_\nu)) = 0.$$

Hence $\tau_{q+1}(\log M_h) = 0$. Then

$$h(z\mathfrak{a}) = e^g \lim_{r \to \infty} \prod_{|y| \leqq r} E\left(\frac{y}{z}, q\right)^{\nu(z,\mathfrak{a})}$$

where

$$g(z) = \sum_{\mu=1}^{q} \frac{1}{\mu!} \frac{d^\mu}{dz^\mu} \log h(z\mathfrak{a}) \big|_{z=0} z^\mu = 0.$$

If $\mathrm{expon}\, \nu < q + 1$, pick any ρ in $\mathrm{expon}\, \nu < \rho \leqq q + 1$. Then $\tau_\rho(n_\nu) = \tau_\rho(N_\nu) = \tau_\rho(K_q(\cdot, n_\nu)) = 0$. Therefore $q + 1 \in C(K_q(\cdot, n_\nu) + N_\nu + n_\nu)$. Therefore $q + 1 \in C(\log M_h)$ and $q + 1 \in C(\log M_{h|\mathfrak{a}})$. Hence $q + 1 \in C(N_{\nu[\mathfrak{a}]})$ which implies that $q + 1$ is an exponent of convergence of $\nu[\mathfrak{a}]$. Consequently, the product

$$\prod_{y \in C} E\left(\frac{z}{y}, q\right)^{\nu(z,\mathfrak{a})}$$

converges absolutely. q.e.d.

If $q + 1$ is a positive, integral exponent of convergence of the divisors ν_1 and ν_2 with $\nu_1(0) = \nu_2(0) = 0$, and if h_λ is the canonical function for ν_λ and q, then $h_1 \cdot h_2$ is the canonical function to $\nu_1 + \nu_2$.

5. **The Lindelöf theorem.** The Lindelöf theorem[25] is a variation of Theorem 4.6. For its proof, several preparations are necessary.

LEMMA 5.1. *Let f be a holomorphic function for $|z| < R$. Define $M_f(r) = \operatorname{Max}_{|z|=r}|f(z)|$. If $0 < \theta < 1$ and $|z| \leqq \theta r < R$, then*

$$\left|\frac{1}{(m-1)!}\frac{d^{m-1}}{d z^{m-1}} f(z)\right| \leqq \frac{1}{(1-\theta)^m}\frac{M_f(r)}{r^{m-1}} \quad \text{if} \quad |z| \leqq \theta r.$$

LEMMA 5.2.[26] (a) *If $q > 0$ is an integer then*

$$\log|E(z,q)| \leqq (\kappa/(q+1))(|z|^{q+1}/(1+|z|)) \quad \text{for} \quad z \in \boldsymbol{C}.$$

(b) *If $0 \leqq m \in Z$, $p \in N$, $0 < \theta < 1$, $0 < r < 1$, $|z| \leqq \theta r$; then*

$$\left|\frac{1}{m!}\frac{d^m}{dz^m}\{z^{p-1}\log E(z,q)\}\right| \leqq \frac{1}{(1-\theta)^{m+1}}\frac{r^{q+p-m}}{1-r}.$$

(c) $(1/m!)(d^m/dz^m)\{z^{p-1}\log E(z,q)\}\big|_{z=0} = 0$ if $0 \leqq m < p + q$.

(d) *If $0 \leqq q \in Z$ and $|z| < 1$, then $|\log E(z,q)| \leqq (|z|^{q+1}/(1-|z|))$.*

LEMMA 5.3.[27] *Let f be a holomorphic function on $B(R) \subseteq \boldsymbol{C}^p$ with $0 < R \leqq \infty$ and with $p > 1$. If $\mathfrak{z} \in B(R)$, then*

$$(p-1)\int_0^1 (1-\tau)^{p-2} f(\mathfrak{z}\tau)\,d\tau = \frac{1}{\Phi(1)}\int_{S(1)} f((\mathfrak{z}|\mathfrak{a})\mathfrak{a})\,\sigma_1(\mathfrak{a}).$$

LEMMA 5.4.[27] *Let f be a holomorphic function on $B(R)$ with $0 < R \leqq \infty$ and with $p > 1$. Suppose that $f(0) \neq 0$. If $\mathfrak{z} \in B(R)$, then*

$$(p-1)\int_0^1 (1-\tau)^{p-2}\log|f(\mathfrak{z}\tau)|\,d\tau \leqq \frac{1}{\Phi(1)}\int_{S(1)} \log|f((\mathfrak{z}|\mathfrak{a})\mathfrak{a})|\,\sigma_1(\mathfrak{a}).$$

LEMMA 5.5. *Let f be a holomorphic function on $B(R)$ with $0 < R \leqq \infty$. Then*

$$(m-1)!\int_0^z\int_0^{x_1}\cdots\int_0^{x_{m-2}} f(x_{m-1}\mathfrak{z})\,dx_{m-1}\cdots dx_1$$

$$= z^{m-1}(m-1)\int_0^1 (1-\tau)^{m-2} f(\tau z \mathfrak{z})\,d\tau.$$

---

25 See Lindelöf [4], p. 375 and Levin [3], Chapter I, Theorem 15.

26 See Nevanlinna [5], p. 225 and Stoll [9], Hilfssatz 3.

27 See Stoll [9], Hilfssatz 6.

LEMMA 5.6.[28] *Let f be an entire function on C^p with $p>1$. Let $m>1$ be an integer. Suppose that $f(0)=1$. For $r>0$ define*

$$H(r) = \operatorname*{Max}_{|\mathfrak{z}| \leqq r} (m-1) \int_0^1 (1-\tau)^{m-2} \log |f(\tau\mathfrak{z})| \, d\tau.$$

Define $v = v_f$. Then

$$\log M_f(r) \leqq 8^{m+2p-1}[H(4r) + N_v(4r) + n_v(4r)]$$

for $r>0$.

LEMMA 5.7. *For $0 \leqq p \in Z$ and $0 \leqq q \in Z$ is*

$$(p+1) \int_0^1 (1-t)^p t^q dt = 1 \Big/ \binom{p+q+1}{q}.$$

LEMMA 5.8.[29] *If P is a homogeneous polynomial of degree q on C^p with $p>1$, then*

$$P(\mathfrak{z}) = \binom{q+p-1}{q} \frac{1}{\Phi(1)} \int_{S(1)} P((\mathfrak{z} \mid \mathfrak{a})\mathfrak{a}) \sigma_1(\mathfrak{a}).$$

LEMMA 5.9. *Let v be a positive divisor on C with $v(0)=0$. Let q be a non-negative integer. For $r>0$, define*

$$f_r(z) = \prod_{|y| \leqq 2r} E\left(\frac{z}{y}, q\right) v(y).$$

Then

$$\log |f_r(z)| \leqq \kappa n_v(2r) + K_q^1(2r; n_v)$$

if $|z| \leqq r$.

PROOF. First, consider the case $q>0$. If $|z|=r$, then

$$\begin{aligned}
\log |f_r(z)| &= \sum_{|y| \leqq 2r} v(y) \log \left| E\left(\frac{z}{y}, q\right)\right| \\
&\leqq \sum_{|y| \leqq 2r} v(y) \frac{\kappa}{q+1} \left(\frac{r}{|y|}\right)^{q+1} \frac{1}{1 + r/|y|} \\
&\leqq \frac{\kappa r^q}{q+1} \sum_{|y| \leqq 2r} \frac{v(y)}{|y|^q} \\
&= \frac{\kappa}{q+1} 2^{-q} n_v(2r) + \kappa r^q \int_0^{2r} \frac{n_v(t)}{t^{q+1}} dt \\
&\leqq \kappa n_v(2r) + K_q^1(2r; n_v).
\end{aligned}$$

28 See Stoll [9], Hilfssatz 7.

29 See Kneser [2], V. This follows immediately from Lemma 5.3 and Lemma 5.7.

Now, consider the case $q = 0$. If $|z| = r$, then

$$\begin{aligned}\log|f_r(z)| &= \sum_{|y|\leqq 2r} \nu(y)\log\left|1 - \frac{z}{y}\right| \\ &\leqq \sum_{|y|\leqq 2r} \nu(y)\log\left(1 + \frac{r}{|y|}\right) \\ &= n_\nu(2r)\log\left(1 + \frac{1}{2}\right) + \int_0^{2r} n_\nu(t)\frac{r}{(1 + r/t)}\frac{dt}{t^2} \\ &\leqq \kappa n_\nu(2r) + \kappa\int_0^{2r} n_\nu(t)\frac{dt}{t} \\ &= \kappa n_\nu(2r) + K_0^1(2r; n_\nu).\end{aligned}$$

Now, the maximum principle proves the lemma. q.e.d.

LEMMA 5.10. *Let ν be a nonnegative divisor on C^p with $p > 1$. Suppose that the positive integer $q + 1$ is an exponent of convergence of ν. Then*

$$P_r(\mathfrak{z}) = \frac{1}{W}\int_{\mathfrak{M}(\nu)\cap R(2r,\infty)} \nu(\mathfrak{y})\log E\left(\frac{(\mathfrak{z}\,|\,\mathfrak{y})}{(\mathfrak{y}\,|\,\mathfrak{y})}, q\right)\omega_{p-1}$$

exists for every $\mathfrak{z}\in B(2r)$ and defines a holomorphic function on $B(2r)$ whose derivatives are obtained by differentiation under the integral. For $\mathfrak{z}\in\overline{B(r)}$,

$$|P_r(\mathfrak{z})| \leqq K_q^2(2r; n_\nu).$$

PROOF. If $0 < \theta < 1$, if $\mathfrak{z}\in B(\theta 2r)$ and $\mathfrak{y}\in\mathfrak{M}(\nu)\cap R(2r,\infty)$, then

$$|(\mathfrak{z}\,|\,\mathfrak{y})/(\mathfrak{y}\,|\,\mathfrak{y})| \leqq |\mathfrak{z}|/|\mathfrak{y}| \leqq 2r\theta/|\mathfrak{y}| \leqq \theta < 1$$

and

$$\begin{aligned}|\log E((\mathfrak{z}\,|\,\mathfrak{y})/(\mathfrak{y}\,|\,\mathfrak{y}), q)| &\leqq |(\mathfrak{z}\,|\,\mathfrak{y})/(\mathfrak{y}\,|\,\mathfrak{y})|^{q+1}/(1 - |(\mathfrak{z}\,|\,\mathfrak{y})/(\mathfrak{y}\,|\,\mathfrak{y})|) \\ &\leqq ((2r\theta)^{q+1}/(1-\theta))(1/|\mathfrak{y}|^{q+1}).\end{aligned}$$

Hence, the integral exists and is holomorphic on $B(2r)$, and the derivatives are obtained by differentiation under the integral. Moreover, the choice $\theta = \frac{1}{2}$ implies

$$\begin{aligned}|P_r(\mathfrak{z})| &\leqq 2r^{q+1}\frac{1}{W}\int_{\mathfrak{M}(\nu)\cap R(2r,\infty)} \frac{\nu(\mathfrak{y})}{|\mathfrak{y}|^{q+1}}\omega_{p-1} \\ &= -\frac{n_\nu(2r)}{2^q} + 2(q+1)r^{q+1}\int_{2r}^{\infty} n_\nu(t)\,t^{-q-2}dt \\ &\leqq K_q^2(2r; n_\nu).\end{aligned}$$

q.e.d.

LEMMA 5.11. *Let A_μ be a set of $m+1$ different points in the complex plane for $\mu = 1, \cdots, p$. Define*

$$A = A_1 \times \cdots \times A_p \subseteq C^p.$$

Take $r_0 \in R$. Suppose that for every $r \geqq r_0$ a polynomial P_r is given such that either $P_r \equiv 0$ or $\deg P_r \leqq m$. Suppose that for every $\mathfrak{z} \in A$ the limit of $P_r(\mathfrak{z})$ for $r \to \infty$ exists. Then

$$P_r(\mathfrak{z}) \to P(\mathfrak{z}) \quad \text{for} \quad r \to \infty$$

exists for every $\mathfrak{z} \in C^p$ and the convergence is uniform on every compact subset of C^p. The limit P is a polynomial such that either $P \equiv 0$ or $\deg P \leqq m$.

PROOF.[30] The lemma will be proved by induction for p. If $p = 0$, it is trivially true, because a polynomial which depends on no variable is constant. Suppose the lemma is already proved for $p-1$. Let $a_0, \cdots, a_m$ be the points of A_p. Define

$$g_\mu(z) = \prod_{\lambda=0;\, \lambda \neq \mu}^{m} (z - a_\lambda),$$

$$h_\mu(z) = g_\mu(z)/g_\mu(a_\mu).$$

Then h_μ is a polynomial of degree m on C with

$$h_\mu(a_\nu) = 0 \quad \text{if } \mu \neq \nu,$$
$$= 1 \quad \text{if } \mu = \nu.$$

Moreover, define a polynomial $Q_{\mu r}$ on C^{p-1} by

$$Q_{\mu r}(z_1, \cdots, z_{p-1}) = P_r(z_1, \cdots, z_{p-1}, a_\mu)$$

for $\mu = 0, \cdots, m$ and all $r \geqq r_0$. Then each $Q_{\mu r}$ is identically zero or has a degree not larger than m. Because $Q_{\mu r}(\mathfrak{z})$ converges for $r \to \infty$ if $\mathfrak{z} \in A_1 \times \cdots \times A_{p-1}$, the polynomial $Q_{\mu r}$converges to a polynomial Q_μ for $r \to \infty$. The convergence is uniform on every compact subset of C^{p-1}. The limit Q_μ is identical zero or has at most degree m. The Lagrange interpolation formula

$$P_r(z_1, \cdots, z_p) = \sum_{\mu=0}^{m} Q_{\mu r}(z_1, \cdots, z_{p-1}) h_\mu(z_p)$$

implies that P_r converges, namely

$$P_r(z_1, \cdots, z_p) \to \sum_{\mu=0}^{m} Q_\mu(z_1, \cdots, z_{p-1}) h_\mu(z_p) = P(z_1, \cdots, z_p)$$

30 My original proof was different and required the convergence on all of C^p. Moreover, it used the algebraic closure of C^p. R. Otter suggested the use of the Lagrange interpolation formula to me.

for $r \to \infty$. The convergence is uniform on every compact subset of C^p. If $P \not\equiv 0$ then $\deg P \leqq m$. q.e.d.

Now, a generalization of a theorem of Lindelöf to $p > 1$ variables can be proved.

THEOREM 5.12.[31] *Let ν be a positive divisor on C^p with $p > 1$. Suppose that $\nu(0) = 0$. Let q be a positive integer and assume that*

$$n_\nu(r) r^{-q} \to 0 \quad \textit{for } r \to \infty \qquad (\textit{i.e.}, \ \tau_q(n_\nu) = 0)$$

and that for every $\mathfrak{z} \in C^p$ the limit

$$P(\mathfrak{z}) = \lim_{r \to \infty} \frac{1}{W} \int_{\mathfrak{M}(\nu) \cap B(r)} \nu(\mathfrak{y}) \left(\frac{(\mathfrak{z} \mid \mathfrak{y})}{(\mathfrak{y} \mid \mathfrak{y})} \right)^q \omega_{p-1}(\mathfrak{y})$$

exists. Define

$$\eta = \inf\{r \mid B(r) \cap \mathfrak{M}(\nu) \neq \varnothing\}.$$

Then $0 < \eta < \infty$. The limit

$$G(\mathfrak{z}) = \lim_{r \to \infty} \frac{1}{W} \int_{\mathfrak{M}(\nu) \cap B(r)} \nu(\mathfrak{y}) e_p \left(\frac{(\mathfrak{z} \mid \mathfrak{y})}{(\mathfrak{y} \mid \mathfrak{y})}, q - 1 \right) \omega_{p-1}$$

exists for every $\mathfrak{z} \in B(\eta)$. The limit is uniform on every compact subset of $B(\eta)$. The function G is holomorphic on $B(\eta)$. One and only one entire function g, called the Lindelöf function to ν and q, exists on C^p such that $g = e^G$ on $B(\eta)$. On C^p is $\nu_g = \nu$. If $M_g(r) = \operatorname{Max}_{\mathfrak{z} \in S(r)} |g(\mathfrak{z})|$ then

$$\log M_g(r)/r^q \to 0 \quad \text{for } r \to \infty$$

i.e., $\tau_q(\log M_g) = 0$ which means $\tau_q[g] = 0$.

PROOF. Because $\tau_q(n_\nu) = 0$, it follows that

$$\operatorname{expon} \nu \leqq q < q + 1.$$

Hence $q + 1$ is a positive integer which is an exponent of convergence of ν. Therefore,

$$H(\mathfrak{z}) = \frac{1}{W} \int_{\mathfrak{M}(\nu)} \nu(\mathfrak{z}) \, e_p \left(\frac{(\mathfrak{z} \mid \mathfrak{y})}{(\mathfrak{y} \mid \mathfrak{y})}, q \right) \omega_{p-1}$$

exists for every $\mathfrak{z} \in B(\eta)$. Moreover, H is a holomorphic function on $B(\eta)$ and the limit

---

[31] For $p = 1$ see Lindelöf [4] page 375 and Levin [3] Chapter I, Theorem 15. If $\nu \equiv 0$, define the constant 1 as the Lindelöf function to ν and q.

$$H(\mathfrak{z}) = \lim_{r\to\infty} \frac{1}{W} \int_{\mathfrak{M}(v)\cap B(r)} v(\mathfrak{z})\, e_p\left(\frac{(\mathfrak{z} \mid \mathfrak{y})}{(\mathfrak{y} \mid \mathfrak{y})}, q\right) \omega_{p-1}.$$

is uniform on every compact subset of $B(\eta)$. If h is the canonical function to v and q then $h = e^H$ on $B(\eta)$. Moreover, $v_h = v$ on C^p. Define

$$Q_r(\mathfrak{z}) = \frac{1}{q}\binom{q+p-1}{q}\frac{1}{W}\int_{\mathfrak{M}(v)\cap B(r)} v(\mathfrak{y})\left(\frac{(\mathfrak{z}\mid\mathfrak{y})}{(\mathfrak{y}\mid\mathfrak{y})}\right)^q \omega_{p-1}.$$

Then Q_r is identically zero or a homogeneous polynomial of degree q. Because

$$Q_r(\mathfrak{z}) \to Q(\mathfrak{z}) = \frac{1}{q}\binom{q+p-1}{q} P(\mathfrak{z}) \quad \text{for} \quad r\to\infty$$

if $\mathfrak{z}\in C^p$, this limit is uniform on every compact subset of C^p and Q is either zero or a homogeneous polynomial of degree q. Define $G = H - Q$ on $B(\eta)$. Then G is holomorphic on $B(\eta)$. The identity

$$e_p(z,q) = e_p(z, q-1) + \frac{1}{q}\binom{q+p-1}{q} z^q$$

implies

$$G(\mathfrak{z}) = \lim_{r\to\infty} \frac{1}{W} \int_{\mathfrak{M}(v)\cap B(r)} v(\mathfrak{y})\, e_p\left(\frac{(\mathfrak{z} \mid \mathfrak{y})}{(\mathfrak{y} \mid \mathfrak{y})}, q-1\right) \omega_{p-1}$$

for $\mathfrak{z}\in B(\eta)$, where the convergence is uniform on every compact subset of $B(\eta)$.

Define the entire function g by

$$g = he^{-Q}.$$

Then $v_g = v_h = v$ on C^p. Moreover,

$$g = e^H e^{-Q} = e^G \quad \text{on} \quad B(\eta).$$

Now, $\log M$, has to be estimated. Because $0\notin\mathfrak{M}(v)$, the divisor v is restrictable to every $\mathfrak{a}\in S(1)$. According to Theorem 4.6, Assertion 5,

$$h(z\mathfrak{a}) = \prod_{y\in C} E\left(\frac{z}{y}, q\right)^{v(y,\mathfrak{a})}.$$

Define

$$h_r(z;\mathfrak{a}) = \prod_{|y|\leqq 2r} E\left(\frac{z}{y}, q-1\right)^{v(y,\mathfrak{a})},$$

$$j_r(z;\mathfrak{a}) = \frac{1}{q}\prod_{|y|\leqq 2r} v(y,\mathfrak{a})\left(\frac{z}{y}\right)^q,$$

$$f_r(z;\mathfrak{a}) = \prod_{|y|>2r} E\left(\frac{z}{y}, q\right)^{v(y,\mathfrak{a})}.$$

Then

$$h(z\mathfrak{a}) = h_r(z;\mathfrak{a})f_r(z;\mathfrak{a})\exp(j_r(z;\mathfrak{a})).$$

For $|z| < 2r$, $\log f_r$ is defined as a holomorphic function of z by setting $\log f_r(0;\mathfrak{a}) = 0$. If $0 < \theta < 1$ and $\mathfrak{z} \in B(\theta 2r)$ then

$$\left| \log E\left(\frac{(\mathfrak{z} \mid \mathfrak{y})}{(\mathfrak{y} \mid \mathfrak{y})}, q\right) \right| \leqq \frac{(2r\theta)^{q+1}}{(1-\theta)} \frac{1}{|\mathfrak{y}|^{q+1}} \quad \text{for } |\mathfrak{y}| \geqq 2r.$$

Hence Lemma 2.1 and Lemma 5.10 imply

$$\begin{aligned} F_r(\mathfrak{z}) &= \frac{1}{W} \int\limits_{\mathfrak{M}(v) \cap R(2r,\infty)} v(\mathfrak{y}) \log E\left(\frac{(\mathfrak{z} \mid \mathfrak{y})}{(\mathfrak{y} \mid \mathfrak{y})}, q\right) \omega_{p-1} \\ &= \frac{1}{\Phi(1)} \int_{S(1)} \log f_r((\mathfrak{z} \mid \mathfrak{a}); \mathfrak{a}) \sigma_1(\mathfrak{a}) \end{aligned}$$

where F_r is holomorphic on $B(2r)$ and where

$$|F_r(\mathfrak{z})| \leqq K_q^2(2r; n_v) \quad \text{for } |\mathfrak{z}| \leqq r.$$

According to Lemma 2.1 is

$$\begin{aligned} \frac{1}{\Phi(1)} \int_{S(1)} j_r((\mathfrak{z} \mid \mathfrak{a}); \mathfrak{a}) \sigma_1(\mathfrak{a}) &= \frac{1}{q} \int\limits_{\mathfrak{M}(v) \cap B(2r)} v(\mathfrak{y}) \left(\frac{(\mathfrak{z} \mid \mathfrak{y})}{(\mathfrak{y} \mid \mathfrak{y})}\right)^q \omega_{p-1}(\mathfrak{y}) \\ &= \left(1 \Big/ \binom{q+p-1}{q}\right) Q_{2r}(\mathfrak{z}). \end{aligned}$$

According to Lemma 5.9,

$$\log |h_r(z;\mathfrak{a})| \leqq \kappa n_{v[\mathfrak{a}]}(2r) + K_{q-1}^1(2r; n_{v[\mathfrak{a}]})$$

for $|z| \leqq r$. Hence

$$\log |h(z\mathfrak{a})| - \Re e\, j_r(z;\mathfrak{a}) - \Re e \log f_r(z;\mathfrak{a}) \leqq \kappa n_{v[\mathfrak{a}]}(2r) + K_{q-1}^1(2r; n_{v[\mathfrak{a}]})$$

for $|z| \leqq r$. This and Lemma 2.5 imply

$$\begin{aligned} \frac{1}{\Phi(1)} \int_{S(1)} \log |h((\mathfrak{z} \mid \mathfrak{a})\mathfrak{a})| \sigma_1(\mathfrak{a}) - \left(1 \Big/ \binom{q+p-1}{q}\right) \Re e\, Q_{2r}(\mathfrak{z}) - \Re e\, F_r(\mathfrak{z}) \\ \leqq \kappa n_v(2r) + K_{q-1}^1(2r; n_v) \end{aligned}$$

if $|\mathfrak{z}| \leqq r$. Therefore

$$\begin{aligned} \frac{1}{\Phi(1)} \int_{S(1)} \log |h((\mathfrak{z} \mid \mathfrak{a})\mathfrak{a})| \sigma_1(\mathfrak{a}) - \left(1 \quad \binom{q+p-1}{q}\right) \Re e\, Q_{2r}(\mathfrak{z}) \\ \leqq \kappa n_v(2r) + K_{q-1}^1(2r; n_v) + K_q^2(2r; n_v). \end{aligned}$$

The identity

$$\log|g((\mathfrak{z}|\mathfrak{a})\mathfrak{a})| = \log|h((\mathfrak{z}|\mathfrak{a})\mathfrak{a})| - \Re e\, Q((\mathfrak{z}|\mathfrak{a})\mathfrak{a})$$

and Lemma 5.8 imply

$$\frac{1}{\Phi(1)}\int_{S(1)} \log|g((\mathfrak{z}|\mathfrak{a})\mathfrak{a})|\,\sigma_1(\mathfrak{a})$$

$$= \frac{1}{\Phi(1)}\int_{S(1)} \log|h((\mathfrak{z}|\mathfrak{a})\mathfrak{a})|\,\sigma_1(\mathfrak{a}) - \left(1\Big/\binom{q+p-1}{q}\right)\Re e\, Q(\mathfrak{z})$$

$$\leqq \left(1\Big/\binom{q+p-1}{q}\right)\Re e\,[Q_{2r}(\mathfrak{z}) - Q(\mathfrak{z})]$$

$$+ \kappa n_\nu(2r) + K^1_{q-1}(2r;n_\nu) + L^2_q(2r;n_\nu)$$

if $|\mathfrak{z}| \leqq r$. Because $Q_{2r}(\mathfrak{z}) \to Q(\mathfrak{z})$ uniformly on $S(1)$ for $r\to\infty$ the maximum

$$L_0(r) = \binom{q+p-1}{q}^{-1} \operatorname*{Max}_{\mathfrak{z}\in S(1)} |Q_{2r}(\mathfrak{z}) - Q(\mathfrak{z})|$$

converges to zero for $r\to\infty$. Hence

$$\binom{q+p-1}{q}^{-1} |Q_{2r}(\mathfrak{z}) - Q(\mathfrak{z})| \leqq L_0(r)r^q \quad \text{if } |\mathfrak{z}| \leqq r.$$

According to Lemma 4.3 and Lemma 4.4

$$0 \leqq \tau_q(K^1_{q-1}(\cdot\,,n_\nu)) \leqq \kappa\tau_q(n_\nu) = 0,$$

$$0 \leqq \tau_q(K^2_q(\cdot\,,n_\nu) \leqq \kappa\tau_q(n_\nu) = 0.$$

Hence

$$\kappa n_\nu(2r) + K^1_{q-1}(2r;n_\nu) + K^2_q(2r;n_\nu) \leqq L_1(r)r^q$$

where $L_1(r)\to 0$ for $r\to\infty$. Define $L_2 = L_0 + L_1$. Then $L_2(r)\to 0$ for $r\to\infty$ and

$$\frac{1}{\Phi(1)}\int_{S(1)} \log|g((\mathfrak{z}|\mathfrak{a})\mathfrak{a})|\sigma_1(\mathfrak{a}) \leqq L_2(r)r^q$$

if $|\mathfrak{z}| \leqq r$. Lemma 5.4 and Lemma 5.6 imply

$$\log M_g(r) \leqq 8^{3p}(L_2(4r)(4r)^q + n_\nu(4r) + N_\nu(4r)).$$

Because

$$0 \leqq \tau_q(N_\nu) \leqq (1/q)\tau_q(n_\nu) = 0$$

this implies

$$\log M_g(r)/r^q \to 0 \quad \text{for } r \to \infty .$$

A similar generalization of a theorem of Lindelöf can be proved for $p > 1$ variables.

THEOREM 5.13[32] *Let ν be a positive divisor on $\mathbf{C}^p$ with $p > 1$. Suppose that $\nu(0) = 0$. Let q be a positive integer and assume that $n_\nu(r)r^{-q}$ is bounded, i.e., $\tau_q(n_\nu) < \infty$. For each $r > 0$ define P_r on $\mathbf{C}^p$ by*

$$P_r(\mathfrak{z}) = \frac{1}{W} \int\limits_{\mathfrak{M}(\nu)\cap B(r)} \nu(\mathfrak{y}) \left(\frac{(\mathfrak{z} \mid \mathfrak{y})}{(\mathfrak{y} \mid \mathfrak{y})}\right)^q \omega_{p-1}(\mathfrak{y}).$$

Suppose that a constant $C > 0$ exists such that

$$|P_r(\mathfrak{z})| \leqq C \quad \text{if } r > 0 \text{ and } \mathfrak{z} \in S(1).$$

Then $q + 1$ is an exponent of convergence of ν. Let h be the canonical function to ν and q. As usual, define

$$M_h(r) = \operatorname*{Max}_{\mathfrak{z} \in S(r)} |h(\mathfrak{z})|.$$

Then $\log M_h(r)r^{-q}$ is bounded for $r \geqq 1$; that means

$$\tau_q[h] < \infty .$$

PROOF. Because $\tau_q(n_\nu) < \infty$, it follows that

$$\operatorname{expon} \nu \leqq q < q + 1 .$$

Hence $q + 1$ is an exponent of convergence of ν. For $\mathfrak{a} \in S(1)$,

$$h(z\mathfrak{a}) = \prod_{y \in C} E\left(\frac{z}{y}, q\right)^{\nu(y,\mathfrak{a})}$$

Define

$$h_r(z;\mathfrak{a}) = \prod_{|y| \leqq 2r} E\left(\frac{z}{y}, q - 1\right)^{\nu(y,\mathfrak{a})},$$

$$j_r(z;\mathfrak{a}) = \frac{1}{q} \sum_{|y| \leqq 2r} \nu(y,\mathfrak{a}) \left(\frac{z}{y}\right)^q,$$

$$f_r(z;\mathfrak{a}) = \prod_{|y| > 2r} E\left(\frac{z}{y}, q\right)^{\nu(z,\mathfrak{a})}.$$

Then

$$h(z\mathfrak{a}) = h_r(z;\mathfrak{a})f_r(z;\mathfrak{a})\exp(j_r(z;\mathfrak{a})).$$

---

[32] For $p = 1$ see Lindelöf [4], p. 375 and Levin [3], Chapter I, Theorem 15.

As in the proof of Theorem 5.12, this implies

$$\frac{1}{\Phi(1)}\int_{S(1)}\log\left|h((\mathfrak{z}\,|\,\mathfrak{a})\mathfrak{a})\right|\sigma_1(\mathfrak{a}) - \frac{1}{q}\,\Re e P_{2r}(\mathfrak{z})$$
$$\leqq \kappa n_\nu(2r) + K^1_{q-1}(2r;n_\nu) + K^2_q(2r;n_\nu).$$

If $r > 0$, then

$$|P_r(\mathfrak{z})| \leqq C|\mathfrak{z}|^q \leqq Cr^q \quad \text{for } |\mathfrak{z}| \leqq r.$$

According to Lemma 4.3 and Lemma 4.4,

$$0 \leqq \tau_q(K^1_{q-1}(\cdot\,,n_\nu)) \leqq \kappa\tau_q(n_\nu) < \infty,$$
$$0 \leqq \tau_q(K^2_q(\cdot\,,n_\nu)) \leqq \kappa\tau_q(n_\nu) < \infty.$$

Hence constants $C_0 > 0$ and C_1 exist such that

$$\kappa n_\nu(2r) + K^1_{q-1}(2r;n_\nu) + K^2_q(2r;n_\nu) \leqq C_0 r^q + C_1$$

if $r > 0$. Define $C_2 = (1/q)C + C_0$. Then

$$\frac{1}{\Phi(1)}\int_{S(1)}\log\left|h((\mathfrak{z}\,|\,\mathfrak{a})\mathfrak{a})\right|\sigma_1(\mathfrak{a}) \leqq C_2 r^q + C_1$$

if $|\mathfrak{z}| \leqq r$. Lemma 5.4 and Lemma 5.6 imply

$$\log M_h(r) \leqq 8^{3p}(C_2(4r)^q + n_\nu(4r) + N_\nu(4r) + C_1).$$

Hence, a constant $C_3 > 0$ exists such that

$$\log M_h(r) \leqq C_3 r^q \quad \text{if } r \geqq 1.$$

Therefore $\tau_q[h] < \infty$. q.e.d.

The last two theorems can be inverted to some extent.

THEOREM 5.14.[33] *Let f be a nonconstant, meromorphic function on $\mathbf{C}^p$ with $p > 1$. Suppose that f is holomorphic at 0 and that $f(0) \neq 0$. Let q be a positive integer. For $r > 0$ and $\mathfrak{z} \in \mathbf{C}^p$ define*

$$P_r(\mathfrak{z}) = \frac{1}{W}\int_{\mathfrak{M}(\nu_f)\cap B(r)} \nu_f(\mathfrak{y})\left(\frac{(\mathfrak{z}\,|\,\mathfrak{y})}{(\mathfrak{y}\,|\,\mathfrak{y})}\right)^q \omega_{p-1}.$$

(a) *If $\tau_q[f] = 0$, then*

$$P_r(\mathfrak{z}) \to P(\mathfrak{z}) \quad \text{for } r \to \infty \text{ if } \mathfrak{z} \in \mathbf{C}^p$$

(the convergence is uniform on every compact subset of $\mathbf{C}^p$ and P is a homogeneous polynomial of degree q or identical to zero).

[33] For (a) see Kneser [2], p. 24.

(b) *If* $\tau_q[f] < \infty$, *and* $s > 0$, *a constant* $K > 0$ *exists such that*

$$|P_r(\mathfrak{z})| \leqq K \quad \textit{for all } r \geqq s \textit{ and } \mathfrak{z} \in S(1).$$

PROOF. Without loss of generality, $f(0) = 1$ can be assumed. If $\mathfrak{M}(v_f) = \varnothing$, the theorem is trivial. Hence, suppose that $\mathfrak{M}(v_f) \neq \varnothing$. Define

$$\eta = \inf\{r \mid \mathfrak{M}(v_f) \cap B(r) \neq \varnothing\}.$$

Then $\eta > 0$ and $B(\eta) \cap \mathfrak{M}(v_f) = \varnothing$. A holomorphic function F with $F(0) = 0$ exists on $B(\eta)$ such that $f = e^F$ on $B(\eta)$. Moreover, $F = \sum_{m=1}^{\infty} F_m$, where the series converges uniformly on every compact subset on $B(\eta)$ and where F_m is either identical zero or a homogeneous polynomial of degree m. For $r > 0$ a $\mathfrak{z} \in C^p$, define

$$U_r(\mathfrak{z}) = \frac{2}{\Phi(r)} \int_{S(r)} \log|f(\mathfrak{y})| \left(\frac{(\mathfrak{z} \mid \mathfrak{y})}{(\mathfrak{y} \mid \mathfrak{y})}\right)^q \sigma_r(\mathfrak{y}),$$

$$Q_r(\mathfrak{z}) = \frac{1}{W} \int_{\mathfrak{M}(v_f) \cap B(r)} v_f(\mathfrak{y}) \frac{(\mathfrak{z} \mid \mathfrak{y})^q}{r^{2q}} \omega_{p-1}(\mathfrak{y}).$$

Now, Proposition 3.3 implies

$$F_q(\mathfrak{z}) = \binom{q+p-1}{p-1} \left[U_r(\mathfrak{z}) - \frac{1}{q} P_r(\mathfrak{z}) + \frac{1}{q} Q_r(\mathfrak{z})\right]$$

for $\mathfrak{z} \in C^p$. Take s such that $0 < s < \eta$. If $\mathfrak{z} \in C^p$, then

$$|U_r(\mathfrak{z})| \leqq \left(\frac{|\mathfrak{z}|}{r}\right)^q \frac{2}{\Phi(r)} \int_{S(r)} |\log|f|| \sigma_r.$$

Because

$$|\log|x|| = \left|\log(1+|x|^2)^{1/2} - \log\frac{(1+|x|^2)^{1/2}}{|x|}\right| \leqq \log(1+|x|^2)^{1/2} + \log\frac{(1+|x|^2)^{1/2}}{|x|}$$

this implies

$$\begin{aligned} |U_r(\mathfrak{z})| &\leqq \left(\frac{|\mathfrak{z}|}{r}\right)^q \frac{2}{\Phi(r)} \int_{S(r)} \log(1+|f|^2)^{1/2} \sigma_r + \left(\frac{|\mathfrak{z}|}{r}\right)^q \frac{2}{\Phi(r)} \int_{S(r)} \log\frac{(1+|f|^2)^{1/2}}{|f|} \sigma_r \\ &= 2(|\mathfrak{z}|/r)^q (m_f(r;0) + m_f(r;\infty)) \\ &\leqq 4(|\mathfrak{z}|/r)^q (T_f(r;s) + m_f(s;0) + m_f(s;\infty)) \end{aligned}$$

where $s > 0$ and $q > 0$. If $\tau_q[f] = 0$, then

$$T_f(r,s) r^{-q} \to 0 \quad \text{for } r \to \infty.$$

Hence

$$U_r(\mathfrak{z}) \to 0 \quad \text{for} \quad r \to \infty \quad \text{if} \quad \mathfrak{z} \in C^p$$

where the convergence is uniform on every compact subset of C^p. If $\tau_q[f] < \infty$, then a constant $K_1 > 0$ exists such that

$$(4/r^q)(T_f(r,s) + m_f(s,0) + m_f(s,\infty)) \leqq K_1$$

for all $r \geqq s > 0$. Hence

$$|U_r(\mathfrak{z})| \leqq K_1 \quad \text{for } r \geqq s \text{ and } \mathfrak{z} \in S(1).$$

If $\mathfrak{z} \in C^p$, then

$$\begin{aligned} |Q_r(\mathfrak{z})| &\leqq \left(\frac{|\mathfrak{z}|}{r}\right)^q \frac{1}{W} \int_{\mathfrak{M}(\nu_f)\cap B(r)} |\nu_f|\, \omega_{p-1} \\ &= (|\mathfrak{z}|/r)^q (n_f(r;0) + n_f(r;\infty)) \\ &\leqq \left(\frac{|\mathfrak{z}|}{r}\right)^q \left(\int_r^{er} n_f(t;0)\frac{dt}{t} + \int_r^{er} n_f(t;\infty)\frac{dt}{t}\right) \\ &\leqq (|\mathfrak{z}|/r)^q (N_f(er,s;0) + N_f(er,s;\infty) \\ &\leqq 2(|\mathfrak{z}|/r)^q (T_f(er,s) + m_f(s;0) + m_f(s;\infty)). \end{aligned}$$

If $\tau_q[f] = 0$, then

$$T_f(r,s)\, r^{-q} \to 0 \quad \text{for } r \to \infty.$$

Hence

$$Q_r(\mathfrak{z}) \to 0 \quad \text{for } r \to \infty \text{ if } \mathfrak{z} \in C^p$$

where the convergence is uniform on every compact subset of C^p. Therefore,

$$P_r(\mathfrak{z}) \to -\left(q\Big/\binom{q+p-1}{p-1}\right) F_q(\mathfrak{z})$$

for $r \to \infty$, where the convergence is uniform on every compact subset of C^p.

If $\tau_q[f] < \infty$, then

$$(2/r^q)(T_f(er,s) + m_f(s;0) + m_f(s;\infty)) \leqq (e^q/2)K_1 = K_2$$

for all $r \geqq s$. Therefore,

$$|Q_r(\mathfrak{z})| \leqq K_2 \quad \text{if } r \geqq s \text{ and } \mathfrak{z} \in S(1).$$

Define

$$K_3 = \left(1\Big/\binom{q+p-1}{p-1}\right) \operatorname*{Max}_{\mathfrak{z}\in S(1)} |F_q(\mathfrak{z})|.$$

Then

$$|P_r(\mathfrak{z})| \leqq qK_3 + qK_1 + K_2 = K$$

for all $r \geqq s$ and $\mathfrak{z} \in S(1)$. If $\mathfrak{z} \in S(1)$ and $0 \leqq r \leqq s < \eta$, then $\mathfrak{M}(v_f) \cap B(r) = \emptyset$ which implies $P_r(\mathfrak{z}) = 0$ for all $\mathfrak{z} \in C^p$. Therefore,

$$|P_r(\mathfrak{z})| \leqq K \quad \text{for } r > 0 \text{ and } \mathfrak{z} \in S(1). \qquad \text{q.e.d.}$$

6. The quotient field theorem. Let p be a positive number. Define

$$\begin{aligned}
\mathfrak{F} &= \{f \mid f \ \textit{meromorphic on } C^p\},\\
\mathfrak{E} &= \{f \mid f \ \textit{holomorphic on } C^p\},\\
\mathfrak{F}_\rho &= \{f \mid f \in \mathfrak{F} \ \textit{with } \tau_\rho[f] < \infty\}, \ \mathfrak{E}_\rho = \mathfrak{F}_\rho \cap \mathfrak{E},\\
\mathfrak{F}_\rho^0 &= \{f \mid f \in \mathfrak{F} \ \textit{with } \tau_\rho[f] = 0\}, \ \mathfrak{E}_\rho^0 = \mathfrak{F}_\rho^0 \cap \mathfrak{E}.
\end{aligned}$$

Obviously, $\mathfrak{F}$, $\mathfrak{F}_\rho$ and $\mathfrak{F}_\rho^0$ are fields, and $\mathfrak{E}$, $\mathfrak{E}_\rho$, $\mathfrak{E}_\rho^0$ are integral domains such that

$$\begin{array}{ccccccc}
\mathfrak{F}_\rho^0 & \subseteq & \mathfrak{F}_\rho & \subseteq & \mathfrak{F} & & \mathfrak{F}_\rho \subseteq \mathfrak{F}^0{}_\sigma \\
\cup\!| & & \cup\!| & & \cup\!| & & \text{if } \ \rho < \sigma. \\
\mathfrak{E}_\rho^0 & \subseteq & \mathfrak{E}_\rho & \subseteq & \mathfrak{E} & & \mathfrak{E}_\rho \subseteq \mathfrak{E}^0{}_\sigma
\end{array}$$

Denote the field of quotients of a ring R by $\mathfrak{Q}(R)$. Rubel's and Taylor's Theorem asserts that $\mathfrak{Q}(\mathfrak{E}_\rho) = \mathfrak{F}_\rho$ if $p = 1$. Now, the relations

$$\mathfrak{Q}(\mathfrak{E}_\rho^0) = \mathfrak{F}_\rho^0 \quad \text{and} \quad \mathfrak{Q}(\mathfrak{E}_\rho) = \mathfrak{F}_\rho$$

shall be proved if $p > 1$.

Proposition 6.1. *Suppose that $f \in \mathfrak{F}_\rho$ and that $0 \notin \mathfrak{M}(v_f)$. Assume that $\rho > 0$ is not an integer. Then $f \in \mathfrak{Q}(\mathfrak{E}_\rho)$. Moreover, if $f \in \mathfrak{F}_\rho^0$ then $f \in \mathfrak{Q}(\mathfrak{F}_\rho^0)$.*

Proof. Define the integer q such that $q < \rho < q + 1$. Then $q \geqq 0$. If f is constant, then $f \in \mathfrak{Q}(\mathfrak{E}_\rho^0)$. Hence, suppose f is not constant. Because $\tau_\rho[f] < \infty$, it follows that

$$\operatorname{expon} v_f^a \leqq \operatorname{ord} f \leqq \rho < q+1 \quad \text{for } a \in C \cup \{\infty\}.$$

Therefore $q + 1$ is an exponent of convergence for v_f^a. Because $0 \notin \mathfrak{M}(v_f) = \mathfrak{M}(v_f^0) \cup \mathfrak{M}(v_f^\infty)$, the canonical function h_a exists for v_f^a and q if $a \in \{0, \infty\}$. Theorem 4.6, Lemma 4.5 and Proposition 4.1 imply

$$\tau_\rho(\log M_{h_a}) \leqq 8^{3p}4^p\left[\left(1 + \frac{\kappa}{\rho - q} + \frac{\kappa}{q + 1 - \rho}\right)\tau_\rho(n_f^a) + \tau_\rho(N_f^a)\right] \tag{6.1}$$

where $N_f^a(r) = N_f(r, s; a)$ and $n_f^a(r) = n_f(r; a)$. Define

(6.2) $$K = 8^{3p}4^p[(1 + \kappa/(\rho - q) + \kappa/(q + 1 - \rho))e^p + 1].$$

Then Proposition 4.1 and (6.1) and (6.2) imply

$$\tau_\rho[h_a] \leqq K\tau_\rho[f] \quad \text{if} \quad a \in \{0, \infty\}.$$

Hence $h_a \in \mathfrak{E}_\rho$. An entire function P exists such that

$$e^P = f \cdot h_\infty/h_0 \in \mathfrak{E} \cap \mathfrak{F}_\rho = \mathfrak{E}_\rho.$$

Therefore $g = e^P h_0 \in \mathfrak{E}_\rho$ and $h = h_\infty \in \mathfrak{E}_\rho$ and

$$f = g/h \in \mathfrak{Q}(\mathfrak{E}_\rho).$$

If $f \in \mathfrak{F}_\rho^0$, then $h_a \in \mathfrak{f}_\rho^0$ for $a \in \{0, \infty\}$. Therefore, $e^P = f \cdot h_\infty/h_0 \in \mathfrak{E} \cap \mathfrak{F}_\rho^0 = \mathfrak{E}_\rho^0$. Hence $g = e^P h_0 \in \mathfrak{E}_\rho^0$ and $h = h_\infty \in \mathfrak{E}_\rho^0$ which proves

$$f = g/h \in \mathfrak{Q}(\mathfrak{E}_\rho^0).$$ q.e.d.

PROPOSITION 6.2. *Let q be a positive integer. Suppose that $f \in \mathfrak{F}_q$ with $0 \notin \mathfrak{M}(\nu_f)$. Then $f \in \mathfrak{Q}(\mathfrak{E}_q)$.*

PROOF. If f is constant, the assertion is trivial. Hence suppose that f is not constant. For $a \in C \cup \{\infty\}$ define $\mu_a = \nu_f^a$. According to (4.1) and Proposition 4.1,

$$\tau_q(n_{\mu_a}) \leqq e^q \tau_q(N_{\mu_a}) \leqq e^q \tau_q[f] < \infty.$$

Hence,

$$\operatorname{expon} \mu_a \leqq q < q + 1.$$

Consequently, $q + 1$ is an exponent of convergence for μ_a. Because $0 \notin \mathfrak{M}(\nu_f) = \mathfrak{M}(\mu_0) \cup \mathfrak{M}(\mu_\infty)$, the canonical function h_a for μ_a and q exists if $a \in \{0, \infty\}$. Define

$$\zeta = e^{i\pi/q}.$$

A nonnegative divisor μ on C^p is defined by $\mu(\mathfrak{z}) = \mu_\infty(\zeta^{-1}\mathfrak{z})$ for $\mathfrak{z} \in C^p$. Define $\rho_a = \mu_a + \mu$ for $a \in \{0, \infty\}$. Then

(6.3) $$\begin{aligned} \tau_q(n_\mu) &= \tau_q(n_{\mu_\infty}) < \infty, \\ \tau_q(n_{\rho_a}) &\leqq \tau_q(n_{\mu_a}) + \tau_q(n_\mu) < \infty \end{aligned}$$

if $a \in \{0, \infty\}$. Therefore $q + 1$ is an exponent of convergence of μ, ρ_0 and ρ_∞ Let u be the canonical function to μ and q; then $g_a = h_a \cdot u$ is the canonical function to ρ_a and q if $a \in \{0, \infty\}$. Define

(6.4) $$P_r^a(\mathfrak{z}) = \frac{1}{W} \int_{\mathfrak{M}(\mu_a) \cap B(r)} \mu_a(\mathfrak{y}) \left(\frac{(\mathfrak{z} \mid \mathfrak{y})}{(\mathfrak{y} \mid \mathfrak{y})}\right)^q \omega_{p-1}(\mathfrak{y}).$$

(6.5) $$Q_r^a(\mathfrak{z}) = \frac{1}{W} \int_{\mathfrak{M}(\rho_a)\cap B(r)} \rho_a(\mathfrak{y}) \left(\frac{(\mathfrak{z}\mid\mathfrak{y})}{(\mathfrak{y}\mid\mathfrak{y})}\right)^q \omega_{p-1}(\mathfrak{y}).$$

(6.6) $$Q_r(\mathfrak{z}) = \frac{1}{W} \int_{\mathfrak{M}(\mu)\cap B(r)} \mu(\mathfrak{y}) \left(\frac{(\mathfrak{z}\mid\mathfrak{y})}{(\mathfrak{y}\mid\mathfrak{y})}\right)^q \omega_{p-1}(\mathfrak{y}).$$

(6.7) $$P_r(\mathfrak{z}) = \frac{1}{W} \int_{\mathfrak{M}(\mu)\cap B(r)} \nu_f(\mathfrak{y}) \left(\frac{(\mathfrak{z}\mid\mathfrak{y})}{(\mathfrak{y}\mid\mathfrak{y})}\right)^q \omega_{p-1}(\mathfrak{y}).$$

Then

$$P_r = P_r^0 - P_r^\infty, \qquad Q_r^a = Q_r + P_r^a$$

and

$$Q_r(\mathfrak{z}) = \frac{1}{W} \int_{\mathfrak{M}(\mu_\infty)} \mu(\zeta\mathfrak{y}) \left(\frac{(\mathfrak{z}\mid\zeta\mathfrak{y})}{(\zeta\mathfrak{y}\mid\zeta\mathfrak{y})}\right)^q \omega_{p-1}(\zeta\mathfrak{y}) = -P_r^\infty(\mathfrak{z}).$$

Hence $Q_r^0 = -P_r^\infty + P_r^0 = P_r$ and $Q_r^\infty = -P_r^\infty + P_r^\infty = 0$.

According to Theorem 5.14b, a constant $K > 0$ exists such that $|P_r(\mathfrak{z})| \leqq K$ for all $\mathfrak{z} \in S(1)$ and $r > 0$. Therefore

$$|Q_r^0(\mathfrak{z})| = |P_r(\mathfrak{z})| \leqq K, \qquad |Q_r^\infty(\mathfrak{z})| = 0 < K$$

for all $r > 0$ and $\mathfrak{z} \in S(1)$. Because of these estimates and of (6.3), Theorem 5.13 can be applied to g_a and implies $\tau_q[g_a] < \infty$ if $a \in \{0, \infty\}$. Hence $g_a \in \mathfrak{E}_q$ for $a \in \{0, \infty\}$.

An entire function P exists such that

$$e^P = f(h_\infty/h_0) = f(g_\infty/g_0) \in \mathfrak{F}_q \cap \mathfrak{E} = \mathfrak{E}_q.$$

Therefore, $g = e^P g_0 \in \mathfrak{E}_q$ and $h = g_\infty \in \mathfrak{E}_q$, which implies

$$f = g/h \in \mathfrak{Q}(\mathfrak{E}_q) \qquad \text{q.e.d.}$$

PROPOSITION 6.3. *Let q be a positive integer. Suppose that $f \in \mathfrak{F}_q^0$ with $0 \notin \mathfrak{M}(\nu_f)$. Then $f \in \mathfrak{Q}(\mathfrak{E}_q^0)$.*

PROOF. The assertion is trivial if f is constant. Hence, suppose that f is not constant. For $a \in C \cup \{\infty\}$ define $\mu_a = \nu_f^a$. According to (4.1) and Proposition 4.1,

$$0 \leqq \tau_q(n_{\mu_a}) \leqq e^q \tau_q(N_{\mu_a}) \leqq e^q \tau_q[f] = 0.$$

Hence

$$\tau_q(n_{\mu_a}) = \tau_q(N_{\mu_a}) = \tau_q[f] = 0, \qquad \operatorname{expon} \mu_a \leqq q < q + 1.$$

Consequently, $q + 1$ is an exponent of convergence for μ_a. Define

$$\zeta = e^{i\pi/q}.$$

A nonnegative divisor μ on C^p is defined by $\mu(\mathfrak{z}) = \mu_\infty(\zeta^{-1}\mathfrak{z})$ for $\mathfrak{z} \in C^p$. Define $\rho_a = \mu_a + \mu$ for $a \in \{0, \infty\}$. Then

$$\tau_q(n_\mu) = \tau_q(n_{\mu\infty}) = 0, \qquad \tau_q(n_{\rho_a}) \leqq \tau_q(n_{\mu_a}) + \tau_q(n_\mu) = 0.$$

Hence

$$\tau_q(n_{\rho_a}) = 0 \quad \text{if } a \in \{0, \infty\}. \tag{6.8}$$

For $a \in \{0, \infty\}$ define P_r^a, Q_r^a, Q_r and P_r as in (6.4), (6.5), (6.6) and (6.7) respectively. Then

$$P_r = P_r^0 - P_r^\infty, \qquad Q_r^a = Q_r + P_r^a,$$
$$Q_r^0 = P_r, \qquad Q_r^\infty = 0.$$

According to Theorem 5.14a,

$$Q_r^0(\mathfrak{z}) = P_r(\mathfrak{z}) \to P(\mathfrak{z}) \quad \text{for } r \to \infty \text{ if } \mathfrak{z} \in C^p \tag{6.9}$$

where the convergence is uniform on every compact subset of C. Obviously

$$Q_r^\infty(\mathfrak{z}) = 0 \to 0 \quad \text{for } r \to \infty \text{ if } \mathfrak{z} \in C^p. \tag{6.10}$$

Because of (6.8), (6.9), (6.10) and because $0 \notin \mathfrak{M}(\rho_a)$, the Lindelöf function g_a to ρ_a and q exists according to Theorem 5.12. Here, $\tau_q[g_a] = 0$ implies $g_a \in \mathfrak{E}_q^0$ for $a \in \{0, \infty\}$. Because $\rho_a = \nu_{g_a}$ and because $\rho_0 - \rho_\infty = \mu_0 - \mu_\infty = \nu_f$, an entire function P exists such that

$$e^P = f g_\infty / g_0 \in \mathfrak{E} \cap \mathfrak{F}_q^0 = \mathfrak{E}_q^0.$$

Therefore, $g = e^P g_0 \in \mathfrak{E}_q^0$ and $h = g_\infty \in \mathfrak{E}_q^0$ and

$$f = g/h \in \mathfrak{Q}(\mathfrak{E}_q^0). \qquad \text{q.e.d.}$$

THEOREM 6.4. (RUBEL AND TAYLOR). *Let ρ be a positive real number, then*

$$\mathfrak{Q}(\mathfrak{E}_\rho) = \mathfrak{F}_\rho \quad \textit{and} \quad \mathfrak{Q}(\mathfrak{E}_\rho^0) = \mathfrak{F}_\rho^0.$$

PROOF. Take $f \in \mathfrak{F}_\rho$. If f is constant then $f \in \mathfrak{Q}(\mathfrak{E}_\rho)$ and $f \in \mathfrak{Q}(\mathfrak{E}_\rho^0)$. Hence, suppose that f is not constant. A point $\mathfrak{c} \in C^p$ exists such that $\mathfrak{c} \notin \mathfrak{M}(\nu_f)$. Define the biholomorphic map $\gamma: C^p \to C^p$ by $\gamma(\mathfrak{z}) = \mathfrak{z} + \mathfrak{c}$. Then $0 \notin \mathfrak{M}(\nu_{f\circ\gamma})$. Because of (4.3) $f\circ\gamma \in \mathfrak{F}_\rho$ (and $f\circ\gamma \in \mathfrak{F}_\rho^0$ if $f \in \mathfrak{F}^0$). Hence $g \in \mathfrak{E}_\rho$ and $h \in \mathfrak{E}_\rho$ (respectively $g \in \mathfrak{E}_\rho^0$ and $h \in \mathfrak{E}_\rho^0$ if $f \in \mathfrak{F}_\rho^0$) exist such that $f\circ\gamma = g/h$, according to Proposition 6.1, 6.2 and 6.3. Because of (4.3), $g\circ\gamma^{-1} \in \mathfrak{E}_\rho$ and $h\circ\gamma^{-1} \in \mathfrak{E}_\rho$ (respectively $g\circ\gamma^{-1} \in \mathfrak{E}_\rho^0$ and $h\circ\gamma^{-1} \in \mathfrak{E}_\rho^0$ if $f \in \mathfrak{F}_\rho^0$). Therefore

$$f = g\circ\gamma^{-1}/h\circ\gamma^{-1} \in \mathfrak{Q}(\mathfrak{E}_\rho)$$

respectively

$$f = g\circ\gamma^{-1}/h\circ\gamma^{-1} \in \mathfrak{Q}(\mathfrak{E}_\rho^0) \quad \text{if } f \in \mathfrak{F}_\rho^0. \qquad \text{q.e.d.}$$

Bibliography

1. H. Kneser, *Ordnung und Nullstellen bei ganzen Funktionen zweier Veränderlichen*. S.-B. Preuss. Akad. Wiss. Kl. **31** (1936), 446–462.

2. ———, *Zur Theorie der gebrochenen Funktionen mehrerer Veränderlichen*. Jber. Deutsch. Math.-Verein. **48** (1938), 1–28.

3. B. Ja. Levin, *Distribution of zeros of entire functions*, Trans. Math. Monographs, Vol. 5, Amer. Math. Soc., Providence, R. I., 1964.

4. E. Lindelöf, *Sur les functions entières d'order entier*, Ann. Sci. École Normal Sup. (2) **22** (1905), 369–395.

5. R. Nevanlinna, *Eindeutige analytische Funktionen*, 2nd ed., Die Grundlehren der math. Wissenschaften vol. **46**, Springer-Verlag, Berlin, 1953.

6. L. A. Rubel, *A Fourier series method for meromorphic functions*, Lecture Notes from the Summer Institute on Entire functions and Related Parts of Analysis, University of California, San Diego, 1966.

7. L. A. Rubel and B. A. Taylor, *A Four ier series method for meromorphic and entire functions*, Bull. Amer. Math. Soc. **72** (1966) 858–860.

8. W. Stoll, *Mehrfache Integrale auf komplexen Mannigfaltigkeiten*, Math. Z. **57** (1952), 116–154.

9. ———, *Ganze Funktionen endlicher Ordnung mit gegebenen Nullstellenflächen*, Math. Z. **57** (1953), 211–237.

10. ———, *Die beiden Hauptsätze der Wertverteilungstheorie bei Funktionen mehrerer Veränderlichen*, I, Acta Math. **90** (1953), 1–115.

11. ———, *Die beiden Hauptsätze der Wertverteilungstheorie bei Funktionen mehrerer Veranderlichen*, II, Acta Math. **92** (1954), 55–169.

12. ———, *Über die neuere Theorie der ganzen und meromorphen Funktionen bei mehreren komplexen Veränderlichen*, Zulassungsarbeit zur wissenschaftlichen Prüfung, Tübingen, 1949, 287 pp. (unpublished).

13. ———, *Normal families of non-negative divisors* Math. Z. **84** (1964), 154–218.

14. R. O. Kujala, *Meromorphic and entire functions of finite -type in several complex variables*, University of Notre Dame Dissertation (1967), 96 pp. (Reference added in proof).

University of Notre Dame
Notre Dame, Ind.

Some Locally Convex Spaces of Entire Functions

B. A. TAYLOR

§0. **Introduction.** The space E of all entire functions in the topology of uniform convergence on compact subsets of the complex plane is a locally convex topological vector space. The dual space of E can be identified with the space E_0 of all entire functions of exponential type, via the Pólya representation of an entire function of exponential type as the Laplace transform of a Borel measure of compact support. The canonical pairing $\langle E, E_0\rangle$ has been a powerful tool in the study of several analytical problems; for example, in proving the existence of solutions to the convolution equation $f * \mu = g$, where μ is a given Borel measure with compact support and f, g are entire functions. In this paper we extend many results regarding the locally convex space E and the pairing $\langle E, E_0\rangle$ to a large class of locally convex spaces of entire functions. We apply our results to generalize some theorems about the convolution equation $f * \mu = g$, and mean periodic entire functions.

The spaces of entire functions we consider are defined as follows. Let K be a family of nonnegative, upper semicontinuous (u.s.c.) functions which vanish at infinity on the complex plane $\boldsymbol{C}$. Suppose also that K contains all such functions of compact support. Then we define $E(K)$ to be the vector space (over $\boldsymbol{C}$) of all entire functions f such that fk vanishes at infinity for each $k \in K$. For $f \in E(K)$ and $k \in K$, let $\|f\|_k = \sup\{|f(z)|k(z)\colon r \in \boldsymbol{C}\}$. Then $\| \ \|_k$ is a seminorm on $E(K)$ and the collection of all such seminorms (with $k \in K$) determines a locally convex topology on $E(K)$. The space E of all entire functions is the space $E(K)$, when K is the family of all nonnegative u.s.c. functions with compact support. (Several other examples are given in §6.)

In order to obtain substantial results in our study of $E(K)$, we have imposed several restrictions on the family K (conditions K1–K4 of §2). The most important of these (K2) implies that the rate of growth of the maximum modulus function of the entire function f completely determines whether $f \in E(K)$. Other restrictions on K imply that $E(K)$ contains the polynomials (K3) and is translation invariant (K4). The basic results about $E(K)$ are proved in §§2, 3. Under suitable conditions, we obtain a representation of the dual space of $E(K)$ as a space $M^{\wedge}(K)$ of entire functions. This is done by identifying a continuous linear functional with its Laplace transform (as in Gelfand-Silov **[10]**, Ehrenpreis **[4—8]**, Malgrange **[16]**). We also show, with a certain additional hypothesis, that this dual space $M^\wedge(K)$ is an algebra of entire functions (under pointwise operations). The equicontinuous subsets of $M^\wedge(K)$ are identified, and this result is used in §4 to obtain a useful sufficient condition that $E(K)$ be barrelled.

A main problem we consider is that of determining when some "natural" topology on the dual space, $M^\wedge(K)$, of $E(K)$ is admissible for the canonical pairing of $E(K)$ with $M^\wedge(K)$. This problem is solved in §5 for topologies on $M^\wedge(K)$ which are given by weighted sup norms. We associate with K the class K^* of all nonnegative u.s.c. functions k^* such that $k^*(w)k(z)\,|\exp(zw)|$ is bounded for each $k \in K$. The family K^* satisfies (roughly) the same conditions as K, so the results of §§2, 3 apply to the locally convex space $E(K^*)$. We then show that $M^\wedge(K)$ is a subspace of $E(K^*)$, and give a necessary and sufficient condition that the relative topology on $M^\wedge(K)$ (as a subset of $E(K^*)$) be admissible for the canonical pairing $\langle E(K), M^\wedge(K)\rangle$. In the example mentioned above, where K is the family of all nonnegative u.s.c. functions with compact support, the family K^* consists of all nonnegative u.s.c. k^* such that $k^*(w) = O(\exp(-A|w|))$ for each $A > 0$. Further, the spaces $E(K^*)$ and $M^\wedge(K)$ both coincide with the vector space E_0 of all entire functions of exponential type, and the topology induced on $E_0 = M^\wedge(K)$ by the seminorms $\|\ \|_{k^*}$ ($k^* \in K^*$) is the strong topology (from the canonical pairing of E_0 with E). This example was first given by Ehrenpreis in **[4]**, **[6]**.

The main question about $E(K)$ which we consider here is whether spectral synthesis is always possible in $E(K)$. That is, is every closed, translation-invariant subspace of $E(K)$ the closure of the linear span of the monomial-exponential functions (i.e. functions of the form $z^j \exp(az)$) it contains? Assuming that $M^\wedge(K)$ is an algebra, it is easily seen (and well known) that this question is equivalent to determining if each weakly closed ideal in $M^\wedge(K)$ is the ideal of all functions in $M^\wedge(K)$ which vanish on some discrete set. In the general case we are unable to make this identification. Accordingly, in §7 we introduce a class of algebras of entire functions, the so called Hadamard algebras, for which this identification is possible. Two sufficient conditions that $M^\wedge(K)$ be an Hadamard algebra are given. We also show, with a slight additional hypothesis, that $E(K)$ is an Hadamard algebra whenever it is an algebra. Consequently, we have identified the closed ideals in $E(K)$ in this case.

We also consider existence theorems for the convolution equation $f * v = g$. It is known (see, for example, Ehrenpreis **[6]**, Malgrange **[16]**, or Muggli **[17]**) that, if v is a Borel measure with compact support which does not annihilate every entire function, then, given any entire function g, there exists an entire function f such that $f * v = g$. In §9, we employ essentially the methods of Ehrenpreis and Malgrange to prove that, under suitable restrictions on K, given $g \in E(K)$, there exists an $f \in E(K)$ such that $f * v = g$. It is, in fact, shown that the measure v can be replaced by any continuous linear functional on $E(K)$.

We remark that the spaces $E(K)$ studied here are closely related to certain locally convex spaces used by Ehrenpreis in his theory of analytically uniform spaces [7], [8]. In Ehrenpreis' terminology, $M^\wedge(K)$ is an analytically uniform space (provided that $E(K)$ is barrelled). However, in many examples it is also true that $E(K)$ is analytically uniform. Theorems 5.1 and 5.2 can be viewed as giving necessary and sufficient conditions for this. Hence, a large class of analytically uniform spaces of entire functions is exhibited by our study.

The methods employed in this paper are from the theory of entire functions of a single complex variable. The one-variable results (and their proofs), however, have straight-forward generalizations to entire functions of several variables. We have, therefore, written the paper in this more general setting. §§7, 8 are notable exceptions; here the proofs are valid *only* for functions of a single variable.

An earlier version of this work constituted the author's Ph.D. dissertation prepared under the direction of Professor L. A. Rubel at the University of Illinois (1965). The author wishes to thank Professor Rubel for his guidance and encouragement.

§1. **Notations and preliminaries.** In this section we give the notation which will be used throughout the paper, and state several miscellaneous facts which will be needed later.

Denote by C^p the usual product space of p copies of the complex plane $C = C^1$, by R^p the real subspace of C^p, and by Z^p the subset of C^p of points with integer coordinates. Let $R^p(+)$ and $Z^p(+)$ be the subsets of R^p and Z^p, respectively, with nonnegative coordinates. For $z = (z_1, \cdots, z_p) \in C^p$ and $m = (m_1, \cdots, m_p) \in Z^p$ we denote by z^m the complex number $z_1^{m_1} \cdots z_p^{m_p}$, and by $|z|$ the point $(|z_1|, \cdots, |z_p|)$ of $R^p(+)$. Also, we write $m! = m_1! \cdots m_p!$, and for $z, w \in C^p$, we write $z \cdot w = z_1 w_1 + \cdots + z_p w_p$. Thus, $\exp(z \cdot w) = \sum(z^m w^m / m!)$, where the sum is over all $m \in Z^p(+)$. The usual notation D^m is used for the differential operator

$$\partial^{m_1 + \cdots + m_p} / \partial z_1^{m_1} \cdots \partial z_p^{m_p},$$

and we write $f^{(m)}(z)$ for $(D^m f)(z)$. In particular, each entire function f on C^p has a representation $f(z) = \sum a_m z^m = \sum (f^{(m)}(0)/m!) z^m$, where the sum is over all $m \in Z^p(+)$. Unless explicitly stated to the contrary, all sums will be over $Z^p(+)$ and will be absolutely convergent, so that the order of summation is irrelevant. When it is necessary to indicate the index of summation, this will be done by the notation $\sum_m$.

For f entire on C^p and $r = (r_1, \cdots, r_p) \in R^p(+)$, let

$$M(r; f) = \sup \{|f(z)| : |z| = r\}$$

be the maximum modulus function of f. We have from the maximum principle that $M(r;f) = \sup\{|f(z)| : |z_1| \leqq r_1, \cdots, |z_p| \leqq r_p\}$. The maximum term of f is denoted by $B(r;f)$; i. e. $B(r;f) = \sup\{|a_m| r^m : m \in Z^p(+)\}$. Here also $r = (r_1, \cdots, r_p) \in R^p(+)$. A closely related function is

$$B^*(r;f) = \sup\{q(m)|a_m| r^m : m \in Z^p(+)\},$$

where $q(m) = \prod_{j=1}^p (m_j + 1)(m_j + 2)$ for $m = (m_1, \cdots, m_p) \in Z^p(+)$. Note that

$$\Sigma[q(m)]^{-1} = \prod_{j=1}^{p} \sum_{m_j=0}^{\infty} \left(\frac{1}{m_j+1} - \frac{1}{m_j+2}\right) = 1.$$

Thus, if $f(z) = \Sigma a_m z^m$, then we have

$$\begin{aligned} |f(z)| &\leqq \Sigma[q(m)]^{-1} |q(m) a_m z^m| \\ &\leqq B^*(|z|;f)\, \Sigma[q(m)]^{-1} = B^*(|z|;f). \end{aligned}$$

Combining this inequality with Cauchy's inequality for the coefficients of entire functions yields

(1.1) $$B(r;f) \leqq M(r;f) \leqq B^*(r;f).$$

Another useful inequality is

(1.2) $$M(r; \partial f/\partial z_j) \leqq M(r + \varepsilon_j; f)$$

where $\varepsilon_j \in R^p(+)$ has all coordinates 0 except the jth, which is equal to one. This follows from the integral representation

$$(\partial f/\partial z_j)(z) = (2\pi i)^{-1} \int [f(z_1, \cdots, z_{j-1}, \mathscr{C}, z_{j+1}, \cdots, z_p)]\, [\mathscr{C} - z_j]^{-2} d\mathscr{C},$$

where the integration is on the contour $|\mathscr{C} - z_j| = 1$.

We shall consider on several occasions in this paper, the linear mapping $f \to Tf$ defined by $(Tf)(z) = \Sigma q(m) a_m z^m$, where $f(z) = \Sigma a_m z^m$ and

$$q(m) = \prod_{j=1}^{p} (m_j + 1)(m_j + 2).$$

If $p(z) = z_1 \cdots z_p$ and δ is the differential operator $\partial^p / \partial z_1 \cdots \partial z_p$, then a short computation shows that $p(z)\,(Tf)\,(z) = \delta(p^2(z))\delta(p(z)f(z))(z)$. Now if $\varepsilon = (1, 1, \cdots, 1) \in R^p(+)$, then we have $M(r;f) \leqq M(r + \varepsilon; pf)$, as is seen by representing h as a p-fold integral over the distinguished boundary of the polydisc $|z_1| \leqq r_1 + 1, \cdots, |z_p| r_p \leqq +1$. Also, p applications of (1.2) show that $M(r; \delta f) \leqq M(r + \varepsilon; f)$. We further have $M(r; pf) = p(r) M(r;f)$. Employing these estimates, we then find that $M(r; Tf) \leqq p(r + 3\varepsilon)[p(r + 2\varepsilon)]^2 M(r + 3\varepsilon; f)$. In particular,

(1.3) $$M(r;\, Tf) \leqq [p(r + 3\varepsilon)]^3 M(r + 3\varepsilon; f).$$

We shall also use several properties of convex functions, especially in §5. Recall that a function h defined on a convex subset D of R^p is convex if for each $x, y \in D$, the function of the real variable t, $g(t) = h(tx + (1\text{-}t)\, y)$ is convex $(0 \leqq t \leqq 1)$. A function $\lambda(r)$ defined for $r \in R^p(+)$ is called a convex function of $\log r$ if the function $h(x) = h(x_1, \cdots, x_p) = \lambda(e^{x_1}, \cdots, e^{x_p})$ is a convex function on R^p.

For a convex function h defined on $R^p(+)$, we denote by $h^*(x)$ the convex function conjugate to h (in the sense of Young; see e.g. [25] or [9]); that is,

$$h^*(x) = \sup\{x \cdot y - h(y) \colon y \in R^p(+)\}$$

(recall that $x \cdot y = x_1 y_1 + \cdots + x_p y_p$). We shall be concerned here only with the case when $h^*(x)$ is finite in some neighborhood of the origin in $R^p(+)$; that is, when $\varepsilon_1 y_1 + \cdots + \varepsilon_p y_p = O(h(y))$ for some $\varepsilon_1, \cdots, \varepsilon_p > 0$. In this case, and with the assumption that h is nondecreasing in each variable, we have that $h^{**} = h$. For a proof, see [9].

Throughout the paper we frequently use the terminology of the theory of locally convex spaces. The reader is referred to the book of Kelley and Namioka [15] for definitions of any unfamiliar terms.

§2. **The space** $E(K)$. In this section we define and give the basic properties (Propositions 2.1–2.8) of the spaces $E(K)$ considered in this paper. We again mention that several examples of the space $E(K)$ are given in §6.

Let K be a family of nonnegative upper semicontinuous (u.s.c.) functions which vanish at infinity on C^p.

DEFINITION 2.1. $E(K)$ is the vector space (over the complex numbers) of all entire functions f on C^p such that $f(z)\,k(z)$ vanishes at infinity for each $k \in K$. That is, $f(z)\,k(z) \to 0$ as $\|z\| \to +\infty$.

$$\text{For } k \in K, \quad \text{let } \|f\| = \sup\{|f(z)\,k(z)| \colon z \in C^p\}.$$

Then $\|\ \|_k$ is a seminorm on $E(K)$, and the family of all such seminorms determines a locally convex topology on $E(K)$. Unless otherwise stated, we shall always suppose that $E(K)$ is equipped with this topology. A net $\{f_\alpha\}$ converges to f in this topology if and only if $f_\alpha k \to fk$ uniformly on C^p, for each $k \in K$.

Throughout this paper we shall always suppose that K satisfies the following four conditions.

K1. If $k, k' \in K$ and $c \geqq 0$, then $ck \in K$ and $\max(k, k') \in K$. Moreover, if $k \in K$ and k' is u.s.c. with $0 \leqq k' \leqq k$, then $k' \in K$. Finally, if $k \geqq 0$ is u.s.c. and has compact support, then $k \in K$.

K2. If $k \in K$, then there exists a radial function $k' \in K$ (i.e. $k'(|z_1|, \cdots, |z_p|) = k'(z_1, \cdots, z_p)$) with $k'(z) \geqq k(z)$ for all $z \in C^p$.

K3. If $k \in K$ and if q is a polynomial, then $|q|k \in K$.

K4. If $k \in K$ and $a = (a_1, \cdots, a_p) \in R^p(+)$, then there exists $k' \in K$ such that $k'(|z_1| + a_1, \cdots, |z_p| + a_p) = k'(|z| + a) \geqq k(z)$.

These four conditions on K imply that:

E1. $E(K)$ contains all polynomials and, moreover, if $f \in E(K)$ and q is a polynomial, then $qf \in E(K)$;

E2. If $f \in E(K)$, if $a = (a_1, \cdots, a_p) \in R^p(+)$, and if g is entire with $M(r; g) \leqq M(r + a; f)$, then $g \in E(K)$.

For most applications it has been necessary to suppose that K satisfies the following stronger conditions.

K3'. Let $D = D(K)$ be the interior of the set $\{a \in C^p: \exp(a \cdot z) \in E(K)\}$. Then D is not empty; in particular, D contains a neighborhood of zero in C^p. In case $D = C^p$, we also assume that $k(z)|\exp(a \cdot z)| \in K$ for each $a \in C^p$ and $k \in K$.

K4'. If $k \in K$ and $\varepsilon > 0$, then $k(\varepsilon z) \in K$.

The corresponding conditions on $E(K)$ are:

E1'. $E(K)$ contains all the exponential functions $e_a(z) = \exp(a \cdot z)$ for $a \in D$.

E2'. If $f \in E(K)$, if $A > 0$, and if g is entire with $M(r; g) \leqq M(Ar; f)$ for all $r \in R^p(+)$, then $g \in E(K)$.

We shall indicate when these stronger assumptions are needed. As a general rule, however, K3' will be assumed when Laplace transforms are needed in our study of the dual space of $E(K)$, and $K4'$ will be assumed when analytical tools from the theory of entire functions are needed (such as estimating the maximum modulus function in terms of the (Nevanlinna) characteristic function).

We wish to point out that the conditions E1, E2 on $E(K)$ are essentially equivalent to the hypotheses K1–K4 on K. For if E is any vector space of entire functions satisfying E1, E2, and if $K = K(E)$ is defined to be the family of all nonnegative u.s.c. functions k such that $k(z)M(|z| : f)$ vanishes at infinity for each $f \in E$, then it is easily checked that K1–K4 hold. A similar remark applies with regard to E1', K3', and E2', K4'. We know of no examples with, in the above notation, $K \neq K(E(K))$ or $E(K(E)) \neq E$.

We also wish to remark that the functions $k \in K$ may be required to be continuous. That is, if one replaces the family K by the smaller class K' of all the $k \in K$ which are continuous, the resulting locally convex space $E(K')$ is the same as $E(K)$; i.e. $E(K) = E(K')$ as sets and also topologically. (The proof of this fact will not be given here, since it is not needed in the following work.) We have chosen to work with the larger class of u.s.c. functions in order to bypass a technical difficulty which arises in the proof of Theorem 5.1.

The most basic facts about $E(K)$ are given in the following propositions, which, for convenience are stated together. The proofs are all routine.

PROPOSITION 2.1. *$E(K)$ is complete.*

PROPOSITION 2.2. *An entire function f belongs to E(K) if and only if* $\sup\{M(r;f)\,k(r): r \in R^p(+)\} < +\infty$ *for each* $k \in K$.

PROPOSITION 2.3. *Translation and partial differentiation are continuous linear mappings of E(K) onto E(K). Multiplication by a polynomial is a continuous linear mapping of E(K) into E(K).*

PROPOSITION 2.4. *Let* $q(m) = \prod_{j=1}^{p}(m_j + 1)(m_j + 2)$. *The mapping T defined by* $T(\sum a_m z^m) = \sum q(m) a_m z^m$ *is a topological isomorphism of E(K) onto E(K). (I.e. a one-one, onto, linear, continuous, and open mapping.)*

PROPOSITION 2.5. *Let f be entire, say* $f(z) = \sum a_m z^m$. *If*

$$B(r;f) = \sup\{|a_m| r^m : m \in Z^p(+)\},$$

then $f \in E(K)$ *if and only if* $k(r)B(r;f)$ *is bounded for each* $k \in K$. *In particular,* $\sum a_m z^m \in E(K)$ *if and only if* $\sum |a_m| z^m \in E(K)$.

PROPOSITION 2.6. *Let B be a set of entire functions. Then the following are equivalent.*

(i) *B is a bounded subset of E(K);*

(ii) *There is a function* $U(r) \geqq 0$ *such that* $M(r;f) \leqq U(r)$ *for all* $f \in B$, *and* $U(r)k(r)$ *is bounded for each* $k \in K$;

(iii) *There is a function* $g(z) = \sum a_m z^m \in E(K)$ *such that, for every* $f(z) = \sum a_m z^m \in B$ *we have* $A_m \geqq q(m)\,|a_m|$, *where q(m) is as in Proposition 2.4.*

PROPOSITION 2.7. *The relative topology on a bounded subset B of E(K) coincides with the topology of uniform convergence on compact subsets of* C^p. *Therefore, the relative topology on B is metrizable. Also, each bounded set is relatively compact.*

PROPOSITION 2.8. *A bounded net* (f_α) *in E(K) converges if and only if* (f_α) *converges pointwise on* C^p. *If* $f \in E(K)$, *then the partial sums of the Taylor series of f form a bounded subset of E(K) and, therefore, converge to f in E(K).*

PROOF OF 2.1. Let $\{f_\alpha\}$ be a Cauchy net in $E(K)$. If D is any compact subset of C^p, then there exists $k \in K$ with $k(z) \geqq 1$ for $z \in D$ (by K1). Therefore, the net $\{f_\alpha\}$ is uniformly Cauchy on each compact $D \subset C^p$, hence, uniformly convergent to an entire function f on each compact subset of C^p. It is then easy to check that $f \in E(K)$ and that $f_\alpha \to f$ in $E(K)$.

PROOF OF 2.2. Follows from K2 and K3.

PROOF OF 2.3. The second assertion follows from K3. That translation is a continuous mapping follows from K2 and K4. Similarly, the mapping $f \to \partial f/\partial z_j$ is an onto mapping, since if $f \in E(K)$, then

$$g(z) = \int_0^{z_j} f(z_1, \cdots, z_{i-1}, \mathscr{C}, z_{i+1}, \cdots, z_p)d\mathscr{C}$$

also belongs to $E(K)$ $(M(r,g) \leqq r_j M(r,f))$, and $f = \partial g/\partial z_j$.

PROOF OF 2.4. The mapping T is clearly 1–1 and linear. That T maps $E(K)$ into $E(K)$ and is continuous follows from K2, K3, K4 and the inequality 1.3 of §1. That T is onto and open follows from K2 and the inequality (1.1) of §1.

PROOF OF 2.5. If T is as in Proposition 2.4, then $M(r; T^{-1}f) \leqq B(r;f)$ by (1.1) of §1. Thus, $T^{-1}f \in E(K)$ by Proposition 2.2 and the hypothesis on $B(r;f)$. Then $f = T(T^{-1}f) \in E(K)$.

The last assertion of the proposition follows from the first, since the two functions $\sum a_m z^m$, $\sum |a_m| z^m$ have the same maximum term function, $B(r;f)$.

PROOF OF 2.6. (i)$\Rightarrow$(ii). Let $U(r) = \sup\{M(r;f): f \in B\}$. Let $k \in K$. We claim $\sup\{U(r)k(r): r \in R^p(+)\} < +\infty$. By K2, it is no loss of generality to suppose that $k(z) = k(|z|)$. Then $U(r)\,k(r) = \sup\{M(r;f)\,k(r): f \in B\} \leqq \sup\{\|f\|_k: f \in B\} = M_k < +\infty$, since B is bounded. Thus, (ii) holds.

(ii)$\Rightarrow$(iii). For $m \in Z^p(+)$, define $A_m = q(m)\,[\inf\{U(r)r^{-m}: r \in R^p(+)\}]$ and $g(z) = \sum A_m z^m$. By Proposition 2.4, $g \in E(K)$ provided that

$$h(z) = \sum [q(m)]^{-1} A_m \in E(K).$$

But $B(r; h) \leqq U(r)$, so $h \in E(K)$ by Proposition 2.5. However, if $f(z) = \sum a_m z^m \in B$, then $|a_m| r^m \leqq M(r,f) \leqq U(r)$, so that (iii) holds.

(iii)$\Rightarrow$(i). To prove B is a bounded subset of $E(K)$, we must show that $\sup\{\|f\|_k: f \in B\} < +\infty$ for each $k \in K$. By K2, it is no loss of generality to suppose that $k(z) = k(|z|)$. If g is as in (iii), we then have $\|f\|_k \leqq \|g\|_k$ for all $f \in B$, so B is a bounded subset of $E(K)$.

PROOF OF 2.7. The topology of $E(K)$ is finer than the topology of uniform convergence on compact subsets of C^p, since K contains all nonnegative u.s.c. functions with compact support. Hence, we have only to prove that each net $\{f_\alpha\}$ in a bounded subset B of $E(K)$, which converges uniformly on compact sets to a function $f \in B$, also converges to f in $E(K)$. To see this, let $U(r)$ be as in (ii) of Proposition 2.6 and let $k \in K$. We may suppose $k(z) = k(|z|)$. For $R > 0$, let $D(R) = \{z \in C^p: |z_j| \leqq R, 1 \leqq j \leqq p\}$. Then $|f_\alpha(z) - f(z)| \leqq 2U(|z|)$ for $z \notin D(R)$. Taking into account the uniform convergence of the f_α on $D(R)$, we therefore have $\limsup \|f_\alpha - f\|_k \leqq \sup\{2k(r)U(r): r \notin D(R)\}$. Now $U(r)\,k(r)$ is bounded for each $k \in K$. From K3, it then follows that $U(r)\,k(r) \to 0$ as $\|r\| \to +\infty$ for each $k \in K$, whence, $\limsup \|f_\alpha - f\|_k = 0$. That is, $f_\alpha \to f$ in $E(K)$.

To prove that the bounded set B is relatively compact, let $\{f_\alpha\}$ be a net in B. Then $M(r;f_\alpha) \leqq U(r)$, so that $\{f_\alpha\}$ is a normal family of entire functions. Hence, a subnet (again denoted $\{f_\alpha\}$) may be selected which converges uniformly on compact sets to an entire function f. Then $M(r;f) \leqq U(r)$ so $f \in E(K)$; moreover, $B \cup \{f\}$ is bounded so $f_\alpha \to f$ in $E(K)$ by the first part of this proposition.

PROOF OF 2.8 Suppose (f_α) converges pointwise to $f \in E(K)$. Then from the theory of normal families and the bound $M(r,f_\alpha) \leqq U(r)$ given by Proposition 2.6, it follows that the convergence is uniform on compact sets, hence in $E(K)$, by Proposition 2.7.

To see that the partial sums of the Taylor series of $f(z) = \sum a_m z^m \in E(K)$ form a bounded set in $E(K)$, note first that $\sum |a_m| z^m \in E(K)$ (by Proposition 2.5). The result then follows from Proposition 2.6 (iii).

Before concluding this section, we wish to point out another natural topology for $E(K)$, an inductive limit topology defined as follows. For $f \in E(K)$, let $E(f) = \{g: g \text{ is entire and } M(r,g) = O(M(r,f))\}$. Then $E(f)$ with the norm $\|g\| = \sup\{M(r,g)/M(r,f): r \in R^p(+)\}$ is a Banach space and $E(K) = \bigcap\{E(f): f \in E(K)\}$. Thus, $E(K)$ can be given a topology as the inductive limit of the Banach spaces $E(f)$. This topology is in some sense more intrinsic, since we need not go outside the space $E(K)$ to define it. Also, it is a barrelled and bornological topology, as the inductive limit of Banach spaces. However, we do not know any examples where this topology is different from the topology determined on $E(K)$ by the seminorms $\| \quad \|_k$ (which is a projective limit topology). Also, it can be shown that in a great many cases these two topologies actually coincide (they coincide if and only if $E(K)$, in the topology from the $\| \quad \|_k$, is barrelled). We have, therefore, not considered this inductive limit topology here.

§3. **The dual space of** $E(K)$. We give in this section two representations of the dual space of $E(K)$, the first as a space of sequences of complex numbers (Proposition 3.1), and the second as a space of analytic functions. For the latter representation, it has been necessary to suppose that K3′ holds.

DEFINITION 3.1. Denote by $\Delta(K)$ the collection of all families of complex numbers (b_m), where m varies over $Z^p(+)$, such that for some $k \in K$ and some constant $A > 0$,

$$|b_m| \leqq A \| z^m \|_k \tag{3.1}$$

for all $m \in Z^p(+)$.

We shall usually suppress the "K" and write Δ for $\Delta(K)$. Also, we shall refer to an element $b = (b_m) \in \Delta$ as a sequence. Following the usual notation from the theory of sequence spaces, Δ^x denotes the set of all sequences (i.e. families (a_m), $m \in Z^p(+)$) of complex numbers such that $\sum |a_m b_m| < +\infty$ for each $(b_m) \in \Delta$. The set $\Delta^{xx} = (\Delta^x)^x$ is defined similarly. Clearly $\Delta^{xx} \supset \Delta$. If $\Delta = \Delta^{xx}$, then Δ is called *perfect*.

Our next proposition shows that Δ may be thought of as the dual space of $E(K)$.

PROPOSITION 3.1. *Let* $(b_m) \in \Delta$ *and*

$$L\left(\sum a_m z^m\right) = \sum a_m b_m \tag{3.2}$$

for $\sum a_m z^m \in E(K)$. *Then the series converges absolutely and defines* L *as a continuous linear functional on* $E(K)$. *Conversely, every continuous linear functional on* $E(K)$ *admits such a representation with* $(b_m) \in \Delta$. *If* $(b_m) \in \Delta^{xx}$, *then* L *is a bounded linear functional on* $E(K)$ *and, conversely, every bounded linear functional on* $E(K)$ *admits such a representation with* $(b_m) \in \Delta^{xx}$.

PROOF. Let $f(z) = \sum a_m z^m$ and $Tf(z) = \sum q(m) a_m z^m$, where $q(m)$ is as in Proposition 2.4. Let $(b_m) \in \Delta$ and choose $k \in K$, $A > 0$ such that (3.1) holds. By K2 we may suppose $k(z) = k(|z|)$. Then

$$\begin{aligned} |L(f)| &\leqq \sum |a_m b_m| \\ &= \sum q(m) |a_m b_m| [q(m)]^{-1} \\ &\leqq [\sup\{q(m) |a_m b_m| : m \in Z^p(+)\}] \sum [q(m)]^{-1}. \end{aligned}$$

From (3.1) and Cauchy's inequality, we have

$$q(m) |a_m b_m| \leqq \sup\{M(r; Tf) k(r) : r \in R^p(+)\} = \| Tf \|_k.$$

Since $\sum [q(m)]^{-1} = 1$, we thus have $|L(f)| \leqq \| Tf \|_k$. By Proposition 2.4, T is continuous, and, consequently, so is L.

Conversely, if L is a continuous linear functional, then for some $A > 0$ and $k \in K$ we have $|L(f)| \leqq A \| f \|_k$. With $b_m = L(z^m)$, this inequality implies (3.1). The representation (3.2) follows from Proposition 2.8 and the continuity of L.

If L is a bounded linear functional on $E(K)$ and if $b_m = L(z^m)$, then it follows easily from the representation of bounded sets given in Proposition 2.6 (iii), that $(b_m) \in \Delta^{xx}$ and (3.2) holds. Conversely, if $(b_m) \in \Delta^{xx}$, then L defined by (3.2) defines a bounded linear functional (again by Proposition 2.6 (iii)).

We shall now represent the dual space of $E(K)$ by a space of analytic functions. For this, it is necessary to suppose that $E(K)$ contains at least some exponential functions. Therefore, throughout the remainder of this section, it is assumed that K3′ holds.

PROPOSITION 3.2. *Suppose that* K3′ *holds and that* L *is a continuous linear functional on* $E(K)$. *Then* $F(w) = L(\exp(w \cdot z))$ *is analytic for* $w \in D$, *where* D *is as in* K3′. *Moreover, for* $m \in Z^p(+)$ *and* $w \in D$, *we have*

$$F^{(m)}(w) = L(z^m \exp(w \cdot z)). \tag{3.3}$$

The correspondence between L *and* F *is* 1–1.

PROOF. Let $h = \Delta w_j$ denote the increment in the jth coordinate of w, and $f_h(z)$ the corresponding difference quotient of $\exp(w \cdot z)$, so that $\lim_{h \to 0} f_h(z) = (\partial / \partial w_j)(\exp(w \cdot z)) = z_j \exp(w \cdot z)$ for each $z \in C^p$. To prove the analyticity of F, it is enough to prove that $f_h(z) \to z_j \exp(w \cdot z)$ in $E(K)$, as $h \to 0$. Standard estimates on the difference quotients, $f_h(z)$, together with Proposition 2 6 (ii), show that the set of all $f_h(z)$ (with h sufficiently small) is a bounded subset of $E(K)$. Since the $f_h(z)$ converge pointwise to $z_j \exp(w \cdot z)$, it then follows from Proposition 2.8 that the $f_h(z)$ also converge to $z_j \exp(w \cdot z)$ in $E(K)$. Thus $(\partial / \partial w_j)\, F(w)$ exists, and (3.3) holds in this special case. Equation (3.4) follows in general by repeated applications of this argument.

To prove the correspondence $L \to F$ is 1–1, we must show that if $F = 0$, then L is the zero functional (the general case reduces to this by linearity). But, if $F(w) = 0$ for all $w \in D$, then $0 = F^{(m)}(0) = L(z^m)$ for all $m \in Z^p(+)$. Therefore, $L(f) = 0$ for all $f \in E(K)$, since by Proposition 2.8, the polynomials are dense in $E(K)$.

The analytic function F defined in the preceding proposition is called the Laplace transform of the continuous linear functional L, and we sometimes write $L^\wedge$ for F.

DEFINITION 3.2. The vector space (under the usual pointwise operations) of all entire functions F which are Laplace transforms of continuous linear functionals on $E(K)$ is denoted $M^\wedge(K)$.

The reason for the notation $M^\wedge(K)$ is the following. It can be shown that each continuous linear functional on $E(K)$ can be represented as an integral, $L(f) = \int fk\, d\mu$, where $k \in K$ and μ is a bounded Borel measure on C^p. Thus, $L^\wedge(w) = \int \exp(w \cdot z)\, k(z)\, d\mu(z)$ is the Laplace transform of the measure $k \cdot \mu$. We do not need this representation, so the proof is omitted.

The results of Propositions 3.1 and 3.2 are summarized in the following theorem.

THEOREM 3.1. *If K3′ holds, then the pairing $\langle E(K), M^\wedge(K)\rangle$ defined by*

$$\langle f, F\rangle = L(f), \tag{3.4}$$

where F is the Laplace transform of the continuous linear functional L, establishes $M^\wedge(K)$ as the dual space of $E(K)$. An entire function $F(w) = \sum (b_m/m!) w^m$ belongs to $M^\wedge(K)$ if and only if $(b_m) \in \Delta$ $(= \Delta(K))$, and in this case

$$\langle f, F\rangle = (F(D)f)\,(0), \tag{3.5}$$

where by $(F(D)f)\,(0)$, we mean $\sum (b_m/m!) f^{(m)}(0)$. This series converges absolutely.

PROOF. The first assertion is a restatement of Proposition 3.2. If $(b_m) \in \Delta$, then $L(\sum a_m z^m) = \sum a_m b_m$ is a continuous linear functional by Proposition 3.1. The Laplace transform of L is easily computed to be $\sum (b_m/m!) w^m = F(w)$. On the other hand, if $F \in M^\wedge(K)$, then $(b_m) = (F^{(m)}(0)) = \langle z^m, F\rangle$ must belong to Δ, by Proposition 3.1. The representation (3.5) follows from the fact that the power series of $f \in E(K)$ converges to f in $E(K)$ (Proposition 2.8).

For some later applications it is convenient to have two other propositions concerning the dual space $M^\wedge(K)$.

Proposition 3.3. *Suppose that* K3′ *holds and that B is a set of functions analytic at the origin. Then B is an equicontinuous subset of* $M^\wedge(K)$ *if and only if there exists* $(b_m) \in \Delta$ *such that* $|F^{(m)}(0)| \leqq |b_m|$ *for all* $m \in Z^p(+)$.

Proof. Recall that a subset of B of the dual space of a locally convex topological vector space E is equicontinuous if and only if there is a neighborhood of zero in E, say N, with $B \supset N^0$. (N^0 is the set of all continuous linear functionals on E which are bounded by 1 on N.) In the present context, this condition is equivalent to $B \subset M^\wedge(K)$, and the existence of a function $k \in K$ such that $|\langle f, F\rangle| \leqq \|f\|_k$ for all $F \in B$ and all $f \in E(K)$. If B is equicontinuous, then $|F^{(m)}(0)| = |<z^m, F>| \leqq \|z^m\|_k = b_m$, so the condition of the proposition is necessary. To prove sufficiency, let $f(z) = \sum a_m z^m$ and $g(z) = \sum q(m) a_m z^m$ (where $q(m)$ is as in Proposition 2.5). Then as was shown in the proof of Proposition 3.2, $|\langle f, F\rangle| = |\sum a_m F^{(m)}(0)| \leqq \sum |a_m F^{(m)}(0)| \leqq A \|g\|_k$, where $|F^{(m)}(0)| \leqq b_m \leqq A \sup\{|z|^m k(z): z \in C^p\}$. By Proposition 2.4, the mapping $f \to g$ is continuous, so there exists a constant A' and a function $k' \in K$ such that $\|g\|_k \leqq A' \|f\|_{k'}$. If $k'' = AA'k' \in K$, then $|\langle f, F\rangle| \leqq \|f\|_{k''}$, so that B is an equicontinuous subset of $M^\wedge(K)$.

Our next theorem establishes that $M^\wedge(K)$ is an algebra of entire functions when $p = 1$ and K3′, K4′ hold. For the cases $p > 1$, our proof requires the additional hypothesis K5 given below. We have stated K5 in such a way that it applies directly in the proof of the theorem; consequently, the statement is quite technical. Roughly speaking, however, some condition like K5 is needed to insure that functions of the form $H(z_1, z_2) = F(z_1)G(z_2) \in M^\wedge(K)$ whenever $F(z_1), G(z_2) \in M^\wedge(K)$. We remark that K5 is satisfied whenever the growth conditions on the $f \in E(K)$ are of the form $M(r; f) \leqq \exp(\lambda_1(r_1) + \cdots + \lambda_p(r_p))$. More precisely, the hypothesis K5 is satisfied whenever, given $k \in K$, there exists a function $k' \in K$, with $k' \geqq k$, and of the form, $k'(z_1, \cdots, z_p) = k_1(|z_1|) \cdots k_p(|z_p|)$. This apparently stronger condition holds in all the examples known to us where K satisfies K5.

Let σ be a subset of $\{1, 2, \cdots, p\}$ and τ the complement of σ relative to this set. For $r \in R^p(+)$, let $r(\sigma) = (\rho_1, \cdots, \rho_p)$, where $\rho_j = r_j$ if $j \in \sigma$, and $\rho_j = 0$ otherwise. Similarly, define $r(\tau)$. Thus $r = r(\sigma) + r(\tau)$.

K5. Given $k \in K$, there exist $k' \in K$ such that

$$k'(r) \geqq \sup\{k(r(\sigma) + x(\tau))\, k(r(\tau) + y(\sigma)): x, y \in R^p(+)\}$$

for all $r \in R^p(+)$ and all subsets σ of $\{1, 2, \cdots, p\}$.

Note that if k is nonincreasing in each variable, the condition reduces to $k'(r) \geqq k(\rho)\, k(r - \rho)$, where ρ ranges over the vertices of the cell

$$0 \leqq x_1 \leqq r_1, \cdots, 0 \leqq x_p \leqq r_p.$$

Theorem 3.2. *Suppose that* $p = 1$ *and that* K *satisfies* K3′, K4′. *Then*

(i) *each* $F \in M^{\wedge}(K)$ *is entire;*

(ii) *if* $F_1, F_2 \in M^{\wedge}(K)$, *then so does* $F_1 \cdot F_2$ (*i.e.* $M^{\wedge}(K)$ *is an algebra under pointwise operations*);

(iii) *if* B_1, B_2 *are equicontinuous subsets of* $M^{\wedge}(K)$, *then*

$$B_1 \cdot B_2 = \{F_1 F_2 \colon F_1 \in B_1,\ F_2 \in B_2\}$$

is an equicontinuous subset of $M^{\wedge}(K)$.

The theorem remains valid for the cases $p > 1$ provided that K5 also holds.

PROOF. The first assertion follows from Proposition 3.2, since $D = C^p$ when K4′ holds (D the open set of K3′). To prove (ii), it suffices to show that $F^2 \in M^{\wedge}(K)$ whenever $F \in M^{\wedge}(K)$, since $FG = \frac{1}{2}(F+G)^2 - \frac{1}{2}F^2 - \frac{1}{2}G^2$. Thus, let $F(w) = \sum (b_m/m!)\, w^m \in M^{\wedge}(K)$, and let $F^2(w) = \sum c_m w^m$. Then $c_m = \sum_j (b_j b_{m-j}/j!(m-j)!)$ where the sum is over all $j \in Z^p(+)$ with $m - j \in Z^p(+)$. We will prove that $(m!\, c_m) \in \Delta$, so that $F^2 \in M^{\wedge}(K)$ by Theorem 3.1.

Let $c'_m = \max\{|b_j b_{m-j}| \colon j, m-j \in Z(+)\}$. Then $|c_m| \leqq c'_m(\sum (1/j!(m-j)!) = (c'_m/m!)\, 2^{m_1 + \cdots + m_p}$. It is a consequence of K4′ that $M^{\wedge}(K)$ is invariant under coordinate-wise multiplication by factors $A^{m_1 + \cdots + m_p}$; hence, we need only prove that $(c'_m) \in \Delta$.

To this end, let $k \in K$ and let $A > 0$ be such that $|b_m| \leqq A \|z^m\|_k$. It is no loss of generality (by K2) to also assume that $k(|z|) = k(z)$. If for $x \in R^p(+)$, we define $h(x) = \sup\{\sum_{j=1}^p x_j \log r_j + \log k(r) \colon r \in R^p(+)\}$, then $h(x)$ is a convex function on $R^p(+)$ (as the upper envelope of a family of linear functions). Moreover, we have $\|z^m\|_k = \exp(h(m))$, so that $c'_m \leqq A^2 \max\{h(j) + h(m-j) \colon j, m-j \in Z^p(+)\} \leqq A^2 \max\{h(x) + h(m-x) \colon x, m-x \in R^p(+)\}$. Now the function $g(x) = h(x) + h(m-x)$ is defined and convex for $0 \leqq x_1 \leqq m_1, \cdots, 0 \leqq x_p \leqq m_p$, and thus assumes its maximum value when x is at a vertex of $0 \leqq x_1 \leqq m_1, \cdots, 0 \leqq x_p \leqq m_p$; i.e., at x such that $x_j = 0$ or $x_j = m_j$ for each j, $1 \leqq j \leqq p$. We are then finished in the case $p = 1$, for we have proved $c'_m \leqq A^2 \exp(h(0) + h(m)) = O(\|z^m\|_k)$.

The case $p > 1$ can be handled using the hypothesis K5. To simplify the notation, it is assumed that $p = 2$; the extension to the cases $p > 2$ will be evident. If the vertex at which $g(x)$ assumes a maximum value is $(0,0)$, then the argument given for $p = 1$ suffices. Hence, we may assume that the maximum of $g(x)$ is equal to $h(s,0) + h(0,t)$, where $m = (s,t) \in Z^2(+)$. We then have $\exp(h(s,0) + h(0,t)) = [\sup\{r_1^s k(r_1, x) \colon r_1 \geqq 0,\ x \geqq 0\}] \cdot [\sup\{r_2^t k(y, r_2) \colon y \geqq 0, r_2 \geqq 0\}] = \sup\{r_1^s r_2^t k(r_1, x)\, k(y, r_2) \colon r_1 \geqq 0, r_2 \geqq 0, x \geqq 0, y \geqq 0\} \leqq \sup\{r_1^s r_2^t k'(r_{1,2}) \colon r_1 \geqq 0, r_2 \geqq 0\}$, where $k' \in K$ is the function whose existence is assumed in K5. Thus, $c'_m \leqq A^2 \|z^m\|_{k'}$. Letting $k'' = \max(k, k') \in K$, we then have $c'_m = O(\|z^m\|_{k''})$ for all $m \in Z^2(+)$, so that $(c'_m) \in \Delta$ as asserted. This completes the proof of (ii).

The final assertion of the theorem now follows easily from (ii) and Proposition

3.3. For, by this proposition, the coefficients of the functions F_j in B_j, $j = 1,2$, are dominated by sequences $(b_m/m!)$, $(c_m/m!)$, respectively, with (b_m), $(c_m) \in \Delta$. Then, the coefficients of the functions in $B_1 \cdot B_2$ are dominated by $d_m/m! = \sum_j (b_j c_{m-j}/j!(m-j!))$. By (ii) of the theorem, $(d_m) \in \Delta$. Thus, $B_1 \cdot B_2$ is equicontinuous by Proposition 3.3. This completes the proof.

REMARK. The above proof does not essentially depend on the assumption K3′. If this hypothesis is omitted, then the above proof establishes that the coefficients $(b_m/m!)$, with $(b_m) \in \Delta$ are the coefficients of a subalgebra of the algebra of formal power series in p indeterminates.

§4. **Sufficient conditions that $E(K)$ be barrelled and bornological.** For certain applications in §§6, 9, it is convenient to know that the locally convex space $E(K)$ is barrelled and bornological (see e.g. [15] or [22] for the definition of these terms). Although this is the case in every example known to us, we have been unable to decide whether it is always true. Therefore, we specify in this section two conditions, each of which is sufficient that $E(K)$ be barrelled and bornological.

DEFINITION 4.1. K is said to have a countable base if there are functions $k_1 \leqq k_2 \leqq \cdots, k_n \in K$, such that each $k \in K$ satisfies $k(z) = O(k_n(z))$ for some $n > 0$.

DEFINITION 4.2. K is said to be countably determined if there are continuous functions $0 < \phi_1(r) < \phi_2(r) < \cdots$ defined for $r \in R^p(+)$ such that

(i) for each $n = 1,2,\cdots,\phi_n(r)$ is nondecreasing in each variable; and

(ii) $K = \{k \in C_0(+) : k(z)\phi_n(|z|)$ is bounded for each $n = 1,2,\cdots\}$

In these definitions, it is tacitly assumed that K1—K4 are satisfied.

THEOREM 4.1. *If K has a countable base, then $E(K)$ is a Frechet space (i.e. complete and metrizable). Therefore, $E(K)$ is bornological and barrelled.*

THEOREM 4.2. *If K is countably determined and if $\phi_1, \phi_2, \cdots$ are as in Definition 4.2., then*

(i) *$E(K) = \{f$ entire: $M(r,f) = O(\phi_n(r))$ for some $n = 1,2,\cdots\}$, and*

(ii) *$E(K)$ is bornological and barrelled.*

The remainder of this section is given to the proofs of these two theorems.

PROOF OF THEOREM 4.1. If K has a countable base and $k_1\, k_2, \cdots$ are as in Definition 4.1, then it is clear that the countable number of seminorms $\| \ \|_{k_n}$, determine the same topology on $E(K)$ as the family of all seminorms $\| \ \|_k$ with $k \in K$. Hence, $E(K)$ is metrizable. By Proposition 2.7 $E(K)$ is complete, so $E(K)$ is a Fréchet space. Every Fréchet space is bornological and barrelled [22, pp. 60–61].

For the proof of Theorem 4.2, we need the following lemma.

LEMMA. *The following are equivalent*:

(i) $\Delta = \Delta^{xx}$;

(ii) *$E(K)$ is barrelled*;

(iii) *$E(K)$ is bornological.*

PROOF. We will show (i) $\Leftrightarrow$ (ii), (iii) $\Rightarrow$ (i), and (i), (ii) $\Rightarrow$ (iii).

(i) $\Rightarrow$ (ii). The space $E(K)$ is barrelled if and only if every weakly bounded subset of Δ, the dual space of $E(K)$ is equicontinuous [22, p. 141]. Also, since $E(K)$ is complete, the weakly bounded subsets of Δ coincide with the strongly bounded subsets of Δ [22, p. 82]. If $B \subset \Delta$ is such a bounded set, and $d_m = \sup\{|b_m|\}$, $m \in Z^p(+)$, where the supremum is over all $b = (b_m) \in \Delta$, then it follows readily using Proposition 2.6 (iii), that $\sup\{|d_m a_m| : m \in Z^p(+)\}$ is bounded for each $\sum a_m z^m \in E(K)$. In view of Proposition 2.4, this implies that $(b_m) \in \Delta^{xx}$. If $\Delta = \Delta^{xx}$, it then follows from Proposition 3.3 that B is equicontinuous.

(ii) $\Rightarrow$ (i). Suppose that $(b_m) \in \Delta^{xx}$, and let L be the associated bounded linear functional on $E(K)$ defined in Proposition 3.1. For $n = 1, 2, \cdots$, define $L_n(\sum a_m z^m) = \sum a_m b_m(n)$, where $b_m(n) = 0$ unless $m_1 \leqq n$, $m_2 \leqq n, \cdots, m_p \leqq n$, in which case $b_m(n) = b_m$. The L_n are continuous linear functionals on $E(K)$, and $L_n(f) \to L(f)$ for each $f \in E(K)$. Since $E(K)$ is barrelled, it follows from the Banach-Steinhaus theorem [22, p. 86] that L is continuous. Hence, $(b_m) \in \Delta$, by Proposition 3.1. Thus, $\Delta = \Delta^{xx}$.

(iii) $\Rightarrow$ (i). On a bornological space, every bounded linear functional is continuous [22, p. 62]. Consequently, $\Delta = \Delta^{xx}$ by Proposition 3.1. We also remark that (iii) $\Rightarrow$ (ii), since every complete bornological space is barrelled [22, p. 63].

(i), (ii) $\Rightarrow$ (iii). Since $\Delta = \Delta^{xx}$, each bounded linear functional on $E(K)$ is continuous. Since $E(K)$ is barrelled, $E(K)$ is a Mackey space [22, p. 132]. These two conditions imply that $E(K)$ is bornological [15, p. 184].

PROOF OF THEOREM 4.2. The first assertion is easy to verify. To prove (ii), it suffices, by the Lemma, to show $\Delta = \Delta^{xx}$. To this end, define

$$a_m(n) = \inf\{r^{-m}\phi_n(r) : r \in R^p(+)\} \tag{4.1}$$

for $m \in Z^p(+)$ and $n = 1, 2, \cdots$. Then $f_n(z) = \sum a_m(n) z^m \in E(K)$, since $B(r; f) \leqq \phi_n(r)$. If $(b_m) \in \Delta^{xx}$, then $\sum_m |a_m(n) b_m| < +\infty$, so there is a positive integer s_n such that $|a_m b_m| < 1$, except for $m \leqq \overline{s_n}$. Here and in the following $\overline{s_n} = (s_n, \cdots, s_n) \in Z^p(+)$ and $m \leqq \overline{s_n}$ means $m_1 \leqq s_n, \cdots, m_p \leqq s_n$. We also choose the s_n with $s_1 < s_2 < \cdots$.

To construct $k \in K$ with $\| z^m \|_k \geqq |b_m|$, we proceed, roughly, as follows. Let $m \in Z^p(+)$. Choose $n = n(m)$ such that $|a_m(n) b_m| < 1$. Then $|b_m| < [a_m(n)]^{-1} = \sup\{r^m(\phi_n(r))^{-1} : r \in R^p(+)\}$. Thus, we try to set $k(r) = [\phi_n(r)]^{-1}$ at some point r where this supremum is attained.

To make the above argument precise, we first suppose that $a_m(n) > 0$ for all m, n. Then choose $r(m, n) \in R^p(+)$ such that

$$r(m, n)^{-m}\phi_n(r(m, n)) \leqq 2a_m(n). \tag{4.2}$$

For each $n = 1, 2, \cdots$, let ρ_n be a positive number so large that $\overline{\rho_n} = (\rho_n, \cdots, \rho_n) \geqq r(m, n)$ (i.e. $\rho_n > r_j(m, n)$ for $1 \leqq j \leqq p$) for the finitely many $m \in Z^p(+)$ with $m \leqq \overline{s_{n+1}}$. We may assume $\rho_1 < \rho_2 < \cdots$, and $\rho_n \to +\infty$. Then let $k(z) = k(|z|)$, where $k(r)$ is any continuous function on $R^p(+)$ such that $[\phi_n(r)]^{-1} \leqq k(r) \leqq [\phi_{n-1}(r)]^{-1}$, for $r \leqq \rho_n$ and not $r \leqq \rho_{n-1}$ (i.e. $\rho_n > r_j$ for all $1 \leqq j \leqq p$ and $r_j > \rho_{n-1}$ for some $1 \leqq j \leqq p$). (The existence of such a function k follows from the Tietze extension theorem.) Then, clearly, $k(z)\phi_n(|z|)$ is bounded for each $n = 1, 2, \cdots$, so that $k \in K$.

To prove $(b_m) \in \Delta$, it is enough to prove $|b_m| \leqq 2\|z^m\|_k$ for all except finitely many $m \in Z^p(+)$. This inequality will be proved except for $m \leqq \overline{s_1}$. Thus, let n be the first positive integer for which $m \leqq \overline{s_{n+1}}$. Such a choice is possible except for $m \leqq \overline{s_1}$. Then $2\|z^m\|_k \geqq 2r(m,n)^{-n}k(r(m,n))$. But $r(m,n) \leqq \overline{\rho_n}$, since $m \leqq s_{n+1}$, so that $k(r(m,n)) \geqq \phi_n(r(m,n))^{-1}$. Combining these inequalities with (4.2), we have $2\|z^m\|_k \geqq [a_m(n)]^{-1}$. However, $[a_m(n)]^{-1} > |b_m|$, since $m \leqq \overline{s_n}$ is false. Thus, $2\|z^m\|_k > |b_m|$ as asserted.

In case some of the $a_m(n)$ are zero, we modify the argument as follows. Replace the $a_m(n)$ which are zero by positive numbers $\varepsilon_m(n)$ so small that $|\varepsilon_m(n)\, b_m| < 1$. Then choose the $r(m,n)$ with $r(m,n)^{-m}\phi_n(r(m,n)) \leqq \varepsilon_m(n)$. The argument can then proceed exactly as above.

§5. **Admissible topologies on** $M^\wedge(K)$. We would like to identify a topology on $M^\wedge(K)$ which is admissible for the pairing $\langle E(K), M^\wedge(K)\rangle$. That is, a locally convex topology $\mathcal{T}$ on $M^\wedge(K)$ with the property that for every $f \in E(K)$, $L_f(F) = \langle f, F\rangle$ is a $\mathcal{T}$-continuous linear functional on $M^\wedge(K)$, and conversely. There are many such topologies on $M^\wedge(K)$ (such as the weak topology or the Mackey topology), but in general these topologies are defined in terms of convergence of the functions in $M^\wedge(K)$ as linear functionals on $E(K)$. We wish, however, to define an admissible topology on $M^\wedge(K)$ in terms of convergence of the functions in $M^\wedge(K)$ as functions on C^p. For suitable classes K, it is possible to do this in a natural way. We associate with K a family K^* of functions k^*, and then define a topology on $M^\wedge(K)$ by seminorms $\|\ \|_{k^*}$ $(k^* \in K^*)$. Theorem 5.1 of this section gives necessary and sufficient conditions that these seminorms determine a topology on $M^\wedge(K)$, which is admissible for the pairing $\langle E(K), M^\wedge K\rangle$.

Throughout this section it is assumed that K3′ holds, and that $D = D(K)$ is as in K3′.

DEFINITION 5.1. Suppose that K satisfies K3′. Then by K^*, the dual class of K, we mean the family of all nonnegative u.s.c. functions k^* on C^p such that $k(z)\,k^*(w)\,|\exp(z\cdot w)|$ is bounded for each $k \in K$.

Now for $F \in M^\wedge(K)$, there exists $k \in K$ and $A > 0$ such that $|\langle f, F\rangle| \leqq A\|f\|_k$ for all $f \in E(K)$. Thus, for $w \in D$, we have $|F(w)\,k^*(w)| = |\langle k^*(w)\exp(z\cdot w), F\rangle| \leqq Ak^*(w)\|\exp(z\cdot w)\|_k \leqq A' < +\infty$, since $k^* \in K^*$, so that $\|F\|_{k^*} = \sup\{|F(w)\,k^*(w)| : w \in D\} < +\infty$ for each $F \in M^\wedge(K)$. Therefore, if $E(K^*)$ is defined to be the vector space of all functions F analytic on D and such that $\|F\|_{k^*} < +\infty$, then $E(K^*)$ is a locally convex topological vector space, and $M^\wedge(K)$ is a subspace of $E(K^*)$. In case $D = C^p$, it is easy to check that K^* satisfies the conditions K1–K4 (and also K3′, with $D(K^*) = C^p$), so that $E(K^*)$ is a vector space of the type studied in §§2–4.

The following definition and theorem establish when the topology induced on $M^\wedge(K)$ by the seminorms, $\| \ \|_{k^*}$, is admissible for the pairing $\langle E(K), M^\wedge(K)\rangle$.

DEFINITION 5.2. Let K satisfy K3′. Let $\lambda(r)$ be a function defined on $R^p(+)$ which is a convex function of $\log r$ (see §1), nondecreasing in each variable, has continuous first partial derivatives, and is such that $k(r)\exp(\lambda(r))$ is bounded for each $k \in K$. The family K is said to be hypoconvex if for each such function $\lambda(r)$, there exists a convex function $\lambda_1(r)$ on $R^p(+)$ such that $\lambda_1(r) \geqq \lambda(r)$, and $k(r)\exp(\lambda_1(r))$ is bounded for each $k \in K$.

An equivalent formulation (the equivalence is a part of the proof of Theorem 5.1) which, however, refers to $E(K)$, is the following.

The family K is hypoconvex if, given any $\lambda(r) = \log M(r;f)$, where $f \in E(K)$, there exists a $\lambda_1(r) \geqq \lambda(r)$ which is convex and such that $k(r)\exp(\lambda_1(r))$ is bounded for each $k \in K$.

Therefore, K is hypoconvex if and only if there exists a family $\{\lambda_\alpha\}$ of convex functions on $R^p(+)$ such that $E(K) = \bigcup_\alpha E(\lambda_\alpha)$, where

$$E(\lambda_\alpha) = \{f \text{ entire}: M(r;f) = O(\exp(\lambda_\alpha(r)))\}.$$

THEOREM 5.1. *Let K satisfy* K3′. *Then the following are equivalent*:

(i) *The relative topology on $M^\wedge(K)$ as a subset of $E(K^*)$ coincides with the Mackey topology and strong topology on $M^\wedge(K)$ from the pairing $\langle E(K), M^\wedge(K)\rangle$ and is therefore admissible for this pairing*;

(ii) *K is hypoconvex.*

PROOF. (i) $\Rightarrow$ (ii). Let $\lambda(r)$ be as in Definition 5.2. We shall first prove that there exists a function $f \in E(K)$ such that $M(r;f) \geqq \exp(\lambda(r))$, and then that that there exists a convex function λ_1 as in Definition 5.2 with $\exp(\lambda_1(r)) \geqq M(r;f)$. Hence, $\lambda_1(r) \geqq \lambda(r)$ so that (ii) follows.

Therefore, we have to prove the existence of such a function $f \in E(K)$. To this end, define $a_m = \inf\{r^{-m}\exp(\lambda(r)): r \in R^p(+)\}$, and $g(z) = \sum a_m z^m$. Then $B(r,g) \leqq \exp(\lambda(r))$, so that $g \in E(K)$. Let S denote the set of all elements $\sigma = (\sigma_1, \cdots, \sigma_p) \in Z^p(+)$ with $\sigma_j = 0$ or $\sigma_j = 1$ for each j, and let $P(z) = \sum z^\sigma$, where the sum is over all $\sigma \in S$. We claim that $f(z) = P(z)g(z)$ is the desired function. To see this, let $h(x) = h(x_1, \cdots, x_p) = \lambda(\exp(x_1), \cdots, \exp(x_p))$. By hypothesis, h is convex. Also, $\log a_m = -h^*(m)$ where $h^*(y) = \sup\{y \cdot x - h(x): x \in R^p\}$ is the convex function conjugate to h (see §1). Now fix $r = (r_1, \cdots, r_p) \in R^p(+)$, with $r_j > 0$, $1 \leqq j \leqq p$. Let $x_j = \log r_j$ and $x = (x_1, \cdots, x_p)$. There exists $y = (y_1, \cdots, y_p) \in R^p(+)$ such that $h^*(y) = y \cdot x - h(x)$ (e.g. $y = (\nabla h)(x) = ((\partial h/\partial x_1)(x), \cdots, (\partial h/\partial x_p)(x))$). Let $m = (m_1, \cdots, m_p) \in Z^p(+)$, where $m_j = [y_j]$ is the greatest integer in y_j. Let $\sigma(r)$ be that element of S with $\sigma_j = 1$ if and only if $r_j \geqq 1$. Then, since h^* is nondecreasing in each variable, $r^{\sigma(r)} a_m r^m$

$= \exp(\sigma(r)\cdot x + m\cdot x - h^*(m)) \geqq \exp((\sigma(r)+m)\cdot x - h^*(y)) \geqq \exp(y\cdot x - h^*(y))$ $= \exp(h(x)) = \exp(\lambda(r))$. Therefore, $M(r;f) = (\sum_\sigma r^\sigma) M(r;g) \geqq r^{\sigma(r)} a_m r^m \geqq \exp(\lambda(r))$, which proves the first assertion.

Now let $f \in E(K)$ be as in the preceding paragraph. By hypothesis, $L(F) = \langle f, F\rangle$ is a continuous linear functional on $M^\frown(K)$ for the K^* topology. Hence, there exists $k^* \in K^*$ and $A > 0$ such that $|L(F)| \leqq A \| F \|_{k^*}$ for all $F \in M^\frown(K)$. In particular, $|f(z)| = |L(\exp(z\cdot w))| \leqq A \sup\{k^*(w)|\exp(z\cdot w)|; w \in D\}$. Since K satisfies $K2$, K^* also satisfies K2; i.e., we may suppose $k^*(|w|) = k^*(w)$. Thus, the above supremum coincides with $\sup\{k^*(x)\exp(r\cdot x): x \in R^p(+) \cap D\}$, where $r = |z| \in R^p(+)$. Letting $\lambda_1(r) = \sup\{r\cdot x + \log k^*(x): x \in R^p(+) \cap D\}$, we then have $|f(z)| \leqq \exp(\lambda_1(|z|))$. Further λ_1 is convex, as the upper envelope of a family of convex functions, and the proof of (i) $\Rightarrow$ (ii) is complete.

PROOF OF (ii) $\Rightarrow$ (i). This implication follows from lemmas 5.1, 5.2 below.

LEMMA 5.1. *The Mackey and strong topologies on $M^\frown(K)$ coincide, and are at least as fine as the topology induced on $M^\frown(K)$ by the seminorms $\|\ \|_{k^*}$ $(k^* \in K^*)$.*

LEMMA 5.2. *Let $\lambda(r)$ be a convex function (on $R^p(+)$) which is nondecreasing in each variable and such that $\sup\{k(z)\exp(\lambda(|z|)): z \in C^p\} < +\infty$ for each $k \in K$. Then there exists a constant $A > 0$ and a function $k^* \in K^*$ such that*

$$|\langle f, F\rangle| \leqq A \| F \|_{k^*}$$

for all $F \in M^\frown(K)$ and all entire functions $f \in E(K)$ such that $M(r;f) \leqq \exp(\lambda(r) - 3\sum_{j=1}^p \log(3 + r_j))$. In particular, if $\psi(r_1, \cdots, r_p) = \lambda(r_1 + 3, \cdots, r_p + 3)$ and if $\psi^*(\rho) = \sup\{\rho\cdot r - \psi(r): r \in R^p(+)\}$ is the convex function conjugate to ψ, then $k^*(w) = \exp(-\psi^*(|w|))$ is such a function.

The implication (ii) $\Rightarrow$ (i) is deduced from these two lemmas as follows. In view of Lemma 5.1, we only have to show that the K^* topology is at least as fine as the strong topology on $M^\frown(K)$. To this end, let N be a strong neighborhood of zero in $M^\frown(K)$. Then $N \supset B^0 = \{F \in M^\frown(K): |\langle f, F\rangle| \leqq 1 \text{ for all } f \in B\}$, for some bounded subset B of $E(K)$. By Proposition 2.6, there exists $g(z) = \sum A_m z^m \in E(K)$ with $A_m \geqq |a_m|$ for all $\sum a_m z^m \in B$. Let $p(z) = z_1 \cdots z_p$, $\varepsilon = (1, 1, \cdots,) \in R^p(+)$, and $h(z) = [p(z + 3\varepsilon)]^2 g(z)$. Then put $\lambda_1(r) = \log M(r;h) = \log h(r)$. Then $\lambda_1(r)$ satisfies the conditions in Definition 5.2. (The fact that $\psi(x) = \log h(e^{x_1}, \cdots, e^{x_p})$ is convex may be seen by employing the following well-known one variable argument. The convexity of ψ is equivalent to the inequality $h(r_1^s\rho_1^{1-s}, \cdots, r_p^s\rho_p^{1-s}) \leqq h(r)^s h(\rho)^{1-s}$ for all $0 \leqq s \leqq 1$ and $r, \rho \in R_p(+)$. To prove this, one considers the function $\alpha(s)$ defined as the quotient of the left side of this inequality by the right side. The function α is analytic for $0 < \operatorname{Re} s < 1$ and continuous for $0 \leqq \operatorname{Re} s \leqq 1$, and further, $|\alpha(\sigma + i\tau)| \leqq \alpha(\sigma)$ for $0 \leqq \sigma \leqq 1$.

The two constants theorem then implies $|\alpha(s)| \leqq \max(\alpha(0), \alpha(1)) = 1$ for all $0 \leqq \operatorname{Re} s \leqq 1$, which proves the desired inequality.) Now choose $\lambda(r) \geqq \lambda_1(r)$ as in Definition 5.2 and let A, k^* be as in Lemma 5.2. Then for $f \in B$, we have $M(r;f) \leqq M(r;g) = \exp(\lambda(r) - 3\log p(r+3\varepsilon))$, hence $|\langle f, F\rangle| \leqq A \| F \|_{k^*}$ for all $F \in M\hat{}(K)$. It follows that B^0 contains the neighborhood of zero $\{F \in M\hat{}(K): \| F \|_{k^*} \leqq A^{-1}\}$, which proves the theorem.

It remains to prove Lemmas 5.1 and 5.2.

PROOF OF LEMMA 5.1. Let N be a neighborhood of zero in $M\hat{}(K)$ for the K^* topology. We may suppose that $N = \{F \in M\hat{}(K): \| F \|_{k^*} \leqq 1\}$. Let

$$B = \{f_w(z): f_w(z) = k^*(w)\exp(w \cdot z),\ w \in D\}.$$

Then B is a subset of $E(K)$, and it follows from Proposition 2.6 (ii) and the definition of K^*, that B is bounded. Moreover, if $F \in B^0$, then $1 \geqq |\langle f_w(z), F\rangle| = |F(w)k^*(w)|$. Hence $B^0 \subset N$, so the strong topology is finer than the K^* topology. The Mackey and strong topologies on $M\hat{}(K)$ always coincide, since every bounded and closed subset of $E(K)$ is compact.

Our proof of Lemma 5.2 is somewhat technical; therefore, we first outline the proof. If $f(z) = \sum a_m z^m \in E(K)$ and $F(w) = \sum B_m w^m \in M\hat{}(K)$, then $|\langle f, F\rangle| = |\sum m! a_m B_m| \leqq \sum m! \, |a_m(B_m)|$ (Theorem 3.1). If $M(r;f) \leqq \exp(\psi(r))$, where $\psi(r)$ is convex on $R^p(+)$, then from Cauchy's inequality

$$|a_m| \leqq \inf\{r^{-m}\exp(\psi(r)): r \in R^p(+)\}.$$

In Lemma 5.3, this infimum is shown to be equal to

$$(m!)^{-1}\sup\{r^m\exp(-\psi^*(r)): r \in R^p(+)\}$$

(except for a nuisance factor), where ψ^* is the convex function conjugate to ψ. The function $k^*(w) = \exp(-\psi^*(|w|))$ belongs to K^*, and the above remarks imply $m! \, |a_m| \leqq \| w^m \|_{k^*}$. Now also $|B_m| \leqq r^{-m}M(r;f)$ for all r; hence, if r is such that $r^m k^*(r) = \| w^m \|_{k^*}$, then $m! \, |a_m B_m| \leqq r^m k^*(r) r^{-m} M(r;F) \leqq \| F \|_{k^*}$. By changing ψ and k^* slightly, we can improve this to $m! \, |a_m B_m| \leqq c_m \| F \|_{k^*}$, where $A = \sum c_m < +\infty$; then $|\langle f, F\rangle| \leqq A \| F \|_{k^*}$ as asserted in the lemma.

It seems reasonable that this lemma should not depend on an estimate of the $|a_m|$. However, the proof given here is the only one known to us.

We now prove the Lemma 5.3 mentioned above

LEMMA 5.3. *Let $\psi(x)$ be a convex function on $R^p(+)$ which is nondecreasing in each variable and such that $\delta \cdot x \leqq \psi(x)$ for some $\delta = (\delta_1, \cdots, \delta_p) \in R^p(+)$ with $\delta_j > 0$, $1 \leqq j \leqq p$. For $y \in R^p(+)$, let*

$$\psi^*(y) = \sup\{y \cdot x - \psi(x): x \in R^p(+)\} \tag{5.1}$$

(ψ^* may be infinite for some values of y). Then for all $u = (u_1, \cdots, u_p) \in R^p(+)$, we have

$$h(u) = \sum_{j=1}^{p} (u_j \log u_j - u_j) + g(u),$$

where

$$h(u) = \sup \left\{ \sum_{j=1}^{p} u_j \log x_j - \psi^*(x) : x \in R^p(+) \right\} \tag{5.2}$$

and

$$g(u) = \inf \left\{ \psi(x) - \sum_{j=1}^{p} u_j \log x_j : x \in R^p(+) \right\}. \tag{5.3}$$

PROOF. For $u, v \in R^p(+)$, denote by uv, $\log u$, u/v, the elements $(u_1 v_1, \cdots, u_p v_p)$, $(\log u_1, \cdots, \log u_p)$, and $(u_1/v_1, \cdots, u_p/v_p)$, respectively, of $R^p(+)$. Fix $u \in R^p(+)$. Since $\delta \cdot x = O(\psi(x))$, we have that $\psi(x) - u \cdot \log x \to +\infty$ as $\|x\| \to +\infty$. Also, since ψ is nondecreasing in each variable and convex, ψ is continuous on $R^p(+)$. Thus, the infimum in (5.3) must be attained at some point $v = (v_1, \cdots, v_p)$. Clearly, $v_j > 0$ provided that $u_j > 0$. We shall show that

$$\psi^*(u/v) = \sum_{j=1}^{p} u_j - \psi(v) \tag{5.4}$$

and that

$$h(u) = u \cdot \log(u/v) - \psi^*(u/v). \tag{5.5}$$

In these equations, the expressions 0/0, and 0 log(0/0), which may occur when $u_j = v_j = 0$, are to be interpreted as 0. The equalities (5.4), (5.5) prove the lemma, for then $h(u) = u \cdot \log u - u \cdot \log v - \sum_{j=1}^{p} u_j + \psi(v) = u \cdot \log u - \sum_{j=1}^{p} u_j + g(u)$ as asserted.

The basic inequality used in proving (5.4) and (5.5) is

$$\psi(x) - \psi(v) \geqq (u/v) \cdot (x - v), \tag{5.6}$$

which is derived as follows. The infimum in (5.3) is attained at v so that

$$\psi(v) - u \cdot \log v \leqq \psi(x) - u \cdot \log x,$$

or

$$\psi(x) - \psi(v) \geqq u \cdot \log(x/v) \tag{5.7}$$

for all $x \in R^p(+)$. For $0 \leqq t \leqq 1$, define $\alpha(t) = \psi(tx + (1-t)v)$. Then from (5.7), $\alpha(t) - \alpha(0) \geqq u \cdot \log[(v/v) + t((x-v)/v)] \sim t[(u/v) \cdot (x-v)]$ for $t \to 0^+$. Now $\alpha(t)$ is convex; therefore, the difference quotient $(\alpha(t) - \alpha(0))/t$ is nondecreasing, so that $\psi(x) - \psi(v) = \alpha(1) - \alpha(0) \geqq \lim_{t\to 0^+}(\alpha(t) - \alpha(0))/t \geqq (u/v)\cdot(x - v)$, and (5.6) is proved.

Now from (5.6), it follows that $(u/v)\cdot x - \psi(x) \leqq (u/v)\cdot v - \psi(v) = \sum_{j=1}^{p} u_j - \psi(v)$, which implies (5.4). Also, $\psi^*(x) - \psi^*((u/v)) = \sup\{x\cdot y - \psi(y): y \in R^p(+)\} - (u/v)\cdot v + \psi(v)$. Substituting $y = v$ yields $\psi^*(x) - \psi^*(u/v) \geqq v\cdot(x - (u/v))$. However, $\log(1+h) \leqq h$ for $-1 < h < +\infty$, so that $v\cdot(x - (u/v)) = \sum_{j=1}^{p} u_j((v_j x_j/u_j) - 1) \geqq \sum_{j=1}^{p} u_j \log(v_j x_j/u_j) = u\cdot(\log x - \log(u/v))$. Therefore, $\psi^*(x) - \psi^*(u/v) \geqq u\cdot(\log x - \log(u/v))$, which implies (5.5).

PROOF OF LEMMA 5.2. Let $\lambda(r)$ and $\psi(r)$ be as in the statement of the lemma. It is no loss of generality to suppose that $\lambda(r) \geqq \delta_1 r_1 + \cdots + \delta_p r_p$ for some $\delta_j > 0$, $1 \leqq j \leqq p$, and all $\| r \|$ sufficiently large. For, if this is not the case, then λ must be constant on some ray in $R^p(+)$. From the fact that λ is nondecreasing in each variable, it then follows that λ is constant in at least one variable, say r_1. Hence, if $M(r;f) \leqq \exp(\lambda(r))$, f must be a function of $z_2, \cdots, z_p$. Therefore, the problem is reduced to one involving functions of $p-1$ variables. A routine induction argument will then complete the proof.

We may, therefore, suppose that λ and, consequently, ψ satisfy the hypotheses of Lemma 5.3. Let ψ^*, h, g be as in Lemma 5.3, and define $k^*(w) = \exp(-\psi^*(|w|))$. The function k^* is u.s.c., since $-\psi^*$ is u.s.c., and on $R^p(+)\cap D$, $k^*(r)$ is nonincreasing in each variable, since $-\psi^*$ has this property. From the known fact $(\psi^*)^* = \psi$ (see e.g. [9]) and the hypothesis that $k(r)\exp(\lambda(r))$ is bounded for each $k \in K$, it then follows that $k^* \in K^*$. We further note that $k^*(w)$ is positive only where $\psi^*(|w|) < +\infty$. Thus, at such a $|w|$, $\lambda(x) \geqq |w|\cdot x - O(1)$ so that necessarily w is a point of the closure of D. Consequently, given $m \in Z^p(+)$, there exists $r \in R^p(+)\cap D$ such that $2r^m k^*(r) \geqq \| w^m \|_{k^*} = \exp(h(m))$.

Now let f satisfy the hypotheses of the lemma. Let $q(m) = \prod_{j=1}^{p}(m_j+1)(m_j+2)$ and $T(\sum a_m z^m) = \sum q(m) a_m z^m$ be as in Proposition 2.4. Then from (1.1) and (1.3),

$$B(r;Tf) \leqq M(r;Tf) \leqq p(r+3\varepsilon)^3 M(r+3\varepsilon;f) \leqq \exp(\lambda(r+3\varepsilon)$$
$$-3\textstyle\sum_{j=1}^{p}\log(6+r_j)(3+r_j)^{-1}) \leqq \exp(\psi(r)).$$

Hence, if $m^{-m}e^{+m} = \exp(-\sum_{j=1}^{p}(m_j\log m_j - m_j))$, and if g, h are as in Lemma 5.3, we have $q(m)\,|a_m| \leqq \inf\{r^{-m}\exp(\psi(r)): r \in R^p(+)\} = \exp(g(m)) = m^{-m}e^m\exp(h(m)) = m^{-m}e^m\| w^m \|_{k^*}$.

Define $c_m = m!\, m^{-m}e^m/q(m)$. An application of Stirling's formula shows that $A = 2\sum c_m < +\infty$. Let $F(w) = \sum(b_m/m!)w^m \in M^\wedge(K)$, and let r be as above, so that $2r^m k^*(r) \geqq \|w^m\|_{k^*}$. Then

$$|a_m b_m| = m!\,|a_m|\,|b_m/m!| \leqq c_m\| w^m \|_{k^*}(r^{-m}M(r;F))$$
$$\leqq 2c_m r^m k^*(r) r^{-m} M(r;F) \leqq 2c_m\| F \|_{k^*}.$$

Therefore, $|\langle f, F\rangle| \leqq \sum |a_m b_m| \leqq A\,\| F \|_{k^*}$ and the lemma is proved. This completes the proof of Theorem 5.1.

We conclude this section by proving, under the hypothesis that K is hypoconvex, that $M^\wedge(K)$ and $E(K^*)$ coincide only when $E(K)$ is a Montel space.

THEOREM 5.2. *Suppose that* K3′ *holds, and that K is hypoconvex. Then $M^\wedge(K) = E(K^*)$ if and only if $E(K)$ is barrelled. Further, if the open set $D(K) = C^p(D(K)$ as in* K3′), *then $M^\wedge(K^*)$ is well defined, and $M^\wedge(K^*) = E(K)$.*

PROOF. This theorem is, for the most part, contained in Lemma 5.2. For, if $f(z) = \sum a_m z^m \in E(K)$, $F(w) = \sum (b_m/m!)w^m$, and k^* are as in the lemma, then, as was established in the proof of the lemma, $\sum |a_m b_m| \leqq A \| F \|_{k^*}$. The only property of F used in the proof was $\| F \|_{k^*} < +\infty$; hence, this inequality also holds for $F \in E(K^*)$.

It follows from this remark and the hypothesis that K is hypoconvex, that $\sum |a_m b_m| < +\infty$ whenever $f \in E(K)$ and $F \in E(K^*)$. In particular, we have $(b_m) \in \Delta^{xx}(\Delta$ and Δ^{xx} as defined in §3). Thus, $M^\wedge(K) = E(K^*)$ if and only if $\Delta = \Delta^{xx}$, which, by the Lemma of §4, is equivalent to $E(K)$ being barrelled.

The last assertion of the theorem follows directly from Lemma 5.2.

§6. **Examples.** We give in this section several examples of the spaces $E(K)$. The first and, in some sense, nicest examples we consider arise as follows. Let $0 < \lambda_1 \leqq \lambda_2 \leqq \cdots$ be a sequence of convex functions on $R^p(+)$ which are nondecreasing in each variable, and with $\lambda_1(r) \geqq \varepsilon_1 r_1 + \cdots + \varepsilon_p r_p$ for some $\varepsilon_1, \cdots, \varepsilon_p > 0$. Then let K be the family of all nonnegative u.s.c. functions k on C^p such that $k(z)\exp(\lambda_n(|z|))$ is bounded for each $n = 1, 2, \cdots$. If we further suppose that, say, $\lambda_n(r) \geqq \lambda_{n-1}(r_1 + 1, \cdots, r_p + 1) + \log[(1 + r_1) \cdots (1 + r_p)]$, then K will satisfy the hypotheses K1—K4 and K3′. Also, K is sequentially determined (in the sense of Definition 4.1) so that $E(K)$ is barrelled and $E(K) = \{f : \log M(r; f) \leqq \lambda_n(r) + O(1)$ for some $n > 0\}$. It follows from the remarks preceding Theorem 5.1 that K is hypoconvex. Consequently, from Theorem 5.1, 5.2, we must have that $M^\wedge(K) = E(K^*)$, and that the topology of $E(K^*)$ coincides with the strong topology on $E(K^*)$ (as the dual space of $E(K)$). That is, $E(K)$ and $E(K^*)$ are dual Montel spaces.

Now if λ_n^* is the convex function conjugate to λ_n, then it is easy to verify (using the fact that $\lambda^{**} = \lambda$) that K^* is the family of all nonnegative u.s.c. functions k^* on C^p such that $k(z)\exp(\lambda_n^*(|z|))$ is bounded for some $n = 1, 2, \cdots$. Thus, $E(K^*) = M^\wedge(K)$ may be decribed as the family of all functions F analytic on the open set $D = D(K)$ (as in K3′) and satisfying $\log M(r; F) \leqq \lambda_n^*(r) + O(1)$ for all $n = 1, 2, \cdots$.

In the above situation, we also have $K^{**} = K$. Thus, we could have given this example by first specifying the λ^*. Notice also, since K^* satisfies K3′ with $D = C^p$, we have $M^\wedge(K^*) = E(K^{**}) = E(K)$.

The specific examples below arise by specializing the λ_n, or λ_n^*, as the case may be. Since the spaces $E(K)$, $E(K^*)$ are dual Montel spaces (and therefore reflexive), we have not distinguished between $E(K)$ and $E(K^*)$; that is, some of the $E(K)$'s in the list below are actually $E(K^*)$'s, in the above terminology. Also, for convenience, we take $p = 1$. The examples have obvious counterparts for $p > 1$. Finally, we do not specify the λ_n explicitly. A suitable choice will be evident.

$E(K)$	$E(K^*)$
1. (Ehrenpreis [4], [6]) All entire functions.	All entire functions of exponential type ($\log M(r;f) = O(r)$).
2. All entire f such that $\log \log M(r;f) = O(r)$.	All entire F such that $\log M(r; F) = o(r \log r)$.
3. All entire f such that $\log \log M(r;f) = o(r)$.	All entire F such that $\log M(r; F) = O(r \log r)$.
4. All entire f of finite order (i.e. $\log M(r;f) = O(r^\rho)$ for some $\rho > 0$).	All entire F of order $\leqq 1$ (i.e. $\log M(r; F) = O(r^{1+\varepsilon})$ for each $\varepsilon > 0$).
5. All entire f of order $\leqq \rho$ ($\rho \geqq 1$) (i.e. $\log M(r;f) = O(r^{\rho+\varepsilon})$ for each $\varepsilon > 0$).	All entire F of order $< q$, where $q^{-1} + \rho^{-1} = 1$ (i.e. $\log M(r; F) = O(r^{q-\varepsilon})$ for some $\varepsilon > 0$).
6. All entire f of order $\leqq \rho$, finite type ($\rho \geqq 1$) (i.e. $\log M(r;f) = O(r^\rho)$).	All entire F of order $\leqq q$, minimal type, where $q^{-1} + \rho^{-1} = 1$ (i.e. $\log M(r; F) = o(r^q)$).
7. All entire f of order $\leqq \rho$, type $< \tau$ ($\rho > 1$) (i.e. $\log M(r;f) \leqq (\tau - \varepsilon)r^\rho + O(1)$ for some $\varepsilon > 0$).	All entire F of order $\leqq q$, type $\leqq \sigma$, where $q^{-1} + \rho^{-1} = 1$ and $\sigma = (\rho\tau)^{1-q}/q$ (i.e. $\log M(r; F) \leqq (\sigma + \varepsilon)r^q + O(1)$ for each $\varepsilon > 0$).
8. All entire f of exponential type $< \tau$ ($\log M(r;f) \leqq (\tau - \varepsilon)r + O(1)$ for some $\varepsilon > 0$).	All F analytic for $\lvert w \rvert < \tau$.

Next consider the case where $\lambda(r)$ is a nondecreasing convex function with $\lambda(r)/r \to +\infty$, and $\lambda_n(r) = \lambda(r + n) + nr$. For simplicity, we retain the assumption $p = 1$. Then $\lambda_n^*(r) = \lambda^*(r - n) - nr$. With $\lambda(r) = (1/\rho)r^\rho$ ($\rho > 1$), we find

9. $$E(K) = \{f: \log M(r;f) = (1/\rho)r^\rho + O(r^{\rho-1})\}$$
$$E(K^*) = \{F: \log M(r;F) = (1/q)r^q - A r^{q-1} + O(1) \text{ for each } A > 0\}. \ (\rho^{-1} + q^{-1} = 1).$$

Along the same lines as examples 1—9, we have the following two examples. Let $K_1 = \{k \geqq 0: k(z) = O(\exp(|z| - n\log(1+|z|)))$ for some $n > 0\}$, and $K_2 = \{k: k(z) = O(\exp(|z| + n\log(1+|z|)))$ for each $n > 0\}$. Then

10. $$E(K_1) = \{f: M(r;f) = O(r^{-n}e^r) \text{ for each } n > 0\}$$
$$M\hat{\ }(K_1) = E(K_1^*) = \{F: F \text{ is analytic for } |w| < 1, \text{ and } M(r;F) = O((1-r)^{-n}) \text{ for some } n > 0\}.$$

11. $$E(K_2) = \{f: M(r;f) = O(r^n e^r) \text{ for some } n > 0\}$$
$$M\hat{\ }(K_2) = A^\infty = \{F: F \text{ is analytic for } |w| < 1, \text{ and } F, F', F'', \cdots \text{ are all uniformly continuous for } |w| < 1\}.$$

To verify that Example 10 is correct, we first show that K_1 is hypoconvex. Let $\lambda(r)$ be a function with $\lambda(r) \leqq r - n\log r + C_n$ for each $n = 1, 2, \cdots$, where the C_n are constants. We have to find a convex function $\lambda_1(r) \geqq \lambda(r)$ with $\lambda_1(r) \leqq r - n \log r + C_n'$ for $n = 1, 2, \cdots$, and some constants C_n'. If $\lambda(r) = r - \varepsilon(r)$, then $\varepsilon(r) \geqq n\log r - C_n$. Thus, there are numbers $0 < r_1 < r_2 < \cdots$ such that $\varepsilon(r) \geqq (n+1)\log r$ for $r \geqq r_n$. We may also suppose $r_{n+1} > 2r_n$. Define $\varepsilon_1(r) = 0$ for $r \leqq r_1$ and $\varepsilon_1(r) = \int_{r_1}^r t^{-1}\alpha(t)\,dt$, where $\alpha(t)$ is an increasing function with $\alpha(r_n) = n$ and $t^{-1}\alpha(t)$ decreasing. Such a choice of $\alpha(t)$ exists; for example, $\alpha(r_n) = n$ and α linear on $[r_n, r_{n+1}]$. Then ε_1 is concave and $\varepsilon_1(r) \leqq \varepsilon(r)$ for $r \geqq r_1$. Hence, $\lambda_1(r) = r - \varepsilon_1(r)$ is convex and $\lambda_1(r) \geqq \lambda(r)$ for $r \geqq r_1$. Since α is nondecreasing and tends to $+\infty$, we also have $n\log r = O(\varepsilon_1(r))$, so that $\lambda_1(r) \leqq r - n\log r + O(1)$ for each $n > 0$. Thus, K is hypoconvex.

A computation shows that $(n + \log n) - n\log(1-r)$ $(0 < r < 1)$ is the convex function conjugate to $r - n\log r$. The assertions in (10) now may be verified in essentially the same way as examples 1–9.

The example 11 does not follow in the same manner as 1—10, since K_2 is not hypoconvex. However, a computation shows that $\sum a_m z^m \in E(K_2)$ if and only if $|a_m| = O(m^A)/m!$ for some $A > 0$. Thus, $F(w) = \sum B_m w^m \in M\hat{\ }(K)$ if and only if $|B_m| = O(m^{-A})$ for each $A > 0$. Hence, the Taylor series for $F(w)$ and all its derivatives converge absolutely and uniformly for $|w| \leqq 1$. Conversely, if F and all its derivatives are uniformly continuous for $|w| < 1$, then it is a consequence of a theorem of Hardy [**12**, p. 70] that $|B_m| = O(m^{-A})$ for each $A > 0$, hence $F \in M\hat{\ }(K)$. We wish to thank D. Williams for calling our attention to this example.

We wish to make three further remarks concerning 11. First, it can be shown that the strong topology on A^∞ from the pairing $\langle E(K_2), A^\infty \rangle$ coincides with the topology induced by the seminorms $\| F \|_n = \sup \{ | F^{(n)}(e^{i\theta}) | : 0 \leqq \theta \leqq 2\pi \}_1$ $n = 1, 2, \cdots$. Secondly, $E(K^*) = H^\infty = \{F : F$ is analytic and bounded for $| w | < 1\}$. In particular, $M^\wedge(K) = A^\infty \neq E(K^*)$. Lastly, the duality between A^∞ and $E(K_2)$ may be used to prove that: given any sequence $\{C_n\}$ of complex numbers, there exists $F \in A^\infty$ such that $F^{(n)}(1) = C_n$, $n = 0, 1, 2, \cdots$. The proof employs the same techniques used in [21].

We conclude this section with the remark that $M^\wedge(K)$ need not be closed under radial minorization. That is, it need not be the case that: given $F \in M^\wedge(K)$ and G entire with $M(r; G) \leqq M(r; F)$, then $G \in M^\wedge(K)$. Of course, 11 gives an example of this. However, $M^\wedge(K)$ can fail to be closed under radial minorization even when K1–K4, K3′, K4′ hold. The examples we know of such K are somewhat involved, and will not be reproduced here. The remarks following the proof of Theorem 7.3 of §7 will make it clear how to construct such examples.

§7. **Hadamard algebras.** Thoughout this section we assume $p = 1$.

It is often the case that $E(K)$ is an algebra under pointwise operations (e.g. examples 1–6 of §6). In this section we identify the closed ideals of $E(K)$ whenever this is the case and K satisfies K4′. Under these conditions, each closed ideal I of $E(K)$ is the ideal of all functions in $E(K)$ which vanish on the common zeros of the functions in I, with multiplicities taken into account ($p = 1$, so there is no difficulty with the notion of multiplicity). The same theorem (and proof) also apply to $M^\wedge(K)$, provided that K satisfies some additional conditions (see e.g. Definition 7.3), and this result will be applied in the next section to obtain series expansions of mean periodic functions in $E(K)$ (for $p = 1$). The main analytical tool we use in our proofs is a theorem (Theorem A below) which is essentially a generalization of Hadamard's factorization theorem for entire functions of finite order.

We now introduce the class of algebras of entire functions to which our analysis applies. Let A be a topological vector space of entire functions of one complex variable, which is also an algebra under pointwise operations. We do not require that multiplication be continuous.

DEFINITION 7.1. ($p = 1$) A is an Hadamard algebra if

(i) A contains all polynomials;

(ii) whenever $f \in A, f(a) = 0$, then $f(z)/(z - a) \in A$;

(iii) given $g \in A$, g not identically zero, there exists a function $h \in A$ and functions $g_n(z) = P_n(z) \cdot e_n(z)$, where $P_n(z)$ is a polynomial and $e_n(z)$ is a unit, such that $gh/g_n \in A$ and $\lim_{n \to \infty} fgh/g_n = f$ for all $f \in A$.

The reason for the name "Hadamard" algebra, is that (iii) is roughly the same as requiring each $g \in A$ to have an infinite product representation, which is consistent with convergence in A. For example, in example 6 of §6, each g has a unique representation in terms of a canonical product; if we choose $h = 1$ and for g_n the partial product over the zeros of g in the disc $|z| \leqq n$, then standard estimates on the rate of growth of canonical products (together with Proposition 2.8) show that (iii) is satisfied.

For an ideal I in A, denote by $Z(I)$ the common zeros of the functions in I (with multiplicities taken into account). For a sequence Z with multiplicities, let $I(Z)$ be the ideal of all functions in A which vanish at least on Z (counting multiplicities, of course).

DEFINITION 7.2. An ideal I in A is saturated if $I = I(Z(I))$.

Evidently, a saturated ideal is the set of all functions vanishing on some discrete subset of C (with multiplicities) and conversely.

THEOREM 7.1. *If A is an Hadamard algebra, then every closed ideal in A is saturated.*

The proof of this theorem is analogous to the proofs used by L. Schwartz [**23**], and Ehrenpreis [**4**]. We first show that "extra" zeros can be divided out one at a time; the condition (iii) then guarantees that this procedure converges in A. Thus, we first prove the following lemma.

Lemma 7.1. ($p = 1$). *Let A be an Hadamard algebra and I an ideal in A. If a has multiplicity m in $Z(I)$ and $f \in I$ has a zero at a of multiplicity $n > m$, then $f(z)/(z-a)^j \in I$, $j = 1, 2, \cdots, n-m$.*

PROOF. It is enough to prove $f(z)/(z-a) \in I$. Since a has multiplicity m in $Z(I)$, there exists $g \in I$ such that $g(z) = (z-a)^m h(z)$ with $h \in A$ and $h(a) \neq 0$. Then $-h(a)f(z)/(z-a) = f(z)[(h(z)-h(a))/(z-a)] + g(z)[f(z)/(z-a)^{m+1}] \in I$ so $f(z)/(z-a) \in I$.

PROOF OF THEOREM 7.1. Let $I' = I(Z(I))$. Clearly, $I \subset I'$, so we have to prove $I' \subset I$. Let $f \in I'$, $g \in I$, g not the zero function (if $I = \{0\}$, the theorem is trivial). Then let $h, g_n = P_n e_n$ be as in (iii) of Definition 7.1. Now $fgh \in I$, since $g \in I$. Further, f and g vanish on I, and gh/P_n is entire, so that $fgh/P_n \in I$ by the lemma. Then $fgh/g_n = e_n^{-1}(fgh/P_n) \in I$, and since I is closed, $f = \lim(fgh/g_n) \in I$. Hence $I = I'$.

Before proving that $E(K)$, $M^\wedge(K)$, under suitable hypotheses, are Hadamard algebras, we need the following lemma on quotients of entire functions.

LEMMA 7.2. ($p = 1$). *Let f, g be entire functions, $g(0) \neq 0$, such that $f(z)/g(z)$ is entire. Then*

$$M(r; f/g) \leqq |g(0)|^{-3}[1 + M(2r; f)]^3[1 + M(2r; g)]^3.$$

PROOF. Let $T(r; h)$ denote the Nevanlinna characteristic function of the meromorphic function h. We refer to [11] for the definition of $T(r; h)$ and proofs of the following inequalities.

(7.1) $$T(r; h_1h_2) \leqq T(r; h_1) + T(r; h_2).$$

(7.2) $$\text{If } h(0) \neq 0, \infty, \text{ then } T(r; h) = T(r; 1/h) + \log|h(0)|.$$

(7.3) $$\text{If } f \text{ is entire, then } T(r; f) \leqq \log^+ M(r; f) \leqq [(R+r)/(R-r)]\, T(R; f) \ (\log^+ x = \max(0, x)).$$

If in (7.3) R is replaced by $2r$, we obtain

(7.4) $$\log M(r; f) \leqq 3T(2r; f).$$

Applying successively (7.4), (7.1), (7.2), and the first part of (7.3), we obtain

$$\log M(r; f/g) \leqq 3\log^+ M(2r; f) + 3\log^+ M(2r; g) - 3\log|g(0)|.$$

The lemma then follows by exponentiating both sides of this last inequality, and using the inequality, $\exp(\log^+ x) \leqq 1 + x\,(x \geqq 0)$.

As mentioned above, we also need the following theorem concerning entire functions.

THEOREM A. [20]. *Suppose that $\lambda(r)$ is a continuous nondecreasing, positive function defined for $r \geqq 0$. Let g be a nonconstant entire function such that $\log M(r; g) \leqq A\lambda(Br)$ for some constants $A, B > 0$. Then there exist constants $C, D \geqq 0$, entire functions $h(z)$, $g_n(z)$, $n = 1, 2, \cdots$, and numbers $R_1 < R_2 < R_3 < \cdots$ such that the following hold*:

(i) *$g_n(z) = P_n(z)e_n(z)$, where $e_n(z)$ is an entire function with no zeros, and $P_n(z)$ is a polynomial whose zeros are those of hg in the disc $|z| \leqq R_n$;*

(ii) *$\lim_{n\to+\infty}(gh/g_n) = 1$ uniformly on compact subsets of the complex plane;*

(iii) *$\log M(r; f) \leqq C\lambda(Dr)$ if f is any of the entire functions g_nh, gh, g_n, gh/g_n.*

We now prove the main theorem of this section.

THEOREM 7.2. ($p = 1$). *Suppose that K satisfies K4′ and that $E(K)$ is an algebra. Then $E(K)$ is an Hadamard algebra. Consequently, $I = I(Z(I))$ for each closed ideal of $E(K)$.*

PROOF. Properties (i), (ii) of Definition 7.1 are clearly satisfied. To prove (iii), let $g(z) = \sum a_m z^m \in E(K)$. We shall apply Theorem A with $\lambda(r) = \log(2 + \sum|a_m|r^m)$. The hypotheses of Theorem A are then satisfied with $A = B = 1$. Let C, D, h, g_n, P_n, e_n be as in Theorem A. Since K4′ holds and $E(K)$ is an algebra, we have from Proposition 2.6 that the set $B(f; C, D) = \{F: F \text{ entire}, \log M(r, F) \leqq \log M(r; f) + C\lambda(Dr)\}$ is a bounded subset of $E(K)$ for any $f \in E(K)$ and any constants C, D. Hence, $fgh/g_n \to f$, in view of (ii), (iii) of Theorem A and Proposition 2.8. Also, from Lemma 7.2 and the hypothesis K4′, we have that any function in $E(K)$ with no zeros is a unit in $E(K)$. In particular, the e_n are units, and this completes the proof.

DEFINITION 7.3. K is hyperconvex if, given $k \in K$, there is a function $k' \in K$ such that $k'(|z|) = k'(z) \geqq k(z)$ and $k'(r) = \exp(-\psi(r))$, where $\psi(r)$ is convex and nondecreasing.

THEOREM 7.3. $(p = 1)$. *Suppose that* K3′, K4′ *hold, and that K is hyperconvex. Then $M^\wedge(K)$, in the weak topology from the pairing $\langle E(K), M^\wedge(K)\rangle$, is an Hadamard algebra.*

PROOF. The theorem is a consequence of the following two assertions.

(1) Suppose that K3′, K4′ hold, and that for each $F \in M^\wedge(K)$, the set $B(F) = \{f : f \text{ entire, and } M(r;f) \leqq M(r;F)\}$ is an equicontinuous subset of $M^\wedge(K)$. Then $M^\wedge(K)$, in the weak topology, is an Hadamard algebra.

(2) If K is hyperconvex, then $B(F)$ $(F \in M^\wedge(K))$ is an equicontinuous subset of $M^\wedge(K)$.

To prove (1), we proceed exactly as in the proof of Theorem 7.2. The only point that remains to be checked is whether the functions $fgh/g_n \to f$ in $M^\wedge(K)$. The hypotheses of (1), and Theorem A show that the fgh/g_n all lie in an equicontinuous subset of $M^\wedge(K)$. Moreover, pointwise convergence of the fgh/g_n implies that $\langle \exp(az), fgh/g_n\rangle \to f(a)$ for each a. Thus, the fgh/g_n, as linear functionals on $E(K)$, form an equicontinuous sequence of continuous functions on $E(K)$ converging on a dense subset of $E(K)$ (namely, the linear span of the exponential functions). Consequently, the fgh/g_n converge to f in the weak topology on $M^\wedge(K)$.

To prove (2) let $F \in M^\wedge(K)$ and $B(F)$ be as in the proposition. There exists $k \in K$ and $A > 0$ such that $|\langle f, F\rangle| \leqq A\|f\|_k$. Thus, if $\exp(-\psi(|z|)) = k'(z) \geqq k(z)$ is as in the definition of hyperconvex, we have $|\langle f, F\rangle| \leqq A\ \|f\|_{k'}$. Let $\psi^*(r) = \sup\{rt - \psi(t) : t \geqq 0\}$ be the convex function conjugate to ψ. Then for each complex number a, $|F(a)| = |\langle \exp(az), F(z)\rangle| \leqq A\,\|\exp(az)\|_{k'} = A\exp(\psi^*(|a|))$; that is, $M(r;F) \leqq A\exp(\psi^*(r))$.

Let $b_m = \|z^m\|_{k'}$. Then $(b_m) \in \Delta(K)$. The mapping $\sum a_m z^m \to \sum (m+1)(m+2)a_m z^m$ is a topological isomorphism of $E(K)$ (Proposition 2.4); hence $((m+1)(m+2)b_m \in \Delta(K)$. We shall prove that $m!\,|a_m| \leqq A'(m+1)(m+2)b_m$ for some $A' > 0$ and each $f(z) = \sum a_m z^m \in B(F)$, so that $B(F)$ is an equicontinuous subset of $M^\wedge(K)$ by Proposition 3.3. Indeed, for such $f \in B(F)$, we have from Cauchy's inequality, $|a_m| r^m \leqq M(r;f) \leqq \exp\psi^*(r))$. Thus

$$|a_m| \leqq \inf\{r^{-m}\exp((\psi^*(r)) : r > 0\}.$$

By Lemma 5.2, this infimum is equal to $m^{-m}e^{+m}[\sup\{r^m \exp(-\psi^{**}(r))\}]$. Since $\psi^{**} = \psi$, this last expression is equal to $m^{-m}e^m\|z^m\|_{k'}$. The desired inequality then follows from Stirling's formula (even $m!\,|a_m| \leqq A' m^{\frac{1}{2}} b_m$).

In connection with (2) of the above proof, we remark (without proof) that the condition, K is hyperconvex, is also necessary in order that (2) hold. Even more, if $B(F) \subset M^\wedge(K)$ for each $F \in M^\wedge(K)$, then K can be shown to be hyperconvex. Thus, to construct a family K such that $M^\wedge(K)$ is not closed under radial minorization (see the remark concluding §6), it suffices to find a family K which is not hyperconvex.

We also wish to remark that K is hyperconvex whenever K satisfies the following two conditions:

(i) K is hypoconvex;

(ii) K is full; i.e. if $M(|z|;f)k(z)$ is bounded for each $f \in E(K)$, then $k \in K$.

In particular, K is hyperconvex in examples 1—10 of §6. To prove the above assertion, note that if K is hypoconvex, then for each $f \in E(K)$ there is a convex function $\lambda(r) \geqq \log M(r;f)$ such that $k(r)\exp(\lambda(r))$ is bounded for each $k \in K$. Fix $k \in K, (k(|z|) = k(z))$, and let $M_\lambda = \sup\{k(r)\exp(\lambda(r))\colon r \geqq 0\}$. Then $k'(z)$ defined by $k'(|z|) = k'(z)$ and $k'(r) = \inf\{M_\lambda \exp(-\lambda(r))\colon \lambda\}$ is a function as in Definition 7.3.

§8. **Spectral synthesis in** $E(K)$. Throughout this section we assume $p = 1$. Following the definition given by L. Schwartz **[23]**, a function f in $E(K)$ is called mean periodic if $T(f)$, the closure in $E(K)$ of the span of the translates of f, is a proper subspace of $E(K)$. A monomial-exponential function is a function of the form $z^j \exp(az)$. The problem of spectral synthesis is that of recovering f from the monomial-exponential functions which lie in $T(f)$; that is, is f the limit of linear combinations of the monomial-exponential functions which lie in $T(f)$? More generally, suppose that V is a variety; that is, a closed, translation-invariant subspace of $E(K)$ (clearly $T(f)$ is a variety). Is V the closed linear span of the monomial-exponential functions which belong to V? If this is the case for every variety V of $E(K)$, we say that spectral synthesis is possible in $E(K)$.

For the space of all entire functions in the topology of uniform convergence on compact subsets of C, it has been shown by Schwartz **[23]**, Ehrenpreis **[4]**, Malgrange **[16]**, and Shapiro **[24]**, that spectral synthesis is possible. Using analogous methods, we show in this section that spectral synthesis is possible in $E(K)$ whenever $M^\wedge(K)$ is an Hadamard algebra in the weak topology from the pairing $\langle E(K), M^\wedge(K)\rangle$. In point of fact, it is possible to use the duality to prove more than this. We show that if $M^\wedge(K)$ is an Hadamard algebra, then to every mean periodic function f there corresponds a unique "canonical" formal series, $\sum d_{n,j} z^j \exp(a_n z)$, the functions $z^j \exp(a_n z)$ being the monomial-exponential functions in $T(f)$. This series need not converge. However, a "summability method" is developed which enables us to represent f as the limit of a sequence of finite sums of appropriately weighted terms of this series.

We remark that an equivalent definition of the mean-periodic functions in $E(K)$ can be given in terms of a convolution equation (see e.g. Schwartz **[23]**).

For, if f is mean periodic, then $T(f) \neq E(K)$. Hence, by the Hahn-Banach theorem, there exists a nonzero continuous linear functional L such that $L(g) = 0$ for all z, where $f_z(t) = f(z+t)$. If L' is the continuous linear functional defined by $L'(g(t)) = L(g(-t))$, then $(f * L')(z) = L(f_z(t)) = 0$ for all z (see §9 for a more complete discussion of $f * L$). These steps are all reversible, so that $f \in E(K)$ is mean periodic if and only if $f * L = 0$ for some nonzero continuous linear functional L on $E(K)$.

The main theorem of spectral synthesis in $E(K)$ is the following.

THEOREM 8.1. *Suppose that $p = 1$, that K satisfies* K3′, K4′, *and that $M^\wedge(K)$ is an Hadamard algebra in the weak topology from the pairing $\langle E(K), M^\wedge(K)\rangle$. Then every variety V of $E(K)$ is the closed linear span of the functions of the form $z^j \exp(az)$ which belong to V.*

PROOF. Let $I = V^\perp = \{F \in M^\wedge(K): \langle f, F\rangle = 0 \text{ for each } f \in V\}$. We claim that I is an ideal in $M^\wedge(K)$. To see this, observe first that V is differentiation invariant, since V is closed and $f'(z) = \lim_{h\to 0} h^{-1}(f(z+h) - f(z))$ for each $f \in E(K)$ (the limit being in $E(K)$). In then follows from the equation $\langle f, zF(z)\rangle = \langle f', F\rangle$ that I is invariant under multiplication by z, hence all polynomials. If $F \in I$ and $G \in M^\wedge(K)$, then by Proposition 3.3 and Theorem 3.2 (iii), the set $\{S_m F\}$, where S_m is the mth partial sum of the Taylor series of G, is an equicontinuous subset of $M^\wedge(K)$. Further, the $S_m F$ converge pointwise to GF. That is, the sequence $S_m F$, as linear functionals on $E(K)$, is an equicontinuous sequence converging on a dense subset of $E(K)$ (the linear span of the exponential functions). Hence, $S_m F \to GF$ in the weak topology. Since $I = V^\perp$ is weakly closed, we must have $GF \in I$, so I is an ideal.

We next note that $z^n \exp(az) \in V$ if and only if a is a zero of order $\geqq n+1$ of each function $F \in I$. To see this, first observe that for $0 \leqq j \leqq n$, $z^j \exp(az)$ is a finite linear combination of translates of $z^n \exp(az)$. Hence, $z^n \exp(az) \in V$ if and only if $z^j \exp(az) \in V, 0 \leqq j \leqq n$. Thus, if $F \in I$, we have that $F^{(j)}(a) = \langle z^j \exp(az), F\rangle = 0$ for $0 \leqq j \leqq n$. Conversely, if $F^{(j)}(a) = 0$ for $0 \leqq j \leqq n$ and each $F \in I$, then $z^j \exp(az) \in I^\perp = V^{\perp\perp}$. But $V^{\perp\perp} = V$ [**22**, p. 126], which proves the assertion.

To complete the proof, let V_0 be the closed linear space of the $z^j \exp(az)$ which belong to V, and let $I_0 = V_0{}^\perp$. Then V_0 is translation invariant, so I_0 is an ideal. Further, $V_0 \subset V$ so $I_0 \supset I$. By the remark in the preceding paragraph, $Z(I_0) = Z(I)$; hence, $I_0 = I$ since $M^\wedge(K)$ is assumed to be an Hadamard algebra. Then $V_0 = V_0{}^{\perp\perp} = I_0{}^\perp = I_0{}^\perp = V^{\perp\perp} = V$, which proves the theorem.

Throughout the remainder of this section we shall be concerned with the definition and properties of the "canonical" series of a function $f \in V$. To define this series, the following lemma about linear functionals on V is needed.

Let V be a variety of $E(K)$, and let $I = V^\perp$. Let $Z = Z(I)$ be the sequence (with multiplicities) of common zeros of the functions in I. Denote by $a_0, a_1, a_2, \cdots$ the distinct elements of Z and by $q_0, q_1, q_2, \cdots$ the multiplicities of the a_n. As in the proof of Theorem 8.1, $z^j\exp(az) \in V$ if and only if for some n, $a = a_n$ and $0 \leqq j < q_n$.

LEMMA 8.1. *Suppose that the hypotheses of Theorem 8.1 are satisfied. Then given any $n = 0, 1, 2, \cdots$ and any j, with $0 \leqq j < q_n$, there exists a unique continuous linear functional on V, $L_{n,j}$ such that $L_{n,j}(z^k \exp(a_m z)) = 0$ unless $m = n$ and $j = k$, in which case $L_{n,j}(z^j\exp(a_n z)) = 1$.*

PROOF. The uniqueness follows from Theorem 8.1. For the existence part, let $F \in V^{\perp} = I$. Then $F^{(j)}(a_n) = 0$ for $0 \leqq j < q_n$, $n = 0, 1, 2, \cdots$. Fix n and j, and consider the functions $F(z)/(z - a_n)^k$, where $0 \leqq k <$ multiplicity of zero of F at a_n. It is routine to verify that a finite linear combination of these functions, say ϕ, can be found with the properties, $\phi^{(j)}(a_n) = 1$ and $\phi^{(k)}(a_m) = 0$, unless $m = n$ and $k = j$. Since $\phi \in M^\wedge(K)$, $L_{n,j}(f) = \langle f, \phi \rangle$ is a continuous linear functional on V. Since $\langle z^k\exp(az), \phi\rangle = \phi^{(k)}(a)$, $L_{n,j}$ is the desired linear functional.

The canonical series of $f \in V$ can now be defined.

DEFINITION 8.1. Suppose that the hypotheses of Theorem 8.1 hold. For an entire function $f \in V$, the canonical series of f is the formal series

$$\sum L_{n,j}(f) z^j \exp(a_n z),$$

where the $L_{n,j}$ are as in Lemma 8.1. If $d_{n,j} = L_{n,j}(f)$, we write $f \sim \sum d_{n,j} z^j \exp(a_n z)$.

We now prove the existence of a "summability method" for the canonical series. The outline of this proof is as follows. Let V be a variety, and $G \in V^{\perp} \subset M^\wedge(K)$. Let $\{G_k\}$ be the functions $Gh/P_k e_k$ of (iii) of the definition of Hadamard algebra, and let L_k be the continuous linear functional corresponding to $G_k(-z) \in M^\wedge(K)$ (i.e. $L_k(g) = \langle g(z), G_k(-z)\rangle$). If $f \in V$, it is true that $f * L_k$ (see §9 for the definition of $*$) is a partial sum of the canonical series of f, with appropriately weighted coefficients. Also, $\langle f * L_k, F\rangle = \langle f, G_k F\rangle$. Since $G_k F \to F$ for all $F \in M^\wedge(K)$, we thus have $\langle f * L_k, F\rangle \to \langle f, F\rangle$ for each $f \in V$. This is enough to imply that $f * L_k \to f$ in $E(K)$. Actually, the proof will be accomplished without ever mentioning the convolution operator, but this is only to simplify the technical details.

To implement this proof, we introduce the following notation. Let Z be as before (i.e. a sequence with multiplicities). A family of complex numbers $c = (c_{n,j})$, $0 \leqq j < q_n$, $n = 0, 1, 2, \cdots$ is called a Z-sequence. That is, if to each pair (n,j) with $0 \leqq j < q_n$, $n = 0, 1, 2, \cdots$, there corresponds a complex number $c_{n,j}$, then this "sequence" is called a Z-sequence. We may consider each Z-sequence as a function defined on Z. In analogy with the usual notation from sequence spaces, we denote by ϕ the collection of all Z-sequences $(c_{n,j})$ with $c_{n,j} = 0$ except for at most finitely many pairs (n,j).

DEFINITION 8.2. Let $c = (c_{n,j}) \in \phi$. If $f \in V$ and f has canonical series $\sum d_{n,j} z^j \exp(a_n z)$, then the c-mean of f is

$$S(f;c) = \sum_{n,j} \left(\left[d_{n,j} \sum_{m=0}^{j} \binom{j}{m} c_{n,m} z^{j-m}\right] \exp(a_n z)\right). \tag{8.1}$$

This sum is well defined, since $c_{n,j} = 0$ except for finitely many (n,j). Note that if all multiplicities q_n are equal to 1, then $S(f;c)$ takes the simple form

$$S(f;c) = \sum_n c_n d_n \exp(a_n z). \tag{8.2}$$

DEFINITION 8.3. By an effective summability method for V, we mean a sequence of elements $c^{(k)}$ of ϕ such that $\lim_{k\to+\infty} S(f;c^{(k)}) = f$ for each $f \in V$.

Under the hypotheses used throughout this section, such a summability method always exists.

THEOREM 8.2. *Suppose that* $p = 1$, *that* K3′, K4′ *hold, and that* $M^\wedge(K)$ *is an Hadamard algebra. Then there exists an effective summability method for each variety* V *of* $E(K)$.

PROOF. Since the closed and bounded subsets of $E(K)$ are compact, every sequence which converges for the weak topology on $E(K)$ (from $\langle E(K), M^\wedge(K)\rangle$) converges in $E(K)$. Thus, it is enough to find $c^{(k)} \in \phi$ such that $\lim_{k\to+\infty} \langle S(f;c^{(k)}), F\rangle = \langle f, F\rangle$ for each $f \in V$ and $F \in M(K)$. To obtain such sequences $c^{(k)}$, let $G \in I (= V^\perp)$, G not the zero function. Then since $M^\wedge(K)$ is an Hadamard algebra, there exists $H \in M^\wedge(K)$, polynomials $P_k(z)$, and units $e_k(z)$ such that $G_k F \to F$ for all $F \in M(K)$, where $G_k = GH/P_k e_k$. Let $c^{(k)} = (c^k_{n\,j})$ be defined by $c^k_{n,j} = G_k^{(j)}(a_n)$. Clearly $c^{(k)} \in \phi$. The theorem follows from the equation,

$$\langle S(f;c^{(k)}), F\rangle = \langle f, G_k F\rangle, \tag{8.3}$$

since $G_k F \to F$ in $M^\wedge(K)$.

To prove equation (8.3), we first prove the following statement: if $f \sim \sum d_{n,j} z^j \exp(a_n z)$, and if $F \in M^\wedge(K)$ has $F^{(j)}(a_n) = 0$ for all but finitely many (n,j), then $\langle f, F\rangle = \sum d_{n,j} F^{(j)}(a_n)$. This may be seen as follows. Let $L_{n,j}$ be the linear functionals of Lemma 8.1. An application of the Hahn-Banach theorem shows that there exist functions $\phi_{n,j} \in M^\wedge(K)$ such that $L_{n,j}(f) = \langle f, \phi_{n,j}\rangle$. Then if $F_1(z) = \sum F^{(j)}(a_n)\phi_{n,j}(z)$, we have $F_1 \in M^\wedge(K)$, since the sum is actually finite. Moreover, the properties of the $L_{n,j}$ and the equation $\langle z^k \exp(az), F\rangle = F^{(k)}(a)$ imply that $F - F_1$ vanishes on $Z(I)$ $(I = V^\perp)$. Since $M^\wedge(K)$ is an Hadamard algebra, $F - F_1 \in I$ (Theorem 6.1). Thus $\langle f, F\rangle = \langle f, F_1\rangle = \sum F^{(j)}(a_n)\langle f, \phi_{n,j}\rangle = \sum F^{(j)}(a_n)\, d_{n,j}$, as asserted.

The equation (8.3) may be seen as follows. Let $F \in M^\wedge(K)$, and apply the statement in the above paragraph to $G_k F$. Then

$$\begin{aligned}\langle S(f,c^{(k)}),F\rangle &= \sum_{n,j}\left[d_{n,j}\left(\sum_{m=0}^{j}\binom{j}{m}c^{k}_{m,j}\langle z^{j-m}\exp(az),F\rangle\right)\right]\\ &= \sum_{n,j}\left[d_{n,j}\left(\sum_{m=0}^{j}\binom{j}{m}G_k^{(j)}(a_n)F^{(j-m)}(a_n)\right)\right]\\ &= \sum_{n,j} d_{n,j}(G_kF)^{(j)}(a_n) = \langle f, G_kF\rangle,\end{aligned}$$

which completes the proof.

As a corollary to the above proof we have

COROLLARY. *A choice of the* $c^{(k)}$ *giving an effective summability method for* V *is*

$$c^{(k)} = (G_k^{(j)}(a_n)).$$

We remark that in many cases (e.g. example 1 of §6) the G_k may be given explicitly in terms of the Hadamard factorization of the function $G \in I$.

As mentioned earlier, we also have the following two corollaries.

COROLLARY. *Suppose that the hypotheses of Theorem* 8.1 *are satisfied, and that* V *is a variety containing the mean periodic function* f. *Then the canonical series of* f *is independent of* V. *That is,* V_1, V_2 *are varieties with* $f \in V_1 \cap V_2$, *and if* $f \sim \sum d_{n,j}z^j\exp(a_nz)$ *in* V_1, $f \sim \sum d'_{m,k}z^j\exp(a_mz)$ *in* V_2, *then the coefficients of* f *corresponding to the monomial-exponentials not belonging to* $V_1 \cap V_2$ *are zero, and the coefficients corresponding to the monomial-exponentials in* $V_1 \cap V_2$ *are the same in both series.*

PROOF. It is enough to prove that if V is a variety with $f \in V$, then the canonical series of f in V is the same as the canonical series of f in $T(f)$. Now $T(f) \subset V$, so the uniqueness part of Lemma 8.1 shows that the coefficients corresponding to the monomial-exponential functions in $T(f)$ are the same in both series. If $z^j\exp(a_nz)$ is in V and not in $T(f)$, then $d_{n,j} = L_{n,j}(f) = 0$, since f is a limit of sums of monomial exponentials from $T(f)$, and $L_{n,j}$ annihilates all of these functions.

COROLLARY. *Suppose that the hypotheses of Theorem* 8.1 *hold and that* f *is mean-periodic, with canonical series* $f \sim \sum d_{n,j}z^j\exp(a_nz)$. *Then* $z^j\exp(a_nz) \in T(f)$ *if and only if* $d_{n,k} \neq 0$ *for some* $k \geqq j$.

PROOF. If $d_{n,k} = 0$ for all $k \geqq j$, then from Theorem 8.2 it follows that f belongs to the variety generated by the monomial-exponentials in $T(f)$ excluding the $z^k\exp(a_nz)$, $k \geqq j$. If $z^j\exp(a_nz) \in T(f)$, then this variety is strictly smaller than $T(f)$, which is a contradiction. Hence, $d_{n,k} \neq 0$ for some $k \geqq j$. On the other hand, if $d_{n,k} \neq 0$, then the preceding corollary implies that $z^k \exp(a_n z) \in T(f)$; therefore also $z^j \exp(a_n z)$ for $0 \leqq j \leqq k$.

§9. **The convolution equation.** If $f \in E(K)$ and L is a continuous linear functional on $E(K)$, then the convolution of f and $L, f * L$, is defined by the equation $(f * L)(z) = L(f_z(t))$, where $f_z(t) = f(z - t)$. If $F(w) = L\hat{}(-w) = L(\exp(-w \cdot z)) = \sum(b_m/m!)w^m$, then $(f * L)(z) = (F(D)f)(z)$, where the right-hand side is equal to $\sum(b_m/m!)f^{(m)}(z)$. This may be seen as follows: $\sum(b_m/m!)f^{(m)}(z) = \sum(1/m!)L((-t)^m)f^{(m)}(z) = L(\sum(f^{(m)}(z)/m!)(-t)^m) = L(f(z - t)) = L(f_z(t))$. The above operations are justified for all values of z, since the power series of each function $f_z(t)$ converges to $f_z(t)$ in $E(K)$.

We note that the equations $f * L = F(D)f$ thus include the differential-difference operators of finite order with constant coefficients. Also, if $k \in K$ and μ is a bounded Borel measure on C^p, then $L(f) = \int fk\,d\mu$ is a continuous linear functional on $E(K)$, and $(f * L)(z) = \int f(z - t)k(t)\,d\mu(t)$. Thus, for example, if $E(K)$ is the entire functions of exponential type, the infinite order difference operator $f \to \sum_{n=0}^{\infty} f(z + n)/n!$ is a convolution operator.

A successful technique for proving existence theorems for the equation $\hat{f} * L = g$ is based on the following general results. Let E be a locally convex topological vector space, whose dual space is E^*. Let T be a linear mapping of E into E and let $T^*: E^* \to E^*$ be defined (if possible) by the equation $\langle Tf, F\rangle = \langle f, T^*F\rangle$ for $f \in E$, $F \in E^*$. (Here $\langle\ ,\ \rangle$ is the canonical bilinear form pairing E and E^*.) Then

(a) T^* is defined on all of E^* and continuous for the weak topology on E^* if and only if T is continuous for the weak topology on E.

(b) $T^{**} = T$ (if T^* exists).

(c) T^* is 1-1 if and only if T has dense range.

(d) If T is continuous for the weak topology on E, then T is continuous for the strong topology on E.

(e) If T is continuous for the weak topology on E, then T has closed range if and only if T^* is open.

(f) If T is continuous (for the original topology on E), then T is weakly continuous.

These results are standard, and can be found (even in a more general form) in ([**15**], Chapter 5, §21) or ([**22**], Chapter 4, §7).

In the case at hand, $E = E(K)$ and T is the convolution operator $Tf = f * L$, for some continuous linear functional L. The adjoint is the multiplication operator $T^*: M\hat{}(K) \to M\hat{}(K)$ defined by $T * G = FG$, where $F(w) = L\hat{}(-w)$. From (a), T is weakly continuous if and only if T^* exists and is weakly continuous. In this case, $FG = 0$ in $M\hat{}(K)$ implies $F = 0$ or $G = 0$. Thus, either $T^* = 0$, or else T^* is 1-1; in this latter case T has dense range by (c). Then by (e), T is onto if and only if $(T^*)^{-1}$ is continuous.

These general remarks are applied to prove the following theorem.

THEOREM 9.1. *Suppose that* K3′, K4′, K5 *hold. For* $f \in E(K)$, *define* Tf *to be the convolution operator* $Tf = f * L$, *or equivalently*, $(Tf)(z) = (F(D)f)(z)$, *where* $F(w) = \sum (b_m/m!)w^m = L(\exp(-w \cdot z)) \in M^\wedge(K)$. *Then*

(i) T *is a linear mapping of* $E(K)$ *into* $E(K)$;

(ii) *if* $E(K)$ *is barrelled, then* T *is a continuous linear mapping of* $E(K)$ *into* $E(K)$;

(iii) T *maps* $M^\wedge(K^*)$ *onto* $M^\wedge(K^*)$; *hence* T *maps* $E(K)$ *onto* $E(K)$ *if* K *is hypoconvex.*

PROOF. We start by considering, instead of T, the adjoint mapping. That is, define $T^*: M^\wedge(K) \to M^\wedge(K)$ by $T^*G = FG$, where F is as in the statement of the theorem. Since K4′, K5 hold, $M^\wedge(K)$ is an algebra and T^* is well defined. Moreover, by Theorem 3.2 (iii), T^* maps equicontinuous sets into equicontinuous sets.

We first prove (ii). Suppose that $M^\wedge(K)$ is equipped with the strong topology from the pairing $\langle E(K), M^\wedge(K)\rangle$. Since $E(K)$ is barrelled, the bounded subsets of $M^\wedge(K)$ coincide with the equicontinuous subsets of $M^\wedge(K)$ ([**22**], p. 141). Hence, T^* is a bounded linear mapping of $M^\wedge(K)$ into $M^\wedge(K)$. From the Lemma of §4, we have that $M^\wedge(K)$ is bornological. On a bornological space, every bounded linear operator is continuous ([**22**], p. 26) so T^* is continuous. Then by (f), T^* is continuous for the weak topology on $M^\wedge(K)$, so that $(T^*)^*: E(K) \to E(K)$ exists and is weakly continuous. Thus, by (d), $(T^*)^*$ is continuous for the strong topology on $E(K)$, which coincides with the usual topology on $E(K)$, since $E(K)$ is barrelled, ([**22**], p. 145). A short computation shows that $T^{**}f = F(D)f = Tf$, which proves (ii).

The assertion (i) can be deduced from (ii) as follows. For each $f \in E(K)$, define $\phi_n(r) = \phi_n(r; f) = nM(nr; f)$, and let $K(f)$ be the class determined by the ϕ_n as in Definition 4.2. Then $E(K(f))$ is barrelled, by Theorem 4.2. Further, $K(f) \subset K$, so that $F \in M^\wedge(K(f))$. Then (ii) applies to show that $T(f) \in E(K(f)) \subset E(K)$, which proves (i).

To prove (iii), we first observe that $(k^*(z))^{\frac{1}{2}} \in K^*$ for each $k^* \in K^*$, whenever K4′ holds. This is because $k(z)\,(k^*(w))^{\frac{1}{2}}|\exp(z \cdot w)| = (k^2(z)k^*(w)|\exp(2z \cdot w)|)^{\frac{1}{2}}$ $= (k^2(\frac{1}{2}(2z)\,k^*(w)|\exp(2z \cdot w)|))^{\frac{1}{2}}$. Now $k^2 = O(k)$, so $k^2 \in K$, and then $k^2(z/2) \in K$ by K4′, so the above expressions are bounded for each $k^* \in K^*$, as asserted.

From the above remark, it follows that multiplication is continuous in $E(K^*)$. Thus, if T^* is extended to be defined on all of $E(K^*)$, then T^* is continuous. Proceeding as before, $T^{**}: M^\wedge(K^*) \to M^\wedge(K^*)$ is well defined, continuous, and $T^{**} = T$.

However, $T^*: E(K^*) \to E(K^*)$ also has a continuous inverse. To see this, we observe first that K^* satisfies K4′, since K does. In the case $p = 1$, we then apply Lemma 7.2 to deduce that $(T^*)^{-1}$ is continuous. The case $p > 1$ follows in the same way, by noting that Lemma 7.2 is also valid in this case. To prove this fact, let f, g be entire functions of p complex variables with $f/g = h$ entire. For

$\mathscr{C}$ a single complex variable, consider $f(\mathscr{C}z_1,\cdots,\mathscr{C}z_p)/g(\mathscr{C}z_1,\cdots,\mathscr{C}z_p)=h(\mathscr{C}z_1,\cdots,\mathscr{C}z_p)$. Applying Lemma 7.2 to this function of a single variable $\mathscr{C}$, with $z_1,\cdots,z_p$ fixed, yields (with $|\mathscr{C}|=1$), $|h(z_1,\cdots,z_p)|\leqq|g(0,\cdots,0)|^{-3}\{1+M(2|z|,f)\}^3\{1+M(2|z|,g)\}^3$, which implies Lemma 7.2 for $p>1$.

Now, since $(T^*)^{-1}$ is continuous, $T=T^{**}$ is an onto mapping of $M^\wedge(K^*)$ by (e). The second part of (iii) follows from Theorem 5.2.

We wish to point out that in the case of a given equation $F(D)f=g$, the hypotheses of the above theorem may quite often be weakened (or even omitted) by using some special properties of the function F. For example, let $E(K)$ be as in example 8 or example 10 of §6, and let

$$(Tf)(z) = \sum_{n=0}^{\infty} \frac{f(z+n)}{n!} = (\exp(\exp(D))f(z)).$$

Thus, $F(z)=\exp(\exp(z))$, and it is easy to check that T^* is a 1-1, onto, continuous, open mapping (i.e. topological isomorphism) of $M^\wedge(K)$ onto itself. Hence, T is a topological isomorphism of $E(K)$ onto $E(K)$. In particular, (Example 8) given any function g of exponential type $<\tau$, there is a unique entire function f of exponential type $<\tau$ such that $g(z)=\sum_{n=0}^{\infty} f(z+n)/n!$.

References

1. R. P. Boas, *Entire functions*, Academic Press, New York, 1954.

2. D. G. Dickson, *Expansion in series of solutions of linear difference-differential and infinite order differential equations with constant coefficients*, Mem. Amer. Math. Soc., No. 23 (1957).

3. P. Dienes, *The Taylor series*, Clarendon Press, Oxford, 1931.

4. L. Ehrenpreis, *Mean periodic functions*. I, Amer. J. Math. **77** (1955), 293–328.

5. ———, *Solutions of some problems of division*. II, Amer. J. Math. **77** (1955), 286–292.

6. ———, *Solutions of some problems of division*. III, Amer. J. Math. **78** (1956), 685–715.

7. ———, *Analytically uniform spaces and some applications*, Trans. Amer. Math. Soc. **101** (1961), 52–74.

8. ———, *Fourier Analysis in Several Complex Variables*, (to appear).

9. W. Fenchel, *Convex cones, sets and functions*, Princeton University notes, 1951.

10. I. M. Gel'fand and G. E. Šilov, *Fourier transforms of rapidly increasing functions and questions of the uniqueness of Cauchy's problem*, Amer. Math. Soc. Transl. (2) **5** (1957), 221–274.

11. W. K. Hayman, *Meromorphic functions*, Oxford University Press, London, 1964.

12. K. Hoffman, *Banach spaces of analytic functions*, Prentice-Hall, Englewood Cliffs, N. J., 1962.

13. L. Hörmander, *La transformation de Legendre et la théorème de Paley-Wiener*, C. R. Acad. Sci. Paris **240** (1955), 392–395.

14. ———, *Linear partial differential operators*, Springer-Verlag and Academic Press, New York, 1963.

15. J. L. Kelley, I. Namioka, et al., *Linear topological spaces*, Van Nostrand, Princeton New Jersey, 1963.

16. B. Malgrange, *Existence et approximation des solutions des équations aux dérivées partielles et des équations de convolution*, Ann. Inst. Fourier, Grenoble **6** (1955–1956), 271–355.

17. H. Muggli, *Differentialgleichungen unendlich hoher Ordnung mit konstanten Koeffizienten*, Comment. Math. Helv. **11** (1938–1939), 151–179.

18. ———, *Differentialgleichungen unendlich hoher Ordnung*, Comment. Math. Helv. **14** (1942), 381–393.

19. L. A. Rubel, *A Fourier series method for entire functions*, Duke Math. J. **30** (1963), 437–442.

20. L. A. Rubel and B. A. Taylor, *A Fourier series method for meromorphic and entire functions* (to appear).

21. ———, *Functional analysis proofs of two theorems in function theory* (to appear).

22. H. Schaefer, *Topological vector spaces*, Macmillan, New York, 1966.

23. L. Schwartz, *Théorie des fonctions moyenne-périodiques*, Ann. of Math. **48** (1947), 857–929.

24. H. S. Shapiro, *The expansion of mean periodic functions in series of exponentials*, Comm. Pure Appl. Math. **11** 1–21 (1958).

25. A. Zygmund, *Trigonometric series*, Vol. I, University Press, Cambridge, 1959.

UNIVERSITY OF ILLINOIS
URBANA, ILLINOIS
UNIVERSITY OF MICHIGAN
ANN ARBOR, MICHIGAN

The Fields of Quotients of Some Rings of Entire Functions

B. A. TAYLOR

1. In [9] it was shown that every meromorphic function (in one complex variable) of a reasonably general rate of growth is the quotient of two entire functions of the same rate of growth.

At this conference, Professor L. Rubel asked if the same result holds true for meromorphic functions of several complex variables. In this note we show that this is the case. Our proof is essentially the same as that given in [9] for the one variable problem. We introduce a characteristic function by averaging certain functions over the distinguished boundary of polydiscs (exactly as for $n = 1$), and apply an extension [7] of the Fourier series method of [9] to prove the theorem.

We point out that Professor W. Stoll has proved in [10] that every meromorphic function in C^n of order $\leqq \rho$, finite type (resp. maximal type, minimal type) is the quotient of two entire functions of the same rate of growth. R. Kujala, in his thesis, has since extended this result to more general classes of meromorphic functions determined by rates of growth that are functions of $r = \| z \| = (\Sigma | z_i |^2)^{1/2}$.

2. For the definition of the characteristic function, we follow essentially the same procedure as Arsove [1]. For brevity the definition is made only general enough for the problem considered here, and several known facts about meromorphic functions are used to simplify some necessary constructions. It seems, however, that the definition can be extended to the class of functions which are differences of plurisubharmonic functions (or even n-subharmonic functions), by using methods analogous to those of [1].

Let f be a function meromorphic on $C^n = C \times \cdots \times C$ (n-times), where C denotes the complex numbers. By a *canonical representation* for f, we mean a pair (g, h) of entire functions such that $f = g/h$ and g, h are locally relatively prime (i.e., at each $z \in C^n$ where $g(z) = h(z) = 0$, the functions g, h are relatively prime in the ring of germs of functions analytic at z). Such representations always exist [2, p. 251]. Then $\log|f| = \log|g| - \log|h|$, and $\log|g|$, $\log|h|$ are plurisubharmonic, so that $\alpha = \max(\log|g|, \log|h|)$ is plurisubharmonic. Let $\bar{r} = (r_1, \cdots, r_n)$, where $r_j > 0$, $1 \leqq j \leqq n$, and

(1) $$L_n(\bar{r},\alpha) = \left(\frac{1}{2\pi}\right)^n \int_{-\pi}^{\pi} \cdots \int_{-\pi}^{\pi} \alpha(r_1 \exp[i\theta_1], \cdots, r_n \exp[i\theta_n])\, d\theta_1 \cdots d\theta_n .$$

Since α is plurisubharmonic and $r_j > 0$ $(1 \leqq j \leqq n)$, $L_n(\bar{r},\alpha)$ exists. We then define the characteristic function of f to be $T(\bar{r};f) = L_n(\bar{r};\alpha)$. Thus, $T(\bar{r};f)$ depends on (g,h). However, if (g_1,h_1) is another canonical representation, then we have $g = g_1 \exp(F), h = h_1 \exp(F)$ for some entire function F. Thus, $\log|g| - \log|g_1| = \log|h| - \log|h_1| = \operatorname{Re} F$ is pluriharmonic and $\max(\log|g_1|, \log|h_1|) = \alpha - \operatorname{Re} F$, so that the characteristic function of f defined with respect to (g_1,h_1) differs from $L_n(\bar{r},\alpha)$ only by a constant. This discrepancy causes us no difficulty.

We wish to derive a representation for $T(\bar{r},f)$ which takes into account the growth of f and the zeros and poles of f. For this, let $E_1 = \{w = (z_2, \cdots, z_n)$: the function $z_1 \to f(z_1,w)$ is not identically 0 or $\infty\}$, and $F_1 = \{w = (z_2,\cdots,z_n)$: $g(z_1,w) = h(z_1,w) = 0$ for some $z_1\}$. If $w \notin E_1 \cup F_1$, then let $n(r,w) = n(r,w;f) = n(r,w;1/h)$ and $N(r,w)$ denote the usual counting function and its average; that is, $n(r,w)$ is the number of poles of the function of one variable, $z_1 \to f(z_1,w)$ in $|z_1| \leqq r$, and $N(r,w) = \int_0^r [n(t,w) - n(0,w)]t^{-1}\, dt + n(0,w) \log r$. Since $w \notin F_1$, $n(r,w)$ is also the number of zeros of $h(\cdot,w)$ in $|z_1| \leqq r$. Now E_1 has zero (Lebesgue) measure in C^{n-1}, since f is not identically zero. Further, it is a consequence of the fact that g,h are locally relatively prime (and a well-known lemma [2, p. 90]) that F_1 has measure zero in C^{n-1}. Consequently, $E_1 \cup F_1$ meets the distinguished boundary of almost all polydiscs $r_j = |z_j|$, $2 \leqq j \leqq n$, in a set of measure zero ($d\theta_2 \cdots d\theta_n$ measure). Therefore, if $(r_2,\cdots,r_n)$ is not in this exceptional set, and if p is the least nonnegative integer for which $\partial^p h(0,w)/\partial z_1^p$ is not identically zero on C^{n-1}, then we have from Jensen's formula that

(2) $$\frac{1}{2\pi}\int_{-\pi}^{\pi} \log|h(re^{i\theta},w)|\, d\theta = N(r,w) + \log\left|\frac{\partial^p h(0,w)}{p!\,\partial z_1^p}\right|$$

for almost all points w on the distinguished boundary of the polydisc $|z_j| = r_j$, $2 \leqq j \leqq p$. (Note that $p = p(w)$; however, $p(w)$ is constant for almost all $w \in C^{n-1}$.) Therefore, taking into account that $\max(\log|g|, \log|h|) = \log^+|f| + \log|h|$, we have

(3) $$\begin{aligned} T(\bar{r};f) = {} & L_n(\bar{r};\log^+|f|) + L_{n-1}((r_2,\cdots,r_n);N(r_1,w)) \\ & + L_{n-1}\left((r_2,\cdots,r_n);\log\left|\frac{\partial^p h(0,w)}{p!\,\partial z_1^p}\right|\right), \end{aligned}$$

for almost all $(r_2,\cdots,r_n)$.

We remark that by iterating this process, a representation for $T(\bar{r};f)$ can be derived which does not depend on the global representation $f = g/h$, but only on $\log^+|f|$ and the behavior of f near the origin.

3. DEFINITION. Let $\lambda(\bar{r}) = \lambda(r_1, \cdots, r_n)$ be a positive continuous function defined for $r_j \geqq 0$, $1 \leqq j \leqq n$, and which is nondecreasing in each variable. The meromorphic function f is said to be of finite λ-type if $T(\bar{r};f) \leqq A\lambda(B\bar{r})$ for some constants $A, B > 0$ and all $r_j \geqq 1$, $1 \leqq j \leqq n$. The class of all meromorphic functions of finite λ-type is denoted by Λ.

We call attention to the fact that the inequality of the above definition is required to hold only for $r_j \geqq 1$, $1 \leqq j \leqq n$. For many growth functions λ, such an inequality implies a similar inequality for all $r_j \geqq 0$, $1 \leqq j \leqq n$ (e.g., any function of $\| r \| = (r_1^2 + \cdots + r_n^2)^{1/2}$). However, there are growth functions for which this is not the case (e.g., $\lambda(r_1, r_2) = \exp[(r_1 + 1)^{(r_2+1)}]$). Throughout the rest of this paper an inequality of the form $\phi(\bar{r}) \leqq A\lambda(B\bar{r})$ will be assumed to hold only for $r_j \geqq 1$, unless explicitly stated to the contrary.

From the inequalities

$$(4) \qquad L_n(\bar{r}; u^+) \leqq \max\{u^+(z) : |z_j| = r_j\} \leqq AL_n(2\bar{r}; u^+)$$

for a plurisubharmonic function u, it follows that an entire function $f \in \Lambda$ if and only if $\log M(\bar{r};f) \leqq A\lambda(B\bar{r})$, where $M(r;f) = \max\{|f(z)| : |z_j| = r_j, 1 \leqq j \leqq n\}$. The inequalities in (4) can be deduced from the corresponding fact for $n = 1$ (even $A \leqq 3^n$), which may be found in [3, p. 18].

We denote the set of entire functions of finite λ-type by Λ_E.

PROPOSITION. *Λ is a field, and Λ_E is a subring of Λ.*

PROOF. This follows from the inequalities
(i) $T(\bar{r}; f + g) \leqq T(\bar{r};f) + T(\bar{r};g) + O(1)$,
(ii) $T(\bar{r}; fg) \leqq T(\bar{r};f) + T(\bar{r};g) + O(1)$,
(iii) $T(\bar{r}; 1/f) = T(\bar{r};f) + O(1)$.
We shall prove only (i), the proof of (ii) being similar, and (iii) is obvious. Thus, write $f + g = G/H$, $f = G_1/H_1$, and $g = G_2/H_2$, where (G, H) etc. are canonical representations of $f + g$, etc. Then $f + g = (H_2G_1 + H_1G_2)/(H_1H_2)$, so that there exists an entire function F such that $H_1H_2 = FH$. If $\alpha_i = \max(\log|G_i|, \log|H_i|)$, $i = 1,2$, then a short computation yields $\max(\log|G|, \log|H|) \leqq \alpha_1 + \alpha_2 + \log 2 - \log|F|$ and (i) follows from this.

Our problem may now be formulated as, "When is Λ the field of quotients of Λ_E?"

THEOREM. *Suppose that $\lambda(\bar{r})$ is slowly increasing in each variable, in the sense that $\lambda(r_1, \cdots, 2r_j, \cdots, r_n) \leqq A_j\lambda(r_1, \cdots, r_n)$ for some $A_j > 0$ $(1 \leqq j \leqq n)$. Then $\Lambda = \Lambda_E/\Lambda_E$.*

We proceed as in the one variable case by first writing $f = g/h$, where g, h are, in some sense, small, and then a suitable symmetrization will complete the proof. Before giving the proof, it is convenient to collect some lemmas.

LEMMA 1. *Suppose that* $f \in \Lambda$ *and let* (g, h) *be a canonical representation of* f. *Let* $\sigma = \Delta(\log|h|)/2\pi$, *where* $\Delta = 4\sum_{j=1}^{n} \partial^2/\partial z_j \partial \bar{z}_j$ *is the Laplacian, the derivatives being taken in the sense of distributions. Then* σ *is a positive measure and*

$$\sigma(\rho) = \int_{\|a\| \leqq \rho} d\sigma(a) \leqq (\rho + 1)^{2n-2} (A\lambda(B\rho, \cdots, B\rho))$$

for some $A, B > 0$. (*Here* $\| a \|^2 = |a_1|^2 + \cdots + |a_n|^2$.)

PROOF. Since $\log|h|$ is plurisubharmonic and therefore subharmonic on R^{2n}, σ is a positive measure. Moreover, it is a theorem of Poincaré (cf. [4], [5]) that σ is surface area on the variety of f. More precisely, $\sigma = \sum_{j=1}^{n} \sigma_j$, where the σ_j are measures defined as follows. For definiteness, suppose $j = 1$. For fixed $w = (z_2, \cdots, z_n) \in C^{n-1}$, let $\mu = \mu(\ \ , w)$ be the measure on C which consists of point masses placed at the zeros of the function $z_1 \to h(z_1, w)$, the weights being equal to the multiplicity of the zero. (Actually

$$\mu(\ \ , w) = (4/2\pi)(\partial^2/\partial z_1 \partial \bar{z}_1)(\log| h(z_1, w) |).)$$

Then $\mu(\ \ , w)$ is defined for almost all $w \in C^{n-1}$. Let $V^{(n-1)}$ denote Lebesgue measure on C^{n-1}, and let ϕ be a C^{∞} function on C^n with compact support. Then

$$\int_{C^n} \phi(z)\, d\sigma_1(z) = \int_{C^{n-1}} \int_{z_1} \phi(z_1, w)\, d\mu(z_1, w)\, dV^{(n-1)}(w). \tag{5}$$

This formula for σ can also be verified directly by applying Fubini's theorem and Green's identities.

Thus, it is enough to prove the lemma for the measures σ_j. For definiteness, again suppose $j = 1$. Let $D^{(n-1)}(\rho)$ denote the polydisc $|z_2| \leqq \rho, \cdots, |z_n| \leqq \rho$. Then for almost all $w \in D^{(n-1)}(\rho)$,

$$N(er, w) \geqq \int_r^{er} [n(t, w) - n(0, w)]\, t^{-1}\, dt + n(0, w) \log(er).$$

Consequently,

$$N(er, w) \geqq n(r, w) \quad \text{for } r \geqq 1. \tag{6}$$

From this inequality, (5) and (3), we thus have for $\rho \geqq 1$,

$$\sigma_1(\rho) \leqq \int_{D^{((n-1)}\rho)} n(\rho, w) dV^{(n-1)}(w) \leqq \int_{D^{((n-1)}\rho)} N(e\rho, w) dV^{(n-1)}(w)$$

$$= O(\rho^{2n-2} T(\bar{\rho}\,; f)),$$

where $\bar{\rho} = (\rho, \cdots, \rho)$. This proves the lemma.

Our next lemma is the many variable analogue of Theorem 4.6 of [9]. In the form below it is implicit in the work of Noverraz [7], where he extended the results of [9] to functions of several variables.

Let $\bar{k} = (k_1, \cdots, k_n)$ be an n-tuple of integers. Denote by $c(\bar{k}, \bar{r}) = c(\bar{k}, \bar{r}, \log|h|)$ the $\bar{k}$th coefficient of $\log|h|$ considered as a function on $|z_j| = r_j$, $1 \leqq j \leqq n$. That is

$$c(\bar{k}, \bar{r}) = \left(\frac{1}{2\pi}\right)^n \int_{-\pi}^{\pi} \cdots \int_{-\pi}^{\pi} (\log|h(r_1 \exp[i\theta_1], \cdots, r_n \exp[i\theta_n])|) \exp(-i\langle \bar{k}, \bar{\theta} \rangle) d\bar{\theta}$$

where $\bar{\theta} = (\theta_1, \cdots, \theta_n)$, $d\bar{\theta} = d\theta_1, \cdots, d\theta_n$, and $\langle \bar{k}, \bar{\theta} \rangle = k_1\theta_1 + \cdots + k_n\theta_n$.

LEMMA 2. *Let h be entire on C^n, and suppose that*

$$|c(\bar{k}, \bar{r})| \leqq A\lambda(Br) \quad \text{for all } \bar{k} \quad \text{and} \quad r_j \geqq 1, \quad 1 \leqq j \leqq n. \tag{7}$$

Then

$$|c(\bar{k}, \bar{r})| \leqq A'\lambda(B'\bar{r}) \left[\prod_{j=1}^{n} (|k_j| + 1) \right]^{-1} \quad \text{for all } \bar{k} \tag{8}$$

and $r_j \geqq 1$, $1 \leqq j \leqq n$. If the last inequality holds, then $h \in \Lambda$.

PROOF. To establish that (7) implies (8) it is sufficient to prove that for each j, $1 \leqq j \leqq n$, (7) implies $|c(\bar{k}, \bar{r})| \leqq A'\lambda(B'\bar{r})(|k_j| + 1)^{-1}$, where A', B' are constant multiples of A, B, the constant depending only on h and λ. However, this follows in exactly the same way as the corresponding part of Theorem 4.6 of [9], except that the estimates used there for $n(r)$, $N(r)$, must now be replaced by the analogous estimates for $L_{n-1}((r_2, \cdots, r_n); n(r_1, w))$, $L_{n-1}((r_2, \cdots, r_n); N(r_1, w))$, etc. The last assertion is a part of Theorem 6 of [7]. It follows from the Schwarz inequality and Parseval's equality. That is, $T(\bar{r}; h) = L_n(\bar{r}; \log^+|h|) \leqq L_n(\bar{r}; (\log|h|)^2) = (\sum |c(\bar{k}, r)|^2)^{1/2} = O(\lambda(O(\bar{r})))$.

PROOF OF THE THEOREM. Let $f \in \Lambda$ and let (g_1, h_1) be a canonical representation of f. That is, $f = g_1/h_1$, and g_1, h_1 are (locally) relatively prime entire functions. Let $\sigma = (2\pi)^{-1}\Delta(\log|h_1|)$ be as in Lemma 1. It follows from the hypothesis that λ is slowly increasing that $\lambda(\rho, \cdots, \rho) = O(\rho^q)$ for some integer $q > 0$. Therefore, from Lemma 1, we have $\sigma(\rho) = O(\rho^q)$ for some $q > 0$. Then from Theorem 5 of [6, p. 394], there exists an entire function h such that h/h_1 and h_1/h are entire, and

$$\log|h(z)| = O(\|z\|^q), \quad \text{where} \quad \|z\|^2 = \Sigma |z_i|^2 \tag{9}$$

for some (possibly larger) $q > 0$.

Since λ is slowly increasing in each variable, there exists an integer $q > 0$ such that

$$(10) \qquad \int_{r_j}^{+\infty} [\lambda(r_1, \cdots, r_{j-1}, t, r_{j+1}, \cdots, r_n)] t^{-(k+1)} dt \leqq A\lambda(r_1, \cdots, r_n)$$

for some $A > 0$, all $k \geqq q$ and all $1 \leqq j \leqq n$ (cf. Proposition 3.5 of [9]). Choose an integer q so large that both (9) and (10) hold, and then define $H(z) = \prod_{j=1}^{q} h(w^j z)$, where w is a primitive $(q+1)$th root of unity. We claim that $H \in \Lambda_E$. This will prove the theorem, since then $G = Hf$ determines a suitable representation of f.

To prove that $H \in \Lambda_E$, we shall apply Lemma 2. That is, we shall prove (7). Now $c(\bar{0}, \bar{r})$ is $(q+1)$ times the $\bar{0}$th coefficient of $\log|h|$, which is equal to $L_n(\bar{r}; \log|h|) \leqq T(\bar{r}; f) + O(1)$, since (hf, h) is again a canonical representation of f. Thus, the appropriate inequality holds in this case. If $\bar{k} = (k_1, \cdots, k_n)$ where $0 < |k_j| \leqq q$ for some $1 \leqq j \leqq n$, then the symmetry in H implies that $c(\bar{k}, \bar{r}) = 0$. Thus, we have only to prove that (7) holds when $\bar{k} = (k_1, \cdots, k_n)$ with every $|k_j| > q$. For the sake of definiteness, we fix our attention on $k = k_1$. It is no loss of generality to assume $k > q > 0$, since $c(\bar{k}, \bar{r}) = \overline{c(-\bar{k}, \bar{r})}$.

Now the coefficients of $\log|H|$ are at most $(q+1)$ times as large as the coefficients of $\log|h|$. Therefore, in proving (7) we may assume that $c(\bar{k}, \bar{r})$ is the coefficient of $\log|h|$. Write $\bar{r} = (r, \bar{\rho})$, where $\bar{\rho} = (r_2, \cdots, r_n)$. Then $c(\bar{k}, \bar{r})$ is the integral over $|z_j| = r_j$ of the function of $w = (r_2 e^{i\theta_2}, \cdots, r_n e^{i\theta_n})$,

$$d(k, r, w) \exp(-ik_2\theta_2 - \cdots - ik_n\theta_n),$$

where

$$d(k, r, w) = \frac{1}{2\pi} \int_{-\pi}^{\pi} (\log|h(re^{i\theta}, w)|) e^{-ik\theta} d\theta.$$

Therefore, it is enough to prove that $L_{n-1}(\bar{\rho}; |d(k, r, w)|) = O(\lambda(O(\bar{r})))$.

For this, write $\{\xi_m(w)\}$ for the zeros of the entire function of one complex variable $\phi(z_1) = h(z_1, w)$, and let $\log(\partial^p h(z_1, w)/p!\,\partial z_1^p) = \sum_{m=0}^{\infty} \alpha_m(w) z_1^m$ near $z_1 = 0$. Here p is as in (2), and the $\xi_m(w)$, $\alpha_m(w)$ are defined for almost all $w \in C^{n-1}$. From 4.2 of [9], we have $d(k, r, w) = a(k, r, w) + b(k, r, w)$, where

$$b(k, r, w) = -\frac{r^{-k}}{2k} \sum_{0 < |\xi_m(w)| \leqq r} \overline{(\xi_m(w))^k},$$

$$a(k, r, w) = \frac{1}{2}\alpha_k(w) r^k + \frac{r^k}{2k} \sum_{0 < |\xi_m(w)| \leqq r} (\xi_m(w))^{-k}.$$

Now $|b(k, r, w)| \leqq n(r, w)/2k$. Also, for $k > q$, it follows exactly as in 1.14 of [9] (see also [8, line 12, p. 440]) that

$$|a(k, r, w)| = \left|\frac{-r^k}{k} \sum_{0 < |\xi_m(w)| \leqq r} (\xi_m(w))^{-k}\right|$$

$$\leqq \frac{r^k}{k} \int_r^{+\infty} t^{-k} dn(t, w),$$

since $\phi(z_1) = h(z_1, w)$ is of order at most q, and $k > q$. After an integration by parts, we have upon applying (6), that $|a(k, r, w)| \leqq N(er, w) + \int_r^{+\infty} N(et, w)t^{-(k+1)}dt$ $(r \geqq 1)$. Therefore,

$$L_{n-1}(\bar{\rho}; |d(r, k, w)|) \leqq 2L_{n-1}(\bar{\rho}; N(er, w)) + \int_r^{+\infty} L_{n-1}(\bar{k}; N(et, w))t^{-(k+1)}\,dt.$$

The first term is $O(\lambda(O(\bar{r})))$ by (3). Also from (3), the last term does not exceed a constant multiple of $\int_r^{+\infty} \lambda(Bt, B\bar{\rho})t^{-(k+1)}dt$. Then it follows from (10) that this term is also $O(\lambda(O(\bar{r})))$, which completes the proof.

References

1. Maynard G. Arsove, *Functions representable as differences of subharmonic functions*, Trans. Amer. Math. Soc. **75** (1953), 327–365.

2. Robert C. Gunning and Hugo Rossi, *Analytic functions of several complex variables*, Prentice-Hall, Englewood Cliffs, N. J., 1965.

3. W. K. Hayman, *Meromorphic functions*, Oxford Univ. Press, London, 1964.

4. Pierre Lelong, *Les fonctions plurisousharmoniques*, Ann. Sci. Ecole Norm. Sup. **69** (1945), 301–338.

5. ———, *Propriétés métriques des variétés analytiques complexes défines par une équation*, Ann Sci. Ecole Norm. Sup. **67** (1950), 393–419.

6. ———, *Fonctions entières (n variables) et fonctions plurisousharmonic d'ordre fini dans C^n*, J. Analyse Math. **12** (1964), 365–407.

7. Philippe Noverraz, *Extension d'une méthode de séries de Fourier aux fonctions sousharmoniques et plurisousharmoniques*, Séminaire P. Lelong 6e année: 1965/66, Analyse n 3.

8. L. A. Rubel, *A Fourier series method for entire functions*, Duke Math. J. **30** (1963), 437–442.

9. L. A. Rubel and B. A. Taylor, *A Fourier series method for meromorphic and entire functions* (to appear).

10. Wilhelm Stoll, *About entire and meromorphic functions of exponential type*, these Proceedings, 392–430.

University of Michigan
Ann Arbor, Michigan

On the Distribution of Asymptotically Neutral Families

MAYNARD THOMPSON

1. **Introduction.** Let D be a bounded Jordan domain in the complex z-plane whose boundary is denoted by Γ. Families of sequences of points on Γ have been useful in problems of approximation and interpolation to functions analytic or harmonic in D (see [2], [5, Chapter VII]). If $\{S^{(n)}\}_{n=1}^{\infty}$, $S^{(n)} = \{z_1^{(n)}, z_2^{(n)}, \cdots, z_n^{(n)}\}$, $z_k^{(n)} \in \Gamma$ for $k = 1, 2, \cdots, n$, and $n = 1, 2, \cdots$, is such a family, then it is frequently necessary to know the relation between the asymptotic distribution of $\{S^{(n)}\}$ on Γ and the convergence properties of the associated polynomials

$$\omega_n(z) = \prod_{k=1}^{n} (z - z_k^{(n)}).$$

For example, if $\{S^{(n)}\}$ is uniformly distributed on Γ in the sense of Weyl, and c is the logarithmic capacity of Γ, then

$$\lim_{n \to \infty} |\omega_n(z)|^{1/n} = c,$$

uniformly on closed subsets of D. It is known [1] that for rectifiable boundary Γ and suitable choice of $S^{(n)}$ one has the stronger convergence

$$\lim_{n \to \infty} \frac{\omega_n(z)}{-c^n} = 1, \tag{1}$$

uniformly on closed subsets of D. Families of sequences for which (1) holds have been called *asymptotically neutral* [3]. The problem of giving necessary and sufficient conditions on a family in order that it be asymptotically neutral is open. The answer apparently depends on both the overall and fine structure of the family. In the special case of a Jordan domain with rectifiable boundary the results given here provide a sufficient condition which among conditions of its kind is best possible. The theorem is based on a simple fact concerning the degree of approximation of Riemann integrals by sums which does not seem to appear in the literature and may be of independent interest.

2. **Approximation of Riemann integrals by sums.** The theorems obtained here are generalizations of known results on the degree of convergence of Riemann sums to the integral of a continuous function.

As a technical convenience we restrict our consideration to the interval $[0,1]$. We will be dealing with families of sequences of points on $[0,1]$ and we always suppose that the families $\{t_k^{(n)}\}_{k=1}^n$, $n=1,2,\cdots$, consist of distinct positive terms and are such that each sequence is ordered according to its lower index $0<t_1^{(n)}<t_2^{(n)}<\cdots<t_n^{(n)}\leqq 1$, $n=1,2,\cdots$. It is a useful convention to adjoin the point 0 to each sequence; we set $t_0^{(n)}=0$ for all n. We refer to such families simply as $\{t_k^{(n)}\}$.

For the applications we have in mind it is appropriate to consider approximating the integral of an absolutely continuous function f. In general we will be measuring the difference

$$E_f^{(n)} = \sum_{k=1}^{n} f(t_k^{(n)})(t_k^{(n)} - t_{k-1}^{(n)}) - \int_0^1 f(t)\,dt.$$

In order to obtain convergence of the desired rapidity the subdivisions defined by $\{t_k^{(n)}\}$ must be quite uniform. We introduce two measures of this uniformity. Set

$$A^{(n)} = \sum_{k=1}^{n} \left| t_k^{(n)} - t_{k-1}^{(n)} - \frac{1}{n} \right|,$$

and

$$M^{(n)} = \underset{1\leqq k\leqq n}{\text{maximum}} \left| t_k^{(n)} - t_{k-1}^{(n)} - \frac{1}{n} \right|.$$

THEOREM 1. *Suppose* $\{t_k^{(n)}\}$ *satisfies* $t_n^{(n)}=1$ *for all* n,

$$M^{(n)} = O(n^{-1}), \tag{2}$$

and f *is any complex valued absolutely continuous function on* $[0,1]$. *Then*

$$E_f^{(n)} = O(n^{-1}). \tag{3}$$

Moreover, for every α, $0<\alpha<1$, *there is a family* $\{t_k^{(n)}\}$, $t_n^{(n)}=1$ *for all* n, *and an absolutely continuous function* f *such that*

$$M^{(n)} = O(n^{-\alpha}) \tag{4}$$

and $nE_f^{(n)}$ *is unbounded as* n *increases to infinity.*

THEOREM 2. *Suppose* $\{t_k^{(n)}\}$ *satisfies* $t_n^{(n)}=1$ *for all* n, *and either*

$$M^{(n)} = o(n^{-1}), \tag{2'}$$

or

$$A^{(n)} = O(n^{-1}). \tag{2''}$$

Then for any absolutely continuous function f *on* $[0,1]$

$$\lim_{n\to\infty} nE_f^{(n)} = \tfrac{1}{2}\{f(1)-f(0)\}. \tag{5}$$

Moreover, examples show that (2) *is not sufficient for* (5).

PROOFS. It is sufficient to give the proofs for a real valued function f. Writing t_k for $t_k^{(n)}$ and Δ_k for $t_k^{(n)} - t_{k-1}^{(n)}$, $k = 1, 2, \cdots, n$, we have

$$f(t_k) - (\Delta_k)^{-1} \int_{t_{k-1}}^{t_k} f(t)\, dt = \int_{t_{k-1}}^{t_k'} f'(t)\, dt, \tag{6}$$

$t_{k-1} < t_k' < t_k$, $k = 1, 2, \cdots, n$. This follows by observing that both sides of (6) are equal to

$$(\Delta_k)^{-1} \int_{t_{k-1}}^{t_k} (t - t_{k-1}) f'(t)\, dt,$$

the left side by an integration by parts, the right side by a mean value theorem for integrals. Therefore

$$\sum_{k=1}^{n} \left| f(t_k) - (\Delta_k)^{-1} \int_{t_{k-1}}^{t_k} f(t)\, dt \right| \leqq \int_0^1 |f'(t)|\, dt = C,$$

and this together with (2) yields

$$\sum_{k=1}^{n} \left| \Delta_k - \frac{1}{n} \right| \left| f(t_k) - (\Delta_k)^{-1} \int_{t_{k-1}}^{t_k} f(t)\, dt \right| = CO(n^{-1}),$$

which gives (3).

In [4] it is shown that for any absolutely continuous function f

$$\lim_{n \to \infty} \left\{ \sum_{k=1}^{n} (\Delta_k)^{-1} \int_{t_{k-1}}^{t_k} (t - t_{k-1}) f'(t)\, dt \right\} = \tfrac{1}{2}\{f(1) - f(0)\}.$$

Using this with (2′) (or (2″)) in the above proof gives (5).

We turn now to the second half of Theorem 1. For a positive real number β let $[\beta]$ denote the integer part of β. Given α, $0 < \alpha < 1$, define $\{t_k^{(n)}\}$ as follows:

$$t_1^{(n)} = [n^{1-\alpha}]/(n + [n^{1-\alpha}]),$$

$$t_k^{(n)} = ((n-1)[n^{1-\alpha}] + n(k-1))/(n-1)(n + [n^{1-\alpha}]), \qquad k = 2, 3, \cdots, n$$

for $n = 1, 2, \cdots$. It is easily verified that (4) is satisfied, indeed it is even true that $A^{(n)} = O(n^{-\alpha})$. Now taking f to be $f(t) = t^{(1-\alpha)/2\alpha}$, $0 \leqq t \leqq 1$, a routine, computation verifies the assertion.

The example for Theorem 2 is given in the proof of Theorem 3.

Next we consider approximating sums for Riemann integrals where the weights correspond to a subdivision of the interval into equal parts but the values of the function need not be computed on equally spaced points. For this discussion it is convenient to define

$$R_f^{(n)} = \sum_{k=1}^{n} f(t_k^{(n)})\, n^{-1} - \int_0^1 f(t)\, dt\, .$$

THEOREM 3. *Suppose* $\{t_k^{(n)}\}$ *satisfies either*

$$A^{(n)} = o(n^{-1}) \tag{8}$$

or

$$A^{(n)} = O(n^{-1}), \tag{8'}$$

and f *is any complex valued absolutely continuous function on* $[0,1]$. *Then respectively*

$$\lim_{n\to\infty} nR_f^{(n)} = \tfrac{1}{2}\{f(1) - f(0)\} \tag{9}$$

or

$$R_f^{(n)} = O(n^{-1}). \tag{9'}$$

Moreover, if t_0 *is any point on* $[0,1]$ *and* f *is any absolutely continuous function on* $[0,1]$, *then there is a family* $\{t_k^{(n)}\}$ *satisfying* (8′) *and*

$$\lim_{n\to\infty} nR_f^{(n)} = \tfrac{1}{2}\{f(1) - f(0)\} - f(t_0).$$

PROOF. The first part of this theorem follows directly from Theorems 1 and 2 since (8) and (8′) imply that

$$\sum_{k=1}^{n} f(t_k)\left(\Delta_k - \frac{1}{n}\right) = \begin{cases} o(n^{-1}) \\ O(n^{-1}) \end{cases}$$

respectively. The restriction in Theorems 1 and 2 that $t_n^{(n)} = 1$ for all n is not important here since (8) and (8′) give

$$|\, t_n^{(n)} - 1 \,| = \begin{cases} o(n^{-1}) \\ O(n^{-1}) \end{cases}$$

respectively.

To obtain the second assertion we select a sequence of rationals m/n, $n = 1, 2, \cdots$, which converges to t_0 as n increases to infinity. For each integer n let m_n, $1 \leqq m_n \leqq n$, be the numerator of the rational selected. Then by Theorem 2

$$\lim_{n\to\infty} n\left\{\sum_{k=1}^{n} f(k/n)n^{-1} - \int_0^1 f(t)\, dt\right\} = \tfrac{1}{2}\{f(1) - f(0)\}.$$

Thus

$$\lim_{n\to\infty}(n-1)\left\{\sum_{k=1;k\neq m_n}^{n} f(k/n)n^{-1} - \int_0^1 f(t)\,dt\right\} = \tfrac{1}{2}\{f(1)-f(0)\} - f(t_0),$$

and the family $\{t_k^{(n)}\}$ defined by

$$t_1^{(n-1)} = 1/n,\ t_2^{(n-1)} = 2/n, \cdots, t_{m_n-1}^{(n-1)} = (m_n-1)/n,$$
$$t_{m_n}^{(n-1)} = (m_n+1)/n, \cdots, t_{n-1}^{(n-1)} = 1,$$

satisfies (8′).

3. **The distribution of asymptotically neutral families.** Let $z = \Phi(w)$ map $|w| > 1$ one-to-one and conformally onto the complement of the closure of D so that the points at infinity correspond and $\Phi'(\infty) > 0$. Under the assumption that D has a rectifiable boundary the function Φ can be extended to give a homeomorphism of $|w| \geqq 1$ onto the complement of D. The function $\phi: \phi(t) = \Phi(e^{2\pi it})$ is a complex valued absolutely continuous function of period 1 which gives a natural parametrization of Γ. Suppose $S^{(n)} = \{z_1^{(n)}, z_2^{(n)}, \cdots, z_n^{(n)}\}$ is so enumerated that $z_k^{(n)} = \phi(t_k^{(n)}), 0 < t_1^{(n)} < t_2^{(n)} < \cdots < t_n^{(n)} \leqq 1$. The family $\{t_k^{(n)}\}$ will be referred to as the parameter family associated with $\{S^{(n)}\}$.

The primary result of this note is

THEOREM 4. *Let D be a bounded Jordan domain with rectifiable boundary and containing the point $z = 0$, and let $\{S^{(n)}\}$ be any family of sequences of points on the boundary of D. If the parameter family associated with $\{S^{(n)}\}$ satisfies (8) then $\{S^{(n)}\}$ is asymptotically neutral on Γ.*

Moreover, there exist families $\{S^{(n)}\}$ whose associated parameter families satisfy (8′) which are not asymptotically neutral on Γ.

The theorem is proved by applying Theorem 3 to the principal value of $\log\{1 - z/\phi(t)\}$. For the details, cf. [**1**], [3a], [**4**].

REFERENCES

1. J. H. Curtiss, *Riemann sums and the fundamental polynomials of Lagrange interpolation*, Duke Math. J. **8** (1941), 525–532.

2. ———, *Interpolation with harmonic and complex polynomials to boundary values*, J. Math. Mech. **9** (1960), 167–192.

3. J. Korevaar, *Asymptotically neutral distributions of electrons and polynomial approximation*, Ann. of Math. **80** (1964), 403–410.

3a. ———, *Limits of polynomials whose zeros lie in a given set*, these preceedings.

4. M. Thompson, *Approximation by polymonials whose zeros lie on a curve*, Duke. Math. J. **37** (1964), 255–266.

5. J. L. Walsh, *Interpolation and approximation*, 3rd ed., Amer. Math. Soc. Colloq. Publ., vol. 20, Amer. Math. Soc., Providence, R. I., 1960.

INDIANA UNIVERSITY
BLOOMINGTON, INDIANA

Mapping of Riemann Surfaces (Nevanlinna Theory)[1]

H. WU

For a holomorphic mapping f from an open Riemann surface V into a compact Riemann surface M, the theory of Nevanlinna is concerned with the quantitative measurement of the pointwise coverage of M by V. In the classical case where $V = \boldsymbol{C}$ and M is the Riemann sphere, Nevanlinna's two main theorems give very precise information in this respect. It has been known for some time that the salient feature about $\boldsymbol{C}$ which makes possible the development of the classical theory is that $\boldsymbol{C}$ can be exhausted by the level lines of $\log r$, which is a harmonic function outside of the origin. One naturally asks for an extension of this theory to the class of harmonically exhaustible Riemann surfaces. See Definition 4.1. The first objective of this paper is to give a complete and coherent account of this extension from a differential geometric point of view. A second objective is to ascertain, in a sense, the most general V for which the theory of Nevanlinna is meaningful. To this end, the analogues of Nevanlinna's two theorems will be established on such an arbitrary open surface, and these theorems seem to point to the conclusion that the harmonically exhaustible surfaces form the only natural domain in which equidistribution theory can flourish.

Our work has been inspired by, and follows in broad outline, Chern's paper of 1960 [7]. It goes without saying that this geometric approach to the two main theorems had its inception in a series of remarkable papers of Ahlfors [1]–[4], which laid down many of the basic ideas exposed below. In a subject so intrinsically geometric as this one, this approach is the most natural and it is futile to try to disguise the many geometric tools that force themselves upon the scene.

The layout of this paper is as follows. Part A (§1 to §8) treats in detail the case of a Riemann surface admitting a harmonic exhaustion. Though this part has been written primarily with the analyst in mind, the function theory used is actually very elementary. In Part B (§9 to §10), the two main theorems on a general domain space will be proved. The presentation here is much more sketchy and is a reflection of our conviction that Nevanlinna theory for a general V is

[1] Work supported partially by the National Science Foundation. A summary of this paper was presented to the 1966 Summer Institute on Entire Functions of the American Mathematical Society.

really not interesting. The only notable result here is perhaps the general First Main Theorem, Theorem 9.3. Incidentally, a general remark about the exposition might be relevant here. Some results in Part A have been included not because they are new, but because on the one hand we wish to have a unified treatment, and on the other we have found that ameliorations in the known proofs are always possible.

In the Appendix, we take up the question of allowing the domain space V to have a nonempty boundary. For reasons explained therein, we believe that this extension of the general theory is indispensable. But since the presence of the boundary does encumber the exposition somewhat, it was thought best to relegate this case to the end.

Part A overlaps some recent work of Sario's **[13]**, but his point of view as well as his method are very different from those adopted here.

PART A.

1. **Preliminaries**. For the basic definitions and facts concerning Riemann surfaces, bordered Riemann surfaces and the differential calculus on them, we refer once and for all to Ahlfors-Sario **[5]**. We restrict ourselves to summarizing the pertinent differential geometrics notions. A good reference for this is **[14]**. *The basic convention is that every object we encounter will be at least of class* C^∞ *(i.e. infinitely differentiable), and that all differential forms (and hence all functions) are complex-valued.* Let ω be a form of degree p $(p = 0,1,2)$; then its exterior derivative $d\omega$ is a uniquely determined differential form of degree $p+1$. We elaborate on this. Let $z = x + iy$ be a local holomorphic coordinate function (i.e. $z \in \Phi$, in the terminology of [**5**]). Then $dz = dx + idy$ and $d\bar{z} = dx - idy$ are defined in this coordinate neighborhood. It is easy to see that every one-form is uniquely expressible as $\mu = pdz + qd\bar{z}$, and every two-form as $\Omega = rdz \wedge d\bar{z}$, where p, q, r are C^∞ complex functions. Let us further define the (local) operators $\partial/\partial z = \frac{1}{2}(\partial/\partial x - i\partial/\partial y)$, $\partial/\partial \bar{z} = \frac{1}{2}(\partial/\partial x + i\partial/\partial y)$. Then one can compute to show: $df = (\partial f/\partial z)dz + (\partial f/\partial \bar{z})d\bar{z}$ for every function f, and $d\mu = ((\partial p/\partial \bar{z})d\bar{z}) \wedge dz + ((\partial q/\partial z)dz) \wedge d\bar{z}$. This motivates the definition: given f, μ, and Ω as above:

$$d'f = (\partial f/\partial z)dz; \qquad d''f = (\partial f/\partial \bar{z})d\bar{z}$$

$$d'\mu = d'q \wedge d\bar{z}; \qquad d''\mu = d''p \wedge dz$$

$$d'\Omega = d''\Omega = 0.$$

d', d'' so defined in terms of local coordinates can easily be shown to be well defined independent of the coordinate function z. Consequently, d', d'' and d are all operators on forms on the whole surface which increase degree by one. One can also prove this fact:

LEMMA 1.1. *Let M_1, M_2 be bordered Riemann surfaces and $f: M_1 \to M_2$ be holomorphic. Let d_1, d_1', d_1'' and d_2, d_2', d_2'' be the various operators associated with M_1, M_2 respectively. Then,*

(1) $f^*d_2 = d_1 f^*$,

(2) $f^*d_2' = d_1'f^*$; $f^*d_2'' = d_1''f^*$,

Here f^* denotes the induced mapping on forms from those of M_2 into those of M_1 (see e.g. [**14**]). Of course (1) is true even for C^∞ mappings between C^∞ manifolds. In general, it is too much work to attach indices to these differential operators to indicate their domain of definition, so one simply writes $f^*d = df^*$, $f^*d' = d'f^*$ and $f^*d'' = d''f^*$ for a holomorphic f.

Now we define another operator which is particularly relevant for the complex structure on a Riemann surface:

$$d^c \equiv i(d'' - d'). \tag{1.1}$$

A few remarks concerning d^c will be important in the sequel. If τ is a harmonic function, then $d^c\tau = *d\tau$, where $*d\tau$ denotes the conjugate differential of $d\tau$. If a function τ is such that its level line $\tau = c$ ($c \in R$) is a submanifold (such is the case if $d\tau$ is nowhere zero on $\tau^{-1}(c)$), then $d^c\tau$ is nowhere zero when restricted to this submanifold. If τ is a real-valued function, then $d^c\tau$ is also a real-valued form in the sense that, if we write $d^c\tau = gdx + hdy$, then g, h are real-valued functions. All these can be verified by simple computation (see (1.3) below).

We collect together a few formal properties of these operators for future reference. They are all trivial. So fix a holomorphic coordinate function $z = x + iy$:

$$\begin{aligned} dx \wedge dy &= (i/2) dz \wedge d\bar{z}, \\ d^c\tau &= -(\partial\tau/\partial y)dx + (\partial\tau/\partial x)dy, \\ d\tau \wedge d^c\tau &= \{(\partial\tau/\partial x)^2 + (\partial\tau/\partial y)^2\} dx \wedge dy. \end{aligned} \tag{1.2}$$

If we let $\bar{\Delta}$ denote the Laplacian in this neighborhood, i.e. $\bar{\Delta} = (\partial^2/\partial x^2 + \partial^2/\partial y^2)$, then we have furthermore:

$$dd^c\tau = (\bar{\Delta}\tau)dx \wedge dy. \tag{1.3}$$

The bar on $\bar{\Delta}$ is to avoid confusion with the general Laplace operator to be introduced in §2.

We shall make extensive use of Stokes' Theorem ([**5**], p. 269, (9)) and it is therefore pertinent to comment on the notion of induced orientation. First recall a special case of this theorem tailored for our application: if D is either (i) a compact subdomain of a Riemann surface whose boundary ∂D is a submanifold except for a finite number of points (the "corners") or (ii) a bordered Riemann surface, and if ω is a C^∞ one-form defined on D, then

$$\int_D d\omega = \int_{\partial D} \omega. \tag{1.4}$$

In either case, D is oriented by the conformal structure, i.e. if $z = x + iy$ is a holomorphic coordinate function of D, then we define the *positive two-forms on* D to be exactly those which are locally the positive multiples of $dx \wedge dy$ (in other words, those $hdx \wedge dy$ with h positive everywhere in the neighborhood). Now the *induced* orientation on ∂D is the following: we have to tell which one-forms on ∂D are positive, and so we choose a *real* coordinate system around a point of ∂D (if we are in case (i), we stay away from the finite exceptional point set), say, $\{\tau, \sigma\}$ such that $\tau =$ constant is exactly ∂D in that neighborhood and such that $d\tau \wedge d\sigma$ is a positive form with respect to the orientation of D. Then all positive multiples of $d\sigma$ are by definition the positive one-forms of ∂D. This definition makes sense because $d\sigma$ is in fact a form nowhere zero when restricted to ∂D, as is trivial to see. This is the orientation of the boundary implicit in the right side of (1.4).

Now we turn to some standard facts about Hermitian (or conformal) metrics. If the metric on the Riemann surface is G, we let Ω denote its volume element. In local coordinates $z = x+iy$:

(1.5) If $G = g(dx^2 + dy^2)$, then $\Omega = gdx \wedge dy$ and the Gaussian curvature of G is $K = (-1/2g)\bar{\Delta} \log g$.

Here as before, $\bar{\Delta} = (\partial/\partial x^2 + \partial/\partial y^2)$. Of course, G is Hermitian means exactly that G assumes this simple form with respect to the real and imaginary parts of a complete complex coordinate function. By the uniformization theorem, every compact Riemann surface is covered by either the complex plane, or the unit disc, or the Riemann sphere. These three spaces can be equipped with a Hermitian metric of constant (resp. zero, negative, positive) Gaussian curvature. We therefore have this well-known fact:

LEMMA 1.2. *Every compact Riemann surface M can be given a Hermitian metric G of constant Gaussian curvature such that $\int_M \Omega = 1$.*

The last assertion is clearly a matter of adjusting constants.

2. **The proximity function.** Let a be a point in a compact Riemann surface M. Our first task is to construct a function u_a with the property that $u_a(p) \to \infty$ as $p \to a$. This function then measures the proximity of p to a. Precisely, we have the following result due to Chern:

THEOREM 2.1. *Let Φ be a C^∞ real two-form on a compact Riemann surface M such that $\int_M \Phi = c > 0$. Let $a \in M$ be a fixed point of M. Then there exists a real-valued function u_a with the following properties:*

(1) *u_a is C^∞ in $M - \{a\}$.*

(2) *$(1/2\pi)dd^c u_a = (1/c)\Phi$ in $M - \{a\}$.*

(3) *If z is any a-centered holomorphic coordinate function in a neighborhood U of a (i.e. $z(a) = 0$), then $u_a(z) + \log |z|$ is C^∞ on U.*

REMARK. Clearly if (3) is true for one holomorphic coordinate function z, it is true for every other. Since $u_a(z) = C^\infty$function $- \log |z|$ in U, it is clear that $u_a(p) \to \infty$ as $p \to a$. We abbreviate (1) and (3) by saying that u_a is C^∞ with a positive logarithmic singularity at a.

The proof of Theorem 2.1 is quite independent of the rest of the paper so that, without loss of continuity, one may simply take for granted this theorem as well as Theorem 2.8 and pass on to §3 — if one so desires. The theorem will be proved by using deRham's theory of currents: we show that if T_a is a solution of $\Delta T_a = (1/c)\Phi - \delta_a$, where δ_a is the Dirac measure, then $u_a = -2\pi * T_a$ is the required function. *All the relevant concepts and definitions will be found in the following.* The excuses for employing the theory of currents are these: (1) It makes possible a complete proof of the fundamental Theorem 2.8. (2) It furnishes the only rigorous and flexible approach to all questions concerning singularities. (3) It is as useful as it is important.

Now we summarize the basic definitions, all specialized to a Riemann surface M. The basic reference here is deRham [**15**]. At first, M need not be compact. Denote by $\mathscr{D} = \bigoplus_{i=0}^{i=2} \mathscr{D}^i$ the space of all complex-valued C^∞ forms with compact support. ($\mathscr{D}^i$ are those of degree i).

DEFINITION 2.1. A *current* on M is a linear functional $T: \mathscr{D} \to C$ which obeys the following continuity condition: if $\{\phi_i\}$ is a sequence in $\mathscr{D}$ with support $\phi_i \subseteq K$ for all i, where K is a compact set in a coordinate neighborhood, and if all the derivatives of the coefficients of the ϕ_i tend uniformly to zero as $i \to \infty$, then $T\phi_i \to 0$. If $T\phi = 0$ for every ϕ of degree $q \neq 2 - p$, we say *T is of degree p*.

EXAMPLE 1. Every locally integrable form α of degree p induces a current T_α of degree p: $T_\alpha(\omega) = \int_M \alpha \wedge \omega$. (Here, we make use of the fact that M is oriented.)

EXAMPLE 2. The Dirac measure δ_a for $a \in M$ is a current of degree 2: $\delta_a(f) = f(a)$, $\forall f \in \mathscr{D}^0$.

We can define the *restriction of a current* T to an open set $E \subseteq M$ by restricting the domain of definition of T to those $\omega \in \mathscr{D}$ such that support $\omega \subseteq E$, i.e. the restriction of T to E gives a current in the Riemann surface E. This is a very important concept. If the restrictions of T and another current T' to E are equal, we say T equals T' in E.

DEFINITION 2.2. The *support of a current* T is the set of points in M which have no neighborhood where T equals zero. We denote it by spt T.

Since the complement of spt T is open, spt T is closed. By a partition of unity, one sees that the restriction of T to the complement of spt T is zero. Thus $(\text{spt } T)^c$ is the largest open set on which $T = 0$.

EXAMPLE 1. Spt $T_\alpha = $ spt α if α is continuous.

EXAMPLE 2. Spt $\delta_a = \{a\}$.

We sometimes write α instead of T_α for locally integrable forms; so every such form can be regarded either as a form or a current.

DEFINITION 2.3. A current T is said to be C^∞ *in an open set* E iff T equals T_α in E for some form α which is C^∞ in E.

If T is a current of degree p and $\phi \in \mathcal{D}$, then by definition, $\bar{T}(\phi) = T(\bar{\phi})$. If $\deg \phi = q$ (here ϕ not necessarily of compact support), then $T \wedge \phi$ is the $(p+q)$-current such that $(T \wedge \phi)(\omega) = T(\phi \wedge \omega)$. Also define $T \wedge \phi = (-1)^{pq}\phi \wedge T$. Thus we have defined the exterior product of a C^∞ current and an arbitrary current; the exterior product of two arbitrary currents is undefinable. The exterior derivative of T is defined by: if $\deg T = p$, $dT(\omega) = (-1)^{p+1}T(d\omega)$. So dT is a current of degree $(p+1)$. Clearly $d^2T = 0$ and $d(T \wedge \phi) = dT \wedge \phi + (-1)^p T \wedge d\phi$. Note that if α is a C^∞ form, $T_\alpha \wedge \phi = T_{\alpha\wedge\phi}$ and $dT_\alpha = T_{d\alpha}$. The last is by Stokes' Theorem.

Now let M be compact, and assume a Hermitian metric G given on M. Let $\mathfrak{A} = \bigoplus_{i=0}^{i=2}\mathfrak{A}^i$ be the space of all C^∞ forms. With respect to G, a homomorphism $*\colon \mathfrak{A}^p \to \mathfrak{A}^{2-p}$ can be defined such that $**\omega = (-1)^p\omega$ for a form ω of degree p. We now define $*$ using local coordinates. If Ω is the volume element of G and f is a C^∞ function, then

$$*(f\Omega) = f, \qquad *f = f\Omega,$$

$$*(p\,dx + q\,dy) = p\,dy - q\,dx.$$

It will be noted that $*$ on one-forms is in fact independent of the Hermitian metric G and coincides with the classical notion of taking the conjugate of a differential. Define now $\delta\colon \mathfrak{A}^p \to \mathfrak{A}^{p-1}$ by $\delta = -*d*$. So $\delta\delta = 0$. Also define $\Delta = d\delta + \delta d\colon \mathfrak{A}^p \to \mathfrak{A}^p$. A form ω is called *harmonic* iff $\Delta\omega = 0$. Clearly:

$$*\Delta = \Delta*, \qquad d\Delta = \Delta d, \qquad \delta\Delta = \Delta\delta. \tag{2.1}$$

We claim that $(-\Delta)$ is an elliptic operator of order two. For this, let $z = x + iy$ be a holomorphic coordinate function and $G = g(dx^2 + dy^2)$ as in (1.5), then an easy computation gives:

$$\begin{aligned} \Delta f &= (-1/g)\bar{\Delta} f, \\ \Delta(p\,dx) &= ((-1/g)\bar{\Delta} p + \text{L.O.T.})\,dx, \\ \Delta(q\,dy) &= ((-1/g)\bar{\Delta} q + \text{L.O.T.})\,dy, \\ \Delta(f\Omega) &= ((-1/g)\bar{\Delta} f)\Omega \end{aligned} \tag{2.2}$$

where $\bar{\Delta} = (\partial/\partial x^2 + \partial/\partial y^2)$ as usual, and L.O.T. means "lower order terms". Another consequence of (2.2) is:

COROLLARY 2.2. *The constants and constant multiples of the volume element are the only harmonic zero-form and two-forms, respectively, in a compact Riemann surface.*

PROOF. (2.2) says that a function f (zero-form) is harmonic in the sense of $\Delta f=0$ iff it is harmonic in the usual sense. So the usual maximum principle applies to show that $\Delta f=0$ iff f is a constant. Same for the two-forms. Q.E.D.

We now extend $*$, δ, and Δ to currents. Given a current T of degree p, define $(*T)(\omega)=(-1)^pT(*\omega)$. $*T$ is a current of degree $2-p$. Consequently, δT has degree $(p-1)$ and $(\delta T)(\omega)=(-1)^pT(\delta\omega)$. Similarly, ΔT is a current of degree p and $(\Delta T)(\omega)=T(\Delta\omega)$. The following theorem states one of the most fundamental properties enjoyed by Δ, and is a consequence of the ellipticity of $(-\Delta)$. First, we must introduce two definitions.

DEFINITION 2.4. A sequence of currents $\{T_i\}$ is said to converge to zero (written $T_i\to 0$) iff for every $\eta\in\mathscr{D}$, $T_i\eta\to 0$.

DEFINITION 2.5. A sequence of C^∞ forms $\{\psi\}$ is said to be convergent to zero in the C^∞ topology (written $\psi_i\rightrightarrows 0$) iff, in every relatively compact coordinate neighborhood, the sequence together with all its derivatives converges uniformly to zero.

THEOREM 2.3. *On an arbitrary Riemann surface, if a sequence of currents $T_i\to 0$ and ΔT_i are C^∞ and $\Delta T_i\rightrightarrows 0$, then T_i are C^∞ and $T_i\rightrightarrows 0$.*

The proof of this is not easy and is given in deRham [**15**, p. 149]. A current T such that $\Delta T=0$ is called a *harmonic current*, and the set of all such is denoted by $\mathscr{H}$. The above theorem clearly implies the famous

WEYL LEMMA (COROLLARY 2.4). *Let S,T be currents such that $\Delta T=S$. Then T is C^∞ in every open set where S is. In particular, $\mathscr{H}$ consists of C^∞ forms.*

Next introduce an inner product on $\mathfrak{A}$ as follows (it should be emphasized that M is compact here):

$$[v_1,v_2]=\int_M v_1\wedge *\bar{v}_2.$$

It is easy to see that [,] is Hermitian positive definite, so that the space of C^∞ forms $\mathfrak{A}$ becomes a pre-Hilbert space. We again have to extend [,] partially to currents. If T is a current and ω is C^∞, then by definition, $[T,\omega]=T(*\bar{\omega})$. Of course if $[T,\omega]=0$, we say T is orthogonal to ω. $[T,S]$ for arbitrary currents T and S is undefined. However, as a consequence of the Weyl Lemma, *it always makes sense to say that a current is orthogonal to $\mathscr{H}$*. Since

$$T(\omega)=[T,(-1)^{\deg\omega}*\bar{\omega}],$$

knowing $[T,\omega]$ for every ω is equivalent to knowing T. This is a useful piece of information. Now let T be a current of degree $(p-1)$ and let ω be C^∞ of degree p, then $0=\{d(T\wedge *\bar{\omega}\}\ (1)=[dT,\omega]-[T,\delta\omega]$, as a simple computation readily shows. Thus $[dT,\omega]=[T,\delta\omega]$. Similarly, $[T,d\omega]=[\delta T,\omega]$. From these it follows that $[\Delta T,\omega]=[T,\Delta\omega]$. In particular, $[\Delta\omega,\omega]=[\delta\omega,\delta\omega]+[d\omega,d\omega]$, so that *for a compact M, ω is harmonic iff $d\omega=\delta\omega=0$.*

We need one more fundamental fact about Δ.

Theorem 2.5. *$\mathscr{H}$ is finite dimensional and for a C^∞ form β, there exists a C^∞ form α such that $\Delta\alpha=\beta$ iff $\beta\perp\mathscr{H}$ ("$\perp$" means "orthogonal to" with respect to $[\ ,\ ]$).*

For a proof of this, see deRham [**15**, p. 154]. On the basis of this theorem, we claim: there exists a unique harmonic projector $H\colon\mathfrak{A}\to\mathscr{H}$ such that $[\phi,\psi]=[H\phi,\psi]$ for all $\psi\in\mathscr{H}$. H is defined by: let $\phi_1,\cdots,\phi_n$ be an orthonormal basis of $\mathscr{H}$ relative to $[\ ,\ ]$, then $H\phi=\sum_i[\phi,\phi_i]\phi_i$. If H' is another such, then $(H-H')\phi\perp\mathscr{H}$ while belonging to $\mathscr{H}$. Hence $(H-H')\phi=0$, or $H=H'$. Clearly $H^2=H$.

We define the Green-deRham operator $\mathfrak{C}\colon\mathfrak{A}\to\mathscr{H}^\perp$ (orthogonal complement of $\mathscr{H}$ in $\mathfrak{A}$) by: $\mathfrak{C}\phi$ is the unique solution of $\Delta\psi=\phi-H\phi$ in $\mathscr{H}^\perp$. Theorem 2.5 says that $\mathfrak{C}\phi$ is indeed well defined. Hence:

$$\phi=H\phi+\Delta\mathfrak{C}\phi\qquad\forall\phi\in\mathfrak{A}. \tag{2.3}$$

Note that $\mathfrak{C}$ is selfadjoint on C^∞ forms with respect to $[\ ,\ ]$: if ϕ,η are C^∞, then $[\mathfrak{C}\phi,\eta]=[\mathfrak{C}\phi,H\eta+\Delta\mathfrak{C}\eta]=[\mathfrak{C}\phi,\Delta\mathfrak{C}\eta]=[\Delta\mathfrak{C}\phi,\mathfrak{C}\eta]=[H\phi+\Delta\mathfrak{C}\phi,\mathfrak{C}\eta]=[\phi,\mathfrak{C}\eta]$.

We collect together all the formal properties of these operators needed in the sequel:

$$\begin{aligned}&H^2=H,\quad \mathfrak{C}H=H\mathfrak{C}=0,\quad H\Delta=\Delta H=0,\\&\mathfrak{C}*=*\mathfrak{C},\quad \mathfrak{C}\delta=\delta\mathfrak{C},\quad \mathfrak{C}d=d\mathfrak{C},\quad \mathfrak{C}\Delta=\Delta\mathfrak{C}.\end{aligned} \tag{2.4}$$

The last set of commutation rules follows from this general fact: if an operator on $\mathfrak{A}$ commutes with Δ, it commutes with H and $\mathfrak{C}$. The latter is easily proved using (2.3). Since $\Delta=d\delta+\delta d$, (2.3) and (2.4) together imply:

$$\begin{aligned}&\phi=H\phi+d(\delta\mathfrak{C}\phi)+\delta(\mathfrak{C}d\phi)\text{ is a unique orthogonal}\\&\text{decomposition of }\phi,\ \forall\phi\in\mathfrak{A}.\end{aligned} \tag{2.5}$$

From this follows trivially the following theorem of Hodge for Riemann surfaces:

THEOREM 2.6. *On a compact Riemann surface M, let $\mathscr{H}^p$ denote the space of harmonic forms of degree p, then*

$$\mathscr{H}^p\cong\textit{closed p-forms/exact p-forms}$$

("$\cong$" denotes isomorphism).

Our task is now to extend (2.5) to currents. First we extend H and $\mathfrak{C}$. Since $\mathscr{H}$ is finite dimensional, [,] is a Hermitian inner product on $\mathscr{H}$, and each current is a linear functional on $\mathscr{H}$, these imply: for each T, there exists a unique $HT \in \mathscr{H}$ such that $[T,\eta] = [HT,\eta]$ for all $\eta \in \mathscr{H}$. Since $\mathscr{H}$ consists of C^∞ forms, HT is a C^∞ form and so certainly a current. Now define $\mathfrak{C}T$ to be that linear functional such that $[\mathfrak{C}T,\phi] = [T,\mathfrak{C}\phi]$ for all $\phi \in \mathscr{D}$. As $\mathfrak{C}$ is selfadjoint on C^∞ forms, this is an honest extension of $\mathfrak{C}$ to currents. But we must check that $\mathfrak{C}T$ is actually a current, i.e. it obeys the continuity condition. So if $\{\phi_i\} \subseteq \mathscr{D}$ such that $\phi_i \rightrightarrows 0$, we will have to prove $(\mathfrak{C}T)(\phi_i) \to 0$. (Since M is compact, the requirement on spt ϕ_i is unnecessary in this case, as is obvious.) First we claim that if $\phi_i \rightrightarrows 0$, then the currents $\mathfrak{C}\phi_i \to 0$. For, if η is C^∞, so is $\mathfrak{C}\eta$, so that $[\mathfrak{C}\phi_i, \eta] = [\phi_i, \mathfrak{C}\eta] \to 0$, proving claim. Next, if $\phi_i \rightrightarrows 0$, then the C^∞ forms $\mathfrak{C}\phi_i \rightrightarrows 0$. For, $\Delta\mathfrak{C}\phi_i = \phi_i - H\phi_i$, and if $\phi \rightrightarrows 0$, $H\phi_i$ certainly $\rightrightarrows 0$ also (since $H\phi_i$ is defined explicitly in terms of an orthonotmal basis of $\mathscr{H}$). This proves $\Delta\mathfrak{C}\phi_i \rightrightarrows 0$ if $\phi_i \rightrightarrows 0$. By Theorem 2.3, $\mathfrak{C}\phi_i \rightrightarrows 0$ as asserted. Finally, let T be an arbitrary current; then $[\mathfrak{C}T, \phi_i] = [T, \mathfrak{C}\phi_i] \to 0$ if $\phi_i \rightrightarrows 0$, which is equivalent to saying that $(\mathfrak{C}T)(\phi_i) \to 0$. Hence *$\mathfrak{C}T$ is a current if T is one.*

If ϕ, η are C^∞, clearly $[H\phi, \eta] = [\phi, H\eta]$ as a consequence of (2.3). We now extend this. Let T be a current, and ϕ be C^∞, then $[HT, \phi] = [HT, H\phi + \Delta\mathfrak{C}\phi] = [HT, H\phi] = [T, H\phi]$. This proves that *$H$ is selfadjoint with respect to [,] even if one of the arguments is a current.* We already know that the same is true of Δ and $\mathfrak{C}$. Having this at hand, we can prove that given any current T, $T = HT + \Delta\mathfrak{C}T$. For, if ϕ is C^∞, $[(H + \Delta\mathfrak{C})T, \phi] = [T, (H + \Delta\mathfrak{C})\phi] = [T, \phi]$ by virtue of (2.3) and (2.4). Since ϕ is arbitrary, our claim is proved. Hence:

THEOREM 2.7. *Given a current S, there exists a current T such that $\Delta T = S$ iff S is orthogonal to $\mathscr{H}$. If S is orthogonal to $\mathscr{H}$, then one such T is provided by $\mathfrak{C}S$.*

On the basis of Theorem 2.7, we can proceed to the proof of Theorem 2.1 along the line already indicated. It should be noted that this method of constructing u_a from the solution of a current equation lends itself to the construction of C^∞ forms with prescribed singularities in general. All the classical existence theorems can be so proved.

PROOF OF THEOREM 2.1. Consider the equation $\Delta T_a = (1/c)\Phi - \delta_a$ of currents of degree two defined on M. We claim:

(A) A solution T_a exists.

(B) $u_a = -2\pi * T_a$ is a function satisfying (1)–(3) of the theorem.

We first attack (A). By Theorem 2.7, it suffices to show that the right side $((1/c)\Phi - \delta_a)$ is orthogonal to $\mathscr{H}$. By Corollary 2.2, every harmonic two-form in a compact M is a constant multiple of the volume element. So if $r \in \boldsymbol{R}$ and Ω denotes the volume element as usual, $[(1/c)\Phi - \delta_a, r\Omega] = ((1/c)\Phi - \delta_a)(\overline{*r\Omega})$

$=((1/c)\Phi-\delta_a)(r)=(1/c)\int_M r\Phi-r=(1/c)\cdot r\cdot c-r=0$. This disposes of (A). (B) is more delicate. First of all, the restriction of $(1/c)\Phi-\delta_a$ to $M-\{a\}$ is $(1/c)\Phi$, which is C^∞. So, $\Delta T_a=(1/c)\phi$ on $M-\{a\}\Rightarrow T_a$ is C^∞ on $M-\{a\}$ by virtue of Weyl's Lemma (Corollary 2.4), $\Rightarrow u_a=-2\pi * T_a$ is a C^∞ function on $M-\{a\}$, thereby proving (1). For (2) observe that in an arbitrary coordinate neighborhood U with coordinate function $z=x+iy$:

$$\begin{aligned}(1/c)\Phi &= \Delta T_a \quad \text{on} \quad M-\{a\}\Rightarrow\\ (1/c)\Phi &= \Delta((-1/2\pi)*u_a) \text{ on } M-\{a\}\\ &= (1/2\pi)*\Delta(-u_a) \qquad (2.1)\\ &= (1/2\pi)*\left(\frac{1}{g}\bar{\Delta}u_a\right) \qquad (2.2)\\ &= (1/2\pi)\left(\frac{1}{g}\bar{\Delta}u_a\right)\Omega \qquad (\text{by definition of } *)\\ &= (1/2\pi)(\bar{\Delta}u_a)\,dx\wedge dy \qquad (1.5)\\ &= (1/2\pi)dd^cu_a \qquad (1.3).\end{aligned}$$

So (2) is proved. Finally, consider $\Delta T_a+\delta_a=(1/c)\Phi$ on U, where U is the coordinate neighborhood of (3) of the theorem. We claim: $\delta_a=\Delta(*(-1/2\pi)\log|z|)$ *on* U. Since $z(a)=0$, we are asked to prove: for every C^∞ function ψ *with support in* U, $\psi(0)=\{\Delta(*(-1/2\pi)\log|z|)\}(\psi)$.

$$\begin{aligned}\text{The right side} &= (*(-1/2\pi)\log|z|)(\Delta\psi) \qquad (2.1)\\ &= ((-1/2\pi)\log|z|)((\Delta\psi)\Omega)\\ &= ((-1/2\pi)\log|z|)(-\bar{\Delta}\psi\,dx\wedge dy) \qquad (2.2)\ \ (1.5)\\ &= (1/2\pi)\int_U \log|z|\cdot\bar{\Delta}\psi\cdot dxdy.\end{aligned}$$

Bur $\psi(0)=(1/2\pi)\int_U \log|z|\cdot\bar{\Delta}\psi\cdot dxdy$ is a classical fact (it is an easy consequence of Stokes' Theorem or, what is the same thing, the classical Green's Identities), so our claim is established. Hence *on* U,

$$(1/c)\Phi=\Delta T_a+\Delta((-1/2\pi)*\log|z|)=\Delta(T_a-(1/2\pi)*\log|z|).$$

By Weyl's Lemma again, $(T_a-(1/2\pi)*\log|z|)$ is C^∞ on U because $(1/c)\Phi$ is. Taking $*$ of it, we see that $-(1/2\pi)(u_a+\log|z|)$ is C^∞ on U. Q.E.D.

To the extent that T_a is only unique up to constant multiples of Ω, u_a is unique only up to additive constants, i.e. the required u_a for each a may be picked from $\{-2\pi*\mathfrak{C}((1/c)\Phi-\delta_a)+r\}$, where $r\in\mathbf{R}$ is arbitrary. (Theorem 2.7). This r can be chosen to accommodate each a. We now uniformize this choice.

THEOREM 2.8. *The u_a may be so chosen that*:

(1) $u_a \geqq 0$ *for all* $a \in M$.

(2) u_a *is a continuous function of a in the following sense*: *given a closed disc neighborhood* $U = \{|z| \leqq 1\}$ (z *is an a-centered coordinate function*) *of a and a compact set K such that* $K \cap \{a\} = \phi$, *there exists a neighborhood* U' *of a such that for all* $a' \in U'$, $(u_a - u_{a'})$ *is* C^∞*-SMALL on K, and* $\{(u_a + \log|z|) - (u_{a'} + \log|z - z(a')|)\}$ *is* C^∞*-SMALL on U*.

By C^∞*-SMALL* we mean that the derivatives of all orders are smaller in absolute value than a pre-assigned positive number.

PROOF. First we note that if functions $\{v_a\}$ for each $a \in M$ have been chosen to satisfy (2), they can be modified to satisfy (1) and (2). For, let $\{v_a\}$ satisfy (2) and let U be a neighborhood of a with an a-centered coordinate function z such that $U = \{|z| \leqq 1\}$. Suppose $v_a \geqq k$ on $M - (\operatorname{Int} U)$. On $\{|z| < 1 + \varepsilon^*\}$, $v_a + \log|z| \equiv \phi_a$ is C^∞. So let $(\phi_a - \log|z|) \geqq k'$ on U. Hence $v_a \geqq k'$ on U. Choose a neighborhood U_a so small that $a' \in U_a \Rightarrow v_{a'} \geqq k - \varepsilon$ on $M - (\operatorname{Int} U)$, and $v_{a'} \geqq k' - \varepsilon'$ on U. Let $k_a \leqq \{k - \varepsilon, k' - \varepsilon'\}$. Altogether, we have: $a' \in U_a \Rightarrow v_{a'} \geqq k_a$. Since M is compact, there exists a constant C such that $v_a \geqq C$ for all $a \in M$. Now define $u_a = v_a - C$; then $\{u_a\}$ clearly enjoys properties (1) and (2).

For the proof of the existence of functions satisfying (2) above, we shall need:

Fact 1. On a compact Riemann surface M, $\phi \rightrightarrows 0 \Rightarrow \mathfrak{C}\phi \rightrightarrows 0$, and if the currents $T_i \to 0$, then $\mathfrak{C}T_i \to 0$ (see Definitions 2.4 and 2.5).

Fact 2. In the unit disc of the complex plane, the *currents* $(\log|z - c_i| - \log|z|) \to 0$ as $c \to 0$ $(c_i \in C)$.

Fact 1 was proved in the process of proving this: T is a current $\Rightarrow \mathfrak{C}T$ is a current. The proof of the intuitively obvious Fact 2 is mildly tricky. Let V_i denote the disc of radius one about the point c_i and let χ_i denote the characteristic function of the intersection of the unit disc and $(V_i - c_i)$ (i.e. the algebraic difference set $\{v - c_i : v \in V_i\}$). If ϕ is a C^∞ function with support in the unit disc, define $\phi_i(z) = \phi(z + c_i)$. Now, we have to prove:

$$\int_{|z|<1} \phi \cdot \log|z - c_i| \, dxdy \to \int_{|z|<1} \phi \cdot \log|z| \, dxdy$$

as $c_i \to 0$. The left side equals $\int_C \phi_i \chi_i \log|z| \, dxdy + \varepsilon_i$, where $\varepsilon_i \to 0$ as $c_i \to 0$ and the integral converges to the right-hand side in view of Lebesgue's Bounded Convergence Theorem. Q.E.D.

Now we prove the existence of u_a satisfying (2). Recall that we have $\Delta T_a = (1/c)\Phi - \delta_a$, $u_a = -2\pi * T_a$ and, in the neighborhood U of a such that $z(a) = 0$ and $U = \{|z| \leqq 1\}$, $\delta_a = \Delta(*(-1/2\pi)\log|z|)$. We let $T_a = \mathfrak{C}((1/c)\phi - \delta_a)$ in the following (Theorme 2.7). So let compact K be given such that $K \cap \{a\} = \varnothing$. We may assume that all a' in the following have been chosen outside of K. Then T_a, $T_{a'}$ are both C^∞ on K. First we show: in a neighborhood of K, $T_{a'} - T_a \rightrightarrows 0$

as $a' \to a$. Since clearly $((1/c)\Phi - \delta_a) - ((1/c)\Phi - \delta_{a'}) \to 0$ as $a' \to a$, Fact 1 implies that $T_a - T_{a'} \to 0$. Moreover, $\Delta T_a - \Delta T_{a'} - 0$ in a neighborhood of K so that by Theorem 2.3, $T_{a'} - T_a \rightrightarrows 0$ in a neighborhood of K. Hence it follows immediately that $u_{a'} - u_a \rightrightarrows 0$ on K as $a' \to a$. Next, on $\{|z| < 1 + \varepsilon\}$, $(T_a - (1/2\pi) * \log|z|) - (T_{a'} - (1/2\pi) * \log|z - z(a')|) = \{\mathfrak{C}((1/c)\Phi - \delta_a) - (1/2\pi) * \log|z|\} - \{\mathfrak{C}((1/c)\Phi - \delta_{a'}) - (1/2\pi) * \log|z - z(a')|\} = \mathfrak{C}(-\delta_a + \delta_{a'}) - (1/2\pi) * (\log|z| - \log|z - z(a')|) \to 0$ because of: $\delta_a - \delta_{a'} \to 0$, Fact 1, and Fact 2. Also, $\Delta(T_a - (1/2\pi) * \log|z|) - \Delta(T_{a'} - (1/2\pi) * \log|z - z(a')|) = ((1/c)\Phi - \delta_a) + \delta_a - \{((1/c)\Phi - \delta_{a'}) + \delta_{a'}\}$ on $\{|z| < 1 + \varepsilon\}$, $= 0$.

Hence, by Theorem 2.3 once more, $(T_a - (1/2\pi) * \log|z|) - (T_{a'} - (1/2\pi) * \log|z - z(a')|) \rightrightarrows 0$ on U, i.e. $(u_a + \log|z|) - (u_{a'} + \log|z - z(a')|) \rightrightarrows 0$ as $a' \to a$. This together with the above fact that $u_{a'} - u_a \rightrightarrows 0$ on K clearly implies that u_a is a continuous function of a. Q.E.D.

3. **The nonintegrated main theorems.** From now on, we assume once and for all that M is a compact Riemann surface with a fixed Hermitian metric G, whose volume element we denote by Ω. *We assume always that G is normalized so that $\int_M \Omega = 1$.* By Theorem 2.1, we can solve for $\Delta T_a = \Omega - \delta_a$ for a fixed $a \in M$, thereby obtaining a function u_a such that:

$$d\gamma_a = \Omega \text{ on } M - \{a\} \text{ where } \lambda_a = (1/2\pi) d^c u_a. \tag{3.1}$$

It is understood that the u_a has been chosen so as to satisfy (1) and (2) of Theorem 2.8.

We now integrate (3.1). Let $f: D \to M$ be a holomorphic mapping from a compact bordered Riemann surface D into M. Since $f^{-1}(a)$ is finite, we let $n(D, a)$ denote the number of pre-images of a in D, *counting multiplicity*. We also define: $v(D) \equiv \int_D f^*\Omega$.

NONINTEGRATED FORM OF THE FIRST MAIN THEOREM. *Let D be a compact bordered Riemann surface and let $f: D \to M$ be a holomorphic mapping. If $a \in M$ is such that $f(\partial D) \cap \{a\} = \varnothing$, then*

$$n(D, a) + \int_{\partial D} f^*\lambda_a = v(D). \tag{3.2}$$

It follows that if $f: M' \to M$ is a holomorphic mapping between compact Riemann surfaces (i.e. $\partial M' = \varnothing$), then $n(M', a) =$ a nonzero constant independent of a for all $a \in M$. For compact Riemann surfaces, this fact will be sharpened in Corollary 3.1. For an improvement in a different direction, see [**19**]. Note that (3.2) is true for certain C^∞ maps: see Chern [**7**].

PROOF. Let $f^{-1}(a) = \{a_1, \cdots, a_n\}$. By hypothesis, $\{a_1, \cdots, a_n\} \subseteq D - \partial D$, so that we can choose a neighborhood U_ε of a with the properties:

(1) $f^{-1}(U_\varepsilon) = U_1 \cup \cdots \cup U_n$, where $a_j \in U_j$, and $U_j \cap U_k = \varnothing$ if $j \neq k$.

(2) A coordinate function z centered at a can be so chosen that $U_\varepsilon = \{|z| < \varepsilon\}$ and $u_a(z) + \log|z|$ is C^∞ in U_ε.
Consequently,

$$\int_D f^*\Omega = \int_{f(D)} \Omega = \lim_{\varepsilon\to 0} \int_{D-U_1-\cdots U_n} f^*\Omega$$

$$= \lim_{\varepsilon\to 0} \int_{D-U_1-\cdots U_n} f^* d\lambda_a \qquad (3.1)$$

$$= \lim_{\varepsilon\to 0} \int_{f(D-U_1-\cdots-U_n)} d\gamma_a = \lim_{\varepsilon\to 0} \int_{D-U_1-\cdots-U_n} df^*\lambda_a$$

(Lemma 1.1)

$$= \int_{\partial D} f^*\lambda_a - \lim_{c\to 0} \sum_{j=1}^{n} \int_{\partial U_j} f^*\lambda_a \qquad (1.4).$$

We are left with proving that the last sum equals $-n(D,a)$. Since $u_a(z) + \log|z|$ is C^∞ in U, $\omega \equiv (1/2\pi)d^c(u_a(z) + \log|z|)$ is also C^∞ there. This implies $\lim_{\varepsilon\to 0}\int_{\partial U_\varepsilon}\omega = 0$. But $\omega = \lambda_a + (1/2\pi)d^c\log|z|$, and also $\partial(f(U_1\cup\cdots\cup U_n)) = \partial U$. Hence,

$$-\lim_{\varepsilon\to 0}\sum_{j=1}^{n}\int_{\partial U_j} f^*\lambda_a = \lim_{\varepsilon\to 0}\sum_{j=1}^{n}\int_{\partial U_j} f^*((1/2\pi)d^c\log|z|).$$

We evaluate each summand of the right side separately. Since f is holomorphic, the integrand can be simplified to $(1/2\pi)d^c\log|f^*(z)|$ ((1.1) and Lemma 1.1). If z_j is a local coordinate function for U_j with $z_j(a_j) = 0$, we may suppose $f^*(z) = z^{m(j)}h(z_j)$, where h is a holomorphic function without zero in U_j and $m(j)$ is a positive integer. Hence

$$\text{Right side integrand over } \partial U_j = \frac{1}{2\pi} d^c(m(j)\log|z_j| + \log|h|)$$

$$= \frac{m(j)}{2\pi} d\theta + \frac{1}{2\pi} d^c\log|h|$$

where $z_j = re^{i\theta}$ is the usual polar decomposition and (1.3) clearly implies $d^c\log|z_j| \equiv d^c\log r = d\theta$. (See discussion after (1.1).) This implies:

$$\lim_{\varepsilon\to 0}\int_{\partial U_j} f^*\left(\frac{1}{2\pi}d^c\log|z|\right) = \frac{m(j)}{2\pi}\int_{\partial U_j} d\theta + \lim_{\varepsilon\to 0}\frac{1}{2\pi}\int_{\partial U_j} d^c\log|h| = m(j)$$

because h is zero-free and so $d^c\log|h|$ is C^∞ in U_j. Finally, $-\lim_{\varepsilon\to 0}\sum_{j=1}^{n}\int_{\partial U_i} f^*\lambda_a = \sum_j m(j) = n(D,a)$ by definition. Q.E.D.

In the same setting $f: D \to M$ as above, f has a finite member of *critical points* $\{\alpha_1, \cdots, \alpha_n\} \subseteq D$, i.e. $df(\alpha_j) = 0$ for all j. As before, assume $\{\alpha_1, \cdots, \alpha_n\} \cap \partial D = \varnothing$.

These α_j's are precisely those points which cover their images $\beta_j \equiv f(\alpha_j)$ with multiplicity greater than one. So if α_j cover β_j $m(j)$ times, it is reasonable to call $(m(j)-1)$ the *stationary index* of f at α_j. The nonintegrated form of the second main theorem gives a formula expressing $\sum_{j=1}^{n}(m(j)-1)$ in terms of geometric and topological invariants of D and M.

This will be done via the Gauss-Bonnet Theorem. Though a well-known theorem, we give here nevertheless a precise definition of the *geodesic curvature* of an oriented (or parametrized) curve γ in an oriented surface with respect to some Riemannian metric $\langle\ ,\ \rangle$ as there is a confusion concerning the sign. We assume γ parametrized by arc-length s, so that ds is the volume element of γ. Let $t = d/ds$ be the unit tangent vector field along γ. Choose a unit normal field n so that $t \wedge n$ defines the same orientation as the surface, i.e. if ω is a positive two-form on the surface, $\omega(t \wedge n) > 0$. By definition, the geodesic curvature of γ is $k_g \equiv \langle D_t t, n\rangle$, where D denotes the covariant differentiation operator. Then $\kappa = k_g ds$ becomes a one-form on the submanifold γ, called the *geodesic curvature form* of γ.

Now we quote a weak form of the Gauss-Bonnet Theorem. Let Δ be a compact bordered Riemann surface equipped with a Hermitian metric $\langle\ ,\ \rangle$. Let $\bar{K}$ and Φ denote the Gaussian curvature and volume element of $\langle\ ,\ \rangle$ respectively, and let $\chi(\Delta)$ *denote the Euler characteristic of* Δ. Then, the theorem states that

$$2\pi\chi(\Delta) - \int_{\partial\Delta} \kappa = \int_{\Delta} \bar{K}\Phi.$$

It should be noted that our convention is such that the Euler characteristic of the Riemann sphere is $+2$.

Return now to the situation $f: D \to M$, we remove from D small discs D_j (to be specified later) about each of the critical points α_j of f and let $D' = D - \bigcup_{i=1}^{n} D_i$. Then f^*G is a Hermitian metric on $D - \{\alpha_1, \cdots, \alpha_n\}$. So f becomes an isometry from $D - \{\alpha_1, \cdots, \alpha_n\}$ equipped with f^*G into M equipped with G. Hence $f^*(K\Omega)$ $\bar{K}\Phi$, where K, Ω are the Gaussian curvature and volume element of G in M as before and $\bar{K}$, Φ are the Gaussian curvature and volume element of f^*G in $D - \{\alpha_1, \cdots, \alpha_n\}$. So retaining the symbol κ for the geodesic curvature of $\partial D'$ in the metric f^*G, we have:

$$2\pi\chi(D') - \int_{\partial D'} \kappa = \int_{D'} f^*(K\Omega).$$

Since clearly $\chi(D') = \chi(D) - n$, where n is the number of critical points of f in D, we have:

$$2\pi\{\chi(D) - n\} - \int_{\partial D'} k = \int_{D'} f^*(K\Omega). \tag{3.3}$$

Now $\int_{\partial D'}\kappa = \int_{\partial D}\kappa - \sum_{j=1}^{n} \int_{\partial D_j}\kappa$, where as usual the minus sign is due to the fact that we orient ∂D_j with respect to D_j. We will evaluate each $\int_{\partial D_j}\kappa$ separately, with D_j still to be chosen. So let z, w be holomorphic coordinate functions around $\alpha_j, \beta_j \equiv f(\alpha_j)$ respectively which vanish at α_j, β_j. Locally $f^*(w) = z^{m(j)}\, h(z)$ where h is holomorphic and zero-free in a neighborhood of α_j, and $1 < m(j) \in Z$. Of course, $(m(j) - 1)$ is just the stationary index of f at α_j. In this neighborhood of α_j, $s = (h)^{1/m(j)}$ is defined. Then clearly $\zeta = z \cdot s$ is also a coordinate function about α_j such that:

$$f^*(w) = \zeta^{m(j)} \text{ and } \zeta(\alpha_j) = 0. \tag{3.4}$$

We now choose D_j to be $\{\zeta : |\zeta| < \varepsilon\}$ where ε is so small that both D_j and $E_j \equiv \{w : |w| < \varepsilon^{m(j)}\}$ lie in the neighborhoods of α_j and β_j respectively where ζ and w are valid as coordinate functions. By (3.4), $f(\partial D_j) = m(j) \cdot \partial E_j$, $f(D_j) = m(j)E_j$. If κ' is the geodesic curvature form of ∂E_j in the metric G, then $\kappa = f^*\kappa'$ because f is an isometry. So,

$$\int_{\partial D_j} \kappa = \int_{\partial D_j} f^*\kappa' = \int_{f(\partial D_j)} \kappa' = m(j) \int_{\partial D_j} \kappa' = m(j)\left\{2\pi - \int_{E_j} K\Omega\right\}$$

(where the last equality is by applying the Gauss-Bonnet Theorem to E_j)

$$= 2\pi m(j) - \int_{m(j)E_j} K\Omega = 2\pi m(j) - \int_{f(D_j)} K\Omega$$

$$= 2\pi m(j) - \int_{D_j} f^*(K\Omega).$$

Thus $\int_{\partial D'}\kappa = \int_{\partial D}\kappa - 2\pi \sum_{j=1}^{n} m(j) - \sum_{j=1}^{n} \int_{D_j} f^*(K\Omega)$. Combining this with (3.3), we get:

$$2\pi\chi(D) - \int_{\partial D} \kappa + 2\pi \sum_{j=1}^{n} (m(j) - 1) = \int_D f^*(K\Omega).$$

It is customary to define $n_1(D) = \sum_{j=1}^{n}(m(j) - 1)$, which is called the *stationary index of f in D*. So we rewrite the above as:

$$2\pi\chi(D) - \int_{\partial D} \kappa + 2\pi n_1(D) = \int_D f^*(K\Omega). \tag{3.5}$$

Suppose now G has constant Gaussian curvature (see Lemma 1.2). Then $\int_D f^*(K\Omega) = K \int_D f^*\Omega = Kv(D)$, in the notation of the nonintegrated first main theorem. But the Gauss-Bonnet Theorem applied to M itself gives $2\pi\chi(M) = \int_M K\Omega = K \int_M \Omega = K$. Hence $\int_D f_1^*(K\Omega) = 2\pi\chi(M)v(D)$. We have proved:

THE NONINTEGRATED FORM OF THE SECOND MAIN THEOREM. *Let f: $D \to M$ be a holomorphic mapping from a compact bordered Riemann surface D into M. Suppose df is nowhere zero along ∂D so that the geodesic curvature form κ of ∂D with respect to the Hermitian metric f^*G in D is defined. Then* (3.5) *is valid. If furthermore G has constant Gaussian curvature, then*

$$\chi(D) - \frac{1}{2\pi}\int_{\partial D} \kappa + n_1(D) = \chi(M)v(D). \tag{3.6}$$

REMARK. (3.2) relates the all-important $v(D)$ to the analytic data while (3.6) expresses $v(D)$ in terms of topological invariants. The validity of (3.6) is crucially dependent on two facts: (α) a Gauss-Bonnet Theorem for bordered Riemann surfaces exists and (β) every Riemann surface carries a metric of constant curvature. These seem to account for all the refined results of Nevanlinna theory on Riemann surfaces which are (as yet) absent in the higher dimensional analogues.

(3.5) has important consequences which are all classical. We list three because we shall need them later on.

COROLLARY 3.1 (HURWITZ FORMULA). *Let $f: D \to M$ be a holomorphic mapping of a bordered Riemann surface D into M. Suppose df is nowhere zero along ∂D and that f is n to one onto its image $D_1 \subseteq M$, counting multiplicities. Then,*

$$\chi(D) = n\chi(D_1) - n_1(D). \tag{3.7}$$

(Note that the assumption that $df|\partial D$ is nowhere zero clearly implies that D_1 is itself a bordered Riemann surface.)

COROLLARY 3.2. *If $f: M' \to M$ is a holomorphic mapping between compact Riemann surfaces, then f is n to one for some n, and*

$$\chi(M') = n\chi(M) - n_1(M'). \tag{3.8}$$

COROLLARY 3.3. *If $0 > \chi(M') > \chi(M)$, then there exists no nonconstant holomorphic mapping $f: M' \to M$.*

Corollary 3.3 follows from (3.8) and the fact that $n_1(M') \geqq 0$. Corollary 3.2 is in turn immediate from (3.7) and the remark after (3.2).

PROOF OF COROLLARY 3.1. We use the notation above. Since f is n to one onto D_1, clearly $\int_D f^*(K\Omega) = \int_{f(D)} K\Omega = n\int_{D_1} K\Omega$ (for a more general result covering this case, see Lemma 6.2). By the Gauss-Bonnet Theorem, this last equals $n(2\pi\chi(D_1) - \int_{\partial D_1}\kappa')$. But $n\int_{\partial D_1}\kappa' = \int_{f(\partial D)}\kappa' = \int_{\partial D} f^*\kappa' = \int_{\partial D}\kappa$ again because f is an isometry and is n to one on ∂D also. Thus: $\int_{\partial D} f^*(K\Omega) = 2\pi n\chi(D_1) - \int_{\partial D}\kappa$. Substitute this into (3.5), we get (3.7). Q.E.D.

4. **Harmonic exhaustion and the integrated main theorems.** In Nevanlinna theory, one is concerned with f: $V \to M$ where V is an open Riemann surface and

M is a compact Riemann surface and f is holomorphic. *This is our standard notation, to be used from now on without further comments.* The object of the study is to make meaningful statements about the coverage of each point of M by points of V under f. One cannot deal with a noncompact V directly. Rather, one restricts the domain of f to a compact subregion D of V where the results of §3 apply, and then one studies the asymtotic behavior of f as D exhausts V. At this point, we impose a strong restriction on V; we want it to be exhaustible by nice subregions in a nice manner.

DEFINITION 4.1: A C^∞ function $\tau: V \to [0, s)$ ($s \leqq \infty$) will be called a *harmonic exhaustion* on the open Riemann surface V iff

(1) τ is onto.

(2) τ is proper, i.e. inverse image of a compact set is compact.

(3) τ is eventually harmonic, i.e. there exists a number $r(\tau)$, $0 \leqq r(\tau) < s$, such that τ is harmonic on $\{p: \tau(p) \geqq r(\tau)\}$.

REMARKS. (A) In view of (1.3), (3) can be rephrased as follows: $dd^c\tau = 0$ on $\{p: \tau(p) \geqq r(\tau)$, where $0 \leqq r(\tau) < s\}$.

(B) It is quite clear that not every Riemann surface can carry a harmonic exhaustion. Specifically, if $s = \infty$, then such a V must be parabolic because the harmonic measure of the ideal boundary is then zero. (See **[5]**, p. 204, 6E and p. 26, 13G. I wish to thank B. Rodin for this remark.) Conversely, a theorem of Nakai **[11]** says that every parabolic Riemann surface in fact carries a harmonic exhaustion with $s = \infty$. For simple spaces such as $M' - \{a_1, \cdots, a_q\}$, where M' is compact and $a_i \in M'$, the exhaustion function is easily constructed: choose holomorphic coordinate function z_i about each a_i, then a C^∞ function which equals $-\log|z_i|$ in $\{|z_i| \leqq 1\} - \{a_i\}$ will do. A similar procedure shows that if the open Riemann surface is obtained from a compact one by removing a finite number of discs, it has a finite harmonic exhaustion (i.e. $s < \infty$).

(C) Note that if V carries a harmonic exhaustion, so does every finite (perhaps ramified) covering surface of V: one simply lifts the exhaustion function up to the covering surface. In this way, one constructs a Riemann surface which carries a finite harmonic exhaustion but has infinite genus, e.g. the Riemann surface of the multi-valued function $(\sin 1/(1-z))^{1/n}$ defined on the unit disc. For further examples, see §8.

(D) We do not as yet distinguish between the cases $s = \infty$ and $s < \infty$; this will be done in §7. It can of course happen that a surface has a harmonic exhaustion of *mixed type*, i.e. going to infinity in one portion of the ideal boundary and approaching a finite limit in another. For instance, a compact surface with a finite number of points *and* discs removed is of this type. This case will for the time being be neglected, but will be dealt with in the Appendix.

(E) It is hardly possible to overemphasize the role of the exhaustion function in the development of the theory. In general, it is not enough to have a nice domain space V but one must choose a nice exhaustion function as well. In the classical case of meromorphic functions, $V = C$ and a correct choice of the exhaustion function there is $\log r$ (outside of some disc). While r, or r^n ($n \in Z$), has the same level lines as $\log r$, it will not lead to any sensible statement about defect values. See Part B, §9. While Nevanlinna ostensibly worked with r, he actually employed $\log r$, e.g. $N(r,a) = \int^r n(t,a)(dt/t) = \int^r n(t,a)\, d\log t$.

Notation. $V[r] = \{p\colon p \in V, \tau(p) \leqq r\}$. Then clearly $\partial V[r] = \{p\colon p \in V, \tau(p) = r\}$ for $r \geqq r(\tau)$.

Since the only interesting part of V is the portion where τ is harmonic, *we shall tacitly assume in the sequel that all the parameter values r, r_0 etc. are greater than $r(\tau)$.*

Let us recall some elementary facts about harmonic functions. Locally, each harmonic function is the real part of a holomorphic function. The following is then obvious.

LEMMA 4.1. *The critical points of a harmonic function τ (i.e. the places where $d\tau$ vanishes) are isolated.*

An $r \in R$ is called a *critical value* of a C^∞ function u if $u^{-1}(r)$ contains a critical point of u. In the case of our τ, Lemma 4.1 and the fact that τ is proper imply that the critical values of τ above $r(\tau)$ are all isolated. Hence except for this isolated set of numbers, $\partial V[r]$ is an imbedded one-dimensional submanifold for all $r \geqq r(\tau)$. (Since $\partial V[r]$ is compact, it has to be a disjoint union of circles.) Thus $V[r]$ is a compact bordered Riemann surface with only an isolated set of r as exceptions. Furthermore, if r belongs to this exceptional isolated set, $\partial V[r]$ is a one-dimensional submanifold of V except for a finite set of points, and these are exactly the critical points of τ on $\partial V[r]$.

Let us be more explicit now about the passage from τ to its associated holomorphic function ζ such that $\operatorname{Re}\zeta = \tau$. Essentially, we want to generalize the situation that in any annulus of C, one can construct from $\log r$ the holomorphic function $\log r + i\theta$ (multivalued) which gives a coordinate system on the annulus-minus-a-radial-slit. So suppose (r_1, r_2) does not contain any critical values of τ, and suppose W is one component of Int $(V[r_2] - V[r_1])$. Let γ be one of the level curves of τ in W, say, $\gamma = W \cap \partial V[r]$ for some $r \in (r_1, r_2)$. We call $\int_\gamma * \, d\tau \equiv \Gamma$ a period of $*d\tau$ and define the conjugate harmonic function ρ of τ by $\rho(p) = \int_{p_0}^{p} *d\tau$, where p_0 is a fixed point of γ. Since $d(*d\tau) = 0$, ρ is defined up to periods, i.e. up to integral multiples of Γ and $d\rho = *d\tau$. Consequently, $\zeta = \tau + i\rho$ is a holomorphic function on W, but is multi-valued. Let $\alpha = \{p \in W\colon \rho(p) = 0 \pmod{\Gamma}\}$ (i.e. $\rho(p) = n\Gamma$, $n \in Z$).

LEMMA *For* 4.2. *$\zeta = \tau + i\rho$ as above $\zeta\colon W - \alpha \to (r_1, r_2) \times (0, \Gamma) \subseteq C$ is biholomorphic (i.e. one-one, onto, and holomorphic).*

PROOF. Clearly ζ is holomorphic, single-valued, onto, and everywhere nonsingular. Suppose $\zeta(p) = \zeta(p')$, then $\tau(p) = \tau(p') = r^*$, say. Thus p and p' lie on the same level curve $\partial V[r^*] \cap W - \alpha \equiv \gamma^*$. If we can check that ρ is one-one on γ^*, then ζ will be one-one on $W - \alpha$ and we are done. Now $\rho(p) = \int_{po}^{p} *d\tau$ $= \int_{po}^{p^*} *d\tau + \int_{p_0}^{p^*} * d\tau$, where p^* is a point of intersection of γ^* and α. The path of integration of the second integral we can choose to be along γ^*. Since $d\tau$ is nowhere zero in W, $*d\tau$ is nowhere zero on γ^*. But $*d\tau$ is actually nowhere zero when restricted to the submanifold γ^* (see discussion after (1.1)) so that the second integral must be either strictly increasing or decreasing as p transverse $\gamma*$. So ρ is one-one on γ^*. Q.E.D.

COROLLARY 4.3. *$\int_{\partial V[r]} *d\tau \equiv L$ is a constant for all $r \geqq r(\tau)$.*

PROOF. It is a continuous function of r, and the lemma says that it is constant in between critical values of τ. Q.E.D.

DEFINITION 4.2. Let (r_1, r_2) contain no critical values of τ. The holomorphic coordinate function of Lemma 4.2 in a component of $\tau^{-1}((r_1, r_2))$ is called a *special coordinate function.*

It is worth remarking that for this special coordinate function $\zeta = \tau + i\rho$ in W, any function μ defined on W has the property:

$$d^c\mu = (\partial\mu/\partial\tau)\, d\rho. \tag{4.1}$$

This is an easy consequence of (1.1) and (1.2) and the remarks after (1.1).

With this preliminary material out of the way, we can now tackle the problem of integrating (3.1) and (3.6). The situation is the following: We have (essentially) a differentiable family of compact bordered Riemann surfaces $V[r]$ in V, and on each of these (3.1) and (3.6) are valid. Naturally, one tries to piece together these informations by integrating with respect to r. The first step in this direction is to transform the line integrals in these formulas. It will be convenient to refer to two standard assumptions on r: given $a \in M$,

($\mathfrak{A}$) $f(\partial V[r]) \cap \{a\} = \varnothing$.

($\mathfrak{B}$) r is not a critical value of τ.

We have $f: V \to M$, of course. The nonintegrated first main theorem implies that under ($\mathfrak{A}$) and ($\mathfrak{B}$):

$$n(r, a) + \int_{\partial V[r]} f^*\lambda_a = v(r)$$

where we have written $n(r, a)$ for $n(V[r], a)$, and $v(r)$ for $v(V[r])$.

LEMMA 4.4. *Under ($\mathfrak{A}$) and ($\mathfrak{B}$), $\int_{\partial V[r]} f^*\lambda_a = (\partial/\partial r)((1/2\pi)\int_{\partial V[r]} f^*u_a *d\tau)$.*

PROOF. Recall that f holomorphic implies $f^*d^c = d^cf^*$. (Lemma 1.1). So by (3.1), $f^*\bar{z}_a = (1/2\pi)d f^*u_a$. Since ($\mathfrak{B}$) holds, we know there exist r_1, r_2 such that $r \in (r_1, r_2)$ and (r_1, r_2) contains no critical value of τ. Working in one component

W of $\tau^{-1}((r_1, r_2))$, we employ special coordinate function $\zeta = \tau + i\rho$, valid on "W minus a strip": $\rho = 0 \pmod{\Gamma}$. In W, $\partial V[r]$ becomes the coordinate curve $\tau = r$. Hence, $f^*\lambda_a = (\partial f^* u_a / \partial\tau)\, d\rho$ by (4.1) and

$$\begin{aligned}\int_{W \cap \partial V[r]} f^*\lambda_a &= \frac{1}{2\pi}\int_{(r,0)}^{(r\,\Gamma)} \frac{\partial f^* u_a}{\partial \tau}\, d\rho \\ &= \frac{\partial}{\partial r}\left(\frac{1}{2\pi}\int_{(r,0)}^{(r,\Gamma)} f^* u_a\, d\rho\right) \\ &= \frac{\partial}{\partial r}\left(\frac{1}{2\pi}\int_{(r,0)}^{(r\,\Gamma)} f^* u_a * d\tau\right) \\ &= \frac{\partial}{\partial r}\left(\frac{1}{2\pi}\int_{W \cap \partial V[r]} f^* u_a * d\tau\right).\end{aligned}$$

Since this is true for each component of $\partial V[r]$, we are done. Q.E.D.

We have proved that under ($\mathfrak{A}$) and ($\mathfrak{B}$),

$$n(r,a) + \frac{\partial}{\partial r}\left(\frac{1}{2\pi}\int_{\partial V[r]} f^* u_a * d\tau\right) = v(r).$$

Suppose (r_1, r_2) is an interval in which ($\mathfrak{A}$) and ($\mathfrak{B}$) hold for all $r \in [r_1, r_2]$. An integration leads us to:

$$\int_{r_1}^{r_2} n(t,a)\,dt + \frac{1}{2\pi}\int_{\partial V[t]} f^* u_a * d\tau \,\Big|_{t=r_1}^{t=r_2} = \int_{r_1}^{r_2} v(t)\, dt \tag{4.2}$$

where we have used the standard notation: $h(t)\,|_{t=r_1}^{t=r_2} \equiv h(r_2) - h(r_1)$. Now, we need a technical lemma whose proof we postpone to §5:

LEMMA 4.5. *$\int_{\partial V[r]} f^* u_a * d\tau$, for a fixed a, is a continuous function of r for ALL r.*

On the basis of Lemma 4.5, we now show that (4.2) is true for an arbitrary finite open interval $(t_0, t_n) \subseteq (r(\tau), s)$ (Definition 4.1). For, by Lemma 4.1, and the fact that f is holomorphic, the points in (t_0, t_n) for which ($\mathfrak{A}$) and ($\mathfrak{B}$) do not hold are finite in number: $t_1, \cdots, t_{n-1}$. So for each of (t_i, t_{i+1}), (4.2) is true and thereby gives n equations of the type (4.2). Use Lemma 4.5 and add; we see that the sum of the middle terms of the left sides of these n equations telescopes and (4.2) becomes true for (t_0, t_n) itself: This therefore forces the definition:

DEFINITION 4.3. Given a number $r_0 \geqq r(\tau)$, we define the *order function* (Nevanlinna characteristic) $T(r)$ to be:

$$T(r) = \int_{r_0}^{r} v(t)\, dt,$$

and the *counting function* $N(r, a)$ to be:

$$N(r, a) = \int_{r_0}^{r} n(t, a)\, dt.$$

The choice of r_0 is immaterial for most of the following discussion except in. §8, so we suppress any reference to it. Restating (4.2), we have:

THEOREM 4.6 (FIRST MAIN THEOREM). *For every* $r \geqq r(\tau)$:

$$N(r, a) + \frac{1}{2\pi} \int_{\partial V[t]} f^* u_a *d\tau \Big|_{t = r_0}^{t = r} = T(r). \tag{4.3}$$

Observe as usual that the terms on the left depend on a whereas $T(r)$ does not. Hence the middle integral compensates for the deficiency of $N(r,a)$ (e.g. $N(r,a) = 0$ if $f(V[r]) \cap \{a\} = 0$). If M is the Riemann sphere equipped with the constant curvature metric, then $u_a(p) = \log 1/\chi(p, a)$ where $\chi(p, a)$ denotes the chordal distance between p and a. See Chern [7]. This retrieves Nevanlinna's original result cast in the form of Ahlfors-Shimizu.

In view of Theorem 2.8 and the fact that $*d\tau$ is a positive one-form on $\partial V[t]$ (see (1.2) and discussion after (1.4)), we have:

$$\int_{\partial V[t]} f^* u_a *d\tau > 0 \quad \text{for all } t. \tag{4.4}$$

Another important fact we need, which is a consequence of Theorem 2.8 also, is the following lemma to be proved in §5 below.

LEMMA 4.7. $\int_{\partial V[r]} f^* u_a * d\tau$, *for a fixed* r, *is a continuous function of* a.

We are now in a position to deduce the fundamental inequality on which Nevanlinna theory is based:

COROLLARY 4.8. $N(r, a) < T(r) + \text{const}$, *where the constant is independent of* r *and* a.

PROOF.

$$\begin{aligned} N(r, a) &= T(r) + \frac{1}{2\pi} \int_{\partial V[r_0]} f^* u_a *d\tau - \frac{1}{2\pi} \int_{\partial V[r]} f^* u_a *d\tau \\ &< T(r) + \frac{1}{2\pi} \int_{\partial V[r_0]} f^* u_a *d\tau \qquad (4.4) \\ &\leqq T(r) + \text{const} \end{aligned}$$

where the constant is chosen to be the maximum of $(1/2\pi) \int_{\partial V[r_0]} f^* u_a *d\tau$ as a varies over the compact M (Lemma 4.7). Q.E.D.

Further discussions of (4.3) and Corollary 4.8 will be found in §6 after Lemma 6.2. For the moment, we turn to the integration of (3.6). First, we set up a

Convention. From now n, the Hermitian metric on M will be assumed to have constant Gaussian curvature. (See Lemma 1.2.) Let us write $\chi(r) = \chi(V[r])$, $n_1(r) = n_1(V[r])$, then under assumption ($\mathfrak{B}$) above and under

($\mathfrak{C}$) df is everywhere nonzero along $\partial V[r]$, we know:

(4.5) $$\chi(r) + n_1(r) - \chi(M)v(r) = (1/2\pi)\int_{\partial V[r]} \kappa$$

where κ is the geodesic curvature form of $\partial V[r]$ in the metric f^*G. To transform $\int \kappa$, we need to define a function h by

(4.6) $$f^*\Omega \equiv h d\tau \wedge *d\tau;$$

h is not defined at the critical points of τ. The important thing to note is that, since $d\tau \wedge *d\tau$ is a positive form on V, and f is holomorphic and hence orientation preserving, *h is a nonnegative function.*

LEMMA 4.9. *Under* ($\mathfrak{B}$) *and* ($\mathfrak{C}$),

$$\int_{\partial V[r]} \kappa = \frac{\partial}{\partial r}\left(\frac{1}{2}\int_{\partial V[r]} (\log h) *d\tau\right).$$

PROOF. As in the proof of Lemma 4.4, we use the special coordinate function $\zeta = \tau + i\rho$ on a component W of some $\tau^{-1}((r_1, r_2))$, $r \in (r_1, r_2)$. Then $f^*\Omega = h d\tau \wedge d\rho$ so that $f^*G = h(d\tau^2 + d\rho^2)$ ((1.5)). It is then well known that the geodesic curvature of the ρ-coordinate curve (i.e. $\partial V[r]$) is $k_g = (1/h^{1/2})(\partial(\log h)/\partial\tau$, and the element of arc-length of the ρ-curve is $ds = h^{1/2} d\rho$, so that $\kappa = \frac{1}{2}(\partial \log h/\partial\tau) d\rho$. For the sake of completeness, we rederive this result here using the above definition of k_g. Now $\langle \partial/\partial\rho, \partial/\partial\rho\rangle = h$ (we use $\langle\ ,\ \rangle$ to denote $f^*G(\ ,\ )$ so that the unit tangent vector field t along $\partial V[r]$ coherent with the orientation of $\partial V[r]$ is $t = (1/h^{1/2})(\partial/\partial\rho)$. Similarly, the unit normal field n along $\partial V[r]$ coherent with the orientation is $n = (-1/h^{1/2})(\partial/\partial\tau)$. (See discussion preceeding the Gauss-Bonnet Formula in §3.) Hence, $\kappa = \langle D_t t, n\rangle d\rho = -\langle D_t t, (1/h^{1/2})(\partial/\partial\tau)\rangle (h^{1/2} d\rho) = -\langle D_t t, \partial/\partial\tau\rangle d\rho$. Using the identity characterizing D (see [10], p. 160, Proposition 2.3), and the fact that $\langle \partial/\partial\rho, \partial/\partial\tau\rangle = 0$, we get:

$$\langle D_t t, \frac{\partial}{\partial\tau}\rangle = \frac{1}{h}\langle D_{\partial/\partial\rho}\frac{\partial}{\partial\rho}, \frac{\partial}{\partial\tau}\rangle = \frac{1}{h}\left\{\frac{-1}{2}\frac{\partial}{\partial\rho}\langle \frac{\partial}{\partial\rho}, \frac{\partial}{\partial\rho}\rangle\right\}$$

$$= \frac{-1}{2h}\frac{\partial h}{\partial\tau} = -\frac{1}{2}\frac{\partial}{\partial\tau}\log h.$$

Hence, $\kappa = \frac{1}{2}(\partial \log h/\partial \tau)d\rho$ as before. So, $\int_{W\cap\partial V[r]}\kappa = (\partial/\partial r)\{\frac{1}{2}\int_{W\cap\partial V[r]}(\log h)*d\tau\}$. This as well as the conclusion of the proof is identical with those of Lemma 4.4.

Q.E.D.

Again we need a technical lemma of the same nature as Lemma 4.5. The proof can be found in §5.

LEMMA 4.10. *$\int_{\partial V[r]}(\log h)*d\tau$ is a continuous function of r.*

The situation here parallels that of the First Main Theorem: we have by (4.5) and Lemma 4.9 that under ($\mathfrak{B}$) and ($\mathfrak{C}$):

$$\chi(r) + n_1(r) - \chi(M)v(r) = \frac{\partial}{\partial r}\left\{\frac{1}{4\pi}\int_{\partial V[r]}(\log h)*d\tau\right\}.$$

Integrate this as before and use Lemma 4.10 to conclude that for an arbitrary $r \geqq r(\tau)$ and r_0 as in Definition 4.3,

$$\int_{r_0}^{r}\chi(t)dt + \int_{r_0}^{r}n_1(t)dt - \chi(M)\int_{r_0}^{r}v(t)dt = \frac{1}{4\pi}\int_{\partial V[t]}(\log h)*d\tau\Big|_{t=r_0}^{t=r}.$$

DEFINITION 4.4. $E(r) = \int_{r_0}^{r}\chi(t)dt, \qquad N_1(r) = \int_{r_0}^{r}n_1(t)\,dt.$

Using these and Definition 4.3, we have arrived at:

THEOREM 4.11 (SECOND MAIN THEOREM). *For all $r \geqq r(\tau)$:*

$$E(r) + N_1(r) - \chi(M)T(r) = \frac{1}{4\pi}\int_{\partial V[t]}\log h*d\tau\Big|_{t=r_0}^{t=r}. \tag{4.7}$$

REMARK. Hermann Weyl seems to be the first one to call a theorem of this type the Second Main Theorem. The significance of this theorem was first recognized by Ahlfors, to whom is also due the magnificent idea of invoking the Gauss-Bonnet Theorem in Nevanlinna theory. What is traditionally called the Second Main Theorem is designated the Third Main Theorem by Weyl; this is Theorem 7.1 below.

5. **Proofs of the technical lemmas.** Before giving the proofs of Lemmas 4.5 and 4.7, let us note that these proofs make no use of the harmonicity of τ and consequently apply to the more general situation considered in §9. The proof of Lemma 4.9, however, uses crucially the fact that τ is eventually harmonic.

PROOF OF LEMMA 4.5. We will first show that the continuity in r of $\int_{\partial V[r]}f^*u_a*d\tau$, for a fixed a, is a consequence of the following assertions:

(α) $\int_{V[r]}d(f^*u_a*d\tau)$ is finite for all r.

(β) If $f(p) = a$ and if W_ε is a holomorphic disc $\{|z| < \varepsilon\}$ about p (i.e. $z(p) = 0$), then $\lim_{\varepsilon\to 0}\int_{\partial W_\varepsilon}f^*u_a*d\tau = 0$.

As always, it is understood that the p of (β) is in the domain of harmonicity of τ, i.e. $\tau(p) \geqq r(\tau)$.

Assume (α) and (β) for the moment, let $f^{-1}(a) \cap V[r] = \{p_1, \cdots, p_l\}$. Enclose each p_i by a holomorphic disc $U_\varepsilon^i = \{|z_i| < \varepsilon\}$ $(z_i(p_i) = 0)$, and let $U_\varepsilon = \bigcup_{i=1}^{i=l} U_\varepsilon^i$. On $V[r] - U_\varepsilon$, f^*u_a is a C^∞ function, and so $f^*u_a *d\tau$ is C^∞ there. Hence Stokes' Theorem applies.

$$\int_{V[r]} d(f^*u_a *d\tau) = \lim_{\varepsilon\to 0} \int_{V[r]-U_\varepsilon} d(f^*u_a *d\tau) \qquad (\alpha)$$

$$= \lim_{\varepsilon\to 0} \left\{ \int_{\partial V_\varepsilon[r]} f^*u_a *d\tau - \sum_{i=1}^{i=l} \int_{\partial U_\varepsilon^i} f^*u_a *d\tau \right\} \qquad (1.4)$$

(where $\partial V_\varepsilon[r]$ denotes the subset of $\partial V[r]$ outside of those U_ε^i which intersect $\partial V[r]$; consequently $\lim_{\varepsilon\to 0} \partial V_\varepsilon[r] = \partial V[r]$)

$$= \int_{\partial V[r]} f^*u_a *d\tau \qquad (\beta).$$

This proves that the last integral is finite for all r. To show continuity in r, it clearly suffices to show that if $r_i \uparrow r$, then

$$\int_{\partial V[r_i]} f^*u_a *d\tau \to \int_{\partial V[r]} f^*u_a *d\tau. \tag{5.1}$$

Let χ_i denote the characteristic function of $V[r_i]$, then left side of (5.1)

$$= \int_{V[r]} d(f^*u_a *d\tau)$$

$$= \int_{V[r]} \chi_i \, d(f^*u_a *d\tau)$$

$$\to \int_{V[r]} d(f^*u_a *d\tau) \qquad \text{(Lebesgue's Bounded Convergence Theorem)}$$

$$= \text{Right side of (5.1)}.$$

The lemma is therefore proved.

We now prove (α). The singularities of $d(f^*u_a *d\tau)$ are located at the isolated set $f^{-1}(a)$, and we only have to examine this two-form at a $p \in V[r] - V[r_0]$ such that $f(p) = a$. So let $\zeta = \sigma + i\rho$ be a p-centered holomorphic coordinate function valid in a neighborhood W of p, and let z be an a-centered holomorphic coordinate function valid in U. For some positive integer m, $f^*(z) = \zeta^m h(\zeta)$, where h is holomorphic and zero free in W (say). In U, $u_a + \log|z| \equiv \phi_a$ is C^∞ and so $f^*u_a = C^\infty$ function $- m \log|\zeta|$, and $*d\tau = (\partial\tau/\partial\sigma)d\rho - (\partial\tau/\partial\rho)d\sigma$ so that

$$d(f^*u_a *d\tau) = (C^\infty \text{ form} - (m/|\zeta|^2)(\sigma d\sigma + \rho d\rho)) \wedge *d\tau$$

$$= C^\infty \text{ form} - m(\partial\tau/\partial\sigma)(\sigma/|\zeta|^2)\, d\sigma \wedge d\rho - m\, \partial\tau/\partial\rho \cdot$$

$$\cdot (\rho/|\zeta|^2) d\sigma \wedge d\rho.$$

Now, $m\,\partial\tau/\partial\sigma$ is C^∞ and $|m(\partial\tau/\partial\sigma)(\sigma/|\zeta|^2)| \leqq |m(\partial\tau/\partial\sigma)|/|\zeta|$, which is clearly locally integrable. Hence so is $m(\partial\tau/\partial\sigma)(\sigma/|\zeta|^2)$. Same for $m(\partial\tau/\partial\rho)(\rho/|\zeta|^2)$. We therefore conclude that (α) is true.

Next, we take up (β). Using the notation of (β), we take a point $q \in \partial W_\varepsilon = \{|\zeta| = \varepsilon\}$. Around q, use polar coordinates $r \equiv |\zeta|$ and θ. Note that $\log r + i\theta$ is then a holomorphic coordinate function at q so that $*d\tau = *((\partial\tau/\partial r) \cdot r\,d\log r + (\partial\tau/\partial\theta)d\theta) = (r(\partial\tau/\partial r)d\theta - (\partial\tau/\partial\theta)d\log r)$. Since we only integrate around ∂W_ε, we neglect terms involving dr in the following computation. Then,

$$\begin{aligned} f^*u_a*d\tau &= (C^\infty \text{ function } - m\log r)\, r(\partial\tau/\partial r)d\theta \\ &= (\partial\tau/\partial r)(C^\infty \text{ function} \cdot r - mr\log r)\, d\theta \\ &\Rightarrow \left|\int_{\partial W_\varepsilon} f^*u_a*d\tau\right| \leqq \text{const}\,(2\pi\varepsilon - m2\pi\varepsilon\log\varepsilon) \to 0 \end{aligned}$$

as $\varepsilon \to 0$. Q.E.D.

PROOF OF LEMMA 4.7. For a fixed r, we will prove $\int_{\partial V[r]} f^*u_a*d\tau$ is a continuous function of a. By the proof of Lemma 4.5, this is equivalent to proving $\int_{V[r]} d(f^*u_a*d\tau)$ is a continuous function of a.

Let U be a neighborhood of a such that $U = \{|z| \leqq 1\}$ with respect to some a-centered coordinate function z and such that $u_{a'} + \log|z - z(a')| \equiv \phi_{a'} \rightrightarrows \phi_a \equiv u_a + \log|z|$ on U as $a' \to a$. (See Theorem 2.8 and Definition 2.5.) We may harmlessly assume that $f^{-1}(U) \cap V[r]$ are disjoint coordinate neighborhoods, say, $W_1, \cdots, W_l$. Let $W = \bigcup_{i=1}^{i=l} W_i$. Then

$$\int_{V[r]} d(f^*u_a*d\tau) = \int_{V[r]-W} d(f^*u_a*d\tau) + \sum_{i=1}^{i=l} \int_{W_i} d(f^*u_a*d\tau).$$

By Theorem 2.8, there is no question of the continuous dependence of the first integral of the right side on a. We need only examine each summand of the last sum separately. Let $a' \in M$. We have to prove that

$$\int_{W_i} d\{(f^*\phi_{a'} - \log|f^*(z) - z(a')|)*d\tau\} \to \int_{W_i} d\{(f^*\phi_a - \log|f^*(z)|)*d\tau\}$$

as $a' \to a$. By Theorem 2.8 again, $\int_{W_i} d\{f^*\phi_{a'}*d\tau\} \to \int_{W_i} d\{f^*\phi_a*d\tau\}$ as $a' \to a$. So we are left with proving:

$$\int_{W_i} d\{\log|f^*(z) - z(a')|*d\tau\} \to \int_{W_i} d\{\log|f^*(z)|*d\tau\} \text{ as } a' \to a. \tag{5.2}$$

So let $p \in W_i$ such that $f(p) = a$. By the usual arguments, we may assume that one branch of $(f^*(z))^{1/m} \equiv \zeta$ gives a local holomorphic coordinate function in W_i for some positive integer m. Note that $\zeta(p) = 0$. Let $\zeta = \sigma + i\rho$. Then

$$d\{\log|f^*(z)|*d\tau\} = m\frac{\partial\tau}{\partial\sigma}\frac{\sigma}{|\zeta|^2}d\sigma\wedge d\rho + m\frac{\partial\tau}{\partial\rho}\frac{\rho}{|\zeta|^2}d\sigma\wedge d\rho.$$

On the other hand, if $\theta_0, \cdots, \theta_{m-1}$ are the mth roots of unity and $\alpha_j = z(a')^{1/m}\cdot\theta_j$, then $\zeta^m = f^*(z)$ implies

$$(f^*(z) - z(a')) = \prod_{i=1}^{m}(\zeta - \alpha_j)$$

or

$$\log|f^*(z) - z(a')| = \sum_{j=1}^{m}\log|\zeta - \alpha_j|.$$

We have therefore,

$$d\{\log|f^*(z) - z(a')|*d\tau\} = \sum_{j=1}^{m}\frac{\partial\tau}{\partial\sigma}\frac{(\sigma-\alpha_j)}{|\zeta-\alpha_j|^2}d\sigma\wedge d\rho$$
$$+\sum_{j=1}^{m}\frac{\partial\tau}{\partial\rho}\frac{(\rho-\alpha_j)}{|\zeta-\alpha_j|^2}d\sigma\wedge d\rho.$$

Note that an equivalent way of saying $a' \to a$ is now: $\alpha_j \to 0$. Hence, to prove (5.2), it is sufficient to prove:

$$\int_{W_i}\frac{\partial\tau}{\partial\sigma}\frac{(\sigma-\alpha_j)}{|\zeta-\alpha_j|^2}d\sigma d\rho \to \int_{W_i}\frac{\partial\tau}{\partial\sigma}\frac{\sigma}{|\zeta|^2}d\sigma d\rho \text{ as } \alpha_j \to 0. \tag{5.3}$$

But the proof of this is entirely analogous to that of Fact 2 used in the proof of Theorem 2.8. Thus we are done. Q.E.D.

PROOF OF LEMMA 4.10. We have to show $\int_{\partial V[r]}(\log h)*d\tau$ is a continuous function of r. If both ($\mathfrak{B}$) and ($\mathfrak{C}$) of §4 are in force, there is nothing to prove. So let both assumptions be violated. Take p such that $df(p) = 0$ and $d\tau(p) = 0$. Let $\sigma = \tau - \tau(p)$ and $\zeta = \sigma + i\rho$ be the usual holomorphic function in a neighborhood of p. ζ is no longer a coordinate function because $d\tau(p) = 0$, but we still choose ζ so that $\zeta(p) = 0$. For some positive integer m, $\eta = (\zeta)^{1/m}$ will then be a valid coordinate function in a neighborhood W of p. Now suppose locally $\Omega = g\,dz \wedge d\bar{z}$ in M, where g is C^∞ and *zero-free*, and suppose $df^*(z)/d\eta = \eta^k l(\eta)$, where $l(\eta)$ is holomorphic and locally *zero-free*. Thus,

$$f^*\Omega = (f^*g)d(f^*(z)) \wedge d\overline{(f^*(z))}$$
$$= |\eta|^{2k}(f^*g)|l|^2 d\eta\wedge d\bar{\eta}.$$

Also, in $W - \{p\}$,

$$
\begin{aligned}
f^*\Omega &= h d\sigma \wedge *d\sigma \quad \text{(definition of } h) \\
&= h d\sigma \wedge d\rho = (ih/2)\, d\zeta \wedge d\bar{\zeta} \\
&= (im^2/2)|\eta|^{2(m-1)}\, h d\eta \wedge d\bar{\eta}.
\end{aligned}
$$

Hence in $W - \{p\}$,

$$
h = |\eta|^{2(k-m-1)}\{2(f^*g)|l|^2/im^2\} \Rightarrow \log h = \{2(k-m-1)\log|\eta| + C^\infty \text{function}\}.
$$

This shows that $\log h$ has the same type of singularities as f^*u_a. The proof of this lemma now proceeds in identical fashion as that of Lemma 4.5. Q.E.D.

6. **Three more technical lemmas.** We bring together here three simple lemmas which will be needed in the following section. The first is always taken for granted in the literature, although it is difficult to see why; the second and third (Lemmas 6.2 and 6.3) are extremely general and useful facts. The proofs in this section can be skipped on first reading.

LEMMA 6.1. *$n(t,a)$ is an integrable function on $[r_1, r_2] \times M$, where $[r_1, r_2]$ is any finite closed subinterval of $(r(\tau), s)$.*

PROOF. Since M is compact, $n(t,a)$ is bounded on $[r_1, r_2] \times M$. So it suffices to prove that it is locally integrable since a partition of unity will do the rest. Let C be the union of the set of all critical values of τ with the set of all r such that $\tau^{-1}(r)$ contains a critical point of f. C is a finite set and so $C \times M$ is a set of measure zero in $[r_1, r_2] \times M$. We will show that on $S \equiv [r_1, r_2] \times M - C \times M$, $n(t,a)$ is locally integrable. It will be understood in the following that each operation is carried out entirely in S.

Let $(t,a) \in S$. We must choose a neighborhood in which the function n is integrable. The choice splits into two cases:

Case 1. $f(\partial V[t]) \cap \{a\} = \varnothing$. Let U be any neighborhood of $a \in M$ such that $f(\partial V[t]) \cap \bar{U} = \varnothing$. We take ε so small that

$$
f(\operatorname{Int} V[t+\varepsilon] - V[t-\varepsilon]) \cap U = \varnothing.
$$

Then $(t-\varepsilon, t+\varepsilon) \times U$ is a neighborhood of (t,a) on which n is constant. This neighborhood will do.

Case 2. $f(\partial V[t]) \cap \{a\} \neq \varnothing$. Let $f^{-1}(a) \cap \partial V[t] = \{p_1, \cdots, p_m\}$. Choose a neighborhood U of a with the properties:

(i) $f^{-1}(U) \cap V[r]$ is a disjoint union $U_1 \cup \cdots \cup U_m \cup U'_1 \cup \cdots \cup U'_k$ where $p_i \in U_i$, $i = 1, \cdots, m$.

(ii) For some ε, $f^{-1}(U) \cap$ closure of $(V[t+\varepsilon] - V[t-\varepsilon])$ consists of exactly $U_1, \cdots, U_m$.

Now we prove that n is integrable in $(t-\varepsilon, t+\varepsilon) \times U$. First, there exist holomorphic maps $g_i: U \to U_i$ such that if $(t', a') \in (t-\varepsilon, t+\varepsilon) \times U$ and $f^{-1}(a')$

$= \{p_1', \cdots, p_m'\}$, then $g_i(a') = p_i'$. Define a sequence of partitions of the cylinder $(t-\varepsilon, t+\varepsilon) \times U$ by:

$$C_i^+ = \{(t', a') \colon \tau(g_i(a')) \leqq t'\},$$

$$C_i^- = \{(t', a') \colon \tau(g_i(a')) > t'\},$$

$i = 1, \cdots, m$. Since τ and each g_i are C^∞, the C_i^+ and C_i^- are Borel sets. Now if $\bar{n} \equiv n(t-\varepsilon, a)$ $(= n(t-\varepsilon, U))$, then clearly $n(t', a') = \bar{n} + \bar{k}$, where k is the total number of C_i^+'s to which (t', a') belong. Thus, $n(t, a)$ is a constant function on the union of all the $\bar{k}$-fold intersections of the C_i^+'s, which is again a Borel set. This implies that $n(t, a)$ defined on $(t-\varepsilon,\ t+\varepsilon) \times U$ is a simple function (Halmos [**9**], p. 84) and so is certainly integrable there. Q.E.D.

LEMMA 6.2. *Suppose $f\colon D \to V$ is a C^∞ map from a compact bordered Riemann surface D into another Riemann surface V and suppose Φ is an integrable form of degree two. Then $\int_D f^*\Phi = \int_V n(D, v)\Phi$, where, as usual, $n(D, v)$ denotes the Algebraic number of preimages of v in D.*

REMARK. (1) The meaning of "Algebraic number of preimages" is as follows. If $f(p) = v$, we say p covers v with multiplicity $+1$ (resp. -1) if f at p preserves (resp. reverses) the orientations of D and V. Higher multiplicity can be defined using local homology groups, but will not be needed as can be seen from the proof. (2) The lemma and the proof are true almost word for word in dimension n.

PROOF. Let $V' = V - \mathscr{C} - f(\partial D)$, where $\mathscr{C}$ is the set of critical values of f, i.e. $\mathscr{C} = \{v \in V\colon$ there exists $p \in f^{-1}(v)$ such that $df(p)$ is singular$\}$. By Sard's Theorem ([**15**], p. 7), $\mathscr{C}$ is a set of measure zero, so that $V - V'$ is a set of measure zero. If $v \in V'$, then $f^{-1}(v)$ is closed, hence isolated (because D is compact) and hence finite. Let $\bar{n}(D, v)$ be the total number of points in $f^{-1}(v) \cap V'$ *without* counting multiplicity. Then each v has a neighborhood *in* V' onto which f maps $\bar{n}(D, v)$ open sets diffeomorphically. Let $\{V_i\}$ be a locally finite covering of V' by such open sets and let $\{\phi_i\}$ be a partition of unity subordinated to $\{V_i\}$. Hence,

$$\begin{aligned}
\int_V n(D, v)\Phi &= \int_{V'} n(D, v)\Phi \\
&= \sum_i \int_{V'} \phi_i n(D, v)\Phi \qquad \text{(by definition of the integral)} \\
&= \sum_i \int_{V_i} n(D, v)(\phi_i\Phi) = \sum_i \int_{f^{-1}(V_i)} f^*(\phi_i\Phi) \\
&= \sum_i \int_{f^{-1}(Vi)} (f^*\phi_i) f^*\Phi = \int_{D - \partial D - \mathscr{P}} f^*\Phi = \int_D f^*\Phi
\end{aligned}$$

where $\mathscr{P}$ is the set of critical points of f in D, and the last equality is because $\int_{\mathscr{P}} f^*\Phi = 0$. Q.E.D.

Before proceeding to the last technical lemma, we pause to show how the above implies an interesting relation between $N(r, a)$ and $T(r)$. Recall that $N(r, a) = \int_{r_0}^{r} n(t, a)dt$. We shall integrate this with respect to a over M:

$$\int_M N(r, a)\Omega = \int_{r_0}^{r} \left(\int_M n(t, a)\Omega \right) dt \qquad \text{(Lemma 6.1 and Fubini)}$$

$$= \int_{r_0}^{r} dt \int_{V[t]} f^*\Omega \qquad \text{(Lemma 6.2)}$$

$$= \int_{r_0}^{r} v(t)dt = T(r), \qquad \text{by definition of } v(t).$$

In other words,

$$T(r) = \int_M N(r, a)\Omega. \tag{6.1}$$

This says that the order function is just the arithmetic mean of the counting function over M. Consequently, by (4.3), the arithmetic mean of the compensating term $((1/2\pi) \int_{\partial V[t]} f^*u_a *d\tau \,|_{t=r_0}^{t=r})$ over M is equal to zero.

By Remark B after Definition 4.1, if V carries an infinite harmonic exhaustion (i.e. $s = \infty$), V has to be parabolic and hence one knows that if $f: V \to M$, then $M - f(V)$ has capacity zero. (Cf. for instance [**13**].) Using (6.1), we now give a simple proof of a weaker result, namely, that under this circumstance $f(V)$ is open dense in M. For if not, $M - f(V)$ has measure $\varepsilon > 0$. We have $N(r, a) < T(r) + \text{const}$ (Corollary 4.8). Since $N(r, a) = 0$ for all $a \in M - f(V)$,

$$\int_{f(V)} N(r, a)\Omega = \int_M N(r, a)\Omega = T(r). \qquad (6.1)$$

On the other hand, $T(r)$ and the constant are both independent of a,

$$\int_{f(V)} (T(r) + \text{const})\Omega = (T(r) + \text{const})(1 - \varepsilon)$$

as $\int_M \Omega = 1$. So, $T(r) < (T(r) + \text{const})(1 - \varepsilon)$, or, $1 < (1 + \text{const}/T(r))(1 - \varepsilon)$. Now $v(t)$ is strictly increasing and positive, so that

$$T(r) = \int_{r_0}^{r} v(t)dt > v(r_0)(r - r_0) \to \infty \text{ as } r \to \infty.$$

(Here $s = \infty$ is used crucially.) Hence as $r \to \infty$, $\text{const}/T(r) \to 0$ and we obtain upon passing to the limit that $1 < (1 - \varepsilon)$, contradiction.

In particular, we retrieve the well-known result that a parabolic Riemann surface admits no bounded holomorphic function (just let M be the Riemann sphere). Also, if V is obtained from a compact Riemann surface M' by removing a set of measure nonzero, V cannot carry an infinite harmonic exhauston because the identity map from V into M' would contradict the above result. In particular, the following special case of a theorem of I.I. Priwalow can be proved ([**17**], p. 210):

There is no proper holomorphic mapping of the unit disc into the complex plane, i.e. a holomorphic function on the disc cannot converge uniformly to infinity towards the boundary.

For if such an f exists, $f^{-1}(0)$ would be a finite set in the disc and outside of this finite set, $\log |f|$ would then give an *infinite* harmonic exhaustion of the disc. Contradiction.

In the above, we integrated $N < T + \text{const}$ over M with respect to Ω to obtain a weak equidistribution statement. One remark we wish to make in this connection is that if one integrates this inequality with respect to more complicated measures, one can obtain various refined results of Nevanlinna theory. This idea of Ahlfors is carried out in the following section for one special case.

LEMMA 6.3 CONCAVITY OF THE LOGARITHM. *Let μ be a positive measure on a topological space R and let E be μ-measurable. Suppose f is a nonnegative and μ-integrable function defined on E, then*

$$\frac{1}{\mu(E)}\int_E (\log f)\,d\mu \leqq \log\left\{\frac{1}{\mu(E)}\int_E f\,d\mu\right\}.$$

PROOF. The following is a trivial rephrasing of the classical proof as given in Nevanlinna [**12** p. 251]. We need this lemma in this general setting because we shall have to let $R = E = \tau^{-1}(r)$, which is a disjoint union of circles.

So let $c = (1/\mu(E)) \int_E f\,d\mu$, and let $\phi = f - c$. Note two things: (i) $\int_E \phi\,d\mu = 0$ and (ii) $\phi/c \geqq -1$. Now for every $t \geqq -1$, $\log(1+t) \leqq t$. Hence by (ii), $\log(1+\phi/c) \leqq \phi/c$. So,

$$\frac{1}{\mu(E)}\int_E (\log f)\,d\mu = \frac{1}{\mu(E)}\left\{\int_E \log c\,d\mu + \int_E \log\left(1+\frac{\phi}{c}\right)d\mu\right\}$$

$$\leqq \log c + \frac{1}{\mu(E)}\int_E \frac{\phi}{c}\,d\mu = \log c \equiv \log\left\{\frac{1}{\mu(E)}\int_E f\,d\mu\right\}. \qquad \text{Q.E.D.}$$

7. **The defect relations.** The two main theorems (4.3) and (4.7) contain all the desirable information we want, with the exception of the two line integrals. What remains to be done is to obtain an estimate for the sum of these line integrals in terms of $T(r)$. Following Ahlfors, we now integrate $N(r,a) < T(r) + \text{const}$ with respect to $\rho\Omega$, where ρ is a function defined on M such that (i) it is integrable and nonnegative and (ii) normalized so that $\int_M \rho\Omega = 1$. We shall specify the explicit form of ρ later on.

Note as usual that the constant in the above inequality is independent of both r and a. (Corollary 4.8.) So,

$$\int_M N(r,a)\rho(a)\Omega < T(r) + \text{const.}$$

But,

$$\begin{aligned}
\int_M N(r,a)\rho(a)\Omega &= \int_{r_0}^{r} dt \int_M n(t,a)\rho(a)\Omega \qquad \text{(Lemma 6.1)}\\
&= \int_{r_0}^{r} dt \int_{V[t]} f^*(\rho\Omega) \qquad \text{(Lemma 6.2)}\\
&\geqq \int_{r_0}^{r} dt \int_{V[t]-V[r_0]} (f^*\rho)(f^*\Omega) \qquad (f^*\rho \text{ is positive})\\
&= \int_{r_0}^{r} dt\left(\int_{V[t]-V[r_0]} h(f^*\rho)d\tau \wedge *d\tau\right) \qquad (4.6)\\
&= \int_{r_0}^{r} dt \int_{r_0}^{t} ds \int_{\partial V[s]} h(f^*\rho) *d\tau.
\end{aligned}$$

The last equality is fairly obvious in general; for us, we can employ special coordinate functions (Definition 4.2) and make use of the fact that there is only a finite number of critical values of τ in $[r_0, t]$. Hence:

$$\int_{r_0}^{r} dt \int_{r_0}^{t} ds \left(\int_{\partial V[s]} h(f^*\rho) *d\tau\right) < T(r) + \text{const.} \tag{7.1}$$

We now apply Lemma 6.3. Using the notation of that lemma, let $R = \partial V[s]$ and $\mu(\gamma) = \int_\gamma *d\tau$ for every open subset γ of $\partial V[s]$. Since $*d\tau$ is a positive form on $\partial V[s]$ (see discussion after (1.4)), μ defines a positive measure on $\partial V[s]$. Also, by Corollary 4.3, $\mu(\partial V[s]) = L$ is a constant independent of s. So Lemma 6.3 implies:

$$\frac{1}{L}\int_{\partial V[s]} \log(hf^*\rho) *d\tau \leqq \log\left\{\frac{1}{L}\int_{\partial V[s]} hf^*\rho *d\tau\right\}$$

or

$$\log L + \frac{1}{L}\left\{\int_{\partial V[s]} \log(f^*\rho) *d\tau + \int_{\partial V[s]} (\log h) *d\tau\right\}$$
$$\leqq \log\int_{\partial V[s]} hf^*\rho *d\tau.$$

Hence, by (7.1),

$$\int_{r_0}^{r} dt \int_{r_0}^{t} \exp\left(\log L + \frac{1}{L}\left\{\int_{\partial V[s]} (\log f^*\rho) *d\tau + \int_{\partial V[s]} (\log h) *d\tau\right\}\right) ds$$
$$< T(r) + \text{const.} \tag{7.2}$$

Of course, we are trying to eliminate the two line integrals in (4.3) and (4.7). Therefore an inspection at (7.2) tells us that there is only one sensible choice of ρ to make: $\log f^*\rho$ should be f^*u_a. More precisely, let $a(1), \cdots, a(q)$ be a finite set of distinct points of M and let

$$(*) \qquad \rho = c_1\left(\prod_{1 \leqq k \leqq q} \exp u_{a(k)}\right)^{2\lambda}, \quad 0 < \lambda < 1$$

where c_1 is so chosen that $\int_M \rho\Omega = 1$. We must first prove ρ is integrable however. The only places that need checking are those near $a(1), \cdots, a(q)$. In fact, near $a(j)$, we only have to inspect the factor $(\exp u_{a(j)})$ of ρ. In the usual notation, $u_{a(j)} = \phi_{a(j)} - \log|z|$, where $\phi_{a(j)}$ is C^∞ in a neighborhood of $a(j)$. Thus, $(\exp u_{a(j)})^{2\lambda} = (\exp \phi_{a(j)})^{2\lambda} \cdot |z|^{-2\lambda} = C^\infty$ function $\cdot |z|^{-2\lambda}$, which is clearly an integrable function near the origin since $\lambda < 1$. So ρ is integrable. Clearly,

$$(7.3) \qquad \log f^*\rho = \text{const} + 2\lambda \sum_{k=1}^{q} f^*u_{a(k)}$$

where the constant is just $\log c_1$. Combining (7.3) and (7.2) and letting $\lambda \uparrow 1$, we have:

$$(7.4) \quad \int_{r_0}^{r} dt \int_{r_0}^{t} \exp\left(\text{const} + \frac{1}{L}\left\{2 \sum_{k=1}^{q} \int_{\partial V[s]} f^*u_{a(k)} *d\tau + \int_{\partial V[s]} \log h *d\tau \right\}\right) ds < T(r) + \text{const}.$$

Now substitute into (7.4) the expressions of the line integrals in (4.3) and (4.7) and make use of Lemma 4.7; it is easy to see that we have proved this: if we define

$$\omega(s) = \frac{4\pi}{L}\left\{\sum_{k=1}^{q} (T(s) - N(a(k), s)) + N_1(s) - \chi(M)T(s) + E(s)\right\} + \text{const}$$

where the constant is independent of s and $\{a(k)\}$, then

$$(7.5) \qquad \int_{r_0}^{r} dt \int_{r_0}^{t} e^{\omega(s)}\, ds < T(r) + \text{const}.$$

We wish to improve on the form of $\omega(s)$. First a definition.

DEFINITION 7.1. $\bar{n}(r, a)$ shall denote the number of pre-images of a in $V[r]$ *without* couting multiplicity. Furthermore

$$\bar{N}(r, a) = \int_{r_0}^{r} \bar{n}(t, a)\, dt.$$

We clearly have: $n(r, a) - \bar{n}(r, a) = \sum_{p \in f^{-1}(a) \cap V[r]}$(stationary index of f at p) $\geqq 0$ so that $n_1(r) \equiv \sum_{p \in V[r]}$(stationary index of f at p) $\geqq \sum_{k=1}^{k=q}\{n(r,a(k)) - \bar{n}(r,a(k))\} \geqq 0$. Hence, $N_1(r) \geqq \sum_{k=1}(N(r,a(k)) - \bar{N}(r,a(k)))$, which implies $\sum_k(T(r) - N(r,a(k)))$

$+ N_1(r) \geqq \Sigma_k(T(r) - \bar{N}(r, a(k)))$. Combining this with (7.5), we have the final result, usually referred to as the second main theorem of the subject:

THEOREM 7.1. *If we define*

$$\omega(r) = \frac{4\pi}{L}\left\{\sum_{k=1}^{k=q} (T(r) - \bar{N}(r, a(k))) - \chi(M)T(r) + E(r)\right\} + \text{const},$$

then

$$\int_{r_0}^{r} dt \int_{r_0}^{t} e^{\omega(s)} ds < T(r) + \text{const} \tag{7.6}$$

where the constants are all independent of r and $\{a(k)\}$.

We are now in a position to derive the defect relations. Let

$$\delta(a) = \liminf_{r \to s}\left(1 - \frac{\bar{N}(r, a)}{T(r)}\right),$$

$$\chi = \limsup_{r \to s}\left(\frac{-E(r)}{T(r)}\right).$$

It is clear that $\delta(a) \leqq 1$ and $\delta(a) = 1$ if a is omitted by f. $\delta(a)$ is called the *defect of* a. In the case that $\limsup_{r \to s} T(r) = \infty$, then it is further true that $0 \leqq \delta(a)$, in view of Corollary 4.8. In Nevanlinna theory this is the only case that one is interested in. All these are well known. We would like to remark that these definitions of $\delta(a)$ and χ actually depend on our choice of the exhaustion function τ. Thus far, the freedom of the choice of τ has not been fully exploited in the literature and the explicit dependence of $\delta(a)$ and χ on τ is not clear. What is known is that "$\delta(a)$ is not invariant under holomorphic transformations" in the case of the complex plane. However, it seems more natural to think of this as the noninvariance of $\delta(a)$ under a different choice of the exhaustion function.

At this point, it becomes crucial to distinguish between the two cases: $s = \infty$ and $s < \infty$. Formally, we make the

DEFINITION 7.2. V is said to admit an *infinite harmonic exhaustion* if $s = \infty$ in Definition 4.1. If $s < \infty$, we say V admits a *finite harmonic exhaustion.*

Case 1. *The infinite case.* (7.5) gives a bound for the doubly integrated $e^{\omega(s)}$, but it is possible to derive from this a bound for $e^{\omega(s)}$ itself. Naturally, we can only expect this bound to hold outside of some exceptional set.

LEMMA 7.2. *Suppose ψ is a once differentiable, positive, increasing function on $[r_0, \infty)$ with $r_0 \geqq 1$. Then for any real number $k > 1$, $\psi' \leqq \psi^k$ on $[r_0, \infty)$ except on an open set $I \subseteq [r_0, \infty)$ for which $\int_I d\log x < \infty$.*

PROOF. Let $I \subseteq [r_0, \infty)$ be the open set on which $\psi' > \nu\psi^k$ where ν is any positive function and $k > 1$ is an arbitrary positive number. Then,

$$v < \frac{1}{\psi^k}\psi' = \frac{1}{(1-k)}(\psi^{1-k})' \qquad \text{on } I$$

$$\Rightarrow \int_I v < \int_I \frac{1}{1-k}(\psi^{1-k})' = \sum_i \left.\frac{1}{(1-k)\psi^{k-1}}\right|_{a_i}^{b_i}$$

$$\leqq \lim_{i \to \infty} \frac{1}{1-k}\left(\frac{1}{\psi^{k-1}(b_i)} - \frac{1}{\psi^{k-1}(a_i)}\right) < \infty$$

where we have written $I = \bigcup_i (a_i, b_i)$ and the last inequality is because ψ is positive and increasing. So outside of I, $\psi' \leqq v\psi^k$. If we let $v = 1/x$, $r_0 \geqq 1$ implies $v \leqq 1$ on $[r_0, \infty)$ and the lemma is now obvious. Q.E.D.

Successive application of Lemma 7.2 to $\psi = \int_{r_0}^t e^{\omega(s)} ds$ and $\psi = \int_{r_0}^r dt \int_{r_0}^t e^{\omega(s)} ds$ give: if $k > 1$, then $e^{\omega(t)} \leqq (\int_{r_0}^t e^{\omega(s)} ds)^k$ for $t \notin I'$ such that $\int_{I'} d\log x < \infty$. $\int_{r_0}^t e^{\omega(s)} ds \leqq (T(t) + \text{const})^k$ for $t \notin I''$ such that $\int_{I''} d\log x < \infty$. Let $I = I' \cup I''$, then

$$e^{\omega(r)} < (T(r) + \text{const})^{k^2}$$

for $r \notin I$ such that $\int_{I''} d\log x < \infty$. In general, we shall let "$\|$" stand in front of an inequality if the latter only holds in $[r_0, \infty) - I$. Thus

$$\|\omega(r) < k^2 \log(T(r) + \text{const}),$$

$$\|\sum_{k=1}^{q} (T(r) - \bar{N}(r, a(k))) \leqq \chi(M)T(r) - E(r)$$

$$+ \frac{Lk^2}{4\pi}\log(T(r) + \text{const}) + \text{const}$$

by virtue of the expression of $\omega(r)$ in Theorem 7.1. Equivalently,

$$\|\sum_k \left(1 - \frac{\bar{N}(r, a(k))}{T(r)}\right)$$

$$\leqq \chi(M) + \frac{-E(r)}{T(r)} + \frac{1}{T(r)}\left\{\frac{Lk^2}{4\pi}\log(T(r) + \text{const}) + \text{const}\right\}.$$

Now the important thing to observe is that, since $\int_{I''} d\log x < \infty$, I does not comprise a neighborhood of ∞, so that the above inequality holds in an unbounded set. Hence,

$$\sum_{k=1}^{q} \delta(a(k)) \leqq \chi(M) + \chi + \limsup_{r \to \infty} \{ \qquad \}.$$

Since $T(r) \to \infty$ as $r \to \infty$ and since the constants in $\{ \quad \}$ are all independent of r and $\{a(k)\}$, l'Hôpital's rule implies that the last term vanishes. Consequently:

THEOREM 7.3 (DEFECT RELATIONS). *If $f: V \to M$ is holomorphic and V admits an infinite harmonic exhaustion, then for any finite set of distinct points $a(1), \cdots, a(q)$ of M:*

$$\sum_{k=1}^{q} \delta(a(k)) \leqq \chi(M) + \chi. \tag{7.7}$$

Case 2. The finite case. The following is the counterpart of Lemma 7.2.

LEMMA 7.4. *Suppose ψ is a once differentiable positive increasing function on $[0, s)$, $s < \infty$. Then for any real number $k > 1$, $\psi'(r) \leqq (1/(s-r))\psi^k(r)$ for all $r \in [0,s) - I$, where I is an open set such that $\int_I d\log(s-r) > -\infty$.*

The proof of this is analogous to that of Lemma 7.2. As before, the fact one should note is that I does not comprise a full neighborhood of s in $[0, s)$ so that $\psi'(r) \leqq (1/(s-r))\psi^k(r)$ is true for numbers arbitrarily close to s. By applying this lemma successively to $\int_{r_0}^{r} e^{\omega(s)}\, ds$ and $\int_{r_0}^{r} dt \int_{r_0}^{t} e^{\omega(s)} ds$, we obtain quite analogously

$$\| \omega(r) \leqq k^2 \log(T(r) + \text{const}) + (k+1) \log(1/(s-r))$$

where "$\|$" means "outside of I". This leads to:

$$\| \quad \sum_k \left(1 - \frac{\bar{N}(r, a(k))}{T(r)}\right)$$

$$\leqq \chi(M) + \frac{-E(r)}{T(r)} + \frac{1}{T(r)} \left\{\frac{Lk^2}{4\pi} \log(T(r)\,\text{const}) + (k+1)\log\frac{1}{s-r} + \text{const}\right\}$$

$$\equiv \chi(M) + \frac{-E(r)}{T(r)} + \varepsilon(r).$$

Suppose we know that

$$\limsup_{r \to s < \infty} \frac{\log(1/(s-r))}{T(r)} = 0. \tag{7.8}$$

Then clearly $\lim_{r \to s} T(r) = \infty$, so that $\limsup_{r \to s} \varepsilon(r) = \lim_{r \to s} \varepsilon(r) = 0$, again by appealing to l'Hôpital's rule. We have therefore proved:

THEOREM 7.4 (DEFECT RELATION). *If $f: V \to M$ is holomorphic and V admits a finite harmonic exhaustion, then for any finite set of distinct points $a(1), \cdots, a(q)$ of M:*

$$\sum_{k=1}^{q} \delta(a(k)) \leqq \chi(M) + \chi + \varepsilon \tag{7.9}$$

where

$$\varepsilon = \limsup_{r\to s} \frac{1}{T(r)}\left\{\frac{L}{4\pi}\log\,(T(r)+\text{const}) + 2\log\frac{1}{s-r} + \text{const}\right\}.$$

Furthermore, $\varepsilon = 0$ *if* (7.8) *is satisfied.*

8. **Additional remarks and examples.** It is not our purpose to draw any conclusions here from the defect relations (7.7) and (7.9) concerning exceptional values and points of ramification. These classical theorems are fully treated in Nevanlinna **[12]** and are true for our general situation with trivial modifications. We will only make such comments about χ and ε of (7.7) and (7.9) as are deemed necessary. We will also compute the right side of the defect relations for two cases explicitly. An application of the general theory will be found at the end of the Appendix.

First, a general remark about the sign of χ. If $\chi(V) < \infty$, then $\chi(V[r])$ must be constant and equal to $\chi(V)$ for all sufficiently large values of r, and *this fact is independent of the exhaustion function* τ *used.* (One of many ways to see this is to employ the homology group argument: $H_*(V,Z) = \lim_{\to} H_*(V[r],Z)$, where $\lim_{\to}$ denotes direct limit and H_* denotes homology.) This means that if we know $\chi(V) < \infty$, we can, and always do, assume r_0 has been so chosen that it is in the constancy range of $\chi(V[r]) \equiv \chi(r)$. Things simplify considerably with this choice. For instance, if $\chi(V) \geqq 0$ (i.e. topologically V is the plane or the punctured plane), then $\chi \leqq 0$. Hence χ can be deleted from the right-hand side of (7.7) and (7.9) in this case. This is the classical Nevanlinna theorem if $M =$ Riemann sphere, except that even here our statement is a bit more general: *the harmonic exhaustion function is arbitrary.*

Now some comments on the right side of (7.7) and (7.9). Again we separate the discussion into two cases.

Case 1. *The infinite case.* A holomorphic mapping $f: V \to M$ is called *transcendental* iff $\limsup_{r\to\infty} r/T(r) = 0$.

LEMMA 8.1. *If f is transcendental and $\chi(V)$ is finite, then* $\chi = 0$.

PROOF. By the above remark, we may assume r_0 to be so large that $\chi(V[t]) = \chi(V)$ for all $t \geqq r_0$. So

$$\chi = \limsup\left(\frac{-E(r)}{T(r)}\right) = \limsup \int_{r_0}^{r} \chi(t)dt/T(r)$$

$$= \limsup \frac{-\chi(V)(r-r_0)}{T(r)} = 0. \qquad \text{Q.E.D.}$$

The following two lemmas explain the meaning of transcendency.

LEMMA 8.2. *If $f: V \to M$ is transcendental, then for every real number r, $f(V - V[r])$ is dense in M.*

PROOF. Suppose false, then there exists $a \in M$ and a neighborhood U of a such that $U \cap f(V - V[r]) = \emptyset$ for some $r \in \boldsymbol{R}$. There is no harm in letting $r = r_0$. So $n(t, a) = n(r_0, a)$ for all $t \geqq r_0$. Hence, $N(r, a) = \int_{r_0}^{r} n(t, a)dt = n(r_0, a)(r - r_0)$. Next, since $t \geqq r_0 \Rightarrow f(\partial V[t]) \cap U = \emptyset$, f^*u_a restricted to $\partial V[t]$ for all such t is bounded above by a constant K.

By (4.3),

$$\begin{aligned} T(r) &= N(r, a) + \frac{1}{2\pi} \int_{\partial V[t]} f^*u_a *d\tau \Big|_{t=r_0}^{t=r} \\ &\leqq N(r, a) + \frac{1}{2\pi} \int_{\partial V[r]} f^*u_a *d\tau \\ &\leqq n(r_0, a)(r - r_0) + \frac{K}{2\pi} \cdot L \qquad \text{(Corollary 4.3)} \\ &= n(r_0, a)r + \text{const.} \end{aligned}$$

So clearly, $\limsup r/T(r) \geqq 1/n(r_0, a) > 0$. Q.E.D.

LEMMA 8.3. *If V is obtained from a compact Riemann surface M' by deleting a finite number of points $m(1), \cdots, m(l)$, then f: $V \to M$ is transcendental iff f is not the restriction of a holomorphic map $\tilde{f}$: $M' \to M$.*

PROOF. This is Chern's theorem and our proof is modelled on his. Lemma 8.2 implies this lemma in one direction; we will prove here that f not transcendental implies f extendable. So let $\lim\sup(r/T(r)) = \beta > 0$, then $\lim\inf(T(r)/r) = 1/\beta < \infty$. Since $N(r, a) < T(r) + \text{const}$ for all $a \in M$, $\liminf(N(r, a)/r) < 1/\beta < \infty$. But

$$\frac{(d/dr)N(r, a)}{(d/dr)r} = n(r, a)$$

and has a limit as $r \to \infty$ because $n(r,a)$ is monotone increasing. So $\lim(N(r,a)/r)$ exists and equals $\lim n(r, a)$ (l'Hôpital's rule again). Hence, $\lim n(r, a) = \lim(N(r,a)/r) = \lim\inf(n(r,a)/r) < 1/\beta < \infty$. But the number β is independent of a, so the number of pre-images of *a for all a* is bounded by a universal constant.

We now realize M as a ramified covering of the Riemann sphere S, i.e. $g: M \to S$ is a holomorphic mapping. Then gf: $V \to S$ clearly also has the property that the cardinality of $(gf)^{-1}(s)$ $(s \in S)$ is universally bounded. If $W_1, \cdots, W_l$ are coordinate neighborhoods of $m(1), \cdots, m(l)$, then the restriction of (gf) to $W_i - \{m(i)\}$ becomes a meromorphic function with a universally bounded number of pre-images. Now, a simple argument based on the Casorati-Weierstrass Theorem and the Baire category theorem implies that if gf had an essential singularity at any of the $m(i)$'s, there would have been at least one $s \in S$ whose pre-image is an infinite set. Since this is not the case, $gf \mid W_i - \{m(i)\}$ can be extended to W_i for all i.

Now fix an i and let $gf(m(i)) = s_i$. Choose an $a_i \in g^{-1}(s_i)$ so that a neighborhood of a_i intersects $f(W_i - \{m(i)\})$. Take a second ramified covering $h: M \to S$ so that $h^{-1}(h(a_i)) = a_i$. Apply the same argument as before to hf; we see that $f(W_i - \{m(i)\})$ is contained in a coordinate neighborhood of a_i. Then $f \mid W_i - \{m(i)\}$ is a holomorphic function, and the extension theorem of Riemann can be applied to give an extension $\tilde{f} \mid W_i$. Do this for each i, and we see that f itself can be extended to $\tilde{f}$ as well. Q.E.D.

We now give an explicit computation of the term χ for an $f: V \to S$ (Riemann sphere), such that V has infinite genus and f is transcendental. This is closely related to an example of B. Rodin ([**15**], Chapter I, §2.11).

Let f be the multi-valued function $(1/(e^{z} - 1))^{1/n}$ on C, and let V be its Riemann surface over C. Then f becomes a holomorphic mapping from V to S. V is clearly an n-fold ramified covering of C and the covering map $\pi: V \to C$ has stationary points at $z = 2m\pi i$, $m \in Z$, each with stationary index $(n-1)$. Let τ' be any C^{∞} function which equals $\log|z|$ outside of some disc of C, and let $\tau = \tau' \circ \pi$ be the harmonic exhaustion function of V. Using the Hurwitz formula (3.7), one sees quite easily that

$$E(r) \leqq ((-(n-1)/\pi)e^{r}) + 0(r) \text{ as } r \to \infty,$$

$$T(r) \geqq (1/\pi)e^{r} + O(r) \text{ as } r \to \infty$$

so that $\chi \leqq (n-1)$. (The details are very similar to those of the example found at the end of this section, and we do not wish to repeat ourselves.) Since $\chi(S) = 2$, we see that $\sum_k \delta(a(k)) \leqq n + 1$. On the other hand, f omits exactly $(n+1)$ points, namely, 0 and the nth roots of -1. Note finally that f is transcendental because

$$0 \leqq \limsup r/T(r) \leqq \limsup r/\left(\frac{1}{\pi}e^{r} + O(r)\right) = 0.$$

Case 2. *The finite case.* First of all, a trivial lemma.

LEMMA 8.4. *If* (7.8) *holds, and* $\chi(V) < \infty$, *then* $\chi = \varepsilon = 0$.

PROOF. If (7.8) is satisfied, we have already noted that $\varepsilon = 0$. Since $T(r) \to \infty$ as $r \to s$, and since $E(r) = \int_{r_0}^{r} \chi(t)dt \to s \cdot \chi(V) < \infty$ as $r \to s$, $\chi = 0$ also. Q.E.D.

Under the assumption of (7.8), the analogue of Lemma 8.2 again holds. But much more is true in this case:

LEMMA 8.5. *If* (7.8) *is true, then for almost all* $a \in M$,

$$\limsup_{r \to s} n(r,a)(s - r) = \infty.$$

COROLLARY 8.6. (7.8) *implies that* $f(V - V[r])$ *is dense in* M *for every real* $r < s$.

For the proof of Lemma 8.5, we first need a generalized form of l'Hôpital's rule whose proof is not too different from the usual version as found in Buck [**6**, p. 92].

LEMMA 8.7. *Let g and h be differentiable on the interval* $[a, b)$, $(b \leqq \infty)$, *such that* h' *exists and is nowhere zero on* $[a,b)$. *If*

$$\lim_{x \uparrow b} g(x) = 0 \text{ and } \lim_{x \uparrow b} h(x) = 0,$$

or if

$$\lim_{x \uparrow b} g(x) = \infty \text{ and } \lim_{x \uparrow b} h(x) = \infty,$$

then

$$\liminf_{x \uparrow b} \frac{g'}{h'} \leqq \liminf_{x \uparrow b} \frac{g}{h} \leqq \limsup_{x \uparrow b} \frac{g}{h} \leqq \limsup_{x \uparrow b} \frac{g'}{h'}. \tag{8.1}$$

PROOF OF LEMMA 8.5. By the remark after (6.1), the integral over M of $\int_{\partial V[t]} f^* u_a * d\tau \Big|_{t=r_0}^{t=r}$ with respect to Ω is zero. It is then clear from (4.4) and Fatou's lemma that

$$\liminf_{r \to s} \int_{\partial V[r]} f^* u_a * d\tau < \infty \text{ for almost all } a \in M. \tag{8.2}$$

Now assume (7.8) is satisfied, and assume also that there is a set E of *positive* measure such that $a \in E \Rightarrow \limsup_{r \to s} n(r, a)(s - r) = \alpha < \infty$. We shall deduce a contradiction. To begin with it is clear that the assumption implies

$$\liminf_{r \to s} \frac{1}{n(r,a)(s-r)} > 0. \tag{8.3}$$

Pick an $a \in E$ such that (8.2) is fulfilled. Then there is a sequence $\{r_i\}$ converging to s such that $\lim_{r_i \to s} \int_{\partial V[r]} f^* u_a * d\tau = \text{const} < \infty$. By (4.3), $T(r_i) \leqq N(r_i, a) + \text{const}$, where the constant does not depend on r_i. Since (7.8) is in force, $T(r_i) \to \infty$, so that

$$\lim_{r_i \to s} (N(r_i, a) + \text{const}) = \infty. \tag{8.4}$$

Hence,

$$\limsup_{r \to s} \frac{\log(1/(s-r))}{T(r)} \geqq \limsup_{r_i \to s} \frac{\log(1/(s-r_i))}{T(r_i)}$$

$$\geqq \limsup_{r_i \to s} \frac{\log(1/(s-r_i))}{N(r_i, a) + \text{const}} \geqq \liminf_{r_i \to s} \frac{\log(1/(s-r_i))}{N(r_i, a) + \text{const}}$$

$$\geqq \liminf_{r \to s} \frac{\log(1/(s-r))}{N(r, a) + \text{const}} \geqq \liminf_{r \to s} \frac{1/(s-r)}{n(r,a)}$$

(by virtue of (8.4) and Lemma 8.7)

$$= \liminf_{r \to s} \frac{1}{(s-r)n(r, a)} > 0 \qquad (8.3).$$

This contradicts (8.7). Q.E.D.

No nice interpretation of (7.8) analogous to Lemma 8.3 is known. We now give an example of a holomorphic function defined in the unit disc which satisfies (7.8) and which has only one essential singularity on the boundary. We also give another example to show that the existence of an essential singularity on the boundary is not quite sufficient to guarantee (7.8). Incidentally, the first example will also show that (7.9) is best possible for holomorphic mappings satisfying (7.8).

So let $D = \{|z| < 1\}$ and consider the multi-valued function

$$f = \{-1 + \exp(2\pi i(1-z)^{-2})\}^{1/n}$$

defined on D. f clearly has $z=1$ as an essential singularity. Let V be the Riemann surface of f. Then $f: V \to S$ (Riemann sphere) is a single-valued holomorphic mapping. Let $\pi: V \to D$ be the natural projection of the ramified covering.

The zeros of f in D are located at $z = (1 - i/n^{1/2})$, $n = 1, 2, 3, \cdots$ and so a circle of radius $r < 1$ encloses at most $1/(1-r)^2$ zeros of f (up to a constant which is clearly independent of r). The inverse images of these zeros are exactly the ramified points of π. Let τ' be the C^∞ function which equals $\log(|z|/\varepsilon)$ outside of a disc of radius 2ε (ε very small compared with 1), so that $\tau': D \to [0, -\log\varepsilon)$ is onto. Let $\tau = \tau' \cdot \pi$. Then τ defines a finite harmonic exhaustion of V. Since the zeros of f on V are simple zeros, it is fairly easy to see that $V[r]$ encloses at most $1/(1-\varepsilon e^r)^2$ zeros of f (up to a constant independent of r). Hence,

$$\begin{aligned} N(r, 0) &= \int_{2\varepsilon}^{r} n(t,0)dt \leqq \int_{2\varepsilon}^{r} \left\{ \frac{1}{(1-\varepsilon e^t)^2} + \text{const} \right\} dt \\ &= \int_{e^{2\varepsilon}}^{e^r} \frac{1}{s} \frac{ds}{(1-\varepsilon s)^2} + O(r) \\ &= \left\{ \frac{1}{(1-\varepsilon s)} - \log \frac{1-\varepsilon s}{s} \right\} \Bigg|_{s=e^{2\varepsilon}}^{s=e^r} + O(r) \\ &\to \infty \text{ as } r \to -\log\varepsilon. \end{aligned}$$

This therefore justifies the use of (8.1) in the following:

$$\begin{aligned} 0 &\leqq \limsup_{r \to -\log\varepsilon} \frac{\log(1/(-\log\varepsilon - r))}{T(r)} \\ &\leqq \limsup_{r \to -\log\varepsilon} \frac{\log(1/(-\log\varepsilon - r))}{N(r,0) + \text{const}} \qquad \text{(Corollary 4.8)} \\ &\leqq \limsup \frac{(-\log\varepsilon - r)^{-1}}{(1-\varepsilon e^r)^{-2} + \text{const}} \qquad (8.1) \\ &\leqq 0 \qquad \text{by repeated application of (8.1).} \end{aligned}$$

This proves that (7.8) is satisfied, and hence $\varepsilon = 0$ in (7.9). We next turn to the computation of χ. We appeal to (3.7), using $\pi: V \to D$ to compute $\chi(V[r])$. Since the zeros of f are the points of ramification of π, and since each such point clearly has stationary index $(n-1)$, we see that the stationary index of π in $V[r]$ is at most

$$\{(n-1)/(1-\varepsilon e^r)^2 + \text{const}\}$$

where the constant is independent of r as usual. So,

$$\chi(r) \geqq n\chi(\pi(V[r])) - \{(n-1)/(1-\varepsilon e^r)^2 + \text{const}\}$$
$$= n - \{(n-1)/(1-\varepsilon e^r)^2 + \text{const}\} = -(n-1)/(1-\varepsilon e^r)^2 + \text{const}.$$

Hence, $E(r) \geqq \int_{2\varepsilon}^{r} \{-(n-1)/(1-\varepsilon e^t)^2 + \text{const}\}\, dt$. By the above result concerning $N(r,0)$, we see that $E(r) \to \infty$ as $r \to -\log \varepsilon$. So (8.1) can be applied, and it is clear that

$$\chi = \limsup_{r \to -\log\varepsilon} \frac{-E(r)}{T(r)} \leqq \limsup \frac{-E(r)}{N(r,0) + \text{const}} \leqq (n-1).$$

Therefore, (7.9) reads in this case

$$\sum_k \delta(a(k)) \leqq \chi(S) + (n-1) = (n+1).$$

But it is clear that f misses exactly $(n+1)$ points: ∞ and the nth roots of -1. Thus (7.9) is optimal at least for meromorphic functions.

Finally, the function $f = -1 + \exp(2\pi i(1-z)^{-1})$ on D also has $z = 1$ as an essential singularity. But in this case:

$$\limsup_{r \to -\log\varepsilon} \frac{\log(1/(-\log\varepsilon - r))}{T(r)} = 1$$

so that (7.8) is not satisfied. The details are very similar to the above and so will be left out.

Part B

9. **Arbitrary exhaustion function and the general theory.** We now pass to the general case, i.e. $f: V \to M$ is a holomorphic mapping from an arbitrary open Riemann surface V into a compact Riemann surface M. V in general will not possess a harmonic exhaustion, but it does possess a nice enough exhaustion function of the following type:

Definition 9.1. A C^∞ function $\tau: V \to [0,s)$ $(s \leqq \infty)$ will be called an *exhaustion function* on the open Riemann surface V iff

(1) τ is onto.

(2) τ is proper, i.e. τ^{-1} (compact) = compact.

(3) The critical points of τ on $\{p: \tau(p) \geqq r(\tau)\}$ for some $r(\tau) < s$ are isolated and *finitely homogeneous*.

We explain (3). Let p be a critical point of τ and let $\zeta = \sigma + i\rho$ be a p-centered holomorphic coordinate function. Then p is *finitely homogeneous* iff $N \equiv (\partial\tau/\partial\sigma)^2 + (\partial\tau/\partial\rho)^2 = |\zeta|^k \psi(\sigma,\rho)$ where k is some positive integer and ψ is a C^∞ function *without zeros*. It is clear that if N has this form relative to one holomorphic coordinate function, it has the same form relative to any other, so the definition makes sense. It is obvious that a finitely homogeneous critical point is automatically isolated. The redundant requirement of isolatedness in (3) is merely for emphasis. Note that if τ is harmonic, all of its critical points are finitely homogeneous. (See the proof of Lemma 4.10 in §5.)

The condition (3) is a bit artificial. Actually the isolatedness is just to insure that Stokes' Theorem (1.4) is applicable and the finite homogeneity is to guarantee the convergence of certain integrals that come up in the Second Main Theorem (Theorem 9.6). On the other hand, by the above remark, every harmonic exhaustion function is an exhaustion function in the above sense. More importantly, every open Riemann surface possesses a *canonical exhaustion function* $\bar{\tau}$ which meets the above requirements and more: (a) $\bar{\tau}$ is onto $[0, \infty)$, (b) $\bar{\tau}$ is everywhere free of critical points and (c) $\bar{\tau}$ is everywhere strictly subharmonic. The simplest way to obtain $\bar{\tau}$ is as follows. By a theorem of Behnke-Stein ([**18**], p. 270, the proof there contains an error), every open Riemann surface is a Stein Manifold. The imbedding theorem of Bishop-Narasimhan ([**18**], p. 224) then implies that every open Riemann surface can be imbedded holomorphically as a closed submanifold of $\boldsymbol{C}^3$. Let f_1, f_2, f_3 be the three functions on V which give this holomorphic imbedding. Then $\bar{\tau} = \sum_i |f_i|^2$ will do.

Finally, we would like to point out that, since we deal with line integrals whose integrands have singularities, the isolatedness of the critical points is probably essential. For a discussion of the difficulties involved, see Whitney [**16**, pp. 100–101].

As before, we let:

$$V[r] = \{p : \tau(p) \leqq r\}.$$

Then $\partial V[r] = \{P : \tau(p) = r\}$ for all $r \geqq r(\tau)$. We will tacitly assume that $r \geqq r(\tau)$ in all that follows. Since the critical points of τ above $r(\tau)$ are isolated, the properness of τ implies again that the critical values of τ are isolated, and that there can be only a finite number of critical points on each $\tau^{-1}(r)$ if $r \geqq r(\tau)$. Consequently, all the $V[r]$ for $r \geqq r(\tau)$ have only a finite number of "corners" and thus (1.4) is valid on them. Writing $n(r, a)$ for $n(V[r], a)$ and $v(r)$ for $v(V[r])$, we translate (3.2) into

$$n(t, a) + \frac{1}{2\pi} \int_{\partial V[t]} f^* d^c u_a = v(t)$$

under the assumptions that

$$(\mathfrak{A}) \qquad f(\partial V[t]) \cap \{a\} = \varnothing,$$

$$(\mathfrak{B}) \qquad t \text{ is not a critical value of } \tau.$$

Integrating from some $r_0 \geqq r(\tau)$ upward, we get:

$$\int_{r_0}^{r} n(t,a)dt + \frac{1}{2\pi}\int_{r_0}^{r} dt \int_{\partial V[t]} f^* d^c u_a = \int_{r_0}^{r} v(t)dt,$$

or, defining $N(r,a)$ and $T(r)$ by this equation, we write:

$$(9.1) \qquad N(r,a) + \frac{1}{2\pi}\int_{r_0}^{r} dt \int_{\partial V[t]} f^* d^c u_a = T(r).$$

This holds if no $t \in [r_0, r]$ violates $(\mathfrak{A})$ or $(\mathfrak{B})$. Clearly, $N(r,a)$ and $T(r)$ are well-defined functions for all r, regardless of $(\mathfrak{A})$ or $(\mathfrak{B})$. We only have to deal carefully with the middle integral.

LEMMA 9.1. *$\int_{V[r]-V[r_0]} d\tau \wedge d^c f^* u_a$ exists for all $r \geqq r(\tau)$.*

This will be proved in §10. Granting this, it is quite obvious that

$$\int_{V[r]-V[r_0]} d\tau \wedge d^c f^* u_a = \int_{r_0}^{r} dt \int_{\partial V[t]} d^c f^* u_a$$

for all $r \geqq r(\tau)$. ((3) of Definition 9.1 makes this assertion almost trivial but, in the presence of (2) of that definition, is actually not needed.) Hence from (9.1):

$$(9.2) \qquad N(r,a) + \frac{1}{2\pi}\int_{V[r]-V[r_0]} d\tau \wedge d^c f^* u_a = T(r), \qquad \text{for all } r \geqq r(\tau).$$

Now suppose $f^{-1}(a) \cap \{V[r] - V[r_0]\} = \{p_1, \cdots, p_l\}$, and let us enclose each p_i by a holomorphic disc U_ε^i of radius ε. Let $U_\varepsilon = \bigcup_i U_\varepsilon^i$. Then,

$$\int_{V[r]-V[r_0]} d\tau \wedge d^c f^* u_a = \lim_{\varepsilon\to 0}\int_{V[r]-V[r_0]-U_\varepsilon} d\tau \wedge d^c f^* u_a .$$

On $V[r] - V[r_0] - U_\varepsilon$, $d\tau \wedge d^c f^* u_a$ is C^∞ and so one may check that: $d\tau \wedge d^c f^* u_a = df^* u_a \wedge d^c\tau = d(f^* u_a d^c \tau) - f^* u_a dd^c \tau$.
Hence,

$$\begin{aligned}\int_{V[r]-V[r_0]} & d\tau \wedge d^c f^* u_a \\ &= \lim_{\varepsilon\to 0}\left\{\int_{V[r]-V[r_0]-U_\varepsilon} d(f^* u_a d^c\tau) - \int_{V[r]-V[r_0]-U_\varepsilon} f^* u_a dd^c\tau\right\} \\ &= \lim_{\varepsilon\to 0}\left\{\int_{\partial V_\varepsilon[t]} f^* u_a d^c\tau\Big|_{t=r_0}^{t=r} - \sum_i \int_{\partial U_\varepsilon^i} f^* u_a d^c\tau \right. \\ &\qquad \left. - \int_{V[r]-V[r_0]-U_\varepsilon} f^* u_a dd^c \tau\right\} \qquad (1.4)\end{aligned}$$

where we have written $\partial V_\varepsilon[t]$ for that portion of $\partial V[t]$ lying outside of those U_ε^i's such that $\partial V[t] \cap U_\varepsilon^i \neq \varnothing$. We now need another lemma whose proof is postponed to §10.

LEMMA 9.2. (a) $\lim_{\varepsilon\to 0}\int_{\partial U_\varepsilon^i} f^*u_a d^c\tau = 0$.

(b) $\int_{\partial V[r]} f^*u_a d^c\tau$ is finite for all $r \geqq r(\tau)$.

(c) $\int_{V[r]} f^*u_a dd^c\tau$ is finite for all r.

Apply this lemma to the last equation and combine it with (9.2); we have proved

THEOREM 9.3 (FIRST MAIN THEOREM). *For all* $r \geqq r(\tau)$,

$$(9.3)\quad N(r,a) + \frac{1}{2\pi}\int_{\partial V[t]} f^*u_a d^c\tau\Big|_{t=r_0}^{t=r} - \frac{1}{2\pi}\int_{V[r]-V[r_0]} f^*u_a dd^c\tau = T(r).$$

REMARK. Recall that if τ is harmonic $dd^c\tau = 0$ and $d^c\tau = *d\tau$. (9.3) therefore generalizes (4.3). These remarks apply to the rest of this section.

The integration of (3.6) proceeds roughly the same as before. Recall that under ($\mathfrak{B}$) and

($\mathfrak{C}$) df is nowhere zero along $\partial V[t]$,

we have $\chi(t) + n_1(t) - \chi(M)v(t) = (1/2\pi)\int_{\partial V[t]} k_g\, ds$, where $\chi(t) \equiv \chi(V[t])$, $n_1(t) = n_1(V[t])$, k_g is the geodesic curvature of $\partial V[t]$ in the metric f^*G, and ds is the element of arc-length of $\partial V[t]$. Integration with respect to t gives, in self-explanatory notation:

$$(9.4)\quad E(r) + N_1(r) - \chi(M)T(r) = \frac{1}{2\pi}\int_{r_0}^{r} dt \int_{\partial V[t]} k_g ds.$$

While the left side makes sense for all r, the right side a priori does not. To deal with this line integral, let

$$f^*\Omega \equiv h d\tau \wedge d^c\tau.$$

Recall from (1.2) that using local coordinates $\zeta = \sigma + i\rho$

$$(9.5)\quad d\tau \wedge d^c\tau = N d\sigma \wedge d\rho$$

where $N = \{(\partial\tau/\partial\sigma)^2 + (\partial\tau/\partial\rho)^2\}$. This means that outside of the set of critical points of τ, $d\tau \wedge d^c\tau$ is a positive form with respect to the orientation of V so that *h is a nonnegative function* defined on V outside of the critical points of τ. From (9.5) and (1.5), the metric f^*G locally assumes the form:

$$f^*G = hN(d\sigma^2 + d\rho^2)$$

so that the (oriented) element of arc-length of $\partial V[t]$ is $ds = h^{1/2}d^c\tau$ (see discussion after (1.1)). Hence outside of the critical points of τ

$$(9.6)\quad k_g d\tau \wedge ds = k_g h^{1/2} N d\sigma \wedge d\rho.$$

The basic fact needed now is the following special case of Beltrami's Theorem which can easily be computed directly (Eisenhart [8], p. 189):

LEMMA 9.4. *Given holomorphic coordinates* $\zeta = \sigma + i\rho$ *and Hermitian metric* $g(d\sigma^2 + d\rho^2)$ *on a Riemann surface, the geodesic curvature of a one-dimensional submanifold defined by a* C^∞ τ *(i.e.* $\tau = \text{const}$*) is*

$$k_g = \frac{1}{(Ng)^{1/2}}\bar{\Delta}\tau + \frac{1}{g}\left(\frac{\partial\tau}{\partial\sigma}\frac{\partial}{\partial\sigma} + \frac{\partial\tau}{\partial\rho}\frac{\partial}{\partial\rho}\right)\left(\frac{g}{N}\right)^{1/2}$$

where N is as in (9.5) *and* $\bar{\Delta} = \partial/\partial\sigma^2 + \partial^2/\partial\rho^2$.

For us, $g = hN$, so that the lemma simplifies to:

$$k_g = \frac{1}{Nh^{1/2}}\left\{\bar{\Delta}\tau + \frac{\partial\tau}{\partial\sigma}\frac{\partial}{\partial\sigma}\log h^{1/2} + \frac{\partial\tau}{\partial\rho}\frac{\partial}{\partial\rho}\log h^{1/2}\right\}$$

Using (9.6), we get

$$k_g d\tau \wedge ds = \bar{\Delta}\tau d\sigma \wedge d\rho + \frac{1}{2}\left\{\frac{\partial\tau}{\partial\sigma}\frac{\partial\log h}{\partial\sigma} + \frac{\partial\tau}{\partial\rho}\frac{\partial\log h}{\partial\rho}\right\} d\sigma \wedge d\rho$$

and so using (1.2) and (1.3), we have arrived at

(9.7) $$k_g d\tau \wedge ds = dd^c\tau + \tfrac{1}{2}\, d\tau \wedge d^c \log h.$$

This formula is valid only outside of the critical points of τ and of f (which correspond to the infinities and zeros of h respectively). Nevertheless, on the basis of this and (3) of Definition 9.1, a proof will be given in §10 of the following:

LEMMA 9.5. $\int_{V[r]-V[r_0]} k_g d\tau \wedge ds$ *is finite for all* $r \geqq r(\tau)$, *and hence* $\int_{r_0}^{r} dt \int_{\partial V[t]} k_g ds = \int_{V[r]-V[r_0]} k_g d\tau \wedge ds$ *for all* $r \geqq r(\tau)$.

Now we *define*: $l(t) = \int_{\partial V[t]} d^c\tau$. Stokes' Theorem implies that $\int_{V[r]-V[r_0]} dd^c\tau = l(r) - l(r_0)$. Combining this with Lemma 9.5, (9.7) and (9.4), we see that

(9.8) $$\begin{aligned} &E(r) + N_1(r) - \chi(M)T(r) \\ &\quad = \frac{1}{2\pi}\{l(r) - l(r_0)\} + \frac{1}{4\pi}\int_{V[r]-V[r_0]} d\tau \wedge d^c \log h. \end{aligned}$$

It will be seen from the proof of Lemma 9.5 that the type of singularity of log h is identical with that of f^*u_a so that the last integral of (9.8) can be treated in the same manner as the middle integral of (9.2). Hence,

$$\int_{V[r]-V[r_0]} d\tau \wedge d^c \log h = \int_{\partial V[t]} (\log h) d^c\tau \Big|_{t=r_0}^{t=r} - \int_{V[r]-V[r_0]} (\log h) dd^c\tau.$$

Combine this with (9.8) and we have proved:

THEOREM 9.6 (SECOND MAIN THEOREM). *For* $r \geqq r(\tau)$ *and* $l(t) = \int_{\partial V[t]} d^c\tau$:

(9.9)
$$E(r) + N_1(r) - \chi(M)(T(r)$$
$$= \frac{1}{2\pi}\{l(r) - l(r_0)\} + \frac{1}{4\pi}\int_{\partial V[t]} (\log h) d^c\tau \Big|_{t=r_0}^{t=r}$$
$$- \frac{1}{4\pi}\int_{V[r]-V[r_0]} (\log h) dd\ \tau.^c$$

To arrive at the defect relations, we again try to obtain a bound for the sum of the two line integrals in (9.3) and (9.9), i.e.

$$\left\{(1/2\pi)\int_{\partial V[r]} f^*u_a d^c\tau + (1/4\pi)\int_{\partial V[r]} (\log h) d^c\tau\right\}.$$

Of course, it would be desirable to be able to have an estimate for the superfluous terms in (9.3) and (9.9) also, namely, $l(r), \int_{V[r]-V[r_0]}(\log h)dd^c\tau$ and $\int_{V[r]-V[r_0]} f^*u_a dd^c\tau$, but we are inclined to believe that this is not always feasible. In any case, we first need an analogue of Corollary 4.8.

LEMMA 9.7. *Both* $\int_{\partial V[t]} f^*u_a d^c\tau$ *and* $\int_{V[t]-V[r_0]} f^*u_a dd^c\tau$ *are continuous functions of a for a fixed t.*

The proof of Lemma 4.7 in §5 carries over to this case without any essential changes, so we omit the proof. By the usual reasoning using the compactness of M, we have from (9.3) that

(9.10)
$$N(r,a) < T(r) + F(r,a) + \text{const}$$

where $F(r,a) \equiv (1/2\pi)\int_{V[r]-V[r_0]} f^*u_a dd^c\tau$ and the constant is independent of r and a. Suppose τ is a superharmonic exhaustion on V, then $F(r,a)$ is nonpositive for all r and a and so can be deleted from (9.10) so that $N(r,a) < T(r) + \text{const}$ again holds. Integrating this inequality in exactly the same way as done in §6, we see that $f(V)$ is open dense in M if $\tau \to \infty$. We summarize this.

THEOREM 9.8. *Let* $f: V \to M$ *be a holomorphic map and let* V *admit an infinite* ($s = \infty$) *exhaustion function which is eventually superharmonic. Then* $f(V)$ *is open dense in* M.

Note that so long as the superharmonic function eventually has isolated critical points, Theorem 9.8 is valid. Finite homogeneity was never used.

Now we integrate (9.10) with respect to $\rho\Omega$ over M, where ρ is again chosen as in (*) of §7. This leads to the generalized version of Theorem 7.1. Since the details are formally the same, we content ourselves with stating the final result:

THEOREM 9.9. *Let* $a(1), \cdots, a(q)$ *be a finite set of distinct points in* M. *Let further*

$$\omega(r) \equiv \frac{4\pi}{l(r)}\left\{\sum_{k=1}^{q} (T(r) - \bar{N}(r, a(k)) - F(r, a(k)) - \chi(M)T(r) + E(r) + G(r) + \text{const}\right\}$$
$$+ \log l(r) + \text{const}.$$

Then

$$\int_{r_0}^{r} dt \int_{r_0}^{t} e^{\omega(s)} ds < T(r) + H(r) + \text{const} \tag{9.11}$$

where $\bar{N}(s,a)$ is as in Definition 7.1, $F(r,a)$ as in (9.10),

$$l(r) = \int_{\partial V[r]} d^c\tau$$

$$H(r) = \int_M F(r,a)\rho(a)\Omega \qquad \textit{(exists because of Lemma 9.7),}$$

$$G(r) = \frac{1}{4\pi}\int_{V[r]-V[r_0]} (\log h) dd^c\tau$$

and the constants are all independent of r and $\{a(k)\}$.

To arrive at the defect relations, one should distinguish between two cases: $s < \infty$ and $s = \infty$. But the infinite case is more interesting, so we treat that case alone. Therefore, *let $\tau: V \to [0, \infty)$ be onto* and define:

$$\delta^*(a) = \liminf_{r\to\infty}\left(1 - \frac{\bar{N}(r,a) - F(r,a)}{T(r)}\right)$$

$$\chi = \limsup_{r\to\infty} \frac{-E(r)}{T(r)}$$

as before, then an application of Lemma 7.2 to (9.11) yields immediately:

$$\sum_{k=1}^{q} \delta^*(a(k)) \leqq \chi(M) + \chi + \varepsilon^* \tag{9.12}$$

where $\varepsilon^* = \limsup_{r\to\infty} [(l(r)/4\pi T(r)) \{\log(T(r) + H(r) + \text{const}) - \log l(r) + \text{const}\} - G(r)/T(r)]$.

Let us examine (9.12) in some detail. To begin with, the generalized defect $\delta^*(a)$ still obeys $0 \leqq \delta^*(a) \leqq 1$ because of (9.10), but it is no longer true that $\bar{n}(r,a) = 0$ for all $r \Rightarrow \delta^*(a) = 1$. This together with the complicated form of ε^* render this generalized defect relation (9.12) almost completely meaningless. A more respectable case is when the exhaustion function is (eventually) superharmonic. In this case, $l(r)$ is monotone decreasing and $H(r)$ is negative. Using (8.1), it is not too difficult to see that

$$\limsup_{r\to\infty} \frac{l(r)}{T(r)} \{\log(T(r) + \text{const}) - \log l(r) + \text{const}\} = 0.$$

Hence, *in the case of an infinite superharmonic exhaustion*

$$\sum_{k=1}^{q} \delta^{\#}(a(k)) \leqq \chi(M) + \chi + \limsup_{r \to \infty} \frac{-G(r)}{T(r)}.$$

There is in general no estimate for $G(r)$ even in this case.

The case of a *subharmonic exhaustion* on V is interesting for different reasons. As remarked earlier (after Definition 9.1), every open Riemann surface carries an exhaustion of this type, so that *this case is actually the most general possible.* Since $F(r, a)$ is now positive, we can replace (9.12) by

$$\sum_{k=1}^{q} \delta(a(k)) \leqq \chi(M) + \chi + \varepsilon^{\#} \tag{9.13}$$

where $\delta(a)$ is the usual defect as defined in §7. (9.13) therefore allows us to count the number of points omitted by f. But now, $l(r)$ and $H(r)$ are strictly monotone increasing functions and an inspection at $\varepsilon^{\#}$ reveals that $\varepsilon^{\#}$ can be highly nonfinite in this case. Therefore (9.13) gives no information about the distribution of values under f unless we can force $\varepsilon^{\#}$ to be finite (as well as χ). Since no nice interpretation is available in general of an f that makes $\varepsilon^{\#}$ finite, this indicates that a meaningful Nevanlinna theory is out of reach when the domain is arbitrary.

Naturally, if τ is a harmonic exhaustion, $F = G = H = 0$ and l stays constant. One then retrieves the results of §7. This seems to be the only case where (9.12) makes any sense and, to a certain extent, this phenomenon explains why the harmonically exhaustible space should occupy a central position in the theory of Nevanlinna.

§10. **Outline of the proofs of the lemmas.**

Proof of Lemma 9.1. We examine the singularities of $d\tau \wedge d^c f^* u_a$, i.e. at $f^{-1}(a)$. So take a p such that $f(p) = a$ and let z be an a-centered holomorphic coordinate function at a; let ζ be a p-centered holomorphic coordinate function at p. As usual, $\phi_a \equiv u_a + \log|z|$ is C^∞ locally. Let $f^*(z) = \zeta^m h(\zeta)$, where h is holomorphic and locally zero free, and m is a positive integer. Then using (1.2) and (1.3):

$$\begin{aligned} f^* u_a &= C^\infty \text{ function} - m \log|\zeta| \\ d^c f^* u_a &= C^\infty \text{ form} + \frac{m}{|\zeta|^2}(\rho\, d\sigma - \sigma\, d\rho) \\ d\tau \wedge d^c f^* u_a &= C^\infty \text{ form} - m\left(\frac{\partial \tau}{\partial \rho} \frac{\rho}{|\zeta|^2} + \frac{\partial \tau}{\partial \sigma} \frac{\sigma}{|\zeta|^2} \right) d\sigma \wedge d\rho \end{aligned}$$

and the coefficient of $d\sigma \wedge d\rho$ in the last equation is clearly a locally integrable function in the plane. Q.E.D.

Proof of Lemma 9.2. The proofs of (a) and (b) are almost identical with the proof of Lemma 4.5 in §5. For (c), it suffices to look at $f^*u_a dd^c\tau$ at a point p such that $f(p) = a$. Using the notation of the preceding proof: $f^*u_a dd^c\tau = C^\infty$ form $- m(\partial^2\tau/\partial\sigma^2 + \partial^2\tau/\partial\rho^2)\log|\zeta|\, d\sigma \wedge d\rho$. This is clearly locally integrable. Q.E.D.

PROOF OF LEMMA 9.5. We need condition (3) of Definition 9.1 here By the formula preceding (9.7)

$$k_g d\tau \wedge ds = C^\infty \text{ form} + \frac{1}{2}\left\{\frac{\partial\tau}{\partial\sigma}\frac{\partial\log h}{\partial\sigma} + \frac{\partial\tau}{\partial\rho}\frac{\partial\log h}{\partial\rho}\right\} d\sigma \wedge d\rho$$

where we have made use of the notation there: $\zeta = \sigma + i\rho$ is a p-centered holomorphic coordinate function. It suffices to show, say,

$$\text{(10.1)} \qquad \frac{\partial\tau}{\partial\sigma}\frac{\partial\log h}{\partial\sigma} \text{ is locally integrable in the } (\sigma,\rho)\text{-plane.}$$

We assume p to be a worst possible point: $d\tau(p) = df(p) = 0$. Recall that $f^*\Omega = hd\tau \wedge d^c\tau = hN\, d\sigma \wedge d\rho$, where $N = \{(\partial\tau/\partial\sigma)^2 + (\partial\tau/\partial\rho)^2\}$ (9.5). Let z be an a-centered coordinate function as usual and let $\Omega = gdz \wedge d\bar{z}$, where g is C^∞ and *zero free*. Let further

$$df^*(z)/d\zeta = \zeta^k\phi(\zeta)$$

where ϕ is holomorphic and *zero free*. Then

$$f^*\Omega = (f^*g)|\zeta|^{2k}|\phi|^2 d\zeta \wedge d\bar{\zeta} = (2/i)(f^*g)|\phi|^2|\zeta|^2 d\sigma \wedge d\rho.$$

Comparison of the two expressions gives

$$\text{(10.2)} \qquad h = |\zeta|^2/N\{^2/i((f^*g)\ |\phi|^2\}.$$

The function in { } is zero free. Hence to prove (10.1), it is sufficient to prove $(\partial\tau/\partial\sigma)\,(\partial/\partial|\sigma|)\,(\log|\zeta|^2/N)$ is locally integrable. But $(\partial\tau/\partial\sigma)\,(\partial/\partial\sigma)\log|\zeta|^2 = (\partial\tau/\partial\sigma)(2\sigma/|\zeta|^2)$ is clearly locally integrable. Thus, we are left with proving:

$$\text{(10.3)} \qquad \partial\tau/\partial\sigma\,\partial\log N/\partial\sigma \text{ is locally integrable in the } (\sigma,\rho)\text{-plane.}$$

Now, $|(\partial\tau/\partial\sigma)\partial\log N/\partial\sigma| \leqq |(N)^{1/2}(\partial\log N/\partial\sigma)|$, and it is likely that, if N has a zero of finite order at the origin (i.e. at least one partial derivative of N is nonzero at the origin), then the right side is an integrable function, so that the left side is also. We do not know how to prove that however, and we have to invoke condition (3) of Definition 9.1. So $N = |\zeta|^k\psi(\sigma,\rho)$ where ψ is C^∞ and zero free. Hence $(\partial\tau/\partial\sigma)(\partial\log N/\partial\sigma) = C^\infty$ function $+ k(\partial\tau/\partial\sigma)(\sigma/|\zeta|^2)$ which is clearly locally integrable. Q.E.D.

APPENDIX

Open Bordered Riemann Surfaces.

So far we have treated the case of $f: V \to M$ where V is an open Riemann surface. In this Appendix, we briefy indicate how to extend the theory of Part A to the case of an open bordered Riemannn surface V. The addition of a boundary to V presents itself in two natural situations. First, very often V has no finite or infinite harmonic exhaustion, but rather admits a harmonic exhaustion of mixed type, i.e., on certain components of the ideal boundary of V, τ approaches infinity while on others τ approaches a finite value. A typical example is a V obtained from a compact M' by removing a finite number of discs $D_1, \cdots, D_l$, and a finite number of points $p_1, \cdots, p_m$. In this case, to analyse a holomorphic mapping $f: V \to M$, we decompose V into two open bordered Riemann surfaces V_1, V_2 such that $V_1 \cap V_2 = \partial V_1 = \partial V_2$ and such that V_1 encloses all the D_i's and V_2 encloses all the p_j's. Then clearly V_1 admits a finite harmonic exhaustion while V_2 admits an infinite one. The study of $f: V \to M$ is reduced to the study of $f: V_1 \to M$ and $f: V_2 \to M$ separately. Second, sometimes one wishes to analyse the local behavior of $f: V \to M$ at one component of the ideal boundary and for this, open bordered Riemann surfaces become a convenient tool. Consider for instance the classical theory of essential singularities of a meromorphic function. Suppose the origin 0 is the singularity of f. Then if we insist on considering Riemann surfaces without boundary, we will have to consider $f: D_\varepsilon \to S$ (Riemann sphere) where D_ε is the deleted open disc of radius ε about 0. This means we will have to consider compact subdomains converging not only towards 0, but also towards the circle $|z| = \varepsilon$ *which is clearly irrelevant for the problem at hand.* The proper thing to do here is to consider $f: B_\varepsilon \to S$ where $B_\varepsilon = \{|z| \leqq \varepsilon\} - \{0\}$ and the exhausting subdomains will be those all having $|z| = \varepsilon$ as a common boundary while the other component converges to 0. This technique is illustrated in the proof of Theorem I.6 below.

So let V^* be a bordered open Riemann surface with boundary ∂V^*, $\tau: V^* \to [0, s)$ is called a harmonic exhaustion if it obeys all the conditions of Definition 4.1 plus this one: $\tau^{-1}(0) = \partial V^*$. This latter is purely a matter of convenience. Most of the proofs of the theorems require only a few formal changes, so we just list the final results. However, we must set up one

Convention. We assume throughout that $f^{-1}(a) \cap \partial V^* = \varnothing$ and df is nowhere zero along ∂V^*.

FIRST MAIN THEOREM I.1. *For $r \geqq r(\tau)$, and λ_a as in* (3.1),

$$N(r, a) + \frac{1}{2\pi} \int_{\partial V^*[t]} f^* u_a * d\tau \Big|_{t=r_0}^{t=r} + (r - r_0) \int_{\partial V^*} f^* \lambda_a = T(r). \tag{I.1}$$

SECOND MAIN THEOREM I.2. *Suppose κ denotes the geodesic curvature of ∂V^* with respect to f^*G, then for $r \geqq r(\tau)$*

$$E(r) + N_1(r) - \chi(M)T(r) = \frac{(r - r_0)}{2\pi} \int_{\partial V^*} \kappa + \frac{1}{4\pi} \int_{\partial V^*[t]} \log h * d\tau \Big|_{t=r_0}^{t=r}. \tag{I.2}$$

Let us denote $\int_{\partial V^*} f^*\lambda_a$ by $C(a)$, then the fundamental inequality of Corollary 4.8 becomes

$$N(r,a) < T(r) + (r - r_0)C(a) + \text{constant}$$

where the constant is independent of a and r. Now all the terms in (I.1), with the exception of $C(a)$, are continuous functions of a, and hence integrable over M. Since (I.1) holds for all $a \in M$ except on $f(\partial V^*)$ — which is a set of measure zero — it is clear that $\int_M C(a)\rho(a)\Omega < \infty$ for ρ as in $(*)$ of §7. Thus an integration of the above inequality with respect to $\rho(a)\Omega$ over M gives

$$\int_{r_0}^{r} dt \int_{r_0}^{t} ds \left(\int_{\partial V[s]} h(f^*\rho) * d\tau\right) < T(r) + \text{constant} \cdot r$$

where the constant is independent of r and a. This leads to

THEOREM I.3. *For distinct points $a(1), \cdots, a(q)$ of M, let*

$$\omega(r) = \frac{4\pi}{L}\left\{\sum_{k=1}^{q} (T(r) - \bar{N}(r, a(k))) - \chi(M)T(r) + E(r)\right\}$$
$$- (r - r_0) \sum_{k=1}^{q} C(a(k)) + \text{constant},$$

then

$$\int_{r_0}^{r} dt \int_{r_0}^{t} e^{\omega(s)}\, ds < T(r) + \text{constant} \cdot r \tag{I.3}$$

where the constants are independent of $\{a(k)\}$ and r.

Define $\delta(a)$ and χ as in §7. From (I.3), one can deduce the defect relations in the usual manner, using Lemmas 7.2 and 7.4.

THEOREM I.4. *Let $f: V^* \to M$ be holomorphic such that df is nowhere zero on ∂V^*, and let $a(1), \cdots, a(r)$ be a finite set of distinct points of M such that $a(k) \notin f(\partial V^*)$ for all k. If V^* admits an infinite harmonic exhaustion ($\tau: V^* \to [0, \infty)$ is onto), then*

$$\sum_{k=1}^{q} \delta(a(k)) \leqq \chi(M) + \chi + \varepsilon_1 \tag{I.4}$$

where $\varepsilon_1 = \limsup_{r\to\infty}\{((r - r_0)/T(r)) \sum_{k=1}^{q} C(a(k))\}$. *If V^* admits a finite harmonic exhaustion, ($\tau: V^* \to [0, s)$, $s < \infty$ is onto), then*

$$\sum_{k=1}^{q} \delta(a(k)) \leqq \chi(M) + \chi + \varepsilon_2 \tag{I.5}$$

where

$$\varepsilon_2 = \limsup_{r\to s} \frac{1}{T(v)}\left\{\frac{L}{4\pi}\log(T(r) + \text{const} \cdot r) + 2\log\frac{1}{s - r} + (r - r_0)\sum_{k=1}^{q} C(a(k))\right\}.$$

It is obvious from (I.4) and (I.5) that if (7.8) holds, then $\varepsilon_2 = 0$ and that if $\limsup_{r\to\infty}(r/T(r)) = 0$, then $\varepsilon_1 = 0$. An f obeying the last condition will be called *transcendental* as in §8. A trivial but important observation is that all the lemmas in §8 remain true if we replace V by V^*; the proofs given in §8 of these carry over to the case of a bordered V^* word for word. We can therefore summarize part of these in the following:

LEMMA I.5. *If V^* is obtained from a compact bordered Riemann surface M' by deleting a finite number of points, then $f: V^* \to M$ is transcendental iff f is not the restriction of a holomorphic map $\tilde{f}: M' \to M$. Furthermore, transcendental f in this case implies $\chi = \varepsilon_1 = 0$ in* (I.4).

As an application of the general theory, we prove the following well-known result:

THEOREM I.6. *Let V be an open Riemann surface and $p \in V$. Let $f: V - \{p\} \to M$ be a holomorphic map into a compact M. Suppose f cannot be extended to an $\tilde{f}: V \to M$. Then $\chi(M) \geqq 0$ and the image under f of every neighborhood of p omits at most $\chi(M)$ points of M.*

PROOF. Let q points $a(1), \cdots, a(q)$ of M be given. Let B_ε be a closed disc neighborhood $\{|z| \leqq \varepsilon\}$ of p relative to some p-centered holomorphic coordinate function such that $f^{-1}(a(i)) \cap \partial B_\varepsilon = \varnothing$ for all i, and that df is nowhere zero along ∂B_ε. Consider now the restriction of f to $f: B_\varepsilon - \{p\} \to M$; by Lemma I.5 and (1.4), we have $\sum_{k=1}^{q} \delta(a(k)) \leqq \chi(M)$. Thus $\chi(M)$ must be $\geqq 0$ in order that this make sense and $f(B_\varepsilon - \{p\})$ omits at most $\chi(M)$ points. Since given any neighborhood of p, we can always find such a B_ε inside it, we are done. Q.E.D.

One final remark is that all the above considerations apply equally well to the case where V^*, instead of being a bordered Riemann surface, is a subdomain of a Riemann surface bounded by piecewise C^∞ curves. In more abstract language, we allow V^* to be a two-dimensional manifold with piecewise C^∞ boundary such that $V^* - \partial V^*$ has a pre-assigned conformal structure. The assertion is self-evident.

REFERENCES

1. L. V. Ahlfors, *Beiträge zur Theorie der meromorphen Funktionen*, Seventh Scand. Congr. Math., Oslo, 1929, pp. 84–88.

2. ———, *Über eine Methoden der Theorie der meromorphen Funktionen*, Soc. Sci. Fenn. Comm. Phys. Math. **8** (1935), no. 10.

3. ———, *Über die Anwendung differentialgeometrischer Methoden zur Untersuchung von Überlagerungsflächen*, Acta. Soc. Sci. Fenn. A (2) **2** (1937), no. 6.

4. ———, *Geometrie der Riemannschen Flächen*, Internat. Congress Math., Oslo, Vol. 1, 1936, pp. 239–248.

5. L. V. Ahlfors and L. Sario, *Riemann surfaces*, Princeton Univ. Press, Princeton, N. J., 1960.

6. R. C. Buck, *Advanced calculus*, 2nd ed., McGraw-Hill, New York, 1965.

7. S. S. Chern, *Complex analytic mappings of Riemann surfaces*. I, Amer. J. Math. **82** (1960), 323–337.

8. L. P. Eisenhart, *An introduction to differential geometry*, Princeton Univ. Press, Princeton, N. J., 1947.

9. P. R. Halmos, *Measure theory*, Van Nostrand, Princeton, N. J., 1950.

10. S. Kobayashi and K. Nomizu, *Foundations of differential geometry*, Interscience, New York, 1963.

11. M. Nakai, *On Evans potential*, Proc. Japan. Acad. **38** (1962), 624–629.

12. R. Nevanlinna, *Eindentige analytische Funktionen*, Springer-Verlag, Berlin, 2*te* Áufl., 1953.

13. K. Noshiro and L. Sario, *Value distribution theory*, Van Nostrand, Princeton, N. J., 1966.

14. B. O'Neill, *Elementary differential geometry*, Academic Press, New York, 1966.

15. G. de Rham, *Variétés différentiables*, Hermann, Paris, 1955.

16. H. Whitney, *Geometric integration theory*, Princeton Univ. Press, Princeton, N. J., 1957.

17. I. I. Priwalow, *Randeigenschaften analytischer Funktionen*, Veb Deutscher, Verlag, Berlin, 1956.

18. R. C. Gunning and H. Rossi, *Analytic functions of several complex variables*, Prentice-Hall, Englewood Cliffs, N. J., 1965.

19. A. Nijenhuis, and R. W. Richardson, *A theorem on maps with nonnegative Jacobians*, Michigan. Math. J. **9** (1962), 173–176.

University of California
Berkeley, California

Problems

LEON EHRENPREIS, editor

1 (G. Piranian). Let $f(z)$ be holomorphic and bounded in $|z| < 1$. Put

$$I(\theta) = \int_0^1 |f'(re^{i\theta})|\,dr = \text{length of image of radius} \quad \arg z = \theta.$$

W. Rudin has given an example of an $f(z)$ with

$$I(\theta) = \infty \text{ almost everywhere.}$$

Is it possible to have $I(\theta) \equiv \infty$?

(W. Rudin, *The radial variation of analytic functions*, Duke Math. J. **22** (1955), 235–242.)

2 (G. Piranian, Ch. Pommerenke). Let $f(z)$ $\sum a_n z^n$ be univalent and holomorphic in $|z| < 1$. If $a_n \to 0$, does $\lim_{r \to 1} f(r)$ exists imply $\sum a_n$ converges?

Is there an example such that $\sum a_n e^{in\theta}$ is divergent for all θ?

3 (C. Rényi). There is an entire function $f(z)$ such that the value 0 is taken infinitely often in every angle, but no other value has this property. Namely, $f(z) = \Phi(z)e^z$, $\Phi(z)$ of order 0 with infinitely many zeros in every angle. In an angle around the negative real axis $f(z) \to 0$ so no other value has this property.

Is there an entire function $f(z)$ such that $f(z) = 0$ and $f(z) = 1$ have roots in every angle, but for other values $c, f(z) \neq c$ in some angle?

4 (C. Rényi). Can an entire function have different exceptional values in different Julia-directions?

5 (J. Clunie). Let $f(z) = \sum a_n z^n$ be holomorphic in $|z| < 1$. Let

$$\rho_n(f) = \inf |\gamma_k|$$

where γ_k runs through the zeros of the nth partial sum $S_n(z)$ of $f(z)$. Is there an $f(z)$ with $\limsup \rho_n(f) = \infty$ $(n \to \infty)$?

Solution by P. Koosis. Observe first that for any entire function $h(z)$ with no zeros , the smallest modulus ρ_n of the zeros of $S_n(z)$ must approach infinity. Thus we can choose n_1 large enough so that the partial sums $S_n(z)$ of e^z for $n \geqq n_1 - 1$ have all their zeros in $|z| > 1$. Note that the first $n_1 - 1$ partial sums of $\exp(z + z^{n_1})$ are the same as those of e^z. We can choose $n_2 > n_1 + 1$ so that the partial sums of $\exp(z + z^{n_1})$ of degree $\geqq n_2 - 1$ have all their zeros in $|z| > 2$. Continuing in this way we obtain a function $\exp(\sum z^{n_j})$ with the desired properties.

Solution by D. J. Newman. One can also get a proof based on the following observation: There can be no implication of the form

$$P(z) \text{ a polynomial with } P(0) = a_0 \neq 0, \quad P'(0) = a_1, \cdots, P^{(k)}(0) = a_k$$

$$\Rightarrow P(z) \text{ has a zero inside } |z| < R.$$

(Indeed, an entire $f(z)$ with no zeros can always be constructed with initial coefficients as given and then P can be chosen as a high partial sum.)

This gives an inductive way of forcing certain partial sums of the series to have all zeros large. Careful estimates can then guarantee that the manufactured series have radius of convergence 1.

6 (E. Strauss) $c_n \uparrow$. Is there an entire function $f(z)$ with $f(x)$ real, $f(x) \uparrow$ such that $f(n) = c_n$?

7 (L. Rubel). Let f_1, f_2, f be assigned entire functions of exponential type. Find necessary and sufficient conditions for the existence of g_1, g_2 of exponential type such that

$$f_1 g_1 + f_2 g_2 = f.$$

In some cases conditions are given in Ehrenpreis, *Fourier analysis in several complex variables*, Interscience, New York, 1968. For $f \equiv 1$ (and one complex variable) a solution was also given by Kelleher and Taylor (see the lecture of Kelleher).

8. (L. Rubel). Is every meromorphic function of exponential type of $n > 1$ complex variables the quotient of two entire functions of exponential type?

9 (C. Rényi). $f(z) = \sum a_n z^n$ entire, $a_n \geqq 0$,

$$\int_0^\infty \frac{f(\lambda x)}{f(x)} dx = \frac{1}{1-\lambda} \qquad (|\lambda| < 1).$$

Do these hypotheses imply $f(z) = Ke^z$?

10. (L. Rubel). Mrs. Rényi has proven: If $\sum a_k z^{n_k} = f(z)$ is periodic and entire and the lower density of the omitted exponents is $> 1/2$, then $f \equiv$ constant. Now, "periodic" means "invariant under $z \to z + 1$". Is it possible to find similar type of gap theorems for f which are holomorphic in $|z| < 1$ and invariant under a discrete group? (Presumably the gap condition will depend on the group.)

11 (Q. I. Rahman). What is the best possible upper bound for

$$\sup_{|z|<1} |p'(z)|$$

when $p(z)$ varies through all polynomials of degree n subject to the conditions

(a) $|p(z)| \leqq 1 \quad (|z| < 1)$

(b) $p(z)$ has m zeros in $|z| \leqq k$.

(For $k = 1$, $m = 0$: $|p'(z)| \leqq n/2$ $(|z| < 1)$: For $m = n$: $|p'(z)| \leqq n$ $(|z| < 1)$. Nothing else is known.

12 (Q. I. Rahman). What is the best possible upper bound of

$$\int_0^{2\pi} |p'(e^{i\theta})|^\delta d\theta \Big/ \int_0^{2\pi} |p(e^{i\theta})|^\delta d\theta \qquad (\delta > 0)$$

when $p(z)$ ranges over all polynomials of degree n with m zeros in $|z| \leqq k$?

13. (H. Wilf). Let us say that the positive number $\tau \in \Omega_p$, if there is no non-constant entire function of exponential type τ such that $f(z)$ and all its derivatives have at least p zeros in $|z| \leqq 1$. Let

$$\omega_p = \sup_{\tau \in \Omega_p} \tau .$$

What is the asymptotic behavior of ω_p as $p \to \infty$? It is known that

$$p/e + \log p/2\pi e + O(1) \leqq \omega_p < 4p/e$$

Whittaker's constant for lacunary entire function, (Wilf, Proc. Amer. Math. Soc. 14 (1963), 238–242.)

14 (D. J. Newman). Let $f(z)$ be an entire function satisfying

$$|f(z)| \leqq |z| e^{|z|} \quad \text{and} \quad |f(re^{i\theta})| \leqq A(\theta) e^r$$

where $A(\theta)$ depends on θ.

Is it true that

$$|f(z)| \leqq \text{constant} \cdot e^{|z|} ?$$

Solution by A. Dvoretzky. Result. Let $\phi(r)$, defined for $0 \leqq r < \infty$, be positive, nondecreasing and satisfy

(1) $$\lim_{r = \infty} \frac{1}{r} \log \phi(r) = 0 .$$

Then there exists an entire function $f(z)$ satisfying

(2) $$M(r) = \max_{|z| = r} |f(z)| < e^r, \qquad (0 \leqq r < \infty),$$

(3) $$\limsup_{r = \infty} e^{-r} M(r) > 0$$

and

(4) $$\lim_{r = \infty} e^{-r} f(re^{i\theta}) \phi(r) = 0$$

for every real θ. (Taking $\phi(r) = r$ and multiplying the constructed f by z shows the answer to Problem 14 is "no".)

Construction. $f(z)$ will be obtained by suitably modifying the canonical product of $\sin z$. We shall make use of the observation that, for every $r > 0$, $|1 - z^2|$ on the circle $|z| = r$ has maxima at $\pm ir$, minima at $\pm r$ and varies monotonically on the arcs between these points.

We shall show that for a suitable sequence of integers

$$(5) \qquad 1 = k_1 < k_2 < \cdots < k_n < \cdots$$

and a suitable sequence of positive numbers

$$(6) \qquad \frac{\pi}{2} > \alpha_1 > \alpha_2 > \cdots > \alpha_n > \cdots, \quad \lim_{n=\infty} \alpha_n = 0,$$

the function

$$(7) \quad f(z) = z \prod_{n=1}^{\infty} P_n(z),\ P_n(z) = \prod_{j=k_{n-1}}^{k_n - 1} \left(1 - \frac{z^2}{\pi^2 j^2 \exp[-2i\alpha_n]}\right), \quad (n=1,2,\cdots)$$

satisfies the requirements.

Indeed,

$$|P_n(z)| \leqq P_n(|z|\exp[i(\pi/2 + \alpha_n)]) = \prod_{j=k_{n-1}}^{k_n-1} (1 + |z|^2/\pi^2 j^2),$$

thus $M(r) \leqq \sin ir$ and (2) holds.

Next we prove (3) by showing that if (5) increases rapidly there exist $r_n \to \infty$ for which $m(r_n) > \frac{1}{8} e^{r_n}$. This is done inductively: Having determined r_{n-1}, k_{n-1} we choose $r_n > 2r_{n-1}$ such that

$$\prod_{j=1}^{k_{n-1}} \left(1 - \frac{r_n^2}{\pi^2 j^2}\right) > \frac{1}{2} \prod_{j=1}^{k_{n-1}} \left(1 + \frac{r_n^2}{\pi^2 j^2}\right)$$

and then $k_n > k_{n-1}$ such that

$$(8) \qquad \prod_{j=k_n}^{\infty} \left(1 + \frac{r_n^2}{\pi^2 j^2}\right) < 2 \prod_{j=k_n}^{\infty} \left(1 - \frac{r_n^2}{\pi^2 j^2}\right).$$

Then

$$|f(r_n \exp[i(\pi/2 + \alpha_n)])| > r_n \cdot \frac{1}{2} \prod_{j=1}^{k_{n-1}-1} \left(1 + \frac{r_n^2}{\pi^2 j^2}\right) \cdot \prod_{j=k_{n-1}}^{k_n - 1} \left(1 + \frac{r_n^2}{\pi^2 j^2}\right)$$

$$\cdot \frac{1}{2} \prod_{j=k_n}^{\infty} \times \left(1 + \frac{r_n^2}{\pi^2 j^2}\right) = \frac{1}{4} \sin ir_n > \frac{1}{8} e^{r_n}.$$

To complete the proof it remains to show that, given the sequence (5), it is possible to choose the sequence (6) so that (4) holds. We first note that (4) certainly holds for $-\pi/2 < \theta < \pi/2$. Indeed, for such θ we have, by the observation made at the beginning of the construction,

$$|P_n(re^{i\theta})| = \prod_{j=k_{n-1}}^{k_n-1} \left|1 - \frac{r^2\exp[2i(\theta-\alpha_n)]}{\pi^2 j^2}\right| < \prod_{j=k_{n-1}}^{k_n-1} \left|1 - \frac{r^2\exp[2i(\pi/2-\delta)]}{\pi^2 j^2}\right|$$

with $\delta = (\pi/2 - |\theta|)/2$ for all n for which $\alpha_n < \delta$. Thus, since $\alpha_n \to 0$ we see from (7) that $|f(re^{i\theta})| < \sin(re^{i(\pi/2-\delta)})$ for all $r > r(\theta)$. This implies (4) by (1). Thus all that remains to complete the proof is to show that (4) is satisfied for $\theta = \pi/2$ for a suitable choice of the sequence (6). But, by the observation and (8),

$$|f(ir)| \leqq r \prod_{j=1}^{k_{n-1}} \left|1 - \frac{r^2\exp[2i(\pi/2-\alpha_n)]}{\pi^2 j^2}\right| \cdot \prod_{j=k_n}^{\infty} \left(1 + \frac{r_n^{\,2}}{\pi^2 j^2}\right)$$

$$< 2\,|\sin r\exp[i(\pi/2-\alpha_n)]| < 4\exp[2r\cos\alpha_n]$$

for all $r \geqq r_n$. Thus, in view of (1), (4) will hold for $\theta = \pi/2$ provided α_n does not decrease too rapidly.

Remarks. (1) By Phragmén-Lindelöf it is impossible to replace (1) by $\limsup_{r=\infty}(1/r)\log\phi(r) < \varepsilon$ for any $\varepsilon > 0$.

(2) We chose for convenience the sine product. This resulted in two directions in the neighborhood of which (4) is not satisfied uniformly. By starting with a canonical product with zeros on one line we could obtain an example with only one such direction.

(3) Similar examples are easily constructed for orders of growth other than the exponential.

Solution by J. Clunie. Let $\{r_n\}_0^\infty$ be an increasing sequence with $r_0 = 1$ and $\{\eta_n\}_0^\infty$ be a decreasing sequence with $\eta_0 = \frac{1}{2}$ satisfying the following conditions:

(i) $\displaystyle\sum_{n=0}^{\infty} \frac{1}{r_n} < 2$;

(ii) $(r_n r_{n+1})^{1/2} > 2^n r_n^2$;

(iii) $re^{-\eta_{n-1}r} < 1 \quad (r \geqq r_n)$;

(iv) $\eta_n (r_n r_{n+1})^{1/2} < 1$;

(v) $4(\log r_n/r_n)^{1/2} \leqq \pi/2^{n+1}$.

Let

$$f(z) = z\sum_{\nu=0}^{\infty} \frac{1}{r_\nu^2}\exp[(1-\eta_\nu)\exp[-i\pi(1-2^{-\nu})]z].$$

For $n \geqq 1$, $r_{n-1} \leqq r \leqq r_{n+1}$ and $|z| = r$,

$$\left|z\sum_{\nu=0}^{n-1} \frac{1}{r_\nu^2}\exp[(1-\eta_\nu)\exp[-i\pi(1-2^{-\nu})]z]\right| \leqq r\sum_{\nu=0}^{n-1}\frac{1}{r_\nu^2}e^r\cdot\exp[-\eta_{n-1}r]$$

$$\leqq 2r\exp[-\eta_{n-1}r]e^r < 2e^r.$$

Under the same conditions,

$$\left|z\sum_{\nu=n+1}^{\infty} \frac{1}{r_\nu^2}\exp[(1-\eta_\nu)\exp[-i\pi(1-2^{-\nu})]z]\right| \leqq re^r\sum_{\nu=n+1}^{\infty}\frac{1}{r_\nu^2} < 2e^r$$

For $z = (r_n r_{n+1})^{1/2} \exp[i\pi(1 - 2^{-n})]$ we have

$$\left| z \frac{1}{r_n^2} \exp[(1-\eta_n)\exp[-i\pi(1-2^{-n})]z] \right| \geqq \frac{(r_n r_{n+1})^{1/2}}{r_n^2} \exp[(1-\eta_n)(r_n r_{n+1})^{1/2}]$$

$$> \frac{2^n}{e} \exp(r_n r_{n+1})^{1/2}.$$

Hence $f(z) \neq O(e^r)$ ($|z| = r; r \to \infty$), but clearly $f(z) = O(re^r)$ ($|z| = r; r \to \infty$).

For small ϕ, $\cos\phi < 1 - \phi^2/4$ and so for such ϕ with $z = r\exp[i\pi(1-2^{-n}+\phi)]$ ($r_n \leqq r \leqq r_{n+1}$),

$$\left| \frac{1}{r_n^2} \exp[(1-\eta_n)\exp[-i\pi(1-2^{-n})]z] \right|$$

$$= \frac{r}{r_n^2} \exp[(1-\eta_n) r\cos\phi] \leqq \frac{r}{r_n^2} \exp[(1-\eta_n)(1-\phi^2/4) r]$$

$$= \frac{r}{r_n^2} e^r \cdot \exp[-(\phi^2 r/4)].$$

Consider $\phi = \phi(r) = 4((\log r)/r)^{1/2}$ and we have, for large r,

$$\left| \frac{z}{r_n^2} \exp[(1-\eta_n)\exp[-i\pi(1-2^{-n})]z] \right| \leqq \frac{e^r}{r_n^2}$$

and so

$$\left| \frac{z}{r_n^2} \exp[(1-\eta_n)\exp[-i\pi(1-2^{-n})]z] \right| < e^r$$

for $z = \exp[i\pi(1-2^{-n}) + i\phi]r$ ($r_n \leqq r \leqq r_{n+1}$) and $4((\log r_n)/r_n)^{1/2} \leqq |\phi| \leqq \pi$. Therefore the above bound is satisfied for $\pi/2^{n+1} \leqq |\phi| \leqq \pi$.

If $\theta \neq \pi$ then clearly $f(re^{i\theta}) = O(e^r)$ ($r \to \infty$); also the same is true from the above for $\theta = \pi$. Hence $|f(re^{i\theta})| \leqq A(\theta)e^r$ for each θ, but $f(z) \neq O(e^{|z|})$ ($|z| \to \infty$).

Solution by L. Ehrenpreis. This is similar to Clunie's solution except that the desired function is constructed as a sum $\sum a_j \exp[ib_j z]$ where $|b_j| = 1$ and $b_j \to 1$ rather than in Clunie's example where the b_j lie in the unit circle.

15 (B. Lepson). Construct a function meromorphic in the whole plane with prescribed zeros, poles, and principal parts.

Solution by D. J. Newman. Let $P(z)$ be a Weierstrass product with zeros at the desired zeros and let $Q(z)$ be a Weierstrass product with zeros with multiplicity at the desired poles. Consider the expression $F(z) = (P(z)/Q(z))e^{f(z)}$. We pick $f(z)$ so that $F(z)$ will have the desired properties. It already has the right zeros and to force the right principal parts we need only force $e^{f(z)}$ to have the correct leading terms in its power series about the poles z_k. Thus we wish

$$e^{f(z)} = a_0 + a_1(z-z_k) + \cdots + a_{n_k}(z-z_k)^{n_k} + \text{higher terms}$$

$(a_0 \neq 0)$. Equivalently we wish

$$f(z) = \log(a_0 + a_1(z-z_k) + \cdots + a_{n_k}(z-z_k)^{n_k}) + \text{higher terms.}$$

Or, in other words, we wish for $f(z)$ to start its power series in a prescribed way at the z_k.

This last is a standard interpolation problem, however, and the problem is solved.

16 (Ch. Pommerenke). Let the Clunie constant be the smallest number γ such that, for all functions $g(z) = z + b_0 + \cdots + b_n z^{-n} + \cdots$ meromorphic and univalent in $|z| > 1$, we have

$$|b_n| \leqq A(\varepsilon) n^{\gamma+\varepsilon-1} \qquad (n = 1, 2, \cdots)$$

for every $\varepsilon > 0$ and some constant $A(\varepsilon)$. Determine or estimate γ. (Note: It is known that $0.139 < \gamma < .497$.)

17 (Ch. Pommerenke). Let

$$f(z) = \sum_{n=1}^{\infty} a_n z^n$$

be analytic, univalent and bounded in $|z| < 1$ and such that

(*) $$|\arg f'(z)| < B \qquad (|z| < 1).$$

Is it true that

(**) $$a_n = O(1/n)?$$

(Note: If (*) is dropped then (**) does not always hold. If (*) is replaced by the stronger assumption that f is close-to-convex then (**) is true.)

18 (Ch. Pommerenke). Let S_p $(p = 1, 2, \cdots)$ be the family of all functions f such that

(i) $f(z) = z + a_2 z^2 + \cdots$ is analytic in $|z| < 1$,

(ii) every value is assumed at most p times,

(iii) $f'(z) \neq 0$ for $|z| < 1$.

Is it true that

$$\max_{f \in S_p} |a_2| = 2p?$$

(Note: If (iii) is dropped then a_2 can become arbitrarily large.)

19 (A. Shields). $f(z)$ holomorphic in $|z| < 1$, $f'(z) \neq 0$ in $|z| < 1$. For which values of c does

$$|f''(z)/f'(z)| \leqq c(1-|z|^2)^{-1}$$

imply that $f(z)$ is univalent? (Known $2(5^{1/2}-2) \leqq c < 2$ works.)

20 (A. Shields). If f is univalent in $|z|<1$, for which α is $f_\alpha(z) = \int_0^z (f'(\zeta))^\alpha \, d\zeta$ univalent? (Known: f_α is always univalent, if $|\alpha| < (5^{1/2}-2)/3$, false for $|\alpha| > \frac{1}{3}$, $\alpha \neq 1$.)

21 (D. J. Newman). Let

$$p(z) = z^2 + 1, \; p_n(z) = p \circ p \circ \cdots \circ p(z) \quad (n \text{ times}).$$

(a) Is there an open, bounded set S such that $p_n(S) \subset S$ for $n = 1, 2, 3, \cdots$? (b) Does the assumption $\{p_n(z)\}$ bounded imply that $p_k(z) = z$ for some k?

Solution by L. Flatto. The answer to both of the above questions is in the negative. We will indeed prove the following result. Let $B = \{z \mid \{p_k(z)\} \text{ is bounded}\}$. B is a perfect nowhere dense set.

The proof is as follows. Choose r so that $(5^{\frac{1}{2}}+1)/2 < r < 2$. Then for $|z| \geqq r$ we have $|z^2+1|/|z| \geqq (|z|^2-1)/|z| \geqq (r^2-1)/r > 1$. It follows that for $|z| \geqq r$, $\lim_{k\to\infty} p_k(z) = \infty$. Let $D_0 = \{z \mid |z| \leqq r\}$ and $D_n = \{z \mid p_0(z) \in D_{n-1}\}$ $(n \geqq 1)$; i.e., $D_n = p_1^{-1}(D_{n-1}) = p_n^{-1}(D_0)$. Since $|p(z)| \geqq ((r^2-1)/r)|z|$ for $|z| \geqq r$ we must have $D_0 \supset D_1$ and thus $D_1 = p_1^{-1}(D_0) \supset p_1^{-1}(D_1) = D_2$. Repeating this argument we obtain $D_0 \supset D_1 \supset \cdots \supset D_n \supset D_{n+1} \supset \cdots$.

We now show that $B = \bigcap_{n=0}^{\infty} D_n$. Suppse $z \in \bigcap_{n=0}^{\infty} D_n$. Then $z \in p_n^{-1}(D_0)$ for all $n \geqq 0$. In particular for any $k > 0$ $p_k(z) \in p_k(p_k^{-1}(D_0)) = D_0$ and so $\{p_k(z)\}$ is a bounded sequence. Hence $\bigcap_{n=0}^{\infty} D_n \subset B$. Conversely let $z \notin \bigcap_{n=0}^{\infty} D_n$ so that $z \notin D_n$ for some n. Then $p_n(z) \notin D_0$ or $|p_n(z)| > r_1$, from which we conclude that $\{p_k(z)\}$ is unbounded. Hence $B = \bigcap_{n=0}^{\infty} D_n$.

We now show that B is a perfect nowhere dense set. We give the following description of D_n. Let C_n denote the boundary of D_n. Let $w = z^2 + 1$ so that

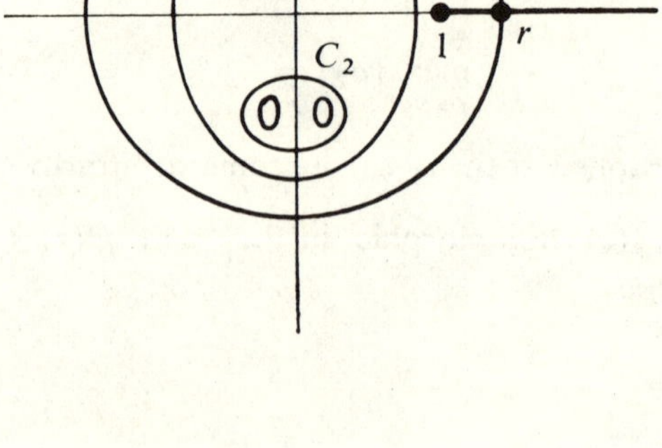

$z=(w-1)^{1/2}$. Since $0<(r-1)^{1/2}<1$, C_1 will not cross the slit $[1,\infty)$. The two branches of $(w-1)^{1/2}$ are analytic functions on the plane minus the slit $[1,\infty)$, one of the branches mapping the w-plane minus $[1,\infty)$ onto the upper z-plane, the other one onto the lower z-plane. It follows that D_2 consists of two components and its boundary C_2 consists of two Jordan curves (see diagram). Repeating this argument we conclude that D_n consists of 2^n components, each of these containing two components of D_{n+1}. Let $d_n=$ greatest diameter of a component of D_n. Our result will be established by proving that $\lim_{n\to\infty} d_n=0$. Consider any component of D_n and call it S_n. Then $S_{n+1}=p_1{}^{-1}(S_n)$ for some S_n. Let the diameter of S_{n+1} be attained by z_1,z_2 and let $z_i=w_i{}^{1/2}-1 (i=1,2)$ where $w_1,w_2\in S_n$. Then $z_2-z_1=\frac{1}{2}\int_{w_1}^{w^2} dw/(w^{1/2}-1)$ and so $|z_2-z_1|$ $\leqq (1/(2|w-1|)^{1/2})|w_2-w_1|$. Now $|1/2|w-1|^{1/2}|\leqq k<1$ on $\Gamma_n \Leftrightarrow |w-1)>\frac{1}{4}$ for $w\in D_n$ and n sufficiently large. Elementary calculations show that this holds for $n\geqq 2$. Thus $d_{n+1}\leqq kd_n$ $(n\geqq 2)$ and so $\lim_{n\to\infty} d_n=0$.

Remark. One could probably also conclude that the 2-dimensional measure of $B=0$. For this we would require the estimate $k<2^{1/2}\Leftrightarrow|w-1|>\frac{1}{2}$ for $w\in D_n$, n sufficiently large.

22 (W. H. J. Fuchs). Problem started with an attempt to show: If f meromorphic, order <1, then deficiency $\delta(0,f'/f)=0$.

This leads to the question of finding refinements of Gauss' theorem that roots of P' lie in convex hull of roots of P: If roots of f are regularly distributed, are those of f'?

23 (W. H. J. Fuchs). If $f(z)=\sum a_k z^{n_k}$, f entire, $\sum(1/n_k)<\infty$, then it is known that every value is taken infinitely often.

Must $f(z)$ have roots in every angle?

24 (W. H. J. Fuchs). Let $f(z)=\sum c_k z^{\lambda_k}$, $\lambda_{k+1}/\lambda_k\geqq q>1$ be analytic in $|z|<1$. Suppose $\sum c_k=\infty$. Does f take every value (for $|z|<1$) infinitely often?

25. (W. H. J. Fuchs). Given a gap series (entire), what is the gap condition to ensure that there are no deficient values? For functions of finite order, the condition is $n_k/k\to\infty$.

26 (H. Wilf). Recall the theorem of *Poulain-Hermite*. Let f,g be polynomials with real zeros only.

$$f(d/dx)\,g(x) \text{ has only real zeros.}$$

(*Proof.* Clear for $f(x)=(x-\alpha)$, then use product.)

Df. f of degree n is *good* if

$$f(d/dx)g(x) \text{ has only real zeros}$$

whenever g (deg n) has only real zeros.

Problem. Which f are good?

By Poulain-Hermite, f having real zeros is good. Also $\sum_0^n (x^j/j!) = f$ is good since $f(d/dx)g = g(x+1)$.

Conjecture. (a) f is good if and only if f extends (by adding $\sum_{j>n} b_j x^j$) to an entire function with only real zeros. (By Poulain-Hermite this condition is sufficient.)

(b) Start with polynomials with real zeros and use the operation $f_1 \square f_2 =$ first n terms of $f_1 f_2$. Is every good polynomial obtained by iteration of this operation?

27 (L. Flatto). Let $f(z) = z + \sum_{n=2}^{\infty} a_n z^n$ converge for $|z| < r$. Let $f_k(z)$ denote the kth iterate of z. Does there exist a z ($|z| < r$) for which either $|f_k(z)| < r$ ($0 \leqq k < \infty$) or $|f_k(z)| < r$ ($-\infty < k \leqq 0$)?

28 (E. F. Collingwood). Consider N_1 in Nevanlinna Theory

$$N_1 = N(r, 1/f) + 2N(R, f) - N(r, f').$$

This is related to algebraic points on the Riemann surface of f (see the paper of Collingwood for notation).

It is known that if $\liminf(N_1(r)/T(r)) < 2$ then there exists at least one center of ramification of unbounded order. What is least condition on $N_1(r)$ to guarantee at least two centers of ramification? *Conjecture*: $\liminf(N_1/T) < 1$.

29 (F. Gross). Does there exist a nonperiodic entire function f such that $f \circ f$ is periodic?

30 (F. Gross). Let f be meromorphic and g transcendental entire. Is it true (as is the case for entire f) that

$$\lim_{\gamma \to \infty} \frac{T(\gamma, f(g))}{T(\gamma, f)} = \infty?$$

31 (F. Gross). Let f be meromorphic and let g be entire. Let p be a polynomial of degree k such that $f(g) = f(p)$. Must g also be a polynomial of degree k?

In case $k = 1$, the answer is "yes" as proven by Neal Goldberg.

32 (F. Gross). If f is entire transcendental and g is nonlinear entire, then can $f(g)$ have only finitely many fix-points?

33 (F. Gross). Do there exist a meromorphic function f and a polynomial of degree other than 2, 3, 4 and 6 such that $f(p)$ is elliptic? For polynomials of degree 2, 3, 4 and 6 the answer is "yes".

34 (F. Gross). A meromorphic function h is said to be prime if it cannot be written as $h = f(g)$, f meromorphic and g entire (or g meromorphic when f is rational) unless one of f or g is linear.

Does there exist a prime elliptic function?

35 (F. Gross). Given a nondenumerable number of measurable subsets of the unit interval of measure $> 1/2$. Can one find for each $\varepsilon > 0$ two of these sets whose intersection is of measure $> \frac{1}{2} - \varepsilon$?

Solution by A. Dvoretzky. Associate with each set s a finite collection of intervals s' with rational end points such that the symmetric difference of s and s' has measure $< \varepsilon/2$. The total number of such finite collections is at most denumerable and so two of them are identical. For these two the original sets could differ by at most ε in measure.

36 (S. M. Shah). Let $f(z)$ be an entire function of order ρ, $0 < \rho < \infty$. Denote by $n_1(r, a)$ the number of simple zeros of $f(z) - a$ in $|z| \leqq r$. Prove (or disprove) that if $a \neq b$ and if

$$n_1(r, a) = O(r^c), \quad n_1(r, b) = O(r^c), \qquad c < \rho$$

then ρ is integer multiple of 1/2.

37 (S. M. Shah). Let $f(z) = \sum_0^\infty a_n z^n$, $a_n \leqq 0$, be an entire function of infinite order and let $\mu(r, f) = \max_n |a_n| r^n$. Prove (or disprove) that if $f(z)$ has a Borel exceptional value (that is $n(r, a) = O(r^c)$ for some a and some positive number c) then

$$\limsup_{r \to \infty} \frac{\mu(r, f')}{M(r, f')} \leqq \frac{1}{2}.$$

38 (B. Lepson). Let E be the set of exponential polynomials, $E = \{\sum_1^n a_k \exp(\alpha_k z)\}$. Prove (or disprove): If $g \in E$, $h \in E$ and g and h have an infinite number of common zeros, then they have a common factor $p \in E$.

Known (Lech). If $g(z) = \sin z$ then the result is true and, in fact, the greatest common divisor of g, h is in E.

39 (B. Lepson). It is known that there exists an $f(x)$ assuming every real value on every interval.

Example. Let the binary expansion of x be $b_1 b_2 \cdots b_k \cdots$. Call $z(n)$ the number of zeros in $b_1 \cdots b_n$. Then $\liminf_{n \to \infty} (z(n)/n) = f(x)$ is an example.

It is known that there is an $f(x)$ assuming every real value on every perfect set. However, the proof is nonconstructive. Can one give an explicit example?

Note: S. Haber has shown that the above example $f = \liminf(z(n)/n)$ does not work.

40 (A. Hyllengren). Let $\sum_1^\infty (b_n/(z + a_n))$ be pointwise convergent on $|z| < 1$ Is the series analytic somewhere?

Solution by B. Lepson.

THEOREM. *If $f_n(z) \to f(z)$ pointwise throughout an open set Ω then $f(z)$ is analytic throughout a dense open subset Ω' of Ω.*

PROOF. Let Δ be any disc inside Ω. A category argument yields a closed non-empty subdisc $\Delta' \subseteq \Delta$ throughout which $f_n(z)$ are bounded. With γ the boundary of Δ' we have $f_n(z) = (1/2\pi i) \int_\gamma (f_n(\zeta)/(\zeta - z))\, d\zeta$. Letting $n \to \infty$ and using the bounded convergence theorem we obtain $f(z) = (1/2\pi i) \int_\gamma (f(\zeta)/(\zeta - z))\, d\zeta$. Hence $f(z)$ is analytic in the interior of Δ' and the proof is complete.

Note: This result is well known and is due to Carathéodory and Osgood.

Question. For the particular series given in this problem, must $f(z)$ be analytic *everywhere* in $|z| < 1$? (Probably not).

41 (R. F. DeMar). Let $G(z, \zeta)$ be analytic for all $z \in$ domain D and all $\zeta \in$ simply connected domain Ω. For what G is the following property P true?

If f is analytic on a simple closed curve $\Gamma \subseteq \Omega$ and

$$\int_\Gamma G(z, \zeta) f(\zeta) = 0 \qquad \text{for all} \quad z \in D,$$

then f is analytic inside Γ.

Examples. 1. $G(z, \zeta) = 1/(\zeta - z)$. G has property P if $D = \bar{\Omega}^c$. It does not if $D = \Omega$.

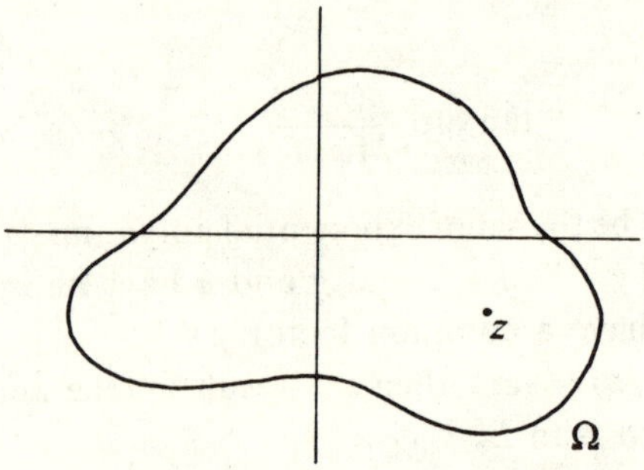

2. $G(z, \zeta) = e^{z\zeta}$. D any domain, G has property P by Pólya's Lemma.

Trivial extension: $G(z, \zeta) = g(z\zeta) = \sum_{n=0}^{\infty} a_n z^n \zeta^n$ convergent on Ω

$$\Rightarrow G(z, \zeta) \text{ has property } P \text{ iff } a_n \neq 0, \; n = 1, 2, \cdots.$$

3. $G(z, \zeta) = e^{z\, w(\zeta)}$; w analytic on Ω.

G has property P iff w is univalent on Ω.

42 (S. Hellerstein). Let f be a nonconstant holomorphic function in the disc $|z| < 1$. Denote $\phi(r) = \phi(r, f)$ the number of points on the circumference $|z| = r$ at which f is real. Find a "good" estimate from above for the rate of growth of ϕ as $r \uparrow 1$ when

(a) f is bounded in $|z| < 1$,

(b) f is univalent in $|z| < 1$.

(It can be shown that for functions of finite order in the unit disc, $\phi(r) = O(\exp\{K/(1-r)\})$ for a suitable constant K (>0).)

References. For entire and meromorphic functions in the plane, cf. H. S. Wilf, *The argument of an entire function*, Bull. Amer. Math. Soc. **67** (1961), 488–489.

S. Hellerstein and J. Korevaar, *The real values of an entire function*, Bull. Amer. Soc. **70** (1964), 608–610.

S. Hellerstein, *The distribution of values of a meromorphic function and a theorem of H. S. Wilf*, Duke Math. J. **32** (1965), 749–764.

43 (S. Hellerstein). Suppose f and g are entire. Let $M^*(r,f,g) = \max\{M(r,f), M(r,g)\}$ where M is the usual maximum modulus function. Assume that as $r \to \infty$,

$$|f(re^{i\theta})| - |g(re^{i\theta})| = o(1/\{M^*(r,f,g)\}^{\varepsilon}) \quad \text{for some } \varepsilon > 0,$$

uniformly in θ $(0 \leqq \theta \leqq 2\pi)$.

Does it follow that f/g is a constant of modulus one?

44 (L. Ehrenpreis). Let $f(z)$ be an entire function satisfying an inequality of the form $|f(z)| \leqq \text{const}(1 + |z|)\exp(c|y|)$ where $z = x + iy$. Can one make any statement about the set of z where $f'(z)/f(z)$ is close to imaginary ("close" being measured by comparing the real part of f'/f to $|f|$)?

For example, if f is of the form $\sum (a_j + b_j z)\exp(ic_j z)$ with real c_j, we see easily that f'/f is close to imaginary when $|y| > A\log(1 + |z|)$ for suitable $A > 0$.

Conjecture. f'/f *is close to imaginary whenever* $|f|$ *is close to* $\text{const}(1 + |z|)\exp(c_0|y|)$ *where* c_0 *is the exponential type of* f.

45 (L. Ehrenpreis). Is the set $\{m + ni\}$, m, n integers sufficient for the topology of H'? In particular, do there exist constants C_1, C_2, C_3, so that, for any entire function f of exponential type 1,

$$\max|f(m + ni)|/\exp(C_1|m + ni|) \leqq C_2 \max|f(z)|/\exp C_2|z|.$$

46 (D. Gaier). Let $\{p_k(z)\}$ $(1 \leqq k < \infty)$ denote a sequence of polynomials with integer coefficients and let $\lim_{k\to\infty} p_k(z) = C$ for $z \in R$, R being an open set. Must C be an integer?

Solution by L. Flatto. No. This is seen as follows. Let $z_0 = (5^{1/2} - 1)/2$ and let $f(z) = 1/(1 + z)$. Note that $|z_0| < 1, f(z_0) = z_0$, and $0 < f'(z_0) = 1/(1+z_0)^2 < 1$. Since $f(z) = \sum_{n=0}^{\infty}(-1)^n z^n$, the equation $p(z) = z$ has a solution z_1 where z_1 is not an integer and $|p'(z_1)| < 1$. z_1 is thus an attractive fixed point of $p(z)$. Let $p_k(z)$ denote the kth iterate of $p(z)$. Then for $\varepsilon > 0$ sufficiently small $\lim_{k\to\infty} p_k(z) = z_1$ for $|z - z_1| < \varepsilon$.

47 (J. Korevaar). Let D be a bounded simply connected region in the z-plane,

and for each n, let $\zeta_{n1},\cdots,\zeta_{nn}$ be a set of equilibrium or Fekete points for D. That is, the points $z_k = \zeta_{nk}$ make the product $\prod_{j \to k} |z_j - z_k|$ an absolute maximum under the condition that $z_1,\cdots,z_n$ lie on the boundary ∂D. Investigate whether if unit masses at the equilibrium points give a small field in D, in the sense that, uniformly on every compact subset of D,

$$\sum_{k=1}^{n} \frac{1}{\zeta_{nk} - z} \to 0$$

as n tends to infinity through a suitable subsequence of the positive integers.

Reference. J. Korevaar, *Limits of polynomials whose zeros lie in a given set*, these Proceedings.

48 (J. Korevaar). Let $A = A(\Omega)$ be the set of monomials $s^m t^n$, where (m,n) runs over a set Ω of points with positive integral coordinates.

(i) If A spans one of the spaces $L^p(I^2)$ $(1 \leqq p < \infty)$ or $C_0(I^2)$, then does it also span the others?

(ii) If $A(\Omega)$ spans $C_0(I^2)$, and the points (m,n) in Ω have $m \geqq 2$, then does $A(\Omega')$ also span $C_0(I^2)$, where Ω' is the set of points $(m-1,n)$ with (m,n) in Ω?

(iii) If the closed span of A in one of the above spaces is not the whole space, then does the closed span contain other than analytic functions?

Reference. J. Korevaar, *Discrete sets of uniqueness for bounded holomorphic functions $f(z,w)$*, these Proceedings.

49 (A. Shields). Let f be analytic in the open unit disc $|z| < 1$ and assume that $f'(z)$ is never zero there.

Is it true that

$$|f''(z)/f'(z)| \leqq 1/(1-|z|) \qquad (|z| < 1)$$

implies f is univalent ?

50 (J. J. Clunie). If f is analytic in $|z| < 1$, is

$$\limsup_{r \to 1} \frac{(1-r)M(r;f')}{M(r,f)} \leqq \frac{1}{e} \ ?$$

Author Index

Roman numbers refer to pages on which a reference is made to an author or a work of an author.

Italic numbers refer to pages on which a complete reference to a work by the author is given.

Boldface numbers indicate the first page of the articles in the book.

Subject Index

rational) unless one of f or g is linear.

Does there exist a prime elliptic function?

whenever g (deg n) has only real zeros.